Inorganic Reaction Mechanisms

Inorganic Reaction Mechanisms

Martin L. Tobe
University College, London

and

John Burgess
University of Leicester

Longman

Addison Wesley Longman Limited
Edinburgh Gate
Harlow
Essex CM20 2JE
England
and Associated Companies throughout the world

Published in the United States of America
by Addison Wesley Longman Inc., New York

Visit Addison Wesley Longman on the world wide web at:
http://www.awl-he.com

First published 1999

ISBN 0 582 23677 0

British Library Cataloguing-in-Publication Data
A catalogue entry for this title is available from the British Library

Library of Congress Cataloging-in-Publication Data
A catalog entry for this title is available from the Library of Congress

Typeset by 40 in Monotype Times
Printed in Malaysia, VVP

Contents

Contents in detail

Preface

The aim of this book is to present an overall view of inorganic reaction mechanisms—substitution, dissociation, addition, electron transfer, intramolecular processes, and their various combinations and extensions—over essentially the whole of the Periodic Table. Although there is the expected dominance of transition metal complexes, compounds of the *sp*-block elements receive appropriate coverage, and occasional mention is made of the lanthanides and actinides. Similarly, the majority of substrates considered are classical inorganic compounds and complexes, but there is full coverage of organometallic systems, including consideration of mechanistic details of homogeneous catalysis, due attention to bioinorganic systems, and a sprinkling of examples from such areas as pharmacology and geochemistry. We restrict ourselves to reactions in solution, covering non-aqueous and mixed aqueous as well as aqueous, and indeed devoting a chapter specifically to medium effects. Although of course we concentrate very much on work published in recent years, we do in places refer to earlier, sometimes much earlier, work in order to sketch the background and context of current research and opinions. The philosophy adopted in this book is that the emphasis is placed on the individual acts of the reaction. The evidence considered will be that gathered, mainly from kinetics but also from stereochemistry and spectroscopy, from studies of systems chosen because they offer the least number of ambiguities and because they throw the most light upon the details of the mechanism. This approach is to be contrasted with many modern reaction mechanism texts, where the emphasis is on mechanism as a guide to synthesis and where reactions that are relevant to the problems of synthesis are given priority.

The breadth of coverage indicated in the preceding paragraph obviously precludes depth in the treatment of many areas of the subject. Obviously we have selected areas of particular importance or interest to us for fuller treatment. Other areas, many of considerable significance, have had to be merely touched on, with references to suitable reviews or recent key papers. In general the text is copiously, though obviously far from comprehensively, referenced—fully comprehensive reviews have been cited wherever possible. We have not hesitated to cite a number of elderly reviews and books. Sometimes seminal or long-established versions provide the best perspectives, especially when written soon after the blossoming of a field by an author centrally involved in the original work. Often the clearest exposition appears before the i-dotting and t-crossing takes over. Fortunately for all interested in inorganic kinetics and mechanisms, the progress of the subject has been methodically and comprehensively chronicled, between January 1969 and mid-1991, in a series of reports. This series comprises the seven *Specialist Periodical Reports on Inorganic Reaction Mechanisms* published by the Chemical Society (latterly the Royal Society of Chemistry), and their continuation, under the title *Mechanisms of Inorganic and Organometallic Reactions* (eight volumes), published by Plenum Press.

The present book grew out of Martin Tobe's 1972 book of the same title, namely *Inorganic Reaction Mechanisms*. A quarter of a century has elapsed since its appearance, during which time the subject of reaction mechanisms in inorganic chemistry has developed from healthy adolescence to mature middle age. The philosophy underlying this book has remained the same, but the size has perforce increased markedly to reflect the enormous increase in data—and perhaps more modest increase in understanding. It was first planned over a decade ago, when Professor Tobe wished to update the text, but felt (quite wrongly!) unable to deal adequately with all the topics to be covered, and therefore co-opted the surviving author to deal with several areas. Tragically, Martin Tobe died early in 1993, and it has been left to the present writer to bring the book to publication. Fortunately Professor Tobe had almost completed the revision and updating of his sections of the manuscript at the time of his death, and had discussed in detail the form and treatment of the other sections of the book with the surviving author. The approach, content, and arrangement of the whole book thus reflect his views, with several sections enshrining his final long-pondered interpretations. It has taken a regrettably long time to finalize the manuscript, but it is hoped that in such sections the original author's aims, approach, and views have been preserved, and that the surviving author's updating has not been detrimental. Of course all errors and omissions, in these and all other parts of the final text, are the surviving author's responsibility.

This book is intended to present the subject matter as outlined above to academics who lecture or research in this and related fields, and to research workers and final-year undergraduates specializing in this topic. It is also hoped that it will be useful to a variety of scientists—academic, industrial, and public service—from other areas who need a detailed introduction to one or more of its main subject areas.

The long and chequered history of the present volume has left us indebted to a large number of people. We cannot hope to attempt to acknowledge them all individually here, but can only mention a selection. Firstly, we would like to acknowledge the influence and inspiration of the University College chemistry school of an earlier generation. In the preface to the 1972 version of *Inorganic Reaction Mechanisms* Martin Tobe singled out Sir Christopher Ingold and Sir Ronald Nyholm; the present writer would like to acknowledge his enormous debt to Reg Prince, himself a product of the Ingold–Hughes school at University College, for introducing him so interestingly, enthusiastically, and enjoyably to inorganic kinetics and mechanisms. We have both been very grateful for the camaraderie of the Inorganic Mechanisms Discussion Group of the Royal Society of Chemistry, which kept us abreast of current developments over many years. Turning to book production, I am very grateful to Alexandra Seabrook, Tina Cadle, Shuet-Kei Cheung, and Pauline Gillett of Addison-Wesley-Longman for their advice, assistance, encouragement, and patience. They deserve all credit and congratulations on finally coaxing a manuscript from an author eternally reluctant to stop 'improving' it. Finally, particular thanks to Rosalie Tobe for her help and support, and especially for her confidence in the surviving author.

John Burgess
Leicester

Abbreviations

Most ligand abbreviations and, where necessary, formulae are defined in the text or in tables at their point of appearance. Abbreviations for frequently encountered ligands in text, tables, figures, and formulae follow.

aq	water—of hydration *or* bulk solvent
bipy [a]	2,2′-bipyridyl
cod	cyclo-octadiene
cp	cyclopentadienyl
*cp	pentamethylcyclopentadienyl
cy	cyclohexyl
cyclam	1,4,8,11-tetraazacyclotetradecane (formula **1** in Chapter 4, p. 148)
cyt	cytochrome
dmf/DMF	*N*, *N*-dimethylformamide (as ligand/solvent)
dmso/DMSO	dimethyl sulphoxide (as ligand/solvent)
edta	ethane-1,2-diamine-tetraacetate (formula **12** in Chapter 7, see p. 325)
etmalt	ethylmaltol(ate) (formula **16** in Chapter 4, see p. 192)
en	ethane-1,2-diamine
fc	ferrocene
imid	imidazole
L	ligand (generally monodentate)
LL (LLL ...)	general bidentate (terdentate ...) ligand
$M^{n+}aq$	hydrated metal ion
nta	nitrilotriacetate
Nu/nucl	nucleophile
OAc	acetate (ethanoate)
OTf	triflate (trifluoromethylsulphonate)

[a] This abbreviation always stands for the 2,2′-compound—its 4,4′-isomer is abbreviated as 4,4′-bipy.

ox	oxalate (ethanedioate)
phen	1,10-phenanthroline
py	pyridine
pz	pyrazine
sep	sepulchrate [b]
terpy	terpyridyl
X^-	halide (Cl^-, Br^-, I^-) unless specified otherwise

[b] Formulae for this and other encapsulating ligands will be found in Chapter 9, as formula **50** in Section 9.5.4.2.

1 Reactions in solution

Chapter contents

1.1 Introduction

The principal aim of the authors of this book is to introduce the reader to the subject of the mechanisms of reactions of inorganic species in solution. Reacting species will include simple ions, compounds, and complexes; we shall deal with organometallic and some bioinorganic compounds as well as those deemed purely inorganic. We shall also cover most of the Periodic Table, though in many cases we shall concentrate on complexes of the *d*-block elements simply because there is so much more known about these. Much of the early development of this subject involved the use of substitution-inert *d*-block complexes, particularly those of the strongly crystal field stabilized ions cobalt(III), chromium(III), and platinum(II). Where possible we shall try to present the overall picture embracing the *s*-, *p*-, *d*-, and *f*-block elements, as in the treatment of general aspects of substitution (Chapter 2), the reactions of solvent exchange and complex formation (Chapter 7), and some redox processes (Chapter 9). Types of reaction covered in the various sections of the book include substitution, electron transfer, intramolecular, insertion, addition, dissociation—and several combinations of pairs of these—as detailed later in this chapter.

It is customary to keep the subject areas of kinetics and mechanisms in the gas phase, in solution, and in the solid state separate. We shall conform to this custom and confine ourselves to reactions in solution, except at a couple of points in the discussions of electron transfer and of medium effects in general. We shall deal with aqueous and with non-aqueous media, though with the expected emphasis on aqueous solutions for reactions of inorganic complexes and on non-aqueous solutions for organometallic compounds. This division is to a considerable extent dictated by solubility patterns; we shall try to bridge the gap where possible. The whole range of solvents will be considered together in the treatment of solvent exchange at simple cations (Chapter 7), and medium effects on reactivity and in the diagnosis of mechanism will form the subject of Chapter 8. One of the main reasons to treat reactions in solution separately from those in the gas or solid phases is the great influence of solvation on reactivity, and sometimes even on reaction mechanism. Simple inorganic cations are generally strongly hydrated in aqueous solution; metal ions are also significantly solvated in organic donor solvents. Simple inorganic anions are also generally strongly hydrated, through hydrogen-bonding with, for example, halide ions or oxygen atoms of oxoanions. On the other hand, such anions are often only very weakly solvated in, e.g., dipolar aprotic solvents such as acetonitrile or dimethyl sulphoxide, and in such media can act as aggressive nucleophiles. Details of structural[1] and thermodynamic aspects of ionic solvation can be found elsewhere.[2]

The majority of inorganic complexes are hydrated, to a greater or lesser extent, in aqueous solution. Ions such as trisoxalatometallates, e.g. $[Cr(ox)_3]^{3-}$, or the hexacyanoferrates are heavily hydrated due to their hydrophilic peripheries; uncharged complexes such as $[Co(NO_2)_3(NH_3)_3]$, $[Fe(CN)_2(bipy)_2]$, or $[NiBr_2(py)_2]$ tend to be only lightly hydrated; while cations such as the tris-1,10-phenanthroline or tris-2,2′-bipyridyl complexes $[M(LL)_3]^{n+}$ are actually lipophilic, like the $PhAs_4^+$ or $^nBu_4N^+$ cations. Organometallic compounds are generally hydrophobic and soluble in organic solvents rather than in water. They are often significantly solvated, but sometimes, as in hydrocarbon solvents, rather small solute–solvent interactions bring us close to gas-phase conditions. Water-insoluble organometallic compounds or coordination complexes which contain aromatic moieties can often be converted into water-soluble derivatives by sulphonation, while the use of R_4N^+ in place of Na^+ or K^+ as counterion may solubilize salts of moderately lipophilic anionic complexes in organic solvents.

Table 1.1 shows some examples of significant and of negligible solvent effects on organometallic and organic reaction rate constants. At one extreme, where solute–solvent interactions are very small, not only is reactivity affected minimally by solvent variation, but reactivity in the gas phase may be very similar to that in solution. Examples of reactions with rate constants in solution very similar to those in the gas phase include:

- thermal decomposition of dinitrogen pentoxide;
- thermal decomposition of di-t-butyl peroxide and of acetyl peroxide;
- intramolecular rotation of the substituted biphenyl (**1**);
- racemization of *d*-pinene (**2**);
- dimerization of cyclopentadiene.

Table 1.1 Examples of significant and negligible solvent effects on rate constants for organometallic and organic reactions.[a]

Reaction	Solvent	Rate constant[b]	Range[c]
ORGANIC			
t-BuCl solvolysis	water	$10^5 k_1 = 3300$	$> 10^6$
	methanol	0.082	
	acetic acid	0.022	
	ethanol	0.009	
Menschutkin: $Et_3N + MeI$	nitrobenzene	$10^5 k_2 = 1380$ (373 *K*)	10^4
	acetone	265	
	benzene	40	
	hexane	0.5	
Cycloadditions:[d]			
[2+2] TCNE + n-butyl vinyl ether	acetonitrile	$10^5 k_2 = 4520$ (303 *K*)	4×10^3
	acetone	882	
	benzene	9.1	
	hexane	1.1	
[4+2] PhNO + 1,3-cyclohexadiene	ethanol	$10^2 k_2 = 1.16$	1.2
	benzene	1.00	
	hexane	1.42	
ORGANOMETALLIC			
$Me_4Sn + Br_2$	acetic acid	$10^3 k_2 = 9600$	5×10^4
	chlorobenzene	117	
	dimethylformamide	14	
	carbon tetrachloride	0.18	
trans-$[IrCl(CO)(PPh_3)_2]$ + methyl iodide	chloroform	$10^3 k_2 = 111$	11
	dimethylformamide	102	
	acetone	72	
	benzene	13	
$[Rh\{P(OMe)_3\}_5]^+$	all solvents	$10^{-2} k_1 = 3.4$	1

[a] Sources for the kinetic data quoted may be traced through J. Burgess and E. Pelizzetti, *Progr. React. Kinet.*, 1992, **17**, 1 (see Table 1).
[b] Rate constants are at 298 K unless stated otherwise; units of k_1 are s^{-1}, of k_2 $dm^3\ mol^{-1}\ s^{-1}$.
[c] This is the total range in all solvents studied, and may be greater than the range of the selected values given in this Table.
[d] The range of rate constants for dimerization of cyclopentadiene in solution (seven solvents) is only from 7 to $19 \times 10^{-7}\ dm^3\ mol^{-1}\ s^{-1}$.

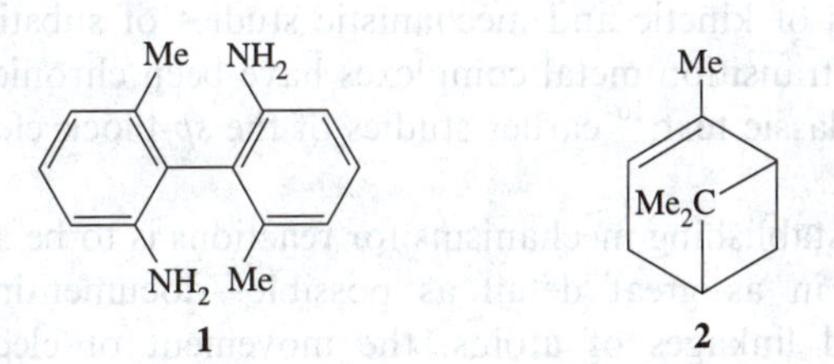

In several of these cases activation parameters were also found to be the same in solution as in the gas phase. The racemization of *d*-pinene is of historical interest not only as the first instance of equality of gas and solution rate constants, but also as the first gas-phase rate constant plus activation energy determination.[3] The whole question of medium effects on reactivities—magnitudes, trends, and explanations—forms the subject of Chapter 8. The effects of solvation can often be quantified through the use of thermodynamic transfer parameters, which can be particularly useful in trying to analyse solvent effects on reactivity into their initial-state and transition-state components.

The quantitative study of chemical kinetics in solution goes right back to Wilhelmy and the inversion of sucrose (1850),[4] with Thénard's observations on the decomposition of hydrogen peroxide arguably forming a qualitative precursor.[5] Recognizable inorganic kinetics, in the sense of establishing dependences of rates on reactant concentrations, goes back to Harcourt and Esson's studies of hydrogen peroxide oxidation of hydrogen iodide and of permanganate oxidation of oxalic acid (1866–7).[7] The published report of the latter system is disappointingly brief and lacks quantitative information—though the authors did report the marked catalysis by the Mn^{2+} product, well known to practitioners of volumetric analysis. The papers on the former reaction, extracts from which were published relatively recently,[8] include experimental details and results. A brief bibliography in one of the standard texts on chemical kinetics[9] facilitates the tracing of other early kinetics work. In practice very little inorganic kinetics work was done for several decades after Harcourt and Esson's experiments.

Systematic study of the kinetics and mechanisms of inorganic substitution reactions really only started after Hughes and Ingold had built their mechanistic framework for nucleophilic substitution in organic chemistry.[10] Indeed Hughes and Ingold did study some inorganic systems towards the end of their lives. These included some organomercury(II) exchange reactions, which were initially published as organic chemistry but later dressed up as a key link between inorganic and organic mechanisms,[11] and ligand exchange and base hydrolysis at octahedral cobalt(III).[12] Mechanisms for electron transfer between inorganic complexes developed, from the early 1950s,[13] from Taube's seminal review[14] of substitution at transition metal centres, while kinetic studies of oxidations by (per)oxoanions and such complexes as hexacyanoferrate(III) have been intensively studied for very many years[15] (cf. Harcourt and Esson's permanganate oxidation above). The early years of kinetic and mechanistic studies of substitution and redox reactions of transition metal complexes have been chronicled in Basolo and Pearson's classic text;[16] earlier studies of the *sp*-block elements are not so easy to track.

The whole idea of establishing mechanisms for reactions is to be able to describe their courses in as great detail as possible—documenting the changing positions and linkages of atoms, the movement of electrons, solvation changes, and so on. To this end it is advisable to gather information from as many sources as possible. Such sources include:

(a) The nature of the *products* of a reaction may give important mechanistic clues. One of the most important examples is the demonstration of the 'inner-sphere' electron-transfer mechanism (see Section 1.5.4 of this chapter, and Chapter 9). Product information has also proved valuable in establishing the different modes of hydrolysis of the trichlorides of nitrogen and phosphorus. Product information could be misleading in the case of nucleophilic substitution at carbon—the S_N1 and S_N2 paths can both give alcohols as products of hydrolysis of alkyl halides. Reactions which give a multitude of products, for example base hydrolysis of difluoramine (where the simple species HNF_2 and OH^- give at least six different products—Section 3.5) may or may not yield mechanistic secrets from product characterization.
(b) Isotopic *labelling*, for example deuterium for hydrogen or the use of radioisotopes, is a special case of product information, permitting monitoring of the destination of selected atoms.
(c) Monitoring the *stereochemical* course of substitution reactions proved very valuable in organic chemistry. Unfortunately stereochemistry is more complicated for octahedral systems than for the tetrahedra of organic chemistry—a great deal of useful information has come out of stereochemical experiments on inorganic systems (Chapter 5), but such information has not played the same central role in inorganic substitution studies as it did in their organic forebears.
(d) The main source of information is usually from *kinetics*. The establishment of a rate law, where possible, is a most important step, while determination of activation parameters (enthalpy, entropy, and volume) can give valuable insights into mechanisms. These two key roles for kinetics in the establishment of mechanisms are discussed further in the following paragraphs/sections. Solvent and salt effects on kinetic parameters, and the mechanistic indications these may give, are dealt with in Chapter 8.

1.2 Ranges of rate constants

For many decades one of the main practical problems in studying inorganic kinetics and mechanisms was the vast range of reactivities encountered. Over the past few decades the seriousness of time-scale problems has decreased enormously, as a variety of fast reaction techniques have been introduced and developed.[17] A certain increase in the range of reactivities that could be monitored was gained by working with very dilute solutions and at low temperatures, but the first technique developed specifically for the study of fast reactions was the continuous-flow method of Hartridge and

Roughton, reported in 1923.[18] Progress during the next thirty years was reviewed in a Faraday Society Discussion;[19] comparable progress was achieved in the following five years.[20] These two review issues chart a great deal of key work on areas of inorganic kinetics and mechanisms previously inaccessible. Thereafter research into fast inorganic reactions continued to develop exponentially in terms of amount of work published, perhaps rather more slowly in terms of innovative science. Technical advances in the field of monitoring fast reactions are still being made. Improvements in optical detection, permitting very small absorbance changes to be accurately monitored, mean that second-order reactions with rate constants as high as 10^9 dm^3 mol^{-1} s^{-1} can be monitored by stopped-flow techniques.[21] Other advances in stopped-flow instrumentation include a multi-mixing arrangement for the generation and study of transients[22] and a pulsed-accelerated-flow apparatus that again permits the measurement of rate constants as high as 10^9 dm^3 mol^{-1} s^{-1}.[23] Time-resolved photoacoustic methods provide another new route to kinetic data,[24] while techniques developed for sequential injection analysis[25] may prove valuable for kinetic applications. In contrast to the great progress on the fast reaction front, there is not much one can do about extremely slow reactions!

Figure 1.1 gives some idea of the ranges of reactivity and rate constants encountered in various areas of inorganic chemistry. The direct relation between rate constant and half-life for first-order reactions means that a half-life scale can be placed alongside the rate constant scale (necessarily logarithmic) to give a clearer idea of the experimental time-scales involved. The left-hand column of Fig. 1.1 shows rate constants and half-lives for water exchange at a variety of metal cations, and immediately emphasizes the range—of about 20 orders of magnitude—between fastest (measured) and slowest (estimated). Details and reasons are given in Chapter 7, for complex formation as well as for solvent exchange. Rate constants for substitution at complexes obviously depend greatly on the nature of the central metal and the leaving group. This is illustrated in Fig. 1.1 for the metal(III) complexes $[M(NH_3)_5X]^{2+}$, which provide the classic examples of octahedral substitution (cf. Chapter 4). Even ligand variation in an apparently rather restricted area can give rise to large changes in reactivity, as shown in Table 1.2 for aquation of a group of dichloro-cobalt(III)-N_4 macrocycle complexes *trans*-$[Co(NNNN)Cl_2]^+$, with the tetraammine analogue for comparison. Table 1.2 also includes activation enthalpies, to show the bond-strength variation which underlies the dramatic rate constant changes.

A similarly wide range of rate constants is known for redox reactions, ranging from the essentially diffusion-controlled limit for reactions of several aquametal cations down to time-scales of several weeks for permanganate or for peroxodisulphate oxidation of water—both thermodynamically very favourable but kinetically very slow. Figure 1.1 also includes reactivity ranges for reduction of cobalt(III) complexes

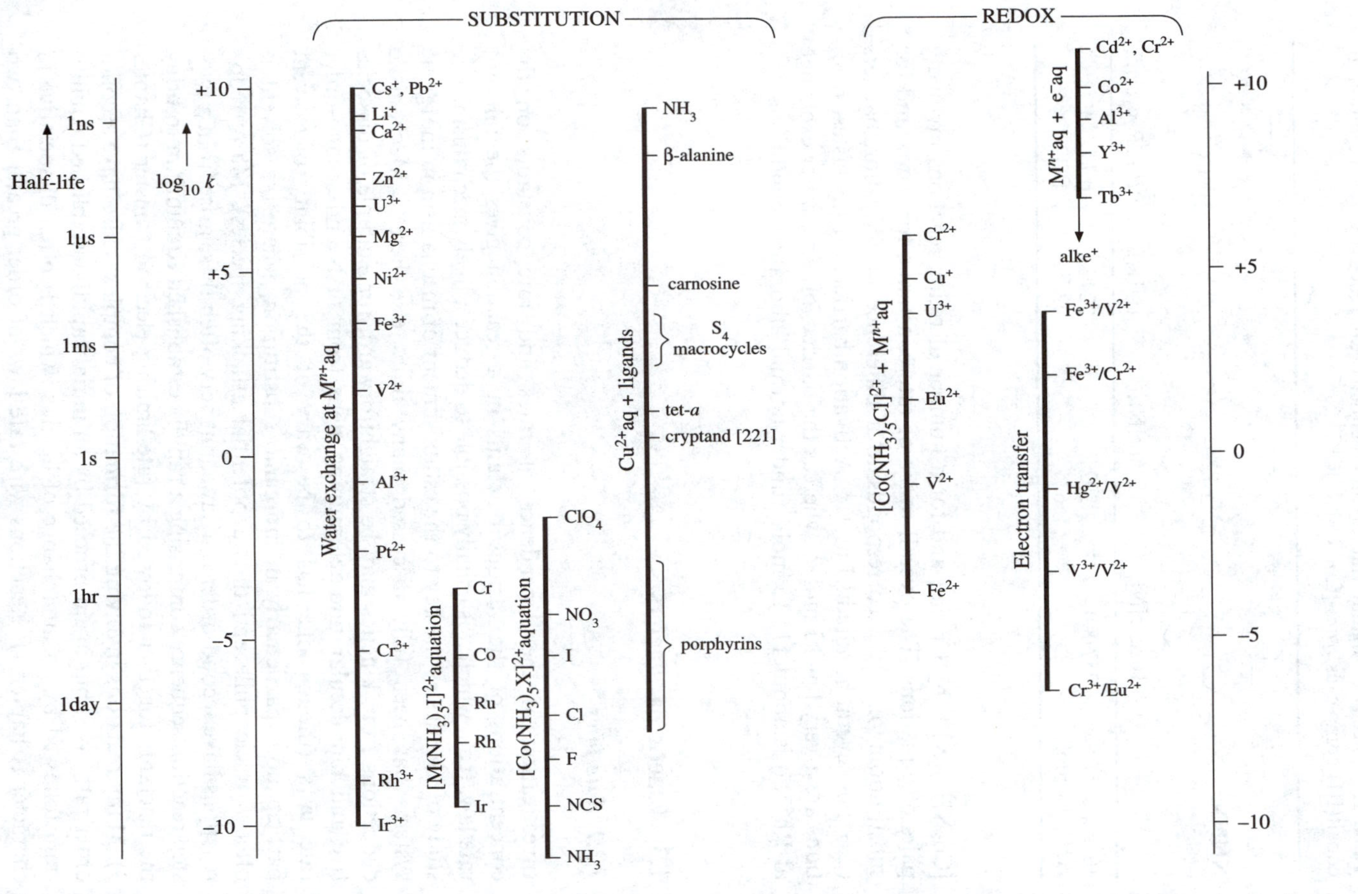

Fig. 1.1 Ranges of rate constants for various types of inorganic reaction (the half-life scale applies to first-order reactions, viz. water exchange and aquation).

Table 1.2 Effects of ligand nature and denticity on reactivity for aquation of cobalt(III) complexes *trans*-$[Co(NNNN)Cl_2]^+$.

NNNN:	16–ane[N]4	15–ane[N]4	$(NH_3)_4$	*RSSR*–cyclam
$10^5 k_{aq}$ (s^{-1})	257 000	992	224	0.11
$\Delta H^\ddagger$ (kJ mol^{-1})	—	62	92	103

$[Co(NH_3)_5X]^{2+}$ by Cr^{2+}aq and for a number of redox reactions involving pairs of aqua-ions. The overall range of rate constants, for redox and for substitution, of $>10^{18}$ corresponds to a difference in activation barriers ($\Delta G^\ddagger$) of several hundred kJ mol^{-1}. Activation barriers for the fastest reactions are a very few kJ mol^{-1}, whereas the barrier for, e.g., water exchange at Rh^{3+}aq is about 115 kJ mol^{-1} under normal laboratory conditions.

1.3 Kinetic parameters

1.3.1 Rate laws

Investigation of the dependence of rates and rate constants on the concentrations of the various reactants in a reaction gives the relevant rate law, from which it is usually possible to deduce a likely mechanism. The study of kinetics as the key to the establishment of mechanism in inorganic systems has been fully discussed many times elsewhere.[26] Mechanistic deductions from kinetics may be straightforward, for example of dissociative and bimolecular rate-determining steps from first- and second-order rate laws (though one has to be wary of the possibility of solvent participation, particularly in inorganic substitution, where the solvent is often a nucleophile and thus a potential attacking species). Many redox reactions have second-order rate laws, but very often these arise from a two-step reaction sequence comprising a rapidly established equilibrium to bring the reactants together followed by rate-limiting electron transfer (Chapter 9). Many reactions involving the formation of complexes also have second-order rate laws, but again proceed by an initial rapidly established equilibrium followed by an interchange of ligands, which is often dissociative in character (Chapter 7). Reactions with rate laws of order greater than two also in general involve one or more pre-equilibria preceding the rate-limiting step. Rate laws consisting of the sum of two terms are not infrequently encountered. They may arise from reactions proceeding to equilibrium rather than to completion, or from the existence of two parallel reaction paths of comparable ease. Examples of the latter are common in substitution in square-planar complexes (Chapter 3), and are also found for some substitution reactions of metal carbonyls (Chapter 6). There are, unfortunately, areas where kinetic information may be ambiguous (see, for example, the classic case of base hydrolysis of cobalt(III)-ammine complexes in Chapter 4); there are also situations where it is impossible to obtain a full rate law. The best example of the latter situation is solvent exchange, since it is clearly impossible to vary the concentration of a solvent in itself in order to obtain its order in the rate law (cf. Chapter 7).

1.3.2 Activation parameters

The temperature dependences of rate constants have been studied assiduously since the earliest days of kinetics. Hood seems to have been the first to report, for chlorate oxidation of Fe^{2+}aq, a linear dependence of logarithms of rate constants against the respective reciprocal temperatures.[27] His preliminary experiments covered a range of only 4 °C, but his second publication listed rate constants at 1 °C intervals between 10 and 25 °C and at 28, 30, and 32 °C—an example of thoroughness all too rarely emulated since. Arrhenius, apparently basing his equation on a suggestion by van't Hoff, confirmed the linearity of plots of logarithms of rate constants against reciprocal temperature for several reactions, and proposed his theory, in 1889.[28] Since then there has been a plethora of activation energies and probability factors, E_a and $\log A$, and of interpretations thereof. In recent decades inorganic solution chemists have tended to favour transition-state theory, and to interpret and discuss temperature dependences of rate constants in terms of enthalpies and entropies of activation, $\Delta H^{\ddagger}$ and $\Delta S^{\ddagger}$. Since the mid-thirties (for organic reactions; mid-fifties for inorganic), there has been a rapidly burgeoning interest (Fig. 1.2) in the pressure dependence of rate constants, which gives volumes of activation, $\Delta V^{\ddagger}$. The interpretation of activation volumes was developed in the main for organic reactions,[29] though the key text[30] does contain a short section on reactions of transition metal complexes. Subsequent reviews[31] have concentrated on inorganic reactions. Activation enthalpies and entropies can be determined relatively easily in almost any laboratory, the only requirement being to be able to thermostat reaction mixtures fairly accurately over at least a 20 °C temperature range, generally around room temperature. The determination of activation volumes requires apparatus that can be thermostated accurately and operated at pressures up to 2 or 3 kbar. Such high-pressure apparatus has been developed from simple vessels only suitable for monitoring rather slow reactions to apparatus which can cope with fast reactions—in particular by the stopped-flow method and by NMR spectrometry—to cover complex formation, redox,

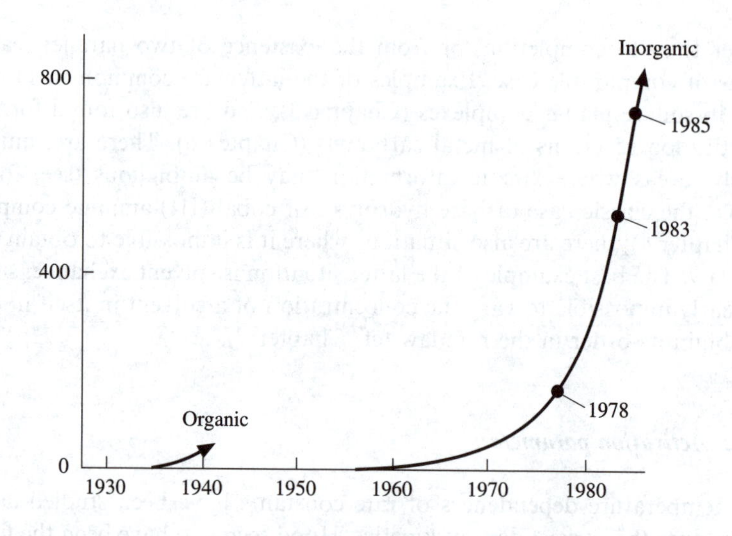

Fig. 1.2 The growth of activation volumes. The curve for inorganic determinations had reached a plateau, possibly even taken a turn downwards, by the mid-1990s.

and solvent-exchange reactions. Unfortunately, such apparatus is still only available in relatively few laboratories.

Activation volumes have the considerable attraction over activation entropies that they seem easier to visualize, although in practice there is often a strong correlation between these two parameters. The increase in volume resulting from bond stretching on forming a dissociative transition state is generally accompanied by an increase in entropy from the greater freedom of the leaving group. Conversely, the decrease in entropy on bringing two reactants together in a bimolecular transition state is associated with a decrease in volume as the more compact transition state is formed. However, it should be added that there are good reasons to prefer $\Delta V^{\ddagger}$ to $\Delta S^{\ddagger}$—the former is derived from the *slope* of a $\log k$ versus pressure plot, whereas the latter is derived from an *intercept* obtained by lengthy extrapolation of a $\log k$ versus reciprocal temperature plot. A further advantage of $\Delta V^{\ddagger}$ is that it can be used, together with ΔV°, to construct volume profiles for reactions and thus track volume changes from initial state to transition state and through to product(s)—an operation that is much less profitable for the parameters H, S, or G. Figure 1.3 shows volume profiles for some much-studied organic reactions. For the Menschutkin reaction[32] of Fig. 1.3(a)—the paucity of partial molar volume data for products of Menschutkin reactions forces us to use this exotic example rather than a standard R_3N+RI—the transition state has properties intermediate between reactants and products. For the Diels–Alder reactions of Fig. 1.3(b)[33] the transition states are, not unexpectedly, much more compact than the reactants, but there is generally only a small increase or

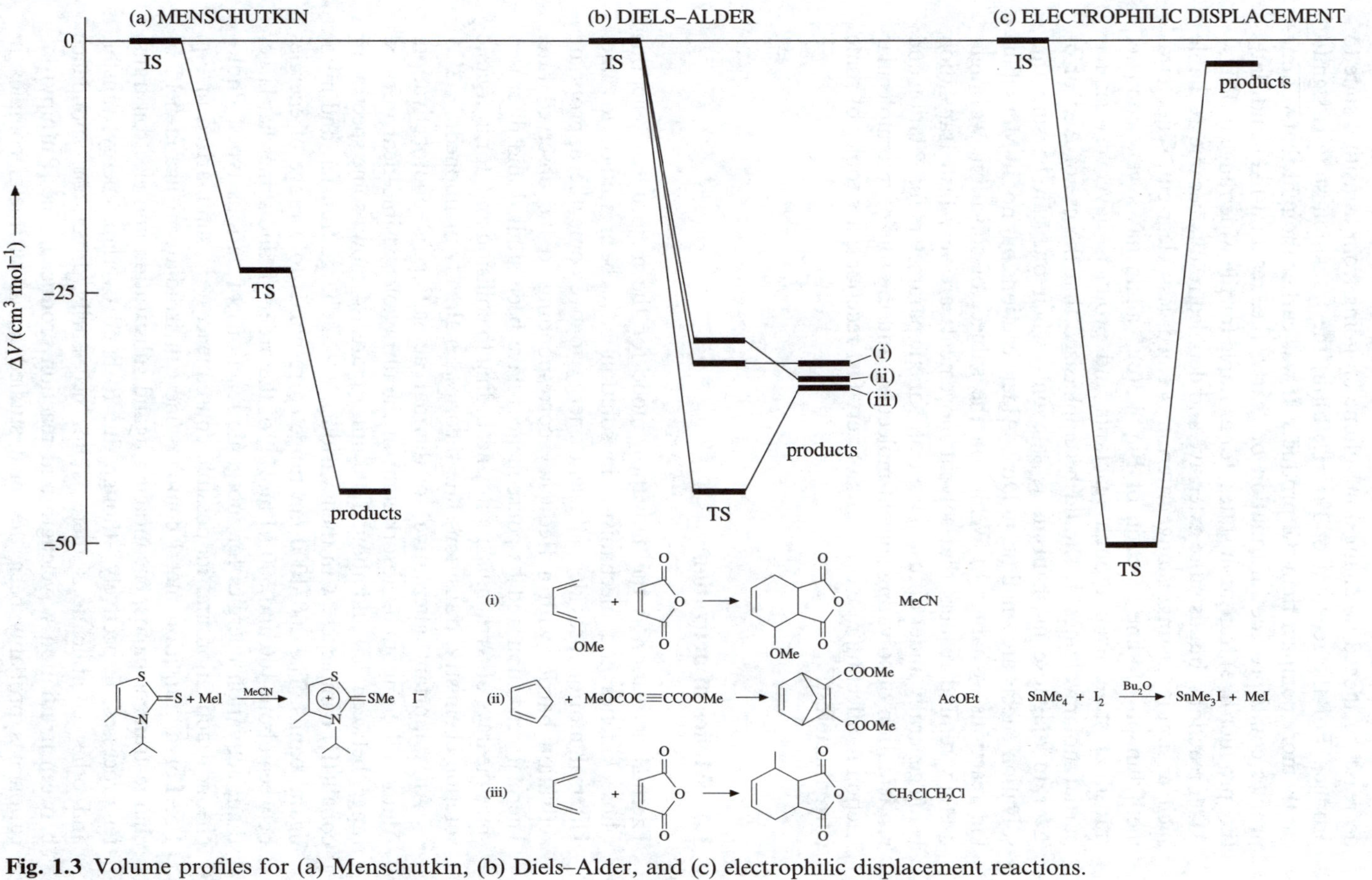

Fig. 1.3 Volume profiles for (a) Menschutkin, (b) Diels–Alder, and (c) electrophilic displacement reactions.

decrease (or indeed no change) in volume on going from transition state to product. For the reaction of Fig. 1.3(b)(iii) the transition state is significantly more compact than the products. It was earlier thought that volume profiles could provide information on 'early' (i.e. reactant-like) and 'late' (i.e. product-like) transition states, for example for the wide range of Diels–Alder reactions, but as more extensive and detailed results became available such a simple distinguishing criterion seemed less clear-cut. The tetramethyltin plus iodine reaction of Fig. 1.3(c)[34] again has a very compact transition state, even though reactants and products have very similar partial molar volumes. It should be emphasized that the use of $\Delta S^{\ddagger}$ or of $\Delta V^{\ddagger}$ to diagnose mechanisms is safe only when one can be sure that solvation effects are small (as in Diels–Alder reactions but not in Menschutkin reactions—Chapter 8). While on the subject of solvation, we might usefully remind readers that solvent movements are very fast, translation being generally under diffusion control. Kinetic parameters for rotation and reorientation of solvating solvent molecules, and the role of intermolecular friction in affecting rate constants of very fast reactions, are topics of some current interest.[35]

1.4 Modes of activation

Thermal activation is the activation mode for the majority of reactions whose kinetics and mechanisms in solution have been studied to date. Thermal motion provides sufficient energy to overcome the appropriate activation barrier with a frequency corresponding to the observed rate. Thermal activation and its consequences have been studied and discussed since the days of Arrhenius, mainly according to collision and, for inorganic solution chemistry, transition-state theories, as already mentioned.

An important alternative to thermal activation is photoactivation. Almost all compounds absorb light in the ultraviolet–visible–near infrared range, between 200 and about 1000 nm. Figure 1.4 shows some spectra for cobalt(III) complexes, with their various $d \rightarrow d$, charge-transfer, and intraligand bands. The 200–1000 nm range corresponds to a range of energies between about 100 and 600 kJ mol^{-1}, of the order of, and somewhat larger than, activation energies for inorganic reactions of various types. Photochemical activation and its possible consequences are summarized in Fig. 1.5. For a transition metal complex one can irradiate in ligand–field or charge–transfer bands, encouraging ligand substitution and electron transfer processes, respectively—though due to band overlap, energy transfer, and other complications there is not necessarily direct correspondence between irradiation wavelength and reaction produced. The $[Ru(bipy)_3]^{2+}$ cation has probably been the most studied transition metal complex in relation to photochemistry, especially photoredox and photoactivated processes.[36]

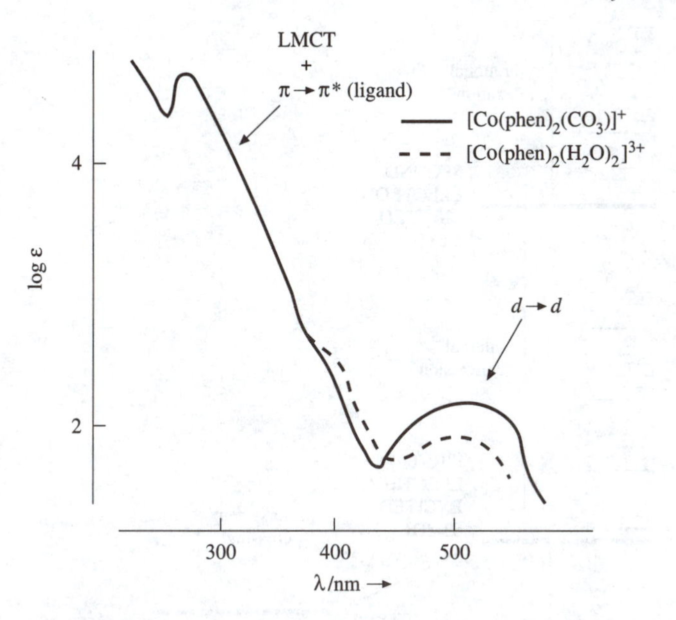

Fig. 1.4 Ultraviolet–visible spectra of two typical cobalt(III) complexes.

One of the most interesting aspects of photoactivation is that products may differ from those of thermal activation. For ternary complexes, different ligands may be replaced in photochemical and in thermal activation. Thus solvolysis of ternary chromium(III)-ammine complexes $[Cr(NH_3)_5X]^{2+}$ under thermal conditions generally results in replacement of X^-, whereas the dominant reaction under photochemical conditions is usually replacement of ammonia. Similarly, very different product isomer ratios may be obtained in thermal and photochemical reaction of complexes of the type *cis*- and *trans*-$[M(NH_3)_4X_2]^+$. Detailed examples will be found in the appropriate sections of Chapter 4, where photochemical substitution at chromium(III) and rhodium(III) complexes will be discussed. It may be commented here that these two metal centres have been by far the most studied in respect to photosubstitution. The problem with photosubstitution at cobalt(III) is the likelihood of significant, or even dominant, reaction via reduction to labile cobalt(II). Such complications are normal for cobalt(III)-ammine and -amine complexes, somewhat less of a problem with cobalt (III)-cyanide complexes. The other very important area is that of photosubstitution at binary and ternary mono- and polynuclear metal carbonyls.

In view of the possibilities of parallel substitution and electron transfer just mentioned, and the greater variety of possible excited states (cf. Fig. 1.5 overleaf), mechanisms tend to be more complicated and less thoroughly understood than in analogous thermally activated sytems. Photochemistry does provide an extra source of evidence in the quantum yield for a reaction, but there are often difficulties in measuring quantum yields accurately. There are also likely to be more complications from solvation effects, with

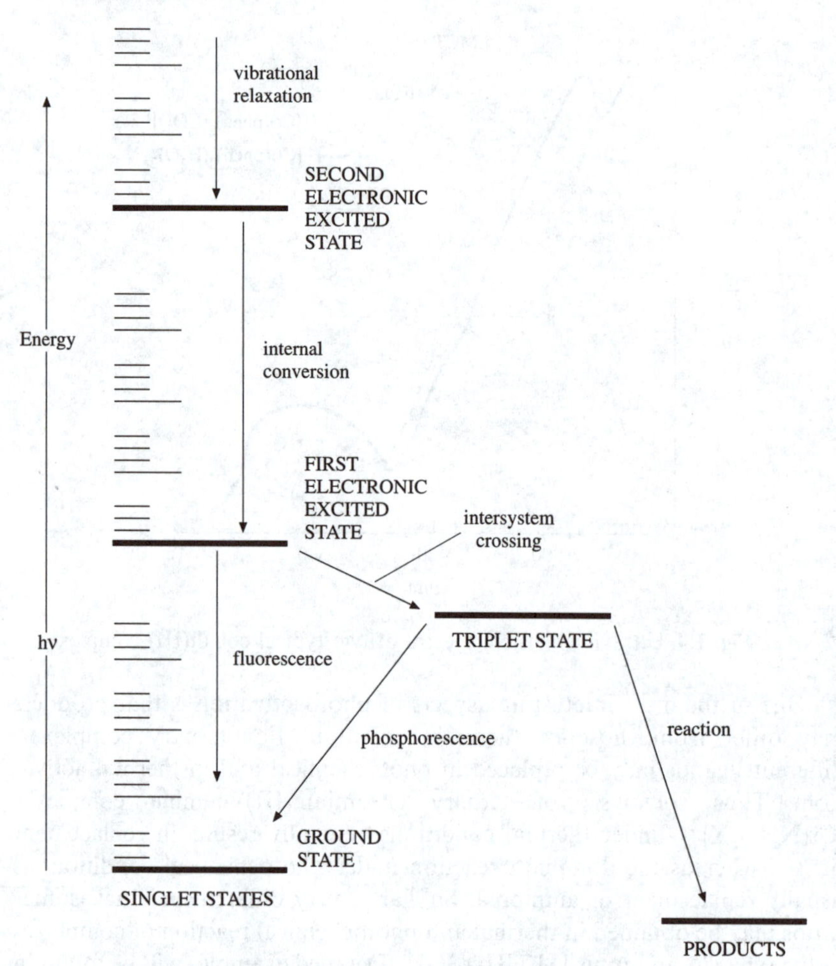

Fig. 1.5 A selection of possible photochemical activation process for a species with a singlet ground state and inanaccessible tripet state.

the trapping of incipient or potential products in solvent cages often being of importance.

Other advantages and attractions of photochemistry include the facile generation of reactive intermediates as an aid to kinetic studies and mechanism elucidation. Thus, for example, complexes of the type $[Cr(CO)_5(LL)]$, $[Fe(CN)_5(LL)]^{n-}$, and $[W(CN)_7(LL)]^{n-}$ containing potentially bidentate ligands LL bonded by only one of the donor atoms can be prepared by photolysis of $[Cr(CO)_6]$, $[Fe(CN)_6]^{n-}$, or $[W(CN)_8]^{n-}$ in the presence of LL. It is then possible to monitor kinetics of chelate ring closure, e.g.

$$[Cr(CO)_5(LL)] \longrightarrow [Cr(CO)_4(LL)] + CO$$

(see Sections 4.7.5.2 and 6.3.1). In similar vein, irradiation of $[Ni(CO)_4]$ in a dinitrogen matrix furnishes $[Ni(CO)_3(N_2)]$ for kinetic studies of the reverse reaction

$$[Ni(CO)_3(N_2)] \longrightarrow [Ni(CO)_4]$$

Photochemical generation of intermediates, actual or postulated, can also be valuable in studies of mechanisms of homogeneous catalysis (Section 10.7). Furthermore, flash photolysis serves as a fast-reaction technique for obtaining kinetic parameters.

There are a number of authoritative and comprehensive books on inorganic photochemistry,[37] including its mathematical aspects.[38] For the reader who needs to go only slightly more deeply than us into photochemical activation in relation to inorganic mechanisms, such treatments as that in Chapter 7 of Jordan's book[39] and in the introduction to this topic in *Comprehensive Coordination Chemistry*[40] should prove sufficient. There are a number of fully comprehensive reviews, each with several hundred references, to provide exhaustive back-up. Early and subsequent fundamental work on photochemistry of transition metal complexes has been documented in depth and detail;[41] the relatively brief treatment of organometallic photochemistry in those reviews is supplemented by two extensive reviews devoted specifically to this area.[42] The photochemistry of electron-transfer processes has also been reviewed,[43] as has the photophysics of the excited states of transition metal complexes[44] and the use of such complexes in photocatalysis and photosensitization.[45] Progress in the areas of metal complexes and organometallic compounds can be followed through the published proceedings of the International Symposia on Photochemistry and Photophysics, where such topics as photochemistry in elucidating organometallic reaction mechanisms[46] and the effects of pressure on photochemical mechanisms for complexes[47] formed the subjects of review articles in the most recently published proceedings.[48]

Sonochemistry provides a third activation mode.[49] Again products and mechanism may differ from those under both thermal and photochemical conditions.[50] The cavitation resulting from ultrasonic activation results in three reaction environments,[51] only one of which is of relevance to the present book. Reactions (generally atom or radical redox processes[52]) may take place in the cavities formed, at the cavity–solvent interface, or in bulk solvent. The first two of these regions are outside our scope. Indeed it has been demonstrated that intra-cavity reactions may be analogous to those in flames.[53] Only processes which take place in the bulk solvent are relevant to us, and in these cases the course of reaction is generally that corresponding to thermal activation. Acoustic activation has proved of considerable interest in several areas of chemistry in recent years, but as far as we can tell has not yet led to any major advances in the area covered by the present book.

Microwave activation has become increasingly popular in recent years, but, as for ultrasonic activation, its value at present lies in the field of preparative rather than mechanistic chemistry.[54]

1.5 Classification of reaction types

1.5.1 Introduction

In any discussion of the mechanism of a chemical reaction, the approach and the emphasis will depend upon the interests of the parties concerned. A synthetic chemist will wish to consider mechanism in so far as it helps to understand the processes involved in the making of specific compounds, the factors that improve yields and the development of new pathways and products. The analytical chemist will be interested in reactions that allow a precise determination of the composition of the material which is being examined. The organometallic chemist may be satisfied with a collection of bows and arrows purporting to indicate the movement of electrons and bonds, and the bioinorganic chemist will wish to be able to understand the sequence of reactions that constitute the biological catalytic cycle and the factors that allow it to be fine-tuned. All of these people realize that a chemical reaction consists of a number of separate steps and those that are wise enough realize that they can do very little until they have an adequate understanding of the mechanisms of these individual stages. This book is concerned with discussion of the main types of reaction processes and will tend to concentrate upon the most direct studies of their mechanism. The secret of success in any study of this sort is to choose the substrate, the techniques, and the conditions used with great care in order to provide the least ambiguous answer to the question that has been asked. To study a system because it is easy to follow or because the compounds happen to be readily available to the researcher may provide many publications but a great deal of luck would be needed if the results were to push back the important frontiers of knowledge. Nevertheless, many of the major advances have sprung from such serendipity but few people are proud to admit this and it would be boorish to draw attention to such occasions.

We propose to classify the various reaction types in a way that is consistent with the usual terminology of coordination chemistry. This, which was once a branch of inorganic chemistry and owed much to the work of one man, Alfred Werner, has long become an approach to chemistry in general. It looks upon a chemical species in terms of a *central atom* (or *ion*) surrounded by a group of *ligands,* which constitute the *coordination shell.* Each ligand, which usually (but not necessarily) can exist as a stable entity when no longer bound to the central atom, can be charged or uncharged and may possess one or more *donor atoms* through which it is bound to the central atom. The *coordination number* (the number of donor atoms bound to the central atom in its coordination shell), the *geometry* (the three-dimensional arrangement of the donor atoms about the central atom), and the *stereochemistry* (the relative position of the ligands within the coordination shell), all go a long way towards describing the immediate environment of the central atom. All of these concepts possess reality in the sense that, if the structure of the complex has been determined, these features can be recognized immediately. The *oxidation state* of the central atom, on the other hand, is a man-made concept. It is defined as 'the charge remaining on the central ion when all the ligands have been removed in their closed-shell configuration'; unfortunately there is often a problem over the meaning of the term 'closed-shell configuration'. While the concept of oxidation state is of considerable use in naming and systematizing the compound, it only approaches reality as the metal–ligand bonding becomes electrovalent.

The concepts of coordination chemistry are extremely useful when applied to situations or processes that relate to a unique centre and are of great value in classifying those mechanisms that take place at a single reaction centre. Since, to a first approximation, coordination chemistry is not dependent upon the ways in which the ligand is bound to the central atom, it can be applied validly to compounds ranging as widely as $[NiCl_4]^{2-}$, $[FeCl_4]^-$, $[TiCl_4]$, even $[CCl_4]$. It starts to lose its usefulness when we consider compounds or processes in which there is no valid single focusing point, such as compounds with metal–metal bonds, clusters such as $Os_3(CO)_{12}$ or $B_{10}H_{14}$, and most organic compounds. However, even in these compounds, if one is discussing a reaction at a single atom, the value of the coordination chemistry approach returns.

For the purpose of classifying their mechanisms, inorganic reactions may be broadly divided into three major categories:

(a) reactions involving changes in the coordination shell;
(b) reactions involving changes in the oxidation state;
(c) reactions involving changes to the ligands while they remain coordinated.

Many well-known reactions are complicated multistage processes that may be made up of a sequence of steps representing two, or all three, of these categories, but generally the task of unscrambling them is not overwhelming. Others may look deceptively simple but are not, and some of the apparently simplest processes consist of two or even more stages. It will be extremely useful to examine each of the categories in turn and consider the complications afterwards.

1.5.2 Reactions involving a change in the composition of the coordination shell

This category can be subdivided into three closely related subdivisions. Addition and dissociation reactions (Sections 1.5.2.1 and 1.5.2.2) are complementary, as for example the forward and reverse reactions involving the familiar Lewis acid–base pair:

$$\underset{\text{Lewis acid}}{BF_3} + \underset{\text{Lewis base}}{NR_3} \rightleftharpoons F_3BNR_3$$

These reactions involve a change in coordination number for the central atom or ion; in the third type, substitution, there is change of composition without change of coordination number (Section 1.5.2.3).

1.5.2.1 Increase in coordination number (addition)

Although the name addition is applied to a large number of important reactions of carbon, the increase in the coordination number at the reaction centre is usually accompanied by a similar change at a related, frequently adjacent, carbon atom. Changes in coordination number in stable carbon compounds usually require a change in the distribution of the electrons between σ- and π-bonding orbitals. When transition metal complexes undergo a change in coordination number there is rarely an accompanying change in bond multiplicity. As a rule, one coordination number is strongly favoured and processes leading towards it usually have very low barriers and are fast.

The most important inorganic centre for addition reactions is three-coordinate boron (see below), where the interest lies in the change of geometry about the boron from planar to tetrahedral as the bond with the nitrogen is formed. Ligand addition to binary halides is another commonly encountered, preparatively if not kinetically, reaction of this type, e.g.

$$PF_5 + F^- = PF_6^-; SnI_4 + 2PPh_3 = SnI_4(PPh_3)_2$$

Addition of alkenes or alkynes to square-planar d^8 complexes plays an important role in some homogeneous catalysis systems:

$$-Ir^{I}- + F_2C{=}CF_2 \rightleftharpoons -Ir(\leftarrow F_2C{=}CF_2)-$$

The iridium in this adduct is still in an oxidation state approximating to +1, but this example is at one end of a spectrum of addition reactions which culminate in the oxidative addition reactions detailed below in which addition is accompanied by an increase in oxidation state to iridium(III) (Section 1.5.4.4).

1.5.2.2 Decrease in coordination number (dissociation)

At least there is no confusion between the name used here and the organic process which is termed *elimination* and requires an increase in the number of π-bonds and a matching decrease in the number of σ-bonds. In order to conform to the philosophy of this book, it is necessary to restrict discussion of addition and dissociation to processes where the species of higher and lower coordination number can exist as long-lived chemical species under suitable conditions. Thus the gas-phase reaction:

$$Me_3BNMe_3 \xrightleftharpoons{\text{dissociation}} Me_3B + NMe_3$$

is a clear-cut example of dissociation (left to right). The species of lower coordination number are able to exist indefinitely when stored separately, though their combination to form the adduct (cf. previous section) is very fast indeed. Slow dissociations that give stable products are very uncommon. One that has been reported is the slow loss of Br^- from the five-coordinate complex $[Au(phen)(CN)_2Br]$, in which the two cyanides are *trans*, to give $[Au(phen)(CN)_2]^+$ (where they are *cis*) plus bromide.[55] The loss of bromide requires extensive rearrangement and it is this that leads to the unusually high barrier to reaction.

1.5.2.3 Ligand substitution

This involves the replacement of one ligand or group by another at a central element, with no change in oxidation state or coordination number of that central atom. Such replacement of one ligand in the coordination shell by another coming from the environment is one of the major areas to be discussed in this book. It is fairly obvious that the act of substitution requires a temporary change in the coordination number of the reaction centre and so the relationship between this section and the previous two sections is important. One of the most important questions that has to be answered when assigning a mechanism to substitution is whether the bond making precedes or follows the act of bond breaking or whether the two processes are synchronous. The development of techniques in recent years to generate species of lower coordination number very rapidly indeed and to study their subsequent recombination or consumption has caused many people to reconsider the concepts developed in the 1950s.

Some examples ranging over the subject areas of this book are:

$$R_3SiCl + OH^- \longrightarrow R_3SiOH + Cl^-$$

$$[Co(NH_3)_5(NCS)]^{2+} + H_2O \longrightarrow [Co(NH_3)_5(H_2O)]^{3+} + NCS^-$$

$$[Pt(py)_2Cl_2] + I^- \longrightarrow [Pt(py)_2ClI] + Cl^-$$

$$[W(CO)_6] + PPh_3 \longrightarrow [W(CO)_5(PPh_3)] + CO$$

There are also a number of special cases of substitution, including:

(a) ligand exchange and solvent exchange:

$$[Cr(H_2O)_6]^{3+} + H_2{}^*O \longrightarrow [Cr(H_2O)_5(H_2{}^*O)]^{3+} + H_2O$$

$$Ni(CO)_4 + {}^*CO \longrightarrow Ni(CO)_3({}^*CO) + CO$$

(b) complex formation:

$$[M(H_2O)_6]^{n+} + L \longrightarrow [M(H_2O)_5L]^{n+} + H_2O$$

(c) isomerization and racemization—when these are intermolecular (intramolecular examples come into Section 1.5.3 below);
(d) insertion reactions—when they are, for instance, methyl migration rather than true insertion (see Chapter 10.5).

All these examples involve nucleophilic substitution at the central metal or ion, since such reaction sites always have an actual or formal positive charge. Examples of electrophilic substitution can be found in coordinated ligand chemistry—again, see below (Section 1.5.5).

1.5.3 Reactions involving change in the positions of the ligands in the coordination shell

This category can be divided into two parts.

1.5.3.1 Changes in geometry

A change in geometry while retaining the same coordination number is rare, but several examples do exist and are readily detected in a number of ways. For example, the existence of allogons (isomers that differ in geometry), is well established in four-coordinate complexes of nickel(II) of the type $[NiL_2X_2]$, where L is a phosphine and X a halogen, particularly bromine. The planar species is diamagnetic and the tetrahedral allogon is paramagnetic; the systems can be readily examined by nuclear magnetic resonance (NMR) spectroscopy. Although the main focus of interest has been the study of equilibria involved and hence the relative stabilities of the two forms, NMR techniques can be applied to the kinetics of the change and hence the nature and magnitude of the barrier to rearrangement.

1.5.3.2 Intramolecular rearrangements

Reactions in which the ligands change their relative positions in the coordination shell are very common. The simplest way that this can take place is through an intramolecular rearrangement in which all the metal–ligand bonds remain intact. In the same way that ligand substitution requires the reaction centre to go through a temporary change of coordination number, an intramolecular rearrangement requires the molecule to pass through an intermediate or a transition state with a different geometry.

Stereochemical change can also take place by the intramolecular rearrangement of a sufficiently long-lived intermediate of lower or higher coordination number, often during a substitution reaction. There are cases where the geometrical consequences of the substitution process itself (the relationship between the stereochemistry of the substrate and the transition state) enforce rearrangement of the coordination shell. All these aspects will be considered in Chapter 5 under the general heading of 'Stereochemical Change'.

1.5.4 Reactions involving a change in the oxidation state

Whereas the concept of coordination number and coordination geometry is real, in the sense that these are related directly to the distribution of the nuclei and electrons in space and can be established by any of the modern techniques of structure determination, the concept of oxidation state is man-made and relates to the analysis of the complex in a way that does not necessarily represent a feasible chemical process. For example, the hypothetical reaction

$$[Co(NH_3)_4(NO_2)Cl]Cl \longrightarrow Co^{3+} + 4NH_3 + NO_2^- + 2Cl^-$$

which is used to define the oxidation state of the cobalt as +III has no reality whatsoever, but the concept of oxidation state remains useful if only as an important part of nomenclature. In the less covalently bound transition metal complexes, where a change in oxidation state does not lead to a change in coordination number, electron transfer to and from essentially non-bonding metal-based orbitals will correspond *pro rata* to a change in oxidation state. Where the bonding is more covalent and stability is governed by considerations of closed shells and 'magic numbers', the coordination number will depend upon the oxidation state. Changes of non-bonding electron pairs to bonding pairs, and vice versa, will lead formally to changes in oxidation state—reactions leading to these changes should also come under the same heading. Reactions leading formally to changes in oxidation state are termed *redox reactions* (since a reduction of one reagent requires the oxidation of another). The following two types of mechanism for such reactions can be identified; they are described in detail in Chapter 9.

1.5.4.1 Intramolecular electron transfer

Here electron transfer takes place within a pre-formed species in which the oxidant and reductant are linked by sharing one or more bridging ligands through which the electron is transferred. This is termed an *inner-sphere* redox reaction. The existence of a bridge between oxidant and reductant can only be demonstrated if the bridge is transferred from one central atom to the other in the actual act of redox, or, more rarely, by direct observation of the bridged intermediate after the act of electron transfer. These observations are the requirements of a convincing demonstration and not the requirements of the mechanism itself.

1.5.4.2 Intermolecular electron transfer

Here the electron is transferred from reductant to oxidant when the two species have made their closest approach and have suitably adjusted their bonding to meet the Franck–Condon conditions for electron transfer. Such a process is generally referred to as an *outer-sphere* redox process. This is the least ambiguous redox process, which takes place with complete retention of the coordination shell of both the oxidant and the reductant. However, it is only easy to establish its operation when the substitutional lability of the reagents is much less than the redox lability. In a classical example such as

$$[Fe^{II}(CN)_6]^{4-} + [Fe^{III}(phen)_3]^{3+} \longrightarrow [Fe^{III}(CN)_6]^{3-} + [Fe^{II}(phen)_3]^{2+}$$

it can be shown by suitable labelling that the coordination shells of the reagents are retained in the products.

1.5.4.2 *Bonding ↔ non-bonding changes*

When the change in the oxidation state is the result of changing the distribution of the electrons in the closed shell between bonding and non-bonding orbitals, problems arise in providing a unique classification of the reaction. Thus, for each of the steps in the reduction of chlorate(V) by sulphite {sulphate(IV)}, one oxygen is transferred from chlorine to sulphur:

$$3[S^{IV}O_3)]^{2-} + [Cl^{V}\,{}^{18}O_3]^{-} \longrightarrow 3[S^{VI}O_3\,{}^{18}O]^{2-} + Cl^{-}$$

so that each step is a two-electron inner-sphere redox process in which the bridging ligand, oxygen, is transferred:

$$[O_3S{:}]^{2-} + [{}^{18}O{-}Cl^{18}O_2]^{-} \longrightarrow [O_3S{:}O^{18}{-}Cl\,{}^{18}O_2]^{3-}$$
$$\longrightarrow [O_3SO^{18}]^{2-} + [Cl^{18}O_2]^{-}$$

This is also, clearly, nucleophilic attack by sulphite on the oxygen of the chlorate ion, leading to the displacement of chlorite {chlorate(III)} (in the subsequent stages the leaving groups are ClO^- and Cl^-). In this process, the non-bonding electron pair on sulphur(IV) becomes bonding in sulphur(VI), while a bonding electron pair on chlorine(V) becomes non-bonding in the chlorine(III) of the leaving group. There is no paradox here; both descriptions are true. The relationship between nucleophilic substitution and electron transfer is pursued further in the next chapter (see Section 2.5).

1.5.4.4 *Oxidative addition and reductive elimination*

Oxidative addition is a combination of addition and change of oxidation state. Examples occur towards the right of the *p*-block, and with low-oxidation, state complexes towards the right of the *d*-block:

$$PCl_3 + Cl_2 = PCl_5$$

$$I^- + Cl_2 = ICl_2{}^-$$

$$[PtBr_4]^{2-} + Br_2 = [PtBr_6]^{2-}$$

$$trans\text{-}[Ir^{I}(CO)Cl(PPh_3)_2] + MeI = [Ir^{III}(CO)MeClI(PPh_3)_2]$$

In the last case, the assignment of the oxidation state of 3+ to the metal follows from the octahedral geometry, inertness, and other properties of the product complex which are more characteristic of the 3+ than of the 1+

state of iridium, though it must be said that there is something a little discomforting about seeming to split methyl iodide into two anions! Oxidative addition is a common and important reaction for complexes of, *inter alia*, rhodium(I) and iridium(I), of platinum(0) and platinum(II), and indeed of a number of square-planar and tetrahedral d^6, d^8, and d^{10} complexes in low oxidation states. For related five-coordinate substrates, the term *oxidative elimination* is more appropriate than oxidative addition, since one of the five original ligands is lost on oxidation to the octahedral product, except in the case of pentacyanocobaltate(II):

$$[Co(CN)_5]^{3-} + \tfrac{1}{2}H_2 = [Co(CN)_5H]^{3-}$$

The converse of oxidative addition, namely reductive elimination, is a combination of dissociation and change in oxidation state, e.g.

$$Me_2SI_2 \longrightarrow Me_2S + I_2$$

$$[PtCl_6]^{2-} \longrightarrow [PtCl_4]^{2-} + Cl_2$$

$$[RhHEtCl(PPh_3)_3] \longrightarrow [RhCl(PPh_3)_3] + C_2H_6$$

This latter type of reaction very often follows oxidative addition in a catalytic cycle (see Chapter 10).

1.5.5 Reactions of coordinated ligands

This category comprises the most heterogeneous collection of phenomena but, in terms of its importance to the world at large, it is probably the most useful way in which we can employ our knowledge of reaction mechanisms. Most coordination chemists concentrate their attention on the metal ions at the centres of their complexes, but substitution and redox reactions can also occur at ligands coordinated to the metal centre. Often, but by no means always, the ligands are organic species and the interest lies in the way in which organic reactions can be modified when the substrate is coordinated to a metal ion. All reactions which either do not occur or are grossly inefficient in the absence of a suitable centre for coordination can be included in this category.

A systematic and detailed treatment of these reactions would merit at least one, and probably two, books dedicated to the subject[56] and is therefore beyond the scope of this one. Nevertheless, it is useful to indicate some of the types of reaction which individually, or in combination, lead to the processes, which range from the use of organometallic compounds as specific catalysts to the highly selective biochemical reactions of metallo-enzymes. Selected aspects, such as template reactions[57] and oxidations of coordinated ligands,[58] will be treated in later chapters, especially in Sections 10.2 and 10.7.

1.6 Chain, oscillating, and clock reactions[59]

These types of reaction are relatively rare in inorganic solution chemistry, though clock and oscillating reactions are occasionally encountered in chemical entertainments. Clock reactions, where visible change in a solution follows a well-defined induction period, are a special case of oscillating reactions. The best-known clock reaction is Landolt's, of iodic acid with sulphite, with added starch to give a sudden appearance of a deep blue coloration after a concentration-dependent induction period.

1.6.1 Chain reactions

These involve the participation of reactive intermediates, generally radicals, which cycle repeatedly. There is a sequence involving an initiation step, chain propagation, and, eventually, termination. A typical example is provided by the classic case of the decomposition of acetaldehyde. Here the initiation step is thermally activated dissociation to give a pair of radicals:

$$CH_3CHO \longrightarrow \cdot CH_3 + \cdot CHO$$

Propagation is through reaction of methyl radicals with acetaldehyde and subsequent regeneration of methyl radicals:

$$CH_3CHO + \cdot CH_3 \longrightarrow CH_3CO\cdot + CH_4$$

$$CH_3CO\cdot \longrightarrow \cdot CH_3 + CO$$

Chain termination is by combination of a pair of methyl radicals:

$$2\cdot CH_3 \longrightarrow C_2H_6$$

Two inorganic examples, one involving oxidation of organochromium species by dioxygen, the other involving peroxodisulphate oxidation of phosphite, will be mentioned in Chapter 9.

1.6.2 Oscillating reactions[60]

Heterogeneous oscillating reactions, generally involving electrodes, have been known for the past 170 years, while a theoretical homogeneous oscillating reaction was described in 1910.[61] But the first report of a real oscillating reaction in homogeneous solution was not published until 1921.[62] That report seems to have aroused little interest, for serious studies of oscillating reactions in solution did not begin until the 1960s. The 1921 report concerned the hydrogen peroxide–iodate reaction, which has become known as the Bray–Liebhafsky reaction. The original system was not really amenable to detailed investigation, but the addition of manganese(II) and malonate (now called the Briggs–Rauscher reaction) improved the kinetics considerably. The other well-known oscillator is the bromate–malonate–

cerium(IV) system, known as the Belousov–Zhabotinsky reaction. This last reaction is probably the best understood,[63] but it should be borne in mind that mechanisms of oscillating reactions are generally very complicated. Analysis of the hydrogen peroxide/thiocyanate/copper(II) oscillator indicated that its description required 30 reactions, containing 26 variables.[64] Study and analysis of chemical oscillators are not made any easier by the fact that most of them have to be studied under conditions well away from their final thermodynamic state. It is often necessary to use continuous rather than batch operation, in other words to operate and monitor the system in continuous flow through a series of reaction vessels. The hydrogen peroxide/thiocyanate/copper(II) system just mentioned is of interest as the first non-halogen oscillator to work in batch mode.

All the early examples of oscillating reactions involved oxoanions of the halogens as their chief oxidants. More recently a number of systems, centred on sulphur oxoanions, has been described, for example hydrogen peroxide (or peroxodisulphate)/thiosulphate/copper(II)[65] and peroxodisulphate/sulphide/silver(I).[66] Other halogen-free examples include the hydrogen peroxide/thiocyanate/copper(II) system mentioned in the preceding paragraph, and permanganate/ninhydrin[67] (ninhydrin is triketohydrindene hydrate, **3**). It has proved very difficult to identify an oscillating reaction based on nitrogen or phosphorus oxoanions.

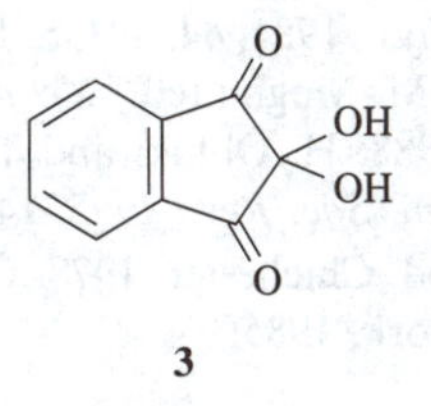

3

As stated at the start of this section, clock reactions are a special case of oscillating reactions, indeed may be regarded as minimalist oscillators. A clock reaction is either an oscillating reaction with such a long time-scale that one only registers the end of the first induction period, or a heavily damped oscillator where only the first cycle is significant (Fig. 1.6). It is possible, though eccentric, to regard ordinary reactions as even more minimalist—infinitely damped—oscillators. Landolt's original clock reaction oscillates, though it oscillates better in the modified form of iodate/malonate/hydrogen peroxide/manganese(II) (plus starch to make the oscillations visible).[68] Such systems as bromate/chlorite/iodide show an even wider range of dynamic behaviour, though the chlorite/iodide mechanism can be analysed in terms of a relatively small number of reactions (between eight and thirteen, depending on author!).[69]

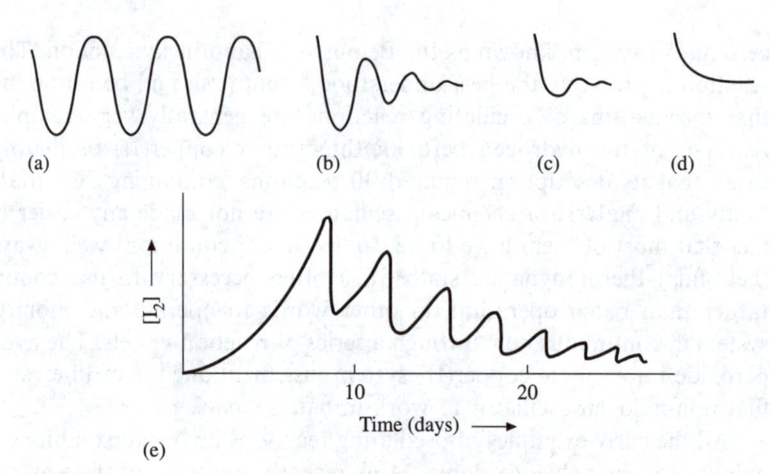

Fig. 1.6 Reactions exhibiting periodicity: sketches (a) to (d) show how increasingly heavy damping converts an oscillating reaction into a single exponential one-direction reaction, while plot (e) shows the variation of iodine concentration with time for the hydrogen peroxide/hydrogen iodide reaction.

References

1. J. E. Enderby and G. W. Nielson, *Rep. Progr. Phys.*, 1981, **44**, 593; S. I. Ishiguro and H. Ohtaki, *J. Coord. Chem.*, 1987, **15**, 237; M. Magini (ed.) *X-Ray Diffraction of Ions in Solution*, CRC Press, Florida, 1988; H. Ohtaki and T. Radnai, *Chem. Rev.*, 1993, **93**, 1157; J. E. Enderby, *Chem. Soc. Rev.*, 1995, **24**, 159.
2. J. Burgess, *Metal Ions in Solution*, Ellis Horwood, Chichester, 1978, Chapters 7 and 8; Y. Marcus, *Ion Solvation*, Wiley, New York, 1985.
3. D. F. Smith, *J. Am. Chem. Soc.*, 1927, **49**, 43.
4. L. Wilhelmy, *Pogg. Ann.*, 1850, **81**, 413.
5. L. J. Thénard, *Ann. Chim. Phys.*, 1818, **9**, 314.
6. A. V. Harcourt and W. Esson, *Phil. Trans.*, 1866, **156**, 193; 1867, **157**, 117.
7. A. V. Harcourt and W. Esson, *Proc. Roy. Soc. London*, 1865, **14**, 470.
8. M. H. Back and K. J. Laidler (eds), *Selected Readings in Chemical Kinetics*, Pergamon, Oxford, 1967, p. 3.
9. K. J. Laidler, *Chemical Kinetics*, 3rd edn, HarperCollins, New York, 1987—see the bibliography at the end of Chapter 1.
10. C. K. Ingold, *Structure and Mechanism in Organic Chemistry*, Bell, London, and Cornell University Press, Ithaca, NY, 1953.
11. For example, H. B. Charman, E. D. Hughes, C. K. Ingold, and H. C. Volger, *J. Chem. Soc.*, 1961, 1142; E. D. Hughes and H. C. Volger, *J. Chem. Soc.*, 1961, 2359; C. K. Ingold, *Helv. Chim. Acta*, 1964, **47**, 1191.
12. For example, D. D. Brown, C. K. Ingold, and R. S. Nyholm, *J. Chem. Soc.*, 1953, 2674; D. D. Brown and C. K. Ingold, *J. Chem. Soc.*, 1953, 2680; C. K. Ingold, R. S. Nyholm, and M. L. Tobe, *J. Chem. Soc.*, 1956, 1691; C. K. Ingold, R. S. Nyholm, and M. L. Tobe, *Nature*, 1960, **187**, 477; C. K. Ingold, R. S. Nyholm, and M. L. Tobe, *Nature*, 1962, **194**, 344; S. Asperger and C. K. Ingold,

J. Chem. Soc., 1956, 2862; B. Bosnich, C. K. Ingold, and M. L. Tobe, *J. Chem. Soc.*, 1965, 4074.

13. H. Taube, H. Myers, and R. L. Rich, *J. Am. Chem. Soc.*, 1953, **75**, 4118; H. Taube and H. Myers, *J. Am. Chem. Soc.*, 1954, **76**, 2103.
14. H. Taube, *Chem. Rev.*, 1952, **50**, 69.
15. T. A. Turney, *Oxidation Mechanisms*, Butterworths, London, 1965; D. A. House, *Chem. Rev.*, 1962, **62**, 185.
16. F. Basolo and R. G. Pearson, *Mechanisms of Inorganic Reactions*, Wiley, New York, 1958, 1967.
17. E. F. Caldin, *Fast Reactions in Solution*, Blackwell, Oxford, 1964; D. N. Hague, *Fast Reactions*, Wiley–Interscience, New York, 1971.
18. H. Hartridge and F. J. W. Roughton, *Proc. Roy. Soc. A*, 1923, **104**, 376; *Proc. Roy. Soc. B*, 1923, **94**, 336.
19. *Disc. Faraday Soc.*, 1954, **17**, 114–234.
20. *Z. Elektrochem.*, 1960, **64**, 1–204.
21. M. Goez, *Ber. Bunsenges. Phys. Chem.*, 1987, **91**, 616.
22. G. C. M. Bourke and R. C. Thompson, *Inorg. Chem.*, 1987, **26**, 903.
23. M. T. Nemeth, K. D. Vogelman, T. Y. Ridley, and D. W. Margerum, *Analyt. Chem.*, 1987, **59**, 283; Yi Lai Wang, J. C. Nagy, and D. W. Margerum, *J. Am. Chem. Soc.*, 1989, **111**, 7838.
24. Review: K. S. Peters, *Angew. Chem. Int. Edn. Engl.*, 1994, **33**, 294.
25. Review: P. J. Baxter and G. D. Christian, *Accounts Chem. Res.*, 1996, **29**, 515.
26. Most recently in R. G. Wilkins, *Kinetics and Mechanism of Reactions of Transition Metal Complexes*, 2nd edn, VCH, Weinheim, 1991.
27. J. J. Hood, *Phil. Mag., Series 5*, 1878, **6**, 371; 1885, **20**, 323.
28. S. Arrhenius, *Z. Phys. Chem.*, 1889, **4**, 226; K. J. Laidler, *Accounts Chem. Res.*, 1995, **28**, 187.
29. W. J. Le Noble, *Progr. Phys. Org. Chem.*, 1967, **5**, 207.
30. N. S. Isaacs, *Liquid Phase High Pressure Chemistry*, Wiley, New York, 1981.
31. R. van Eldik (ed.), *Inorganic High Pressure Chemistry: Kinetics and Mechanisms*, Elsevier, Amsterdam, 1986; R. van Eldik and J. Jonas (eds), *High Pressure Chemistry and Biochemistry (NATO ASI Series C, vol. 197)*, Riedel, Dordrecht, 1987; R. van Eldik, T. Asano, and W. J. Le Noble, *Chem. Rev.*, 1989, **89**, 549; R. van Eldik, *Pure Appl. Chem.*, 1992, **64**, 1439; R. van Eldik and A. E. Merbach, *Comments Inorg. Chem.*, 1992, **12**, 341.
32. T. Asano and W. J. Le Noble, *Chem. Rev.*, 1978, **78**, 407 (see entry 517 in Table II).
33. J. Sauer and R. Sustmann, *Angew. Chem. Int. Edn. Engl.*, 1980, **19**, 779 (Table 20).
34. N. S. Isaacs and K. Javaid, *Tetrahedron Lett.*, 1977, 3073.
35. See, e.g., W. von E. Doering, Yi-qun Shi, and Da-chuan Zhao, *J. Am. Chem. Soc.*, 1992, **114**, 10763.
36. E. Krausz and J. Ferguson, *Progr. Inorg. Chem.*, 1989, **37**, 293.
37. V. Balzani and V. Carassiti, *Photochemistry of Coordination Compounds*, Academic Press, New York, 1970; A. W. Adamson and P. Fleischauer (eds), *Concepts of Inorganic Photochemistry*, Wiley, New York, 1975; H. Yersin and A. Vogler (eds), *Photochemistry and Photophysics of Coordination Compounds*, Springer, Berlin, 1987; O. Horváth and K. L. Stevenson, *Charge Transfer Photochemistry of Coordination Compounds*, VCH, Weinheim, 1993.
38. G. J. Ferraudi, *Elements of Inorganic Photochemistry*, Wiley, New York, 1988.

39. R. B. Jordan, *Reaction Mechanisms of Inorganic and Organometallic Systems*, OUP, Oxford, 1991, Chapter 7.
40. C. Kutal and A. W. Adamson, *Comprehensive Coordination Chemistry*, Vol. 1, eds G. Wilkinson, R. D. Gillard, and J. A. McCleverty, Pergamon, Oxford, 1987, Chapter 7.3.
41. A. W. Adamson, W. L. Waltz, E. Zinato, D. W. Watts, P. D. Fleischauer, and R. D. Lindholm, *Chem. Rev.*, 1968, **68**, 541; B. R. Hollebone, C. H. Langford, and N. Serpone, *Coord. Chem. Rev.*, 1981, **39**, 181; J. Sykora and J. Sima, *Coord. Chem. Rev.*, 1990, **107**, 1.
42. E. A. Koerner von Gustorf, L. H. G. Linders, I. Fischler, and R. N. Perutz, *Adv. Inorg. Chem. Radiochem.*, 1976, **19**, 65; M. S. Wrighton, *Top. Curr. Chem.*, 1976, **65**, 37.
43. J. F. Endicott, K. Kumar, T. Ramasami, and F. P. Rotzinger, *Progr. Inorg. Chem.*, 1983, **30**, 141.
44. G. A. Crosby, *Accounts Chem. Res.*, 1975, **8**, 231.
45. H. Hennig, D. Rehorek, and R. D. Archer, *Coord. Chem. Rev.*, 1985, **61**, 1.
46. W. Boese, K. McFarlane, B. Lee, J. Rabor, and P. C. Ford, *Coord. Chem. Rev.*, 1997, **159**, 135.
47. G. Stochel and R. van Eldik, *Coord. Chem. Rev.*, 1997, **159**, 153.
48. *Proceedings of the 11th International Symposium on Photochemistry and Photophysics, Kraków, 1995*, eds Z. Stasicka and G. Stochel; *Coord. Chem. Rev.*, 1997, **159**.
49. A. Henglein, *Ultrasonics*, 1986, **25**, 6.
50. For example, M. J. Begley, S. G. Puntambekar, and A. H. Wright, *J. Chem. Soc., Chem. Commun.*, 1987, 1251.
51. C. M. Krishna, Y. Lion, T. Kondo, and P. Riesz, *J. Phys. Chem.*, 1987, **91**, 5847.
52. M. Gutiérrez, A. Henglein, and J. K. Dohrmann, *J. Phys. Chem.*, 1987, **91**, 6687.
53. E. J. Hart and A. Henglein, *J. Phys. Chem.*, 1987, **91**, 3654.
54. R. van Eldik and C. D. Hubbard, *New J. Chem.*, 1997, **21**, 825.
55. L. Cattalini, G. Marangoni, G. Paolucci, B. Pitteri, and M. L. Tobe, *Inorg. Chem.*, 1987, **26**, 2450.
56. See, e.g., R. P. Houghton, *Metal Complexes in Organic Chemistry*, Cambridge University Press, Cambridge, 1979; E. C. Constable, *Metals and Ligand Reactivity*, Ellis Horwood, Chichester, 1990; E. C. Constable, *Metals and Ligand Reactivity: An Introduction to the Organic Chemistry of Metal Complexes*, VCH, Weinheim, 1996.
57. L. F. Lindoy, *The Chemistry of Macrocyclic Ligand Complexes*, Cambridge University Press, Cambridge, 1989.
58. O. Mønsted and G. Nord, *Adv. Inorg. Chem.*, 1991, **37**, 381; O. Vollárová and J. Benko, *Current Topics in Solution Chemistry*, 1994, **1**, 107.
59. J. H. Espenson, *Chemical Kinetics and Reaction Mechanisms*, 2nd edn, McGraw-Hill, New York, 1995, Chapter 8.
60. D. O. Cooke, *Inorganic Reaction Mechanisms*, The Chemical Society, London, 1979, Chapter 7; S. K. Scott, *Oscillations, Waves, and Chaos in Chemical Kinetics*, OUP, Oxford, 1994.
61. A. Lotka, *J. Phys. Chem.*, 1910, **14**, 271.
62. W. C. Bray, *J. Am. Chem. Soc.*, 1921, **43**, 1262.
63. Review of mechanisms of inorganic oscillating reactions : R. M. Noyes and R. J. Field, *Accounts Chem. Res.*, 1977, **10**, 273.

64. Yin-Luo, M. Orbán, K. Kustin, and I. R. Epstein, *J. Am. Chem. Soc.*, 1989, **111**, 4541.
65. M. Orbán and I. R. Epstein, *J. Am. Chem. Soc.*, 1987, **109**, 101; M. Orbán and I. R. Epstein, *J. Am. Chem. Soc.*, 1989, **111**, 2891.
66. Q. Ouyang and P. De Kepper, *J. Phys. Chem.*, 1987, **91**, 6040.
67. L. Treindl and A. Nagy, *Chem. Phys. Lett.*, 1987, **138**, 327.
68. T. S. Briggs and W. C. Rauscher, *J. Chem. Ed.*, 1973, **50**, 496.
69. M. T. Beck and G. Rábai, *J. Phys. Chem.*, 1986, **90**, 2204; O. Citri and I. R. Epstein, *J. Phys. Chem.*, 1987, **91**, 6034.

2 Substitution reactions: general considerations

Chapter contents

2.1 Bond making and bond breaking

A process in which a ligand in the coordination shell is replaced by another from the environment is called a *substitution reaction.*[1] If there is no temporary change in the oxidation state of the reaction centre (frequently seen as a change in the number of non-bonding electrons) the process can be called a *simple substitution reaction.* There are many reactions of great importance in which the reaction centre passes through different oxidation states as a part of an overall substitution process; these will be discussed in Chapter 10.

In the act of simple substitution a bond is made and a bond is broken, so the reaction centre must undergo a temporary change in coordination number. There are two important parts to the description of a single act of ligand substitution. The first concerns the changes in the electron distribution that occur as the acts of bond making and bond breaking take place. We shall see shortly that, to some extent, this depends more on the nature of the entering group and the leaving group than it does on the nature of the reaction centre. The second, which will be discussed in the next section, is concerned with the timing of the bond-making and bond-breaking processes and is probably the cause of more semantic arguments than any other aspect of mechanism.

In the seminal work of Ingold and Hughes,[2] substitution reactions at carbon were classified in terms of the mode of the breaking of the bond between the reaction centre and the leaving group, which could be *homolytic* or *heterolytic* and, if the latter, further subdivided according to whether the electrons of the bond remain with the reaction centre or depart with the leaving group.

$$\mathrm{M{:}X \longrightarrow M\cdot + \cdot X} \qquad \text{HOMOLYSIS}$$

$$\mathrm{M{:}X \longrightarrow M^+ + X^-} \qquad \text{NUCLEOPHILIC HETEROLYSIS}$$

$$\mathrm{M{:}X \longrightarrow M{:}^- + X^+} \qquad \text{ELECTROPHILIC HETEROLYSIS}$$

Such processes are termed *electrophilic* and *nucleophilic*, respectively. The seemingly unexplained shift of the nomenclature from the leaving process arises because the terms *electrophilic/nucleophilic substitution* are actually condensed versions of the full nomenclature *substitution by electrophilic/nucleophilic reagents*. The nomenclature arising from this classification, e.g. S_N1, S_N2, S_E1, S_E2, S_H1, S_H2 (the molecularity of the process, see below, is also indicated) has become part of the vocabulary of every young student of organic chemistry, be they sixth-former or first-year undergraduate, and remains, despite recent controversial attempts to replace it,[3] part of the basic framework of organic chemistry. These labels, while providing a convenient but rough description of reagents and reaction types, are insufficient to indicate the detailed information that reaction mechanists are now capable of harvesting from their studies.

While working well for substitution reactions at carbon in a climate where the Lewis concept of electron-pair bonds and the valence-bond view of bonding dominated, such a method of classification presented problems when extended to reactions of inorganic interest. Heterolytic processes are essentially Lewis acid–base reactions with the reaction centre a Lewis base for electrophilic processes and a Lewis acid when they are nucleophilic. In the latter case, the change in the number of nucleophiles (or ligands) about the reaction centre has no effect upon the number of non-bonding electrons and therefore, irrespective of the molecularity of the process, does not lead to any temporary change in the oxidation state of the reaction centre. Such reactions therefore fit readily into the category of simple substitution. The attachment of an electrophile, on the other hand, brings two previously non-bonding electrons into bonding and is equivalent to a two-electron oxidation. If a ligand then leaves as a Lewis acid the electrophilic substitution is complete, but it is also possible to stabilize the higher oxidation state intermediate by the attachment of a Lewis base (ligand) from the environment. Such processes can be synchronous and the Lewis acid and base fragments might be part of the same molecule. They all fall into the category of *oxidative addition*, which is discussed fully in a later chapter (Chapter 10). A dissociative electrophilic process, causing a bonding pair of electrons to become non-bonding, is thus a two-electron reduction. The gain of an electron-pair acceptor restores the oxidation state and completes the act of substitution, whereas a synchronous or subsequent loss of a Lewis base leads to *reductive elimination*, also discussed in Chapter 10. Such behaviour patterns are summarized in Fig. 2.1.

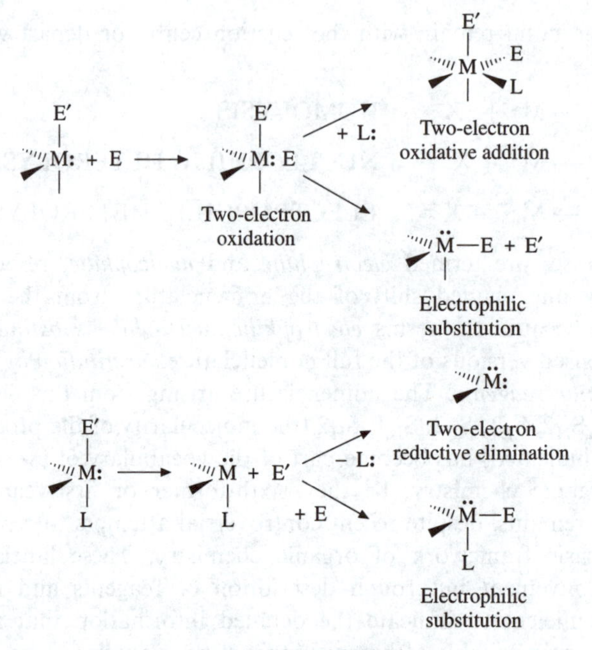

Fig. 2.1 Bond making and bond breaking.

Similar schemes can be written for the various homolytic processes, which can be formulated as one-electron redox processes that can be reversed by the appropriate loss or gain of an odd-electron species. In electrophilic and homolytic processes the legalistic definition of simple substitution is not obeyed and such processes are best considered elsewhere. This section will therefore be restricted to nucleophilic substitution, which, in view of the Lewis-acid character of the metal ion, dominates the substitution reactions of the classic coordination complexes of metal ions and only starts to let us down when the metal–ligand bonding becomes covalent, as is the case in organometallic chemistry.

2.2 The molecularity (or timing) of the substitution process

Ingold defined the molecularity of a reaction stage as 'the number of molecules necessarily undergoing covalency change' and made it clear that any looser definition, such as 'the number of molecules involved in the process', brings too much prominence to the solvation changes involved in going from the ground state to the transition state. Since, with this definition, the molecularity is essentially a description of the transition state, the major investigative probe is *kinetics*. However, although the rate law of a single-stage reaction may allow us to compare the composition of the transition state with that of the reagents' ground state, if the reaction is multistage or if we are not aware of the nature of the reagents' ground state, then the interpretation of the kinetic rate law may be ambiguous. For example, a very common mechanism for ligand substitution involves a rapid and reversible pre-equilibrium association of the reagents followed by a slow rearrangement of the aggregate:

$$\mathrm{AX} + \mathrm{Y} \underset{K}{\overset{\text{fast}}{\rightleftharpoons}} \{\mathrm{A{-}X{-}{-}{-}Y}\} \xrightarrow[k]{\text{slow}} \text{products}$$

This situation leads to the rate law:

$$-\mathrm{d}[\text{substrate}]/\mathrm{d}t = \frac{kK[\text{substrate}][\mathrm{Y}]}{1 + K[\mathrm{Y}]}$$

This is true irrespective of the nature of the interaction that holds the aggregate together (ion-association, solvation, coordination), the mechanism of the rearrangement of the aggregate, or indeed the nature of the reaction of the aggregate. When Y is a catalyst, such as an enzyme, and AX is the substrate, this kinetic form is the classic Michaelis–Menten equation. If the conditions are such that $K[\mathrm{Y}] \gg 1$ over the whole range of [Y] studied, the substrate is always completely in the form of the aggregate and

$$-\mathrm{d}[\text{substrate}]/\mathrm{d}t = k[\text{substrate}]$$

In the absence of evidence to the contrary, one might have concluded that Y was not present in the rate-determining step leading to the transition state and that the process was unimolecular. One might say, quite legitimately, that, since the aggregate can be looked upon as a single entity whose rearrangement was the rate-determining step, the process was indeed unimolecular, but such a description of the act of reaction would be unacceptably incomplete. In this case the rate law provides information about the distribution of the substrate between unreactive and reactive forms and, since the composition of the reactant solution can be investigated independently by non-kinetic techniques, such as conductance or spectroscopy (UV–visible, NMR, etc.), there is no reason to make such a mistake.

The early studies of organic reaction mechanisms did not encounter these problems and there was usually a one-to-one relationship between kinetic order and molecularity—the simple labelling, e.g., S_N1, S_N2, etc. was adequate. Later, it was realized that the labels ought to contain more information and the introduction of labels such as S_N1(lim) and S_N2(lim) to indicate extreme cases made it clear that it was necessary to consider a whole spectrum of possibilities.

2.3 The Langford–Gray nomenclature

2.3.1 Descriptions and definitions

When the first detailed kinetic studies were made of substitution reactions of transition metal complexes, especially in solvents other than water, it was soon realized that kinetic order was not a good criterion of molecularity and that substitution frequently took place within a pre-formed aggregate. Langford and Gray in their excellent monograph on substitution reactions[4] attempted to produce an operational definition by introducing the concepts of a *stoichiometric* mechanism and an *intimate* mechanism.

The stoichiometric mechanism could take one of three forms.

(a) A *dissociative* process with an identifiable intermediate of lower coordination number. This was termed the *D* mechanism.
(b) An *associative* process with an identifiable intermediate of higher coordination number—the *A* mechanism.
(c) An *interchange* process, labelled *I*, in which the acts of bond making and bond breaking were either synchronous or else took place within the pre-formed aggregate.

The definition therefore depends upon the identification of an intermediate species. Such a species does not have to live long enough to be produced in sufficient amount to be isolated and characterized independently (although it can happen in certain unusual circumstances, which will be discussed in the appropriate context); nor does it need to be present in sufficient amount to be directly detectable by electronic (electron spin resonance—ESR) or NMR spectroscopy. The sensitivity of ESR to very low concentrations of suitable odd-electron species makes it an important technique for the detection of transients in the study of reactions involving free radicals, but its application in mechanistic studies of inorganic complexes is rare. An intermediate can be characterized through its influence on the kinetics, the competition characteristics, and/or the stereochemistry of the reaction. These approaches constitute such an important part of the way in which the finer details of the mechanism are elucidated that they will not be discussed in a general fashion now but will be considered in the context of the specific mechanisms discussed in later chapters.

The stoichiometric interchange mechanism was then further classified according to its *intimate* mechanism, which was an indication of its mode of activation.

(a) A *dissociatively activated* intimate mechanism, indicated by subscript $_\mathrm{d}$, has a rate-determining transition state in which there is no direct interaction between the reaction centre and the entering group, while
(b) An *associatively activated* intimate mechanism, subscript $_\mathrm{a}$, has a transition state in which there is bonding between the incoming group and the reaction centre.

The symbols I_a and I_d are thus used to denote associative and dissociative interchange, respectively, with *I* in use to indicate intermediate situations—associative and dissociative are the black and white between which many shades of grey are possible. Since interaction between the entering group and the reaction centre will exert a considerable effect on the energy of the transition state while having no effect upon the ground state,

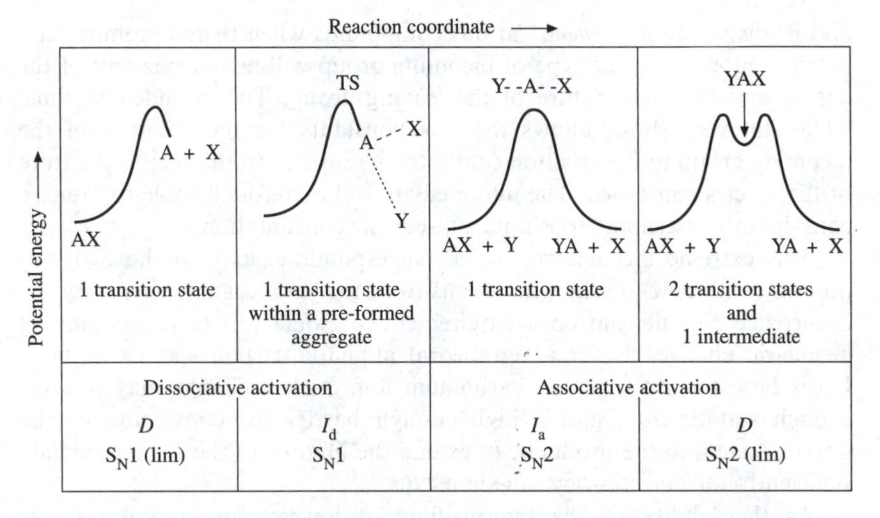

Fig. 2.2 The relationship between the mechanism of substitution, and its energy profile, in terms of the Langford–Gray and Hughes–Ingold nomenclature.

an experimental means of determining the intimate mechanism should, in theory, be available. In a dissociative mechanism, the rate of reaction should be insensitive to the nature of the entering group (except for small effects arising from solvation interactions in the ground state of any aggregate), whereas the rate of an associatively activated process should show considerable sensitivity towards the nature of the entering group. The relationships are summarized in Fig. 2.2. If one avoids being obsessed with borderlines and concentrates on the the area between them, it is possible to think of the mechanism as a spectrum of possibilities ranging between an intermediate of lower coordination number with a long enough lifetime to exert strong nucleophilic discrimination and one of higher coordination number with sufficient lifetime to do something to characterize itself, such as rearrange. For any particular reaction our task is to determine its position within this spectrum.

2.3.1.1 *Dissociative activation—the* D *mechanism*

At the dissociative extreme, we can visualize a situation where the intermediate of lower coordination number lives long enough to fully equilibrate its environment before it captures the entering group. Under these circumstances the factors that relate to the production of the intermediate are completely separated from those involved in its consumption. The rate of consumption of substrate is therefore dependent upon the nature of the leaving group and independent of the nature and the concentration of the incoming group. The stereochemistry of the product and its distribution between various compounds when there is competition between more than one type of incoming group will be independent of the charge and chemical nature of the leaving group. The extended lifetime, while still very short, allows the rate constants for the addition of the incoming group to the reaction centre to be sensitive to the chemical nature of the species concerned. The intermediate will therefore be able to exercise considerable discrimination in its subsequent consumption.

This extreme mechanism, which corresponds exactly to the S_N1(lim) process or to the *D* mechanism, seems to be rare in inorganic chemistry. Its occurrence may depend on a requirement for some sort of accompanying geometric change, as for a tetrahedral aliphatic carbon species losing a Lewis base to give a planar carbonium ion. Such a change may provide enough stabilization, and a high enough barrier for conversion of the intermediate into the products, to extend the lifetime of the *D* intermediate sufficiently for detection by kinetic means.

As the lability of the intermediate of lower coordination number increases, its ability to discriminate between potential entering groups will become less controlled by their chemical natures and more by their concentrations so that the solvent, always present in large molar excess, will begin to dominate the substitution process. The consumption of the intermediate will become more and more likely to occur while the leaving group is in close proximity to it, and the products of the reaction, and their stereochemistry, will be strongly affected by the nature of the leaving group.

The behaviour of extremely short-lived dissociative intermediates can sometimes be studied by picosecond laser flash photolysis. $Mo(CO)_5$, thus generated by photolysing $Mo(CO)_6$, can be so lacking in discrimination that it will combine with solvents such as chlorobenzene and even cyclohexane to form solvento species. The least stable species will dissociate the fastest and the final distribution of product, in accordance with the thermodynamic properties of the species concerned, is the result of a very large number of successive dissociations and captures. A great deal of emphasis has to be placed on the discriminating ability of this intermediate, which can be measured in relative terms by allowing two nucleophiles to compete for it. The extent of nucleophilic discrimination can then be taken as a measure of the relative lifetime of the intermediate. An ultra-short-life intermediate will have a very small barrier to the attachment of the incoming group, whether it be the long-term partner in a stable product or a transient visitor. If the barrier is small (comparable to kT), even for the transient visitor, its height cannot be sensitive to the nature of the entering group and hence there will be little or no discrimination.

2.3.1.2 *Dissociative activation—the* I_d *mechanism*

When discrimination is poor, the only way that a reagent, whose concentration is much less than that of the solvent, can improve its chances of competing with the solvent for the reactive intermediate is for it to be waiting in the inner solvation shell when the act of dissociation takes place. This is rather akin to the need to be in the queue at the bus stop in the rush hour in order to stand a chance of getting on the bus. It is for this reason that reaction within a pre-formed aggregate becomes so important. Once pre-association becomes necessary the mechanism shifts from D to I_d, as in the important area of formation of complexes (Chapter 7). The important point to make is that if we wish to retain the original definition of a dissociative intimate mechanism, the term I_d cannot be applied to any process in which there is direct bonding between the entering group and the reaction centre. The term 'partly dissociative' is about as meaningful as the term 'slightly pregnant'.

2.3.1.3 Associative activation—the I_a *mechanism*

Once there is need for direct bonding between the entering group and the reaction centre in the transition state the intimate mechanism becomes associative. A substitution in which bond making and bond breaking are synchronous leads to a single transition state. The change from dissociative to associative activation occurs as soon as there is direct interaction between the reaction centre and both the entering and leaving groups, but there will be a grey area in which the techniques at our disposal do not allow us to make this distinction. When the substitution is symmetrical, i.e. the entering and leaving groups are chemically identical, the Law of Microscopic Reversibility requires a symmetrical transition state for a synchronous process and so the extent of bonding with the reaction centre and the entering group must be identical to that between the reaction centre and the leaving group. When the two groups differ, the degrees of bond breaking and bond making can differ. We can consider a range of binding in the transition state from the fully open (indistinguishable from the dissociative mechanism) to the fully closed. Restrictions may be introduced by the nature of the reacting system, e.g. nucleophilic substitution at tetrahedral carbon is unlikely to allow the sum of the bond orders of the entering and leaving group to be greater than unity. All of these processes are associative interchanges, I_a.

2.3.1.4 Associative activation—the A *mechanism*

The changeover from the I_a mechanism to the A mechanism occurs when the single transition state is replaced by an intermediate and two transition states, one in which bond making is much more advanced than bond breaking (*the bond-making transition state*) and the other in which bond breaking is more advanced than bond making (*the bond-breaking transition state*). When the reaction is a symmetrical exchange, i.e. X and Y are chemically identical, the two transition states (bond making and bond breaking) are energetically degenerate and differ only in the identity of the entering and leaving groups. But when there is net chemical change, one of the two will lie at higher energy. The deeper the energy well in which the

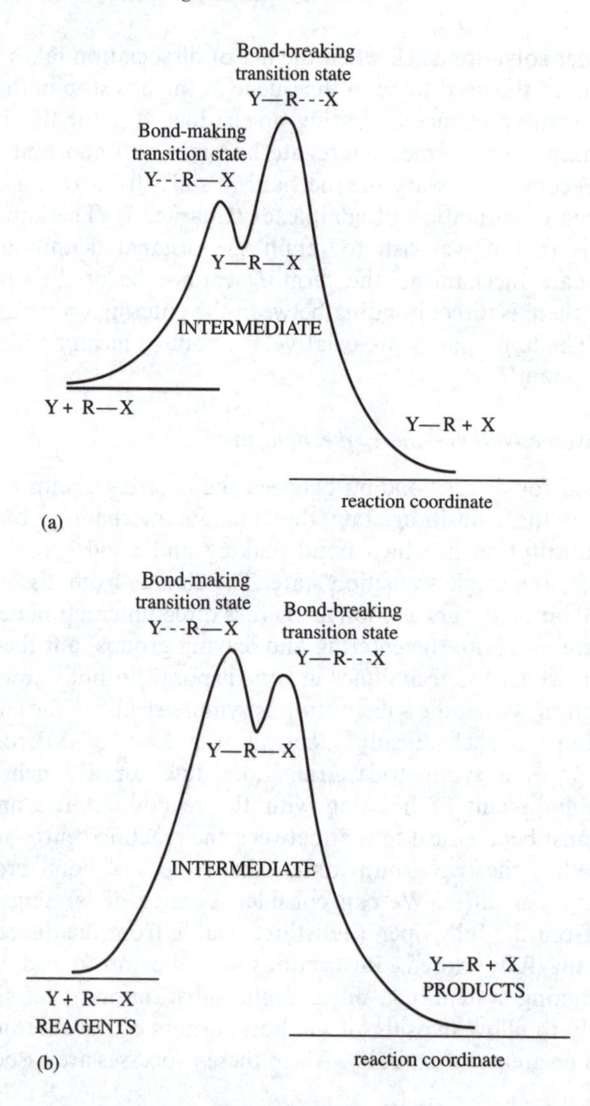

Fig. 2.3 Energy profiles for the *A* mechanism for substitution, showing the relationship between the intermediate and the bond-breaking transition states: (a) the bond-breaking transition state at higher energy; (b) the bond-making transition state at higher energy.

intermediate sits, i.e. the more stable the intermediate, the more likely are the two transition states to differ. There is no reason why one of the two should always be higher than the other and it is possible to visualize the two cases, as shown in Fig. 2.3.

Although there is a tendency to refer to them as 'bond-breaking rate-limiting' and 'bond-making rate-limiting' it should be realised that bond making plays an important part in both. A reaction that goes to completion can be written as

$$AX + Y \underset{k_{-a}}{\overset{k_a}{\rightleftharpoons}} X{-}A{-}Y \overset{k_b}{\longrightarrow} X + AY$$

If the concentration of the intermediate remains small we can use the stationary state approximation and show that

$$-d[AX]/dt = d[AY]/dt = \frac{k_a k_b}{k_{-a} + k_b}[AX][Y] = k_2[AX][Y]$$

i.e. $$k_2 = \frac{k_a k_b}{k_{-a} + k_b}$$

In case 1, $k_b \ll k_{-a}$ and so the second-order rate constant simplifies to

$$k_2 = (k_a k_b)/k_{-a}$$

so that, even though the bond-breaking transition state lies at higher energy and this is often referred to as a bond-breaking rate-limiting case, the second-order rate constant not only depends upon the breaking of the bond with the leaving group X, k_b, but is also very sensitive to the nature of Y and reflects the relative stability of the intermediate of higher coordination number (k_a/k_{-a}).

In case 2, $k_b \gg k_{-a}$ and the second-order rate constant reduces to

$$k_2 = k_a$$

This is truly a case of rate-limiting bond formation.

2.3.1 Weaknesses in the Langford–Gray nomenclature

In recent years it has become apparent that the advantages possessed by the Langford–Gray approach over the Hughes–Ingold nomenclature are more illusory than real and that, as has been indicated above, much of the middle ground remains poorly defined. But at least the range of interchange mechanisms, $I_d \rightarrow I \rightarrow I_a$, recognizes that there is a middle ground, something which is less easily accommodated within the S_N1/S_N2 system. The weak point in the Langford–Gray nomenclature is that the same word, 'interchange', has been used for the dissociative rearrangement of a pre-formed aggregate, the associative rearrangement of a pre-formed aggregate, and the synchronous substitution of an encounter complex. As a result the nomenclature has been hijacked from time to time and given new meanings. For example, it has become common practice to divide solvent exchange reactions into I_d and I_a mechanisms according to whether the volume of activation is positive or negative.

A little thought will show that the associative and dissociative interchange are not opposite sides of the same coin. The I_d mechanism, as defined above in terms of interchange within a pre-formed aggregate, represents a two-stage process consisting of the rapid and reversible formation of the aggregate and the subsequent interchange in which the group that is to enter is a bystander. The I_a mechanism is a synchronous process with a transition state in which the entering group is partially bound. The idea of an associatively activated interchange within a pre-formed aggregate which loomed large in the original concept of the I_a mechanism is not invalid, but cases where the rate of substitution obeys saturation kinetics, i.e. $K[Y]$ as defined above becomes much greater than 1 and the limiting rate constant, k, is very sensitive to the nature of Y (a requirement for associative interchange within a pre-formed aggregate), are vanishingly rare. The reaction

$$[Pt(Me_4en)(Me_2SO)X]^+ + X^- \longrightarrow [Pt(Me_4en)X_2]^+ + Me_2SO$$

where Me_4en= N,N,N',N'-tetramethyl-1,2-diaminoethane and X = Cl, Br, I, takes place readily in the non-polar solvent dichloromethane and the kinetics indicate interchange within the pre-formed ion-pair $[Pt(Me_4en)(Me_2SO)X]^+X^-$ or ion-triplet $[Pt(Me_4en)(Me_2SO)X]^+2X^-$.[5] There is surprisingly little discrimination between the three halides.

There is still an active debate about how free an intermediate of lower coordination number must become before the I_d mechanism can be considered to have changed to D.[6] Much elegant work has been (and is still being) done to resolve this question and to estimate the stage of development that the intermediate of lower coordination number reaches before it is captured by the entering nucleophile. The bulk of the work concerns the base-catalysed substitution reactions of octahedral d^6 cobalt (III)-acido-amine complexes, a system that possesses many unique features of interest and which is discussed in detail in Section 4.7. The generation of intermediates of lower coordination number by laser flash photolysis and the study of their subsequent reactions have already been briefly mentioned (Section 2.3.1.1). It is probably as a result of the systematic study of volumes of activation in systems where electrostriction effects (changes in solvation caused by changes in charge distribution) are negligible that the apparently lucid distinctions between D, I_d, I_a and A have been shown to be illusory. Of all the activation parameters, $\Delta V^\ddagger$ can probably be interpreted with the least ambiguity. Claims that it is also the easiest to determine precisely are also probably true in the sense that, unlike $\Delta S^\ddagger$ and $\Delta H^\ddagger$, it is an independent variable and does not rely on a long extrapolation for its determination. On the other hand, even with a range of pressure as large as 2000 atmospheres, the change in the magnitude of the rate constant is usually far less than that found for a change of 30 °C, the minimum range of temperature that a serious kineticist would use to determine $\Delta H^\ddagger$ and $\Delta S^\ddagger$. The $\Delta V^\ddagger$ values for symmetrical solvent-exchange processes of the type

$$[M(solv)_N]^{n+} + {}^*solv \longrightarrow [M(solv)_{(N-1)}({}^*solv)]^{n+} + solv$$

have been studied at great length by Merbach and others and are discussed in more detail in Chapter 7. For a liquid of such open structure as water, it can be assumed that the act of coordination is equivalent to a considerable compression. A calculation based on the molar volume of pure water and the molar volume of coordinated water derived from crystallographic data for aqua-cations forecasts a decrease in volume of about -10 cm^3 mol^{-1}. Consequently, a fully dissociative process should have a $\Delta V^\ddagger$ value of $+10$ cm^3 mol^{-1}, while one in which the bond with the incoming water is fully formed in the transition state should have a $\Delta V^\ddagger$ value of -10 cm^3 mol^{-1}. Of course, one would expect the other bound ligands to change their positions to mitigate this effect. Experimental values for $\Delta V^\ddagger$ for water exchange with $[M(H_2O)_6]^{n+}$ ($n = 2, 3$) are generally smaller in magnitude than these limits and vary from reaction centre to reaction centre (see Chapter 7 for details). It has been suggested that, depending mainly upon the size and coordination number of the reaction centre, there is a spectrum of modes of activation ranging from fully associative to fully dissociative. But this, unfortunately, merely gives new meanings to old words and adds little to our understanding of the situation.

Ligand- and solvent-exchange processes are symmetrical and therefore the entering and leaving groups must be bound to the metal to the same extent in the transition state of a synchronous process or in the intermediate. This is shown in Fig. 2.4, where the possibilities for a synchronous process are set out.[7] The diagram indicates where the *D* and the *A* mechanisms would fit into this scheme. All the pathways shown may be treated as associatively activated; they differ in the extent to which bond formation has developed in the transition state. This is represented by the diagonal running from bottom left to top right. It should be clear from Fig. 2.4 that associatively activated processes can have positive volumes of activation as well as negative ones—all they tell us is the degree of openness or compactness of the transition state. There is no justification for the postulation that $\Delta V^\ddagger = 0$ should be taken as the dividing line between I_d and I_a mechanisms; $\Delta V^\ddagger = 0$ simply tells us that, in a symmetrical transition state, the incoming and leaving groups share a single bond. The application of these principles to water exchange at metal cations is taken up in Chapter 7 (see Section 7.4.1).

2.4 Coordination number and substitution mechanism

One might imagine that there should be some connection between coordination number and substitution mechanism, in that the more atoms or ligands there are around a central atom, the harder it might be to bind a further group. Such reasoning would suggest a favouring of dissociative substitution at centres with higher coordination numbers, associative at those with lower coordination numbers. But of course ground-state geometry is already determined by a combination of steric and electronic factors, both of the central atom and of its ligands, so its favoured change on going to the

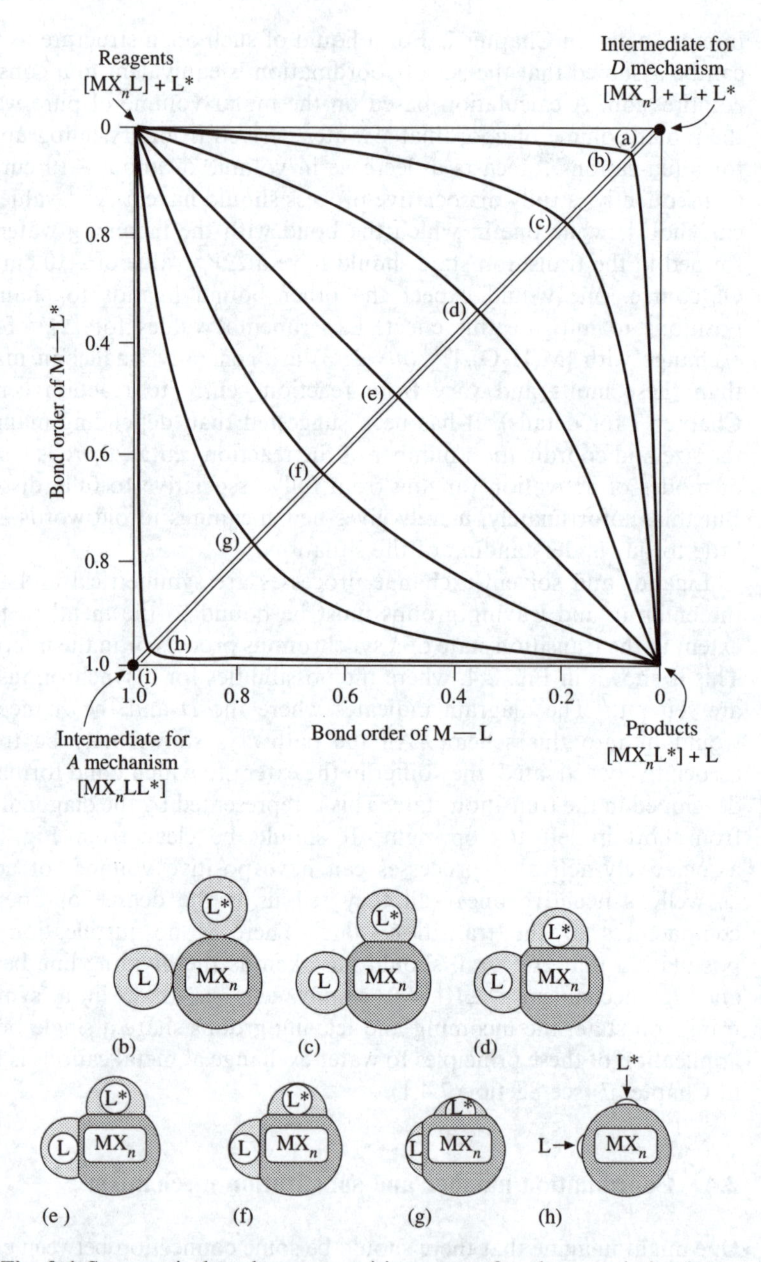

Fig. 2.4 Symmetrical exchange transition states for the associative interchange, $MX_NL + L^* \rightleftharpoons MX_NL^* + L$, showing the smooth decrease from a transition state with the equivalent of N bonds in the fully dissociative mechanism (pathway through top right-hand corner) to the fully associative mechanism (pathway through the bottom left-hand corner) with the equivalent of N + 2 full bonds in the transition state. The transition states are depicted diagrammatically in the lower part of the figure; the change in the value of $\Delta V^{\ddagger}$ as the transition state becomes less open can be seen.

transition state will also be determined by the resultant of a number of factors. So we should not be surprised that tetrahedral *sp*-block elements undergo substitution by S_N1 and by S_N2 (I_d and I_a) pathways, with the balance changing as one descends a Periodic Table group, or that the same range ($I_a \rightarrow I \rightarrow I_d$) is exhibited as one traverses the first-row *d*-block M^{2+} and M^{3+} centres. The small differences in energy between various geometries available to seven- and eight-coordinate species permit *f*-block elements to react variously by associative and dissociative paths. The sole area where there appears to be a considerable degree of consistency is that of square-planar *d*-block complexes, though even here the establishment of a few examples of undoubtedly dissociative substitution (Section 3.3.11) has generated a significant number of exceptions to the general rule of associative substitution at these centres.

2.5 Single-electron transfer mechanisms

In recent years an enormous amount of interest has developed in organic reactions that appear from their stoichiometry to be nucleophilic substitutions but which, on closer examination, can be shown to involve two single-electron transfers, as set out in Fig. 2.5. Indeed, it has been suggested that, since electrons usually transfer one at a time, all heterolytic processes should

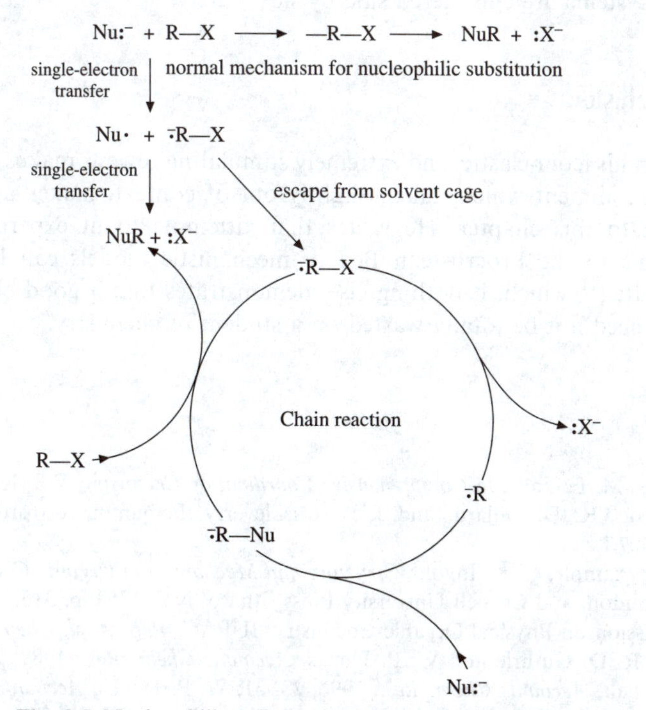

Fig. 2.5 Nucleophilic substitution via single-electron transfer.

be considered in this light. When the substrate and the nucleophile come together in the encounter complex, a single-electron transfer from the nucleophile to the substrate would lead to the formation of a trapped radical pair. If the radicals then recombined rapidly in the cage with the expulsion of X^-, the kinetics and products would be the same as those of a classical nucleophilic substitution. Attempts have been made to distinguish between the two concepts by examining the extent to which the rate depends upon the Lewis basicity (a measure of the ability to act as an electron-pair donor) or the redox potential (a measure of its ability to donate a single electron) of the 'nucleophile'. If the radicals can escape from the solvent cage the distinction can be made clearly and the process takes on the characteristics of a free-radical chain reaction. This mechanism has been labelled $S_{RN}1$[8] and was first observed in the reaction between 4-nitrobenzyl chloride and the anion of 2-nitropropane. This reaction gave a 92% yield of *C*-alkylated product, whereas with other 4-substituted benzylic chlorides the expected *O*-alkylated product was obtained exclusively.[9]

Reactions of this sort have received considerable attention in recent years from organic chemists, but the general concept of electron-transfer catalysis has been well known in the reactions of transition metal complexes since the early 1950s. Redox-catalysed substitution will be mentioned again in Chapter 4 (Section 4.7.2.1), but the attention of the reader may here be drawn to two provocative articles in which the organic, organometallic, and inorganic systems are considered side by side.[10]

2.6 Conclusion

Swaddle, in his iconoclastic and extremely stimulating review, makes a most perceptive comment which, taken slightly out of context, makes a fitting conclusion to this chapter. He writes that 'attempts to fit experimental observations to the Procrustean Bed of mechanistic models can lead to sterile debate';[11] which, if nothing else, demonstrates that a good classical education need not be totally wasted on a student of chemistry.

References

1. Reviews: M. L. Tobe, in *Comprehensive Coordination Chemistry*, Vol. 1, eds. G. Wilkinson, R. D. Gillard, and J. A. McCleverty, Pergamon, Oxford, 1987, Chapter 7.1.
2. See, for example, C. K. Ingold, *Structure and Mechanism in Organic Chemistry*, Bell, London, and Cornell University Press, Ithaca, NY, 1953, p. 315.
3. Commission on Physical Organic Chemistry, IUPAC, *Pure Appl. Chem.*, 1989, **61**, 23; R. D. Guthrie and W. P. Jencks, *Accounts Chem. Res.*, 1989, **22**, 343; G. A. Olah, *Accounts Chem. Res.*, 1990, **23**, 31; W. P. Jencks, *Accounts Chem. Res.*, 1990, **23**, 32; R. D. Guthrie, *Accounts Chem. Res.*, 1990, **23**, 33.

4. C. H. Langford and H. B. Gray, *Ligand Substitution Processes*, Benjamin, New York, 1965.
5. G. Alibrandi, R. Romeo, L. M. Scolaro, and M. L. Tobe, *Inorg. Chem.*, 1992, **31**, 5061.
6. P. Lay, *Comments Inorg. Chem.*, 1991, **11**, 235.
7. R. A. More O'Ferrall, *J. Chem. Soc. (B)*, 1970, 274; D. A. Jencks and W. P. Jencks, *J. Am. Chem. Soc.*, 1977, **99**, 7948; T. W. Swaddle, *Adv. Inorg. Bioinorg. Chem.*, 1983, **2**, 95 (see pp. 108–113).
8. J. F. Bunnett, *Accounts Chem. Res.*, 1978, **11**, 413.
9. H. B. Bass and M. L. Bender, *J. Am. Chem. Soc.*, 1949, **71**, 3842.
10. M. Chanon and M. L. Tobe, *Angew. Chem. Int. Edn. Engl.*, 1982, **21**, 1; M. Chanon, *Accounts Chem. Res.*, 1987, **20**, 214.
11. T. W. Swaddle, *Adv. Inorg. Bioinorg. Chem.*, 1983, **2**, 95 (specifically p. 97).

3 Substitution at two-, three-, four-, and five-coordinate reaction centres

Chapter contents

3.1 Introduction

Tetrahedral stereochemistry is one of the most common geometries, square-planar structures are very important for d^8 centres, five-coordination is relatively less common, while two- and three-coordination are rare. These types of reaction centre are considered together in this chapter since their kinetic and mechanistic patterns, particularly for compounds of the *p*-block elements, are often informatively similar or complementary. Thus, for example, it is convenient not to separate hydrolysis of organoboron halides from substitution at organosilicon halides, and it seems sensible to consider oxygen exchange at $MO_3{}^{n-}$ oxoanions alongside that at $MO_4{}^{n-}$. We shall deal with the two most common geometries, tetrahedral and square-planar, first, then consider five-, three-, and two-coordination in that order.

Octahedral complexes are so common and so important, and show such a range of behaviour, that it seems best to allot them a separate chapter (Chapter 4). Kinetic studies of complexes with coordination numbers higher than six are still quite uncommon. They most frequently deal with solvent exchange or complex formation, and will therefore be found in the chapter (Chapter 7) devoted to these special and linked topics.

3.2 Tetrahedral compounds and complexes

3.2.1 Occurrence

Tetrahedral geometry is extremely common and is found in a wide range of chemical species, in the *s*-, *p*-, and *d*-blocks. The main areas are shown in

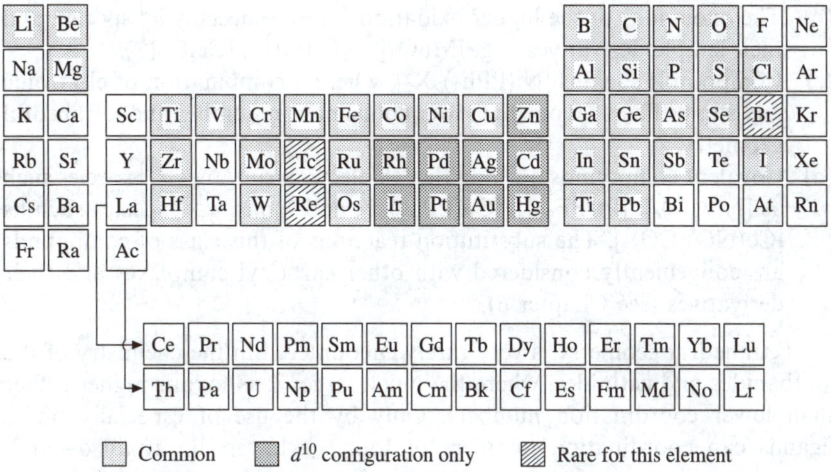

Li	Be											B	C	N	O	F	Ne
Na	Mg											Al	Si	P	S	Cl	Ar
K	Ca	Sc	Ti	V	Cr	Mn	Fe	Co	Ni	Cu	Zn	Ga	Ge	As	Se	Br	Kr
Rb	Sr	Y	Zr	Nb	Mo	Tc	Ru	Rh	Pd	Ag	Cd	In	Sn	Sb	Te	I	Xe
Cs	Ba	La	Hf	Ta	W	Re	Os	Ir	Pt	Au	Hg	Ti	Pb	Bi	Po	At	Rn
Fr	Ra	Ac															

Ce	Pr	Nd	Pm	Sm	Eu	Gd	Tb	Dy	Ho	Er	Tm	Yb	Lu
Th	Pa	U	Np	Pu	Am	Cm	Bk	Cf	Es	Fm	Md	No	Lr

Common d^{10} configuration only   Rare for this element

Fig. 3.1 Occurrence of tetrahedral four-coordination.

Fig. 3.1. In terms of the number of representative compounds, the field is dominated by those of the aliphatic carbon atom. However, this is but one prolific example of the characteristic behaviour of those light elements which can use a valence shell with four pairs of bonding electrons. Tetrahedral geometry is common for essentially covalent boron and nitrogen compounds, e.g. $[BF_4]^-$ and $[NMe_4]^+$, although both atoms can adopt lower coordination numbers with or without making use of multiple bonding. Carbon, on the other hand, can only adopt a stable lower coordination number when there is multiple bonding. Lithium and beryllium, likewise, are characterized by tetrahedral coordination, but the bonding may take on a measure of electrovalence, especially in their organometallic compounds.

On going down the *p*-block to the heavier members of the groups there is scope for higher coordination numbers, and these eventually tend to be favoured. Nevertheless, with appropriate choice of ligand, Al, Ga, Si, Ge, Sn, P, As, Cl and I can all form tetrahedral compounds and substitution reactions at some of these centres, especially at silicon, have been studied in great detail.

In transition metal chemistry, tetrahedral geometry appears in a number of situations.

(a) Species of the type $[MX_4]^{n-}$, in which X is usually a monodentate anionic ligand such as chloride or bromide and the bonding can be treated as mainly electrovalent. The central ion, M, is either truly or approximately spherical. Examples include $TiCl_4$, $[MnCl_4]^{2-}$, $[FeCl_4]^-$, $[CoCl_4]^{2-}$, $[NiCl_4]^{2-}$ and, on the fringe of transition metal chemistry, $[HgBr_4]^{2-}$. This type of complex is rare for second- and third-row elements. Conditions of size and charge are important here, the smaller F^- forming octahedral six-coordinate species, e.g. $[TiF_6]^{2-}$, $[FeF_6]^{3-}$.

(b) The oxoanions of the higher oxidation states, especially d^0 species, and their simple derivatives, e.g. $[MnO_4]^-$, $[CrO_4]^{2-}$, $[Cr_2O_7]^{2-}$.
(c) Compounds such as $[Ni(PPh_3)_2X_2]$, where a combination of electronic and steric effects favour the lower coordination number and tetrahedral geometry.
(d) Covalent compounds obeying the '18-electron rule', i.e. d^{10} species such as $[Ni(CO)_4]$, $[Pt(PF_3)_4]$, $[Cu(PPh_3)_3I]$, and 'quasi d^{10}' species such as $[Co(NO)(CO)_3]$. The substitution reactions of this class of compounds are conveniently considered with other carbonyl complexes and their derivatives (see Chapter 6).

Tetrahedral geometry is very rarely encountered in the chemistry of the lanthanides and actinides, where the much larger ions favour higher rather than lower coordination numbers. Only by the use of especially bulky ligands can coordination numbers of four, or lower, be attained—and, once made, these compounds are generally not examined in relation to ligand substitution kinetics and mechanisms.

3.2.2 *General features of substitution*

It is not possible in this context to talk about a typical mechanism for tetrahedral substitution, and each of the groups listed above in Section 3.2.1 ought to be treated separately. It is possible to find entire textbooks devoted to very specialized aspects of substitution reactions at carbon and it is not usual to restrict the term 'substitution' in the way we have done for the inorganic systems. Electrophilic and homolytic substitution processes are at least as important as nucleophilic ones. The same is true, although on a much smaller scale, for other reaction centres. It is important, in this book, to discuss the organic mechanisms to the same level of sophistication as the others because a comparison of the behaviour of carbon and other reaction centres will show many similarities. It soon becomes apparent why the element carbon has such an extensive chemistry.

3.2.3 *Tetrahedral centres: the light elements*

3.2.3.1 *Beryllium*

The substitution reactions of beryllium are typified by solvent exchange with, and complex formation from, tetrahedral solvento-species $[BeS_4]^{2+}$, and are therefore discussed in detail in Chapter 7. Suffice it to say here that mechanisms of these substitutions range from clearly associative—for $S = H_2O$ the exchange is with one of the largest negative volumes of activation found for water exchange (-13.6 cm^3 mol^{-1})[1]—to equally clearly dissociative. The mode of activation depends on the size of S; for bulkier molecules the mechanism changes to dissociative and the $\Delta V^\ddagger$ values become positive. It would appear that, as in other cases where the bonding has a reasonable electrovalent content and the central metal ion is spherical, the mechanism is governed by considerations of size and charge.

3.2.3.2 Boron

Both tetrahedral four-coordinate geometry and three-coordinate planar six-electron species are well represented in boron chemistry. A good treatment of kinetics and mechanisms of reactions of boron compounds, both tetrahedral and trigonal, is available.[2] In its hydride chemistry, the problems of electron deficiency are solved in non-classical ways and the boranes and their derivatives offer a wealth of reaction centres, whose discussion is beyond the scope of this book.

The planar three-coordinate species, while being strong Lewis acids, are often capable of independent existence in the absence of suitable Lewis bases. In comparison with carbon and the carbenium ion, three-coordinate species are much less unstable and one might thus predict that a dissociative mechanism could be the dominant pathway for substitution at tetrahedral boron centres. A typical reaction,

$$Me_3BNMe_3 + {}^*NMe_3 \longrightarrow Me_3B^*NMe_3 + NMe_3$$

takes place at a rate that is independent of the concentration of NMe_3 and with all the other characteristics of a D mechanism (e.g. $\Delta S^\ddagger = +65\ \mathrm{J\,K^{-1}\,mol^{-1}}$; $\Delta H^\ddagger$ approximates to the B–N bond dissociation enthalpy).[3] In the cases where the three-coordinate boron intermediate is especially unstable, an associative, I_a, process is observed. Purcell and Kotz[2] show very well how the transition from a purely I_a mechanism in the displacement of NR_3 from H_3BNR_3 by phosphines, through parallel D and I_a pathways for the displacement of amines from R_3NBH_2R', to a dominant D pathway is related to the electronic and steric factors that stabilize the three-coordinate intermediate and destabilize the five-coordinate transition state. Table 3.1 indicates this progress from first-order to second-order rate law via the composite rate law

$$\text{rate} = \{k_1 + k_2[\text{L}]\}\ [\text{adduct}]$$

and shows how the signs of the determined activation entropies support mechanistic assignments on the basis of order with respect to the attacking amine or phosphine L. High activation enthalpies, typically in the range 100–150 kJ $\mathrm{mol^{-1}}$ for the k_1 path, are consistent with stretching a strong boron–nitrogen bond in generating a dissociative transition state.[4] The importance of steric factors can be illustrated by comparing k_1 and k_2 for reaction of Me_3NBH_2R with ${}^nBu_3P$ for R = mesityl or t-butyl on the one hand, and simple alkyl groups on the other. The much higher rate constants for the mesityl and t-butyl compounds suggest that steric bulk does not significantly impede nucleophile approach to the boron, but that its weakening effect on the boron–nitrogen bond is important in both dissociative and associative activation.

Dissociation of F_3BPF_3 obeys simple first-order kinetics, with a high

Table 3.1 Activation entropies for reaction of borane adducts with amines and phosphines, in diglyme or 1,2-dichlorobenzene.

		$\Delta S^{\ddagger}$ ($J\,K^{-1}\,mol^{-1}$)	
Adduct	Reagent	k_1 term	k_2 term
H_3BNMe_3	$^{n}Bu_3P$	—	−21
$^{i}BuH_2BNMe_3$	$^{n}Bu_3P$	+75	−38
$^{t}BuH_2BNMe_3$	$^{n}Bu_3P$	+102	−71
$(4-XC_6H_4)_2HBNH_2Me$[a]	DPE[b]	+46 to +150	—

[a] X = H, F, Cl, Br, Me, or OMe.
[b] DPE = 2,2′-diphenylethylamine.

activation enthalpy ($100\,kJ\,mol^{-1}$) and positive activation entropy, corresponding to the expected rate-limiting dissociative boron–phosphorus bond breaking. Analogous borane adducts H_3BL, where L = an amine, a phosphine, or carbon monoxide, perforce undergo two-step dissociation in view of the non-existence of BH_3:

$$\text{e.g.}\quad H_3BCO \underset{k_2}{\overset{k_1}{\rightleftharpoons}} BH_3 + CO$$

$$BH_3 + H_3BCO \xrightarrow{k_3} B_2H_6 + CO$$

The rate law approximates to

$$-d[H_3BCO]/dt = k_1[H_3BCO]$$

(there is also a small contribution from the k_2 back reaction), indicating that rate-limiting boron–carbon bond breaking is followed by rapid formation of diborane (and a little competitive reformation of H_3BCO) in this reaction, and almost certainly in analogous reactions of other compounds H_3BL.

Formation reactions of adducts H_3BL from diborane and L have high activation enthalpies, since it is necessary to separate the two BH_3 units. This requires some $150\,kJ\,mol^{-1}$. In contrast, formation of adducts of BF_3 takes place extremely rapidly, with activation enthalpies of $10\,kJ\,mol^{-1}$ or less. It seems that the conversion of planar sp^2 boron into the potentially four-coordinate tetrahedral sp^3 configuration is facile.

A variant on the Lewis acid/base reactions discussed above is provided by the following equilibrium (L = e.g. pyridine or quinuclidine):

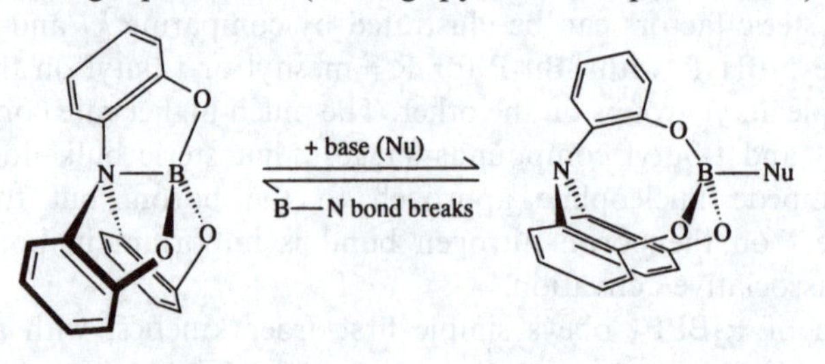

In the direction left $\rightarrow$ right this is nucleophilic attack at four-coordinate (geometry close to tetrahedral) boron; the determined activation entropy indicates an associative (S_N2) mechanism. Bimolecular nucleophilic substitution at an analogous bridgehead carbon is impossible, but the relative ease of breaking the boron–nitrogen bond in the strained substrate permits S_N2 attack here.[5]

3.2.3.3 Carbon

Many man-centuries of effort have been,[6] and indeed still are being, devoted to research on substitution at saturated carbon. Since it would be presumptuous to attempt to condense this into a few brief paragraphs it will only be possible to discuss those aspects that are of interest in their application to the treatment of the reactions at other centres. Nevertheless, we wish to emphasize that, although the title of this book is *Inorganic Reaction Mechanisms*, there is no fundamental difference between organic and inorganic reaction mechanisms and any subdivision of this sort is purely for administrative or political convenience.

The basic rules of the game played by carbon in its chemistry are extremely simple and an adequate understanding of structure, bonding, and mechanism relating to classical organic compounds (hydrocarbons and their derivatives with *p*-block donor atoms) can be based on very elementary valency theory, which would not get us very far if we tried to apply it to transition metal chemistry. We see that this is true if we try to explain the analogous chemistry of the boranes and their derivatives, or even organometallic compounds such as ferrocene and other π-complexes, by the simple methods that work well enough for alkyl halides, ketones, carboxylic acids, and a variety of other organic compounds. The naivety and inherent conservatism of carbon with respect to its bonding habits is also reflected in its reaction patterns. No facile path exists for substitution; indeed, all reaction paths are energetically unfavourable, some more so than others. This is why the substitution reactions of tetrahedral carbon under normal circumstances are very much slower than the rate at which the reagents can come together. In addition, the rates of these reactions are extremely sensitive, not only to the nature of the leaving group but also to the nature of the other ligands attached to the reaction centre and to the nature of the entering group for associatively activated reactions. This combination of kinetic inertness and sensitivity of reaction rate is one of the causes of the uniqueness of the chemistry of carbon and the means whereby the vast range of compounds that constitute organic chemistry, and indeed biology, are able to exist. Platinum(II) and cobalt(III), to cite the two most-studied transition metal centres, match carbon in this respect but lack two other requirements, namely the inertness of the C–C bond and the ability to do chemistry at one part of the molecule while the skeleton remains intact, and the ability to form multiple Pt–Pt or Co–Co bonds.

Both homolytic and heterolytic substitution processes can be demon-

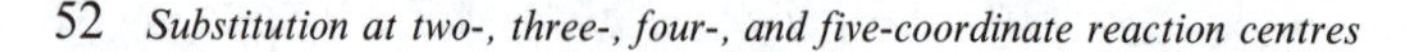

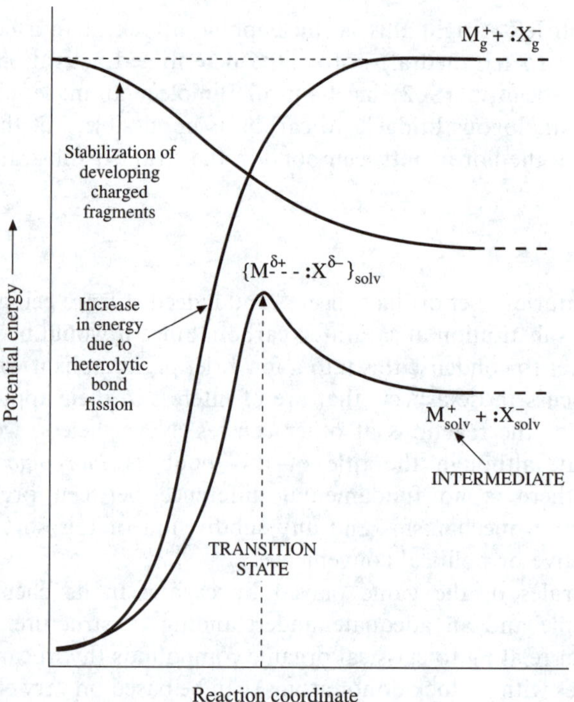

Fig. 3.2 The effect of solvation on the developing fragments in a *D* mechanism—relation between transition state and intermediate of lower coordination number.

strated for carbon. As a general rule, homolytic bond fission in an isolated molecule requires less energy than heterolytic fission since the latter leads to a significant charge separation. Homolysis is therefore most likely in reactions that occur in the gas phase or in non-polar solvents. The extra energy required for heterolysis can be adequately compensated by solvation in polar solvents and it is likely that heterolysis will be favoured in them. It is of interest to note that, because of the combination of bond breaking and fragment stabilization, there may be significant bonding between the leaving group and the reaction centre in the transition state, even in a dissociatively activated process (Fig. 3.2).

There are many examples of heterolytic processes in the gas phase and homolytic reactions in polar solvents, but these constitute the exception rather than the rule, unless the homolytic processes are free-radical chain reactions. Heterolysis can be accomplished either electrophilically or nucleophilically depending upon the requirements of the reactions.

As a rule, the standard substitution reactions of tetrahedral carbon atoms involve reagents that are formally Lewis bases and are therefore nucleophilic. Typical reactions are shown in Table 3.2. There is no preponderant path for nucleophilic substitution. A dissociative (*D*) mech-

Table 3.2 Typical substitution reactions of tetrahedral carbon.

Reaction	Mechanism
Nucleophilic	
$(CH_3)_3CBr + H_2O \longrightarrow (CH_3)_3COH + H^+ + Br^-$	$D(S_N1)$
$(CH_3)_2CHBr + H_2O \longrightarrow (CH_3)_2CHOH + H^+ + Br^-$	$D(S_N1)$
$CH_3CH_2Cl + I^- \longrightarrow CH_3CH_2I + Cl^-$	$I_a(S_N2)$
$(CH_3)_3CCH_2Br + I^- \longrightarrow (CH_3)_3CCH_2I + Br^-$	$I_a(S_N2)$
Electrophilic	
$(sec\text{-}C_4H_{10})_2Hg + HgBr_2 \longrightarrow 2(sec\text{-}C_4H_{10})HgBr$	S_E2
$CH_3HgBr + {}^*HgBr_2 \longrightarrow CH_3{}^*HgBr + HgBr_2$	S_Ei (S_E2 with a cyclic transition state)
$HNC_5H_4^+\text{-}CH_2Mn(CO)_5 + Hg^{2+} \longrightarrow HNC_5H_4^+\text{-}CH_2Hg^+ + Mn(CO)_5^+$	S_E2 (open transition state)
$HN^+C_5H_4\text{-}CH_2Mn(CO)_5 + H_2O \longrightarrow HN^+C_5H_4\text{-}CH_3 + Mn(CO)_5(OH)$	S_E1 (S_N1 at Mn)

anism with abundant evidence for a three-coordinate intermediate (a carbenium ion), which reacts independently of its mode of formation and which exhibits considerable discrimination, has been assigned to a number of reactions. In general, such a mechanism will be favoured by poor nucleophiles, labile leaving groups, and electron-releasing ligands attached to the reaction centre. Carbenium ions can be generated under conditions where they are not subsequently consumed, for example, by protonating alkenes or generating irreversibly labile leaving groups in 'super acids'. Stable, metastable, and even unstable carbenium ions, both of the classical three-coordinate type and the 'non-classical' five-coordinate variety, which account for the rearrangement process that can be observed and which have much in common with the boranes, have been studied, mainly by NMR.

Associatively activated nucleophilic substitution is favoured in reactions with stronger nucleophiles and under conditions that reduce the likelihood of a dissociative mechanism. Bond making and bond breaking both contribute to the energetics of the transition state, but since these usually have opposing requirements the resultant relationship between reactivity and the variables such as electron displacement by the spectator ligands and the strength of the bond with the leaving group will depend entirely upon the reaction being studied, and generalizations are not possible. The mechanism is, at all times, interchange (I_a) and no evidence for a transient five-coordinate intermediate (let alone an isolable one) has been forthcoming.

Electrophilic substitution is encountered when the leaving group is more electropositive than carbon and the entering group is functioning as a Lewis acid (otherwise attack will be at the donor and not at carbon). Sometimes, when the entering group is a strong enough electrophile, the leaving group may be persuaded, at least temporarily, to adopt an electron-deficient role. The bimolecular mechanism has been well established for certain transmetallation reactions. At no stage is the carbon called upon to have more than eight electrons in its valence shell and an *A* mechanism is possible. The dissociative mechanism has also been demonstrated.

The examination of the steric courses of substitution requires some signposting in order to relate macrochemical observation to the molecular process. The high symmetry of the tetrahedron makes this rather difficult. If the reaction centre is attached to four monodentate ligands the only mode of signposting available is to make all of these different so that the reaction centre becomes asymmetric.

Much of the early work on the stereochemistry of substitution at tetrahedral carbon hinged on the relationship between the rate of mutarotation or racemization of an optically active substrate and the rate of the accompanying substitution reaction. In the absence of adequate knowledge of the relationship between the sign of rotation and the relative configuration, a particularly ingenious approach was used by Hughes in which he compared the rates of exchange of radioactive iodide with the rates of racemization of optically active *sec*-octyl iodide in the presence of an excess of sodium iodide.[7] Both processes were first-order in both substrate and iodide and the rate constant for racemization was twice that for exchange indicating that each act of exchange inverted the configuration. This was one of the first applications of radioisotopes to the problems of reaction mechanism. In cyclic systems, inversion at one centre can be related to a signpost on an adjacent atom and inversion will lead to a diastereomeric product (Fig. 3.3). The process can readily be followed by NMR.

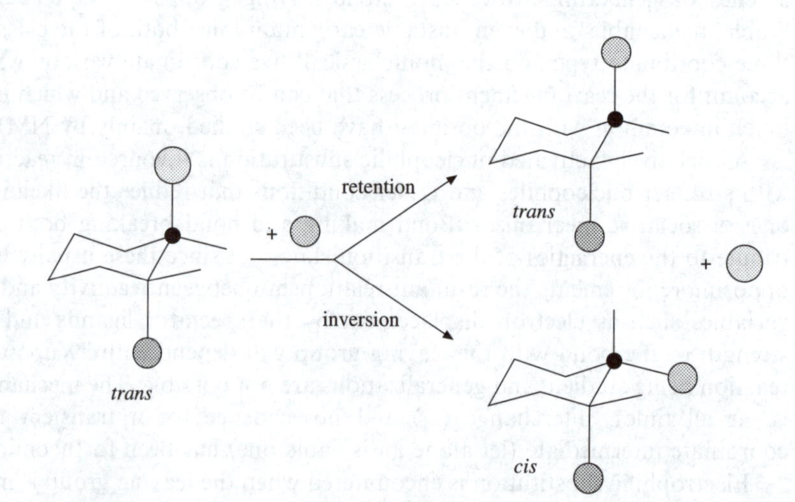

Fig. 3.3 Signposting a cyclic or chelated tetrahedral system by ● on an adjacent atom. The *cis* and *trans* forms can be identified by their different NMR spectra.

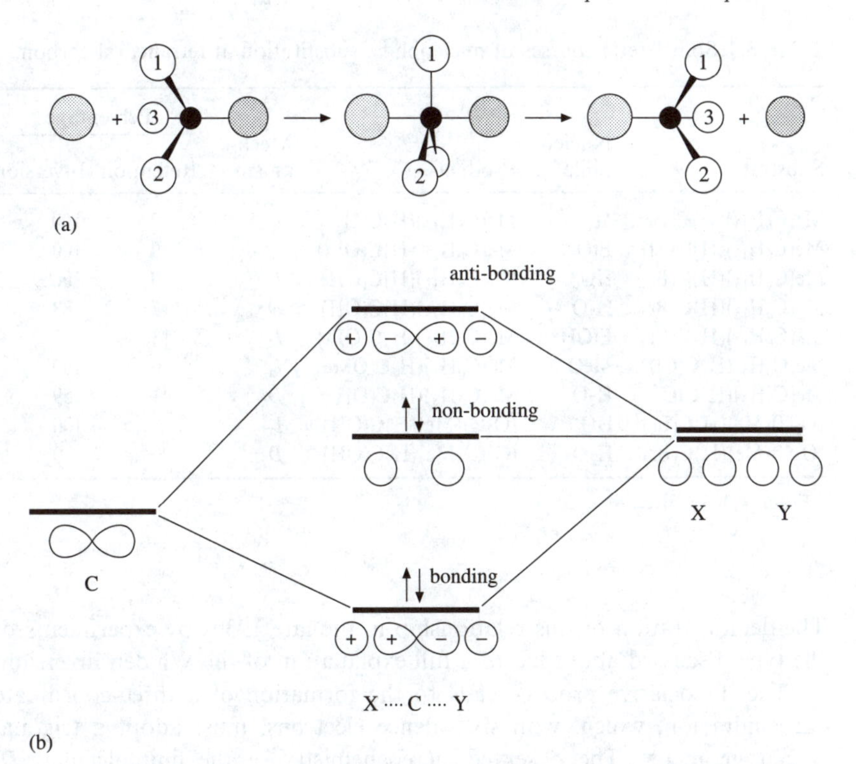

Fig. 3.4 (a) $I_a(S_N2)$ substitution at carbon leading to inversion of configuration. (b) Molecular orbital diagram showing the bonding of the entering and leaving groups in the trigonal bipyramidal I_a transition state.

The relative simplicity of the mechanistic pathways available to tetrahedral carbon in its nucleophilic substitution reactions is also reflected in the stereochemistry of substitution. The bimolecular nucleophilic process is clearly defined. In order to accommodate ten electrons in the five-coordinate transition state, a trigonal bipyramidal geometry is required with the entering and leaving groups in the axial positions (Fig. 3.4(a)). The three spectator ligands take up the equatorial positions in the trigonal plane and their bonding (fig. 3.4 (b)) can be treated in terms of σ-interaction with the sp^2 hybrid orbitals on the carbon. The remaining, p_z, orbital combines with the σ-symmetry orbitals on the entering and leaving groups to give three molecular orbitals, two of which are filled. The orbital associated with the last two electrons (HOMO) is located almost entirely on the entering and leaving groups and so the valence shell of the carbon atom does not contain more than eight electrons. This constraint means that each act of associative, I_a, substitution at a tetrahedral centre requires attack at the face that is remote from the leaving group and leads to an inversion of configuration.

Table 3.3 Some steric courses of nucleophilic substitution at tetrahedral carbon.[a]

Substrate	Nucleophile	Product	Mechanism	Percentage Retention	Percentage Inversion
$Me(C_6H_{13})(H)\mathbf{C}(I)$	$^*I^-$	$Me(C_6H_{13})(H)\mathbf{C}(^*I)$	I_a	0	100
$Me(C_6H_{13})(H)\mathbf{C}(Cl)$	EtO^-	$Me(C_6H_{13})(H)\mathbf{C}(OEt)$	I_a	0	100
$Me(C_6H_{13})(H)\mathbf{C}(Br)$	HO^-	$Me(C_6H_{13})(H)\mathbf{C}(OH)$	I_a	0	100
$Me(C_6H_{13})(H)\mathbf{C}(Br)$	H_2O	$Me(C_6H_{13})(H)\mathbf{C}(OH)$	D	17	83
$Me(C_6H_{13})(H)\mathbf{C}(Br)$	EtOH	$Me(C_6H_{13})(H)\mathbf{C}(OEt)$	D	13	87
$Me(C_6H_5)(H)\mathbf{C}(Cl)$	MeO^-	$Me(C_6H_5)(H)\mathbf{C}(OMe)$	I_a	0	100
$Me(C_6H_5)(H)\mathbf{C}(Cl)$	H_2O	$Me(C_6H_5)(H)\mathbf{C}(OH)$	D	41	59
$(O_2C)(Me)(H)\mathbf{C}(Br)$	HO^-	$(O_2C)(Me)(H)\mathbf{C}(OH)^-$	I_a	0	100
$(O_2C)(Me)(H)\mathbf{C}(Br)$	H_2O	$(O_2C)(Me)(H)\mathbf{C}(OH)^-$	D	100	0

[a] From p. 388 of Ref. 6.

The demonstration of this relationship in the late 1930s by experiments of the type discussed above led to a full explanation of the Walden inversion.

The dissociative process leads to the formation of a three-coordinate carbenium ion, which, with six valence electrons, must adopt a trigonal planar geometry. The observed stereochemistry of the unimolecular, *D*, substitution should reflect this planar intermediate. The two pathways for recombination, from the front or back, should be equivalent once the leaving group has left the solvation shell, and the product ought to be racemic. A random selection of steric courses for aliphatic substitution is given in Table 3.3. These data show that the I_a mechanism is invariably accompanied by inversion of configuration, but that the dissociatively activated process rarely leads to complete racemization in the act of substitution. When the aliphatic secondary carbon atom in the 2-octyl halides, for example, undergoes dissociatively activated substitution there is predominant but not exclusive inversion of configuration. A phenyl substituent, on the other hand, while not affecting the complete inversion in the associative substitution, causes the product of the dissociatively activated process to be very nearly racemic. It is reasonable to believe that the three-coordinate intermediate is consumed before it can properly equilibrate its environment and that the leaving group is still close enough to partially block approach adjacent to itself. To a purist this might mean that the mechanism is more properly labelled I_d rather than *D*, but it should be remembered that in solvolytic processes the solvent is always close at hand. The phenyl group can stabilize the carbenium ion sufficiently for its lifetime to increase to the point where the environment has been equilibrated. A configuration-holding substituent such as $-CO_2^-$ is electrostatically attracted to the developing carbenium ion before the leaving group has gone and, while not being able to bind to the vacated site, can block entry from the back of the carbenium ion. The product is then formed mainly with retention of configuration.

3.2.4 Tetrahedral centres: the heavier p-block elements

3.2.4.1 Silicon, germanium, and tin

The congeners of carbon in Group 14, Si, Ge, and Sn, present many aspects of chemistry superficially similar to that of carbon, though there is the usual gradation of properties as one descends the group. The organoelement chemistry on which so much mechanistic interpretation depends has been comprehensively reviewed, at least for silicon[8] and tin.[9] In particular, Si, Ge, and Sn behave rather differently from carbon in so far as coordination numbers are concerned. Coordination numbers less than four are very rare for Si, Ge, Sn, and Pb, requiring specially designed compounds for their existence. On the other hand, coordination numbers of five or six, unknown for carbon, are quite common (as in SiF_6^{2-}, $Ge(OH)_6^{2-}$, and SnX_4L_2 adducts) and indeed preferred as the group is descended.[10] Their possession of accessible *d*-orbitals permits additional $d\pi$-bonding interactions (π-donor and/or π-acceptor) in initial and/or transition states. Their lower ionization energies (first ionization energies are 1086, 786, 762, 709, and 716 kJ mol^{-1} for C, Si, Ge, Sn, and Pb, respectively) are reflected in the existence of such species as the cations $[Si(acac)]^+$, $[Si(C_6H_5)_3(bipy)]^+$, and even '$[Si(bipy)_3]^{4+}$'.[11] This factor should make dissociative pathways via R_3M^+ transition states easier for these elements than for carbon, though, as will emerge in the following paragraphs, this effect seems to be negligible in practice. The elements silicon to lead are all considerably less electronegative than carbon, with the result that they almost invariably function (in their highest oxidation state at least) as the Lewis acid component in any heterolytic process. It is not immediately apparent from these trends whether, or how, one should expect the relative importance of the two nucleophilic substitution mechanisms, namely S_N1 and S_N2, to change as one descends this group of the Periodic Table. Indeed, it could be that interchange mechanisms of intermediate character—*I* rather than I_d (S_N1) or I_a(S_N2)—might assume considerable importance for lower members of this group. One further variation is that the site of attack might change, and indeed there is considerable evidence for substitution at certain organolead compounds involving initial attack at α-carbon rather than at the lead. Thus, whereas reaction of R_4E compounds with I_2 takes place by iodide attack at the central element E when this is Si, Ge, or Sn, when E = Pb the mechanism is electrophilic attack (S_E2) of I^+ at carbon adjacent to the lead.

There proves to be a profound change in the kinetic and mechanistic pattern of substitution reactions at the lower members of Group 14, from silicon downwards.[12] The most obvious change is the enormous increase in lability, so much so that adequate comparison requires the use of fast-reaction techniques for these lower elements. Where kinetic order can be determined it is reasonably clear that the substitutions are associatively activated, even under the conditions that were the most favourable for dissociative activation in the corresponding carbon case. For example, Ph_3CCl undergoes substitution by a *D* mechanism and the planar triphenylmethyl (trityl) carbenium ion can be isolated and characterized. Ph_3SiCl, on the other hand, undergoes substitution in a typically associative fashion. There are a large number of reactivity trends, for variation in central atom, variation in substituent electronic or steric effects, and so on, which have almost without exception provided supporting evidence for S_N2 rather than S_N1 reaction at silicon, germanium, and tin. Several workers spent much effort over many years in seeking a three-coordinate intermediate in substitution at silicon. A review of the situation in the mid-1970s stated that there then existed no *proof* of an R_3Si^+ ion as intermediate or transition state, but that the possibility of the existence of such species could not be ruled out.[13] More of the background can be found in a later paper claiming the first demonstration of the solvolytic generation of an R_3Si^+ ion.[14] However, the few reactions at silicon that, at one time or another, have been thought to be dissociatively activated have been shown to involve rate-determining associative solvolysis. At the time of writing, there is still no evidence for the transient existence of a planar R_3Si^+ species in solution, but several species with 'considerable silylium-ion character' have been established. These are $R_3Si \cdots\cdots L^+$ species with a weakly bonded group or solvent molecule in a fourth coordination site.[15]

The evidence from steric and electronic effects provides interesting illustrations of how the tailoring of substituents can assist in the diagnosis of mechanism. We shall digress to this topic for a couple of paragraphs, since the principles have proved equally useful in a variety of transition metal and organometallic systems. Before setting out on an extensive series of kinetic experiments it is, of course, a good idea to make sure that the results should provide an unequivocal answer to the question being asked, or hypothesis being tested. In the case of using substituent steric effects to probe mechanism, the first requirement is to examine whether rate constant trends are in opposite directions for S_N1 and S_N2 mechanisms. To do this properly, one needs to forecast effects on both the initial and the transition states. For dissociative activation, any steric strain will decrease as the initial state moves towards the transition state. Therefore activation barriers will decrease, and rate constants increase, as steric constraints increase. On the other hand, the key feature of associative attack is the incipient bonding to the incoming nucleophile in the transition state. Now increasing steric crowding around the reaction centre will hinder approach of the nucleophile and lead to decreasing rate constants. With these forecasts in mind, it is apparent that the observed reactivity trends for hydrolysis of organogermanium and organotin chlorides R_3MCl, where rate constants decrease in the order:

R = methyl > R = phenyl > R = 1-naphthyl

are consistent with an S_N2 mechanism, but not with an S_N1 mechanism.

In the case of substituent electronic effects, an electron-releasing group will discourage nucleophilic approach and reduce reactivity in an S_N2 mechanism, but would facilitate generation of an R_3M^+ moiety in an S_N1 mechanism. An electron-withdrawing group will have the opposite effects. In practice it has been established that, for example, rate constants for nucleophilic attack at $(4\text{-X-}C_6H_4)_3MCl$, where M = Si, Ge, or Sn, decrease as electron release by the *para*-substituent X increases. This again supports the operation of an S_N2 mechanism. These predictions for substituent steric and electronic effects are summarized in Fig. 3.5.

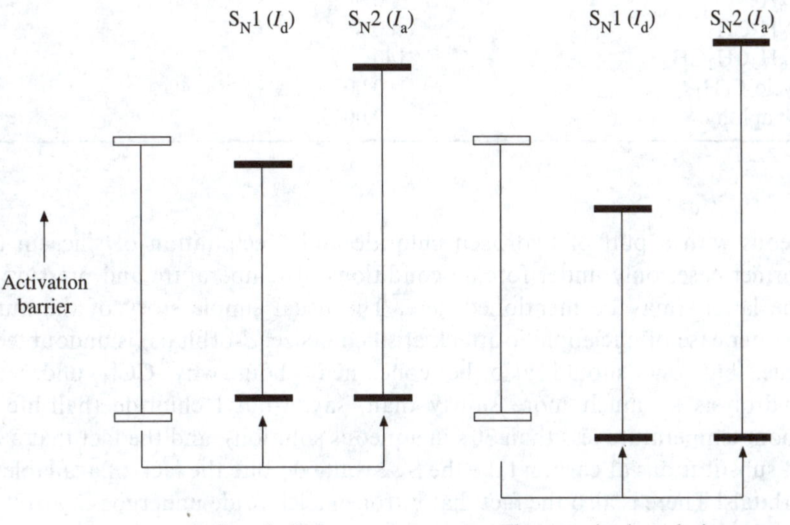

Fig. 3.5 Steric and electronic effects of substituents on activation barriers for S_N1 and S_N2 substitution.

With the profound mechanistic difference between carbon and the lower Group 14 elements, direct comparison of the rate constants for substitution is meaningless, beyond the qualitative observation made above that stopped-flow techniques are almost always needed for the lower members, but very rarely for analogous organic substitutions. However, one can compare k_1 for the hydrolysis of Ph_3MF in 40% aqueous acetone at 25 °C—2.1×10^{-4} s^{-1} when M = C and 1.5×10^{-6} s^{-1} when M = Si. This is one of the few examples where the silicon compound is significantly less reactive than its carbon analogue. However, whereas added base has no effect upon the rate of hydrolysis of the trityl fluoride, it can increase the rate of hydrolysis of the silicon analogue by at least six orders of magnitude. The hoary old chestnut of the enormously greater enthusiasm of silicon tetrachloride than of carbon tetrachloride to undergo hydrolysis—instanta-

Table 3.4 First-order rate constants for the reaction

$$R_3MCl + {}^{36}Cl^- \longrightarrow R_3M^{36}Cl + Cl^-$$

in 2:5 acetone/dioxan at 25 °C ([LiCl] = 10^{-3} mol dm^{-3}).

R	M:	$10^3 k_{exch}/s^{-1}$		
		Si	Ge	Sn
C_2H_5		38		
n-C_3H_8		18		
n-C_4H_{10}		11	1300	
n-C_6H_{14}		9	920	
C_6H_5		0.9	150	> 4000
$C_6H_5CH_2$		400		
$C_6H_5CH_2CH_2$		140		
cyclo-C_6H_{12}		0.007	40	
1-naphthyl		0.0013		

neous with a puff of hydrogen chloride and precipitation of silica in the former case, only under forcing conditions of temperature and pressure in the latter—may be mentioned here. The usual simple story of the much greater ease of nucleophilic attack at silicon (size; *d*-orbitals) is undoubtedly true, but one should also be concerned about why CCl_4 undergoes hydrolysis so much more slowly than, say, t-butyl chloride (half-life at room temperature less than 30 s in aqueous solution), and the fact that a lot of substitutions at carbon take the S_N2 route despite the lack of available *d*-orbitals! There is also the fact that nitrogen trichloride undergoes hydrolysis not by attack at the nitrogen but rather by interaction at the chlorine (cf. Section 3.5.2 below).

Silicon and germanium display similar mechanistic patterns, but there are reactivity differences depending on the nature of the reagents. The data in Table 3.4 for the exchange of labelled chloride ion with R_3MCl in acetone–dioxan[16] indicate a reactivity sequence Si < Ge ≪ Sn, as do the reactions of M_2R_6 with iodine, mentioned below. On the other hand, many examples are known where the reactivity of the silicon species is greater than that of its germanium analogue. For example, for base-catalysed solvolysis of the hydrides RMH, the reactivity order is Si > Ge ≫ Sn, and rate constants for the hydrolysis of Ph_3MCl in acetone containing 4.0 mol dm^{-3} water are 4.0 s^{-1} for M = Si and 0.034 s^{-1} for M = Ge. There are no easy generalizations and explanations of relative reactivities in this area !

The distinction between the I_a and A mechanisms for these associatively activated substitutions hinges, in the absence of any convincing kinetic evidence for transient five-coordinate intermediates, on the existence of stereochemical evidence for the presence of a reactive five-coordinate intermediate. The report that 1-naphthylphenylmethylfluorosilane racemizes in n-pentane when methanol is added, at a rate that depends upon the concentration of methanol, suggests that a symmetrical (or rapidly fluxional) five-coordinate intermediate is formed although, in this case, cleavage of the Si–F bond does not follow.[17] A similar reaction has been observed for the corresponding fluorogermane.

The stereochemical rules associated with substitution at tetrahedral silicon[18] and at germanium are much less rigid than those relating to substitution at tetrahedral carbon, and indeed also at tetrahedral phosphorus. Inversion of configuration is observed in a number of cases and, although all processes are associatively activated, retention of configuration is also common.[19] The rate of chloride exchange with 1-naphthylphenylmethylchlorosilane in 5:1 dioxan/acetone is equal to the rate of racemization (not half the rate as would be required for substitution with inversion) and a symmetrical intermediate was suggested. However, a fluxional trigonal bipyramidal intermediate that lives long enough to racemize intramolecularly would give the same results—we are back to the I versus I_a problem! The stereochemical behaviour seems to depend upon the requirements of the leaving group, X. A good leaving group, such as Cl, will usually be replaced with inversion of configuration, and a trigonal bipyramidal transition state with entering and leaving group in the axial position, of the type encountered in the I_a substitutions of carbon, is indicated. On the other hand, a poor leaving group, such as OMe or H, is replaced with retention of configuration. It has been suggested that the entering and leaving groups are adjacent in the transition state or intermediate because the leaving group requires some electrophilic assistance that it can gain from a cyclic transition state (Fig. 3.6). Other factors play some part. For example, when the situation is otherwise fairly balanced, a more polar solvent will tip it in favour of path (a), where the charge separation in the transition state is greater than in path (b). However, the view is widely held at present that while the entering group approaches at a face and takes up an axial position in a trigonal bipyramid and the departing group leaves from an axial position of a trigonal bipyramid, the intermediate lives long enough to undergo a large number of pseudorotations (see Chapter 5) and the stereochemistry of the reaction will depend upon the relative stabilities of the various isomers of the five-coordinate intermediate. All of this is consistent with a typical A mechanism. Evidence has been presented to suggest that the stereochemical course, at least for compounds containing Si–OR or Si–F groups, depends on the hard or soft character of the entering group, with the former attacking equatorially, the latter axially.[20] Complete inversion of configuration with each act of associative substitution is not necessarily proof of a synchronous (I_a) mechanism. If the entering and leaving groups are the most electronegative of the set of five ligands involved, then the trigonal bipyramid in which they both occupy axial positions will be the most stable and inversion of configuration will result. The alternative pathway, whereby the entering group approaches an

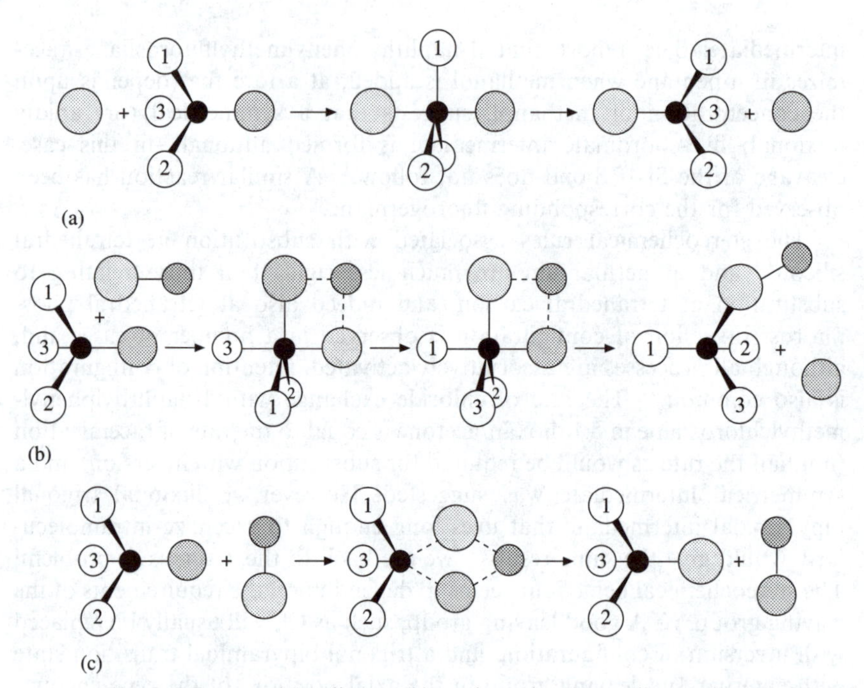

Fig. 3.6 Remote (a), adjacent (b), and synchronous adjacent (c) attack at a tetrahedral reaction centre. For a good leaving group, substitution takes place with inversion by way of a trigonal bipyramid with entering and leaving groups occupying axial positions (a). For a poor leaving group, the entering group may attack at a face adjacent to the leaving group and a remote site may provide electrophilic assistance (b), or, less likely, the mechanism may be synchronous, involving a four-centre ten-electron transition state (c). In (b), the rearrangement shown as the second stage is needed to get the leaving group into an axial position appropriate for departure.

edge of the tetrahedron and is immediately in the trigonal plane (and the possibility of the leaving group departing from the trigonal plane) is largely discounted in favour of a fluxional intermediate. Nevertheless, it has been suggested that the acid-catalysed ring opening of the tris(2-phenoxy)-aminosilane (**1**) must take place with equatorial ligand departure.[21]

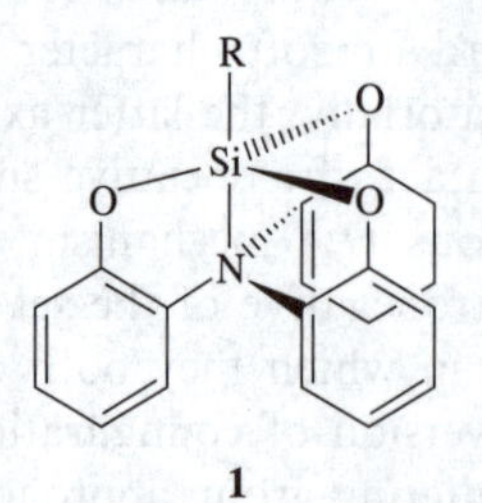

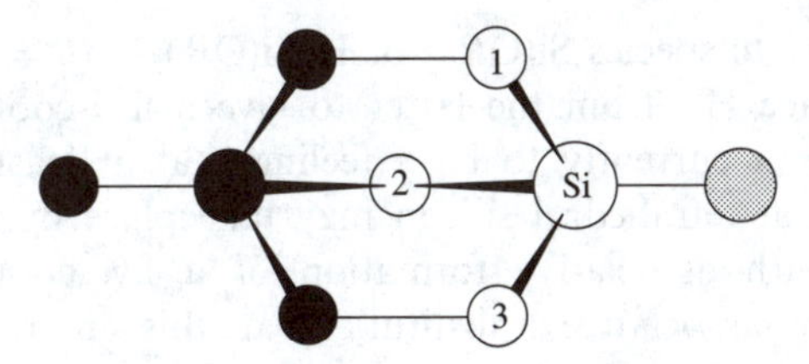

Fig. 3.7 Reaction centre at a bridgehead—bimolecular attack from the rear is prevented, and the donor groups 1, 2, and 3 are unable to become coplanar with the reaction centre.

Table 3.5 Approximate relative rates of base-catalysed hydride solvolysis at bridgehead silicon.[a]

Silane	Relative rate	Silane	Relative rate	Silane	Relative rate
Si CH_3 H	10^4–10^5				
Si—H	10^3	Et_3SiH	1	Si CH_3 H	0.1
Si—H	10				

[a] From L. H. Sommer, O. F. Bennett, P. G. Campbell, and D. R. Weyenberg, *J. Am. Chem. Soc.*, 1957, **79**, 3295; L. H. Sommer and O. F. Bennett, *J. Am. Chem. Soc.*, 1959, **81**, 251.

The lower stringency of the stereochemical rules is seen in the substitutional lability of a 'bridgehead' atom (Fig. 3.7). The three 'chelate rings' block associative attack from the rear of the reaction centre and prevent the formation of a trigonal bipyramidal transition state (or intermediate) with the entering and leaving groups occupying axial positions (groups 1, 2, and 3 cannot become coplanar with the reaction centre). While the reactivity of a bridgehead carbon is very greatly reduced, that of an adjacent silicon can be reduced, enhanced, or even unaffected depending on the nature of the reactions studied (Table 3.5).

In view of the fact that even trigonal bipyramidal silicon is coordinatively unsaturated when surrounded by electronegative ligands, it is not surprising that it has been suggested that it is sometimes easier to displace an anionic ligand associatively from an anionic five-coordinate silicate than for it to occur dissociatively. Thus, it has been shown that the four-coordinate $Si(OR)_4$ or $HSi(OR)_3$ are far less susceptible to alcoholysis than the corresponding five-coordinate species $Si(OR)_5$ or $H_nSi(OR)_{5-n}$ ($n = 1$ or 2) and that RO^- will displace H^- from the latter to give a five-coordinate product.[22] There does seem currently to be a feeling that at least some nucleophilic substitutions at tetrahedral silicon may take place by a fully fledged *A* mechanism, with associative formation of a five-coordinate intermediate followed by associative substitution at this intermediate through a six-coordinate transition state.

Electrophilic reactions of tetra-alkyl and tetra-aryl compounds SnR_4 and PbR_4—and of ternary compounds of the type R_3SnR' or R_3SnCl—with halogens and interhalogen compounds such as ICl are a feature of tin(IV) and lead(IV) chemistry.[23] SnR_4 and PbR_4 also react with mercury(II) halides, where the mechanism is always bimolecular, but may involve a four-centre (cyclic) or open transition state,[24] depending on the nature of the metal, the groups R, the attacking species, and the solvent.[25] Steric effects on reactivity in reactions of $SnEt_4$ and of Sn^nBu_4 with a series of mercury(II) carboxylates indicate open transition states.[26] In all cases the key step in determining reactivity appears to be the breaking of a tin–(lead–) carbon bond; the stereochemistry, kinetics, and mechanisms of tin–carbon bonds have been conveniently reviewed.[27] Analogous reactions of Sn_2R_6 and Pb_2R_6 may involve a significant contribution from metal–metal bond breaking, at least under certain conditions of reactants and solvent.[28] Reactions of the series of compounds M_2Me_6 with iodine provide a convenient measure of reactivity going down the group from M = Si to Sn. In chlorobenzene solution at 293 K rate constants are 0.97×10^{-3} (Si), 9.2×10^{-2} (Ge), and too fast to measure (Sn); Sn_2Ph_6 reacts much more slowly than Sn_2Me_6.[29]

3.2.4.2 Phosphorus and sulphur

Phosphorus and sulphur, as the second-row representatives of Groups 15 and 16, can offer a range of tetrahedral species that can undergo substitution relatively slowly. However, they also introduce some new features of structure, and of kinetic and mechanistic behaviour. Many of their tetrahedral species contain multiple bonds, for example $(C_6H_5)_3P{=}O$, indicating that the 3*d*-orbitals have been brought into use in the ground state. The kinetics of substitution often reveal a strong dependence of rate upon the concentration and nature of the entering group and therefore indicate associatively activated mechanisms. Thus, the reactivity trend $NR_2 < OR < R < X$ (R = Me or Et, X = halide) is consistent with associative activation for substitution at compounds L_2PXE (L = groups cited above, E = O or S), as also are solvent effects on complexes of the type L′L″P(O)Cl, where L′, L″ = NMe_2, NEt_2, OEt, and Me variously.[30] The occurrence of stable five-coordination (and indeed of stable six-coordination, as in species such as PF_6^-) is a characteristic feature of the highest oxidation state and is much more prevalent than in the corresponding (no lone pair) oxidation state in Group 14. As for silicon (cf. previous section), there are suggestions of nucleophilic substitution taking place by an *A* mechanism, with both formation of the *A* intermediate and its onward reaction involving transition states of increased coordinate number, namely six for the second stage.[31] It has been suggested that anionic species of the type $[Cl_2PO_2]^-$ hydrolyse dissociatively, but this mode of activation is rare.

As in the case of silicon, the stereochemistry of substitution depends upon the nature of the entering and leaving groups.[30,32] The rate constant for the reaction

$$\text{EtPhPO(OMe)} + {}^*\text{OMe}^- \rightleftharpoons \text{EtPhPO}({}^*\text{OMe}) + \text{OMe}^-$$

is exactly one half of that of the racemization of the optically active substrate under identical conditions, indicating that each act of methoxide exchange leads to a product with inverted configuration and thereby cancels out two units of optical activity.[33] This was the first demonstration of a Walden inversion at an element other than carbon. In other reactions, substitution takes place with complete retention of configuration and frequently the stereochemistry is non-specific. In this sense the behaviour resembles that of silicon and probably can be explained in the same way. In general, it is thought that approach of the nucleophile and departure of the leaving group takes place at or from a face, i.e. to or from the axial position of the trigonal bipyramidal intermediate.[34] Strongly electronegative groups like to be in axial sites in trigonal bipyramidal intermediates or transition states. Inversion and retention are finely balanced, as illustrated by the case of the reaction of compounds (**2**) with EtS^-, for inversion is observed for the reaction when X = Cl, but retention when X = Br.[35]

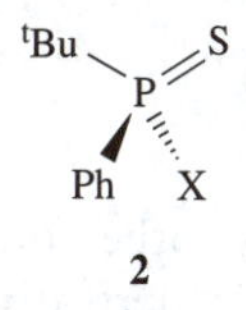

Many cases are to be found in the substitution reactions of cyclic phosphorus compounds, where the problems of ring strain control the form of the five-coordinate intermediate and hence the steric course of substitution.[36] Since inversion of configuration requires that the entering and leaving group take up axial positions in the trigonal bipyramidal transition state or intermediate, the chelate (assuming that we are not studying a ring-opening reaction) would have to span equatorial sites and a 'bite angle' of 120° would be preferred. This is most readily achieved with a larger (six- or more-membered) ring and a smaller ring would only be able to link an axial and an equatorial position. The occurrence of rapid intramolecular rearrangement in five-coordinate phosphorus species will be discussed at length in Chapter 5; it is generally assumed that the five-coordinate intermediates of the *A* mechanism can live long enough in many instances to rearrange in the ways described above for silicon, but with greater facility.

The main body of information on kinetics and mechanisms of nucleophilic substitution at sulphur refers to four species types, covering the oxidation state range 2+, 4+, and 6+ and stereochemistries from two- to four-coordinate (though all could be considered tetrahedral if lone pairs are counted as attached groups). These four types of reaction centre are sulphenyl, **3**, sulphinyl, **4**, sulphonium, **5**, and sulphonyl, **6**.[37] The last type is the most relevant in this section, with a typical set of nucleophilic substitution reactions at **7** showing both activation entropies and solvent effects characteristic of the S_N2 mechanism.[38]

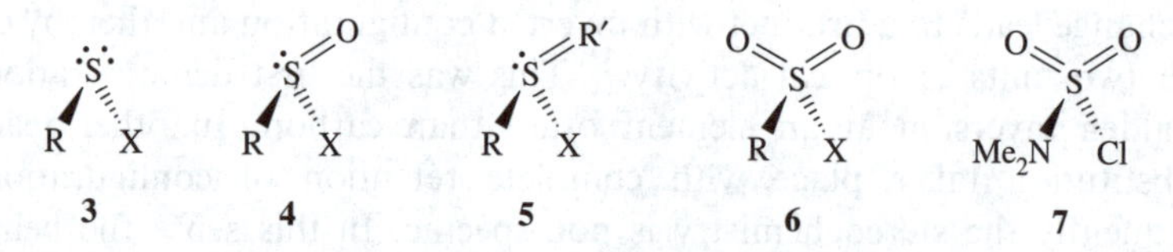

Sulphur is also tetrahedrally disposed in Lewis acid–base adducts of the type R_3NSO_3. Here base hydrolysis also appears to be an S_N2 process, in view of the established rate law

$$-d[R_3NSO_3]/dt = k_2[R_3NSO_3][OH^-]$$

The difference between S_N2 attack at sulphur here, in contrast to S_N1 substitution in the boron analogues R_3NBF_3, may be attributed to the availability of vacant *d*-orbitals of appropriate energy only in the case of sulphur, to the greater size of sulphur than boron—or both.[39]

3.2.5 *Complexes of* d*-block elements*

The tetrahedral geometry in transition metal chemistry, other than in the case of the high oxidation state oxoanions and their derivatives (next section), is essentially a 'default' geometry in cases where either the distribution of the non-bonding electrons is spherically symmetrical or else the interactions between the low field strength ligands and a metal ion that is not fussy will lead to a structure that reflects the most efficient way of packing ligands whose bulk or charge may cause the system to adopt the lower coordination number of four. Therefore, this geometry (or a closely related squashed (D_{2d}) geometry) is found in anionic complexes $[MX_4]^{n-}$ and uncharged complexes $[MX_2L_2]$ with bulky ligands L (often a phosphine or pyridine ligand) and X (Cl, Br, I), with M = Fe^{III}, Fe^{II}, Co^{II}, Ni^{II}, and in Cu^I complexes $[Cu(LL)_2]^+$. Of these M^N, Co^{II}, with its d^7 electron configuration, is the centre where tetrahedral geometry is least disfavoured relative to octahedral. Solvolytic interference can be enormous—substitution in $[Cu(biquinolyl)_2]^+$ by 2,9-dimethyl-1,10-phenanthroline is an associative process in acetone but goes mainly by a solvolytic pathway in methanol.[40] The existence of equilibria between different stereochemistries and coordination numbers[41] adds further challenges and can lead to confusion:

$$[CoCl_4]^{2-} + H_2O \rightleftharpoons [CoCl_3(H_2O)]^- + Cl^-$$

$$[CoCl_3(H_2O)]^- + H_2O \rightleftharpoons [CoCl_2(H_2O)_2] + Cl^-$$

$$\underset{\text{tetrahedral}}{[CoCl_2(H_2O)_2]} + 2H_2O \rightleftharpoons \underset{\text{octahedral}}{[CoCl_2(H_2O)_4]}$$

Such reactions are generally fast, so it is more convenient to study the exchange of neutral ligands in non-aqueous solvents. This means that independent examination of parameters such as leaving group, entering group, and even spectator ligands may be considerably restricted. Nonetheless, rate laws and kinetic parameters have been determined, by variable-temperature NMR spectroscopy, for a range of ligand-exchange reactions of the type

$$[M(PR_3)_2X_2] + {}^*PR_3 \rightleftharpoons [M(PR_3)({}^*PR_3)X_2] + PR_3$$

with M = Fe, Co, or Ni; $PR_3 = PPh_3$, $P(4\text{-tolyl})_3$, or P^nBuPh_2; and X = Cl, Br, or I. These reactions follow second-order rate laws (in $CDCl_3$), and almost all have activation entropies within the range between −70 and $-120\,J\,K^{-1}\,mol^{-1}$, indicating an associative mechanism. Similar comments apply to the analogous reaction of the nitrogen-donor complex [Co(2-picoline)$_2$Cl$_2$], in d_6-acetone ($\Delta S^\ddagger = -61\,J\,K^{-1}\,mol^{-1}$). The nickel complexes react about ten times faster, and the iron complexes several hundred times faster, than their cobalt analogues. These reactivity differences are mainly determined by $\Delta H^\ddagger$ differences.[42] The exchange reaction

$$[Co(PPh_3)_2Br_2] + {}^*PPh_3 \rightleftharpoons [Co(PPh_3)({}^*PPh_3)Br_2] + PPh_3$$

in $CDCl_3$ again obeys a simple second-order rate law, with this time both a negative $\Delta V^\ddagger$($-12\,cm^3\,mol^{-1}$, from pressure-dependent NMR spectra) and a negative $\Delta S^\ddagger$($-79\,J\,K^{-1}\,mol^{-1}$).[43] Everything points to associative activation for substitution at tetrahedral transition metal complexes, except for the odd case of nickel carbonyl, $Ni(CO)_4$ ($\Delta V^\ddagger$ for whose reaction with PPh_3 is $+8\,cm^3\,mol^{-1}$;[44] see Chapter 6). However, in very unusual circumstances, there are deviations from the simple bimolecular mechanism. One of the best-documented involves exchange of the particularly bulky ligand hexamethylphosphoramide, hmpa, at cobalt(II). Here the rate law is

$$\text{rate} = \{k_1 + k_2[\text{hmpa}]\}[Co(\text{hmpa})_2X_2]$$

The k_2 term corresponds to the normal associative mechanism, but the k_1 term is assigned to a dissociative process, favoured by the steric constraints introduced by the bulk of the hmpa. The fact that the activation energy for the k_1 term is about three times larger than that for the k_2 term adds support to this assignment.[45] Conversely, complexes with the d^8 electron configuration, some of which can exist in both (high-spin) tetrahedral and (low-spin) square-planar forms, can also form five-coordinate species in the presence of an excess of ligand L, provided that L is not bulky. When $L = PMe_3$ the trigonal bipyramidal species $[NiL_3X_2]$ becomes the stable form and the rate-determining step for the exchange is dissociation of the 18-electron five-coordinate species.[46]

3.2.6 Oxoanions and their derivatives

3.2.6.1 Oxygen exchange

Kinetics of oxygen exchange at a number of tetrahedral oxoanions have been established.[47] The pH-dependence of rates gives information on the reactivities of variously protonated forms of the respective anions, and about possible base hydrolysis. Thus, for example, oxygen exchange between labelled water and various halate anions (chlorate, bromate, iodate) appears to involve water attack at XO_3^-, HXO_3, and $H_2XO_3^+$, and hydroxide attack at XO_3^-, in various proportions depending on substrate and conditions. Activation entropies for most pathways are negative, but in view of the composite character of many of the values, often containing entropy contributions from protonation equilibria, it would be unwise to proffer this as strong supporting evidence for S_N2 mechanisms. In practice there are several more lines of evidence which suggest associative activation for this group of reactions.

The situation with respect to Group 6 (16) oxoanions is generally complicated, although oxygen exchange between sulphate and water, and between thiosulphate and water, follow simple rate laws in acidic media. These, respectively, are

$$\text{rate} = k_2[HSO_4^-][H^+] \qquad \text{and} \qquad \text{rate} = k'_2[S_2O_3^{2-}][H^+]$$

However, in the case of sulphate it is believed that the rate-limiting step is dissociative ($H_2SO_4 \rightarrow H_2O + SO_3$), whereas for thiosulphate associative attack by water at the central sulphur of the protonated form $HS_2O_3^-$ was proposed.

Complicated rate laws are again the norm for oxoanions of Group 5 (15) elements, in their various oxidation states. The rate law and rate/pH profile (Fig. 3.8) for oxygen exchange at phosphate reflect the respective reactivities of the variously protonated forms $H_4PO_4^+$, H_3PO_4, and $H_2PO_4^-$, while for phosphite the four species $H_4PO_3^+$, H_3PO_3, $H_2PO_3^-$, and HPO_3^{2-} all seem to be required to rationalize the rate/pH profile and rate law. Distinguishing between the various possible mechanisms, including the usual S_N1/S_N2 question, proved difficult. The situation appears more straightforward for some arsenic species, for markedly positive and negative activation entropies for oxygen exchange at arsinates, $R_2AsO_2^-$, and arsonates, $RAsO_3^{2-}$, indicate the operation of S_N1 and S_N2 mechanisms, respectively.[48] It is perhaps worthwhile to document the more complicated situation for inorganic arsenate, where for oxygen exchange

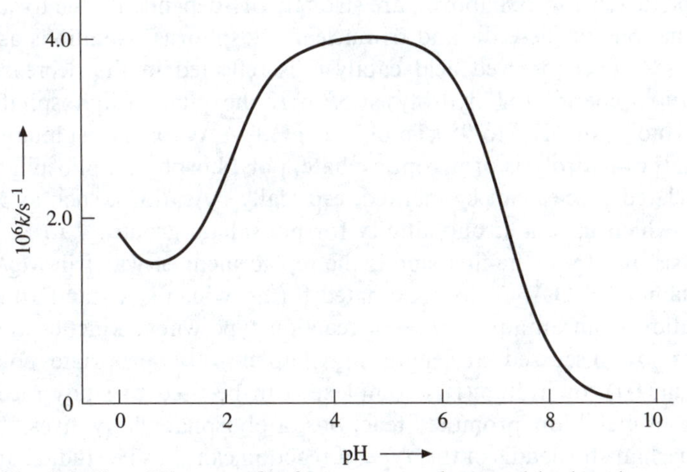

Fig. 3.8 The pH/rate profile for oxygen exchange with phosphate.

$$\text{rate} = k_0[\text{AsO}_4{}^{3-}] + k_1[\text{HAsO}_4{}^{2-}] + k_2[\text{H}_2\text{AsO}_4{}^-] + k_3[\text{H}_2\text{AsO}_4{}^-]^2 + k_4[\text{H}_2\text{AsO}_4{}^-][\text{HAsO}_4{}^{2-}] + k_5[\text{H}_2\text{AsO}_4{}^-]^2$$

This impressive (or perhaps oppressive!) array of terms serves to emphasize both the importance of (de)protonation equilibria, as already invoked in several systems mentioned above, and the possibility of contributions from binuclear species, which leads us into the next paragraph.

Hydrolysis of bi-, tri-, and polynuclear oxoanions is a process of intrinsic interest, and also of great importance in relation to the role of, for example, polyphosphate anions in biochemistry. A good example is that of dichromate:

$$[\text{Cr}_2\text{O}_7]^{2-} + \text{H}_2\text{O} \rightleftharpoons 2[\text{HCrO}_4]^-$$

Hydrolysis of dichromate is acid-catalysed, probably via protonation of the bridging oxygen,[49] and base-catalysed. The rate law for base hydrolysis is simply:

$$\text{rate} = k_2[\text{Cr}_2\text{O}_7{}^{2-}][\text{base}]$$

The rate constants for hydrolysis by water, hydroxide, ammonia, and lutidine correlate with their basicities, once allowance has been made for steric (lutidine) and electrostatic (hydroxide) effects. The base-catalysed reactions have remarkably low activation enthalpies; their activation entropies are in the range -120 to $-240\,\text{J}\,\text{K}^{-1}\,\text{mol}^{-1}$ The various pieces of kinetic evidence all point to an S_N2 mechanism.[50] The mechanism of hydrolysis of the *sp*-block analogue pyrosulphate, $S_2O_7{}^{2-}$, is probably the same.[51] The most interesting feature of kinetic studies of hydrolysis of this anion is the extensive information on cation effects on reactivity (Chapter 8).[52] Hydrolysis of $S_2O_7{}^{2-}$ occurs at the same rate throughout the pH range 1–10; rates of hydrolysis of $P_2O_7{}^{4-}$ and of $P_3O_{10}{}^{5-}$, as of mononuclear phosphate (cf. Fig. 3.8 above), are strongly pH-dependent, due to successive protonations of these di- and tri-nuclear phosphorus oxoanions as the pH decreases. The observed acid catalysis is reflected in the decrease in the activation energy for hydrolysis of pyrophosphate (diphosphate) from 125 kJ mol^{-1} at pH 7 to 95 kJ mol^{-1} at pH 1.[53] As for pyrosulphate, cation catalysis of hydrolysis of pyrophosphate, polyphosphates, phosphate esters, and related species can be marked, especially by cations, such as Mg^{2+} or Zn^{2+}, which have a strong affinity for phosphate groups.[54] In the case of catalysis by Mg^{2+}, the first step is the replacement of water in Mg^{2+}aq by, for example, $P_2O_7^{4-}$ or its protonated forms, which is a standard complex formation from an aqua-ion[55]—a reaction type whose kinetics and mechanisms are discussed at length in Chapter 7. Appropriate cobalt(III), rhodium(III), or iridium(III) complexes can be used to study mechanistic details of metal-ion–promoted reactions of phosphate derivatives.[56] Kinetic and mechanistic details of this type of reaction can thus be studied at leisure, since these metal(III) centres are inert, and the results extrapolated to reactions promoted by substitution-labile M^{n+}aq cations.

3.2.6.2 Reactions with ligands

Many oxoanions, especially those of the early first-row transition (3*d*) elements, are strong oxidizing agents and thus their reactions studied kinetically are often redox processes (see Chapter 9). However, it can be shown that the rate-determining step is frequently a substitution reaction in which the species to be oxidized enters the coordination shell of the metal. In general, although these reactions are relatively fast, rates are dependent upon the concentration and nature of the entering group and have all the characteristics of associatively activated processes with high nucleophilic discrimination.[57]

3.3 Substitution in square-planar complexes

3.3.1 Introduction[58]

Since the planar geometry does not represent the lowest energy arrangement of four ligands in an electrostatically bonded system, it is necessary to have special circumstances before this geometry is favoured. In square-planar complexes of the main-group elements, such as XeF_4, $[ClF_4]^-$, $[ICl_4]^-$, there are also two non-bonding pairs of electrons and such species are formally pseudo-octahedral compounds. In general, the substances in question are highly reactive materials towards processes other than substitution, and the replacement of the electronegative ligands required for their stabilization results in general decomposition. In the area of transition metals, electronic factors provide exceptional stabilization of square-planar geometry specifically for d^8 ions and especially for the second and third rows of the Periodic

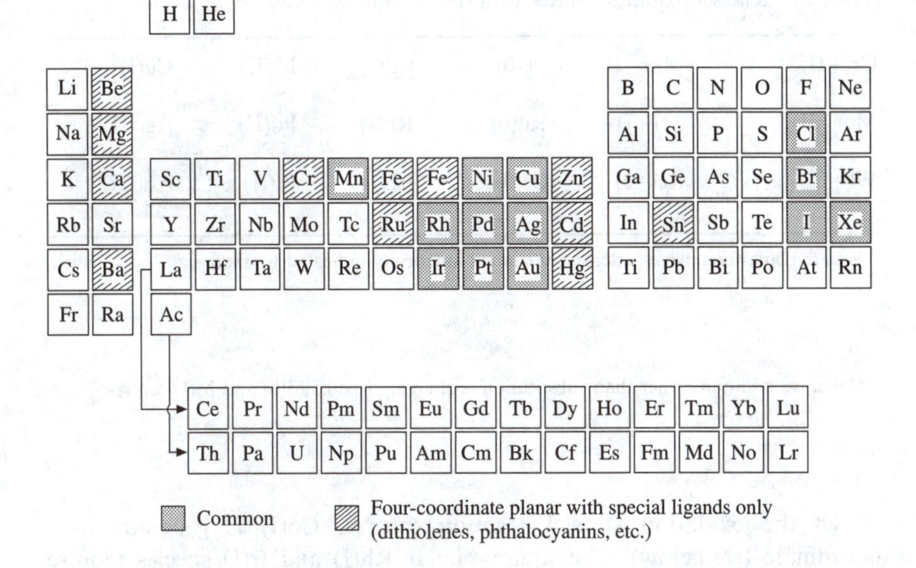

Fig. 3.9 The occurrence of planar four-coordination.

Table. Here and elsewhere in the *d*-block, complexes with highly conjugated bidentate or tetradentate ligands, such as those of the type $[M(L\cap L)_2]$, where L∩L is generally, but not necessarily, a two-sulphur donor (e.g. the maleonitriledithiolate anion, mnt^{2-}), are of considerable interest because of the extensive electronic delocalization over the metal–ligand system.[59] Electrons can be added or subtracted from the appropriate molecular orbitals of this type of complex without significantly changing its general shape; here the question of the oxidation state of the metal is best left unanswered. The four-coordinate planar geometry is a requirement of the ligand system and ligand substitution reactions usually lead to a general breakdown. The overall occurrence of planar four-coordination is summarized in Fig. 3.9.

The bulk of four-coordinate planar complexes whose ligand-substitution reactions have been studied in a systematic fashion are drawn from complexes whose central ion has the low-spin d^8 electron configuration. They are highlighted in Table 3.6. The lower oxidation states (0, –I, –II) are generally found in carbonyl and substituted carbonyl complexes and their organometallic derivatives. Their metal centres either achieve the predicted 18-electron valence shell with an appropriate coordination number,[60] e.g. 6 in $Cr(CO)_6$, 5 in $Fe(CO)_5$, or 4 in $Ni(CO)_4$, or else they form a wide variety of metal–metal bonded or carbonyl-bridged cluster species such as $Os_3(CO)_{12}$. Ligand-substitution reactions of such compounds are discussed elsewhere in this book (see Chapter 6).

Table 3.6 Known oxidation states with the d^8 configuration.[a,b]

Cr(−II)	Mn(−I)	Fe(0)	Co(I)	Ni(II)	Cu(III)
Mo(−II)	Tc(−I)	Ru(0)	Rh(I)	Pd(II)	Ag(III)
W(−II)	Re(−I)	Os(0)	Ir(I)	Pt(II)	Au(III)

[a] Square-planar complexes extensively studied kinetically indicated by shading:

M(*N*)

[b] Oxidative addition rather than substitution is characteristic of M(0) and M(I) centres.

Of the oxidation state I reaction centres, Co(I) is generally five-coordinate (cf. below). The square-planar Rh(I) and Ir(I) species require special ligands for their stabilization and even then are very sensitive to oxidative addition. Such reactions and their involvement in catalysed organometallic processes (Chapter 10) have dominated the interest of researchers involved in mechanistic studies in this area. Nevertheless, substitution reactions of such complexes, for example of rhodium(I)[61] and iridium(I)[62] species, have been studied, although the number of examples is small.

Cu(III) and Ag(III) are highly oxidizing states and, while some redox reactions of the $[Ag(OH)_4]^-$ ion are thought to be initiated by a rate-determining substitution and allow a little insight into behaviour patterns,[63] only Au(III) gives a sufficient range of four-coordinate planar complexes to allow systematic studies[64] to be carried out. Even here redox processes can often interfere, albeit in a way that generally can be recognized.

The divalent d^8 metal complexes are the most characteristic of the four-coordinate planar species. The bulk of the information relating to four-coordinate planar substitution mechanisms comes from the reactions of Pt(II) substrates, where the combination of redox stability and relatively low reactivity serves to make the synthesis of specifically designed substrates and the study of their kinetics simple enough to attract a large number of research groups over the years. Complexes of Pd(II) are in most respects similar to those of Pt(II), but the much greater reactivity (an increase of rate constant by a factor of 10^5 is quite usual) has made them less attractive for study. Ni(II) presents a wide range of coordination numbers, geometries and spin multiplicities, with square-planar complexes relatively uncommon. In treating substitution in four-coordinate planar complexes the main discussion will therefore frequently revolve about Pt(II) substrates, but the other reaction centres will be considered in the appropriate contexts.

3.3.2 General considerations for d^8 *complexes*

The elements whose oxidation states come within our interest lie on the right-hand side of the *d*-block and their bonding is reasonably covalent. As a result, the electrostatically important factors, such as charge and size, which are so important in determining mechanism in many octahedral systems, play only a secondary role and the concept of '*magic numbers*' or electron counting becomes important. The eighteen-electron valence shell tends to dominate the requirements for stability and so the four-coordinate d^8 metal complexes, with a sixteen-electron valence shell, are coordinatively unsaturated.

It therefore seems reasonable to expect that associatively activated substitution with its eighteen-electron valence shell transition states and intermediates would be energetically more feasible than dissociatively activated substitution with its fourteen-electron valence shell, three-coordinate intermediates. Long before Tolman[65] set down these rules it was known that the kinetics of the substitution reactions of four-coordinate, square-planar, d^8 complexes were consistent with associative activation and, until recently, substitution reactions where the preferred reaction path was dissociative were unknown. In this sense, there is a very close resemblance to the substitution reactions of the tetrahedral main-group elements of the second and subsequent rows, e.g. Si, P, which have accessible empty orbitals to accommodate the increase in coordination number (Section 3.2.4).

Systematic studies of the kinetics of substitution at platinum(II) began in the 1950s.[66] As was the case in the other systems discussed in the previous chapters, the need to work in coordinating solvents gave rise to kinetic ambiguities. It is usually necessary to use non-coordinating and, where possible, non-polar solvents if solvent interference is to be avoided. Solubility considerations and the need to avoid pre-equilibrium ion association would restrict such studies to the displacement of neutral ligands from neutral complexes by neutral nucleophiles. Classic examples are provided by two closely related reactions, *cis* ↔ *trans* isomerization of $[PtCl_2(Me_2S)_2]$, which is catalysed by added Me_2S, with the rate law, in dichloromethane solution:

$$\text{rate} = k_2[PtCl_2(Me_2S)_2][Me_2S]$$

and ligand exchange at the *trans* isomer of the palladium analogue:

$$trans\text{-}[PdCl_2(Me_2S)_2] + {}^*Me_2S \longrightarrow [PdCl_2(Me_2S)({}^*Me_2S)] + Me_2S$$

The latter reaction is, at least in chloroform, not complicated by any significant isomerization, and again obeys the simple second-order rate law

$$\text{rate} = k_2[PdCl_2(Me_2S)_2][Me_2S]$$

These reactions were studied by NMR line broadening.[67] The main problem with this technique is that labile systems are required, and independent variation of the entering group and leaving group in order to separate the factors that determine reactivity is either impossible or else very difficult. The use of NMR as a simple spectrophotometric monitoring technique under slow- or even stopped-flow conditions is making this type of study much more common. Indeed, the best example of a dissociatively activated mechanism in square-planar substitution (see below) was first studied by this technique.[68]

3.3.3 The kinetics and mechanisms of substitution

3.3.3.1 General features

Since most of the kinetic studies[69] that have been published were carried out in interfering (i.e. coordinating) solvents it is necessary to investigate the nature of this interference before the mechanistic assignment is secure. In a coordinating solvent, such as methanol or water, under the conditions where the concentration of entering nucleophile, Y, is large enough compared with that of the complex, (a) to ensure that the kinetics remain first-order, and (b) to drive the reaction to completion, the process

$$[L_3M{-}X + Y] \longrightarrow [L_3M{-}Y] + X \qquad \text{(charges omitted)}$$

usually obeys the rate law (Fig. 3.10):

$$-d[L_3M{-}X]/dt = k_{obs}[L_3M{-}X] = \{k_1 + k_2[Y]\}[L_3M{-}X]$$

Thus the (pseudo) first-order rate constant, k_{obs}, is given by

$$k_{obs} = k_1 + k_2[Y]$$

This is frequently referred to as 'the typical rate law for square-planar substitution'. It is, of course, only typical in coordinating solvents under the conditions described above.

Quite often, usually because the substrate undergoes reversible solvolysis:

$$[L_3M{-}X] + S \rightleftharpoons [L_3M{-}S] + X$$

and it is necessary to add an excess of X to suppress this, the kinetics are studied under conditions where $[X] \gg [L_3M{-}X]$ and then the rate law may take the form

$$k_{obs} = \frac{k_1 k_Y[Y]}{k_X[X] + k_Y[Y]} + k_2[Y]$$

The simpler expression is just a limiting form of this when $k_Y[Y] \gg k_X[X]$. At the other limit, when $k_X[X] \gg k_Y[Y]$, the general expression reduces to

$$k_{obs} = \left\{\frac{k_1 k_Y}{k_X[X]} + k_2\right\}[Y]$$

i.e. a simple second-order rate law with mass-law retardation by X. Care must be taken in discussing limiting cases in this way. The conditions of this approximation require that $k_1 \gg k_2$ if the limiting form is to be anything other than $k_{obs} = k_2[Y]$. Often, data taken in the intermediate region, where

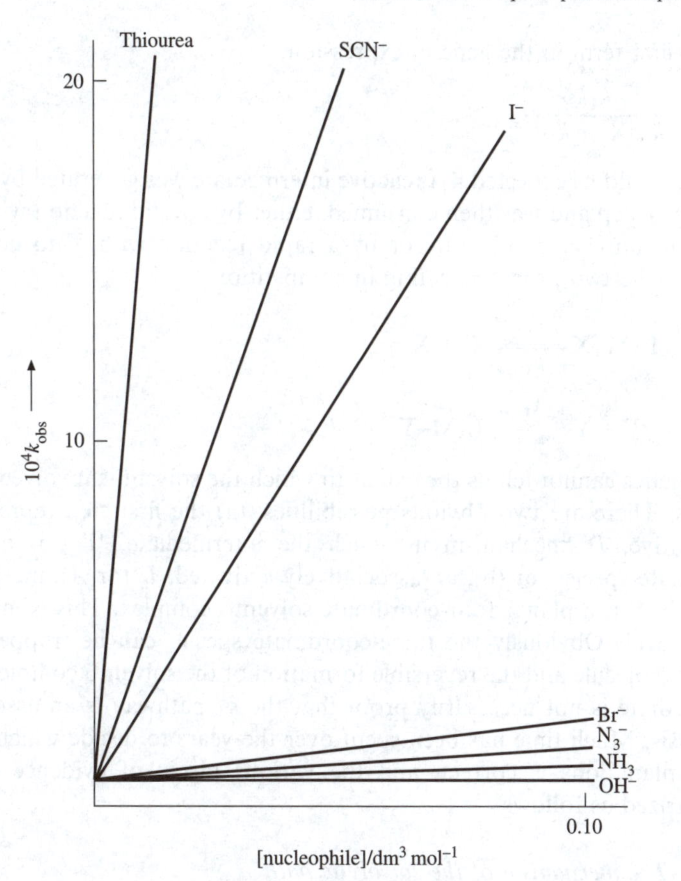

Fig. 3.10 Kinetic pattern for substitution at $[PtCl_2(en)]$, in aqueous solution at 308 K {[nucleophile] ≫ $[Pt^{II}]$}.

the dependence of k_{obs} on [Y] is not linear, are treated as if it were—especially if linear regression programs without graphics are used and the curvature is not spotted. The mechanistic explanations can then be quite bizarre (no references are given to avoid embarrassment). Other rate laws have been encountered and will be discussed in their appropriate place.

In all cases, k_2 is sensitive to the nature of the entering group, Y, and the extent of this sensitivity is of considerable importance in any discussion of the finer details of the act of substitution. It is therefore reasonably certain that this k_2 pathway corresponds to that of the simple second-order rate law that is observed in reactions in non-interfering solvents and is also due to an associatively activated substitution. Negative volumes of activation for k_2 pathways,[70] as for example the values of $-12.6\,cm^3\,mol^{-1}$, $-13.1\,cm^3\,mol^{-1}$, and $-14.5\,cm^3\,mol^{-1}$ for nitrite substitution in $[Pt(dien)X]^+$ for X = Cl, Br, and I, respectively,[71] support associative substitution.

The first term in the general expression,

$$\frac{k_1 k_Y}{k_X[X] + k_Y[Y]}$$

is what would be expected if a reactive intermediate was generated by a first-order (k_1) step and was then consumed, either by a rapid reaction with X to return to unreacted substrate or by a rapid reaction with Y to go on to product, the two processes acting in competition.

$$L_3M\text{–}X \underset{k_X}{\overset{k_1}{\rightleftharpoons}} \text{'I'} + X$$

$$\text{'I'} + Y \xrightarrow{k_Y} L_3M\text{–}Y$$

The kinetics cannot tell us the extent to which the solvent is involved in this process. There are two obvious possibilities: (a) the first step represents a dissociative, *D*, mechanism in which the intermediate, 'I' , is a three-coordinate species; or (b), an associatively activated, I_a (or *A*), mechanism in which 'I' is a planar four-coordinate solvento complex. This is indicated in Fig. 3.11. Obviously the three-coordinate species can be trapped by a solvent molecule and the reversible formation of the solvento complex as an intermediate is not necessarily proof that the k_1 pathway is an associative solvolysis. Much time has been spent over the years to decide which of the two explanations is correct, and the various pieces of evidence can be summarized as follows.

3.3.3.2 The mechanism of the solvolytic path

Five approaches to this question have, fortunately, all produced the same answer.

(i) *The dependence of the rate law upon the nature of the solvent.* Providing the solvent can coordinate to the metal, the k_1 pathway is usually present and can sometimes dominate the kinetics of substitution. Cases where a k_1 pathway is found when the reactions are carried out in a non-coordinating solvent are very few at the moment. They are of considerable interest and will be discussed separately below.

(ii) *The kinetics of substitution reactions of complexes containing bond-weakening (high* trans-*influence) ligands.* When the bond to the leaving group is weakened in the ground state, the reactivity of the substrate is increased. This is just as true for k_2 as for k_1, although it is true that k_1 is affected much more than k_2 and in extreme cases only the stronger nucleophiles can be seen to enter by way of the k_2 path. Taken by itself such information is ambiguous and might have been explained in terms of an enhancement of the dissociative character of the substitution were it not for other evidence.

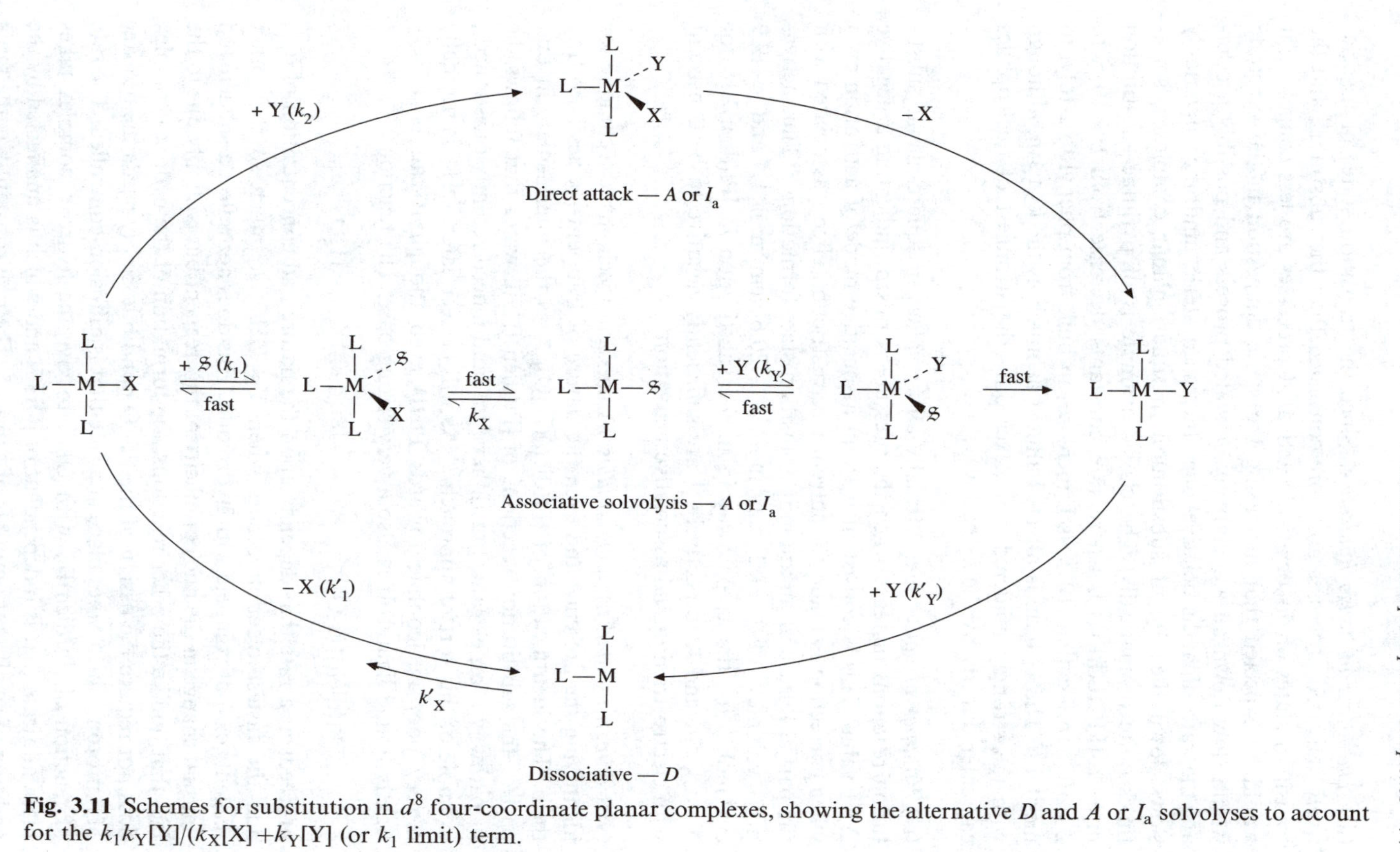

Fig. 3.11 Schemes for substitution in d^8 four-coordinate planar complexes, showing the alternative D and A or I_a solvolyses to account for the $k_1k_Y[Y]/(k_X[X]+k_Y[Y]$ (or k_1 limit) term.

(iii) *The effect of steric hindrance.* Since the transition states of associatively activated processes are more congested than the substrates, and the transition states of dissociatively activated processes are less congested, it would be expected that increasing the steric congestion in the substrate would slow down an associatively activated process and possibly accelerate the rate of a dissociative one. Increased steric hindrance invariably slows down the rate of substitution in square-planar complexes, sometimes quite dramatically. The classic example is of pyridine substitution into *cis*-$[PtClL(PEt_3)_2]$, where rate constants decrease from 8×10^{-2} to 2×10^{-4} to less than 10^{-6} as L increases in bulk from phenyl to 2-tolyl to mesityl.[72] The k_1 term is affected almost as much as the k_2 term, although with the strongly hindered complexes only the strongest of nucleophiles can enter directly by the k_2 pathway.

(iv) *Volumes of activation.* Virtually all k_1 values in square-planar substitution reactions increase when the pressure is raised, indicating a negative $\Delta V^‡$ value. This is consistent with an associative mode of activation even when correction is made for solvent electrostriction effects associated with the partial release of an anion in transition-state formation.[73] Thus activation volumes of -21.2 $cm^3\,mol^{-1}$, -20.6 $cm^3\,mol^{-1}$, and -19.9 $cm^3\,mol^{-1}$ for the k_1 pathway in nitrite substitution in $[Pt(dien)X]^+$ for X = Cl, Br, and I, respectively,[71] are sufficiently negative to be consistent with electrostriction and associative activation.

(v) *Competition and trapping studies.* A classical mechanistic technique for identifying the presence and even the nature of transient species is to trap them. The identification of $[Pt(dien)(H_2O)]^{2+}$ as the intermediate in that part of the substitution reactions of $[Pt(dien)I]^+$ in water that follows the k_1 pathway is an elegant, but rather rare (and indeed recently questioned), example of this sort of approach.[74] As will be seen later, OH^- is normally a very poor nucleophile towards Pt(II) even though it can be a very tightly bound ligand. It is also a very strong base. The reaction

$$[Pt(dien)(H_2O)]^{2+} + Y^- \longrightarrow [Pt(dien)X]^+ + H_2O$$

proceeds at a rate that depends upon the nature and concentration of Y^-, and the fastest reaction occurs when $Y = OH^-$, so much so that other reagents cannot compete with hydroxide. This does not reflect a sudden and unusual change in the nucleophilicity of this reagent but is simply due to the fact that, unlike the other reactions, the formation of the hydroxo complex does not require the fission of the Pt–O bond but simply needs the transfer of a proton, which takes place at a rate that is diffusion-controlled. The rate of reaction of $[Pt(dien)I]^+$ with OH^- to give the hydroxo complex takes place at a rate that is independent of $[OH^-]$ and which is unaffected by the addition of significant amounts of extra I^-. The rate constant has the same value as k_1 for this substrate determined for the entry of other nucleophiles in the absence of hydroxide. If the intermediate were a three-coordinate species, e.g. $[Pt(dien)]^{2+}$, a retardation of rate by I^- would be expected because of competition between I^- and H_2O (and possibly even OH^-) for this labile intermediate. If, on the other hand, $[Pt(dien)(H_2O)]^{2+}$ was formed directly by associative solvolysis, deprotonation in the presence of hydroxide would be much faster than the reverse reaction with I^- (whatever its concentration) and the rate would be independent of $[I^-]$. This technique can only be applied to the detection of solvent species that are stabilized by deprotonation and has not been used extensively. Nevertheless, it is quite usual to determine the value of k_1 in a protonic solvent by studying the solvolysis in the presence of a non-coordinating base, or hydroxide.

The conclusion to be drawn from all the kinetic evidence is that the associative mode of activation is dominant in the solvolytic (k_1) as in the k_2 path. Dissociative activation has now been identified and fully characterized in certain cases; it will be considered separately below (Section 3.3.11). Nowadays it is less necessary to recite all the above evidence in favour of assigning an associative mechanism to the solvolytic reaction, as the special requirements of the dissociative mechanism clearly allow it to be ruled out in most cases.

3.3.3.3 Non-platinum centres

Palladium(II) complexes undergo substitution in mechanistically similar ways to their platinum(II) analogues. However, they are very much more labile, rate constants usually being some five orders of magnitude greater than those of analogous Pt(II) substrates.[69] Complexes of gold(III) also undergo substitution by associatively activated processes, as does $[Rh^I(CO)(PPh_3)_2(ClO)_4]$, one of the few complexes of rhodium(I) whose substitution kinetics have been studied.[75] At the non-Pt centres the solvolytic pathway frequently contributes little or nothing to the overall reaction rate. Unusual rate laws, with higher orders with respect to the entering nucleophile, have also been reported. Thus, for example, there is a term in $[Br^-]^3$ in the rate law for displacement of the 2,2′-bipyridyl from $[Ag(bipy)Br_2]^+$. Often such results[76] are associated with the displacement of multidentate ligands, sometimes they cannot be reproduced. On occasions when only monodentate ligands are involved, satisfactory explanations have yet to be offered.

3.3.3.4 Evidence for five-coordinate intermediates

There has for decades been a debate as to whether associative substitution at square-planar centres is I_a or A in nature, in other words whether it is a one-stage or two-stage process. There have, therefore, been many attempts to seek evidence for the existence, even transiently, of five-coordinate intermediates.[77] Such attempts have sometimes involved kinetic tests, at other times the detection or even isolation of five-coordinate species. There are, of course, a good number of stable five-coordinate platinum(II) complexes to provide role models for the elusive kinetic intermediates. Strong kinetic evidence in favour of five-coordinate intermediates has been presented for such reactions as substitution at the $[Pd(dien)(NH_3)]^{2+}$ and $[Pd(dien)(py)]^{2+}$ cations and displacement of triphenyl antimony from $[RhCl(cod)(SbPh_3)]$ by amines.[78] Substitution kinetics at iridium(I) are extremely rarely studied, but it is interesting to note that although phosphine replacement in Vaska's compound, *trans*-$[IrCl(CO)(PPh_3)_2]$, is too fast to monitor by conventional techniques there is non-kinetic evidence for an *A* mechanism, i.e. for a transient five-coordinate intermediate.

3.3.3.5 *Bioinorganic and pharmacological relevance*

We shall now step out of our ivory tower for a brief excursion into an area of kinetic and mechanistic studies of considerable relevance and importance. The success of drugs such as cisplatin[79] and its second- and third-generation successors in the treatment of certain types of cancer has led to a number of studies of kinetics and mechanisms of substitution at such complexes.[80,81] Investigations of substitution at cisplatin, *cis*-$[PtCl_2(NH_3)_2]$, itself are complicated by its ready hydrolysis to $[PtCl(NH_3)_2(H_2O)]^+$ and the possibility of formation of some $[Pt(NH_3)_2(H_2O)_2]^{2+}$, by the possibility of *cis*$\rightarrow$*trans* isomerization, and by pH-dependent coordinated water/coordinated hydroxide equilibria.[82] The kinetics of nucleophilic substitution at carboplatin, $[Pt(NH_3)_2(cbdc)]$, where cbdc = 1,1′-cyclobutane-dicarboxylate,[83] and of formation of the palladium analogue of carboplatin from $[Pd(NH_3)_4]^{2+}$,[84] are more straightforward. Ring opening is the first step in the former reaction when conducted in acid (not HCl) solution, with the second-step proton-promoted opening of the chelate ring. In the presence of chloride, there is a simple second-order kinetic pattern ($k_{obs} = k[Cl^-]$). In the latter reaction there is no indication of any reaction through an aqua-intermediate $[Pd(NH_3)_3(H_2O)]^{2+}$, but there is NMR evidence for the intermediacy of transient $[Pd(NH_3)_3(\text{monodentate cbdc})]^{n+}$. The kinetics of reaction of cisplatin and of carboplatin with the platinum-loving entering group thiosulphate indicate, as expected, reaction predominantly through the k_2 pathway.[85] Carboplatin is resistant to thermal hydrolysis, but can be photolysed quite readily; kinetics and quantum yields have been established.[86] These studies are linked to *in vivo* situations through kinetic studies of interactions with phosphate entering groups. A simple-model system of this type is provided by reactions of $[Pt(en)Cl_2]$ and of $[Pt(dien)Cl]^+$ with inorganic phosphates—ortho-, pyro-, and tri-phosphates.[87] A better link is afforded by reactions of $[Pt(dien)Cl]^+$ with the 5′-monophosphates of adenosine (AMP), guanosine (GMP), cytidine (CMP), and thymidine (TMP). NMR studies of competition between these phosphates and methionine revealed, especially in the case of 5′-GMP, the balance between kinetic and thermodynamic control. Reaction with methionine is relatively rapid, but the final product is the 5′-GMP complex.[88]

3.3.4 The stereochemistry of substitution

In the dominant substitution pathway, which is associatively activated, the transition state(s) must be at least five-coordinate and, if the stoichiometric mechanism is associative, there must be an intermediate of higher coordination number between two transition states. Under normal circumstances substitution proceeds with complete retention of configuration and it can be inferred that the three non-participating groups must retain the same positions relative to one another throughout the course of the reaction. The fact that the rate of reaction is very sensitive to the electron displacement properties of the ligand *trans* to the leaving group (see the section on the *trans* effect below) while responding mainly to the bulkiness of the *cis* ligands (see the section on the *cis* effect below) suggests strongly that the three ligands also retain their symmetry inequivalence throughout and that the transition state(s) and intermediate resemble trigonal bipyramids with the *cis* ligands occupying the axial sites (Fig. 3.12).

The square-pyramidal geometry frequently found in five-coordinate d^8 metal-ion complexes may be traversed during the act of substitution but it represents an arrangement below the energy maxima of the transition states. No attempt has been made to include the effects of solvation and desolvation in these processes even though one might expect to find that

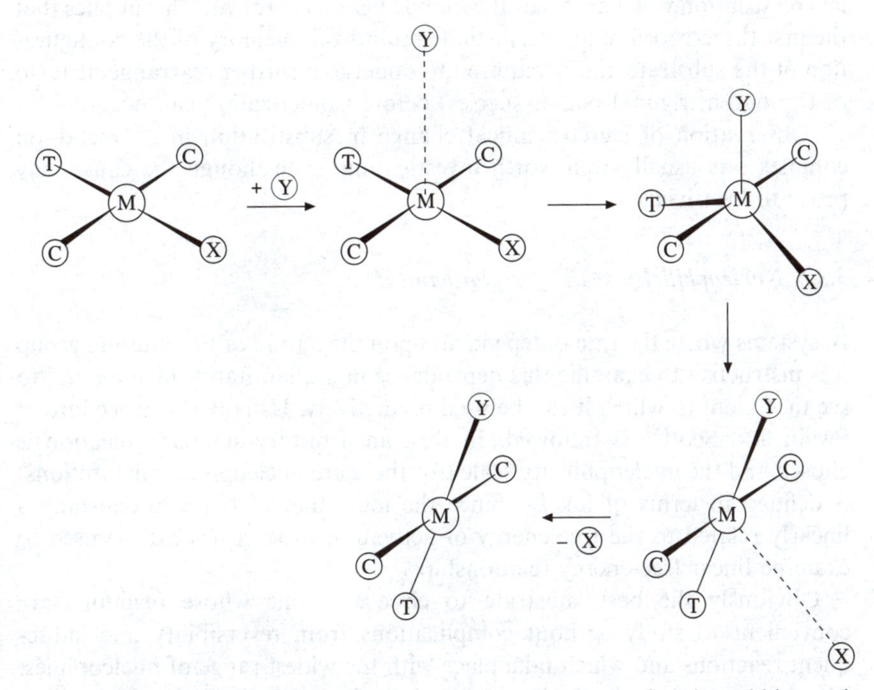

Fig. 3.12 Substitution in a planar substrate by way of a trigonal bipyramidal intermediate.

the interaction between the metal ion and the solvent molecules in the axial positions above and below the coordination plane is far from negligible. Much was made of this in the early days of these studies and it was suggested that the nucleophile and the solvent competed for these axial sites, with the act of substitution resembling a dissociative collapse of the pseudo-octahedral species.[89] This mechanism was soon discarded, and the demonstration that the four coordinated waters in $[M(H_2O)_4]^{2+}$ (M = Pd, Pt) remain distinct from any others that may be solvating the complex in aqueous solution and that water exchange is mechanistically identical to any other act of substitution allows it to be ruled out completely.[90] Nevertheless, the role of the solvent is still a major one and is far from being well understood. A detailed discussion of this aspect would be premature at this time. The preponderance of cases of retention of configuration indicates that the five-coordinate intermediate species usually does not live long enough to undergo the intramolecular rearrangements that are so facile in such systems. There are a few examples of substitution reactions in which there is stereochemical change, and the number is increasing. It is plausible to believe that some of these are cases where, for one reason or another, the five-coordinate intermediate does live long enough to rearrange.

The three-coordinate intermediate in the rare dissociatively activated substitution mechanism (Section 3.3.11) is probably T-shaped. Isomerization that takes place through a dissociative mechanism is sensitive to mass-law retardation when excess of the mobile ligand is present. This implies that the first three-coordinate intermediate retains full memory of the configuration of the substrate and is required to undergo a further rearrangement (to or through a trigonal-planar species) before isomerization can occur.

Observation of stereochemical change in substitution in d^8 metal-ion complexes is usually well worth investigating, even though the cause may prove to be trivial.

3.3.5 Nucleophilicity scales for platinum(II)

In systems where the rate is dependent upon the nature of the entering group it is instructive to examine this dependence in a quantitative fashion and to see the extent to which it can be used predictively. Usually the procedure of Swain and Scott[91] is followed. In this, an arbitrary standard reaction is chosen and the nucleophilicity scale (for these are nucleophilic substitutions) is defined in terms of log k_2. Since the logarithm of the rate constant is linearly related to the free energy of activation these scales can be used to examine linear free-energy relationships.

Obviously the best substrate to choose is one whose reactions are convenient to study without complications from reversibility and subsequent reactions and which take place with the widest range of nucleophiles. The substrate should also have a moderately good discriminating ability. These criteria were not actually applied to the chosen substrate.[92] It just so happened that, at the time the first survey was made, much of the data had already been obtained for other purposes.

The original n_{Pt} scale was defined by

$$n_{Pt} = \log (k_2/k_1)$$

using kinetic data for the reaction

$$trans\text{-}[Pt(py)_2Cl_2] + Y^{n-} \longrightarrow trans\text{-}[Pt(py)_2Cl(Y)]^{(n-1)-} + Cl^-$$

in methanol at 30.0 °C (py = pyridine). Some data for *trans*-[Pt(piperidine)$_2$Cl$_2$] were mixed in because the difference in the behaviour of the two substrates appeared to be negligible. Since k_2 and k_1 differ in their dimensions, the logarithmic representation is strictly out of order and, once it was accepted that the first-order pathway was an associative solvolysis, k_1 was replaced by a second-order rate constant, so that

$$n_{Pt}{}^0 = \log (k_2/k_1{}^0)$$

where $k_1{}^0 = k_1[\mathrm{MeOH}]$, and, since 1 dm^3 of methanol at 30 °C contains 24.3 moles, $k_1{}^0 = k_1/24.3$. Therefore

$$n_{Pt}{}^0 = n_{Pt} + \log 24.3 = n_{Pt} + 1.39$$

This distinction has not always been maintained, and a further complication, for which one of us must take the blame, has been the tendency to use $n_{Pt}{}^0$ to indicate the nucleophilicity scale appropriate to uncharged substrates (as distinct from $n_{Pt}{}^-$, $n_{Pt}{}^+$, $n_{Pt}{}^{2+}$, etc.; cf. below). The reader is advised to be wary when reading the original literature.

A number of additions have been made to the original compilation (mainly sulphur-donor nucleophiles) and the most recent values are collected in Table 3.7. The most valuable use that can be made of this compilation is to examine the factors that contribute towards the nucleophilicity towards Pt(II) of a particular reagent. The following points are worth remembering:

(a) The scale covers many orders of magnitude in reactivity (i.e. the rate of reaction is very sensitive to the nature of the entering group, an observation that is extremely rare in coordination chemistry outside the substitution reactions of d^8 metal complexes).

(b) The nucleophilicities follow typical *Class b*, or *Soft*, acid–base behaviour, i.e. compounds with second-row donor atoms are much more effective than the compounds of the corresponding light (first-row) element:

$$I^- > Br^- > Cl^- \gg F^-$$

(this is reflected in the thermodynamic stability; F^- complexes of Pt(II) are not easy to make and are unstable in coordinating solvents, other than HF);

$$Se > S > Te \gg O,$$

Table 3.7 A selection of n_{Pt}^0 values[a] listed according to the donor atoms.

Carbon		**Nitrogen**			
$C_6H_{11}NC$	6.34	NH_3	3.07	NO_2^-	3.22
CN^-	7.14	$C_5H_{10}NH$	3.13	imidazole	3.44
		$PhNH_2$	3.16	N_3^-	3.48
		C_5H_5N	3.19	NH_2OH	3.85
				NH_2NH_2	3.86
Oxygen		**Halogens**		**Phosphorus**	
MeOH	0.00	F^-	<2.2[b]	$(Et_2N)_3P$	4.57
MeO^-	<2.4[b]	Cl^-	3.04	$(MeO)_2PO^-$	5.01
OH^-	<2.4[b]	Br^-	4.18	$(MeO)_3P$	7.23
$MeCO_2^-$	<2.4[b]	I^-	5.46	Ph_3P	8.93
				$(Bu^n)_3P$	8.96
				Et_3P	8.99
		Sulphur			
Me_2SO	2.56[c]	$(4\text{-}MeC_6H_4)_2S$	3.68[e]	$(CH_2)_5S$	5.02
$(4\text{-}MeC_6H_4)MeSO$	2.60[d]	$(4\text{-}MeOC_6H_4)_2S$	3.73[e]	$(CH_2)_4S$	5.14
(Ph)MeSO	2.90[d]	$(4\text{-}HOC_6H_4)_2S$	3.85[e]	SCN^-	5.75
$(4\text{-}MeOC_6H_4)MeSO$	3.16[d]	PhMeS	3.99[d]	SO_3^{2-}	5.79
$(4\text{-}ClC_6H_4)_2S$	3.21[e]	$(4\text{-}ClC_6H_4)MeS$	4.04[d]	$(Me_2N)_2CS$	5.83[c]
Ph_2S	3.22[e]	PhSH	4.15	Me_4thiaz[f]	6.07[c]
$(4\text{-}ClC_6H_4)PhS$	3.25[e]	$(4\text{-}MeC_6H_4)MeS$	4.24[d]	thiaz[f]	6.39[c]
$(4\text{-}FC_6H_4)_2S$	3.30[e]	$(4\text{-}H_2NC_6H_4)_2S$	4.27[e]	$(NH_2)_2CS$	7.17
$(4\text{-}NH_2C_6H_4)\text{-}(4\text{-}NO_2C_6H_4)S$	3.31[e]	Et_2S	4.52	PhS^-	7.17
$(PhCH_2)_2S$	3.43	$(4\text{-}MeOC_6H_4)MeS$	4.55[d]	$S_2O_3^{2-}$	7.34
$(4\text{-}MeOC_6H_4)PhS$	3.64[e]	Me_2S	4.87		
Arsenic		**Selenium**		**Tin**	
Ph_3As	6.89	$(PhCH_2)_2Se$	5.53	$SnCl_3^-$	5.44
Et_3As	7.68	Me_2Se	5.70		
		$SeCN^-$	7.11		
		Antimony			
		Ph_3Sb	6.79		

[a] Data from R. G. Pearson, H. Sobel, and J. Songstad, *J. Am. Chem. Soc.*, 1968, **90**, 319.
[b] No k_2 term could be observed.
[c] L. Cattalini, M. Bonivento, G. Michelon, M. L. Tobe, and A. T. Treadgold, *Gazz. Chim. Ital.*, 1988, **118**, 725; this paper contains data for 18 substituted thioureas and similar compounds.
[d] T. Hoover and P. Zipp, *Inorg. Chim. Acta*, 1982, **63**, 9.
[e] J. R. Gaylor and C. V. Senoff, *Can. J. Chem.*, 1971, **49**, 2390.
[f]

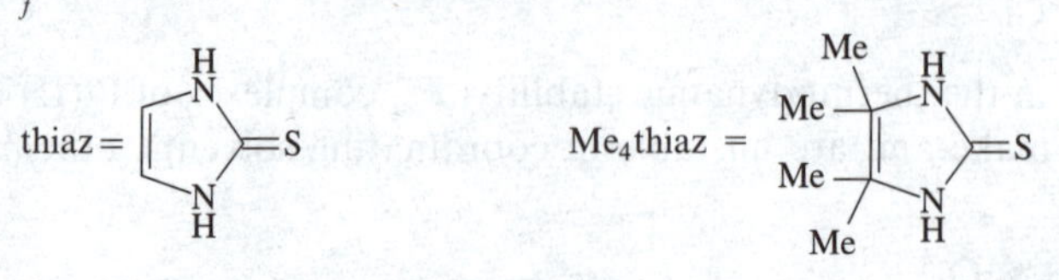

but in this case the kinetic and thermodynamic aspects do not match since OH^- and OR^- complexes can be very stable indeed:

$$P > As > Sb \gg N.$$

Although the number examined is small, carbon ligands that function as σ-donors are generally very good nucleophiles.

(c) The nucleophilicities of ligands containing a *soft* donor are extremely sensitive to the nature of the substituents and the extent to which its other electrons are involved in bonding. This is seen at its best in the range of $n_{Pt}{}^0$ values for sulphur-donor ligands. The nucleophilicities of ligands with *hard* donors (only nitrogen offers the range of nucleophiles to support this statement) do not vary much with major variation of the substituents, or the 'oxidation state' of the donor. The effects of systematic variation of the basicity of amines and other heterocyclic nitrogen bases has not been studied using the standard reaction and so the variation in the $n_{Pt}{}^0$ value with basicity is not known. However, this type of study has been made elsewhere in Pt(II) chemistry many times and, while plots of log k_2 against the pK_a of the protonated nitrogen base are frequently linear, the slopes are very small and occasionally are negative.[93]

(d) Some nucleophiles are less reactive than expected, e.g. Me_2SO (binding finally through S), OH^-, and other anionic oxygen-donor ligands.

3.3.6 Applicability of the platinum(II) nucleophilicity scale

It is of little use to introduce the concept of nucleophilicity scales if the order (qualitative or quantitative) is not maintained throughout a wide range of substrates. For many compounds it has been found that for a specific substrate, a plot of log k_2 for a particular nucleophile against its $n_{Pt}{}^0$ value is linear, i.e.

$$\log k_2 = S n_{Pt}{}^0 + C$$

The slope, S, is called the *nucleophilic discrimination factor* and the intercept, C, is the *intrinsic reactivity*. The concept of nucleophilic discrimination finds wide application in discussions relating to the intimate mechanism of the reaction and is often related to the extent of bond making in the rate-determining transition state. Considerable use has been made of the concept of n_{Pt}, not only in predicting and explaining reactivity, but also in application to ground-state phenomena. An early example was Sacconi's attempt to use the sum of the n_{Pt} values of the donor atoms of the ligands in the coordination shell as a measure of their 'softness' in order to explain the factors that determine the spin multiplicity of the metal ion.[94]

In the many years that have followed the introduction of the $n_{Pt}{}^0$ scale it has been realized that its application is somewhat restricted. It was shown quite early on[95] that certain nucleophiles, e.g. $SeCN^-$, $(NH_2)_2CS$, and NO_2^-, did not always provide points that sat on the line and it was suggested that the ability of the entering group to act as a π-acceptor might enhance its reactivity (it would be wrong to use the word nucleophilicity for what is essentially an electrophilic assistance) by lowering the energy of the transition state. Since this contribution would also depend upon the π-donor ability of the substrate, it would make the reagents more reactive than predicted by their $n_{Pt}{}^0$ value towards those substrates that were better π-donors than *trans*-$[Pt(py)_2Cl_2]$, e.g. $[PtCl_4]^{2-}$, and less reactive towards those that are worse, e.g. $[Pt(dien)Br]^+$. Such reagents were said to be *biphilic*.

More recently, as the amount of published data has increased, it has become clear that the $n_{Pt}{}^0$ scale is best applied to uncharged nucleophiles and that charge plays a much greater part in determining reactivity than was originally thought.[96] Steric crowding, on the other hand, has only a minor effect.[97] Thus, anionic reagents such as $[PtCl_4]^{2-}$ and $[Pt(L)Cl_3]^-$ (where $L = SMe_2$, SEt_2, Me_2SO, PMe_3, PEt_3, PPh_3, $AsEt_3$, etc.) are far less reactive than neutral reagents with similar $n_{Pt}{}^0$ values when the substrate is anionic,[98] and the reverse is true, but less markedly so, for cationic substrates.[99] By choosing suitable standard reactions, it has become customary to develop a nucleophilicity scale for each charge type of substrate, and then the plots of log k_2 against the appropriate nucleophilicity are substantially linear. However, in none of these scales is there the extensive data collection that can be found in the $n_{Pt}{}^0$ scale. No doubt in time worthy and ingenious researchers will be able to persuade some funding body that a knowledge of nucleophilicity scales of this sort will lead to a cure for cancer, or a diminution in the greenhouse effect, or some other politically favoured aim or objective.

3.3.7 Nucleophilicity scales for palladium(II) substrates

The greater reactivity of palladium(II) makes an analogous quantification of an $n_{Pd}{}^0$ scale somewhat more difficult. If we assume that the usually quoted factor of 10^5 for the increase in reactivity on going from Pt(II) to Pd(II) applies throughout, then strong nucleophiles, such as PEt_3, might react with *trans*-$[Pd(py)_2Cl_2]$ with a second-order rate constant in the region of $10^7\ dm^3\ mol^{-1}\ s^{-1}$ in methanol at 30 °C. Such rate constants are inaccessible to non-reversible techniques unless means can be found to buffer extremely low concentrations of the nucleophile. Attempts have been made to overcome this problem by using less reactive substrates for the standard reaction. For example, choosing *trans*-$[Pd(P^iPr_3)_2(NO_2)_2]$ as standard reagent and measuring the rate constants for the displacement of the first nitrite in methanol at 30 °C, values of $n_{Pd}{}^0$ (defined in the same way as $n_{Pt}{}^0$) were found to be almost numerically identical to the $n_{Pt}{}^0$ values of the same reagents. As in the case of Pt(II), nucleophilicities of amines and other nitrogen bases are not very sensitive to the proton basicity of these species, while the dependence of nucleophilicity on the inductive effects of substituents bound to sulphur can be quite marked but depends upon the nature of the substrate. Nothing has yet been done to see the effect of the charge of the substrate on nucleophilicity scales towards palladium(II). Much more needs to be done in this area, but the technical difficulties are considerable.

3.3.8 Nucleophilicities towards gold(III) substrates

Although much has been published on the kinetics of substitution at four-coordinate planar Au(III),[58] there has been no systematic attempt to establish nucleophilicity scales. Because of the different oxidation states representing the d^8 configuration, it is not possible to make a direct comparison between Pt(II) and Au(III). If the same ligands are used, the substrates will differ in charge and if complexes of the same charge are compared, the coordination shells will have to differ. In addition to this, complexes with the same coordination shell frequently react differently. For instance contrast

$$[Pt(py)Cl_3]^- + N_3^- \longrightarrow cis\text{-}[Pt(py)(N_3)Cl_2]$$

with

$$[Au(py)Cl_3] + N_3^- \longrightarrow [AuCl_3(N_3)]^- + py$$

even though both reactions are carried out in methanol. A collection of rate constants for the reactions of nucleophiles with $[AuCl_4]^-$ will be found in Table 3.8, where they are compared with data for the corresponding reactions of $[PtCl_4]^{2-}$. The very much greater reactivity of Au^{III} compared with Pt^{II} is apparent.

One of the problems encountered in studies of the substitution reactions of Au(III) is that the 'softer' nucleophiles are reducing agents and Au(III) is readily reduced to Au(I). Frequently, the ligand-substitution process can be observed as a rapid initial reaction and the subsequent reduction is slow. Thus the reaction of $[AuCl_4]^-$ with Me_2S consists of a rapid sequence of substitutions, some of which are reversible, to give neutral $[Au(Me_2S)Cl_3]$ and cationic $[Au(Me_2S)_2Cl_2]^+$. The reduction goes by way of an inner-sphere two-electron process in which the chloride transfers from Au to S. The direct reduction becomes more facile as the number of Me_2S ligands in the complex increases (probably due to the change in the charge). There is no indication that direct reduction of $[AuCl_4]^-$ makes a significant contribution to the reaction.

Table 3.8 Comparison of the rate constants for the reaction $[MCl_4]^{n-} + Y^{m-} \rightarrow [MCl_3(Y)]^{(n+m-1)-} + Cl^-$ for M = Au, Pt in water at 25 °C.

M	Au ($n = 1$)	Pt ($n = 2$)	
Y	$k_2/dm^3\,mol^{-1}\,s^{-1}$	$10^5 k_2/dm^3\,mol^{-1}\,s^{-1}$	k_{Au}/k_{Pt}
H_2O	0.020[a]	5.0[a]	400
Cl^-	1.5	[b]	
Br^-	63	4.8	1 310 000
Pyridine	1.6	340	470
SCN^-	13 000	1000	1 300 000
I^-	< 85 000[c]	1960	< 4 340 000

[a] First-order rate constant (s^{-1}).
[b] No chloride-dependent path detected.

$$[AuCl_4]^- + Me_2S \rightleftharpoons [AuCl_3(Me_2S)] + Cl^-$$

$$[AuCl_3(Me_2S)] + Me_2S \rightleftharpoons [AuCl_2(Me_2S)_2]^+ + Cl^-$$

$$[L_3Au\text{–}Cl] + SMe_2 \longrightarrow [L_3Au\cdots Cl\cdots SMe_2]$$

$$\longrightarrow L + [L_2Au] + ClSMe_2{}^+$$

$$ClSMe_2{}^+ + H_2O \longrightarrow OSMe_2 + 2H^+ + Cl^-$$

The analogous reaction with phosphines eventually leads to reduction of the gold and the formation of the phosphine oxide, presumably by a similar mechanism.

In the reaction with I^-, the product usually contains $[AuI_2]^-$ and $I_3{}^-$ and the reaction appears to take place in a single stage. This poses the question as to whether the rate-determining step is a substitution (followed by a fast redox process) or a two-electron inner-sphere redox process.

$$[Cl_3AuCl]^- + I^- \rightarrow [Cl_3Au\cdots Cl\cdots I]^{2-} \rightarrow Cl^- + [AuCl_2]^- + ICl$$

$$ICl + 2I^- = I_3{}^- + Cl^-$$

The rate constant for the reaction of $[AuCl_4]^-$ with I^- is so much greater than those for the reaction with Br^- and Cl^- that a rate-determining redox process is indicated. This is not always the case and in methanol the discrepancy in the rate constants for the same set of reactions is much less, although iodide is still comparatively too reactive for substitution to be rate-limiting. The reaction of I^- with *trans*-$[Au(C_6H_5)(MeOH)Cl_2]^-$ in methanol appears to be a rate-limiting substitution followed by reduction.

In general, the reactions of normal Au(III) substrates have a smaller contribution from the solvolytic pathway than do those of Pd(II) and Pt(II). This is true even for anionic reagents such as $[AuCl_4]^-$, $[AuBr_4]^-$, and *trans*-$[Au(CN)_2X_2]^-$ (X = Cl, Br), which do not appear to have the aversion towards anionic nucleophiles possessed by $[PtCl_4]^{2-}$. An exception can be found in the reactions of $[Au(C_6H_5)Cl_3]^-$ in methanol, where the solvolytic pathway dominates the reactions with anionic nucleophiles and there is no evidence for direct reaction between two anionic species.[101]

A major difficulty that is likely to arise if any attempt is made to apply nucleophilicity scales to the reactions of Au(III) complexes is that reactivity sequences appear to change with minor changes in the nature of the leaving group. It is thought that the substitution reactions of Au(III) have much more I_a character than those of Pd(II) and Pt(II) and thus the separation of entering-group and leaving-group effects is far less possible.[102] This is discussed further in the next section.

3.3.9 The dependence of reactivity on the nature of the leaving group

This is by far the most difficult effect to systematize, because not only is it very dependent upon the nature of the reaction centre but it can also be intimately connected to the nature of the entering nucleophile and the nature of the *trans* ligand. As a rule, the study must be limited to the most replaceable ligand in the complex (although in favourable circumstances ligand-exchange studies at equilibrium can provide rate constants for more than one site exchange). The nucleofugacity (leaving ability) of a particular ligand must be a combination of its intrinsic lability (assuming that there is such a thing, and not to be confused with intrinsic reactivity as defined in Section 3.3.7) and the labilizing effects exerted upon it by the other ligands in the complex. An example can be found in the substitution reactions of *cis*-$[Pt(Me_2S)_2(NH_3)_2]^{2+}$, where the first ligand to be displaced is always NH_3. Although usually considered to be a very poor leaving group, the ammonia is under the relatively strong *trans* effect (see below) of Me_2S, whereas the better nucleofuge, Me_2S, is under the very weak *trans* effect of NH_3. However, the second ligand to be displaced depends upon the nature of the first group that entered; Cl^-, Br^-, and I^- will cause the second ammonia to be displaced, while stronger *trans*-effect ligands such as NO_2^-, N_3^-, and SCN^- cause the displacement of the Me_2S lying *trans* to them.[103] When *trans* to the same ligand, Me_2S is always more labile than NH_3 and the *trans*-isomer of $[Pt(Me_2S)_2(NH_3)_2]^{2+}$ is much less labile than the *cis*, always losing the dimethyl sulphides first.

Sometimes the group that leaves depends upon the nature and the conditions of the reaction. Thus, *cis*-$[Pt(Me_2S)_2Cl_2]$ loses Me_2S when reacted with neutral nucleophiles such as amines in a non-polar solvent such as dimethoxyethane, whereas chloride, under the *trans* effect of the Me_2S, is readily replaced in polar solvents.[104] It was suggested that it is the inability of a non-polar solvent to support the ionogenic process that controls this behaviour, but one should not rule out the possibility that the chloride ion-pair is a highly reactive transient intermediate in a two-stage process.

The use of substrates containing a strong *trans*-effect ligand can therefore extensively widen the range of leaving groups that can be studied under otherwise identical conditions. Sometimes it is possible to secure the three remaining sites by linking them with a terdentate ligand which forms stable complexes. Ring opening, if it does occur, is likely to be reversible and not interfere with the study of the displacement of the monodentate ligand. One of the most extensive series of reactions used to study the leaving-group effect has been the reaction

$$[Pt(dien)X]^{(2-n)+} + Nu^{x-} \longrightarrow [Pt(dien)(Nu)]^{(2-x)+} + X^{n-}$$

where for the entry of a particular nucleophile, the rate constants for the displacement of X decrease in the order[105]

$$H_2O \gg Cl^- > Br^- > I^- > N_3^- > \underline{S}CN^- > NO_2^- > CN^-$$

At one time it was thought that the slopes of the plots of log k_2 against $n_{Pt}{}^0$ were independent of the nature of the leaving group and that the leaving group effect resided solely in the intrinsic reactivity. While this may be true for many Pt(II) substrates, there are a large number of exceptions, and once away from Pt(II) even the nucleophilicity sequences may depend upon the nature of the leaving group. This is true for the displacement of heterocyclic nitrogen bases (L) from $[AuCl_3L]$[106] or from *trans*-$[Au(CN)_2ClL]$,[107] where the order of nucleophilicity of Cl^-, Br^-, N_3^- and NO_2^- depends upon the basicity of L.

In a dissociatively activated substitution, the lability of a substrate will depend upon the factors that promote bond dissociation, including not only bond strength (for heterolytic fission) but also any stabilization of the intermediate of lower coordination number and solvation of the developing fragments. In an associatively activated process the importance of bond breaking will depend entirely upon the intimate mechanism and may range from insignificance to the same importance as is encountered in a dissociatively activated process. It has already been shown in Chapter 2 that an *A* mechanism has two transition states and that the general expression

$$k_2 = \frac{k_a k_b}{k_{-a} + k_b}$$

can have two limiting forms. Here k_a is the rate constant for forming the intermediate, k_{-a} is the rate constant for the dissociation of the intermediate back to substrate plus nucleophile, and k_b is the rate constant for the dissociation of the intermediate to give products. When the bond-forming transition state has the higher energy, $k_2 = k_a$, but, even though the rate constant for bond breaking does not enter the expression for k_2, the leaving group still forms part of the coordination shell of the reaction centre and therefore must exert some effect upon its electrophilicity and hence upon k_a. Quite obviously, when the bond-breaking transition state lies higher and

$$k_2 = \frac{k_a k_b}{k_{-a}}$$

the rate constant must be sensitive to the factors influencing bond breaking.

Frequently, the rate constants for the entry of a common nucleophile into a series of substrates differing only in the nature of the the leaving group tend to parallel the equilibrium constants, i.e. in a reaction of the type

$$[L_3MX] + Y \underset{k_{-2}}{\overset{k_2}{\rightleftharpoons}} [L_3MY] + X$$

a plot of log k_2 against log K (the equilibrium constant $= k_2/k_{-2}$) is linear. This might be expected as a typical linear free-energy relationship, in this case between the free energy of activation and the standard free energy change for the reaction. In their interpretation of Leffler and Grunwald's original proposal,[108] Langford and Gray[109] suggested that the slope of such a plot will give some indication of the mode of activation, slopes of < 0.5 indicating an associative mode of activation—an argument used extensively in the assignment of mechanism to solvolytic reactions of octahedral complexes (see Chapter 4). However, in reactions involving amine displacement of the type

$$trans\text{-}[Pt(L)(am)Cl_2] + Cl^- \overset{k_2}{\rightleftharpoons} [Pt(L)Cl_3]^- + am$$

$$\{\text{equilibrium constant } K\}$$

which have been studied extensively, the plot of log k_2 against log K has, for instance, a slope of 1.0 in the case where L = dimethyl sulphoxide.[110] This might suggest that the process is dissociatively activated, but the kinetics and all the other evidence say otherwise. The Leffler–Grunwald relationship is still valid, because this is a case of an *A* mechanism where the bond-breaking transition state lies at the higher energy and, to that extent, the process must be considered to be dissociatively activated. Great care must be taken when these non-synchronous substitution processes are being considered.

When there is more than one equivalent leaving group, their relative reactivities depend upon a number of factors. Obviously each act of substitution will cause some change in the substrate, either by altering the labilizing roles of the spectator ligands, as in the case of *cis*-$[Pt(Me_2S)_2(NH_3)_2]^{2+}$, or by changing the charge type. In systems where there are two equivalently located potential leaving groups and the change in the spectator ligand effects as a result of the first act of substitution is small, the two successive rate constants are statistically related, i.e. the value of the first is approximately twice that of the second. One notable exception to this rule is the mutual labilization of a pair of *cis*-dimethyl sulphoxides. For example, while the single sulphoxide in $[Pt(dien)(Me_2SO)]^{2+}$ is very difficult to replace (at least 10^4 times more difficult than the displacement of H_2O from the equivalent complex), the rate constant for the displacement of the first dimethyl sulphoxide from *cis*-$[Pt(en)(Me_2SO)_2]^{2+}$ is anything from 15 to 400 times greater than that for the displacement of water from *cis*-$[Pt(en)(Me_2SO)(H_2O)]^{2+}$, depending upon the strength of the entering nucleophile. Similar effects are found in other *cis*-bis(dimethyl sulphoxide) complexes, such as *cis*-$[PtCl_2(Me_2SO)_2]$,[111] and it has been suggested that the two sulphoxides lower the energy of the five-coordinate intermediate and allow it to rearrange. The absence of examples of this type of mutual labilization involving other ligands prevents these suggestions from being tested at the present time.

This section is linked to the next by a series of recent studies prompted by the thought that asynchronous substitution, a characteristic of square-planar complexes, provides an opportunity to separate the energetic contributions of bond formation and bond rupture to free energies of activation. This approach has been applied to forward and reverse reactions of equilibria involving complexes of the type $[Pt(LLL)(L)]^{2+}$, $[Pt(LL)(L)X]^{+}$, $[Pt(LLL)Cl]^{+}$, and $[Pt(LL)XY]$, where L is an amine or pyridine and LL, LLL are bidentate or terdentate nitrogen/sulphur-donor ligands. The effects of such factors as ligand basicity and σ- and π-bonding for the leaving and entering groups, and effects of non-leaving ligands, were probed. Comparisons of L = ammonia versus L = substituted pyridines, and slopes of reactivity/basicity plots, confirm the importance of platinum–nitrogen and platinum–sulphur π-bonding interactions, where these are feasible. The various effects can be summarized in terms of chemical potentials of initial and transition states, as affected by electronic and, where relevant, steric effects.[112]

3.3.10 The effect of the non-participating ligands on the lability of the complex

3.3.10.1 Introduction

In an associatively activated substitution of a four-coordinate planar d^8 metal complex, the intermediate is a trigonal bipyramid with the entering group, the leaving group, and the *trans* ligand in the trigonal plane. The two ligands that were *cis* to the leaving group occupy the axial positions. As a consequence, the T-shaped relationship between the *trans* ligand and the two *cis* ligands remains intact throughout the act of substitution (Fig. 3.12 above). Not only does this lead to substitution with retention of configuration, but it also means that the the two *cis* ligands remain symmetrically distinct from the *trans* ligand throughout the act of substitution. As a result, the effect on the reactivity of varying the nature of the ligand *trans* to the leaving group is totally different from the corresponding *cis* effect.

3.3.10.2 The history of the trans *effect*

In view of the fact that this is one of the few mythical beasts of inorganic chemistry that can stand up to a quantitative examination, a brief historical introduction is not out of place. The ability of certain ligands to promote substitution *trans* to themselves was first noticed and made use of in synthetic chemistry by Alfred Werner at the turn of the century,[113] but the naming and systematic application of the *trans* effect had to wait another quarter of a century. The work of Chernayev and his colleagues[114] in the preparation of square-planar complexes of platinum(II) showed that ligands could be placed in a sequence of increasing ability to direct substitution *trans* to themselves and that their position in this sequence was qualitatively retained on going from one complex to another. One consequence of this was that some measure of planning could be introduced into the synthesis of inorganic complexes, which until then had been of the 'make now—characterise later' variety. We might comment that this is not to say that such an approach is no longer encountered in other parts of synthetic inorganic chemistry—as, for example, in the case of technetium pharmaceuticals, where diagnostic efficacy was demonstrated several years before oxidation state, stoichiometry, or structure were established. To return to substitution at platinum(II), the *trans*-effect sequence of the classic ligands, i.e. those known in the mid-1920s, is:

$$H_2O < OH^- < NH_3 \sim RNH_2 < \text{pyridine} < Cl^- < Br^- < SCN^- \sim I^- \sim NO_2{}^- \sim SO_3{}^{2-} \sim PR_3 \sim SR_2 \sim SC(NH_2)_2 < CO \sim C_2H_4 \sim CN^-$$

A familiar example of its application is to the preparations of the *cis*- and *trans*-isomers of $[Pt(NH_3)_2Cl_2]$ by reacting $[PtCl_4]^{2-}$ with NH_3 and $[Pt(NH_3)_4]^{2+}$ with concentrated HCl, respectively, substitution *trans* to Cl being more facile than substitution *trans* to NH_3. This is just conceivably bogus, because a careful analysis of the reactivities of complexes of the type $[Pt(NH_3)_{(4-n)}Cl_n]^{(2-n)+}$ suggests that a greater *cis* labilizing effect of NH_3 may be a more plausible explanation. A more complicated example, spelt out in detail elsewhere, involves the synthesis of the three isomers of $[Pt(NH_3)(py)BrCl]$.[115]

When attention was first paid to the mechanism of the *trans* effect, inorganic chemists, in general, were ignorant of the distinction between lability and instability and only bond-weakening models were considered. An early review that gives a good summary of the preparative applications and the first explanations of the mechanism is that of Quagliano and Schubert.[116] It was realized that the *trans* effect frequently predicted products that were not the most stable isomer of a particular ligand combination and therefore had to be accounted for in kinetic and hence mechanistic terms. This came at a time when interest in the mechanism of ligand substitution was just dawning, and the view is well summarized in the first edition of Basolo and Pearson (published in 1958).[117]

In the early 1950s, in the wake of the success of the bonding models for M–C≡O in metal carbonyls and the binding of ethene in Zeise's salt, $K[Pt(C_2H_4)Cl_3]$, in which back-bonding from filled orbitals on the metal

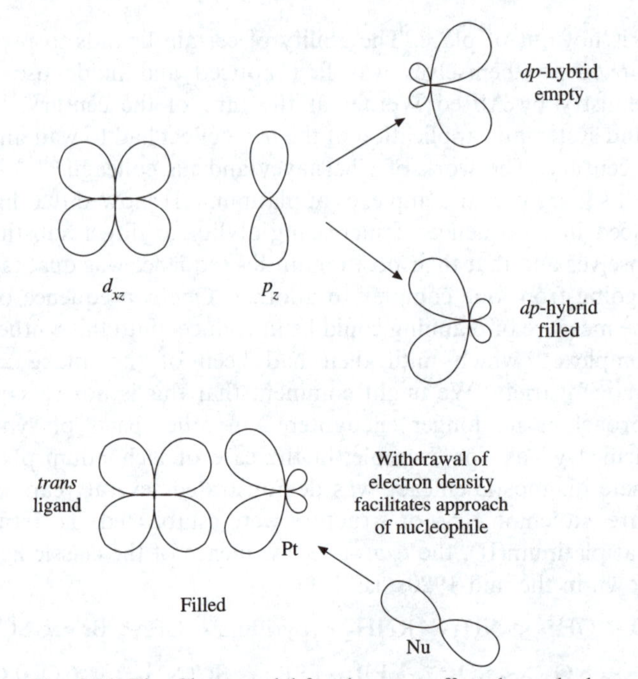

Fig. 3.13 The Chatt model for the *trans* effect through electron withdrawal by a π-acceptor ligand.

into empty orbitals of π-symmetry on the ligand played a major part, Chatt applied the same reasoning to offer a mechanism for the *trans* effect. He pointed out that the strong *trans*-effect ligands then known, C_2H_4, CN^-, CO, PR_3, SR_2, all possessed empty orbitals of π-symmetry that could withdraw charge from the metal and minimize the disturbance produced by the binding of the fifth ligand.[118] Chatt's model, Fig. 3.13, involved a hybridization of the d_{xz} and p_z orbitals on the metal to produce a pair of *dp*-hybrid orbitals, one of which was filled and overlapped with the empty *p*-orbital on the *trans* ligand, while the other was ready to accept the electrons from the incoming nucleophile. An analogous π-acceptor mechanism was proposed by Orgel,[119] who used a molecular orbital-based argument to show that the presence of a π-acceptor in the trigonal plane would lower the energy of the five-coordinate transition state/intermediate. It is interesting to observe that both mechanisms assumed an associative mode of activation some five years before systematic kinetic studies showed that this was indeed true. Previous explanations in terms of bond weakening had assumed a dissociative mechanism. Some years later, compounds of the type *trans*-$[Pt(PR_3)_2RX]$ (R = H, alkyl, aryl; X = Cl, Br) were synthesized and the *trans* effect of ligands such as H and CH_3 was shown to be considerable.[65] Since these could not function as π-acceptors, an alternative explanation was called for. This work was carried out at a time that coincided with the period when inorganic chemists had woken up to the importance of applying new and not so new physico-chemical techniques to the problem of structural assignment. Starting with magnetism and vibrational spectroscopy, then nuclear magnetic resonance and, once the problems of data collection and processing were eased by the use of modern computers, X-ray diffraction, they were frequently in possession of information about bond strength and bond length. Considerable interest developed in the effects that one ligand had upon the bonding of others in the complex and it was soon observed that there was a very specific effect of a ligand upon its *trans* partner in both square-planar and octahedral complexes. Since it was much more 'modern' to make these measurements instead of the tedious and old-fashioned kinetic studies, a great deal was learned about these ground-state effects and comparatively few systematic kinetic studies were carried out during the same period. After a period of semantic confusion when the term '*trans* effect' was used for any phenomenon that changed the properties of the ligand *trans* to the source of perturbation, it was decided to reserve the term '*trans* effect' for the kinetic aspects, i.e. *the effect of a coordinated group upon the rate of substitution reactions of the ligand* trans *to itself,*[120] and to use the term '*trans* influence' for any non-kinetic ground-state effect in which there was a *trans*-specific perturbation.

3.3.10.3 Quantitative aspects of the trans *influence*[121]

In so far as the amount of interest is indicated by the number of publications, there has been a much greater emphasis on *trans* influence than there has been upon *trans* effect. This probably reflects the small number of people that have the patience to carry out reliable kinetic studies. The non-kinetic effects that are *trans*-specific cover a range of properties that are related directly or indirectly to the strength of the metal–ligand bond.[122] It is generally assumed that, in the absence of bond-multiplicity changes, the length of a bond between a particular pair of atoms is a measure of the strength of that bond—the longer the bond, the weaker it has become. In a similar way, bond-stretching force constants will decrease with bond weakening. However, a proper normal coordinate analysis of the vibrational spectrum of a substrate molecule is often not possible and there is a temptation to use a vibrational frequency 'assigned' to a particular bond instead. This might be useful in a qualitative sense, but the coupling of vibrations and other factors that separate vibrational frequencies in a polyatomic molecule from the individual force constants can frequently lead one into error. In the days when metal–ligand stretching frequencies were not easy to measure, the ν_{N-H} frequencies were used as a measure of the M–N bond strength in metal–amine complexes, the idea being that the weakening of the M–N bond would lead to a strengthening of the N–H bond.[123] Trends in Pt–Cl stretching frequencies and Pt–Cl bond lengths according to the nature of the *trans* ligand are shown in Table 3.9. A *trans*-influence sequence based on bond lengthening and a judicious choice of bond-stretching frequencies is

Table 3.9 The *trans* influence on Pt–Cl bond distances and stretching frequencies.

L	PR_3	Pt–Cl (Å)[a]	ν_{Pt-Cl} (cm^{-1})[b]
(a) Complexes of the type *trans*-[Pt(PR$_3$)$_2$LCl]			
Cl^-	PEt_3	2.294	
$C_2F_5^-$	$PMePh_2$	2.361	
$C_6H_5^-$	PPh_3	2.408	
CH_3^-	$PMePh_2$	2.412	
$(CH_3)_3SiCH_2^-$	PMe_2Ph	2.415	
H^-	$PEtPh_2$	2.422	
Ph_2MeSi^-	PMe_2Ph	2.45	
(b) Complexes of the type [PtLCl$_3$]$^-$ [c]			
CO_2		2.289	322
Cl^-		2.317	305–330[d]
Me_2SO		2.309	309
NH_3		2.321	
C_2H_4		2.340	309
Me_2S; Et_2S			310; 307
$AsMe_3$; $AsEt_3$			272; 280
PPh_3; PMe_3			279; 275
PEt_3		2.382	271

[a] Sources can be traced through bond distance tabulations in S. A. Cotton, *Chemistry of Precious Metals*, Blackie, London, 1997; further examples of the *trans* influence on Pt–Cl bond distances can be found in this text and in F. R. Hartley, *Chem. Soc. Rev.*, 1973, **2**, 163.
[b] From R. J. Goodfellow, P. L. Goggin, and D. A. Duddell, *J. Chem. Soc. (A)*, 1968, 504.
[c] Pt–Cl bond distances for Cl *trans* to L.
[d] For details see P. L. Goggin and J. Mink, *J. Chem. Soc., Dalton Trans.*, 1974, 1479.

$$CO < H_2O < NH_3 \sim NH_2R < C_2H_4 < Cl^- \sim Me_2SO < Br^- < Me_2S < Ph_3As < Ph_3P < CH_3^- < H^- < Si(CH_3)_3^-$$

The spin–spin coupling between ^{195}Pt and a suitable magnetic nucleus in the ligand, preferably that of the donor atom, e.g. ^{31}P, ^{13}C or even ^{1}H (unfortunately the lack of interest that Pt shows for F rules out the use of ^{19}F unless one is interested in coupling through more than one bond) allows NMR to be used to probe the *trans* influence. In general, metal–ligand coupling constants are very sensitive to variation in the nature of the *trans* ligand but are hardly affected by a change in the nature of the *cis* neighbour. For example, in *cis*-[Pt(PEt$_3$)$_2$(CH$_3$)Cl], $^1J_{^{195}Pt-^{31}P}$ for P *trans* to CH_3 (1719 Hz) is very much less for the other P, which is *trans* to Cl (4179 Hz).[124] The results of a study of an extensive series of complexes of the type [Pt(dppe)(CH$_3$)X] or [Pt(dppe)(CH$_3$)L]$^+$ (dppe = 1,2-bis(diphenylphosphino)ethane, the chelate being chosen to prevent the *cis* isomer changing into the *trans* form; X is an anionic ligand and L is neutral) are collected in Table 3.10. In all the compounds examined, the coupling between Pt and the P *trans* to CH_3 is independent of the nature of the *cis*

Table 3.10 *Trans* influence for neutral (L) and anionic (X) ligands as measured by $^{1}J_{^{195}Pt-^{31}P}$ coupling constants in $[Pt(dppe)(CH_3)X]$ and $[Pt(dppe)(CH_3)L]^+$ (dppe = 1,2-bis(diphenylphosphino)ethane).[a]

L	$^{1}J_{^{195}Pt-^{31}P}$/Hz	L	$^{1}J_{^{195}Pt-^{31}P}$/Hz
O_2NO^-	4510	O_2N^-	3345
CH_3CN	4370	Ph_3As	3300
Cl^-	4224	OC	3212
$CH_3CO_2^-$	4122	HS^-	3201
I^-	4050	NC^-	2870
$4\text{-}CH_3OC_6H_4O^-$	3840	Ph_3P	2743
C_5H_5N	3738	$(PhO)_3P$	2718
NCS^-	3719	$O_2NCH_2^-$	2580
HO^-	3546	$NCCH_2^-$	2422
Ph_3Sb	3472	$CH_3COCH_2^-$	2346
PhS^-	3380	H_3C^-	1794

[a] Coupling constants for the P *trans* to L or X; data from T. G. Appleton and M. A. Bennett, *Inorg. Chem.*, 1978, **17**, 738.

ligand (X or L), but the coupling between Pt and the other P changes significantly.[125] ^{35}Cl NQR and Mössbauer (on iodo complexes) spectroscopies have provided parallel information on the *trans* effect and the *trans* influence.[126]

The bulk of the evidence, including extended Hückel molecular orbital calculations on *trans*-$[Pt(NH_3)LCl_2]$,[127] suggests that the *trans*-specific bond-weakening effect operates through the σ-framework and that a strong σ-donor will enhance its bonding at the expense of its *trans* partner. Langford and Gray developed a simple and easily visualized model (Burdett and Albright provide a more thorough analysis, applicable to the *trans* influence in octahedral as well as square-planar sytems)[128] which is frequently quoted and in which the pair of *trans* ligands compete for the use of the appropriate *p*-orbital on the platinum (Fig. 3.14). Calculations of the overlap integrals between the donor orbital of the *trans*-influencing ligand and the 6*p*-orbital on the platinum are consistent with the observed *trans*-influence sequence. However, if the *trans* influence on the spin–spin coupling constants arises from the same source as the *trans* bond weakening, a description in terms of competition for π-orbitals cannot be used and it has been suggested that a strong *trans*-influence ligand reduces the 6*s* character of the Pt(II) contribution to the orbital binding its *trans* partner. If, instead of using the p_x- and p_y-orbitals of the Langford and Gray model, we use $5d_{x^2-y^2}6s$-hybrids, the two models can be reconciled.[129]

The bond-length *trans*-influence sequence cannot be totally reconciled with the spin–spin coupling sequence, partly because bond lengths are not a sensitive enough measure of relatively low *trans* influences and partly because multiple bonding, i.e. contributions from π-bonding back donation,

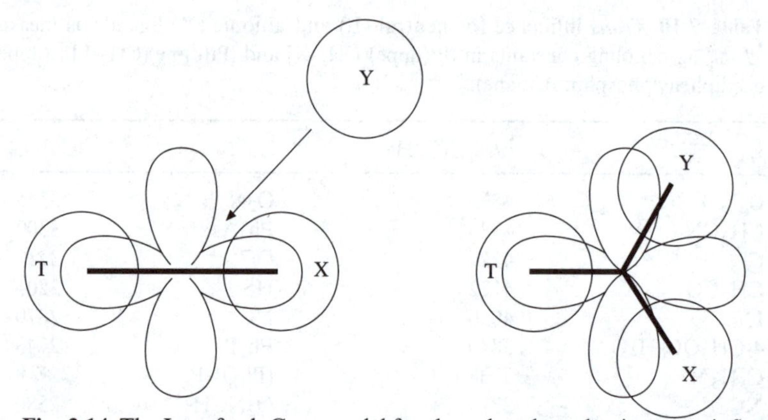

Fig. 3.14 The Langford–Gray model for the σ-bond-weakening *trans* influence.

in the ground state may affect the bond length. The case of CO is probably the most relevant in this context. In $[Pt(CO)Cl_3]^-$, the Pt–Cl bond *trans* to CO is shorter than those *cis*, while the coupling constant reported in Table 3.10 above suggests a moderate *trans* influence. It has been suggested that the back donation in the ground state, which is an essential part of the binding of CO, counteracts the σ *trans* influence. But it is always a risky business to rely upon arguments that involve the fortuitous cancelling out of opposing effects.

3.3.10.4 Quantitative aspects of the trans *effect*

Although is is possible to collect sufficient data from the literature to cobble together a systematic presentation of the quantitative kinetic *trans* effect, there are few extensive studies that examine a large group of substrates which differ from one another only in the nature of the ligand *trans* to the leaving group, and fewer still where there is an extensive study of the effect of changing the leaving group as well. Rate constants for reactions of the substrates *trans*-$[PtL(am)Cl_2]$ (displacement of am) and $[PtLCl_3]^{-\text{ or }2-}$ (displacement of Cl^- *trans* to L) are collected in Table 3.11, and those for the displacement of Cl^- from *trans*-$[Pt(PEt_3)_2YCl]$, where Y is an anionic ligand, are collected in Table 3.12.

In the solvolytic reactions of the first set of complexes, the first-order rate constant increases along the sequence

$$H_2O < NH_3 < Cl^- < Br^- < Me_2S < Et_2S < Me_2SO < AsEt_3$$
$$< PPh_3 < PMe_3 < P^nBu_3 < P(OMe)_3 < C_2H_4$$

the extremes spanning more than 11 orders of magnitude.[130] Changing the reaction may change the size of this range but, as yet, it does not change the order of effectiveness. The displacement of am = pyridine allows a quantitative ordering of the strong *trans*-effect ligands:

Table 3.11 The kinetic *trans* effect of neutral ligands.

	Rate constants, $10^3 k_1 (s^{-1})$, for the reaction *trans*-$[Pt(L)(am)Cl_2] + MeOH \rightarrow$ *trans*-$[Pt(L)(MeOH)Cl_2] + am$		Rate constants, $k_1 (s^{-1})$, for the solvolytic reaction $[Pt(L)Cl_3]^- + S \rightarrow$ *trans*-$[Pt(L)(S)Cl_2] + Cl^-$	
L	am = py[a]	am = 3-Cl-py	S = MeOH[b]	S = H_2O[c]
C_2H_4	6800[d]			10 000
CO	1210[d]			
$P(OMe)_3$	365	5240	10.7	
PPh_3	27	335	3.1	
PEt_3	26	495	6.6	
P^nBu_3	23	380	5.2	
PMe_3	15	275	4.8	
$AsEt_3$	1.7	39	0.67	
Me_2SO	0.11	2.8	0.0081	0.098
Et_2S		0.13	0.0024	
Me_2S		0.039	0.0014	
Br				0.000 084
Cl				0.000 009 2
NH_3				0.000 006 3
H_2O				0.000 000 028

[a] py = pyridine.
[b] In methanol at 30.0 °C.
[c] In water at 25.0 °C.
[d] In 95:5 methanol:water at 25.0 °C.

Table 3.12 The *trans* effect of anionic ligands: rate constants for the reaction *trans*-$[Pt(PEt_3)_2RCl] + Y \rightarrow$ *trans*-$[Pt(PEt_3)_2RY]^+ + Cl^-$ in methanol at 25 °C.

R	Y: MeOH $10^3\ k_1\ (s^{-1})$	$(NH_2)_2CS$ $k_2\ (dm^3\ mol^{-1}\ s^{-1})$
H	560	6 310
C_2H_5	48	15.5
CH_3	40[a]	≥1.5[a]
C_6H_5	10.2[a]	6.3[a]
$4\text{-}FC_6H_4$	5.0[a]	5.9[a]
$3\text{-}FC_6H_4$	4.0[a]	4.0[a]
$3\text{-}CF_3C_6H_4$	3.0	3.2
$2\text{-}CH_3\text{-}C_6H_4$	2.0[a]	0.65[a]
$2,4,6\text{-}(CH_3)_3C_6H_2$	0.44[a]	0.049[a]
C_6F_5	0.020	0.35
Cl	0.00015[a]	47

[a] At 30.0 °C.

$$Me_2SO < AsEt_3 < PMe_3 < P(OMe)_3 < CO < C_2H_4$$

but pyridine is bound too strongly to allow weaker *trans*-effect ligands to be studied, and even the more weakly bound 3-chloropyridine only allows the study to be extended to $L = Me_2S$.[131] The rate constants for the sequence

$$Cl^- < C_6F_5{}^- < C_6H_5{}^- < CH_3{}^- < C_2H_5{}^- < H^-$$

cover nearly seven orders of magnitude in the solvolytic reactions of *trans*-$[Pt(PEt_3)_2LCl]$ in methanol. In this series a great deal of work has been carried out to investigate the ways in which the *trans* effect could be quantified in terms of the nucleophilic discrimination of the substrate, and it was once suggested that the *trans* effect might be quantified in terms of the nucleophilic discrimination factor, which becomes smaller as the *trans* effect increases. Indeed, in many of these complexes, the nucleophilic discrimination is so small that most normal nucleophiles enter by way of the solvolytic rate-determining step and only the strongest have a measurable direct pathway. In these systems, plots of log k_2 (where known) against $n_{Pt}{}^0$ are markedly curved, suggesting a co-operativity on the part of the strongest nucleophiles.[132] Because of this there is a complete change in the reactivity sequence of the above series of complexes when the entering group is changed from the solvent to the strong nucleophile thiourea, chloride moving up to second place:

$$C_6F_5{}^- < 3\text{-}CF_3C_6H_5{}^- < C_6H_5{}^- < CH_3^- < C_2H_5{}^- < Cl^- < H^-$$

This is a consequence of the dramatic change of nucleophilic discrimination and shows the need for care when trying to make sweeping generalizations.

3.3.10.5 *Current views on the mechanisms of the* trans *effect*

In any study of substituent effect, electronic or steric, irrespective of the geometry of the reaction centre, or even the nature of the reaction, it is convenient to guide the discussion by reference to the very simple but fundamental diagram depicted in Fig. 3.15. One would use a diagram of this sort to distinguish between thermodynamic stability relating to ΔG° (the difference between the standard free energy of the reagents and products) and the kinetic inertness, as measured by $\Delta G^\ddagger$ (the difference between the standard free energy of the reagents and that of the transition state—assuming one accepts the concept that transition states in bulk can be assigned standard free energy changes). Clearly $\Delta G^\ddagger$ can be affected by changes in the ground state and/or the transition state. It is therefore convenient to separate substituent effects into ground-state effects and transition-state effects, bearing in mind that a change in substituent can contribute to both. This was well recognized by Ingold,[133] who talked in terms of polar effects, divided into inductive and conjugative displacements, and polarization effects, divided into inductomeric and electromeric displacements. These concepts remain even though the nomenclature is different. In modern inorganic terms these would be ground-state effects,

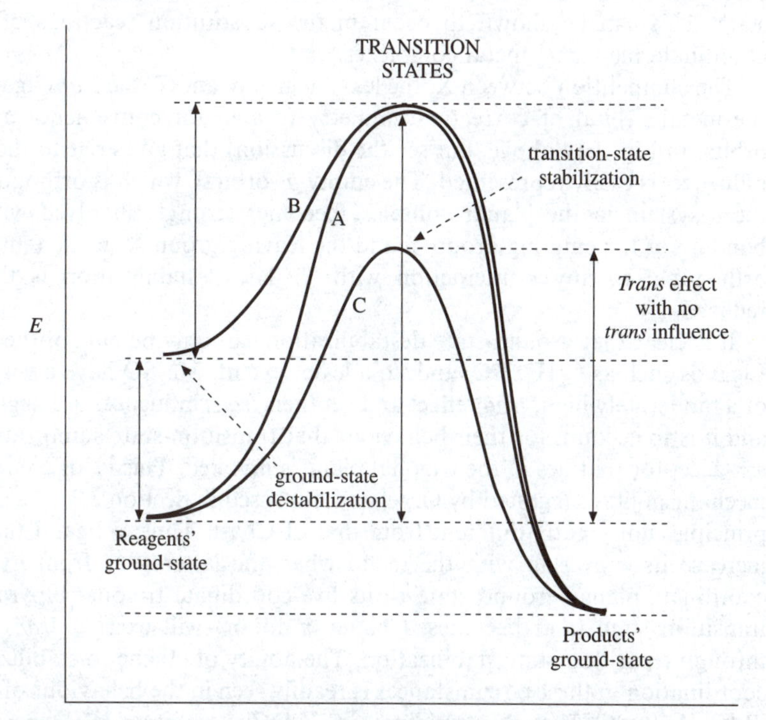

Fig. 3.15 Labilization arising from B: ground-state destabilization (*trans* effect plus *trans* influence) and C: transition-state stabilization (*trans* effect alone) compared with the reference reaction A.

divided into σ- and π-displacements and transition states, similarly subdivided.

Any change that reduces the stability of the ground state more than it does that of the transition state will increase the rate constant of the reaction. A change that destabilizes the transition state more than it does the ground state will reduce the rate constant of the reaction. An example of this is to be found in steric hindrance of an associatively activated reaction, where congestion in the ground state is exacerbated when the incoming group increases the coordination number of the reaction centre in the transition state. Any change that takes place in the ground state is potentially observable by other, non-kinetic, probes, whereas a change restricted to the transition state is seen only as a kinetic effect.

The *trans* influence is therefore a ground-state destabilization seen as a weakening of the bond between the metal and the leaving group that is *trans* to the perturbing ligand. If the process was dissociatively activated, the fact that there is little or no bonding between the leaving group and the reaction centre in the transition state would mean that ground-state bond weakening must lead to an increase in lability. However, in an associatively activated process there is no necessary relationship between ground-state bond weakening and increased lability, and it must be demonstrated that the perturbation that destabilizes the ground state is mitigated in the transition state. This can be shown to occur in the substitution reactions of four-coordinate planar d^8 metal complexes.

The competition between X, the leaving group, and T, the *trans* ligand for the metal orbital of correct σ-symmetry (drawn for convenience as a *p*-orbital in Fig. 3.14 above, but see the discussion) that gives rise to the *trans* influence, is easily represented. The empty p_z-orbital, which is orthogonal to the σ-system in the planar complex, becomes strongly involved with the bonding of the entering group Y and the leaving group X, while remaining orthogonal to any σ-interaction with T. The destabilization is thereby reduced.

It is clear that ground-state destabilization can only be part of the story. Ligands such as C_2H_4, CO, and, to a lesser extent, Me_2SO have a very high or a moderately high, *trans* effect and yet their *trans* influences are negligible, and it is to account for their behaviour that transition-state stabilization by a π-acceptor that lies in the trigonal plane is invoked. This is, of course, the mechanism first suggested by Orgel and discussed in Section 3.3.10.2 and, in principle, not greatly different from that of Chatt. Thus, a ligand that can increase its π-overlap with the metal when the latter goes from its four-coordinate planar ground state to its five-coordinate trigonal bipyramidal transition state (and becomes a better π-donor) will exert a *trans* effect through transition-state stabilization. The ability of ethene to stabilize five-coordination in these circumstances is readily seen in the behaviour of *trans*-$[Pt(C_2H_4)(py)Cl_2]$ with pyridine in $CDCl_3$ solution,[134] where the ^1H NMR spectrum not only indicates very rapid exchange between free and coordinated pyridine at all temperatures studied but also signals for a five-coordinate species, $[Pt(C_2H_4)(py)_2Cl_2]$. With chelating nitrogen donors (LL) the exchange is prevented, and five-coordinate species of the type $[Pt(C_2H_4)(LL)Cl_2]$, **8**, have been isolated. Moreover, in the case where LL is butane-2,3-dionebisphenylhydrazone, the trigonal bipyramidal nature of the complex has been confirmed by single-crystal X-ray diffraction.[135]

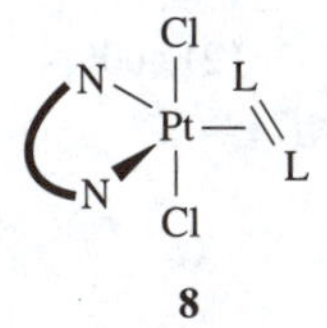

8

While there is no difficulty in accounting for the *trans* effect of ethene in terms of its π-acidity and consequent transition-state stabilization, and the combined *trans* effect and *trans* influence of ligands such as H and CH_3 in terms of their σ-basicity and consequent ground-state stabilization, problems arise when we want to assign the relative contributions for ligands that have demonstrable *trans* influence and also possess empty orbitals of the correct symmetry to function as π-acids. Since it is not possible to make measurements on the transition state directly it is not easy to devise an objective way of determining this. It is a well-known fact that calculation can always provide the answer expected. The greater *trans* effect of $P(OMe)_3$ compared with that of PMe_3, and of Me_2SO when compared with Me_2S, indicate that π-interactions must contribute to the *trans* effects of the phosphite and sulphoxide (the electronegative oxygens substantially reduce the σ-basicity and increase the π-acidity). Attempts are being made to use the idea that bond-weakening effects lead to open transition states and hence small nucleophilic discriminations, while stabilized transition states would be more compact and lead to larger nucleophilic discriminations. The low discrimination is evident in complexes with high *trans*-influence ligands, but complexes containing ethene and carbon monoxide undergo rapid reactions with nucleophiles that frequently involve π–σ transformations and addition of the nucleophile to the ethene.

3.3.10.6 The effects of varying the cis *ligands*

Whereas reactivity is very sensitive to the electron-displacement properties of the ligand *trans* to the leaving group and not very sensitive to variations in its bulk, the reverse is true for the *cis* ligands. It is easy to see how steric hindrance will exert a much greater effect from the *cis* position by building models of *cis* and *trans* isomers containing a selection of bulky ligands. An increase in the cone angle of the *trans* ligand or the introduction of bulky ligands adjacent to the donor atoms will increase the non-bonding repulsions between this ligand and its *cis* partners, but this will not change greatly on going from the ground state to the transition state. On the other hand, any increase in the bulk of the ligands *cis* to the leaving group will result in

Table 3.13 The steric *cis* effect of *ortho*-methylation of an aryl ligand. Rate constants for the reactions $[Pt(PEt_3)_2RX] + MeOH \rightarrow [Pt(PEt_3)_2R(MeOH)]^- + Cl^-$ and $[Pt(PEt_3)_2RX] + CN^- \rightarrow [Pt(PEt_3)_2R(CN)] + Cl^-$ are k_1 and k_2, respectively.

R	X	$10^3k_1(s^{-1})^a$	$k_2(dm^3\,mol^{-1}\,s^{-1})^b$
	*cis*-Br	6000	$21\,400^c$
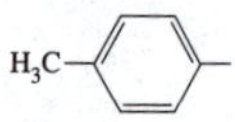 H_3C	*cis*-Br	4280	
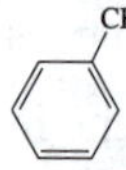 CH_3	*cis*-Br	54.4	2460^c
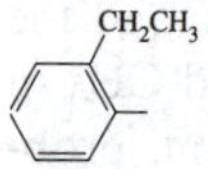 CH_2CH_3	*cis*-Br	16.2	
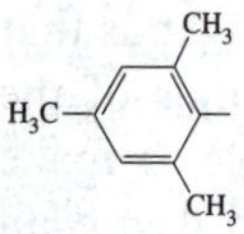 CH_3, H_3C, CH_3	*cis*-Br	0.19	0.312
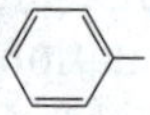	*trans*-Cl	10.0	34.8
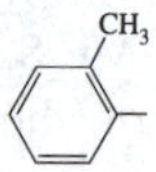 CH_3	*trans*-Cl	2.29	0.90
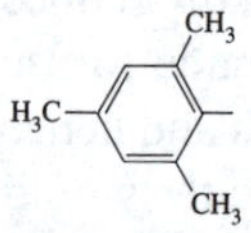 CH_3, H_3C, CH_3	*trans*-Cl	0.405	0.043

[a] In methanol at 30.0 °C.
[b] In propan-2-ol at 30.0 °C.
[c] X = Cl.

further crowding when the number of *cis* partners changes from two to three in the trigonal-bipyramidal transition state.

A classic example of steric hindrance is the way in which the reactivity of complexes of the type $[Pt(PEt_3)_2RCl]$ changes with increasing *ortho*-methyl substitution in the aryl group, R, along the sequence, R = phenyl, 2-methylphenyl, 2,4,6-trimethylphenyl (Table 3.13).[136] The absence of any significant response to changes in the nature of the substituent in the 4-position shows that this effect is steric in origin and the much greater response when R is *cis* to the leaving group is fully evident in both the values for k_1, the solvolytic pathway and k_2 for the reactions of the strong nucleophile CN^-. This was one of the earliest indications that the transition states for substitution were trigonal bipyramidal.

Although the effects of varying the electron-displacement properties of the *cis* ligands upon the reactivities of the substrates are small when compared with *trans* effects, they can still be seen. Because they are small they are difficult to systematize and it is not easy to distinguish a site-specific phenomenon from a general effect. Proper studies of *cis* effects require some means of anchoring the ligand *trans* to the site of variation, otherwise one has no control of the choice of leaving groups. Variation of the *cis* ligands, L, in complexes of the type *trans*-$[PtL_2Cl_2]$, where Cl^- is the leaving group, affects the nucleophilic discrimination and, in the sequence L = piperidine, SMe_2, PEt_3, the nucleophilic discrimination factor increases quite considerably.[137] The PEt_3 complex has a much lower intrinsic reactivity than the other two, probably because of the steric effects, but the corresponding *trans*-$[Pt(PMe_3)_2Cl_2]$ complex readily isomerizes to the *cis* species and the appropriate measurements have not yet been made. The rate constants for the replacement of Cl^- from $[Pt(en)LCl]^+$ are not greatly sensitive to the nature of L (the intrinsic reactivities are virtually identical provided there is no steric hindrance), but there is a noticeable increase in the nucleophilic discrimination along the sequence $L = NH_3 < SMe_2 < Me_2SO \sim PMe_3 \sim P(OMe)_3$.[138] It has been observed that variation in the basicity of amines *cis* to the leaving group affects the reactivity quite significantly, the least basic amine producing the most reactive substrate. There have been particularly thorough studies, kinetic and theoretical, of both the *cis* and the *trans* effects operating in the rather simple and important system Pt^{II}–Cl^-–H_2O–DMSO.[139]

3.3.11 The dissociative mechanism for substitution in four-coordinate planar metal complexes[140]

3.3.11.1 General considerations

Because of the ready availability of the pathway for associative activation it has long been of considerable interest to find out just how dominant this pathway can be in the substitution reactions of four-coordinate planar d^8 metal complexes and, more to the point, what has to be done to tip the balance to the point where the dissociatively activated mechanism is observed. In many respects this resembles the problems encountered in attempting to persuade tetrahedral silicon and other *p*-block elements of the second and subsequent rows to adopt a dissociative mode of activation.

The methodology of the approach must involve either the suppression of the dominant associative pathway or a promotion of the dissociatively activated pathway or, if one is very smart, a combination of both. In principle, the following approaches are worth considering:

(a) Use a poor nucleophile to suppress the associative pathway.
(b) Suppress the associative pathway by using sterically hindered substrates.

(c) Enhance the act of bond breaking by incorporating a sufficiently strong *trans*-influence ligand to the point where no assistance is needed from bonding with the incoming group.

The use of weak nucleophiles has not been successful, because in coordinating solvents a point is always reached where the solvolytic process takes over and, except in very special circumstances, these have always been shown to be associatively activated. With very weak nucleophiles there is always the problem that the stability of the product may be too low for reaction to take place. In principle, if a suitable fluoro complex of Pt(II) could be synthesized and dissolved in a non-coordinating solvent, such as liquid HF, which could also dissolve a source of fluoride ions, then fluoride exchange might be observed to be dissociative. The nearest approach to this has been the exchange of H_2O between solvent and $[M(H_2O)_4]^{2+}$ (M = Pd, Pt), where, after some initial excitement,[141] it was shown unambiguously that the $\Delta V^{\ddagger}$ values were fully consistent with associative activation.[142]

3.3.11.2 *Steric retardation*

Most of the effort to identify a dissociatively activated mechanism has centred upon the use of steric hindrance to disfavour the associative process. The replacement of three, four, or all five of the amine protons of the ligand 1,5-diamino-3-azapentane (dien) by methyl or ethyl groups leads to considerable congestion above and below the coordination plane in complexes of the type $[Pt(LLL)X]^{n+}$ (where LLL represents the terdentate-substituted dien). Such complexes were indeed called *pseudo-octahedral*. The increased steric hindrance led to an enormous decrease in reactivity, so much so that, rather than work at temperatures greater than 80 °C (or get involved with research-student-unfriendly runs lasting several months!), the reaction centre was changed from Pt(II) to Pd(II) (Table 3.14). In addition to the great reduction in intrinsic reactivity, the nucleophilic discrimination is also reduced to the point where all but the strongest nucleophiles enter by way of the solvolytic pathway. Nowadays we realize that low nucleophilic discrimination indicates an open transition state, which is to be expected when there is congestion, but, at the time this work was done, the absence of a k_2 pathway was taken as evidence that a dissociative mechanism had taken over. Although it is true that ligands like Et_4dien prevent the formation of six-coordinate complexes $[M(Et_4dien)X_2]$, an enormous number of five-coordinate complexes of the type $[M(Et_4dien)X]^{n+}$ have been described in the literature and indeed this ligand is a prime favourite for the generation of five-coordination. The volumes and entropies of activation of the reactions of the type listed in Table 3.14 are negative and similar in magnitude. For instance for the reactions

$$[Pd(R_n dien)Cl]^+ + H_2O \longrightarrow [Pd(R_n dien)(H_2O)]^{2+} + Cl^-$$

Table 3.14 The effect of *N*-substitution on the rate of replacement of Cl^- from $[Pd(LLL)Cl]^+$ (LLL = $R_2NCH_2CH_2N(R')CH_2CH_2NR_2$).[157]

R	R′	$k_1 (s^{-1})$	$k_2 (dm^3\ mol^{-1}\ s^{-1})$
H	H	37.8	a wide range, dependent upon the nature of the nucleophile
CH_3	H	0.90	small for most nucleophiles
			OH^- 4.53
			I^- 0.28
CH_3	CH_3	0.24	small for most nucleophiles
			$(NH_2)_2CS$ 1.4
			$S_2O_3^{2-}$ 12.6
CH_3CH_2	H	0.0021	very small for most nucleophiles
			OH^- 0.0376
			I^- 0.0008
			$S_2O_3^{2-}$ 5.9
CH_3CH_2	CH_3	0.00065	zero for all nucleophiles except
			$S_2O_3^{2-}$ 0.052

where the ligands $R_n$dien comprise a number of tetra- and penta-methyl and -ethyl *N*-substituted diethylenetriamine ligands, activation volumes and entropies lie between -9 and $-15\,cm^3\,mol^{-1}$ and -67 and $-106\,J\,K^{-1}\,mol^{-1}$, respectively.[143] It could be argued that electrostriction around the leaving chloride was the major contributor to these negative values—after all, the values for benzyl chloride hydrolysis are $-9\,cm^3\,mol^{-1}$ and $-55\,J\,K^{-1}\,mol^{-1}$—but activation volumes and entropies for reaction of $[Pd(R_n dien)(H_2O)]^{2+}$ with thiourea and its methyl-substituted derivatives of -9 to $-15\,cm^3\,mol^{-1}$ and -56 to $-85\,J\,K^{-1}\,mol^{-1}$ cannot be attributed to electrostriction. Indeed, although the order of reactivity for a series of nucleophiles (iodide, pyridine, tetramethylurea, triphenyl phosphine) with $[Pd(LLL)Cl]^+$, where LLL is the sterically demanding ligand **9**, is in the opposite order to that expected from the n_{Pt} scale: activation entropies and volumes are all negative, indicating associative solvolysis.[75] Rate constants in fact correlate with ligand bulk, which is perfectly reasonable in view of their likely role in hindering approach of the entering solvent molecule. The norbornyl substituent in **10** also provides steric hindrance, and could conceivably promote chloride dissociation through nucleophilic assistance. However, the activation volumes and volume profile for aquation and the reverse anation reaction for **10** clearly indicate associative activation, indeed possibly with a transient five-coordinate intermediate, here too.[144]

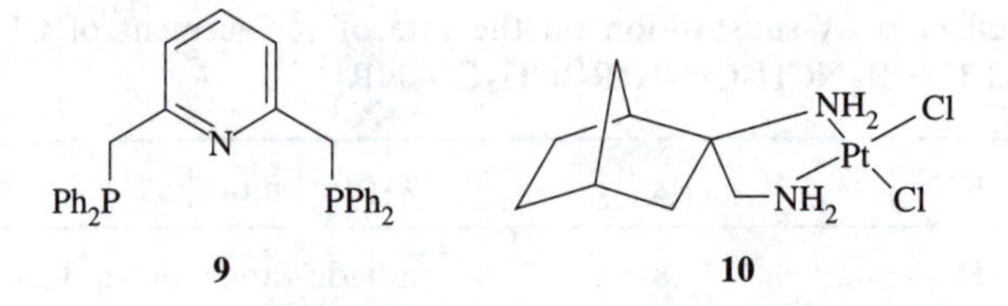

These results indicate quite clearly that *N*-substitution may slow down the reaction and reduce the nucleophilic discrimination, but that the mode of activation remains associative. These systems have been extensively studied in recent years, mainly with a view to confirming the associative mode of activation, and it was observed that OH^-, usually a very poor nucleophile, is less inhibited by steric hindrance than the others and that in the reaction with $[Pd(Et_4dien)Cl]^+$, it behaves as if it were a better nucleophile than iodide. However, when there is no amine proton present, the activity of OH^- vanishes. Furthermore, the value of k_2 for OH^- is much enhanced when there is an available proton on the nitrogen *cis* to the leaving group, as in $[Pd(Me_3dien)Cl]^+$.

It has already been shown (Section 3.3.10.6) that *ortho*-methylation of the phenyl ligand (R) in $[Pt(PEt_3)_2(R)X]$ (X = Cl, Br) leads to a strong retardation when R is *cis* to X. As in the case of the *N*-substituted dien complexes, this reduction in reactivity is accompanied by a large decrease in nucleophilic discrimination, but in this case also, $\Delta V^{\ddagger}$ values indicate associative activation throughout. However, in polar solvents such as methanol and in the absence of added nucleophile other than Cl^- or Br^-, the *cis* complex changes to the *trans* isomer at a rate that is retarded by added halide ion. The mechanism of this isomerization is discussed in Chapter 5 and so will not be repeated here other than to say that it involves the rearrangement of a three-coordinate T-shaped intermediate formed by the dissociation of the coordinated halide. Such intermediates have also been invoked in, for example, reductive elimination (Chapter 10) of alkanes from dimethylpalladium(II) and trialkylgold(III) complexes.[145] Since the *cis* → *trans* isomerizations are accompanied by ligand exchange, such exchanges are dissociatively activated, but in every case there is an alternative pathway that is considerably faster and associatively activated. In the case where R = 2,4,6-trimethylphenyl (mesityl) the steric hindrance is sufficient to reduce the rate constant for substitution down to that for isomerization and it was once thought that here, at last, was dissociatively activated substitution, but, as is discussed in Chapter 5, the $\Delta S^{\ddagger}$ and $\Delta V^{\ddagger}$ values indicate associative activation in this case. It is now generally agreed that the isomerization of *cis*-$[Pt(PEt_3)_2(mesityl)X]$ (X = Cl, Br) goes by way of the fast dissociation of the methanolo complex formed, in the rate-determining step, by an associative solvolysis. Direct studies of the isomerization of the methanolo complexes show that, in all cases, including R = mesityl, the activation parameters for isomerization are fully consistent with dissociative activation. In principle, this system should only be used as an example of dissociatively activated substitution in the absence of a better

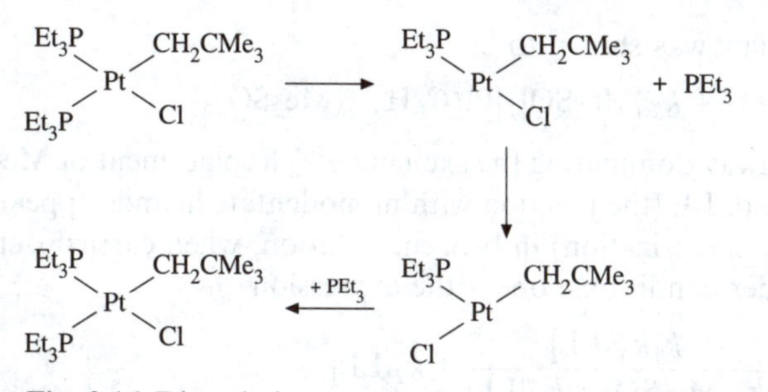

Fig. 3.16 Dissociative mechanism for isomerization of *cis*-$[Pt(PEt_3)_2$(neopentyl)Cl].

one. It nevertheless serves to show just how much the associative mode of activation is preferred in square-planar d^8 systems. The combination of steric hindrance and bond weakening is nearly enough to tip the balance, but not quite. Were it not for the fact that the three-coordinate intermediate is capable of rapid rearrangement and leads to isomerization, the small, background, dissociatively activated pathway would be missed.

In less polar solvents, the dissociation of an anionic leaving group is less favoured and, in the case of the isomerization of *cis*-$[Pt(PEt_3)_2$(neopentyl)Cl] (neopentyl = Me_3CCH_2–) in 2-propanol, when the normal mode is suppressed by adding chloride, an alternative pathway, involving the dissociation of the phosphine, is observed (Fig. 3.16).[146] This is also seen in dichloromethane solution, and for (uncatalysed) *cis* ↔ *trans* isomerization of $[Pd(C_6F_5)_2(tht)_2]$ in chloroform. In the latter case, the assignment of a dissociative mechanism rests on the observed mass-law retardation effect of added tetrahydrothiophen (tht), on the large positive activation entropy, and on the high value of the activation enthalpy, indicative of a dominant role for bond breaking in transition-state formation.[147]

3.3.11.3 Ground-state bond weakening

The promotion of a dissociative mechanism, as distinct from the suppression of an associative mechanism, does not appear to work when there is only a single *trans*-influence ligand present, even one as strong as hydride. Thus all substitution reactions of *trans*-$[Pt(PEt_3)_2(H)Cl]$ are still associatively activated in spite of the enormously enhanced lability. It is true that the transition state has been opened up and the nucleophilic discrimination is small, but there is no doubt that the entering group is bound to the platinum in the transition state.

However, when two *cis* σ-bonded organic groups are present, the mechanism changes. Although assumed, without any evidence, by organometallic chemists, the dissociative mechanism was first established kinetically in the exchange of Me_2SO with *cis*-$[Pt(C_6H_5)_2(Me_2SO)_2]$ in $CDCl_3$, where the rate law was shown to be

$$\text{rate} = \{k_1 + k_2\,[\mathrm{Me_2SO}]\}[\mathrm{Pt(C_6H_5)_2(Me_2SO)_2}]$$

with the k_1 pathway dominating the exchange.[148] Replacement of Me_2SO by a chelating ligand, LL (the reaction with monodentate ligands appears to be complicated by isomerization) in benzene solution, when carried out under pseudo-first-order conditions, obeys the expression

$$k_{\mathrm{obs}} = \frac{k_1 k_{\mathrm{Y}}[\mathrm{LL}]}{k_{\mathrm{X}}[\mathrm{Me_2SO}] + k_{\mathrm{Y}}[\mathrm{LL}]} + k_2[\mathrm{LL}]$$

with the k_2 pathway only appearing for the strong nucleophiles such as $Ph_2P(CH_2)_nPPh_2$ ($n = 2$ or 3) and $PhS(CH_2)_2SPh$.

The first term clearly indicates the presence of a labile intermediate and the problems of deciding whether it is a three-coordinate species or a four-coordinate solvento species are precisely those discussed in Section 3.3, but the chances of benzene or chloroform binding to platinum(II) are very slim and the similarity of the magnitudes of k_1 in the two solvents made it very likely that this was the elusive dissociatively activated substitution mechanism promoted above the normal associative process. It has since been shown that a wide range of complexes of the type *cis*-$[PtR_2L_2]$ (R = Me, Ph; L = Me_2SO, Me_2S) undergo substitution with dissociative activation[149] (Table 3.15). High activation enthalpies, up to $100\,\mathrm{kJ\,mol^{-1}}$, are entirely consistent with dissociative activation. It is of interest to note that, when the entering group has a sufficiently high nucleophilicity, a parallel associative pathway can be observed and this is most apparent when Me_2SO is the leaving group. This is a further indication of the ability of Me_2SO to stabilize the associative transition state. The volumes of activation for the exchange of free and coordinated Me_2S in *cis*-$[PtPh_2(Me_2S)_2]$ in benzene, and of Me_2SO in *cis*-$[PtPh_2(Me_2SO)_2]$ in chloroform and in *cis*-$[PtMe_2(Me_2SO)_2]$ in benzene are $+4.7 \pm 0.5, +5.5 \pm 0.8$, and $+4.9 \pm 0.5\,\mathrm{cm^3\,mol^{-1}}$, respectively. All these values are fully consistent with dissociative activation, prompted by the dominant effect of the Pt–C bonding to the phenyl or methyl ligand.[150] It is of interest to note that very small amounts of *cis*-$[PtMe_2(Me_2SO)_2]$ catalyse the *cis* → *trans* isomerization of $[Pt(PEt_3)_2(\text{neopentyl})Cl]$ in dichloromethane, presumably because this species acts as a reversible sponge for the PEt_3 that is released in the generation of the stereochemically labile three-coordinate intermediate (cf. Fig. 3.16 above). Although detailed studies have yet to be carried out, it is likely that the analogous phosphine and arsine complexes react similarly. The *trans* influences of the two *cis* ligands do not seem to reinforce one another, and it is likely that the presence of two σ-bonded organic groups increases the stability of the three-coordinate intermediate. Replacement of one of the neutral ligands by carbonyl completely removes the dissociative mechanism. Substitution at *cis*-$[PtPh_2(Et_2S)_2]$ is I_d, but at *cis*-$[PtPh_2(CO)(Et_2S)]$ is associative (I_a) with high nucleophilic discrimination.[151]

Table 3.15 Derived rate constants for the reaction *cis*-$[Pt(R)_2(L)_2] + LL \rightarrow [Pt(R)_2(LL)] + 2L$.[a]

LL[b]	*cis*-$[Pt(CH_3)_2(Me_2SO)_2]$		*cis*-$[Pt(CH_3)_2(Me_2SO)_2]$		*cis*-$[Pt(C_6H_5)_2(Me_2SO)_2]$		*cis*-$[Pt(C_6H_5)_2(Me_2S)_2]$	
	$10^2k_1(s^{-1})$	$k_2(M^{-1}s^{-1})$	$10^2k_1(s^{-1})$	$k_2(M^{-1}s^{-1})$	$10^2k_1(s^{-1})$	$k_2(M^{-1}s^{-1})$	$10^2k_1(s^{-1})$	$k_2(M^{-1}s^{-1})$
bipy	1.12	0.006	2.34	0.00	1.40	0.00	0.53	0.00
bpym	1.35	0.03	2.59	0.02	2.30	0.12	0.51	0.00
phen	1.07	0.06	2.75	0.00	2.00	0.00	0.53	0.00
dpte	1.16	0.16	2.88	0.05	2.50	0.26	0.60	0.05
dppe		226		35	2.00	10.0	0.42	3.87
dppp		148		28	1.90	9.3	0.59	2.46

[a] In benzene at 30.0 °C; data taken from Ref. 149.
[b] bipy = 2,2′-bipyridyl; phen = 1,10-phenanthroline; bpym = bipyrimidine; dpte = 1,2-bis(phenylthio)ethane; dppe = 1,2-bis(diphenylphosphino)ethane; dpte = 1,3-bis(diphenylphosphino)ethane.

It has been suggested that phosphine redistribution at iridium(I):

$$[IrCl(CO)(PR_3)_2] + [IrCl(CO)(PR'_3)_2] \longrightarrow 2[IrCl(CO)(PR_3)(PR'_3)]$$

takes place by a dissociative mechanism. In view of the proposal, mentioned above, that reactions of the type

$$[IrCl(CO)(PR_3)_2] + PR'_3 \longrightarrow [IrCl(CO)(PR'_3)] + PR_3$$

take place by the *A* mechanism, it may be that the redistribution reaction takes place by dissociation of a phosphine from one complex followed by associative attack by this released phosphine at the iridium of the other complex![152]

3.4 Five-coordinate compounds and complexes

3.4.1 Occurrence

The subject of five-coordinate complexes[153] was considered sufficiently important to merit a separate, if brief, chapter in this book's previous incarnation, since five-coordination is fairly common in a number of areas of the Periodic Table.[154] However, as there has been so little further study of substitution kinetics at five-coordinate complexes, and so much for other coordination numbers, we have perforce downgraded this group of complexes to a brief mention in the current book. Five-coordination is manifested in the forms of the trigonal bipyramid and of the square-based pyramid—and in various intermediate geometries. The widespread occurrence and range of geometries are documented, for binary species with five identical ligands, ML_5, in Table 3.16.[155] The angles specified in this table to characterize ideal and intermediate geometries are defined in Fig. 3.17. For an ideal trigonal bipyramid θ_1 and $\theta_2 = 90$ and $120°$; for a square-based pyramid $\theta_1 = \theta_2$ (for ML_5 with five equal bond distances to the central atom or ion). It should be borne in mind that the exact geometry may be very dependent on crystal packing; the nature of the counterion is important for

Table 3.16 Examples of compounds and complexes ML_5 with coordination number five (five identical unidentate ligands).[a]

	TRIGONAL BIPYRAMID (D_{3h})	steady change in geometry →				SQUARE PYRAMID (C_{4v})
$\Delta^b =$	30	27	21	7	7	2
	$CdCl_5^{3-}$	PPh_5	$Co(C_6H_7NO)_5^{2+}$	$Nb(NMe_2)_5$	$SbPh_5$	$InCl_5^{3-}$

[a] $Ni(CN)_5^{3-}$ exists in both trigonal bipyramidal and square pyramidal geometries.
[b] Δ = the difference between θ_1 and θ_2—see Fig. 3.17.

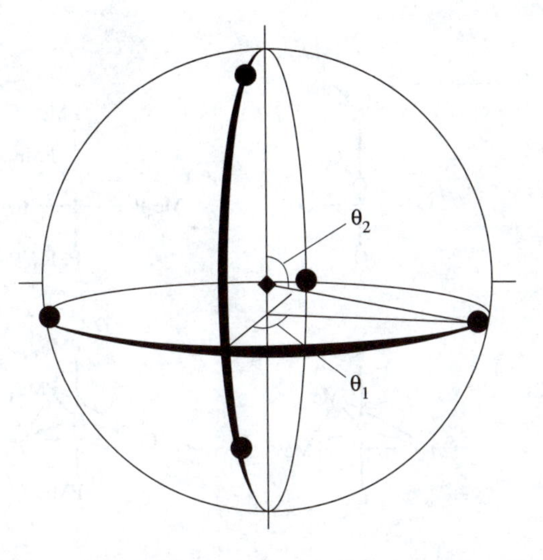

Fig. 3.17 Definition of angles characterizing five-coordinate geometries.

charged species, while the presence of solvent of crystallization may also affect the geometry. The most striking example of the latter effect is the comparison between $AsPh_5$ and its solvate $AsPh_5 \frac{1}{2}C_6H_{12}$, whose values (as defined in Table 3.16/Fig. 3.17) are 7 and 30°, respectively. The trigonal bipyramid is the more common form for ML_5 species; other examples conforming, more or less, to this geometry include $Fe(CO)_5$, $[Pt(SnCl_3)_5]^{3-}$, and a second form of $[Ni(CN)_5]^{3-}$. For complexes with square-based pyramidal geometry there is the possibility that the five ligands may not all be at the same distance. Thus, for instance, $[Cr(CN)_5]^{2-}$ has Cr–CN = 1.88 and 2.10 Å for the basal and axial ligands, respectively. Ternary complexes ML_4L' and $ML_3L'_2$ are perhaps more relevant than symmetrical ML_5 species to associative substitution mechanisms of four-coordinate compounds and complexes. These too exist in the forms of the trigonal bipyramid and square-based pyramid, and in various intermediate geometries. Various geometries and ligand arrangements for ML_4L' and $ML_3L'_2$ are shown in Fig. 3.18. There are also a fair number of five-coordinate ternary complexes containing bidentate ligands, $M(LL)_2L'$, which are generally square pyramidal in shape. In these the bidentate ligands define the square base, with the ligand L′ axial. Such is the case for several d^8 complexes, e.g. $[Pt(diars)_2I]^+$, and for vanadium(IV) and technetium(V) complexes $[MO(LL)_2]^{n+}$. Ligand-field stabilization for five-coordinate species of trigonal bipyramidal, square pyramidal, and various intermediate geometries, has been discussed.[156] Ligand-field stabilization energies of the four- and six-coordinate transition states for fully dissociative and associative activation are, at least for the likely common geometries, a matter of common knowledge!

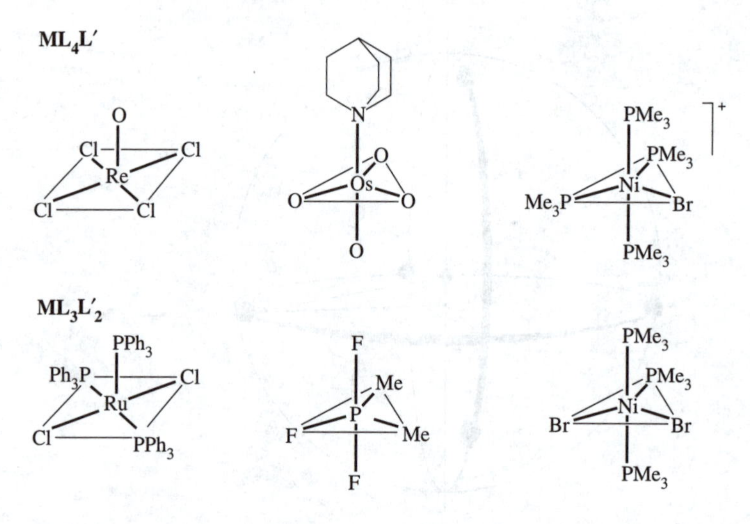

Fig. 3.18 Geometries and ligand arrangements for a selection of five-coordinate species ML_4L' and $ML_3L'_2$.

3.4.2 *Kinetics and mechanisms*

The importance of coordination number five in relation to transition states for associative substitution at four-coordinate substrates has been dealt with above—the mechanistic importance of intramolecular processes *within* five-coordinate species is detailed in Chapter 5. However, very little is known about kinetics and mechanism of substitution at five-coordinate species of either geometry. The few kinetic studies which have been made on five-coordinate complexes mainly deal with square-based pyramidal complexes of d^8 configuration, which are closely related to square-planar complexes of this same configuration.

Tetradentate tripod ligands of the type shown as **11** stabilize five-

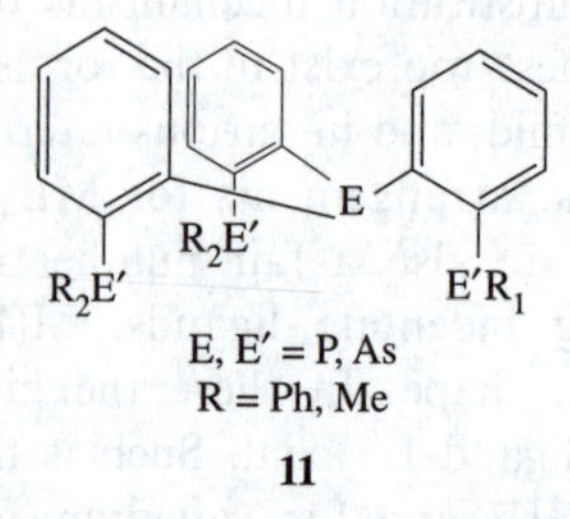

coordinate geometry at the d^8 centres Ni^{II}, Pd^{II}, and Pt^{II}, permitting study of the kinetics of substitution processes of the type

$$[M(\mathbf{11})X]^+ + Y = [M(\mathbf{11})Y]^+ + X$$

which are in fact relatively slow. The steric demands of the tripod ligands force X to be in an axial position in square pyramidal geometry, and prevent the $[M(\mathbf{11})]^{2+}$ moiety from assuming a planar (or tetrahedral) configuration. The rate law is reminiscent of square-planar substitution, namely,

$$\text{rate} = \{k_1 + k_2 [\text{Y}]\}[\text{M}(\mathbf{11})\text{X}^+]$$

but now the k_1 term is strongly dependent on the nature of the incoming nucleophile Y. Hence a simple dissociative mechanism for this pathway is ruled out. The mechanism may involve a rapid pre-equilibrium involving reversible loss of one of the donor arsenic atoms followed by bimolecular attack at the four-coordinate species of exotic geometry thus produced. An alternative suggestion invokes a weird ion-pairing mechanism. The k_2 term can at least have a simple interpretation, that of associative attack via a six-coordinate transition state. Phosphine substitution in a series of bis-dithiolato-complexes $[M(S_2C_2Ph_2)_2(PR_3)]$ appears to proceed by a simple associative mechanism.[157]

3.5 Three-coordinate compounds and complexes

3.5.1 Occurrence

The coordination number of three is fairly rare. There are a number of very important *p*-block compounds in Groups 13 and 15 of the Periodic Table—trigonal planar and pyramidal, respectively—and a few *d*- and *f*-block complexes where this low coordination number is forced on an unwilling ion by particularly bulky ligands. This miscellaneous collection of species hardly warrants a chapter devoted to this coordination number, but some of the few kinetic and mechanistic studies of such species provide useful supplementary information or illustrations to our discussion of substitution at tetrahedral centres. Moreover, some of the simple three-coordinate compounds of Groups 13 and 15—boron halides, ammonia, nitrate, phosphines—are of considerable importance.

3.5.2 Kinetics and mechanisms

Differences between substitution mechanisms at tetrahedral and trigonal centres are likely to arise from the greater ease of forming a transition state, or an intermediate, of higher coordination number at the latter. In other words, the balance between associative and dissociative activation should be tipped towards the former, favouring I_a and A substitution. Substitution at trigonal organoboron halides by bases, e.g.[158]

$$R_2BCl + L = R_2BL^+ + Cl^-$$

could proceed by any of four mechanisms, the simple S_N1 or S_N2 mechanisms already encountered frequently for tetrahedral species in this chapter, or, a realistic third possibility here, an A mechanism:

$$R_2BCl + L \underset{}{\overset{\text{fast}}{\rightleftharpoons}} R_2BClL \xrightarrow{\text{slow}} R_2BL^+ + Cl^-$$

or, fourthly, a four-centre transition state or intermediate:

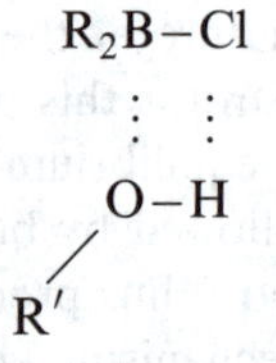

The fact that orange Ph_2B^+ can be generated from diphenylboron chloride quite easily, by the addition of $AlCl_3$ to a nitrobenzene solution, favours an S_N1 pathway. But competition experiments on diphenylboron chloride, bromide, fluoride, thiocyanate, and analogous organic derivatives reacting with water, butanol, or phenol were not consistent with the operation of an S_N1 mechanism. The particular stability of tetrahedral boron species, and the ease of their formation from three-coordinate boron compounds, would seem to favour the third route. In fact there seems to be no good evidence for the intermediacy of tetrahedral species R_2BClL of significant lifetime in reactions of this type. Thus the kinetics of reaction of phenylboron dichloride with a series of nitroanilines conformed to a simple second-order rate law, with reactivity increasing with increasing basicity of the incoming group. These observations are consistent with the S_N2 or *A* mechanisms, as were solvent effects on reactivity. The time profiles for reactants and products of these reactions conform to the pattern shown as Fig. 3.19(a), corresponding to S_N2 attack, rather than Fig. 3.19(b), characteristic of an *A* mechanism. Although displacement of chloride from trigonal boron seems in general to be S_N2 in character, the situation with respect to leaving bromide is less clear-cut. One might well deduce from the activation parameters in Table 3.17 that bromide replacement was S_N1 in mechanism, from both the positive activation entropies and the much higher activation energies. However, the fact that rate constants show significant dependence on the nature and concentration of the incoming

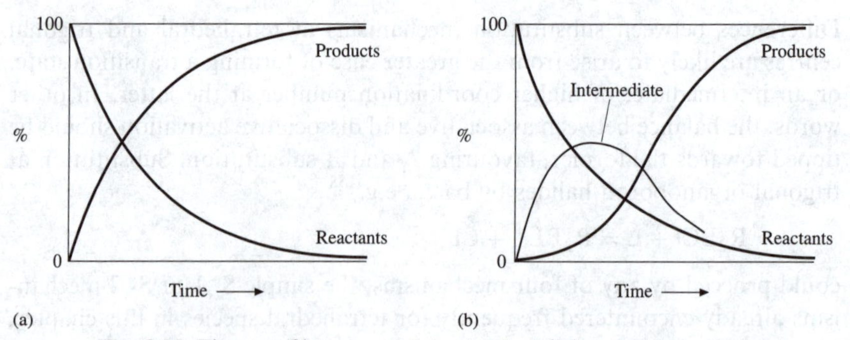

Fig. 3.19 Time profiles expected for (a) S_N2 (I_a) and (b) *A* substitution.

Table 3.17 Activation parameters for base attack at boron trihalides, in acetonitrile solution.

	$\Delta S^{\ddagger}$ (J K^{-1} mol^{-1})		$\Delta H^{\ddagger}$ (kJ mol^{-1})	
Incoming base	BCl_3	BBr_3	BCl_3	BBr_3
Dinitroaniline	−77	+43	40	80
Dinitronaphthylamine	−65	+53	40	85

group suggests that the mechanism of bromide replacement is on the S_N1/S_N2 border—perhaps a Langford–Gray *I* mechanism here, I_a for chloride replacement. Finally, we should add that exchange of alkoxy groups at trigonal boron occurs in yet another manner—probably by the four-centre route stated fourth in the list of mechanistic possibilities above.

The hydrolyses of the trichlorides of nitrogen and of phosphorus provide a particularly good example of product information giving clear indications of mechanism. These reactions also provide a further illustration to those cited above of the important mechanistic differences between substitution mechanisms for first-row elements and their lower-row analogues (cf. e.g. silicon, germanium, and tin versus carbon in Section 3.2.4.1). Hydrolysis of nitrogen trichloride gives, providing this dangerous system does not explode, ammonia and hypochlorous acid:

$$NCl_3 + 3H_2O = NH_3 + 3HClO$$

whereas hydrolysis of phosphorus trichloride gives hydrogen chloride and phosphorous acid:

$$PCl_3 + 3H_2O = HPO(OH)_2 + HCl$$

In this latter case, the availability of vacant *d*-orbitals on the phosphorus facilitates nucleophilic attack at that atom by water-oxygen. In the case of nitrogen trichloride this pathway is not feasible; rather dipolar interactions result in approach of water-hydrogen to the nitrogen, hydroxide to the chlorine.

Base hydrolysis of difluoramine[159] nicely illustrates a kinetic ambiguity that will assume considerable importance later in relation to cobalt(III)–ammine complexes (Section 4.7). The rate law is simple second-order:

$$-d[HNF_2]/dt = k_2[HNF_2][OH^-]$$

the simplest explanation of which is S_N2 attack at the nitrogen. However, as we have just seen, nucleophilic attack by oxygen at pyramidal nitrogen does not seem to be a favoured pathway. An alternative mechanism, termed S_N1 CB (≡ unimolecular nucleophilic substitution at a conjugate base), is consistent with the rate law and avoids this problem, though at the expense of a small increase in complexity:

Table 3.18 Derived second-order rate constants for reaction of difluoramine, HNF_2, with a variety of bases.

Base	k_2(dm^3 mol^{-1} s^{-1})	Base	k_2(dm^3 mol^{-1} s^{-1})	Base	k_2(dm^3 mol^{-1} s^{-1})
OH^-	6.9×10^2	Br^-	$\approx 8 \times 10^{-3}$	$CH_3CO_2^-$	4.7×10^{-5}
CN^-	4.6×10^{-2}	$H_2PO_4^-$	5.8×10^{-3}	F^-	very slow
NCS^-	1.8×10^{-2}	Cl^-	$\approx 5 \times 10^{-4}$		

$$HNF_2 + OH^- \xrightleftharpoons{\text{fast}} NF_2^- + H_2O$$

$$NF_2^- \xrightarrow{\text{slow}} \text{products}$$

If the equilibrium constant for the initial very rapidly established (de)protonation equilibrium is K, and the rate constant for the second, rate-determining, step is k, then the rate law forecast for this mechanism is:

$$-d[HNF_2]/dt = Kk[HNF_2][OH^-]$$

This is of a form identical to the experimental rate law stated above, with the composite constant Kk replacing the experimental rate constant k_2. The S_N1 CB mechanism must involve the generation of a dissociative transition state, $[F\text{–}N\cdots F^-]^\ddagger$, which reacts by itself and/or with water. In contrast to the trichloride hydrolyses discussed in the preceding paragraph, product nature is of no assistance here, since no fewer than six products were identified, including N_2F_2 and N_2 as major products and N_2F_4, N_2O, NO_3^- and NO_2^- in small amounts. We should add that the formation of a number of products from an S_N1 transition state should not cause concern—hydrolysis of t-butyl chloride produces varying ratios of t-butyl alcohol and butenes depending on the solvent, all from the same S_N1 transition state. Although the original investigators of this system favoured reaction via the conjugate base NF_2^-, subsequently a simple S_N2 mechanism was preferred on the grounds of analogy to a claim that hydroxide reacted with NF_3 by an S_N2 mechanism. The route was also felt to accommodate the relative reactivities for reaction of HNF_2 with a variety of nucleophiles (Table 3.18).

Sulphur provides a link between this section and the next, between these two sections and the discussion of tetrahedral compounds, and between inorganic and organic chemistry, in that kinetic and mechanistic studies have ranged over all these reaction types.[160]

3.6 Two-coordinate compounds and complexes

3.6.1 Occurrence

The most likely place to encounter two-coordinate linear complexes is at the right-hand edge of the *d*-block, i.e. at centres with d^n configurations of high *n*. Indeed, practically all known two-coordinate centres are d^{10}, including such long-known species as $[Ag(NH_3)_2]^+$, $[Au(CN)_2]^-$, $[CuCl_2]^-$, and $[Hg(CN)_2]$, the important alkylmercury(II) species RHgX, and the recently discovered complexes $[M(CO)_2]^+$ (with M = Ag, Au). In other regions of the Periodic Table there are some isolated examples of two-coordination imposed by, for example, stringent steric demands (e.g. $[Co\{N(SiMe_3)_2\}_2]$) or stereochemically active lone pairs (e.g. ICl_2^-, XeF_2).

3.6.2 Kinetics and mechanisms

3.6.2.1 Silver[161] and gold

Forty years ago it was shown that exchange of ammonia with the $[Ag(NH_3)_2]^+$ cation was fast at −33.5 °C; more recently it has been established that Ag^+ reacts more rapidly than most other M^+ cations with cryptands in DMSO solution. Ag^+aq (which may or may not be linear!) reacts very rapidly with thiosulphate and with 1,10-phenanthroline—in both cases with rate constants greater than 10^9 dm^3 mol^{-1} s^{-1}. Indeed, most complex formation reactions of Ag^+aq are fast and believed to be associative in character.[162]

There is slightly more information on gold(I). Gold(I) complexes seem to be rather less reactive than silver(I) or mercury(II), though thiol exchange at gold(I) is fast on the NMR time-scale. Gold(I) complexes probably generally undergo substitution associatively.[163] Thus, for example, the 1-methylpyridine-2-thione (mpt, **12**) complex [Au(CN)(mpt)] reacts very much more rapidly with CN^- than with HCN. Substitution at the anti-arthritis drug auranofin by, e.g., thiols, chloride, or $AuCl(PEt_3)$, is again fast, typically on the stopped-flow time-scale. For these reactions activation enthalpies are reported to be relatively low, in the 20–30 kJ mol^{-1} range. However, mechanisms of exchange of tertiary phosphines with gold(I)-halide compounds $AuX(PR_3)$ have been claimed to range from associative to dissociative, depending on the nature of the tertiary phosphine PR_3.[164] In contrast to these standard nucleophilic substitutions, it appears that methylgold(I) complexes, MeAuL, react with thiols by a free-radical chain mechanism.[165]

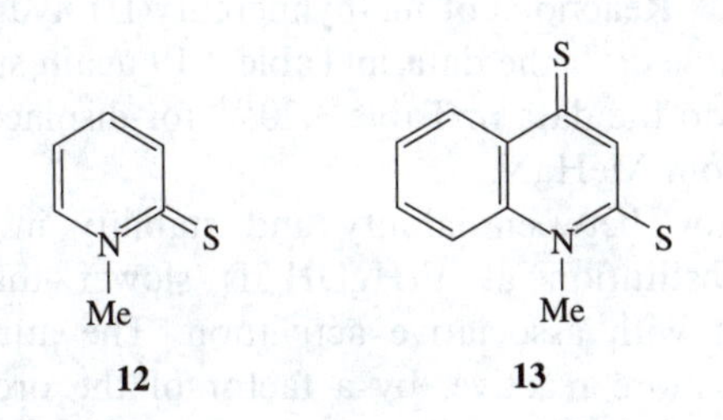

Table 3.19 Second-order rate constants for reactions of methylmercury(II) hydroxide with ligands, in aqueous solution at 293.2 K.

	$k_f/dm^3\ mol^{-1}\ s^{-1}$		$k_f/dm^3\ mol^{-1}\ s^{-1}$
MeHgOH + py	8×10^4	MeHgOH + Cl^-	1×10^4
bipy	6×10^5	SCN^-	2×10^5
pada	6×10^5	Br^-	2×10^5
terpy	8×10^5	SO_3^{2-}	2×10^5
phen	8×10^6	I^-	7×10^6

Table 3.20 Second-order rate constants for the displacement of 4-nitro-2-sulphonato-thiophenolate, nst, or 4-nitro-3-carboxylato-thiophenolate, nct, from their methylmercury(II) derivatives by the incoming groups shown, at 293.2 K.

MeHg(nst)		MeHg(nct)	
		Cl^-	7.9×10^2
Br^-	2.8×10^3		
		SCN^-	1.8×10^3
imid	1.8×10^4		
SO_3^{2-}	2.5×10^4		
I^-	*ca* 2×10^5		
OH^-	3.3×10^5		
$S_2O_3^{2-}$	2.1×10^7	$S_2O_3^{2-}$	6.7×10^7
CN^-	3.2×10^8	CN^-	1.2×10^9
$HOCH_2CH_2S^-$	1.4×10^9	$HOCH_2CH_2S^-$	1.7×10^9

3.6.2.2 Mercury

There have been a number of kinetic investigations of alkylmercury(II) compounds, RHgX. Thus, for example, kinetic studies of alkylmercury thione derivatives $MeHgL^+$ (L = 1-methylpyridine-2-thione, **12**, or 1-methylquinaldine-4-thione, **13**) with cyanide, thiones, and phosphorus donor ligands revealed almost diffusion-controlled reactions, believed to be associative in mechanism.[166] Reactions of methylmercury(II) hydroxide have been quite extensively studied;[167] the data in Table 3.19 again suggest an associative mechanism, as do the data in Table 3.20[168] for displacement of more exotic groups, X^-, from MeHgX.

There is a good correlation between lability and stability in these reactions. That bromide substitution at EtHgOH is slower than at MeHgOH is again consistent with associative activation. The aqua-ion $[MeHg(H_2O)]^+$ is very much more reactive, by a factor of the order of 10^4, for reaction with pada than the much-studied hydroxo-compound

MeHgOH, but MeHgCl can be very unreactive. Thus the symmetrization reaction

$$2MeHgCl \longrightarrow Me_2Hg + HgCl_2$$

which can be promoted by the addition of a Lewis acid to scavenge the $HgCl_2$ formed, is very slow.[169]

In the course of this chapter we have progressed from the very common and important four-coordinate systems to the rather unusual coordination numbers of three and two. In the following chapter we return to the mainstream, dealing with the most common coordination geometry of all, octahedral.

References

1. P.-A. Pittet, G. Elbaze, L. Helm, and A. E. Merbach, *Inorg. Chem.*, 1990, **29**, 1936.
2. K. F. Purcell and J. C. Kotz, *Inorganic Chemistry*, W. B. Saunders Co., Philadelphia, 1977, Chapter 7, pp. 393–401.
3. A. H. Cowley and J. H. Mills, *J. Am. Chem. Soc.*, 1969, **91**, 2910.
4. W. L. Budde and M. F. Hawthorne, *J. Am. Chem. Soc.*, 1971, **93**, 3147; D. E. Walmsley, W. L. Budde, and M. F. Hawthorne, *J. Am. Chem. Soc.*, 1971, **93**, 3150.
5. E. Müller and H.-B. Bürgi, *Helv. Chim. Acta*, 1987, **70**, 499.
6. C. K. Ingold, *Structure and Mechanism in Organic Chemistry*, Cornell University Press, Ithaca, NY, 1953, brings together the first few decades.
7. E. D. Hughes, F. Juliusberger, S. Masterman, B. Topley, and J. Weiss, *J. Chem. Soc.*, 1935, 1525.
8. C. Eaborn, *Organosilicon Chemistry*, Butterworths, London, 1960.
9. W. P. Neumann, *The Organic Chemistry of Tin*, Wiley–Interscience, New York, 1970; A. G. Davies, *Organotin Chemistry*, Wiley–VCH, Chichester, 1997.
10. See, e.g., C. Chuit, R. J. P. Corriu, C. Reye, and J. C. Young, *Chem. Rev.*, 1993, **93**, 1371.
11. D. Kummer and H. Köster, *Z. Anorg. Allg. Chem.*, 1973, **402**, 297.
12. Review: R. H. Prince, in *MTP International Review of Science*, Vol. 9, ed. M. L. Tobe, Butterworths, London, 1972.
13. R. J. P. Corriu and M. Henner, *J. Organomet. Chem.*, 1974, **74**, 1.
14. Y. Apeloig and A. Stanger, *J. Am. Chem. Soc.*, 1987, **109**, 272.
15. J. B. Lambert, L. Kania, and Shizhong Zhang, *Chem. Rev.*, 1995, **95**, 1191.
16. M. W. Grant and R. H. Prince, *J. Chem. Soc. (A)*, 1969, 1138.
17. L. H. Sommer, W. D. Korte, and P. G. Rodewald, *J. Am. Chem. Soc.*, 1967, **89**, 862.
18. Reviews: R. J. P. Corriu and C. Guérin, *J. Organomet. Chem.*, 1980, **198**, 231; R. J. P. Corriu, C. Guérin, and J. J. E. Moreau, *Top. Stereochem.*, 1984, **15**, 43.
19. L. H. Sommer and J. D. Citron, *J. Am. Chem. Soc.*, 1967, **89**, 5797; L. H. Sommer and J. McLick, *J. Am. Chem. Soc.*, 1967, **89**, 5802; R. R. Holmes, *Accounts Chem. Res.*, 1979, **12**, 257.
20. R. J. P. Corriu, C. Guérin, and J. Masse, *J. Chem. Res. (S)*, 1977, 160; R. J. P. Corriu and C. Guérin, *J. Chem. Soc., Chem. Commun.*, 1977, 74; *J. Organomet. Chem.*, 1978, **144**, 165.

21. J. Woning, L. M. Daniels, and J. G. Verkade, *J. Am. Chem. Soc.*, 1990, **112**, 4601.
22. R. J. P. Corriu, C. Guérin, B. J. L. Henner, and Q. Wang, *Organometallics*, 1991, **10**, 3200.
23. M. Gielen and J. Nasielski, *J. Organomet. Chem.*, 1967, **7**, 273; S. Boué, M. Gielen, and J. Nasielski, *J. Organomet. Chem.*, 1967, **9**, 443; P. Verbiest, L. Verdonck, and G. P. van der Kelen, *Int. J. Chem. Kinet.*, 1993, **25**, 107.
24. M. H. Abraham and G. F. Johnston, *J. Chem. Soc. (A)*, 1970, 193; 1971, 1610; M. H. Abraham, R. J. Irving, and G. F. Johnston, *J. Chem. Soc. (A)*, 1970, 199; M. H. Abraham and F. J. Dorrell, *J. Chem. Soc., Perkin Trans. II*, 1973, 444.
25. M. H. Abraham and F. Behbahany, *J. Chem. Soc. (A)*, 1971, 1469; M. H. Abraham, *J. Chem. Soc., Perkin Trans. II*, 1972, 1343.
26. M. H. Abraham, D. F. Dadjour, and C. J. Holloway, *J. Organomet. Chem.*, 1973, **52**, C27.
27. M. Gielen, *Accounts Chem. Res.*, 1973, **6**, 198.
28. D. N. Hague and R. H. Prince, *J. Inorg. Nucl. Chem.*, 1966, **28**, 1039.
29. M. Gielen, J. Nasielski, and G. Vandendungen, *Bull. Soc. Chim. Belges*, 1971, **80**, 165.
30. J. C. Lockhart, *Introduction to Inorganic Reaction Mechanisms*, Butterworths, London, 1966, Chapter 3.
31. R. R. Holmes, *Accounts Chem. Res.*, 1979, **12**, 257.
32. R. F. Hudson and M. Green, *Angew. Chem. Int. Edn. Engl.*, 1963, **2**, 11.
33. M. Green and R. F. Hudson, *J. Chem. Soc.*, 1963, 540.
34. F. H. Westheimer, *Accounts Chem. Res.*, 1968, **1**, 70; J. Emsley and D. Hall, *The Chemistry of Phosphorus*, Harper & Row, London, 1976.
35. J. Omelanczuk and M. Mikolajczyk, *J. Chem. Soc., Chem. Commun.*, 1994, 2223.
36. J. Burgess, in *Inorganic Reaction Mechanisms*, Vol. 1, The Chemical Society, London, 1971, pp. 119–120 (and many of the early volumes in the sister Specialist Periodical Report series, *Organophosphorus Chemistry*, Vol. 1, ed. S. Trippett, The Chemical Society, London, 1970; subsequent volumes published by the Chemical Society, then by the Royal Society of Chemistry); B. J. Walker, *Organophosphorus Chemistry*, Penguin, Harmondsworth, 1972, Chapter 3.
37. Reviews: F. Ciuffarin and A. Fava, *Progr. Phys. Org. Chem.*, 1968, **6**, 81; J. L. Kice, *Progr. Inorg. Chem.*, 1972, **17**, 147.
38. O. Rogne, *J. Chem. Soc.(B)*, 1969, 663; 1970, 727; 1970, 1056.
39. B. E. Fleischfresser and I. Lauder, *Aust. J. Chem.*, 1962, **15**, 251; I. G. Ryss, L. P. Bogdanova, S. L. Idel's, and T. N. Kotlyar, *Russ J. Inorg. Chem.*, 1969, **14**, 1577.
40. U. M. Frei and G. Geier, *Inorg. Chem.*, 1992, **31**, 187.
41. Review: J. K. Beattie, *Adv. Inorg. Chem.*, 1988, **32**, 1; J. K. Beattie and W. E. Moody, *J. Coord. Chem.*, 1994, **32**, 155.
42. L. H. Pignolet and W. de W. Horrocks, *J. Am. Chem. Soc.*, 1968, **90**, 922; L. H. Pignolet, D. Forster, and W. de W. Horrocks, *Inorg. Chem.*, 1968, **7**, 828; S. S. Zumdahl and R. S. Drago, *J. Am. Chem. Soc.*, 1967, **89**, 4319.
43. F. K. Meyer, W. L. Earl, and A. E. Merbach, *Inorg. Chem.*, 1979, **18**, 888.
44. K. R. Brower and T.-S. Chen, *Inorg. Chem.*, 1973, **12**, 2198.
45. S. S. Zumdahl and R. S. Drago, *Inorg. Chem.*, 1968, **7**, 2162;

46. P. F. Meier, A. E. Merbach, M. Dartiquenave, and T. Dartiquenave, *Inorg. Chem.*, 1979, **18**, 610.
47. H. Gamsjäger and R. K. Murmann, *Adv. Inorg. Bioinorg. Mech.*, 1983, **2**, 317.
48. A. Okumura, M. Fushimi, H. Hanaki, A. Yamada, Y. Gofuku, M. Kagawa, N. Niwa, M. Yamamoto, and M. Iida, *Bull. Chem. Soc. Jpn.*, 1992, **65**, 1397.
49. R. Baharad, B. Perlmutter-Hayman, and M. A. Wolff, *J. Phys. Chem.*, 1969, **73**, 4391.
50. P. Moore, S. F. A. Kettle, and R. G. Wilkins, *Inorg. Chem.*, 1966, **5**, 220.
51. H. K. Hofmeister and J. R. Van Wazer, *Inorg. Chem.*, 1962, **1**, 811.
52. E. Thilo and F. V. Lampe, *Z. Anorg. Allg. Chem.*, 1963, **319**, 387; *Chem. Ber.*, 1964, **97**, 1775.
53. J. R. Van Wazer, E. J. Griffith, and J. F. McCullough, *J. Am. Chem. Soc.*, 1955, **77**, 287.
54. M. Kawabe, O. Ohashi, and I. Yamaguchi, *Bull. Chem. Soc. Jpn.*, 1970, **43**, 3705; S. Ueda and Y. Sasaki, *Bull. Chem. Soc. Jpn.*, 1971, **44**, 1972; J. R. Blackburn and M. M. Jones, *J. Inorg. Nucl. Chem.*, 1973, **35**, 1597, 1605, 2421.
55. R. C. Patel and R. S. Taylor, *J. Phys. Chem.*, 1973, **77**, 2318.
56. P. Hendry and A. M. Sargeson, *Progr. Inorg. Chem.*, 1990, **38**, 201.
57. Review: K. Saito and Y. Sasaki, *Adv. Inorg. Bioinorg. Mech.*, 1982, **1**, 179.
58. Reviews: L. Cattalini, *Progr. Inorg. Chem.*, 1970, **13**, 263; M. L. Tobe, in *Comprehensive Coordination Chemistry*, Vol. 1, eds G. Wilkinson, R. D. Gillard, and J. A. McCleverty, Pergamon, Oxford, 1987, Chapter 2.2; R. J. Cross, *Adv. Inorg. Chem.*, 1989, **34**, 219.
59. J. A. McCleverty, *Progr. Inorg. Chem.*, 1968, **10**, 49; U. T. Mueller-Westerhoff and B. Vance, in *Comprehensive Coordination Chemistry*, Vol. 2, eds G. Wilkinson, R. D. Gillard, and J. A. McCleverty, Pergamon, Oxford, 1987, Chapter 16.5.
60. N. V. Sidgwick and R. W. Bailey, *Proc. Roy. Soc. A*, 1934, **144**, 521; N. V. Sidgwick and H. M. Powell, *Proc. Roy. Soc. A*, 1940, **176**, 153.
61. Chong Shik Chin, Sung-Nak Choi, and Kye-Duck Kim, *Inorg. Chem.*, 1990, **29**, 145.
62. J. S. Thompson and J. D. Atwood, *J. Am. Chem. Soc.*, 1991, **113**, 7429.
63. E. T. Borish and L. J. Kirschenbaum, *Inorg. Chem.*, 1984, **23**, 2355.
64. L. K. Skibsted, *Adv. Inorg. Bioinorg. Mech.*, 1986, **4**, 137.
65. C. A. Tolman, *Chem. Soc. Rev.*, 1972, **1**, 337.
66. F. Basolo, *Adv. Chem. Ser.*, 1965, **49**, 81.
67. R. Roulet and C. Barbey, *Helv. Chim. Acta,* 1973, **56**, 2179.
68. S. Lanza, D. Minniti, P. Moore, J. Sachinidis, R. Romeo, and M. L. Tobe, *Inorg. Chem.*, 1984, **23**, 4228.
69. Review: A. Peloso, *Coord. Chem. Rev.*, 1973, **10**, 123.
70. R. van Eldik, T. Asano, and W. J. Le Noble, *Chem. Rev.*, 1989, **89**, 549.
71. R. van Eldik, *Mech. Inorg. Organomet. Reactions*, 1994, **8**, 399 (see p. 406).
72. F. Basolo, J. Chatt, H. B. Gray, R. G. Pearson, and B. L. Shaw, *J. Chem. Soc.*, 1961, 2207.
73. D. A. Palmer and H. Kelm, *Inorg. Chim. Acta*, 1976, **19**, 117.
74. H. B. Gray and R. J. Olcott, *Inorg. Chem.*, 1962, **1**, 481.
75. Chong Shik Chin, Sung-Nak Choi, and Kye-Duck Kim, *Inorg. Chem.*, 1990, **29**, 145.
76. G. Annibale, L. Cattalini, A. A. El-Awady, and G. Natile, *J. Chem. Soc., Dalton Trans.*, 1974, 802.

77. J. Burgess, *Ann. Rep. Chem. Soc.*, 1968, **65**, 395 (see p. 406 for early references).
78. L. Canovese, L. Cattalini, P. Uguagliati, and M. L. Tobe, *J. Chem. Soc., Dalton Trans.*, 1990, 867 and references therein.
79. Reviews: M. Green and M. Green, *Transition Met. Chem.*, 1985, **10**, 196; M. Green, *Transition Met. Chem.*, 1987, **12**, 186; M. Green, M. Garner, and D. M. Orton, *Transition Met. Chem.*, 1992, **17**, 164.
80. Reviews: J. Reedijk, A. M. J. Fichtinger-Schepman, A. T. van Oosterom, and P. van de Putte, *Structure Bonding*, 1987, **67**, 53.
81. Review: A. Pasini and F. Zunino, *Angew. Chem. Int. Edn. Engl.*, 1987, **26**, 615.
82. J. Reedijk, A. M. J. Fichtinger–Schepman, A. T. van Oosterom, and P. van de Putte, *Structure Bonding*, 1987, **67**, 62.
83. L. Canovese, L. Cattalini, G. Chessa, and M. L. Tobe, *J. Chem. Soc., Dalton Trans.*, 1988, 2135.
84. K. J. Barnham, M. I. Djuran, U. Frey, M. Mazid, and P. J. Sadler, *J. Chem. Soc., Chem. Commun.*, 1994, 65.
85. F. Elferink, W. J. F. van der Vijgh, I. Klein, and H. M. Pimedo, *Clin. Chem.*, 1986, **32**, 641.
86. M. Pujol, J. Part, M. Trillas, and X. Domènech, *Monatsh. Chem.*, 1993, **124**, 1077.
87. R. N. Bose, N. Goswami, and S. Moghaddas, *Inorg. Chem.*, 1990, **29**, 3461.
88. K. J. Barnham, M. I. Djuran, P. del S. Murdoch, and P. J. Sadler, *J. Chem. Soc., Chem. Commun.*, 1994, 721.
89. D. Banerjea, F. Basolo, and R. G. Pearson, *J. Am. Chem. Soc.*, 1957, **79**, 4055.
90. L. Helm, L. I. Elding, and A. E. Merbach, *Helv. Chim. Acta*, 1984, **67**, 1453; *Inorg. Chem.*, 1985, **24**, 1719.
91. C. G. Swain and C. B. Scott, *J. Am. Chem. Soc.*, 1953, **85**, 141.
92. U. Belluco, L. Cattalini, F. Basolo, R. G. Pearson, and A. Turco, *J. Am. Chem. Soc.*, 1965, **87**, 241.
93. Full documentation can be found in M. L. Tobe, *Comprehensive Coordination Chemistry*, Vol. 1, Chapter 7.1; see p. 314 (Table 15).
94. L. Sacconi, *J. Chem. Soc.(A)*, 1970, 248.
95. L. Cattalini, A. Orio, and M. Nicolini, *J. Am. Chem. Soc.*, 1966, **88**, 5734.
96. D. Banerjea, F. Basolo, and R. G. Pearson, *J. Am. Chem. Soc.*, 1957, **79**, 4055.
97. H. Krüger and R. van Eldik, *J. Chem. Soc., Chem. Commun.*, 1990, 330.
98. M. L. Tobe, A. T. Treadgold, and L. Cattalini, *J. Chem. Soc., Dalton Trans.*, 1988, 2347 and references therein.
99. R. Romeo and M. Cusumano, *Inorg. Chim. Acta*, 1981, **49**, 167; M. Bonivento, L. Canovese, L. Cattalini, G. Marangoni, G. Michelon, and M. L. Tobe, *Inorg. Chem.*, 1981, **20**, 3728.
100. M. Cusumano, G. Faraone, V. Ricevuto, R. Romeo, and M. Trozzi, *J. Chem. Soc., Dalton Trans.*, 1974, 490.
101. E. Ahmed, R. J. H. Clark, M. L. Tobe, and L. Cattalini, *J. Chem. Soc., Dalton Trans.*, 2701, 1990.
102. L. Cattalini, A. Orio, and M. L. Tobe, *J. Am. Chem. Soc.*, 1967, **89**, 3130; L. Cattalini, V. Ricevuto, A. Orio, and M. L. Tobe, *Inorg. Chem.*, 1968, **7**, 51; K. M. Ibne-Rasa, J. O. Edwards, and J. L. Rogers, *J. Solution Chem.*, 1975, **4**, 609.
103. G. Annibale, L. Canovese, L. Cattalini, G. Marangoni, G. Michelon, and M. L. Tobe, *Inorg. Chem.*, 1984, **23**, 2705.

104. L. Canovese, L. Cattalini, G. Marangoni, G. Michelon, and M. L. Tobe, *Inorg. Chem.*, 1981, **20**, 4166.
105. F. Basolo, H. B. Gray, and R. G. Pearson, *J. Am. Chem. Soc.*, 1960, **82**, 4200.
106. L. Cattalini and M. L. Tobe, *Inorg. Chem.*, 1966, **5**, 1145.
107. L. Cattalini, A. Orio, and M. L. Tobe, *Inorg. Chem.*, 1967, **6**, 75.
108. J. E. Leffler and E. Grunwald, *Rates and Equilibria of Organic Reactions*, Wiley, New York, 1963, p.156.
109. C. H. Langford and H. B. Gray, *Ligand Substitution Processes*, Benjamin, New York, 1965, p. 61.
110. R. Romeo and M. L. Tobe, *Inorg. Chem.*, 1974, **13**, 1991.
111. M. Bonivento, L. Canovese, L. Cattalini, G. Marangoni, G. Michelon, and M. L. Tobe, *Inorg. Chem.*, 1981, **20**, 1493.
112. B. Pitteri, G. Marangoni, L. Cattalini, and L. Canovese, *J. Chem. Soc., Dalton Trans.*, 1994, 169; M. Bellicini, L. Cattalini, G. Marangoni, and B. Pitteri, *J. Chem. Soc., Dalton Trans.*, 1994, 1805; B. Pitteri, G. Marangoni, and L. Cattalini, *J. Chem. Soc., Dalton Trans.*, 1994, 3539; B. Pitteri, G. Marangoni, and L. Cattalini, *Polyhedron*, 1995, **14**, 2331; B. Pitteri, G. Marangoni, L. Cattalini, and T. Bobbo, *J. Chem. Soc., Dalton Trans.*, 1995, 3853; M. Bonivento, L. Cattalini, G. Marangoni, B. Pitteri, and T. Bobbo, *Transition Met. Chem.*, 1997, **22**, 46.
113. A. Werner, *Z. Anorg. Allg. Chem.*, 1893, **3**, 367.
114. I. I. Chernayev, *Ann. Inst. Platine USSR*, 1926, **4**, 246, 261.
115. A. D. Hel'man, E. F. Karandashova and L. N. Essen, *Dokl. Akad. Nauk. S.S.S.R.*, 1948, **63**, 47; F. Basolo and R. G. Pearson, *Mechanisms of Inorganic Reactions*, 2nd edn, Wiley, New York, 1967, p. 354.
116. J. V. Quagliano and L. Schubert, *Chem. Rev.*, 1952, **50**, 201.
117. F. Basolo and R. G. Pearson, *Mechanisms of Inorganic Reactions*, Wiley, New York, 1958, p. 190.
118. J. Chatt, L. A. Duncanson, and L. M. Venanzi, *J. Chem. Soc.*, 1955, 4456.
119. L. E. Orgel, *J. Inorg. Nucl. Chem.*, 1956, **2**, 137.
120. F. Basolo and R. G. Pearson, *Progr. Inorg. Chem.*, 1962, **4**, 381.
121. Reviews: F. R. Hartley, *Chem. Soc. Rev.*, 1973, **2**, 163 (good simple introduction); S. A. Cotton, *Chemistry of Precious Metals*, Blackie, London, 1997.
122. Reviews: R. McWeeny, R. Mason, and A. D. C. Towl, *Disc. Faraday Soc.*, 1969, **47**, 20; T. G. Appleton, H. C. Clark, and L. E. Manzer, *Coord. Chem. Rev.*, 1973, **10**, 335.
123. R. J. Goodfellow, P. L. Goggin, and D. A. Duddell, *J. Chem. Soc. (A)*, 1968, 504.
124. F. H. Allen and A. Pidcock, *J. Chem. Soc. (A)*, 1968, 2700.
125. T. G. Appleton and M. A. Bennett, *Inorg. Chem.*, 1978, **17**, 738.
126. C. W. Fryer and J. A. S. Smith, *J. Chem. Soc. (A)*, 1970, 1029; R. V. Parish, *Coord. Chem. Rev.*, 1982, **42**, 1.
127. S. S. Zumdahl and R. S. Drago, *J. Am. Chem. Soc.*, 1968, **90**, 6669.
128. C. H. Langford and H. B. Gray, *Ligand Substitution Processes*, Benjamin, New York, 1965, p. 27; J. K. Burdett and T. A. Albright, *Inorg. Chem.*, 1979, **18**, 2112.
129. T. G. Appleton, H. C. Clark, and L. E. Manzer, *Coord. Chem. Rev.*, 1973, **10**, 335.
130. R. Gosling and M. L. Tobe, *Inorg. Chem.*, 1983, **22**, 1235 and references therein.

131. L. Canovese, M. L. Tobe, and L. Cattalini, *J. Chem. Soc., Dalton Trans.*, 1985, 27.
132. M. Cusumano, P. Marricchi, R. Romeo, V. Ricevuto, and U. Belluco, *Inorg. Chim. Acta*, 1979, **34**, 169.
133. C. K. Ingold, *Structure and Mechanism in Organic Chemistry*, Cornell University Press, Ithaca, NY, 1953, Chapter 2, Section 7.
134. G. Natile, L. Maresca, and L. Cattalini, *J. Chem. Soc., Dalton Trans.*, 1977, 651.
135. L. Maresca, G. Natile, M. Calligaris, P. Delise, and L. Randaccio, *J. Chem. Soc., Dalton Trans.*, 1976, 2386.
136. R. Romeo, D. Minniti, and M. Trozzi, *Inorg. Chem.*, 1976, **15**, 1134; G. Faraone, V. Ricevuto, R. Romeo, and M. Trozzi, *J. Chem. Soc., Dalton Trans.*, 1974, 1377.
137. U. Belluco, L. Cattalini, F. Basolo, R. G. Pearson, and A. Turco, *J. Am. Chem. Soc.*, 1965, **87**, 241.
138. M. Bonivento, L. Canovese, L. Cattalini, G. Marangoni, G. Michelon, and M. L. Tobe, *Inorg. Chem.*, 1981, **20**, 3728.
139. L. I. Elding and Ö. Gröning, *Inorg. Chem.*, 1978, **17**, 1872; J. K. Burdett, *Inorg. Chem.*, 1977, **16**, 3013.
140. Review: R. Romeo, *Comments Inorg. Chem.*, 1990, **11**, 21.
141. Ö. Gröning, T. Drakensberg, and L. I. Elding, *Inorg. Chem.*, 1982, **21**, 1820.
142. L. Helm, L. I. Elding, and A. E. Merbach, *Inorg. Chem.*, 1985, **24**, 1719.
143. R. van Eldik, in *High Pressure Chemistry and Biochemistry*, eds R. van Eldik and J. Jonas, Reidel, Dordrecht, 1987, p. 333 (see especially p. 338); M. Kotowski, S. Begum, J. G. Leipoldt, and R. van Eldik, *Inorg. Chem.*, 1988, **27**, 4472.
144. J.-L. Jestin, J.-C. Chottard, U. Frey, G. Laurenczy, and A. E. Merbach, *Inorg. Chem.*, 1994, **33**, 4277.
145. S. Komiya, T. A. Albright, R. Hoffmann, and J. K. Kochi, *J. Am. Chem. Soc.*, 1976, **98**, 7255; A. Gillie and J. K. Stille, *J. Am. Chem. Soc.*, 1980, **102**, 4933.
146. G. Alibrandi, L. Monsù Scolaro, and R. Romeo, *Inorg. Chem.*, 1991, **30**, 4007.
147. D. Minniti, *Inorg. Chem.*, 1994, **33**, 2631.
148. S. Lanza, D. Minniti, P. Moore, J. Sachinides, R. Romeo, and M. L. Tobe, *Inorg. Chem.*, 1984, **23**, 4428.
149. D. Minniti, G. Alibrandi, M. L. Tobe, and R. Romeo, *Inorg. Chem.*, 1987, **26**, 3956.
150. U. Frey, L. Helm, A. E. Merbach, and R. Romeo, *J. Am. Chem. Soc.*, 1989, **111**, 8161.
151. R. Romeo, A. Grassi, and L. Monsù Scolaro, *Inorg. Chem.*, 1992, **31**, 4383.
152. J. S. Thompson and J. D. Atwood, *J. Am. Chem. Soc.*, 1991, **113**, 7429; R. L. Rominger, J. M. McFarland, J. R. Jeitler, J. S. Thompson, and J. D. Atwood, *J. Coord. Chem.*, 1994, **31**, 7.
153. R. S. Nyholm and M. L. Tobe, *Experientia*, Suppl. 9, 1964, 112; *The Stabilization, Stereochemistry, and Reactivity of Five-coordinate Complexes*, in *Essays in Coordination Chemistry*, Birkenhäusen, Basel, 1964, pp. 112–127.
154. E. L. Muetterties and R. A. Schunn, *Quart. Rev.*, 1966, **20**, 245; P. L. Orioli, *Coord. Chem. Rev.*, 1971, **6**, 285; J. S. Wood, *Progr. Inorg. Chem.*, 1972, **16**, 227; M. C. Favas and D. L. Kepert, *Progr. Inorg. Chem.*, 1980, **26**, 325; R. R. Holmes, *Progr. Inorg. Chem.*, 1984, **32**, 119.
155. E. L. Muetterties and L. J. Guggenberger, *J. Am. Chem. Soc.*, 1974, **96**, 1751.

156. Review: C. Furlani, *Coord. Chem. Rev.*, 1968, **3**, 141.
157. D. A. Sweigart, D. E. Cooper, and J. M. Millican, *Inorg. Chem.*, 1974, **13**, 1272.
158. J. C. Lockhart, *J. Chem. Soc. (A)*, 1962, 1197; 1966, 809; R. Heyes and J. C. Lockhart, *J. Chem. Soc. (A)*, 1968, 326; J. R. Blackborow and J. C. Lockhart, *J. Chem. Soc. (A)*, 1968, 3015.
159. A. D. Craig and G. A. Ward, *J. Am. Chem. Soc.*, 1966, **88**, 4526.
160. Review: S. Oae, *Organic Sulfur Chemistry: Structure and Mechanism*, CRC Press, Boca Raton, Fla., 1991.
161. See, e.g., H. U. D. Wiesendanger, W. H. Jones, and C. S. Garner, *J. Chem. Phys.*, 1957, **27**, 668; M. M. Farrow, N. Purdie, and E. M. Eyring, *Inorg. Chem.*, 1975, **14**, 1584; B. G. Cox, J. Garcia-Rosas, H. Schneider, and Ng van Truong, *Inorg. Chem.*, 1986, **25**, 1165.
162. M. M. Farrow, N. Purdie, and E. M. Eyring, *Inorg. Chem.*, 1975, **14**, 1584.
163. P. N. Dickson, A. Wehrli, and G. Geier, *Inorg. Chem.*, 1988, **27**, 2921.
164. C. B. Colburn, W. E. Hill, C. A. McAuliffe, and R. V. Parish, *J. Chem. Soc., Chem. Commun.*, 1979, 218; M. J. Mays and P. A. Vergagno, *J. Chem. Soc., Dalton Trans.*, 1979, 1112.
165. A. Johnson and R. J. Puddephat, *J. Chem. Soc., Dalton Trans.*, 1975, 115.
166. I. Erni and G. Geier, *Helv. Chim. Acta*, 1979, **62**, 1007.
167. G. Geier, I. Erni, and R. Steiner, *Helv. Chim. Acta*, 1977, **60**, 9.
168. G. Geier and H. Gross, *Inorg. Chim. Acta*, 1989, **156**, 91.
169. K. Stanley, J. Martin, J. Schnittker, R. Smith, and M. C. Baird, *Inorg. Chim. Acta*, 1978, **27**, L111.

4 Substitution in octahedral complexes

Chapter contents

4.1 Overview

Six-coordination, in the octahedral geometry with which it is predominantly associated, is by far the most commonly occurring ligand arrangement in coordination chemistry. It is favoured by the packing of ligands about a spherical metal ion, and it is the favoured arrangement for the d^3 and low-spin d^6 configurations in their relatively covalent compounds. Six-coordination is very common for other configurations of *d*-block metal ions, though the octahedron may be distorted, either tetragonally (e.g. Jahn–Teller distortion in d^4 and d^9 complexes) or towards a trigonal prism (e.g. some bidentate oxygen and, more commonly, bidentate sulphur donor ligands).[1] Such tetragonal or trigonal distortions, especially the former, can exert a profound effect upon the reactivity of the system. The important areas are summarized in Table 4.1, which is by no means comprehensive.

Because this geometry covers such a wide range of oxidation states and electron configurations as well as modes of bonding, one might have expected to find examples of the whole range of modes of activation but, with some exceptions, the bulk of the substitution processes are interchanges. At one time it was thought that the I_d mechanism dominated. There was talk of this being the ‘typical mechanism for octahedral substitution’—exemplified by the Eigen–Wilkins mechanism for complex formation[2] (discussed in depth in Chapter 7). More recently it has emerged that the associative mode of activation is perhaps just as common—with a

Table 4.1 Examples of the occurrence of octahedral geometry.

Electrovalent bonding	*s*-block	$[Mg(H_2O)_6]^{2+}$
		NaCl in the solid state
	d-block	$[Mn(H_2O)_6]^{2+}$
Electrovalent/covalent bonding	*p*-block	$[SiF_6]^{2-}$
	d-block	$[Ni(H_2O)_6]^{2+}$ (regular)
		$[Cr(H_2O)_6]^{2+}$ (distorted)
Covalent bonding	*p*-block	SF_6
	d-block	$Cr(CO)_6$

The low-spin d^6 configuration particularly favours six-coordinate octahedral complexes; it is represented by the following oxidation states:

V(−I)	Cr(0)	Mn(I)	Fe(II)	Co(III)	Ni(IV)
Nb(−I)	Mo(0)	Tc(I)	Ru(II)	Rh(III)	Pd(IV)
Ta(−I)	W(0)	Re(I)	Os(II)	Ir(III)	Pt(IV)

range of geometries[3] possible for such transition states.[4] It remains true, nevertheless, that no octahedral substrate has yet been shown to possess a nucleophilic discrimination that equals that found in the associative reactions of square-planar d^8 metal complexes or of carbon. It is by no means unlikely that the reason such systems are lacking is the reluctance of researchers in this area to make the necessary detailed studies, or to have the good fortune to light on an appropriate system.

In principle, the two simplest groups of reactions to handle in a text on kinetics and mechanisms in solution should be solvent exchange and substitution in organometallic compounds—in both cases observable for a variety of octahedral species. In the case of the former the simplicity stems from the fact that the solvent is both nucleophile and medium—a simplicity which makes diagnosis of mechanism through rate-law determination almost impossible, and a special character and relation with complex formation that makes it more appropriate to deal with such reactions in a separate chapter (see Chapter 7). Another good reason for dealing with solvent exchange separately is that the overall mechanistic picture is best built up from kinetic results on solvento ions covering the whole range of solvation numbers and stereochemistries encountered—from tetrahedral and square-planar (Be^{2+}; Pd^{2+}, Pt^{2+}) through to nine-coordinate *f*-block solvento cations. Consideration of octahedral species separately would cause an artificial and unhelpful division of results and discussion, while the fact that solvation numbers are not known with certainty for a significant number of metal ions also plays havoc with any attempt to treat this area strictly according to coordination number and geometry.

In the case of organometallic compounds, the prevalence of low oxidation states and the dominance of the eighteen-electron rule, along with the different type of ligands, groups, and reactions involved, all make it more convenient to deal with their reaction mechanisms separately (Chapters 6 and 10). In organometallic chemistry, solute–solvent interactions rarely cause the complications they do when dealing with substitution at classical inorganic complexes in donor solvents (especially water!), so it is with an area lying between these and classical Werner complexes that we start our detailed consideration of mechanisms. Fortunately, there have been some kinetic studies of uncharged octahedral species where solute–solvent interactions also cause minimal difficulty, so we deal with these sytems first, before embarking on the centrally important theme of substitution at the most studied octahedral transition metal centre, cobalt(III). The field is still dominated, for a variety of reasons, by information relating to low-spin d^6 Co(III) complexes. However, this dominance has been much diminished in recent years as people have been forced to tackle more difficult systems to maintain some semblance of novelty, fast-reaction techniques have become more widely available, and the preparative coordination chemistry of the other elements has been developed. Although Co(III) has proved to be a non-typical reaction centre, we will discuss the patterns that are found in its reactions first, and then we will see how the behaviour of the other reaction centres differs.

4.2 General theoretical considerations

4.2.1 The valence bond approach

In his landmark review, Taube,[5] using a valence bond/hybridization model for bonding, suggested that if one of the *d*-orbitals not involved in the hybrid orbitals used to bond the ligands, d^2sp^3 in the case of octahedral complexes, was empty, the complex would be labile. Note the incipient Tolman's rules, the hint of coordination unsaturation and the consequent facile pathway for associative activation, all as long ago as 1952! If the non-bonding electrons could not all be accommodated in these orbitals, the theory at that time required that the *d*-orbitals of the next quantum shell were in the bonding, which was now weak ('ionic') and the complex was consequently labile—the hint was that the mechanism was dissociatively activated. It was then assumed that reaction centres that lacked the electron configurations that did not lead to these facile pathways for substitution were inert to substitution. Thus for an octahedral substrate, slow substitution reactions were predicted for the d^3, d^4 low-spin, d^5 low-spin, and d^6 low-spin configurations, while d^0, d^1, and d^2 complexes were labile by way of associative substitution and d^4 high-spin, d^5 high-spin, d^6 high-spin, d^7, d^8, d^9, and d^{10} complexes were labile by way of dissociative activation. To the extent that the areas where substitutionally inert complexes existed were correctly identified, this assignment was successful; but it rested on shaky foundations and was not even semi-quantitative. The distinction between labile and inert was purely arbitrary and subsequent work was to show that there was no natural break in reactivity from the very slow to the very fast. Nevertheless, this review pointed the way and by making clear the distinction between thermodynamic stability and kinetic inertness its contribution to the education of inorganic chemists was of major importance.

4.2.2 The crystal field activation energy approach

Basolo and Pearson, who probably provided the first readable account of crystal field theory in the first edition of their textbook,[6] used as the basis for their model the assumption that the change in crystal field stabilization energy (CFSE) on going from the ground state to the transition state made a significant contribution to the activation energy of the substitution reaction. The crystal field model was an electrostatically based bonding theory in which the ligands removed some of the degeneracy of the *d*-orbitals of an isolated atom and the resultant distribution of the non-bonding electrons among these orbitals resulted in a decrease in the total energy, except where all five orbitals were empty, d^0, half-filled, d^5 high-spin, or filled, d^{10}. The orbital splitting patterns for an octahedral substrate and the possible geometries of the transition states (there was always some confusion between transition states and intermediates) were calculated in terms of $10Dq$ (or Δ), the difference in energy between the lower and higher levels in an octahedral field. These are shown in Fig. 4.1.[7] The crystal field activation energy (CFAE) was then taken to be the loss of crystal field stabilization energy on going from the ground (initial) state to the transition state. When there was any possible ambiguity it was assumed that there is no change in spin multiplicity on going from the ground state to the transition state. The values of the CFAE thus estimated are collected in Table 4.2, with the minimum loss of crystal field stabilization energy (or maximum gain) indicated for each electron configuration. It is clear that the d^3, d^6 low-spin, and d^8 configurations should yield inert reaction centres. It is also clear that other configurations, notably high-spin d^4, d^7, and d^9, actually gain CFSE on going to another coordination number and it is these that exhibit Jahn–Teller distortions. Reduction of symmetry, in these cases by a tetragonal distortion, when the higher-symmetry ground state is degenerate, leads to a non-degenerate ground state at lower energy. It is once again clear that the trigonal bipyramidal five-coordinate geometry will always be less favoured than the square-based pyramid in a dissociative process, and that the octahedral wedge (C_{2v}) geometry is usually favoured over the pentagonal bipyramid (D_{5h}) in the seven-coordinate intermediate. The increase in $10Dq$ for a particular set of ligands on going from a $3d$ to a $4d$ to a $5d$ reaction centre not only ensures that the high-spin configuration becomes a rarity beyond the first row, but it also means that the CFAE barriers increase and leads to the prediction that, for any one configuration, the reactivity increases on going down a group. Although the CFAE approach looks quantitative, it is not. The surprise is that it has any predictive power

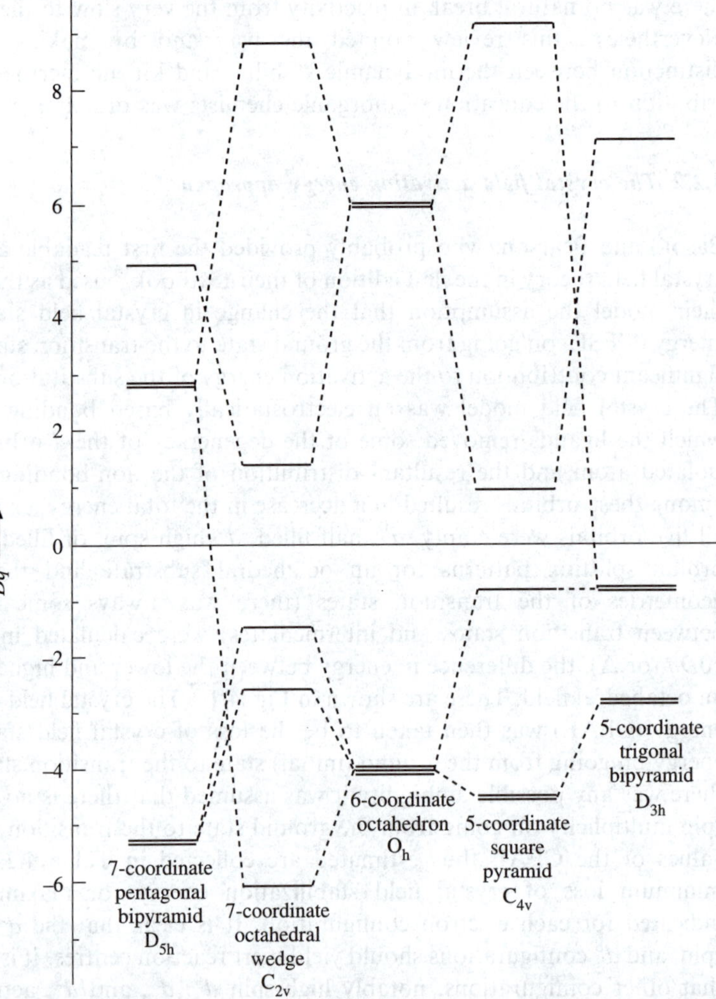

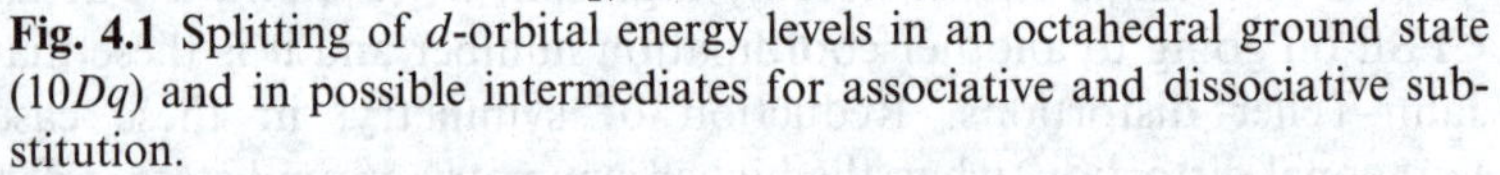

Fig. 4.1 Splitting of d-orbital energy levels in an octahedral ground state ($10Dq$) and in possible intermediates for associative and dissociative substitution.

at all, with the right predictions usually made for the wrong reasons. Despite these negative comments, we must admit that the CFAE approach has actually proved very fruitful in encouraging rationalization and prediction in this area of substitution kinetics, right from early tentative applications to reactions of trisoxalatometallates[8] up to recent strong strictures.[9]

4.2.3 Molecular orbital approaches

Attempts have been made from time to time to apply the more versatile molecular orbital model to the problems of reactivity and mechanism of

Table 4.2 Crystal field activation energies (CFAE)/Dq for ligand substitution in octahedral substrates as a function of their electron configuration and mode of activation. The smallest barrier for each configuration is shaded.

Mode of activation	Dissociative		Associative	
Symmetry of transition state	D_{3h}	C_{4v}	C_{2v}	D_{5h}
d^0	0	0	0	0
d^1	1.28	−0.57	−2.08	−1.28
d^2	2.56	−1.14	−0.68	−2.56
d^3	5.75	2.00	1.80	4.26
d^4 low spin	7.02	1.43	−0.26	2.98
d^4 high spin	−1.08	−3.14	−2.79	1.07
d^5 low spin	8.30	0.86	1.14	1.70
d^5 high spin	0	0	0	0
d^6 low spin	11.48	4.00	3.63	8.52
d^6 high spin	0.27	−0.57	−2.08	−1.28
d^7 low spin	4.66	−1.14	−0.98	5.34
d^7 high spin	2.55	−1.14	−0.68	−2.56
d^8	5.73	2.00	1.80	4.26
d^9	−1.08	−3.14	−2.79	1.07
d^{10}	0	0	0	0

substitution. In their excellent textbook, which provided first-class insights into many aspects of inorganic chemistry, and which would be thoroughly recommended were it not out of print, Purcell and Kotz apply the angular overlap method to this question.[10] In this approach, the extent to which ligand-based orbitals are stabilized and metal-based d-orbitals are destabilized as a result of overlap is estimated for the octahedral complex substrate and the appropriate transition state. In the absence of π-interaction with appropriate ligand orbitals, occupation of the metal-based orbitals with π-symmetry does not affect the extent of σ-bonding, while occupation of the metal-based orbitals of σ-symmetry will reduce the stability of the complex because these orbitals are antibonding. This model has also been discussed by Burdett[11] and can be extended to take account of π-interactions. The difference between the total stabilization of the octahedral ground state and the appropriate transition state (intermediate) is then assumed to make a contribution towards the activation energy and in this way it resembles the CFAE approach. It is a pity that transition states of higher coordination number (i.e. associative activation) were not considered, but at that time volumes of activation for solvent exchange that clearly indicated an associative mechanism had not been measured. As it stands, this approach also resembles that of CFAE in so far as it depends upon the application of symmetry arguments and does not attempt to calculate overlap for specific orbitals. Refinements that deal with specific orbitals are not uncommon and are usually more successful when the facts to be predicted are already known. Some that appear in the literature achieve good agreement between prediction and experimental fact by the simple expedient of omitting all facts that do not agree with prediction (references are omitted for obvious reasons!). In 1994 detailed criticisms of the role of ligand-field theory and molecular orbital theory, as well as of crystal field theory, in describing bonding in transition metal complexes appeared in a mainstream textbook on transition metal chemistry,[9] and cast further doubt on the use of these theories, applied to transition or initial states, in rationalizing kinetic behaviour.

4.2.4 *Inferences drawn from a knowledge of the factors leading to the occurrence of stable five- and seven-coordination*

Octahedral geometry is far too widespread for the arguments that were used successfully to predict the dominance of the associative mode of activation in the special case of substitution at square-planar d^8 metal complexes to be applied here. Nevertheless, the eighteen-electron rule would predict that octahedral complexes of metal ions with d^0, d^1, and d^2 electron configurations are coordinatively unsaturated and therefore might prefer an associatively activated pathway for substitution (d^3 and d^4 complexes would require a change in spin multiplicity on going to the transition state). Although complexes of such metal ions with coordination numbers greater than 6 are by no means uncommon, they are most frequently found with second- and third-row elements. These and other examples of coordination numbers of 5, 7, or 8 in *d*-block complexes may be found in standard undergraduate textbooks.[12] Examples particularly relevant to the present discussion include Ti(IV), with a d^0 configuration, which forms eight-coordinate $[Ti(diars)_2Cl_4]$ (diars = *o*-phenylene-bis(dimethylarsine)) and the seven-coordinate $[Ti(Me_2dtc)_3Cl]$ ($Me_2dtc = NN'$-dimethyl-dithiocarbamate), while V(III) with its d^2 configuration forms the pentagonal bipyramidal $[V(CN)_7]^{4-}$ ion. This ability to form stable complexes with coordination numbers greater than six is matched by the ability to form stable complexes with low coordination numbers, e.g. the d^2 V(III) $[V(NMe_3)_2Cl_3]$ and the d^1 Ti(III) $[Ti(NMe_3)_2Cl_3]$, not to mention the well-known and numerous d^0, d^1, and d^2 tetrahedral tetraoxo anions, $[MO_4]^{n-}$, where the low coordination number must be stabilized by further donation from the ligand to the metal. In the more electrovalent complexes size and packing may become a dominant factor when crystal field effects are small. Thus, the high-spin d^5 Fe(III) complex [Fe(edta)(H_2O)] is seven-coordinate, and of the $d^0/f^n M^{3+}$ ions Sc(III), Y(III), and the tervalent lanthanides, only Sc^{3+} perhaps forms a six-coordinate aqua-ion[13] (though binuclear $[Sc_2(OH)_2(H_2O)_{10}]^{4-}$ and its picolinato derivative contain seven-coordinate Sc^{3+},[14] and $Sc(H_2O)_9{}^{3+}$ has recently been established in hydrated scandium triflate[15]), the rest having coordination numbers up to nine.

At the other end of the *d*-block, where six-coordination is usually found, five-coordination can be generated with a suitable choice of ligand. According to Purcell and Kotz,[16] the extent to which six-coordination is preferred over five is relatively small for the elements of the first row of the *d*-block. This would reinforce suggestions that facile dissociatively activated pathways exist for their substitution reactions. Tris-(2-methylaminoethyl) amine (Me_6tren) is a quadridentate ligand with trigonal symmetry that forms trigonal bipyramidal complexes of the type $[M(Me_6tren)X]^+$ (M = Cr(II), Mn(II), Fe(II), Co(II), Ni(II), Cu(II), or Zn(II)). Ni(II) has an extensive five- and four-coordinate chemistry even for the high-spin d^8 configuration, and tetrahedral $[MX_4]^{n-}$ (X = halide) anions are well known for M = Fe(II), Fe(III), Co(II), Ni(II), Cu(II), and Zn(II). The negative charge density at the metal in $[MX_6]^{4-}$ seems to be too much for the divalent metal ions and is only found rarely, e.g. in $[Fe(CN)_6]^{4-}$. Lower coordination numbers are far less common in the chemistry of the second- and third-row *d*-block elements until the d^8 configuration is reached. The greater covalency in the bonding of these elements (and also $3d^n$ elements when the ligands are soft enough) requires a greater respect for the eighteen-electron rule. The low-spin d^6 configuration does not easily give rise to lower coordination numbers unless there is an extremely strong *trans*-influence ligand present. Five-coordinate Rh(III) and Ir(III) complexes of this type seem to be more accessible, but to make stable five-coordinate Co(III) complexes a conjugated planar macrocycle or pseudo-macrocycle seems to be necessary.

4.2.5 Conclusions

The general conclusion is that the dissociatively activated pathways should become more common as the *d*-shell fills and be less common for reactions of the second- and third-row elements of the *d*-block. The mechanisms of the substitution reactions of complexes with poorer σ-donors should be determined more by electrostatic and packing considerations (charge and size). Associative activation can be readily swung over to the dissociative mode if a strong *trans*-influence ligand is present—remember that more than one such ligand is required to cause the switch in square-planar d^8 metal complexes.

These statements can now be made from behind the protective barricades of established fact. At one time it was thought that the associative pathway would be a rare occurrence for octahedral substitution, but now it seems to be quite commonplace, almost to the extent that there are some who would propose, with all the enthusiasm of recent converts, that a dissociatively activated mechanism is now the rarity, confined mainly to Co(III). What is abundantly clear is that the interchange mechanism is dominant in this area and that the labels I_d and I_a mean different things to different people.

4.3 Substitution reactions in non-coordinating solvents

To people coming fresh to the study of the mechanism of inorganic substitution reactions it might seem strange that we are devoting only a short section of this chapter to substitution reactions in non-interfering, i.e. non-coordinating, solvents. 'Why', they may well ask, 'did the old researchers spend so much of their time studying reactions in solvents that did interfere?' There were at least three reasons for this. The first, and probably the most important, was the difficulty in finding suitable substrates and reagents that would dissolve in the non-interfering solvents and yet would remain independent until the act of substitution took place. Secondly, although many feasible systems were known, for example the exchange of neutral ligands with neutral complexes, they were generally too fast for the techniques available at the time and their study had to wait until dynamic NMR was adequately refined. Thirdly, solvent interference was occasionally not identified as such and, in the absence of strong nucleophilic discrimination, rate-determining solvolysis could be mistaken for dissociative activation. Recent work on the fate of intermediates of lower coordination number generated by laser flash photolysis makes one wonder whether there is such a thing as a 'non-interfering' solvent. However, interference by trapping an intermediate of lower coordination number is essentially irrelevant in dissociatively activated processes because the solvent has to leave again before the ultimate substitution takes place and the incoming nucleophile has to trap the 'bare' intermediate.

The earliest systems of this sort to be examined systematically involved the exchange and replacement of carbon monoxide in metal carbonyl complexes. Here the reactions were slow enough (frequently too slow except at very high temperatures) to be followed by classical labelling techniques, either with radioactive ^{14}CO, or by using infrared spectroscopy together with heavy-isotope-enriched CO. Because the mechanistic features of these reactions depend upon closed-shell configurations (the compounds being strongly covalent) and follow the same patterns, it is convenient to treat them all together (Chapter 6), in accordance with the policy outlined in Section 4.1 above.

Probably one of the most significant sets of octahedral systems to have been studied in non-interfering solvents, such as CD_2Cl_2 and $CDCl_3$, is the exchange between free and coordinated neutral ligand L in uncharged complexes of the type $[MX_{(6-n)}L_n]$. The first studies[17] related to M(V) reaction centres, involving exchange of L with complexes of the type $[MX_5L]$ (M = Sb, Nb, Ta; X = Cl, Br). The kinetics were followed by NMR line broadening of signals assigned to free and coordinated L, and the ability to make these measurements at high pressure has allowed Merbach and his co-workers to determine volumes of activation as well as the usual enthalpies and entropies of activation. The complexes of the two *d*-block elements Nb and Ta (formally d^0 reaction centres) exchange with rate laws that depend upon the nature of the donor atom in L. With O and N donors, e.g. L = Me_2O, Et_2O, MeCN, Me_3CCN, $(MeO)Cl_2PO$ (but also with $(Me_2N)_3PS$ as the exception to the rule), the rate of exchange is independent of [L], but sensitive to the nature of L (as leaving group), X, and of course M. The Et_2O complex is more labile than that of the less bulky Me_2O and the volumes of activation are positive. All the evidence points to a classical *D* mechanism. The five-coordinate intermediate MX_5, while a strong Lewis acid, is much more discriminating than, for example, $Mo(CO)_5$. The softer S, Se, and Te donors, on the other hand, exchange at rates that are first-order in [L], the exchange of Et_2S being much slower than that of the less bulky Me_2S, and the volumes of activation are all negative. All in all, an associative mode of exchange is indicated. The exchange of Me_2S with $[TaBr_5(Me_2S)]$ takes place by both exchange pathways in parallel. Only N and O donors have been studied with the analogous Sb(V) complexes, and all of these exchanges are dissociatively activated.

An octahedral complex with five identical ligands whose exchange is not being monitored does not give us the opportunity to examine the steric course of the substitution process. The analogous M(IV) complexes $[MX_4L_2]$ (M = Ti, Zr, Hf, Sn; L is a neutral oxygen or sulphur donor ligand) exist in interconvertible *cis* and *trans* forms and the exchange of neutral oxygen donor ligands has been studied, together with the *cis* ↔ *trans* isomerization. In the cases of M = Sn[18] the exchange is first-order in complex and independent of the concentration of L; a positive volume of activation is consistent with the postulated dissociative activation. The mechanistic changeover on going from Me_2O to Me_2S is not found here and the high positive volume of activation ($+38.4\ cm^3\ mol^{-1}$) found for *cis*-$[SnCl_4(Me_2S)_2]$ confirms the assignment of a dissociative mechanism. Although both the *cis* and *trans* isomers are present the exchange takes place in the *cis* isomer only and at a rate that is much greater than that of *cis* → *trans* isomerization. Although the evidence is not conclusive, it is believed that the isomerization is intramolecular. The analogous Ti(IV) complexes behave in the same way—all exchanges take place in the *cis* complex and the activation is dissociative.[19] The kinetics of ligand exchange in the analogous Zr and Hf complexes are second-order and depend upon the nature and concentration of L; the volumes of activation are negative, and thus associative activation is indicated here.[20] The bulky hexamethylphosphoramide is an exception and the rate is independent of its concentration. This steric effect and the general increase in the importance of associative activation when we go to reaction centres that are in the second and third row of the Periodic Table is consistent with the idea that size plays an important part in determining the mechanism of substitution at these spherical d^0 or d^{10} reaction centres (as also at d^6 carbonyls—see Chapter 6). It is of interest to note that the *cis* ↔ *trans* isomerization of the Zr and Hf complexes is much faster than ligand exchange and must therefore be intramolecular.

4.4 Substitution reactions in coordinating solvents

When a poorly discriminating substrate undergoes ligand substitution in a coordinating solvent the part played by the solvent in the substitution process is dominant simply by virtue of its high concentration. This is particularly true for substitution reactions of octahedral complexes in water. A typical example is the reaction

$$[Co(NH_3)_5(NO_3)]^{2+} + NCS^- \longrightarrow [Co(NH_3)_5(NCS)]^{2+} + NO_3^-$$

where a detailed examination of the changes in absorbance with time showed that the process occurred in two stages, namely

$$\text{(i)} \quad [Co(NH_3)_5(NO_3)]^{2+} + H_2O \longrightarrow [Co(NH_3)_5(H_2O)]^{3+} + NO_3^-$$

$$\text{(ii)} \quad [Co(NH_3)_5(H_2O)]^{3+} + NCS^- \longrightarrow [Co(NH_3)_5(NCS)]^{2+} + H_2O$$

and there was no evidence for the direct replacement of the nitrate by the thiocyanate.[21] The ability to detect such a two-stage reaction will depend upon the ratio of the rate constants for the two stages and the absorbance of the reagent, intermediate and product. It is of considerable interest that a re-investigation of this reaction showed that the initial NCS product contains 75% of the *N*-bonded isomer and 25% of the unstable *S*-bonded isomer, which then rearranges.[22] In aqueous solution many simple acido-amine cobalt(III) complexes can be shown to undergo ligand substitution in this way, but with the second- and third-row elements, such as Rh(III) and Ir(III), and even with certain Co(III) complexes, where the desire to coordinate to water is low, the maximum concentration of the aqua intermediate can be too small to be detected. Under these circumstances one observes a smooth conversion of the reagent to the final product in a single stage and the rate is either independent of the concentration of the entering and leaving groups (it being assumed that the reaction goes to completion) or else follows the usual mass-law retardation already discussed for the analogous substitution process in the substitution reactions of planar d^8 metal complexes (Chapter 3). In solvents with a poorer coordinating ability than water this phenomenon, which may be called *cryptosolvolysis*, is the normal behaviour.

The usual substitution reactions encountered in these systems are therefore the replacement of a ligand by the solvent, *solvolysis*, and the replacement of coordinated solvent by another ligand. This process is usually termed *anation*, because, in the early work, the entering ligands were anionic. However, it applies equally well to the entry of uncharged ligands and is identical to the processes termed *complexation* or *complex formation* in the substitution reactions of labile systems. One of the major differences in the study of labile and inert systems is that, in the former, where the rate at which the system returns to an equilibrium that has been disturbed is being measured, the kinetic analysis gives the rate constants for both the solvolysis and the complex formation. In inert systems it is possible to study the solvolysis and the anation process separately under non-reversible conditions. As mentioned earlier in this chapter, it is convenient to consider these processes together, but separately from solvolysis. Consequently, solvent exchange and complex formation are covered in Chapter 7, where various hydration (solvation) numbers are treated in a unified manner.

Since the major part of the information available comes from studies of solvolysis in aqueous solution, i.e. *aquation*, the bulk of the discussion in this chapter will relate to aqueous solution. Apart from the greater chance of finding reversibility, the mechanistic features of solvolysis in other solvents, are the same. A major exception is to be found in the reactions of cobalt(III) acido-amine complexes in liquid ammonia and other basic solvents, where the ability of the solvent to remove a proton from the coordinated amine is sufficient to activate the base-catalysed solvolytic pathway. This will be discussed in Section 4.6.

As in the case of the solvent-exchange reactions, the study of solvolytic reactions is greatly disadvantaged by the fact that it is not possible to vary the concentration of the solvent. It is therefore not possible to assign the order with respect to the entering group, nor is it possible to vary the nature of the entering group without, at the same time, changing the nature of the solvent. Volumes of activation can be used as a criterion of mechanism but the correction for electrostriction effects associated with the displacement of anionic ligands and the resultant charge separation on going to the transition state is extremely important as this contribution may dominate the sign and magnitude of $\Delta V^{\ddagger}$. Chapter 8 details this difficulty—suffice it to point out here that for the undoubtedly dissociative solvolyses of t-butyl chloride, $\Delta V^{\ddagger}$ values range from -2 to -32 dm^3 mol^{-1}, and most certainly do not have values close to the $+10$ dm^3 mol^{-1} expected for a simple dissociative process. Nevertheless, the solvolytic reaction is ideal for any study of the way in which reactivity, as against molecularity or nucleophilic discrimination, depends upon the nature of the leaving group and the nature and positions of the other ligands in the coordination shell. A major concern will be the way in which the reactivity depends upon the nature of the leaving group and the nature and position of the spectator ligands and, perhaps more important still, the ways in which these responses depend upon the nature of the central atom.

Apart from the magnitudes of the rate constants and the ways in which the kinetic data are collected, there is no difference in the solvolytic behaviour of labile and non-labile complexes. Since the ability to separate the variables is so much easier when inert substrates are studied, it will be convenient, in this section, to concentrate upon the reactions of such complexes. There are several major collections of kinetic and stereochemical information on substitution reactions at inert-metal centres. One deals very comprehensively (451 references and numerous tabulations) with complexes of cobalt(III) and chromium(III).[23] Others range rather more widely, if slightly less comprehensively.[24] Mechanisms are discussed in these, and other, review articles,[25] one of which usefully outlines the role of activation volumes in establishing ligand-substitution mechanisms.[26]

4.5 Cobalt(III) complexes: aquation

Provided that a sufficiently weakly bound ligand, e.g. Cl^-, Br^-, NO_3^-, is being displaced, the position of solvolytic equilibrium in aqueous solution at the concentrations of complex generally used for spectrophotometry ($<10^{-2}$ mol dm^{-3}) lies well over to the side of the aqua complex and so it is simplicity itself to study the kinetics of the solvolysis. All that is necessary is to examine the changing spectrum of an acidified solution of the substrate. For this reason the literature is filled with rate constants for the aquation of cobalt(III) complexes. In general, the data can be considered in three categories.

4.5.1 The effect of the leaving group[27]

Using the criteria of maximum information and minimum complication, the best set of substrates is provided by the $[Co(NH_3)_5X]^{n+}$ series, for which relevant data[23–25,28], are collected in Table 4.3. The most labile members included in the list, i.e. when X is $CF_3SO_3^-$ or ClO_4^-, are determined by the time taken to get the reaction started. Indeed, the ease and rapidity of loss of trifluorosulphonate ligands (cf. Table 4.9 in Section 4.7.1 below) is of great value in preparative coordination chemistry, especially when dealing with very inert centres such as osmium(II).[29] The least labile X in Table 4.3 is determined by the extent to which the alternative loss of ammonia becomes important. The values of the rate constants to be assigned to the least labile species have been a matter of controversy. Studies in the presence of added ammonia plus a non-coordinating ammonium salt to ensure that the pH is kept low enough to prevent base catalysis have been necessary to obtain the

Table 4.3 The dependence of the rate constants for the aquation of $[Co(NH_3)_5X]^{n+}$ upon the nature of the leaving group, X.

X	$10^7 k_{aq}(s^{-1})$	X	$10^7 k_{aq}(s^{-1})$	X	$10^7 k_{aq}(s^{-1})$
ClO_4^-	810 000	Me_2SO	180	$CF_3CO_2^-$	1.7
$CF_3SO_3^-$	270 000	I^-	83	$CH_3CO_2^-$	0.27
$4\text{-}NO_2C_6H_4SO_3^-$	6300	H_2O	59	NO_2^-	0.12
ReO_4^-	3120	Br^-	39	HCO_2^-	0.026
$(MeO)_3PO$	2500	Cl^-	18	N_3^-	0.021
$MeSO_3^-$	2000	$HCONMe_2$	15	NCS^-	0.0037
$(NH_2)_2CO$	510	SO_4^{2-}	8.9	PO_4^{3-}	0.0033
NO_3^-	240	$CCl_3CO_2^-$	5.8	NH_3	0.000 058

true value for the displacement of NO_2^- from the $[Co(NH_3)_5(NO_2)]^{2+}$ cation. In some of the early work insufficient care was taken to make provision for the reversibility of the solvolysis of the more tightly bound ligands.

The rate constants in Table 4.3 cover a range of many orders of magnitude, limited only by the conditions outlined above. There is a good correlation—a very good correlation if one sticks to mononegative ligands—between aquation rate constants and stability constants for formation of the respective complexes (Fig. 4.2).[7,30] This is hardly surprising since, as we shall see in a later chapter, the reverse (anation) process takes place at rates that are essentially insensitive to the nature of the entering ligand, other than its charge.

The halides are rather poor leaving groups, but their removal can be facilitated by the addition of halogenophiles,[31] such as Ag^+ or Hg^{2+} for Cl^-, Br^-, I^-, or a hard cation such as Be^{2+}, Al^{3+}, or Th^{4+} for the removal of F^-. Some typical kinetic studies, including some non-cobalt(III) systems, are cited in Table 4.4—more can be found associated with a discussion of the applicability of the hard and soft acids and bases concept to this type of reaction.[32] Conversion of, for instance, the poor leaving group Cl^- into the good leaving group $HgCl^+$, e.g.

$$trans\text{-}[Co(en)_2Cl_2]^+ + Hg^{2+} \xrightarrow{\text{fast}} trans\text{-}[Co(en)_2Cl(ClHg)]^{3+}$$

$$trans\text{-}[Co(en)_2Cl(ClHg)]^{3+} \longrightarrow cis\text{-} + trans\text{-}[Co(en)_2Cl(H_2O)]^{2+} + HgCl^+$$

is also a good example of the reaction of a coordinated ligand—when viewed as electrophilic attack of the Hg^{2+} on the coordinated Cl^- (Chapter 10.2). An extensive correlation of rate constants for catalysed aquation with appropriate stability constants has been demonstrated for M^{n+}-catalysed aquation not only of chloro-cobalt(III) complexes but also chloro-chromium(III), chloro-rhodium(III), and even t-butyl chloride.[33] There are two important special cases of M^{n+}-catalysed aquation. One is acid-catalysed

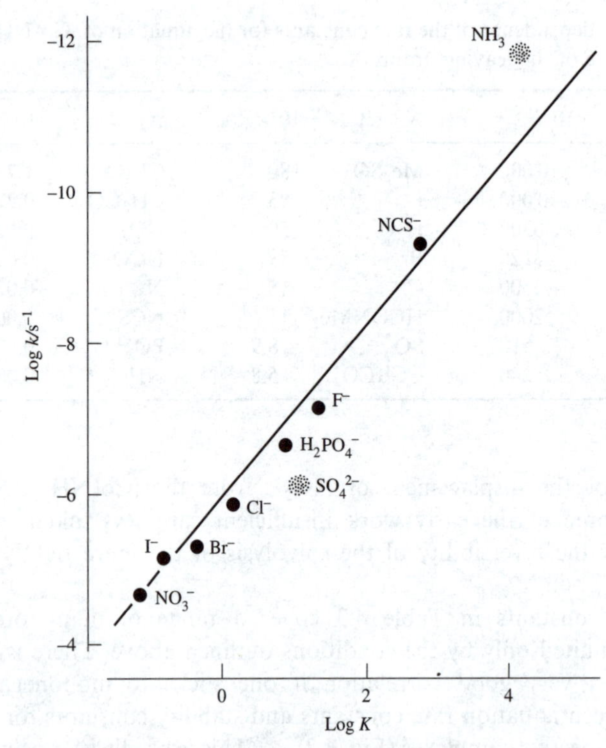

Fig. 4.2 Correlation between rate constants for aquation and stability constants for formation of complexes $[Co(NH_3)_5X]^{n+}$.

Table 4.4 Examples of kinetic studies of metal-catalysed aquation of halogenocobalt(III) complexes.

Soft/Soft	Ref.	Hard/Hard	Ref.
$[Co(NH_3)_5Cl]^{2+}/Ag^+, Hg^{2+}, Tl^{3+}$	*a*	$[Co(NH_3)_5F]^{2+}/Al^{3+}$	*g*
cis-$[Co(en)_2Cl_2]^+/Hg^{2+}$	*b*	*trans*-$[Co(en)_2F_2]^+/Be^{2+}, Al^{3+}, Sc^{3+}, Th^{4+}$	*h*
$[Co(hedta)Br]^-/Ag^+, Hg^{2+}, Pb^{2+}$	*c*	$[Co(NH_3)_5(ntaH_2)]^{2+}/Ca^{2+}, Ni^{2+}, Zn^{2+}, Fe^{3+}$	*i*
$[Cr(H_2O)_5Cl]^{2+}/ Ag^+, Hg^{2+}$	*d*		
$[Co(NH_3)_5(NCS)]^{2+}/Ag^+$	*e*		
$[Rh(NH_3)_5I]^{2+}/Ag^+$	*f*		

[a] F. A. Posey and H. Taube, *J. Am. Chem. Soc.*, 1957, **79**, 255.
[b] A. M. Sargeson, *Aust. J. Chem.*, 1964, **17**, 385.
[c] G. Schwarzenbach, *Helv. Chim. Acta*, 1949, **32**, 845; W. C. E. Higginson and M. P. Hill, *J. Chem. Soc.*, 1959, 1620.
[d] J. H. Espenson and J. P. Birk, *Inorg. Chem.*, 1965, **4**, 527.
[e] H. Miller, *Polyhedron*, 1986, **5**, 1965.
[f] C. St. E. Boyce, G. C. Lalor, and H. Miller, *J. Inorg. Nucl. Chem.*, 1979, **41**, 857.
[g] M. Iida, E. Kai, K. Nishimoto, and A. Yamatera, *Bull. Chem. Soc. Jpn*, 1981, **54**, 1818.
[h] I. V. Kozhevnikov, E. S. Rudakov, T. V. Vasserman, and V. A. Bigeeva, *Russ. J. Inorg. Chem.*, 1973, **18**, 1780.
[i] R. D. Cannon and J. Gardiner, *Inorg. Chem.*, 1974, **13**, 390.

aquation, i.e. $M^{n+} = H^+$, important for complexes with leaving groups that can readily be protonated, e.g. azide, pyrazine, and many oxoanions. The second is redox catalysis by, e.g. Fe^{2+}—a special case of the M^{2+}-catalysed aquation of $[Co(NH_3)_5(ntaH_2)]^{2+}$ mentioned in Table 4.4. This will be taken up in Chapter 9.

Alternative approaches to converting a reluctant leaving group into an eager leaving group involve the reaction of coordinated azide with nitrous acid, e.g.[34]

$$[Co(NH_3)_5(N_3)]^{2+} + HNO_2 + H^+ \longrightarrow [Co(NH_3)_5(N_4O)]^{3+} + H_2O$$

$$[Co(NH_3)_5(N_4O)]^{3+} + H_2O \longrightarrow [Co(NH_3)_5(H_2O)]^{3+} + N_2 + N_2O$$

and permanganate oxidation of coordinated dimethyl sulphoxide, as for example in $[Co(NH_3)_5(dmso)]^{3+}$, to dimethyl sulphone.[35] There has been much argument as to whether M^{n+}-catalysed aquation, HNO_2-catalysed aquation, and perhaps also base hydrolysis (see below), proceed through a common five-coordinate intermediate or transition state.[36] Essentially such arguments hinge on distinctions between intermediates and transition states, and between D, I_d, and I mechanisms. One has to be watchful for minor pathways in M^{n+} and in HNO_2-catalysed aquation, since there may be small contributions from, for example, NOCl, $HgCl^+$ and even $HgCl_2$ catalysis.[37]

Another rather unusual case of rapid aquation is provided by those oxoanion ligands MO_n^{x-} in which the M–O bond is sufficiently weak for it to break rather than the Co–O bond. This time we have an intramolecular reaction of a coordinated ligand. A recent example is provided by aquation of the $[Co(NH_3)_5(OMoO_3)]^+$ cation.[38] This aquation pathway is more commonly encountered for chromium(III) complexes.

4.5.2 *The effect of the nature and position of the other ligands in the complex*

The discussion under this heading was, for a long time, dominated by the behaviour patterns of the series of complexes pioneered by Werner, *cis*- and *trans*-$[Co(en)_2LX]^{n+}$, where much is known about the way in which the rates of displacement of X (usually Cl^-, Br^- or NO_3^-) by H_2O and the steric course of the process depend upon the nature and position of ligand L with respect to the leaving group. It is now possible to compare a whole range of species in which the two diaminoethanes are replaced by other nitrogen donors and even to make extensive comparisons of complexes with five nitrogen donors (usually amine groups from combinations of multidentate or macrocyclic systems).

A selection of data for the $[Co(en)_2LCl]^{n+}$ aquations is presented in Table 4.5. It is immediately clear that the reactivity is much affected by the nature of ligand L and, to a much lesser extent, its position with respect to the leaving group. There is a great difference between the non-participating

Table 4.5 Rate constants, activation parameters, and steric courses for the spontaneous aquation of *cis*–and *trans*-$[Co(en)_2LCl]^{n+}$ at 25 °C.

	cis					*trans*[a]				
L	$10^4 k_{aq}(s^{-1})$	$\Delta H^‡$ (kJ mol^{-1})	$\Delta S^‡$ (J K^{-1} mol^{-1})	% *cis*	% *trans*	$10^4 k_{aq}(s^{-1})$	$\Delta H^‡$ (kJ mol^{-1})	$\Delta S^‡$ (J K^{-1} mol^{-1})	% *cis*	% *trans*
OH	120	94.1	+33	84	16	16	107	+61	25	75
Cl	2.4	115	+72	76	24	0.42	09	+36	74	26
Br	1.4	94.1	−2.9	>95	<5	0.45	103	+17	50	50
$MeCO_2$	–	–	–	–	–	0.031	112	+28	75	25
$PhCO_2$	–	–	–	–	–	0.012	123	+55	75	25
−NCS	0.11	84.4	−56	100	0	0.000 45	125	+30	40	60
CO_3	–	–	–	–	–	1.0	130	+97	0	100
N_3	2.4	89	−17	86	14	2.5	87	−25	91	9
NH_3	0.0050	102	−25	100	0	0.0034	97	−46	100	0
$MeNH_2$	0.023	102	−29	100	0	–	–	–	–	–
NH_2OH	0.050	82	−70	–	–	–	–	–	–	–
CN	0.0062	103	−13	100	0	0.82	94	−8	100	0
$-NO_2$	1.1	91	−13	100	0	9.8	87	−8	100	0
S_2O_3	–	–	–	–	–	40	–	–	100	0
SO_3[b]	–	–	–	–	–	130 000	67	+1	100	0

[a] Rate constants for aquation of analogous series of complexes *trans*-$[Co(cyclam)XCl]^{n+}$ and *trans*-[Co*trans*-[14]-diene)XCl]$^{n+}$ can be found in Ref. 24.
[b] Data for the dissociation of water from *trans*-$[Co(en)_2(SO_3)(H_2O)]^+$.

effects observed here and those discussed in connection with substitution in four-coordinate planar d^8 metal complexes (Chapter 3.3), where only the ligand *trans* to the leaving group exerts any marked influence on the reactivity that is not steric in origin. With some of these cobalt(III) complexes, labilization from the *cis* position can be greater than that from the *trans*.

Interest in the labilizing effects of these spectator ligands has been strong since the early 1950s and, mainly because of restrictions in the types of ligand that were represented in this type of complex, the classification has followed different lines from that used in the d^8 systems. Emphasis was placed on the potential π-interactions between the metal and the ligand. Ligands that were π-donors, such as OH^-, Cl^-, Br^-, SCN^-, and RCO_2^-, generated a different behaviour pattern from those that were π-acceptors, such as NO_2^- and CN^-, or that had no potential for π-involvement at all, e.g. NH_3, H_2O. Originally it was thought that the π-donors facilitated a dissociatively activated substitution by stabilizing the five-coordinate intermediate, while the π-acceptors facilitated an associatively activated substitution, but the evidence in favour of a dissociative intimate mechanism throughout became overwhelming and alternative explanations were sought for the duality of behaviour.

Aquation reactions of the complexes containing a potential π-donor ligand were found to be faster when this ligand was *cis* to the leaving group, and the *trans* isomers aquated with stereochemical change. In recent years it has been shown that the *cis* isomers, initially believed, with some puzzlement, to aquate with complete retention of configuration, also aquate with stereochemical change. The steric courses are also given in Table 4.5. Although the kinetics and steric courses of the aquations of these $[Co(en)_2LX]^{n+}$ complexes have been studied extensively, one should not leap to the conclusion that this is normal behaviour. Indeed, it is far from normal and one of the conclusions that we will come to is that cobalt(III), far from being a typical octahedral reaction centre, is highly unusual, though the reason is still not clear. With few exceptions this type of stereochemical change is restricted to these amine complexes of Co(III) and is not found in substitution at other reaction centres. As will be seen in the next section, suitable choice of steric restraint in the coordination shell of the substrate can prevent stereochemical change, and the lability of some Co(III) complexes can be very sensitive to the steric requirements of the ligand system. In the bis(1,2-diaminoethane) series, the labilizing sequence is $OH > N_3 > CO_3 > Cl \approx Br > RCO_2 > NCS$ when *trans* to the leaving group and $OH > Cl \approx N_3 > Br > NCS$ when it is *cis*. In all the cases where both isomers can be compared, the *cis* isomer is more labile than the *trans*. This is especially marked in the case of the isothiocyanato complexes, where the factor is about 250; a satisfactory explanation for this has yet to be offered.

Complexes which have no spectator ligands that can function as π-donors undergo aquation with retention of configuration and when the spectator ligand can be formulated as a π-acceptor the labilization is much more effective in the *trans* isomer. We start to see the beginnings of the classical *trans* effect combined with a *trans* influence. In the bis(1,2-diaminoethane) series, the biggest effect is to be found in the case of $L = SO_3^{2-}$. In the *trans* isomer the lengthening of the Co–Cl bond *trans* to Cl is considerable and the complex is extremely labile. Although the data for the displacement of chloride are not available, it has been estimated that the ratio of the rate constants for the displacement of water from the two isomers of $[Co(en)_2(SO_3)(H_2O)]^+$, $k_{trans}/k_{cis} \geqslant 7 \times 10^6$.[39] The absence of complexes with very strong σ-donor ligands (e.g. PR_3, CH_3) prevents a proper study of this *trans* effect, but in more suitable Co(III) complexes, usually with macrocyclic or quasi-macrocyclic ligands, there is a plethora of complexes with strong σ-donor ligands. For these there is ample evidence, e.g. from X-ray structure determinations,[40] that a very important *trans* bond-weakening effect can operate. Unlike the corresponding reactions of the four-coordinate planar d^8 metal ion complexes, which are associatively activated, bond weakening in the ground state here must, of necessity, lead to greater lability when the mode of activation is essentially dissociative. Indeed, the particularly strong *trans* effect of *S*-bonded sulphite leads not only to enhanced reactivity but also to a change of mechanism from I_d to D.

To complement the above discussion of electronic and bonding effects, we should add that steric effects may also be important, though it is not always easy to separate steric and electronic contributions to reactivity trends. One of the classic demonstrations of the operation of a dissociative mechanism for aquation of cobalt(III) complexes involved a series of substituted ethane-1,2-diamine complexes *trans*-$[Co(R_n en)_2Cl_2]^+$. As shown in Table 4.6, increasing the bulk of the $R_n$en ligands leads to marked increases in aquation rates. The simplest explanation is that of relief of steric strain on forming a dissociative transition state, but methyl substitution leads to an inductive increase in electron density at the cobalt as well as steric congestion.[30] It was then shown that the electronic effects were probably much smaller than the steric effects by establishing very small substituent effects on rate constants for aquation of complexes [Co(en)$_2$(X-

Table 4.6 Effect of methyl substitution on aquation rate constants for cobalt(III) complexes $[Co(LL)_2Cl_2]^+$; all values are for aqueous solution at 298 K.[a]

LL	$10^5k(s^{-1})$	LL	$10^5k(s^{-1})$
$H_2NCH_2CH_2NH_2$	3.2		
$H_2NCHMeCH_2NH_2$	6.2	$H_2NCH_2CH_2NHMe$	1.7
$H_2NCMe_2CH_2NH_2$	22	$H_2NCH_2CH_2NHEt$	6.0
$H_2NCHMeCHMeNH_2$	42	$H_2NCH_2CH_2NH^nPr$	12.0
$H_2NCMe_2CMe_2NH_2$	3200		

[a] Data selected from p. 162 of Ref. 7.

Table 4.7 Substituent effects on rate constants for aquation of 4-substituted pyridine complexes $[Co(en)_2(4\text{-X-py})Cl]^{2+}$ and *trans*-$[Co(4\text{-X-py})_4Cl_2]^+$ in aqueous solution.

	$[Co(en)_2(4\text{-X-py})Cl]^{2+}$	*trans*-$[Co(4\text{-X-py})_4Cl_2]^+$
	$10^5 k_{aq}$ (s^{-1}; 323 K)	$10^5 k_{aq}$ (s^{-1}; 298 K)
H	1.1	0.82
4-Me	1.4	1.5
4-OMe	1.5	

py)Cl]$^{2+}$ and *trans*-$[Co(X\text{-py})_4Cl_2]^+$, where the substituents X were in the 4-position and thus ideally situated to transmit their electronic effects to the nitrogen and thence to the metal without causing any steric crowding (Table 4.7).[41] The replacement of the ammonias in $[Co(NH_3)_5Cl]^{2+}$ or $[Co(NH_3)_5L]^{3+}$ (L = an uncharged leaving group such as urea or DMSO) by primary amines has a marked effect on aquation rate constants. The relief of steric crowding in $[Co(NH_2R)_5Cl]^{2+}$ or $[Co(NH_2R)_5L]^{3+}$ on going to the I_d transition state results in an acceleration of around 100 times in aquation rates for these primary amine complexes compared with their respective uncrowded pentaammine parents. More positive activation entropies and volumes for $MeNH_2$ complexes than for their NH_3 analogues may be interpreted in terms of a lesser role for the approaching incoming water in the formation of the I_d transition state for the more crowded complexes.[42]

Regardless of the nature of the leaving and non-leaving ligands, all aquations of cobalt(III) complexes are dissociative in character. However, their precise nature ranges from D through I_d towards I. A true D mechanism probably only operates in rare cases, for example in sulphito complexes (cf. above), and perhaps in pentacyanocobaltates, $[Co(CN)_5L]^{n-}$. The extent of interaction with the incoming water is small, presumably depending on the nature of the complex, and can be speculated upon at length. Volume profiles for aquation of $[Co(NH_3)_5X]^{2+}$ and *cis*- and *trans*-$[Co(en)_2X_2]^+$, with X variously Cl, Br, or NO_3, provide perhaps the best support for a small but significant role for incoming water in the respective transition states.[43]

4.5.3 The effect of geometric constraint produced by multidentate ligands

This section is intimately bound up with the two previous ones and to a great extent provides information about the mechanisms of labilization exerted by the spectator ligands. It is possible, especially in the case of cobalt(III) complexes, to synthesize and characterize an enormous number of species derived from $[Co(NH_3)_5X]^{n+}$ and $[Co(en)_2LX]^{n+}$ by replacing the simple nitrogen donors by more complicated combinations of multidentate and macrocyclic amine ligands, whereby stereochemical restrictions are placed upon the system without, at the same time, varying the electron displacement of the spectator ligands, by a large amount. It is also possible to use multidentate and macrocyclic ligands to introduce donors that otherwise would not bind to the cobalt in monodentate ligands, e.g. R_2O and R_2S. Long before macrocycles were big business, 1,4,8,11-tetraazacyclotetradecane (cyclam), **1**, was used to provide a restraint on the

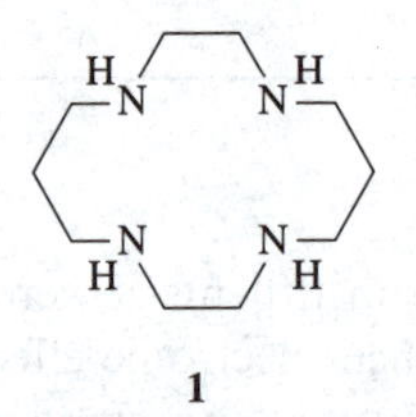

formation of a trigonal bipyramidal intermediate. Each of the four nitrogen donors, when coordinated, is tetrahedral and chiral and so there is scope for configurational isomerism. Curtis[44] pointed out an elegant way to represent this in two dimensions, using + and – signs to indicate whether the amine protons lie above or below the mean plane of the ligand. There are five possibilities, as shown, with estimates of their relative strain energies, in Fig. 4.3. In order for the ligand to be able to fold about a diagonal and

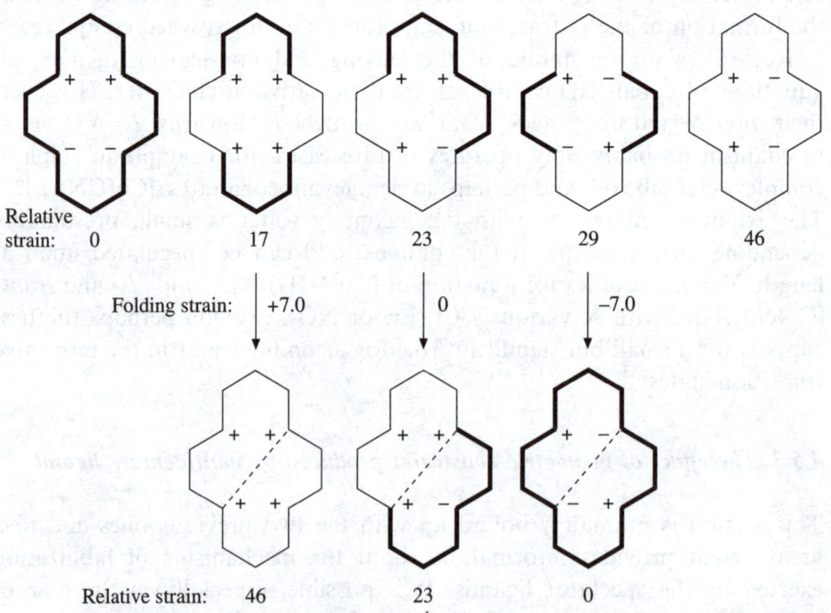

Fig. 4.3 Relative strains (kJ mol^{-1}) in various macrocycle conformations, and folding strains in interconversions.

thereby generate a trigonal bipyramid or even form a *cis* complex, the hydrogens at each end of a diagonal have to be pointing in the same direction, because folding must take place away from hydrogen. Fig. 4.3 includes estimates for the energies required to fold those forms which can be folded.

The *R, R, S, S* configuration will only form a *trans* di-acido complex, whereas the *R, R, R, R* form allows folding and is, indeed, found in both a *cis* and a *trans* di-acido complex. Aquation of *trans*-[Co(*R, R, S, S*-cyclam) Cl_2]$^+$, which cannot fold and form a trigonal bipyramid, takes place with complete retention of configuration. The rate constant is $1.1 \times 10^{-6}\ s^{-1}$ at 25 °C, with $\Delta H^{\ddagger} = 102.8\ kJ\ mol^{-1}$ and $\Delta S^{\ddagger} = -12.5\ J\ K^{-1}\ mol^{-1}$. The *trans* isomer of [Co(*R, R, R, R*-cyclam)Cl_2]$^+$ aquates with stereochemical change and is much more labile, with $k_{aq} = 1.75 \times 10^{-3}\ s^{-1}$ at 25 °C, $\Delta H^{\ddagger} = 101.2\ kJ\ mol^{-1}$, and $\Delta S^{\ddagger} = +41.8\ J\ K^{-1}\ mol^{-1}$.[45] No one has yet set out to make a comparison of the reactivity and steric course of a series of such pairs of the general type *trans*-[Co(cyclam)LCl]$^{n+}$, but the very large increase in solvolytic lability reported on going from *trans*-[Co(en)$_2$Cl$_2$]$^+$ to the corresponding *trans*-[Co(tn)$_2$Cl$_2$]$^+$ (tn = 1,3-diaminopropane), first observed by Alfred Werner at the beginning of the century, is found only in complexes of the type *trans*-[Co(tn)$_2$LCl]$^{n+}$ (displacement of Cl^-) for those ligands A that are said to promote the formation of the trigonal bipyramidal intermediate, e.g. OH, Cl, NCS. The enhancement of the lability of the other complexes (L = CN, NO_2, NH_3, and SO_3)[46] is much more modest. It has been suggested that the steric hindrance that results from the six-membered chelate rings is relieved when the five-coordinate intermediate is trigonal bipyramidal.

A similar, but less marked, effect is found in differences in the lability of the two configurational isomers, *trans*-[Co(*R, S*-2,3,2-tet)Cl_2]$^+$ and *trans*-[Co(*R, R*-2,3,2-tet)Cl_2]$^+$ (2,3,2-tet = 1,9-diamino-3,7-diazanonane, **2**). Here

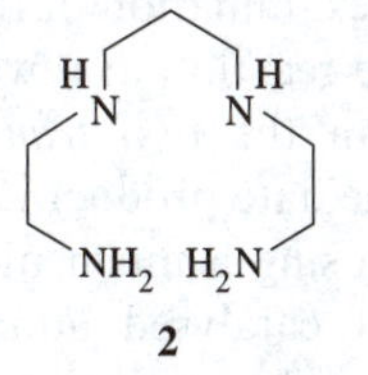

only two of the four nitrogens are anchored but the folding is restricted in the *R, S* isomer but not in the other, which is some 20 times more reactive at 25 °C.[47] The more labile isomer aquates with stereochemical change and a positive entropy of activation.

4.5.4 The steric course of aquation[48]

Apart from the reactions of certain cobalt(III) complexes, substitution in octahedral systems generally proceeds with complete retention of configuration. However, so much work has been done on these Co(III) systems that one might be forgiven for thinking that substitution with stereochemical change was normal in octahedral chemistry. One of the reasons why Werner was able to collect pairs of *cis* and *trans* isomers of the complexes of the type $[Co(en)_2LX]^{n+}$ was the facile way in which the cobalt could be isomerized; the other reason was the sensitivity of these systems to base-catalysed substitution.

These Werner complexes provided the bulk of the early information about the steric course of ligand substitution in octahedral systems. The geometric (*cis–trans*) isomerism and the chirality of the *cis* isomer provided a reasonable measure of signposting for such a simple system. The *trans* isomers fall into two categories, depending upon the nature of ligand L. For L = OH, Cl, Br, $\underline{N}CS$, RCO_2, CO_3, and N_3 aquation is accompanied by considerable stereochemical change, while with L = NH_3, NO_2, CN or SO_3 aquation takes place with complete retention of configuration. Since exchange of water for water is also a solvolytic process, it is not surprising that the aqua complexes, *cis*- and *trans*-$[Co(en)_2L(H_2O)]^{(n+1)+}$, readily isomerize in solution when L is a ligand that promotes stereochemical change. This can make a quantitative determination of the steric course difficult unless the lability of $[Co(en)_2LX]^{n+}$ is significantly greater than that of the corresponding aqua complex. For X = Cl, the most commonly studied leaving group, the situation is borderline. In the case of *trans*-$[Co(en)_2Cl_2]^+$, the rate constant for the aquation at 25 °C is three times larger than that for the subsequent isomerization.

$$trans\text{-}[Co(en)_2Cl_2]^+ + H_2O \longrightarrow 28\%\,cis\text{-} + 72\%\,trans\text{-}[Co(en)_2(H_2O)Cl]^{2+}$$

$$cis\text{-}[Co(en)_2(H_2O)Cl]^{2+} \rightleftharpoons trans\text{-}[Co(en)_2(H_2O)Cl]^{2+}$$
$$[75\%\ cis + 25\%\ trans \text{ at equilibrium}]$$

The analysis must therefore be made by a careful examination of the change in absorbance with time in the early stages of the reaction at a wavelength where the *cis* product absorbs much more than the two *trans* species. Fortunately, the *cis–trans* distribution of the immediate product is independent of the nature of the leaving group. By choosing a much more labile leaving group, usually by taking advantage of catalysed aquation (cf. Section 4.5.1 above) the product composition can be studied under conditions where the rate of formation is considerably greater than that of any subsequent isomerization. For example,

$$trans\text{-}[Co(en)_2Cl(N_3)]^+ + HNO_2 + H^+ \xrightarrow{\text{fast}} trans\text{-}[C(en)_2Cl(N_4O)]^{2+} + H_2O$$

$$trans\text{-}[Co(en)_2Cl(N_4O)]^{2+} + H_2O \xrightarrow{\text{fast}} cis\text{-} + trans\text{-}[Co(en)_2Cl(H_2O)]^{2+} + N_2 + N_2O$$

and

$$trans\text{-}[Co(en)_2Cl_2]^+ + Hg^{2+} \xrightarrow{\text{fast}} trans\text{-}[Co(en)_2Cl(ClHg)]^{3+}$$

$$trans\text{-}[Co(en)_2Cl(ClHg)]^{3+} \longrightarrow cis\text{-} + trans\text{-}[Co(en)_2Cl(H_2O)]^{2+} + HgCl^+$$

Detailed discussion of the often complicated (and occasionally revised!) steric courses of these and other 'assisted' aquations, and their relation to those of the corresponding spontaneous aquations, would take us well beyond the scope of this book. The interested reader could start investigating with the reports on Hg^{2+} and Ag^+ catalysis of aquation of *trans*-$[Co(NH_3)_4(ND_3)X]^{2+}$,[49] or on Hg^{2+} and NO^+ catalysis of aquation of $[Co(tren)(NH_3)X]^{2+}$, with X = Cl and N_3, respectively,[36] or with Jackson's comprehensively documented comparison of catalysed and spontaneous aquations[50] or Worrell's useful short introduction to the key early work on such reactions of tetraazamacrocyclic complexes.[51]

Care has to be taken to ensure that the catalysis of the aquation does not introduce a new mechanism. The acid displacement of NO_2 can be very fast in a strong enough acid, such as CF_3SO_3H, and has been shown to take place by way of an intramolecular nitro → nitrito interconversion. Final bond breaking takes place at N, and the oxygen of the aqua ligand has been shown to come from the original nitro group. Since the whole process takes place without the cobalt becoming five-coordinate, it has to be stereoretentive (Fig. 4.4).

For a long while it was thought that stereochemical change was confined to the *trans* members of this series, the *cis* isomers aquating with complete retention of configuration. In spite of the difficulties in rationalizing this with a simple mechanistic argument, nobody thought it necessary to re-examine the early data until a few years ago. It has now been shown that, in all cases where the *trans* isomer aquates with stereochemical change, the corresponding *cis* isomer does likewise, albeit with a smaller amount of rearrangement and with no loss of chirality. In other words, the *cis* part of the aqua product has not undergone any racemization. An interesting exception is found in the $[Co(en)_2(N_3)X]^+$ series, where the *trans* isomer

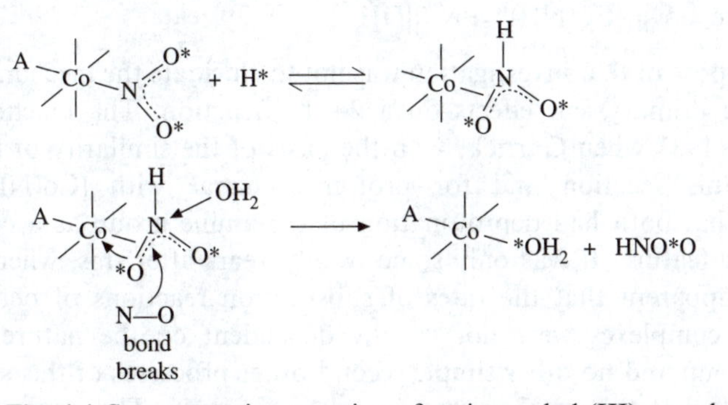

Fig. 4.4 Stereoretentive aquation of a nitro-cobalt(III) complex.

aquates with complete retention of configuration while the *cis* isomer yields some *trans* product.

Stereochemical change can indicate the existence of a trigonal bipyramidal transition state or intermediate, and the question then arises as to whether the water enters the trigonal bipyramidal intermediate by one of three possible pathways or whether the intermediate is square-planar but fluxional, the water entering the empty axial position (Fig. 4.5).

The subtleties of this problem are considered in Chapter 5, where the conclusion will be reached that these π-donor ligands can stabilize the trigonal bipyramid intermediate when they lie in its trigonal plane. The distribution of product between the isomers is independent of the nature of the leaving group, and the *cis* part of the product from a single enantiomer of a *cis* substrate retains its chirality. Nevertheless, the product ratio from a *trans* species is not the same as that from its *cis* isomer.

4.6 Cobalt(III) complexes: base-catalysed substitution reactions[52]

4.6.1 General

A class of reactions that has been studied extensively and which deserves a section of its own is typified by the reactions of cobalt(III) acido-amine complexes with hydroxide in aqueous solution. Base hydrolysis of this group of complexes is very much faster than aquation, as can be seen from Table 4.8.[24,52] Although the general features are well understood and there are a number of first-class papers that examine the details of the mechanism to a depth that exceeds anything found elsewhere in the study of transition metal substitution reactions, some of the basic questions still remain unanswered.

The first kinetic study, carried out more than sixty years ago, was of the reaction

$$[Co(NH_3)_5Br]^{2+} + OH^- \longrightarrow [Co(NH_3)_5(OH)]^{2+} + Br^-$$

This was shown to have a simple second-order rate-law:

$$\text{rate} = k_{OH}[Co(NH_3)_5Br^{2+}][OH^-]$$

but the purpose of that investigation was not to elucidate the mechanism but to examine primary salt effects on a 2+/1− reaction. The reaction next surfaced in 1937 when Garrick,[53] on the basis of the similarity of the rate laws for this reaction and for proton exchange with $[Co(NH_3)_6]^{3+}$, suggested that both had deprotonation of the amine group as a common mechanistic feature. It was only some twenty years after this, when it was becoming apparent that the rates of substitution reactions of octahedral cobalt(III) complexes were not usually dependent on the nature of the entering group and no other simple second-order processes of this sort had been found, that this mechanism was taken seriously. There was a lively

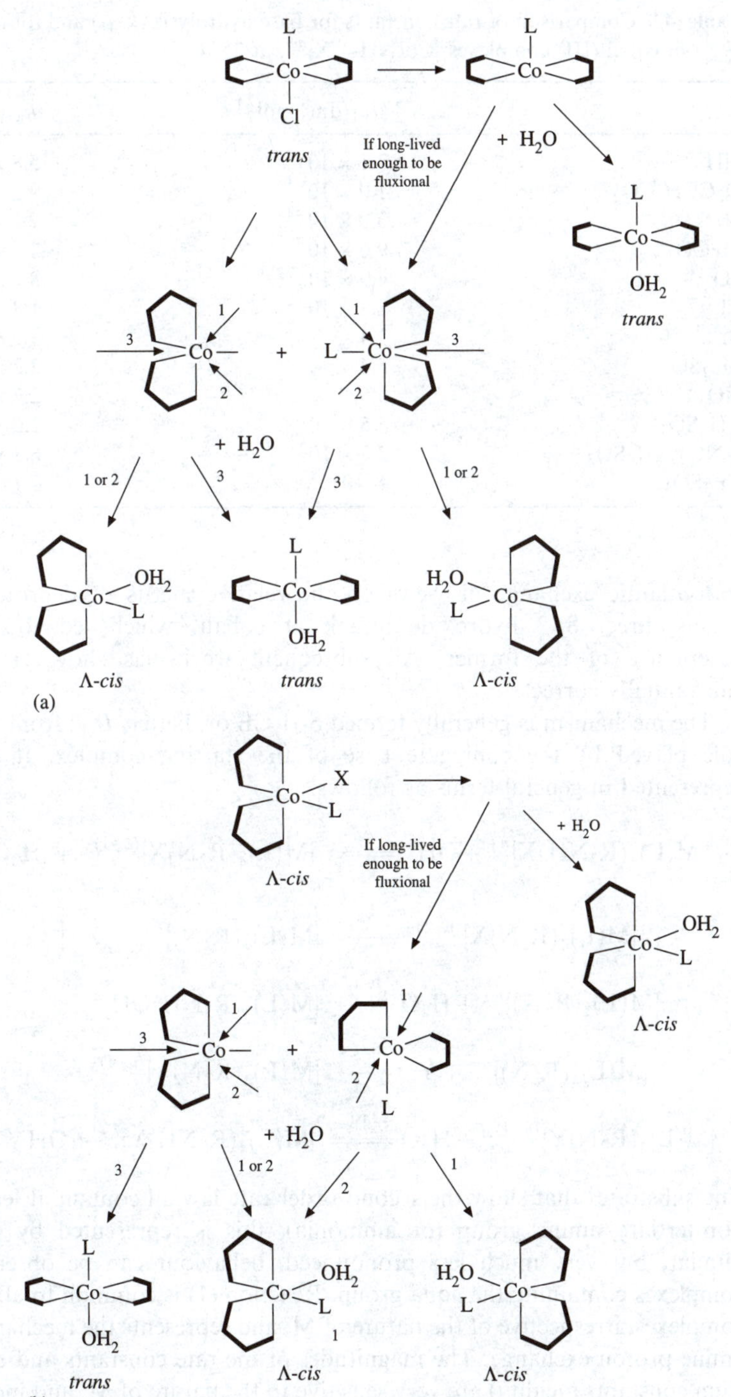

Fig. 4.5 Possible steric courses of aquation of (a) *trans*- and (b) *cis*-$[Co(en)_2LX]^{n+}$.

Table 4.8 Comparison of rate constants for base hydrolysis (k_{OH}) and for aquation (k_{aq}) of cobalt(III) complexes $[Co(NH_3)_5X]^{n+}$, at 25 °C.

X	$k_{OH}(dm^3\ mol^{-1}\ s^{-1})$	$k_{aq}(s^{-1})$
NH_3	7.1×10^{-7}	5.8×10^{-12}
$O_2CCH_2CO_2^-$	1.0×10^{-5}	9.8×10^{-9}
N_3^-	3.0×10^{-4}	2.1×10^{-9}
$MeCO_2^-$	9.6×10^{-4}	2.7×10^{-8}
SO_4^{2-}	4.9×10^{-2}	8.9×10^{-7}
Cl^-	2.3×10^{-1}	1.8×10^{-6}
Br^-	1.4	3.9×10^{-6}
Me_2SO	5.4	2.2×10^{-5}
NO_3^-	5.5	2.7×10^{-5}
$CH_3SO_3^-$	5.5×10^1	2.0×10^{-4}
$4\text{-}NO_2C_6H_4SO_3^-$	2.7×10^2	6.3×10^{-4}
$CF_3SO_3^-$	$> 10^4$	2.7×10^{-2}

transatlantic exchange of views on the relative merits of deprotonation versus direct S_N2 hydroxide attack at cobalt, which led to general acceptance of the former. All subsequent work has shown it to be substantially correct.

The mechanism is generally termed S_N1 CB or, better, D_{cb}, from the key role played by the conjugate base of the starting complex. It can be represented in general terms as follows:

$$M(L)_4(R_2NH)X]^{n+} + OH^- \underset{k_{-1}}{\overset{k_1}{\rightleftharpoons}} [M(L)_4(R_2N)X]^{(n-1)+} + [H_2O] \quad (1)$$

$$[M(L)_4(R_2N)X]^{(n-1)+} \xrightarrow{k_2} [M(L)_4(R_2N)]^{n+} + [X^-] \quad (2)$$

$$[M(L)_4(R_2N)]^{n+} + H_2O \xrightarrow{\text{fast}} [M(L)_4(R_2NH)OH]^{n+} \quad (3)$$

$$[M(L)_4(R_2N)]^{n+} + Y^- \xrightarrow{\text{fast}} [M(L)_4(R_2N)Y]^{(n-1)+} \quad (4)$$

$$[M(L)_4(R_2N)Y]^{(n-1)+} + H_2O \overset{\text{fast}}{\rightleftharpoons} [M(L)_4(R_2NH)Y]^{n+} + OH^- \quad (5)$$

The substrates that show the second-order rate law all contain at least one non-tertiary amine group (or ammonia); this is represented by R_2NH. Similar, but very much less pronounced, behaviour can be observed in complexes containing the aqua group. Reaction (1) is common to all amine complexes, irrespective of the nature of M, and represents the mechanism of amine proton exchange. The magnitudes of the rate constants and equilibrium constants (acidity) are very sensitive to the nature of M, and indeed to the nature and position of the other ligands in the complex. Activation entropies, $\Delta S^{\ddagger}$, give an indication of the stereochemistries of transition states, in that stereochemical change is associated with a higher activation entropy than is retention of configuration, and can indicate whether a trigonal bipyramidal or a square pyramidal intermediate is more likely.[55] This will be discussed in more detail later.

The lability of the amido species towards ligand substitution is extremely dependent upon the nature of M. Of the compounds adequately studied, only those of Co(III) and Ru(III) show any marked sensitivity to base hydrolysis, although the effect is observable in the reactions of Cr(III) and Rh(III) complexes at high enough hydroxide concentration. The dissociative nature of the process as indicated in equation (2) is demonstrated by the existence of base-catalysed substitution (1) + (2) + (4) + (5) in competition with base hydrolysis, (1) + (2) + (3). Practically all the studies (many of them extremely elegant probes of the finer details of the reaction) have been made on cobalt(III)-ammine species. The big problem with these is their reluctance to lose a proton, which makes it very difficult to detect or characterize their conjugate bases. In the special case of base hydrolysis of $[Co(NH_3)_3(NO_2)_3]$ the dependence of k_{obs} on hydroxide concentration is sufficiently curved for a double reciprocal plot of $1/k_{obs}$ versus $1/[OH^-]$ to give estimates of the equilibrium constant for conjugate base formation and the rate constant for onward reaction. The particularly favourable features of this system are the uncharged nature of the complex, minimizing ion-association complications, and the strongly electron-withdrawing properties of the nitro ligands.[56] In principle, ammine complexes of platinum(IV) should be more suitable than those of cobalt(III), since the formal 4+ charge on the metal promotes proton loss, but these complexes are extremely stable and inert. However, gold(III) has a more marked polarizing effect on coordinated ammonia than cobalt(III)—the pK of $[Au(NH_3)_4]^{3+}$ is 7.5—and it has been shown that the main path in ammonia exchange with this complex is via a bimolecular transition state consisting of $[Au(NH_2)(NH_3)_3]^{2+}$ and ammonia.[57]

Although aqueous solution studies dominate the literature in this area, there is some information on non-aqueous systems. Thus there is sufficient evidence to show that CH_3O^- in methanol will catalyse solvolysis and ligand substitution, for example at the *cis*-$[Co(en)_2Cl_2]^+$ cation.[54,58] The most interesting non-aqueous solution studies of this type have been those carried out by Balt in liquid ammonia.[59] Here the solvent acts as the base and the equivalent acid–base process to that in equation (1) above is

$$[M(L)_4(R_2NH)X]^{n+} + NH_3 \longrightarrow [M(L)_4(R_2N)X]^{(n-1)+} + NH_4^+$$

These studies require considerable care, because water in liquid ammonia is a strong acid and will retard the reaction. It has proved possible for a number of complexes to split experimentally determined rate constants into equilibrium constants for conjugate base formation and rate constants for dissociation of the respective conjugate bases. The two-term rate law established for proton exchange with these complexes in liquid ammonia indicates that both amino and amido forms are reacting[60] (cf. the gold-ammine system in the previous paragraph).

4.6.2 The rate law

Provided the concentration of the amido complex remains small enough, application of the stationary-state approximation gives the rate law

$$\text{rate} = k_{OH}[\text{complex}][OH^-]$$

with

$$k_{OH} = k_1k_2/(k_{-1}+k_2)$$

There are few, if any, substrates containing a single, secondary amine ligand, especially with the range of isomers that would allow a simple test of some of the postulates that will be discussed below. There are also usually, especially in the case of a relatively simple complex, several equivalent protons (in, for instance, *trans*-$[Co(en)_2Cl_2]^+$ there are eight). Under these circumstances the expression includes a statistical term:

$$k_{OH} = nk_1k_2/(k_{-1}+k_2)$$

k_1 remaining the rate constant for proton exchange. In many of the substrates examined, for example, *sym*-[Co(trenen)Cl], **3**, there are a number

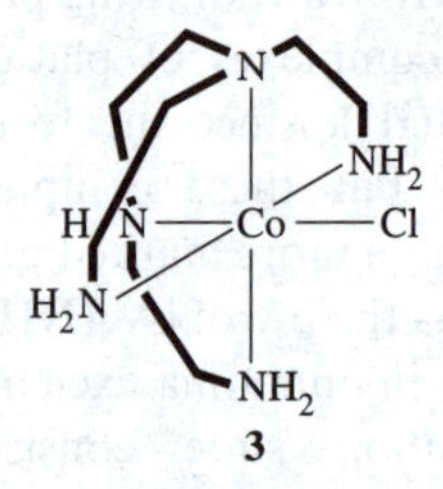

3

of sets of equivalent protons, and there is a possibility that more than one of the amido bases is capable of causing dissociative substitution. Under these circumstances the expression becomes

$$k_{OH} = \sum n_ik_1{}^ik_2{}^i/(k_{-1}{}^i+k_2{}^i)$$

and the task of analysing the system becomes extremely difficult, although it is possible to use NMR to evaluate all the $k_1{}^i$ terms.

Usually the proton transfer processes are much faster than the dissociation of the amido base, i.e. $k_{-1} \gg k_2$, and the system exhibits specific base catalysis. The expression for k_{OH} reduces to

$$k_{OH} = \sum n_ik_1{}^ik_2{}^i/k_{-1}{}^i$$

Since $nk_1{}^i/k_{-1}{}^i = K_{hyd}{}^i$, the equilibrium constant for equation (1) on p. 154, and $K_{hyd}{}^i = K_a^i/K_w$, the ratio of the acid dissociation constants for the *i*th proton and the ionic product of water, respectively, the expression can be written as

$$k_{\mathrm{OH}} = K_{\mathrm{w}}{}^{-1} \sum n_i K_{\mathrm{a}}{}^{i} k_2{}^{i}$$

Things get even more out of hand when the conjugate base becomes labile enough for the above approximation not to hold, and when $k_2 \gg k_{-1}$, the expression changes to

$$k_{\mathrm{OH}} = \sum n_i k_1{}^{i}$$

Thus deprotonation has become rate-limiting and, since such a system exhibits general base catalysis, the full rate expression becomes

$$\text{rate} = \sum n_i k_1{}^{i}[\mathrm{OH^-}] + \sum\left(\sum k_{\mathrm{B}}{}^{i}[\mathrm{B}]\right)$$

where each base, B, present in solution can make an independent contribution to the deprotonation process. Needless to say, nobody has yet attempted a full analysis of such a complicated system. If proton transfer is rapid and reversible, the reaction will take place in a single stage even when the concentration of the amido base is significant. Provided that $[\mathrm{OH^-}]$ is held constant in any kinetic run, the reaction will be first-order in complex, with

$$k_{\mathrm{obs}} = k_2[\mathrm{M(L)_4(R_2N)X}]^{(n-1)+} = k_2 K_{\mathrm{hyd}}[\mathrm{OH^-}]/(1 + K_{\mathrm{hyd}}[\mathrm{OH^-}]$$

In other words, there will be a departure from the simple first-order dependence on $[\mathrm{OH^-}]$ when $K_{\mathrm{hyd}}[\mathrm{OH^-}]$ is no longer $\ll 1$, and if it is possible to reach a point where $K_{\mathrm{hyd}}[\mathrm{OH^-}]$ is no longer $\ll 1$, k_{obs} becomes independent of $[\mathrm{OH^-}]$. Although this could be a way of determining k_2, it is only unambiguous if there is just one type of proton in the complex.

The general expression takes the form

$$k_{\mathrm{obs}} = \sum K_{\mathrm{hyd}}{}^{i} k_2{}^{i}[\mathrm{OH^-}]/(1 + \sum K_{\mathrm{hyd}}{}^{i}[\mathrm{OH^-}])$$

and only in the cases where the removal of the most acidic proton leads to the formation of the most labilizing amido group does the limiting rate constant correspond to that for the dissociation of the amido base.[61] Although there are many examples of base hydrolysis that exhibit saturation kinetics, in most cases it can be shown that the derived deprotonation equilibrium constant does not relate to the formation of the labilizing amido base. Indeed, in some cases, for example the displacement of urea from $[\mathrm{Co(NH_3)_5\{OC(NH_2)_2\}}]^{3+}$,[62] the most acidic proton is in the leaving group and its removal forms a complex that is less labile than the original substrate.

4.6.3 Evidence for the involvement of the amido conjugate base

The fact that the rate law for proton exchange[63] is the same as that for hydroxide substitution might have been enough to make Garrick suggest a common step but it is far from enough to constitute a proof. The demonstration that the base catalysis disappears when the substrate contains no amine protons constitutes a negative proof, but its importance is much diminished by the scarcity of suitable substrates.

Indirect evidence comes from several sources. Thus, for example, the very much faster (of the order of 10^5 times) base hydrolysis of $[Co(NH_2R)_5Cl]^{2+}$ (R = Me, ^{n}Pr, ^{i}Pr) than of $[Co(NH_3)_5Cl]^{2+}$ strongly suggests relief of steric strain in forming a five-coordinate amido conjugate-base intermediate.[42] Capture studies, in which base hydrolysis is carried out in the presence of a suitable nucleophilic anion, have been held to support the intermediacy of a transient five-coordinate conjugate base.[64] The argument is based on product distributions, particularly the ratio of linkage isomers produced when an ambidentate ligand such as thiocyanate, thiosulphate, or nitrite is used as added anion.

The direct proof comes from the study of complexes where the reprotonation of the conjugate base occurs at a rate that is comparable with, or less than, that of its subsequent dissociation, i.e. when $k_2 \geqslant k_{-1}$. Under these circumstances, it is possible to put an isotopic label on the proton concerned and follow its progress through the act of substitution—when $k_{-1} \gg k_2$ any label will be lost long before there is ligand substitution. Complexes of the type *trans*-$[Co(L_4)Cl_2]^+$, where L_4 is a combination of four nitrogen donors consisting of two bidentates, e.g. $(en)_2$, or one quadridentate, either linear such as 2,3,2-tet (formula **1** in Section 4.5.3, p.148)[65] or cyclic such as cyclam (formula **2** in Section 4.5.3, p.149),[66] fit into this category. If a sample in which all amine protons are replaced by deuterium is placed in a suitable buffer solution and allowed to react for a specific time before the mixture is acidified and all reaction stopped, it is usually possible to separate the unreacted substrate from the reaction product and examine the extent of proton exchange in each independently. With *trans*-$[Co(en)_2Cl_2]^+$, where there is only one N–H signal, it can be shown that there is exchange in the recovered unreacted material and also in the recovered product. It is possible to estimate how much exchange took place in the product after it was formed. When this is taken into account it can be shown that the recovered product has exchanged one more proton than the substrate from which it was formed, and it is reasonable to infer that this indicates that the reaction passed through the deprotonated conjugate base. Analysis of the data can give a value for the ratio k_{-1}/k_2. In addition to showing that the act of losing chloride is associated with the exchange of one proton, it can also be shown that this takes place at a secondary nitrogen. With the 2,3,2-tet complex, similar experiments are more informative, as the amido conjugate base is at least 20 times more likely to undergo dissociative activation than to be protonated. It can also be shown that the site of the amido group is on one of the secondary nitrogens. Furthermore, it can be shown that both the *RR*(*SS*) and *RS* isomers are converted into the *RS* product in the act of base hydrolysis. This is also clear evidence that the deprotonated nitrogen became planar in the act of base-catalysed hydrolysis. The effects of steric constraints on base hydrolysis rates has been fully discussed in relation to tetraazamacrocycle complexes of the $[Co(LLLL)Cl_2]^+$ type.[67]

Thermochemical evidence for the intermediacy of a common conjugate base species in a series of base hydrolyses of complexes $[Co(NH_3)_5X]^{n+}$ has been provided by the establishment of enthalpy profiles from kinetic and calorimetric measurements. These indicated a constant enthalpy difference between intermediate (or transition state) and products, thereby implicating reaction through a conjugate base common to all members of the series.[68] Activation volumes[69] and volume profiles[70] for base hydrolysis of cobalt(III)-ammine complexes are consistent with the $S_N1\,CB$ mechanism but do not provide proof. Nonetheless, one obtains convincingly positive activation volumes within the short range of +5 to +10 $cm^3\,mol^{-1}$ for dissociation of the conjugate base if one analyses the experimental results into detailed volume profiles for a series of cobalt(III) complexes of charge 1+, 2+, and 3+, allowing for contributions from conjugate base formation, conjugate base dissociation, and solvation effects.

4.6.4 Base-catalysed substitution

The published studies of base-catalysed substitution (all but one relating to Co(III) systems) not only provide evidence that the base-catalysed process is dissociatively activated but also lead us into areas where we have to question very closely what we mean by the 'a dissociative mechanism'. In essence, all the published material relates to studies of (i) competition between two nucleophiles, one of which is usually the solvent, for the intermediate species, or (ii), provided adequate signposting is present, the relationship between the stereochemistry of the reagents and the products and the key test becomes the way in which the behaviour depends upon the nature of the leaving group. Small differences in the product ratios are of utmost importance when it come to describing the finer details of the mechanism and so the need for very precise techniques is obvious. The scope of this book does not permit a detailed discussion of these results, but an interested reader is recommended to go to the cited literature.[24,52]

The types of question that can be answered are:

(a) Is the product composition independent of the nature of the leaving group? (Answer: no, there is a slight dependence upon its charge.)
(b) Is the amido group reprotonated before or after the five-coordinate intermediate is consumed? (Answer: after.)
(c) Is there proton transfer within the five-coordinate amido species? (Answer: there is, as yet, no evidence to suggest this, but keep looking).

The bulk of the evidence is consistent with the existence of a five-coordinate intermediate whose lifetime is so short that it usually has insufficient time to equilibrate its solvation environment and can only combine with whatever is present before the act of dissociation. Most of the studies have been carried out using substrates which are not unduly sensitive to base hydrolysis. A comparison of the effect of ionic strength upon the competition between N_3^- and H_2O for the five-coordinate intermediates derived from the $[Co(NH_3)_5X]^{n+}$ cations with $X = NO_3^-$ or DMSO suggests that $[Co(NH_3)_4(NH_2)]^{2+}$ does not have time to equilibrate, whereas bulkier $[Co(CH_3NH_2)_4(CH_3NH)]^{2+}$ does.[71]

4.6.5 The lability of the amido complex

Although there are a number of cases where the rate law for base hydrolysis departs from the simple first-order dependence on $[OH^-]$, such as that of $[Co(NH_3)_3(NO_2)_3]$ mentioned in Section 4.6.1 above, there is no system where it can be shown unambiguously that the removal of the most acidic proton generates the labile conjugate base. As a consequence, it is not possible to assign with confidence a value to the rate constant for the dissociation of the conjugate base. However, it can easily be shown that the value of the limiting rate constant obtained from the analysis of these plots must be smaller than the real value of k_2. Base hydrolysis of complexes containing the bis-pyridyl triamine ligands picdien (1,9-bis(2′-pyridyl)-2,5,8-triazanonane; **4**, **5**) or picditn (picditn = 1,11-bis(2′-pyridyl)-2,6,10-triazaundecane; **6**, **7**) is very fast, suggesting that they might be suitable substrates for establishing rate constants for conjugate base dissociation. The rate laws

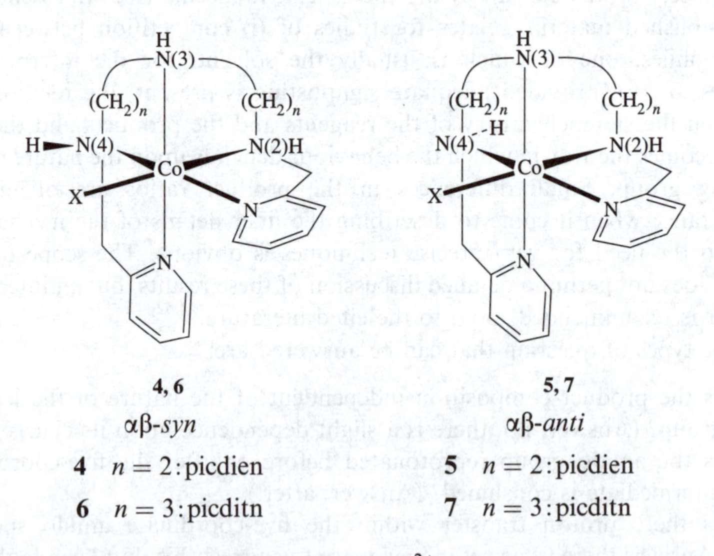

4, 6
αβ-*syn*
4 $n = 2$: picdien
6 $n = 3$: picditn

5, 7
αβ-*anti*
5 $n = 2$: picdien
7 $n = 3$: picditn

for base hydrolysis of $[Co(picditn)X]^{2+}$ complexes, and of *anti*-αβ-$[Co(picdien)(N_3)]^{2+}$, are simple first-order in hydroxide throughout the concentration range studied, but base hydrolysis of *anti*-

αβ-[Co(picdien)(NCS)]$^{2+}$ and of *syn*-αβ-[Co(picdien)(NO_2)]$^{2+}$ give curved plots. Unfortunately the presence of non-equivalent protons complicates interpretation, but it has been estimated that the effect of deprotonating *anti*-αβ-[Co(picdien)(NCS)]$^{2+}$ increases lability by as much as 15 orders of magnitude.[72] There are a number of further examples of *anti*-αβ-*syn*-αβ unusual labilization by heterocyclic nitrogen-donor ligands[73], but there does not seem to be any appreciable amount of quantitative information on reactivity of their respective conjugate bases.

4.6.6 The E2 mechanism

The S_N1 CB mechanism appears to operate for base hydrolysis of the vast majority of cobalt(III)-ammine and -amine complexes, but there is some evidence that certain base hydrolyses might occur by a synchronous deprotonation/dissociation process—in other words, the lifetime of the conjugate base decreases to a negligible value, though the five-coordinate intermediate still has a significant lifetime.[74] This process (Fig. 4.6) bears some resemblance to bimolecular elimination at carbon, hence the E2 label. Although evidence has been presented for the operation of such a mechanism for certain macrocyclic amine complexes [Co(LLLL)X_2]$^+$, it is often possible to rationalize the experimental results by more than one mechanism,[75] and further examination of some systems where the E2 mechanism was, or might be expected to be, operative have cast doubts upon its validity.[76]

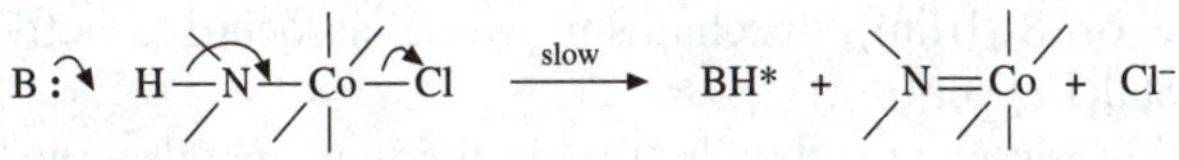

$$\mathrm{B{:}}\ \ \mathrm{H{-}N{-}Co{-}Cl} \xrightarrow{\text{slow}} \mathrm{BH^{*}} + \mathrm{N{=}Co} + \mathrm{Cl^{-}}$$

Fig. 4.6 The E2 mechanism for base hydrolysis of cobalt(III)-amine-halide complexes.

4.7 Substitution at other metal centres

4.7.1 General

So far, much of this chapter has concentrated on reactions of cobalt(III) complexes, for the good historical reason that the main body of the early research into stereochemistry, kinetics, and mechanisms of substitution at octahedral centres was carried out on these complexes. Thanks to Werner, Jørgensen, and their students, a vast amount of experience in methods of preparation was on hand, so that preparation and characterization of starting materials and products was relatively straightforward. Moreover, the substitution time-scale for these very inert complexes was appropriate for the kinetic techniques available at that time. For many years, the relative

Table 4.9 Rate constants for aquation (k_{aq}) and for base hydrolysis (k_{OH}) of trifluoromethanesulphonate-pentaammine metal(III) complexes, $[M(NH_3)_5(OSO_2CF_3)]^{2+}$, in aqueous solution at 298 K.

M^{III}	$k_{aq}(s^{-1})$	M^{III}	$k_{aq}(s^{-1})$	M^{III}	$k_{aq}(s^{-1})$	$k_{OH}(dm^3\ mol^{-1}\ s^{-1})$
Cr	1.1×10^{-2}			Co	2.7×10^{-2}	1.0×10^{4}
		Ru	9.3×10^{-2}	Rh	1.9×10^{-2}	4.6×10^{1}
		Os	8.8×10^{-4}	Ir	2.3×10^{-4}	1.4×10^{-2}

ease with which the wide range of cobalt(III) substrates could be prepared tended to obscure the fact that Co(III) is unique and that apart from a great sensitivity to *trans* bond weakening, other reaction centres show far less sensitivity to the nature of the other ligands in the complex, whether their role is electronic or steric. In the latter part of this chapter we shall deal fairly briefly with substitution at several other octahedral transition metal centres, highlighting some facets of octahedral substitution that are relatively unimportant at cobalt(III). Thus photochemical substitution[77] is much more important at chromium(III) and at rhodium(III) than at cobalt(III). Comparison of rhodium(III) and iridium(III) complexes with their cobalt (III) analogues illustrates changes in reactivity and mechanism down the Periodic Table (cf. solvent exchange, Chapter 7, and also substitution in metal carbonyls, Chapter 6). Substitution at low-spin iron(II) complexes, including pentacyanoferrates and a number of bioinorganic model compounds, provides examples of substitution by the limiting dissociative, *D* or S_N1(lim), mechanism, while associative activation is common at metal(IV) centres.

Before getting down to details for individual metals, the relative reactivities of a number of metal centres should be considered.[78] Table 4.9 shows trends for aquation and for base hydrolysis for a number of pentaammine complexes of the good leaving ligand trifluoromethanesulphonate.[79] Changes in aquation rate constants on going from element to element along a row are small, but there are marked decreases in reactivity on going down a group—with a much bigger change for base hydrolysis than for aquation. Table 4.10[80] shows reactivities which parallel the ligand field stabilizations of the respective metal centres. Decreasing reactivity with increasing ligand field strength is also reflected in activation energies and enthalpies. These generally, apart from a few idiosyncratic values for cobalt(III), increase with increasing ligand field strength, though there are a number of exceptions. Close correlations between rate constants and activation parameters are, however, not always observed. This is not altogether surprising, in view of the change of mechanism from dissociative to associative as one goes away from cobalt(III), either down the Periodic Table to rhodium and iridium, or back across the first series of transition metals. These trends are discussed in Chapter 7, since the best illustrations

Table 4.10 Relative reactivities of various transition metal centres in their complexes $[M^{III}(NH_3)_5X]^{2+}$, X = Cl or Br.

Metal	$10^7 k_{298}$ (s^{-1})	$\Delta H^{\ddagger}$ (kJ mol^{-1})
Chloro-complexes (X = Cl)		
Cr	95	87
Co	18	94
Rh	0.48	100
Ir	0.011	
Ru	7.1	91
Bromo-complexes (X = Br)		
Cr	950	98
Co	39	98
Rh	0.34	102
Ir	0.011	110

and evidence come from kinetics of solvent exchange and complex formation.

4.7.2 *Chromium(III)*

Chromium(III), like cobalt(III), forms extensive series of pentaammine complexes, $[Cr(NH_3)_5X]^{n+}$, ethane-1,2-diamine complexes, *cis*- and *trans*-$[Cr(en)_2XY]^{n+}$, and so on. But chromium(III) also, in contrast to cobalt (III), forms penta-aqua, $[Cr(OH_2)_5X]^{n+}$, and related series, thanks to its very much less marked oxidizing tendencies. However, the synthesis and characterization of chromium(III) complexes seems to be often rather more difficult than that of cobalt(III) complexes. As a result, while there is a generous supply of kinetic data for substitution at chromium(III) centres, there is significantly less kinetic and mechanistic information than for cobalt(III). There are excellent treatments of chromium(III) complexes in Pascal's treatise,[81] and specifically of chromium(III)-ammine and -amine complexes in Garner and House's review.[82] The latter, running to over 200 pages, gives a comprehensive picture of the chemistry of these complexes, including 65 pages on the kinetics (several extensive tables of kinetic parameters) and mechanisms of their reactions.

4.7.2.1 *Reactivity and mechanism*

Substitution at chromium(III) complexes tends to be significantly less slow than at their cobalt(III) analogues. This difference stems from generally lower activation enthalpies, which can be interpreted as a consequence of the smaller ligand field stabilization of chromium(III), d^3, than of cobalt(III), d^6—as low-spin $t_{2g}{}^6$ in all its complexes of kinetic (substitution)

Table 4.11 Comparison of rate constants and activation enthalpies for aquation of pentaammine complexes of chromium(III) and cobalt(III), $[Cr(NH_3)_5X]^{2+}$ and $[Co(NH_3)_5X]^{2+}$.

	$10^7 k_{298}(s^{-1})$		$\Delta H^{\ddagger}$ (kJ mol^{-1})	
	Cr	Co	Cr	Co
$[M(NH_3)_5F]^{2+}$	2.5	0.86	103	102
$[M(NH_3)_5Cl]^{2+}$	95	18	91	97
$[M(NH_3)_5Br]^{2+}$	950	39	91	97
$[M(NH_3)_5I]^{2+}$	10 340	83		
$[M(NH_3)_5(N_3)]^{2+}$	0.36	0.021		
$[M(NH_3)_5(NCS)]^{2+}$	0.92	0.0037	103	126
$[M(NH_3)_5(CF_3CO_2)]^{2+}$	87	1.7	85	109
$[M(NH_3)_5(SO_4)]^{+}$	12	8.9[b]		95[b]
$[M(NH_3)_5(OH_2)]^{3+}$ [a]	520	57	97	111

[a] For water exchange.
[b] Acid-independent term.

interest. Comparisons are shown in Table 4.11[7,24,83] but need to be interpreted or rationalized with care, for there is considerable evidence for a marked difference in mechanisms between chromium(III) and cobalt(III) substitutions. Table 4.12[7,24,83,84] compares activation volumes for aquation of several pairs of pentaammine complexes, showing a clear distinction between negative values for chromium(III), positive for cobalt(III). These reactions have been selected to involve only uncharged leaving groups, to

Table 4.12 Activation volumes for aquation reactions of chromium(III)- and cobalt(III)-pentaammine complexes:

$$[M(NH_3)_5L]^{3+} + H_2O \rightarrow [M(NH_3)_5(H_2O)]^{3+} + L$$

involving uncharged leaving groups, L.

	Activation volumes $\Delta V^{\ddagger}$ (cm^3 mol^{-1})	
Leaving group (L)	$M = Cr^{III}$	$M = Co^{III}$
OH_2	−5.8	+1.2
$OCHNH_2$	−4.8	+1.1
$OCHNMe_2$	−7.4	+2.6
$OC(NH_2)_2$	−8.2	+1.3
$OC(NMe_2)_2$	−3.8	+1.5
$OSMe_2$	−3.2	+2.0
$OP(OMe)_3$	−8.7	

Table 4.13 Comparison of rate constants and activation enthalpies for aquation of penta-aqua and pentaammine complexes of chromium(III), $[Cr(H_2O)_5X]^{2+}$ and $[Cr(NH_3)_5X]^{2+}$.

	$[Cr(H_2O)_5X]^{2+a}$		$[Cr(NH_3)_5X]^{2+}$	
X	$10^7 k_{298}$ (s^{-1})	$\Delta H^\ddagger$ (kJ mol^{-1})	$10^7 k_{298}$ (s^{-1})	$\Delta H^\ddagger$ (kJ mol^{-1})
F	0.006	120	1	111
Cl	2.8	102	73	91
Br	39	100	680	91
I	830	96	1000	90
NCS	1.0	115	16	103
NO_3	720	90	7000	78

[a] Data given apply to the acid-independent path.

minimize electrostriction complications (cf. Chapter 8). It is thus fairly safe to interpret the results in terms of I_a and I_d mechanisms for these chromium(III) and cobalt(III) systems, respectively.[85]

For chromium(III), unlike cobalt(III), it is possible to have water as spectator ligand, so one can compare kinetic parameters for $[Cr(H_2O)_5X]^{2+}$ with those for $[Cr(NH_3)_5X]^{2+}$. This is done in Table 4.13,[82,83,86] which shows that the aqua-complexes have rate constants for aquation which are one to two powers of ten higher than those for the respective ammine complexes. These reactivity differences are reflected in the activation energies, which are of the order of 10 kJ mol^{-1} lower for the more reactive aqua complexes. Activation enthalpies are small and negative, and essentially equal for a given $[Cr(H_2O)_5X]^{2+}$/ $[Cr(NH_3)_5X]^{2+}$ pair. Tables 4.11 and 4.13 also illustrate the variation in kinetic parameters with variation of leaving group.

Some similarities and differences between aquation at chromium(III) and cobalt(III) can be nicely illustrated in the aquation of the isomers of the $[Cr(en)_2Br_2]^+$ cation. Figure 4.7, which is a simplified version of a more comprehensive scheme,[82] shows that the pattern for chromium is very similar to that for the analogous cobalt(III) complexes for the first two stages. In both cases the two bromide ligands are sequentially lost, with the second stage more than one order of magnitude slower than the first. The *cis* isomers are considerably more reactive than their *trans* analogues for both metals. Isomerization processes are less important in the chromium system. The major difference is the further reaction of $[Cr(en)_2(H_2O)_2]^{3+}$, eventually to the aqua-ion $[Cr(H_2O)_6]^{3+}$. There may be complications arising from a significant amount of chromium–nitrogen bond breaking in parallel with the chromium–X bond in aquation of complexes of the type $[Cr(NH_3)_5X]^{2+}$ and $[Cr(en)_2X_2]^+$, especially for large ligands. There is also the possibility of acid catalysis, even for weakly basic leaving groups, giving a two-term rate law

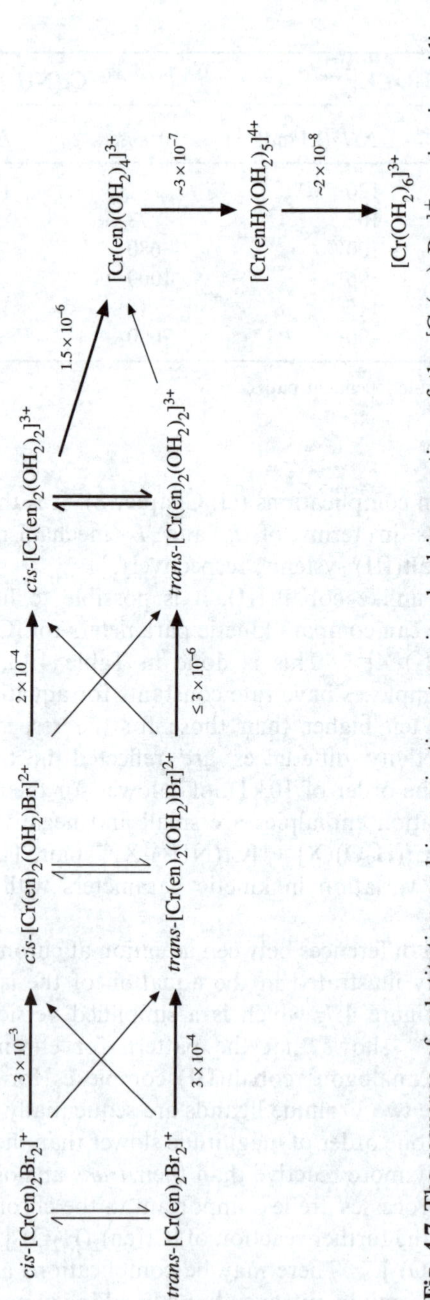

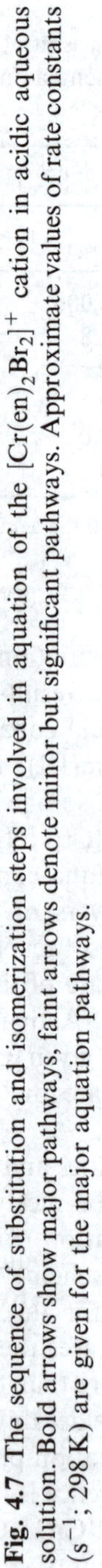

Fig. 4.7 The sequence of substitution and isomerization steps involved in aquation of the $[Cr(en)_2Br_2]^+$ cation in acidic aqueous solution. Bold arrows show major pathways, faint arrows denote minor but significant pathways. Approximate values of rate constants (s^{-1}; 298 K) are given for the major aquation pathways.

with $k_{obs} = k_1 + k_2[H^+]$. Aquation of *cis*-$[Cr(en)_2F_2]^+$ exhibits both these features—under photochemical conditions it undergoes 100% Cr–N bond fission to give $[Cr(en)(monodentate\text{-}enH)(OH_2)F]^+$ as primary product.[87]

One of the most important differences between chromium(III)- and cobalt(III)-ammine complexes lies in the field of base hydrolysis, with the former much less susceptible than the latter.[88] Thus, for example, the disappearance of $[Cr(NH_3)_5Cl]^{2+}$ in 0.1 mol dm^{-3} NaOH is only about ten times faster than in dilute acid. This is in marked contrast to the acid catalysis of the preceding paragraph. As described in Section 4.6 above, hydroxide attack at cobalt(III)-ammine and -amine complexes normally proceeds through a conjugate base which is very much—10^5–10^{13} times—more reactive than the complex itself. In the case of chromium(III), the conjugate base is more modestly, perhaps around 10^3–10^4 times more, reactive than the complex.[89] For a given complex, the activation energy for base hydrolysis is usually 20 kJ mol^{-1} or more higher than that for aquation. Further, there is no stereochemical change associated with base hydrolysis at chromium(III)-ammine complexes, whereas base hydrolysis of cobal-t(III)-ammine complexes often results in significant stereochemical change.

It would be natural to assume rate-limiting dissociation of the conjugate base, as for cobalt, and indeed comparisons of kinetic data for base hydrolysis, aquation, and mercury-catalysed dissociation for a varied range of chromium(III)- and cobalt(III)-ammine and -amine complexes can in the main be accommodated within a general S_N1 CB/D_{cb} mechanism.[90] There are nonetheless a number of experimental results which do not conform but rather suggest at least a degree of associative character to base hydrolysis for at least some chromium(III) complexes.[91] Thus, for example, activation volumes for the *NN*-dimethylacetamide complexes $[M(NH_3)_5(dma)]^{3+}$ are +25 cm^3 mol^{-1} for M = Cr, +43 cm^3 mol^{-1} for M = Co, with the large difference suggesting perhaps S_N2 CB in the former case. Similar activation volume differences have been established for $[M(NH_3)_5Cl]^{2+}$ and $[M(NH_3)_5I]^{2+}$, but for $[M(NH_2Me)_5Cl]^{2+}$ essentially equal activation volumes for the Cr and Co complexes suggest that the greater steric bulk of the methylamine ligands is forcing dissociative behaviour on the chromium complex as well as the cobalt (as in aquation[92]). It is probably best to view these base hydrolysis mechanisms as an interchange process, I_{cb}, with the ligands present controlling the precise associative/dissociative balance.[93]

This is perhaps the appropriate point to mention Gillard's electron transfer theory[94] for base catalysis of ammine complexes, introduced originally to circumvent some of the then extant difficulties with the rival S_N2 and S_N1 CB mechanisms. Electron transfer to cobalt(III) to give labile cobalt(II) is the key step in this hypothesis. This could be facile for cobalt(III) but would be very difficult for chromium(III), which is eminently compatible with the much greater base hydrolysis reactivity of the cobal-t(III) complexes than their chromium(III) analogues. Although this mechanism has been strongly criticized on energetic grounds,[95] it does have analogues in organic chemistry.[96]

Mercury(II), and other 'soft' cations, catalyse dissociation of chromium(III) complexes of the $[Cr(NH_3)_5X]^{2+}$ and $[Cr(en)_2X_2]^+$ (X = Cl, Br, I) type, just as they catalyse dissociation of their cobalt(III) analogues (Section 4.5.1). The extra feature for chromium in this respect is catalysis of dissociation by Cr^{2+}aq. In this case, however, catalysis operates by an electron-transfer cycle (inner-sphere—cf. Chapter 9) and is reliant on the extreme lability of chromium(II), rather than on the generation of good leaving groups of the HgX^+ type.

4.7.2.2 Photochemistry[97]

As already mentioned (Chapter 1, Section 1.5.2, and in the previous section), the photochemistry of chromium(III) complexes is often markedly different from their chemistry under normal thermal-activation conditions. Thus, for example, thermal aquation of $[Cr(NH_3)_5(NCS)]^{2+}$ gives exclusively $[Cr(NH_3)_5(H_2O)]^{3+}$, while photoaquation gives almost entirely $[Cr(NH_3)_4(H_2O)(NCS)]^{2+}$. In contrast to cobalt(III), there is not the problem of possible reduced products, though irradiation at charge-transfer frequencies may induce redox pathways for dissociation. Photochemical substitution at chromium(III) complexes has been widely studied for a long time, and there are numerous reviews on the relevant photochemistry[98,99] and photophysics.[100]

Many years ago, Adamson proposed a set of rules to correlate and rationalize photosubstitution at chromium(III).[101] His main guides to products of photoaquation were:

(a) Ligand loss will take place from the axis with the lowest value of $10Dq$ (summed over the two ligands).
(b) If the two ligands on this axis differ, the one with the larger $10Dq$ will be lost in photoaquation (photosubstitution).
(c) The quantum yield will be approximately the same as that for the hexakisligand complex of the ligand with the lower (lowest) $10Dq$ value.

Thus for photoaquation of $[Cr(NH_3)_5Cl]^{2+}$, given the spectroscopically based $10Dq$ values for ammonia and chloride in chromium(III) complexes, one would expect predominant loss of ammonia, rather than of chloride as in thermally activated aquation. Ammonia loss is indeed predominant, with a quantum yield over 70 times that for loss of chloride. This stereospecific pathway of the replacing group entering *trans* to the leaving ligand may be viewed as an edge-displacement process. If prevented by spectator ligand constraints, then the complex is usually photochemically inert, being reluctant to follow the alternative thermal route. Adamson's guidelines have proved very useful, though from time to time significant exceptions, e.g. some aqua-ammine species and complexes containing fluoride, have

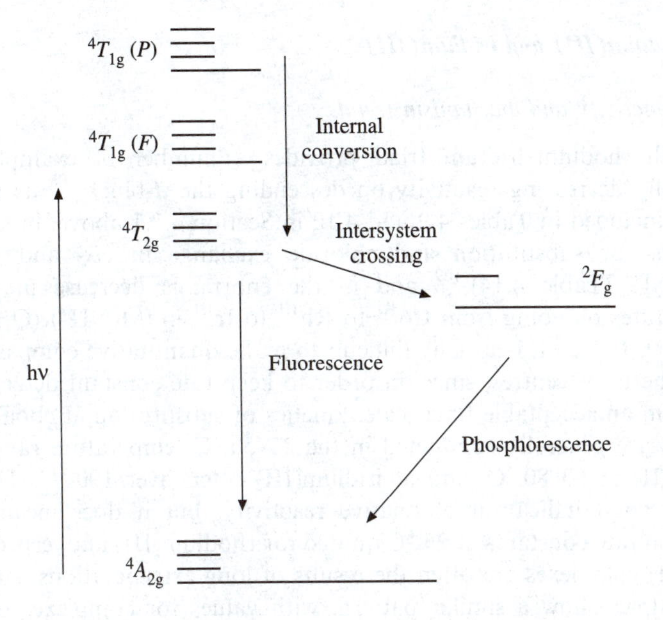

Fig. 4.8 Photoactivation of chromium(III) complexes.

been demonstrated.[102] The main problem in understanding photosubstitution mechanisms for chromium(III) complexes has been establishing which of the possible excited states is involved.[82,103] Figure 4.8 puts the states in question—$^4T_{2g}$, $^4T_{1g}$, and ^2E_g—in the context of the general photoactivation scheme given in Fig. 1.5. The photomechanism is indeed believed to be different for, e.g., ammonia and halide loss from complexes $[Cr(NH_3)_5X]^{2+}$. There is a large amount of information on excited state lifetimes, with one tabulation listing values for nearly sixty chromium(III) complexes.[104]

Until the mid-1980s, photodissociation of chromium(III) complexes was generally assumed to be dissociative in character. Indeed even at the start of the present decade the first *ab initio* study of such processes was centred on the fluxional behaviour of the five-coordinate intermediate in a limiting dissociative process.[105] However, investigations of product distributions, as in photo-induced water exchange at the various chromium(III)-water-ammine complexes, began to suggest associative *trans* attack.[106] Moreover, it was established many years ago that activation volumes for photochemical loss of ammonia from chromium(III)-ammine-anion complexes are significantly negative, generally in the range -6 to -9 cm^3 mol^{-1}, suggesting associative activation.[107] Photochemical loss of anionic ligands, where relevant, usually has a somewhat more negative value, suggesting an electrostrictive hydration contribution in addition to the intrinsic associative activation contribution. Activation is thus associative under photochemical conditions as under thermal.

4.7.3 Rhodium(III) and iridium(III)

4.7.3.1 Reactivity and mechanism trends

The cobalt–rhodium–iridium triad provides a number of examples of dramatically decreasing reactivity on descending the *d*-block,[108] as in the examples included in Tables 4.9 and 4.10 in Section 4.7.1 above, in kinetic parameters for substitution and chloride exchange in *cis*- and *trans*-$[M(en)_2Cl_2]^+$ (Table 4.14),[109] and in the enormous decrease in water exchange rates on going from Co^{III} to Rh^{III} to Ir^{III} in $[M(NH_3)_5(OH_2)]^{3+}$ (Chapter 7). Indeed it is usually difficult to make quantitative comparisons between the three centres, since, in order to keep rate constant determinations within an acceptable time-scale, kinetics of substitution at cobalt(III) complexes are generally conducted in the 25–50 °C temperature range, at rhodium(III) at 60–80 °C, and at iridium(III) often over 100 °C. This in itself is a good indication of relative reactivity, but it does mean that substitution rate constants at 25 °C quoted for rhodium(III) and, especially, iridium(III) complexes are often the results of long extrapolations. Activation enthalpies show a similar pattern, with values for complexes of the $[ML_4Cl_2]^+$ type ranging from 95 to 110 kJ mol^{-1} for M = Co, from 100 to 120 for M = Rh, and from 110 to 130 for M = Ir.

A number of different approaches to the mechanism of substitution at rhodium(III) give somewhat different views of the precise mechanism of such reactions. Activation volumes for solvolysis of $[Rh(NH_3)_5(OSO_2CF_3)]^{2+}$ in water, methanol, and acetonitrile are −4.1, −6.6, and −7.8 cm^3 mol^{-1}, respectively, which tend to suggest associative activation.[110] Electrostriction around leaving trifluoromethanesulphonate may contribute to the negative values for water and methanol but is unlikely to contribute much in acetonitrile. The activation volume for aquation of tris-oxalatorhodium(III) is −8 cm^3 mol^{-1}, again suggesting associative activation, though an activation entropy of +22 J K^{-1} mol^{-1} for this aquation is uncomfortably inconsistent. However, the fact that $\Delta V^‡$ and $\Delta S^‡$ for aquation of $[Rh(ox)_3]^{3-}$ lie between positive values for both parameters for $[Co(ox)_3]^{3-}$ and negative values for both for $[Cr(ox)_3]^{3-}$ suggests that substitution at rhodium here is an interchange process of character intermediate between I_a and I_d.[111] Such an intermediate character was also deduced from the relative slopes of the LFER plots for aquation of

Table 4.14 Kinetic parameters for chloride exchange with *trans*-$[M(en)_2Cl_2]^+$ in aqueous solution.

	$k(s^{-1})$ at 353 K	$\Delta H^‡$ (kJ mol^{-1})
trans-$[Co(en)_2Cl_2]^+$	8×10^{-2}	98
trans-$[Rh(en)_2Cl_2]^+$	2.5×10^{-5}	102
trans-$[Ir(en)_2Cl_2]^+$	2.5×10^{-7}	106

a number of rhodium(III), cobalt(III), and chromium(III) complexes of the $[M(NH_3)_5X]^{2+}$ type.[112] Reactivity comparisons for aquation of such complexes with their pentaamine analogues $[M(RNH_2)_5X]^{2+}$ lean towards the I_a mechanism.[113] Doubtless the relative timing of bond breaking and bond making depends somewhat on the individual complex and entering group, but the general conclusion, from the preceding observations and a number of other studies, would seem to be that substitution at rhodium(III) complexes of the ammine- or amine-anion type involves interchange of a modestly associative nature. This conclusion is compatible with evidence from activation volumes for water exchange (Chapter 7). In all probability, interchange at iridium(III) has more associative character than at rhodium(III).

If we move from the classical ammine and amine complexes to dimethylglyoximates, for example $[Rh(dmgH)_2(py)R]$, then the mechanism of replacement, for instance of the pyridine by triphenylphosphine, becomes dissociative, as for analogous cobalt(III) systems. Also as for cobalt(III), there is a good correlation of rate constants with the σ-donor properties of the group R, over a range of five powers of 10 in reactivity. There is also a good correlation of reactivity with bond length, as revealed by parallel X-ray diffraction studies.[114]

The most interesting feature of base hydrolysis of rhodium(III)-ammine- or -amine-anion complexes is the smallness of the contribution from the hydroxide-dependent term in the rate law:[115]

$$-\mathrm{d}[\mathrm{Rh^{III}\ complex}]/\mathrm{d}t = \{k_1 + k_2[\mathrm{OH^-}]\}[\mathrm{Rh^{III}\ complex}]$$

Indeed, the $k_2[OH^-]$ term was not at first detected for some complexes. Subsequent studies using higher hydroxide concentrations revealed its small but significant contribution, as, for instance, in the study of base hydrolysis kinetics for *trans*-$[Rh(en)_2X_2]^+$, X = Cl, Br, I.[116] This study also established the identity of the respective k_1 terms, and their corresponding $\Delta H^\ddagger$ and $\Delta S^\ddagger$ values, with those for aquation in acid solution. Kinetic parameters, including positive activation volumes for $[M(NH_3)_5X]^{2+}$, stereochemical courses of reaction, and steric effects on reactivity, all point to a dissociative mode of activation, D_{cb} or S_N1CB, for base hydrolysis of this type of rhodium complex. An S_N1CB mechanism also operates in liquid ammonia solution, where, as for cobalt, the equilibrium constant for conjugate base formation and the rate constant for its dissociation can each be established.[117]

4.7.3.2 *Photochemical substitution in rhodium(III) complexes*

Rhodium(III) complexes, like chromium(III) complexes (*v.s.*), have been intensively studied as substrates for photochemical substitution.[118,119] Recent development of interest in relation to photochemotherapeutic applications arises from their marked thermal inertness but relative photochemical lability.[120] The reluctance of both Rh^{III} and Cr^{III} to undergo

Table 4.15 Quantum yields (Φ) for photosolvolysis of rhodium-ammine-halide complexes, in aqueous solution.

	$\Phi(X^-)$	$\Phi(NH_3)$	$\Phi(X^-)/(NH_3)$
$[Rh(NH_3)_6]^{3+}$	—	0.07	—
$[Rh(NH_3)_5Cl]^{2+}$	0.14	<0.001	>140
$[Rh(NH_3)_5Br]^{2+}$	0.02	0.17	0.12
$[Rh(NH_3)_5I]^{2+}$	0.01	0.87	0.012
trans-$[Rh(NH_3)_4Cl_2]^+$	0.13	<0.002	>65

reduction to the 2+ state helps to simplify their photochemistry, certainly by comparison with cobalt(III) complexes. The photochemistry of ligand substitution at rhodium(III) is probably better established and understood than that of chromium(III).[121] Photosolvolysis of complexes $[Rh(NH_3)_5X]^{n+}$ results in parallel loss of ammonia or of ligand X. Product ratios (Table 4.15)[122] depend on temperature, pressure, solvent, and the nature of X, since activation parameters and solvation effects differ, particularly when X is a hydrophilic anion. In aqueous solution, chloride loss dominates for X = Cl, ammonia and bromide replacement are of comparable importance for X = Br, whereas for X = I reaction proceeds almost entirely by ammonia loss. The dependence of quantum yield for ammonia and for chloride loss from $[Rh(NH_3)_5Cl]^{2+}$ on solvent reflects the very different solvation properties of ammonia and of chloride in water and in organic media. Quantum yields for halide loss are much smaller in organic solvents than in water, due to less favourable solvation in organic media.[99,118]

The dependence of activation volumes on leaving ligand and solvent, including hydrogen/deuterium isotope effects, has been fully studied, analysed, and discussed in relation to photochemistry and photophysics.[123] The key result is that activation volumes for leaving halide are markedly negative, for loss of ammonia of similar magnitude but opposite sign. Both results can be interpreted in terms of dissociative ligand loss from the photoexcited state, with marked solvation of leaving halide giving the overall negative values for their loss. Labelling the ammonia ligands with ^{15}N permitted the demonstration that photoaquation of $[Rh(NH_3)_5Cl]^{2+}$ and of its bromo-analogue involved parallel loss of ammonia *cis* and *trans* to the halide.[124] A number of detailed studies of product distributions has tended to confirm that selection rules based on the angular overlap model,[125] like the original photosubstitution rules of Adamson,[101] generally give a correct overall prediction, but that they are not infallible, and often cannot account for minor products.

In summary, most of the evidence suggests that photosubstitution at rhodium(III) takes place dissociatively, with the transient five-coordinate intermediate and its possible stereochemical rearrangements playing a key role.[126] This is the case not only for complexes containing strongly labilizing ligands such as sulphite[127] but also for standard ternary ammine complexes containing ligands such as halide or water. However, there have been hints of significant associative character to photosubstitution at just a few rhodium(III) complexes.[128]

The photochemistry of iridium(III) complexes has been considered alongside that of analogous rhodium and cobalt complexes.[99]

4.7.4 Ruthenium and osmium

4.7.4.1 Reactivity and mechanism trends

Perhaps the most interesting feature of substitution at ruthenium(II) and ruthenium(III) is that both oxidation states are inert, though the former is much less inert—often with the time-scale for substitution only a few seconds—than the latter. This double inertness is an enormous advantage to the study of electron transfer. Indeed, ruthenium is the only metal in the Periodic Table with two substitution-inert oxidation states, which can conveniently be used to study one-electron exchange (two-electron exchange, e.g. platinum(II)/(IV), is complicated by the difference in coordination numbers of the two oxidation states) with no fear of dissociation in one or both states on the relevant electron transfer time-scale. The lability of most cobalt(II) complexes is a problem in studying electron transfer kinetics, and indeed also in establishing cobalt(II)/(III) redox potentials, for instance by polarography. Only by the use of encapsulating ligands can cobalt(II) be rendered inert. The situation is worse for iron, as the great majority of both iron(II) and iron(III) complexes are substitution-labile, though the small group of low-spin iron(II) complexes—with ligands such as cyanide, 2,2′-bipyridyls, and 1,10-phenanthrolines—are very inert and have played an important role in the study of redox reactions. Osmium(II) and osmium(III) complexes are also substitution-inert,[129] in practice sometimes so inert that considerable difficulties may arise in synthesizing target complexes. There is also the problem that osmium(II) is very strongly reducing.

After this redox-related digression it is time to refer readers to Table 4.16, which contains kinetic data for aquation of a selection of pentaammine complexes of ruthenium and osmium[130,131] in oxidation states (II) and (III). The lack of any clear trends immediately signals that reactivity is determined by a number of factors. In particular, π-bonding can be an important factor in determining metal–ligand bond strengths in the 2+ oxidation state. The most dramatic comparisons of 2+ and 3+ complexes are those of the chloride and dinitrogen complexes of ruthenium(II) and (III), but there is a similar reversal of reactivity for the osmium(II)/osmium(III)-benzene complexes. Substitution rate constants for the osmium(II) complexes $[Os(NH_3)_5(\eta^2\text{-}C_6H_5X)]^{2+}$ increase in the order

Table 4.16 Rate constants for substitution at a selection of ruthenium- and osmium-pentaammine complexes in aqueous solution at 298 K.

	Ru^{II}	Ru^{III}	Os^{II}	Os^{III}
$[M(NH_3)_5(N_2)]^{2+,3+}$	2×10^{-6}	70	$< 10^{-8}$	2×10^{-2}
$[M(NH_3)_5Cl]^{+,2+}$	6.3	9.3×10^{-7}	0.2	$\sim 10^{8}$
$[M(NH_3)_5Br]^{+}$	5.4			
$[M(NH_3)_5(OSO_2CF_3)]^{2+}$		9.3×10^{-2}		8.8×10^{-4}
$[M(NH_3)_5(O_2CH)]^{+}$	1.3			
$[M(NH_3)_5(O_2CCF_3)]^{+}$	5.3			
$[M(NH_3)_5(O_2CCH_3)]^{+}$	<5[a]			
$[M(NH_3)_5(HO_2CCH_3)]^{2+}$	17.5			
$[M(NH_3)_5(imid\text{-}2\text{-}CO_2)]^{+}$		2×10^{-4}		
$[M(NH_3)_5(Me_4\text{-}ascorbate)]^{2+}$		3.3×10^{-5}		
$[M(NH_3)_5(\eta^2\text{-}C_6H_6)]^{2+,3+}$			3.5×10^{-5}	3.5×10^{2}
$[M(NH_3)_5)(\eta^2\text{-}C_6H_5CF_3)]^{2+,3+}$			1.1×10^{-6}	6×10^{3}
$[M(NH_3)_5(\eta^2\text{-}Me_2CO)]^{2+,3+}$	*b*		$\sim 10^{-1}$	

[a] $k = 25.5\,s^{-1}$ when $-Cr(OH_2)_5$ is coordinated to the acetate to form a binuclear μ-acetato-$Ru^{II}Cr^{III}$ complex.
[b] Labile on the preparative time-scale.

$$X = CF_3 < NMe_2 < CMe_3 < H$$

with the η^2-C_6HMe_4 analogue reacting more quickly, whereas for the osmium(III) analogues the CF_3-substituted complex reacts much more rapidly than the parent $[Os(NH_3)_5(\eta^2\text{-}C_6H_6)]^{3+}$. This reversal of effects of electron release or withdrawal by substituents is consistent with only a minor contribution from π-bonding for the 3+ complexes, but a major contribution for the 2+ complexes. The balance between σ- and π-bonding for the latter results in the interesting situation that both electron-withdrawing and electron-releasing substituents can stabilize $[Os(NH_3)_5(\eta^2\text{-}C_6H_5X)]^{2+}$ in comparison with the parent $[Os(NH_3)_5(\eta^2\text{-}C_6H_6)]^{2+}$ cation. In the case of the 3+ complexes, reactivity parallels arene basicity.[131]

Reactivities of ruthenium(II) and ruthenium(III) complexes can usefully be compared with those of cobalt(III) analogues by considering substitution at *trans*-$[M(LLLL)X_2]^{+,2+}$ (Table 4.17).[132] Such comparisons suggest that d^6 Ru^{II} behaves rather more like d^5 Ru^{III} than like d^6 Co^{III}. Reactivity trends for the macrocyclic ligand entries for ruthenium(II) and for ruthenium(III) in Table 4.17 show steric effects consistent with dissociative activation. This conclusion is consistent with a limited LFER plot for $[Ru^{III}(NH_3)_5X]^{2+}$ cations but unfortunately is in stark contrast to the strong indication of associative activation, at least at ruthenium(III), from the measured activation volume of $-30\ cm^3\ mol^{-1}$.[133] Non-leaving ligand effects on reactivity at ruthenium(II) have been assessed for *cis*- and *trans*-$[M(NH_3)_4LX]^{+}$ (Table 4.18).[134] Here *cis* effects are dominated by π-donor properties, *trans* effects by σ-donor properties. Substitution in phthalo-

Table 4.17 Comparison of substitution rate constants[a] for ruthenium(II)-, ruthenium(III)-, and cobalt(III)-ammine and -amine-halide complexes of the type *trans*-$[M(LLLL)X_2]^+$, in aqueous solution at 298.2 K.

	Ru^{II}		Ru^{III}		Co^{III}	
X_2:	Cl_2	Br_2	Cl_2	Br_2	Cl_2	Br_2
$(NH_3)_4$	1.0		1.7×10^{-6} [b]	3.0×10^{-6} [b]	3.9×10^{-6}	3×10^{-6}
$(en)_2$	0.35	0.44	4.2×10^{-6}	5.6×10^{-6}	3.5×10^{-5}	1.4×10^{-4}
2,3,2-tet[c]	0.066	0.094	4.8×10^{-7}		1.5×10^{-5}	
cyclam[c]	0.023		$\ll 10^{-7}$		1.1×10^{-6}	2.2×10^{-5}
tet-a[c]	*ca* 5		6×10^{-9}	1.1×10^{-7}		
tet-b[c]				19×10^{-6}		

[a] Activation parameters corresponding to the rate constants tabulated are available in most cases. It is interesting to note the marked increase in activation enthalpies on going from ammonia to chelates to macrocyclic complexes, for example

Complex:	*trans*-$[Ru(NH_3)_4Cl_2]^+$	$[Ru(en)_2Cl_2]^+$	$[Ru(2,3,2\text{-tet})Cl_2]^+$	$[Ru(\text{tet-a})Cl_2]^+$
$\Delta H^{\ddagger}/\text{kJ mol}^{-1}$:	91	98	103	141

[b] Rate constants for substitution at *trans*-$[Ru(NH_3)_4\{P(OEt_3)\}_2]^{3+}$ are remarkably similar, despite the difference in nature and charge of the leaving group here (S. E. Mazzetto, E. Rodrigues, and D. W. Franco, *Polyhedron*, 1993, **12**, 971).

[c] Ligand formulae:

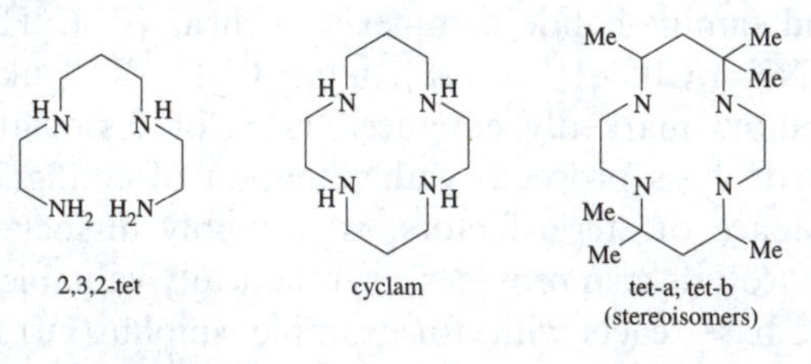

cyanine complexes $[M(pc)L_2]^{n+}$ is limiting dissociative (*D*) in mechanism, for M = Ru just as for M = Fe, with the five-coordinate intermediates showing very little discrimination. The ruthenium complexes react, as one would expect, much more slowly than their iron analogues.[135] Substitution in complexes of the *cis*- and *trans*-$[M(NH_3)_4X_2]^+$ type is stereoretentive.[136]

Fortunately for the study of electron transfer, the ruthenium–nitrogen bonds in ammine complexes are extremely reluctant to break. The half-life for aquation of $[Ru(NH_3)_6]^{2+}$ is much more than an hour,[137] for $[Ru(NH_3)_6]^{3+}$ longer still—both for thermal and for photochemical aquation. The $[Ru(NH_3)_6]^{3+/2+}$ couple can therefore be used in outer-sphere electron transfer studies (Section 9.5), and $[Ru(NH_3)_5]$-modified metalloproteins in the study of intramolecular electron transfer (Section 9.6.5), without fear of complications from dissociation. Similarly great reluctance to undergo dissociation on the part of $[Ru(bipy)_3]^{2+}$ is again a great advantage in the use of this complex in photochemistry. When photolysed sufficiently aggressively, $[Ru(bipy)_3]^{2+}$ does lose one bipy, replaced by solvent or, in the presence of chloride, one chloride and one solvent

Table 4.18 Rate constants[a] for a selection of ruthenium(II)-ammine-halide complexes $[Ru^{II}(NH_3)_4LX]^{n+}$ in aqueous solution at 298 K.

	X = Cl^-		X = Br^-		X = I^-	
	cis	*trans*	*cis*	*trans*	*cis*	*trans*
NH_3	5.1		3.4			
Cl^-	~32	0.9				
py[b]	0.4	1.5	0.40	2.1	0.75	0.9
isn[b]	0.13	0.5	0.18	0.43	0.14	0.5
MeCN	0.18	0.10				

[a] Activation enthalpies are all 70 kJ mol^{-1} within the stated uncertainties of ± 12 kJ mol^{-1}; activation entropies are in the range 0 to −35 J K^{-1} mol^{-1}, again all the same within the stated uncertainties. The actual values, where known, are detailed in Ref. 108.
[b] py = pyridine; isn = isonicotinamide.

molecule. Activation volumes for such reactions lie in the range +8 to +17 cm^3 mol^{-1}, indicating a dissociative mechanism.[138]

The hexaammine complexes $[Ru(NH_3)_6]^{2+}$ and $[Ru(NH_3)_6]^{3+}$ are also very inert in alkaline solution, with the former having a half-life of days in molar NaOH.[137] However, proton exchange at $[Ru(NH_3)_6]^{3+}$ is facile.[139] Ruthenium-ammine and -amine-halide complexes such as $[Ru(NH_3)_5X]^+$ (X = Cl, Br, I),[139] $[Ru(NH_3)_5Cl]^{2+}$,[140] or *cis*-$[Ru(en)_2Cl_2]^+$,[139,141] like their cobalt(III) analogues, show markedly enhanced rates of dissociation in alkaline media. Base hydrolysis proceeds with retention of configuration, and, based on the evidence of steric factors, is probably dissociative in character.[132,141] Base hydrolysis also provides a synthetically valuable route, for the amido-conjugate base reacts with, for example, sulphur(IV) to give coordinated sulphamate or nitric oxide to give coordinated dinitrogen, and can be oxidized to give coordinated nitrosyl.[142] These reactions will have interesting mechanisms, albeit not then elucidated.

Ruthenium-ammine or -amine-halide complexes, again like those of cobalt (Section 4.5.1) and rhodium, undergo aquation more rapidly in the presence of halophilic cations such as mercury(II), but this effect is decreasingly less marked in the order $Co^{III} > Rh^{III} > Ru^{III}$.[143] It is not clear whether the smallness of this effect in the case of Ru^{III} is attributable to a smaller equilibrium constant for formation of the Ru^{III}–X–Hg intermediate or to HgX^+ being a poorer leaving group from Ru^{III}.

The kinetics and mechanisms of reactions of osmium complexes have been reviewed;[144] hexahalide and a few chlorosolvento complexes of ruthenium and osmium appear in Section 4.7.6 below.

4.7.4.2 Photosubstitution in ruthenium complexes

The photochemistry of ruthenium complexes has been much investigated, though the main interest has been in redox chemistry, especially involving the excited-state complex *$[Ru(bipy)_3]^{2+}$. However, the high level of interest in this species has ensured that its photosubstitution behaviour has been studied and reviewed,[145] while a more general review of photosubstitution at ruthenium complexes places their photochemistry in the context of other transition metal complexes.[99] The intermediacy of a monodentate-bipy species in dissociation of $[Ru(bipy)_3]^{2+}$, for long a matter for discussion and investigation,[146] has now been established.[147] Such evidence, and that from determination of activation volumes of +12 and +9 $cm^3 mol^{-1}$ (at 288 K, in acetonitrile) for photodissociation of $[Ru(bipy)_3]^{2+}$ and $[Ru(phen)_3]^{2+}$, respectively,[148] strongly supports dissociative activation. On the other hand, it has been claimed, from added ion effects on reactivity, that these reactions are associative in character.[149] This latter assignment is somewhat less convincing, both intrinsically and also in view of other observations of added ion effects which can be ascribed to an important role for ion pairing in dissociative solvolysis. Early photochemical studies on ruthenium-ammine and -amine-ligand complexes often showed parallel loss of both ligands, as for example in irradiation of $[Ru(NH_3)_5(py)]^{3+}$; quantum yields depend on the irradiation wavelength.[150] More recent work can be traced through the report on the comparative photochemistry of complexes *trans*-$[Ru^{II}(NH_3)_4LCl]^+$ and their cyclam analogues.[151]

4.7.5 *Iron(II)*

High-spin iron(II) complexes are labile, but low-spin iron(II) complexes have the $t_{2g}{}^6$ electronic configuration, which confers high ligand field stabilization and high crystal field activation energies. The most familiar of these low-spin iron(II) complexes is hexacyanoferrate(II), whose inertness is well exemplified by its particularly long half-life, of more than a week, for cyanide exchange. Complexes with 2,2′-bipyridyl, 1,10-phenanthroline, related diimine ligands, and a range of their substituted derivatives, are generally rather less inert, and have been extensively studied in respect of their kinetics of dissociation in acid and in basic solution, and indeed in mixed and non-aqueous media, in salt solutions, and in micelles and microemulsions. The pentacyanoferrates(II), $[Fe^{II}(CN)_5L]^{n-}$, provide another group of moderately inert complexes whose dissociation kinetics have been much studied, while there are also a considerable number of substitution-inert bioinorganic complexes and models, generally with the iron(II) firmly coordinated by a tetraazamacrocyclic ligand but undergoing kinetically monitorable substitution at the axial positions.

4.7.5.1 *Iron(II)-diimine complexes*

The iron(II)-diimine complexes $[Fe(LL)_3]^{2+}$, where LL = 1,10-phenanthroline (phen, **8**), 2,2′-bipyridyl (bipy, **9**), a diazabutadiene (**10**), or indeed any ligand containing the delocalized chelating moiety **11** (sometimes called the 'ferroin group'), have been popular with inorganic chemists for many decades, ever since Blau reported the synthesis of 2,2′-bipyridyl and the intense red colour of its Fe^{2+} complex (though perhaps the complex he prepared actually contained **12**) in the 1890s.[152] They are very stable and

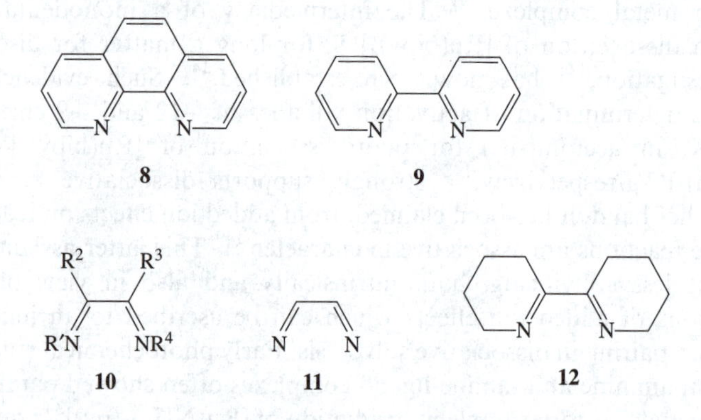

inert, intensely coloured, low-spin (t_{2g}^{6}) complexes. Their stability, charge-transfer spectra, redox potentials, solvation, and many other properties can be tailored by varying substituents on the parent ligands. They aquate in the presence of acid, they undergo base hydrolysis, and they react with cyanide (to give low-spin $[Fe(CN)_2(LL)_2]$, then $[Fe(CN)_4(LL)]^{2-}$)—all at rates which can usually be monitored by conventional spectrophotometric techniques, at atmospheric pressure and at elevated pressures.

Aquation of $[Fe(phen)_3]^{2+}$, and of most of its ligand-substituted derivatives, follows first-order kinetics:

$$-\mathrm{d}[\mathrm{Fe(phen)_3}^{2+}]/\mathrm{d}t = k[\mathrm{Fe(phen)_3}^{2+}]$$

There is no dependence on acid concentration, since the added acid protonates the released ligand *after* rate-determining breaking of iron–nitrogen bonds:

$$[\mathrm{Fe(phen)_3}]^{2+} \underset{k_s}{\overset{k_d}{\rightleftharpoons}} [\mathrm{Fe(phen)_2}]^{2+} + \mathrm{phen} \qquad \text{SLOW}$$

$$[\mathrm{Fe(phen)_2}]^{2+} \rightleftharpoons [\mathrm{Fe(phen)}]^{2+} + \mathrm{phen} \qquad \text{FAST } (K_2)$$

$$[\mathrm{Fe(phen)}]^{2+} \rightleftharpoons \mathrm{Fe}^{2+} + \mathrm{phen} \qquad \text{FAST } (K_1)$$

The low-spin↔high-spin change takes place in the first of the three equations above. Loss of the first phen ligand from $[Fe(phen)_3]^{2+}$ is thus rate-limiting, with protons scavenging released phen very rapidly and preventing its recombination with the bis-phen complex. This scavenging can also be achieved by other metal ions, such as Zn^{2+} or Ni^{2+}, or by oxidants such as hydrogen peroxide (which presumably convert the phen to its *N*-oxide and prevent recombination). Hydroxide can probably scavenge released Fe^{2+}, since it is usually added in aerobic conditions, where

Table 4.19 Rate constants for dissociation of $[Fe(X{-}phen)_3]^{2+}$ cations in acidic aqueous solution at 298 K, together with respective ligand pK_a values, stability constants for complex formation ($\log_{10} \beta_3$), and redox potentials ($\varepsilon°$).

Complex	$10^5 k_{aq}(s^{-1})$	pK_a	$\log_{10} \beta_3$	$\varepsilon°$(V)
$[Fe(5NO_2\text{-phen})_3]^{2+}$	49	3.25	17.8	1.26
$[Fe(5Cl\text{-phen})_3]^{2+}$	24	4.26	19.7	1.11
$[Fe(5Ph\text{-phen})_3]^{2+}$	8.0	4.80	21.1	1.08
$[Fe(phen)_3]^{2+}$	7.3	4.97	21.2	1.07
$[Fe(5Me\text{-phen})_3]^{2+}$	3.5	5.28	21.9	1.02
$[Fe(5,6Me_2\text{-phen})_3]^{2+}$	1.1	5.60	23.0	0.97

dissolved oxygen can readily convert it into hydroxo-iron(III) species (see below). The above sequence of equilibria explains why the formation of tris-diimine complexes obeys fourth-order kinetics[153]:

$$+d[Fe(phen)_3]^{2+}/dt = k_4[Fe^{2+}][phen]^3$$

Rate-determining formation of the tris-complex from the bis-complex, preceded by the pre-equilibria generating the bis-complex, means that k_f in the first equation in the sequence above is composite ($k_4 = K_1 K_2 k_f$).

To return to dissociation, rate constants are markedly affected by substitution in the 1,10-phenanthroline ring, as is documented for a series of 5- and 6-substituted complexes in Table 4.19. These rate constants mirror the electron-releasing and -withdrawing properties of the substituents, correlating with ligand pK values and with stability complexes.[154,155] Dissociation rate constants are markedly sensitive to pressure, with a range of activation volumes between +12 and +22 $cm^3\,mol^{-1}$ (again at 298.2 K).[156–158] Activation volumes do not seem to correlate neatly with ligand-substituent properties, but they indicate that the mechanism of dissociation is strongly dissociative—probably limiting *D*. Space-filling models suggest that *both* iron–nitrogen bonds to the leaving phen ligand have to stretch to form the transition state, with a bond extension estimated to be from 1.97 Å in the initial state to 2.6 Å, corresponding to about 40% of that required for bond fission.[156] The observed activation volumes here are probably determined by a combination of ligand bulk, bond extension, and solvation. High activation enthalpies, which, like activation volumes but in contrast to rate constants, do not correlate well with ligand properties, lie generally within the range 80–120 $kJ\,mol^{-1}$.[155] Such values are consistent both with this steric difficulty and with the high ligand–field stabilization in these low-spin $t_{2g}{}^6$ complexes.

Complexes of 2,2′-bipyridyl are often taken to be very similar to those of 1,10-phenanthroline, but in this area of chemistry there is one important difference, the flexibility of 2,2′-bipyridyl in contrast to the rigidity of 1,10-phenanthroline. The kinetic consequences of the flexibility of bipy are that

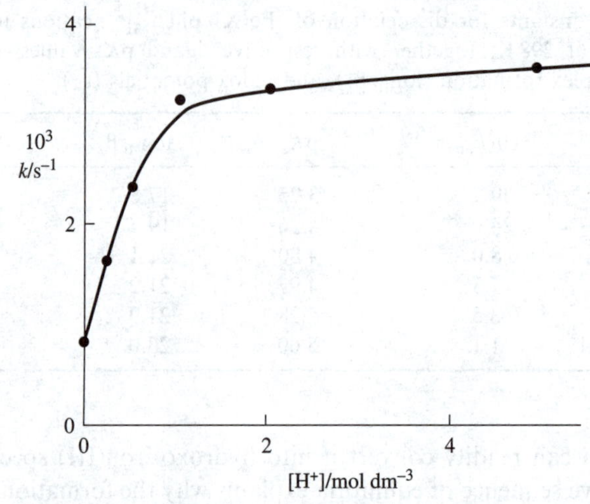

Fig. 4.9 The dependence of aquation rate constants on pH for the $[Fe(bipy)_3]^{2+}$ cation.

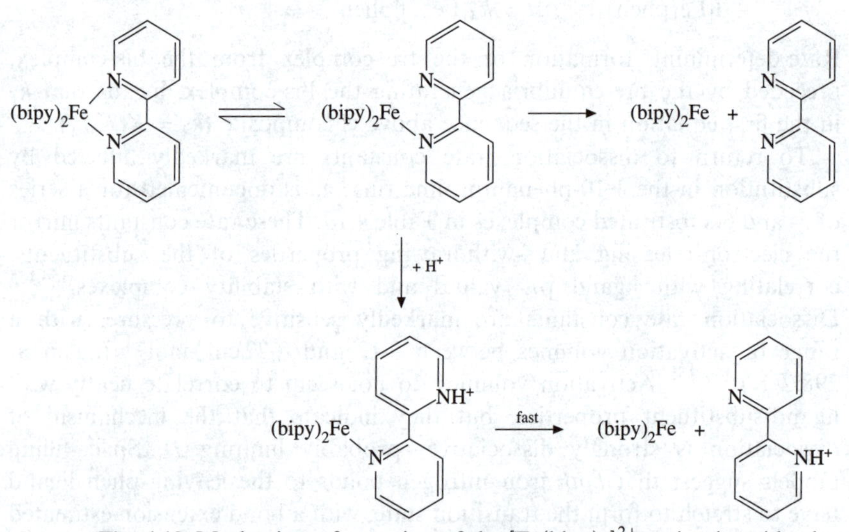

Fig. 4.10 Mechanism of aquation of the $[Fe(bipy)_3]^{2+}$ cation in acid solution.

aquation of $[Fe(bipy)_3]^{2+}$ in acid solution obeys a more complicated rate law than aquation of $[Fe(phen)_3]^{2+}$. The pH dependence of aquation rate constants for $[Fe(bipy)_3]^{2+}$ is shown in Fig. 4.9 and may be explained by invoking equilibrium protonation of the uncoordinated nitrogen in an intermediate containing one monodentate bipy ligand[159] (Fig. 4.10), analogous to that mentioned in relation to the photochemistry of $[Ru(bipy)_3]^{2+}$ (Section 4.7.4.2 above). Similar kinetic (rate law) and mechanistic considerations apply to dissociation, isomerization, and racemization of a variety of complexes of flexible bidentate ligands. The case of $[Cr(en)_2F_2]^+$

was mentioned in Section 4.7.2.1; the complex $[Co(NH_3)_5(enH)]^{4+}$, in which the en is constrained to be monodentate by virtue of protonating one end, was characterized some time ago.[160] Other examples include reactions of oxalato complexes $[M(ox)_3]^{3-}$ and racemization of $[Ge(acac)_3]^+$. Reaction schemes of this type are classified under the general title of 'one-end-off' mechanisms.

A number of substituted 1,10-phenanthrolines, and a few substituted 2,2′-bipyridyls, are readily available, but it can be a major problem in organic synthesis to produce derivatives bearing just the required substituents for specific investigations.[161] Likewise, terpyridyl is readily, if expensively, available, but quaterpyridyl and higher dentate polypyridyls are not.

However, it is a much simpler matter to make Schiff-base analogues (**13**) containing the same diimine chelating moiety (**11** above) from pyridine 2-carboxaldehyde or 2-acetyl or 2-benzoyl pyridine and appropriate primary amines. As the primary amines can be aliphatic or aromatic, with a very large choice of substituted anilines readily available, a wide range of ligands and thus iron(II) complexes are accessible.[162] Terpyridyl analogues can be made from 2,6-diacetyl pyridine; linear hexadentate ligands from, for example, trien and pyridine 2-carboxaldehyde (**14**).[163] To maximize stability, and minimize dissociation rates, complexes of encapsulating ligands such as **15** can be synthesized fairly easily.[164]

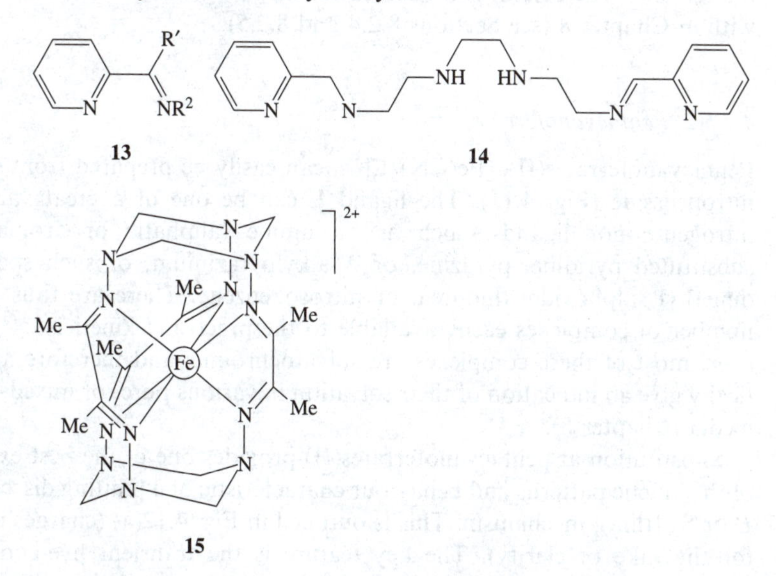

Base hydrolysis of, and cyanide attack at, these tris-diimine complexes generally obey two-term rate laws, for example[165]

$$-d[Fe(phen)_3]^{2+}/dt = \{k_1 + k_2[OH^-]\}[Fe(phen)_3]^{2+}$$

The k_1 term may, by comparison with rate constants for dissociation (*v.s.*), be assigned to rate-limiting dissociation and rapid subsequent scavenging (to give '$Fe(OH)_3$' with hydroxide, $[Fe(CN)_6]^{4-}$ with cyanide). There has been a lengthy and at times emotional argument[166,167] over the mechanism of the k_2 term. This has been attributed to nucleophilic attack at the central iron or, more interestingly, to attack at the ligand.[168] Despite the valiant marshalling of a range of evidence, kinetic, stereochemical, and spectroscopic, in favour of the latter,[167,169] this matter has yet to be resolved to the satisfaction of all interested parties. Regardless of the *point* of attack, bimolecular reaction should be accompanied by negative entropies and volumes of activation. In practice, activation volumes are surprisingly positive, generally between about +10 and +20 $cm^3 mol^{-1}$.[157] The most likely explanation is desolvation of the heavily hydrated hydroxide or cyanide on entering the transition state, though the observed values suggest the de-electrostriction of a disturbingly large number of water molecules—the difference in volume on going from hydration shell to bulk solvent is only about 2 or 3 $cm^3 mol^{-1}$ per water. However, the fact that activation volumes for cyanide attack at the fairly closely related complex $[Mo(CO)_4(bipy)]$ (employed here for solubility reasons) are +4 $cm^3 mol^{-1}$ in methanol, in which cyanide will be much less solvated than in water, and −9 $cm^3 mol^{-1}$ in dimethyl sulphoxide, where cyanide is effectively unsolvated, support this hypothesis,[170] as do trends in binary aqueous solvent mixtures.[171] The role of solvation in the kinetics of these reactions is dealt with in Chapter 8 (see Sections 8.2.4 and 8.2.5).

4.7.5.2 *Pentacyanoferrates*[172]

Pentacyanoferrates(II), $[Fe(CN)_5L]^{3-}$, can easily be prepared from sodium nitroprusside (Fig. 4.11). The ligand L can be one of a great variety of nitrogen-donor ligands, such as an amine (aliphatic or aromatic), a substituted pyridine, pyrazine, or *N*-alkylpyrazinium, or such species as dimethyl sulphoxide, thiourea, or nitrosobenzene. There are thus a large number of complexes easily available to the practising kineticist.[173] Moreover, most of these complexes are solvatochromic, and therefore automatically give an indication of their solvation in various pure- or mixed-solvent media (Chapter 8).

Substitution at pentacyanoferrates(II) provides one of the best examples of the kinetic patterns and behaviour characteristic of a limiting dissociative, *D* or S_N1(lim), mechanism. This is outlined in Fig. 4.12(a) (charges omitted for the sake of clarity). The key feature is the transient five-coordinate intermediate, $[Fe(CN)_5]^{3-}$, which persists long enough to discriminate between the leaving group, L, the incoming ligand, L′, and the solvent. $[Fe(CN)_5]^{3-}$ has little interest in water as a sixth ligand, so the actual discrimination is simply L versus L′. The normal pattern of rate constants for a given complex reacting with a series of nucleophiles is shown in Fig. 4.12(b). All the curves tend to the same limiting rate constant, $k_{lim} \equiv k_1$ of

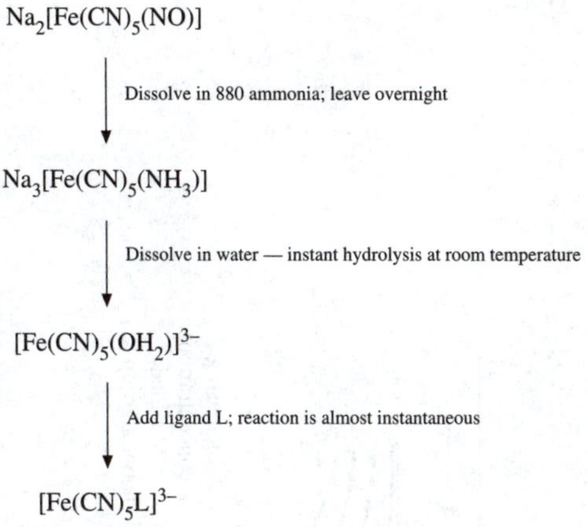

$Na_2[Fe(CN)_5(NO)]$

↓ Dissolve in 880 ammonia; leave overnight

$Na_3[Fe(CN)_5(NH_3)]$

↓ Dissolve in water — instant hydrolysis at room temperature

$[Fe(CN)_5(OH_2)]^{3-}$

↓ Add ligand L; reaction is almost instantaneous

$[Fe(CN)_5L]^{3-}$

Fig. 4.11 The preparation of pentacyanoferrate(II) complexes.

Fig. 4.12(a), at high nucleophile concentration. The curvature of the k_{obs} versus nucleophile concentration plots reflects the relative affinities of the incoming ligands (L′) for the $[Fe(CN)_5]^{3-}$ intermediate. If L′ is a very weak nucleophile towards $[Fe(CN)_5]^{3-}$, then k_{lim} would only be reached at impossibly high concentrations. In such cases, the experimentally accessible portion of the k_{obs} versus nucleophile concentration plot may appear to be a straight line, as shown in Fig. 4.12(b).[174] There is another possibility for a weak incoming ligand, and that is that substitution may not go to completion. In such a situation the observed rate constant is the *sum* of the forward and reverse rate constants, and will therefore be higher than k_{lim}. The experimentally observed dependence of k_{obs} on nucleophile concentration is then as shown in the top curve of Fig. 4.12(b).[175] The complete range of behaviour has been demonstrated for substitution at the $[Fe(CN)_5(4\text{-CNpy})]^{3-}$ anion, though the families of curves do not give quite the same tidy and well-spaced arrangement as the idealized Fig. 4.12(b)![176] If we ignore the two special cases of particularly weak incoming groups and concentrate on the reactions giving the 'normal' pattern, namely the bold curves in Fig. 4.12(b), then we find that the double reciprocal plot of $1/k_{obs}$ versus 1/[nucleophile] is of the form shown in Fig. 4.12(c). In such a plot, k_1 may be obtained from the intercept on the *y*-axis, while the slope gives k_{-1}/k'_L. If one carries out series of runs with increasing amounts of leaving ligand added, then the graph of k_{obs} against nucleophile concentration is as shown in Fig. 4.12(d).[177] The decrease in observed rate of product formation in the presence of added leaving ligand is often termed 'mass-law retardation'. If for substitution in a given complex one obtains plots of the form shown in Fig. 4.12(a), (b), and (c), this constitutes pretty well conclusive evidence for the operation of the *D* mechanism. Often the experimental results are less extensive, with resultant lower confidence in the assignment of mechanism.

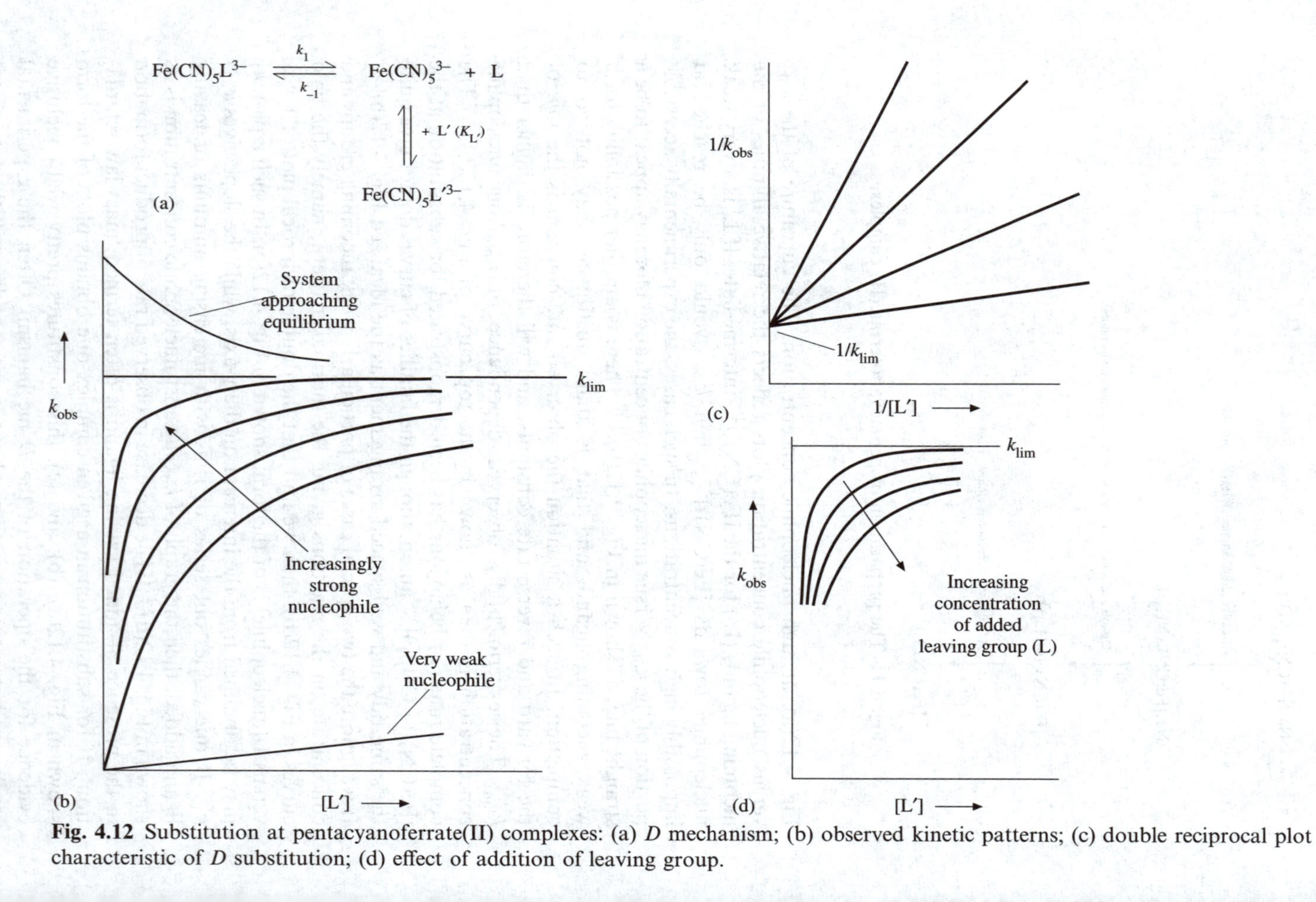

Fig. 4.12 Substitution at pentacyanoferrate(II) complexes: (a) *D* mechanism; (b) observed kinetic patterns; (c) double reciprocal plot characteristic of *D* substitution; (d) effect of addition of leaving group.

Kinetic parameters for dissociation of a number of pentacyanoferrates(II) are listed in Table 4.20.[178] Activation enthalpies for ligand dissociation are high, as would be expected for low-spin iron(II), with its $t_{2g}{}^6$ configuration; activation entropies and volumes are positive, as expected for dissociative activation. What is surprising about the activation volumes is the lack of correlation with any measure of ligand bulk. As for the diimine complexes discussed in the previous section, these must reflect the volume and cone angle of the ligand, and bond extension in the transition state (i.e. volume swept out by the leaving group on going from the initial to the transition state), as well as solvation (electrostriction) changes. The picture is much simpler for pentacyanoferrates(III), where activation volumes for dissociation of the cytosine, cytidine, and cytidine monophosphate complexes vary, from +2 to +16 cm^3 mol^{-1}, in proportion to the bulk of the leaving group.[179]

There are a number of other groups of complexes where the *D* mechanism operates, or might well operate. The most closely related are pentacyanoruthenates(II)[180] and pentacyanocobaltates(III). In the case of the latter, early indications[181] of *D* substitution have been followed by more extensive and detailed studies, which appear to support a range of dissociative mechanisms from limiting *D*[182] to substitutions with at least some interchange character.[183] The latter would parallel pentacyanoferrates(III), with a contrast of mechanism between pentacyanoferrates of these two 3+ metal centres and the 2+ of pentacyanoferrates(II) with their limiting *D* mechanism. The question of *D* versus I_d for substitution at pentacyanometallates crops up again in Chapter 7 in relation to anation at $[M(CN)_5(H_2O)]^{n-}$. The most studied set of *D*-substitution substrates comprises Group VI metal(0)-carbonyl derivatives $M(CO)_5L$, where kinetic patterns, mass-law retardation, and activation entropies and volumes are all consistent with *D* substitution (cf. Chapter 6).[176,184] Limiting dissociative activation is encouraged, though not necessarily always attained, by the presence of a strongly labilizing group such as sulphite, as in substitution in complexes of the type $[M(CN)_4(SO_3)L]^{n-}$[185]. Moving away from pentacyanides and pentacarbonyls, *D* activation has also been established for complexes of the type [Fe(N_4-macrocycle)LL′], discussed in the following section, and for reaction of a bis-aqua-porphyrin-cobalt(III) complex with thiocyanate.[186]

Returning to pentacyanocobaltates, in the specific case of thiocyanate substitution at $[Co^{III}(CN)_5Br]^{3-}$ it has been shown that $[Co^{III}(CN)_5(NCS)]^{3-}$ and $[Co^{III}(CN)_5(SCN)]^{3-}$ are produced by two parallel pathways. One is the *D* route, with marked cation catalysis attributed to M^{n+} mediating the approach of the thiocyanate anion to the anionic transient $[Co(CN)_5]^{2-}$. The other, pH-dependent, path is through $[Co(CN)_5(H_2O)]^{2-}$ or $[Co(CN)_5(OH)]^{3-}$. This competitive situation obtains under both thermal and photochemical activation.[187] Under photochemical conditions it is often possible to replace either L or cyanide from pentacyano complexes, $[M(CN)_5L]^{n-}$, while photolysis facilitates the cyanide departure

Table 4.20 Kinetic parameters[a] for substitution at pentacyanoferrates(II), $[Fe(CN)_5L]^{n-}$, in aqueous solution at 298.2 K.

L	$10^3k(s^{-1})$	$\Delta H^{\ddagger}(kJ\,mol^{-1})$	$\Delta S^{\ddagger}(J\,K^{-1}\,mol^{-1})$	$\Delta V^{\ddagger}(cm^3\,mol^{-1})$
GENERAL SEQUENCE				
Quinoxaline	330	82	+17	+17.9
Ammonia	22	94 or 102	+33 or +63	+16.4
Thiourea	13	88	+21	+20.9
Nitro	10			+20.1
Aliphatic amines	2.8 to 22.7	92 to 107	+29 to +71	+16 to +24
Substituted pyridines	~0.2 to 2	93 to 105	+16 to +50	~ +20
Dimethyl sulphoxide	0.075	110	+46	
***N*-HETEROCYCLES**				
Pyridines:				
unsubstituted	1.1	104	+46	
4-cyano	1.0	105	+50	+20
4-t-butyl	0.92			+11.4
4-butylpentyl	0.37			+16[b]
4-phenyl	0.7			+0.4
4-(4′-pyridyl)[c]	0.62	111	+67	+13.5
Piperidine	7.5	91	+15	
Pyrazine	0.42	110	+59	+12.5
2-Methylpyrazine	0.77	114	+44	+19.4
N-Methylpyrazinium	0.28	115	+75	+0.9
N-n-Pentylpyrazinium	0.20			+9.6
BINUCLEAR COMPLEXES				
$-NC-Fe(CN)_4(4CNpy)$	1.02			
$-N$(pyrazine)$N-M(NH_3)_5$ $M=Rh^{III}$	1.11			
Ru^{III}	2.9			
Ru^{II}	0.72			
$-N$(4,4′-bipyridine)$N-Ru^{II}(NH_3)_5$	2.47			

[a] In some cases the values quoted are means of two or more reported values.
[b] In 20% methanol.
[c] Rate constants for dissociation of the related pentacyanoferrates with leaving groups

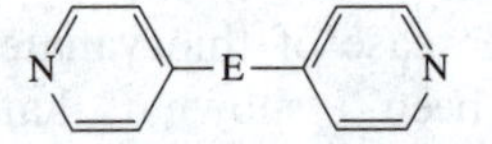

where $E=CH_2$, CH_2CH_2, $CH_2CH_2CH_2$, $CH{=}CH$, $C{\equiv}C$, etc. are all in the range 0.6–$0.8\times10^{-3}\,s^{-1}$; while rate constants for dissociation of the pentaammine-cobalt(III) derivatives

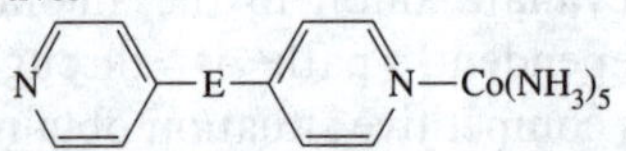

with similar linking groups E are in the range 1–$4\times10^{-3}\,s^{-1}$.

needed for ring closure in the case where L represents a chelating ligand monodentately bound.[188] Such behaviour provides a link with the photochemical initiation of the $[Fe(CN)_5(LL)]^{n-} \rightarrow [Fe(CN)_4(LL)]^{(n-1)-}$ and $[M(CO)_5(LL)] \rightarrow [M(CO)_4(LL)]$ (Section 6.3.1) chelate ring-closure reactions.

Finally, it should be emphasized that the presence of several cyanide or carbonyl ligands in a Werner complex or an organometallic compound may have one of two very different mechanistic consequences. For the pentacyano and pentacarbonyl derivatives discussed in this section, the five cyanide or carbonyl ligands stabilize the $M(CN)_5$ or $M(CO)_5$ entities sufficiently for them to act as transient discriminating intermediates in limiting dissociative substitution, whereas in the case of the Group VI carbonyls π-bonding withdraws electron density from the vicinity of the metal, making nucleophilic approach easier and resulting in a significant, or even dominant, associative pathway (cf. Section 6.2.2).

4.7.5.3 Bioinorganic and model systems

There are a large number of actual and model bioinorganic systems containing a planar four-nitrogen-donor macrocyclic ligand and a pair of *trans* monodentate ligands. Often these two *trans* ligands can be replaced by two other ligands in a two-step sequence exhibiting simple kinetics, usually by the *D* mechanism. There has been a sequence of investigations, from simple model systems, such as iron(II)-phthalocyanine-DMSO-piperidine,[189] through biomimetic complexes[190,191] to the real thing, as in studies of kinetic and equilibrium parameters for oxygen and carbon monoxide complexes of picket-fence and basket-handle porphyrins and of myoglobin and haemoglobin.[192] Further extensions include other oxygen carriers, such as deoxyhaemerythrin, the kinetics of reaction of whose iron(II) sites with oxygen, nitric oxide, fluoride, cyanate, and azide have been reported.[193] Photochemical substitution can be very much faster than thermal.[194]

4.7.6 Hexahalogenometallates

These complexes provide a convenient link between the 2+ and 3+ oxidation states considered so far in this chapter and the final section devoted to 4+ complexes. Kinetic studies have so far been limited, for good chemical reasons, to a selection of chloride and bromide complexes of the second and third rows of the *d*-block, some of which are very inert to substitution. Rate constants for aquation, in some cases estimated from halide exchange data when these appear to involve rate-controlling aquation, have been collected together in Table 4.21. For ease of comparison, this table contains estimated rate constants at 298 K. In view of the long extrapolations often required in their calculation, significant specific cation effects, and the possibility of complications from redox-catalysed pathways (e.g. for $RhCl_6^{3-}$, $IrCl_6^{2-}$) or photochemical activation (e.g.

Table 4.21 Kinetic parameters for aquation of hexachloro- and hexabromo-metallates.*

	$M^{IV}X_6^{2-}$			$M^{III}X_6^{3-}$				$M^{II}X_6^{4-}$
	k_{25} (s^{-1})	$\Delta H^{\ddagger}$ ($kJ\,mol^{-1}$)	$\Delta S^{\ddagger}$ ($J\,K^{-1}\,mol^{-1}$)	k_{25} (s^{-1})	$\Delta H^{\ddagger}$ ($kJ\,mol^{-1}$)	$\Delta S^{\ddagger}$ ($J\,K^{-1}\,mol^{-1}$)	$\Delta V^{\ddagger}$ ($dm^3\,mol^{-1}$)	k_{25} (s^{-1})
$TcCl_6^{n-}$	$\sim 1 \times 10^{-7}$	133						
$TcBr_6^{n-}$	*$\sim 4 \times 10^{-4}$*	*117*						
$ReCl_6^{n-}$	$\sim 2 \times 10^{-9}$	109						
$ReBr_6^{n-}$	*5×10^{-7}*	*109*	−25					
$RuCl_6^{n-}$	$\sim 10^{-3}$			~ 1				$\sim 10^{-3}$
$OsCl_6^{n-}$	$\sim 5 \times 10^{-9}$	136	+42					
$OsBr_6^{n-}$	*$\sim 10^{-6}$*							
$RhCl_6^{n-}$				1.8×10^{-3}	102	+42	+21.5	
$IrCl_6^{n-}$	$\sim 10^{-9}$			$\sim 10^{-5}$	115	+59		
$IrBr_6^{n-}$				*4×10^{-5}*	*96*	*−13*		
$PtCl_6^{n-}$	4×10^{-7}							
cf. $PtCl_4^{n-}$	3.7×10^5	89	−30 to −60					

* Values for the hexobromometallates are italicised.

$PtCl_6^{2-}$, $TcCl_6^{2-}$), these should be viewed as order of magnitude values. Despite this, Table 4.21 does show semi-quantitatively how the nature of the metal and its oxidation state, and of the halide, control reactivity and kinetic parameters. Inspection of, for example, the reactivity sequence $RuCl_6^{2-} \rightarrow RuCl_6^{3-} \rightarrow RuCl_6^{4-}$ shows that the dependence on oxidation state is not simple, though there are the expected decreases in reactivity on going down the Periodic Table (for the metal) and from bromide complexes to their chloro- analogues. Some of the rate constants in Table 4.21 are of considerable importance in relation to redox kinetic studies, for several of these complexes—especially $IrCl_6^{2-}$ and $IrCl_6^{3-}$—are frequently involved in such studies, and it is important that they are so slow to aquate.

In view of the reluctance of many members of this group of complexes to undergo aquation, it is hardly surprising that there have been several studies of metal-ion catalysis of aquation. As ever, mercury(II) is the popular cation for this purpose for the hexachloro- and hexabromo-cations, while for hexafluoro- analogues such as OsF_6^{2-}, ReF_6^{2-}, and PtF_6^{2-} (and indeed PF_6^{-}) zirconium(IV) has proved effective. This set of reactions has the added interest of rate constants being independent of the nature of the central metal, presumably arising from rate-limiting dissociation of inactive polynuclear zirconium species to catalytically active monomers.[195] The $IrCl_6^{3-}/In^{3+}$ system[196] provides a link between this catalysed aquation and the specific cation effects mentioned in the previous paragraph. It is not immediately apparent whether this soft cation is actually forming an intermediate, to give $InCl^{2+}$ as a good leaving group, or whether its 3+ charge is merely providing strong and effective ion pairing.

Very inert centres, such as iridium(III) or osmium(IV), permit the isolation of intermediates in halide-exchange reactions of the type $MX_6^{n-} + 6Y^- \rightarrow MY_6^{n-} + 6X^-$, and thus the establishment of the complete series of rate constants or stereochemical courses of reaction. The pattern of rate constants for successive replacement of the bromides in $OsBr_6^{2-}$ by chlorides is shown in Figure 4.13(a).[197] For simplicity, only forward rate constants are shown. To complement this kinetic picture, the results of a study of the full stereochemical course of substitution in the oxalato-halide system $[Os(ox)I_4]^{2-} + 4Br^- \rightarrow [Os(ox)Br_4]^{2-} + 4I^-$ is shown in Figure 4.13(b).[198]

It would be interesting and useful to have sets of kinetic parameters for the six successive stages, if possible including data for *cis* and *trans* or *mer* and *fac* isomers where relevant, in hydrolysis reactions of some hexahalogenometallates. The only approximation to this is provided by the $[Ru^{III}Cl_6]^{3-}$ sequence shown in Table 4.22.[199] Other kinetic data on ternary aqua-halide complexes are very fragmented—the nearest comparable sequence is that for aquation of pentacyanonitrosylchromate(III), where the use of ESR spectroscopy permitted characterization of intermediates as well as the establishment of rate constants and Arrhenius parameters for two of the stages.[200] There has been some interest in ternary dimethyl sulphoxide-chloride-ruthenium species in connection with their anti-tumour activity,

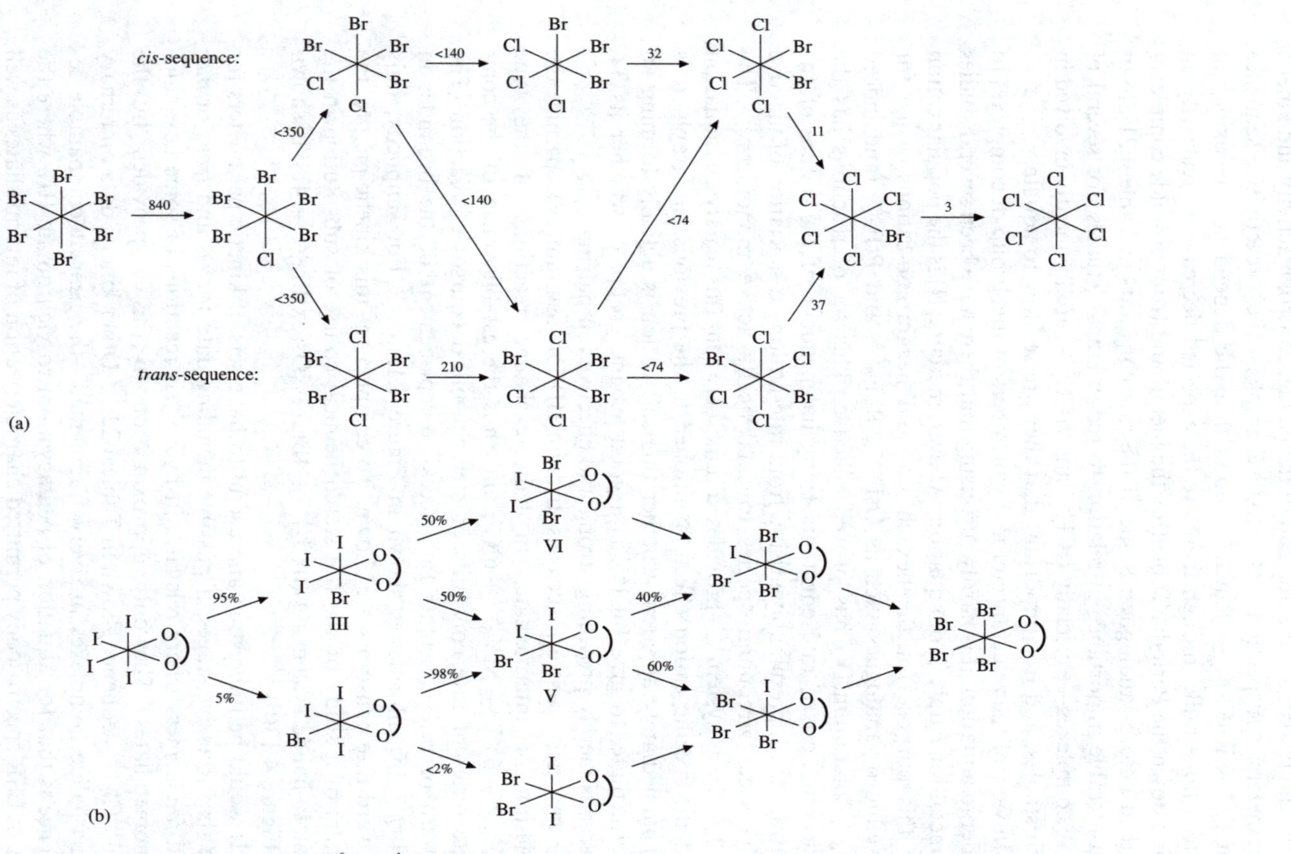

Fig. 4.13 (a) Rate constants ($10^6 k/s^{-1}$; at 353 K) for the successive replacement of the bromide ligands of $[OsBr_6]^{2-}$ by chloride. For simplicity we show only rate constants for the forward reactions. (b) The stereochemical course of successive substitution steps in the replacement of the iodide ligands in the ternary oxalato complex $[Os(ox)I_4]^{2-}$ by bromide.

Table 4.22 Kinetic parameters, at 298 K, for successive steps in aquation of $[Ru^{III}Cl_6]^{3-}$.

	k (s^{-1})	$\Delta H^{\ddagger}$ (kJ mol^{-1})
$[RuCl_6]^{3-}$	~1	
$[RuCl_5(OH_2)]^{3-}$	[a]	
$[RuCl_4(OH)_2]^{3-}$	9×10^{-5}	96
$[RuCl_3(OH_2)_3]^{3-}$	3×10^{-5}	99
$[RuCl_2(OH_2)_4]^{3-}$	5×10^{-6}	103
$[RuCl(OH_2)_5]^{3-}$	$< 10^{-8}$	

[a] No value available.

but kinetic and mechanistic information on these species is limited and mainly qualitative.[201]

4.7.7 *Metal(IV) centres*

4.7.7.1 *Platinum(IV) complexes*

One might well expect that substitution at platinum(IV), a 4+ low-spin d^6 centre, would be extremely slow. It is, however, rather hard to find examples to supplement the aquation rate constant given for $PtCl_6^{2-}$ in the preceding section. The usual mechanism for substitution at platinum(IV) involves redox catalysis by platinum(II), often with significant contributions from other ions present.[202] Typically, the rate law for such a reaction conducted in the presence of a potential ligand such as chloride is:

$$\text{rate} = k[Pt^{IV}][Pt^{II}][Cl^-]$$

For many years it was believed that base hydrolysis of $[Pt(NH_3)_5Cl]^{3+}$ and of *cis*-$[Pt(NH_3)_4Cl_2]^{2+}$ represented two of very few non-redox-catalysed nucleophilic substitutions at platinum(IV). The respective amido-conjugate bases were the predominant species present, but mechanistic interpretation was clouded by the non-integral orders in hydroxide.[203] Base hydrolysis of $[Pt(CN)_4Br_2]^{2-}$ has been another candidate; here there cannot be an S_NCB mechanism.[204] More recent examples, involving ternary dihydroxy complexes, are clouded by possible isomerization complications.[205] The normally very good leaving group $CF_3SO_3^-$ is replaced remarkably slowly by water. The rate constant for aquation of $[Pt(NH_3)_5(OSO_2CF_3)]^{3+}$ at 353 K is only about 10^{-5} s^{-1}; determination of this rate constant clearly caused the authors considerable trouble.[79] Substitution at $[Pt(S_5)_3]^{2-}$ by sulphide, sulphite, hydroxide, or triphenyl phosphine is another non-simple case, for here the initial attack by the nucleophile is at sulphur α to the metal atom.[206]

4.7.7.2 Ternary metal(IV) complexes

Ligand exchange at some metal(IV) centres was briefly mentioned in Section 4.3, where its mechanism was indicated to depend on the nature of the metal centre and of the ligands. Metal(IV) complexes of formula $M(LL)_2X_2$ are known for a variety of metals, M, of monoanionic bidentate ligands, LL^-, and of anions, X^-. Signs that some complexes of this type, especially budotitane, *cis*-$[Ti(bzac)_2(OEt)_2]$, may have significant anti-tumour activity[207] have inspired some interest in their substitution kinetics. Their formal similarity to cisplatin, *cis*-$[Pt(NH_3)_2Cl_2]$, prompted thoughts that partial aquation of a species $M(LL)_2X_2$, particularly where X = halide, may well be important in relation to their possible use in medicine. If partially aquated species are indeed important in this respect, then clearly it is necessary to know something about their substitution kinetics, if only to match time-scales for aquation against those of bodily processes.

Studies of substitution kinetics at bis-cyclopentadienyltitanium dichloride[208] have subsequently been extended to several series of compounds $M(LL)_2Cl_2$ to cover a range of metal(IV) centres and spectator ligands LL^-. Kinetics of substitution at $M(cp)_2Cl_2$ and at $M(LL)_2Cl_2$ with LLH = ethylmaltol, **16**, follow the same two-term rate law as substitution at square-planar complexes, namely:

$$-\mathrm{d}[M(LL)_2Cl_2]/\mathrm{d}t = \{k_1 + k_2[\text{nucleophile}]\}[M(LL)_2Cl_2]$$

This is illustrated, for the titanium-cyclopentadienyl and titanium-ethylmaltol systems, in Fig. 4.14(a) and (b). Figure 4.15 shows the effects of varying the non-leaving ligands (LL^-) and the metal(IV) centre, for thiocyanate substitution at $M(etmalt)_2Cl_2$.[209] It is not yet altogether clear what interpretation should be placed on the relatively small k_1 pathway, but the

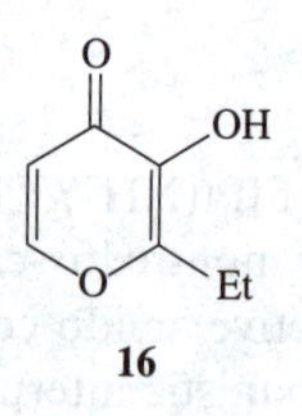

16

relative values of k_2 provide an illustration of factors influencing reactivity at metal(IV) centres around the Periodic Table. Thus the *p*-block elements of Group 14 show rapidly increasing reactivity on descending the group, whereas their *d*-block analogues titanium, zirconium, and hafnium show the opposite trend. Ligand field effects are also illustrated by the considerably lower reactivity of the molybdenum(IV) complex (d^2) than of the d^0 analogue zirconium. The results also demonstrate the possibility of tuning reactivity by varying the ligand LL^-. Such variation may also be used to optimize solvation properties, i.e. the partition coefficient and hydrophilic–lipophilic balance (HLB), for membrane crossing.

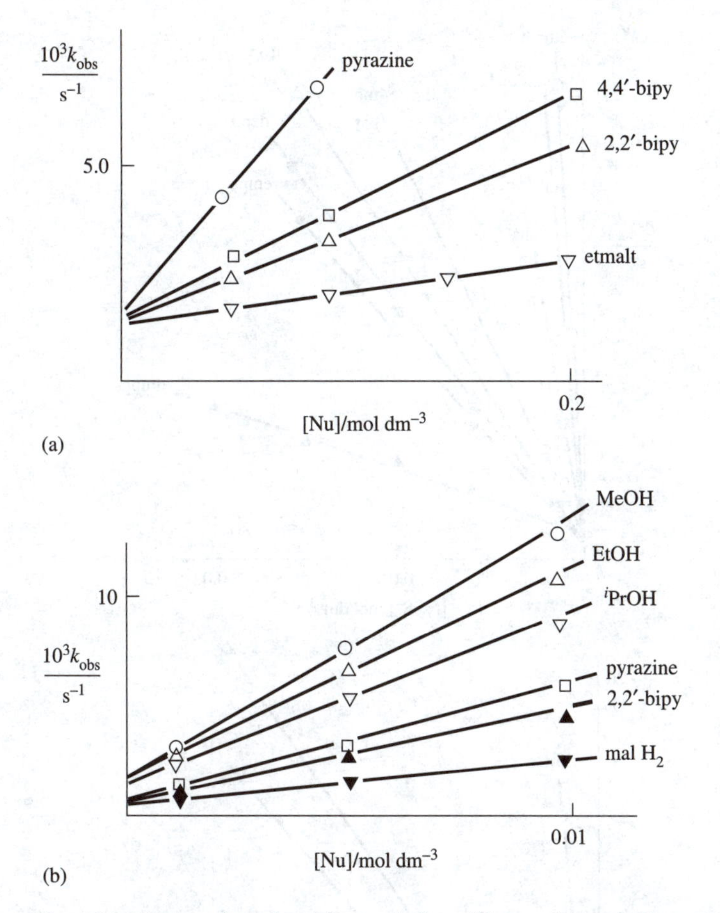

Fig. 4.14 Effects of nucleophile variation. Observed first-order rate constants for the first stage in nucleophilic attack at (a) $[Ti(cp)_2Cl_2]$ and (b) $[Ti(etmalt)_2Cl_2]$ in acetonitrile solution, at 298.2 K.

There are a number of ternary octahedral complexes of metal(IV), and metal(V) and metal(VI), centres based on the *trans*-dioxo unit O=M=O. The tetracyano derivatives actually exist as $O{=}M{-}OH_2$ in acid, and O=M−OH at higher pHs. Kinetics of substitution at *trans*-$[M^{IV}O_2(CN)_4]^{4-}$, *trans*-$[M^{IV}O(OH)(CN)_4]^{3-}$, and *trans*-$[M^{IV}O(OH_2)(CN)_4]^{2-}$ by, for instance, 2,2′-bipyridyl, have been fairly extensively studied,[210,211] with bond distance information from X-ray diffraction studies of crystals containing these anions providing valuable insights into mechanisms.[211] Activation entropies for oxygen exchange at *trans*-$[W^{IV}O(OH_2)(CN)_4]^{2-}$ and at *trans*-$[Re^{V}O(OH_2)(CN)_4$ are +64 and $+17\ J\ K^{-1}\ mol^{-1}$, respectively, indicating dissociative activation. Activation entropies for oxygen exchange at *trans*-$[W^{IV}O(OH)(CN)_4]^{3-}$ and at *trans*-$[Re^{V}O(OH)(CN)_4]^{2-}$ are −40 and $-3\ J\ K^{-1}\ mol^{-1}$, respectively, suggesting an interchange mechanism with significant associative character.[212] The

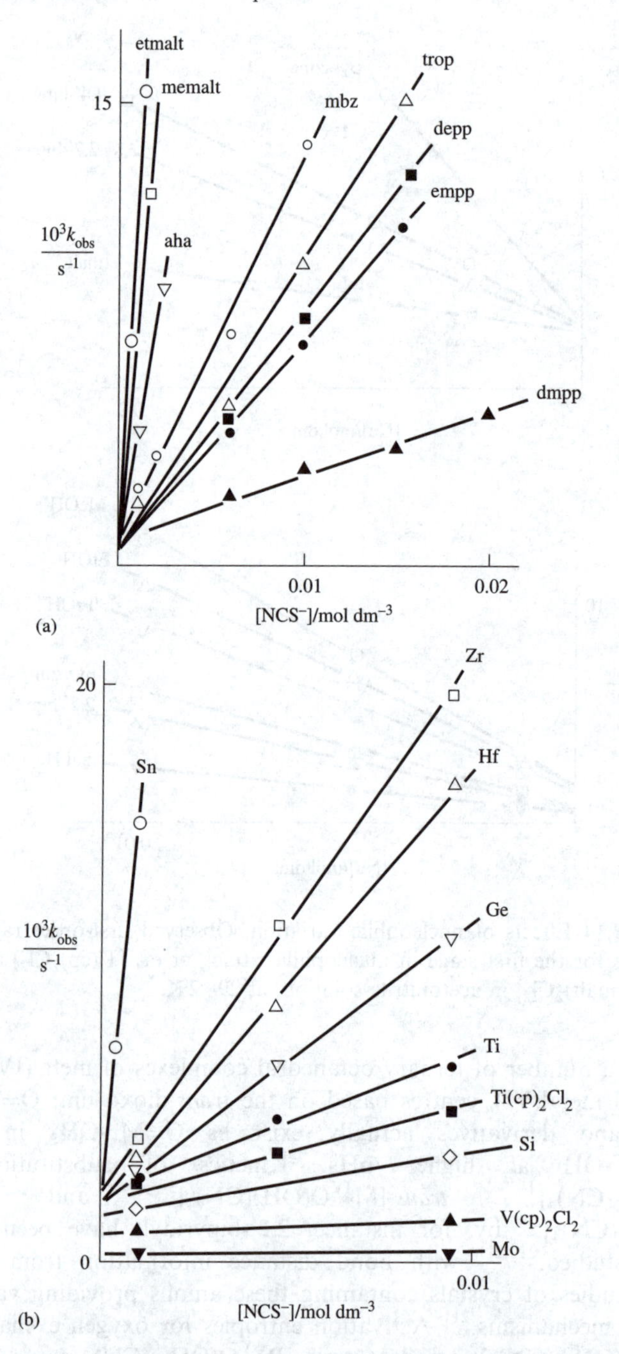

Fig. 4.15 Effects of varying non-leaving ligands and metal. (a) Observed first-order rate constants for reactions of $[Sn(LL)_2(NCS)Cl]$ with thiocyanate in acetonitrile solution, at 298.2 K. (b) Observed first-order rate constants for reactions of $[M(LL)_2(NCS)Cl]$, LL = etmalt except where cp specified, with thiocyanate in acetonitrile solution, at 298.2 K.

main point of interest here is the operation of a dissociative mechanism at unhindered metal(IV) and even metal(V) centres. Activation entropies again indicate dissociative activation at metal(V) centres in pyridine exchange at *trans*-$[M^{V}O_2(py)_4]^+$, M = Tc or Re, and in solvolysis of *trans*-$[M^{V}O_2(py)_4]^+$.[213]

4.8 Substitution at eight-coordinate complexes

A brief mention of kinetics and mechanism at eight-coordinate metal centres can conveniently form a footnote to octahedral substitution. There have been rather few such studies. Many have involved exchange or substitution at solvento-cations and are therefore to be found in Chapter 7, but the group of octacyanometallates deserves at least cursory mention here. Their photochemistry has been examined over many years, as may be seen, for example, from published comparisons of the results of early fundamental work with those for hexacyanometallates $[M(CN)_6]^{3-}$ (M = Cr, Mn, Fe, Co)[214] and more recent comparisons of thermal and photochemical substitution, in cyano complexes of transition metals in general[215] and in particular in solvolysis of $[W(CN)_8]^{4-}$ as a function of pH.[216] There have been several investigations of photolysis of $[W(CN)_8]^{4-}$ and of $[Mo(CN)_8]^{4-}$ in the presence of ligands such as 2,2′-bipyridyl, 1,10-phenanthroline, or ethanolamine. Here the usual sequence of events is thermal substitution in species such as $[M(CN)_7(OH)]^-$ and $[M(CN)_6(CNH)(OH_2)]^{2-}$ produced in the initial photolytic step.[217] Such studies are closely related to those of substitution at *trans*-$[M^{IV}O_2(CN)_4]^{4-}$ and its protonated derivatives, as mentioned in the preceding section. Metal-ion catalysis of dissociation of $[M(CN)_8]^{n-}$ anions can be achieved with chromium(III) or titanium(IV),[218] and can be linked to similar catalysis of aquation of octahedral halogenocobalt(III) complexes (Section 4.5.1) through Hg^{2+} catalysis of aquation of hexacyanoferrate(II).[219]

References

1. D. L. Kepert, *Progr. Inorg. Chem.*, 1977, **23**, 1.
2. M. Eigen and R. G. Wilkins, *Adv. Chem. Ser.*, 1965, **49**, 55.
3. D. L. Kepert, *Progr. Inorg. Chem.*, 1979, **25**, 41; *Comprehensive Coordination Chemistry*, Vol. 1, eds G. Wilkinson, R. D. Gillard, and J. A. McCleverty, Pergamon Press, Oxford, 1987, Section 2.4.
4. References to possible geometries for five-coordination, relevant to fully dissociative substitution, have already been cited in Chapter 3.
5. H. Taube, *Chem. Rev.*, 1952, **50**, 69.
6. F. Basolo and R. G. Pearson, *Mechanisms of Inorganic Reactions*, Wiley, New York, 1958.
7. Data from F. Basolo and R. G. Pearson, *Mechanisms of Inorganic Reactions*, 2nd edn, Wiley, New York, 1967, p. 69.

8. R. W. Oliff and A. L. Odell, *J. Chem. Soc.*, 1964, 2417.
9. M. Gerloch and E. C. Constable, *Transition Metal Chemistry*, VCH, Cambridge, 1994.
10. K. F. Purcell and J. C. Kotz, *Inorganic Chemistry*, Saunders, Philadelphia, 1977, p. 720.
11. J. K. Burdett, *J. Chem. Soc., Dalton Trans.*, 1976, 1725.
12. See Chapter 10 of Ref. 10; F. A. Cotton and G. Wilkinson, *Advanced Inorganic Chemistry*, 5th edn, Wiley–Interscience, 1988—see the detailed tables of coordination numbers and stereochemistries given for individual *d*-block elements.
13. See H. Kanno, T. Yamaguchi, and H. Ohtaki, *J. Phys. Chem.*, 1989, **93**, 1695 for the best evidence that Sc^{3+}aq is not a simple octahedral species, D. T. Richens, *The Chemistry of Aqua Ions*, VCH, Weinheim, 1996, for a detailed discussion of Sc^{3+}aq, and ref. 10 of Chapter 7 (p327).
14. F. Matsumoto, Y. Ohki, Y. Susuki, and A. Ouchi, *Bull. Chem. Soc. Jpn*, 1989, **62**, 2081; Ma Jian-Fang, Jin Zhong-Sheng, and Ni Jia-Zuan, *Polyhedron*, 1995, **14**, 563.
15. A. B. Ilyukhin and S. P. Petrosyants, *Russ. J. Inorg. Chem.*, 1994, **39**, 1449 (*Zh. Neorg. Khim.*, 1994, **39**, 1517).
16. See p. 596 of Ref. 10.
17. A. E. Merbach and J. C. Bünzli, *Helv. Chim. Acta*, 1972, **55**, 1903; R. Good and A. E. Merbach, *Inorg. Chem.*, 1975, **14**, 1030; C. P. M. Favez and A. E. Merbach, *Inorg. Chim. Acta*, 1977, **60**, 2695; H. Vanni and A. E. Merbach, *Inorg. Chem.*, 1979, **18**, 2758; A. E. Merbach, *Pure Appl. Chem.*, 1982, **54**, 1479.
18. C. T. G. Knight and A. E. Merbach, *J. Am. Chem. Soc.*, 1985, **106**, 804; *Inorg. Chem.*, 1985, **24**, 576.
19. E. Turin, R. Nielsen, and A. E. Merbach, *Inorg. Chim. Acta*, 1987, **134**, 79.
20. M. Turin-Rossier, D. Hugi-Cleari, U. Frey, and A. E. Merbach, *Inorg. Chem.*, 1990, **29**, 1374.
21. J. W. Moore and R. G. Pearson, *Inorg. Chem.*, 1964, **3**, 1334.
22. W. G. Jackson, S. S. Jurisson, and B. C. McGregor, *Inorg. Chem.*, 1985, **24**, 1788.
23. D. A. House, *Coord. Chem. Rev.*, 1977, **23**, 223.
24. J. O. Edwards, F. Monacelli, and G. Ortaggi, *Inorg. Chim. Acta*, 1974, **11**, 47; M. L. Tobe, Substitution Reactions, in *Comprehensive Coordination Chemistry*, Vol. 1, eds G. Wilkinson, R. D. Gillard, and J. A. McCleverty, Pergamon Press, Oxford, 1987, Chapter 7. 1, p. 281 (see Table 4).
25. For example, T. W. Swaddle, *Adv. Inorg. Bioinorg. Chem.*, 1983, **2**, 95; T. W. Swaddle, *Comments Inorg. Chem.*, 1991, **12**, 237; P. A. Lay, *Coord. Chem. Rev.*, 1991, **110**, 213.
26. R. van Eldik, *Pure Appl. Chem.*, 1992, **64**, 1439.
27. G. A. Lawrance, *Adv. Inorg. Chem.*, 1989, **34**, 145.
28. J. L. Uzice and R. Lopez de la Vega, *Inorg. Chem.*, 1990, **29**, 382.
29. For example, Zai-Wei Li, W. D. Harman, P. A. Lay, and H. Taube, *Inorg. Chem.*, 1994, **33**, 3635.
30. G. E. Rodgers, *Introduction to Coordination, Solid State, and Descriptive Inorganic Chemistry*, McGraw-Hill, New York, 1994, pp. 101–102.
31. Review: R. Banerjee, *Coord. Chem. Rev.*, 1985, **68**, 145.
32. M. M. Jones and H. R. Clark, *J. Inorg. Nucl. Chem.*, 1971, **33**, 413.
33. E. S. Rudakov and I. V. Kozhevnikov, *Tetrahedron Lett.*, 1971, 1333; I. V. Kozhevnikov and E. S. Rudakov, *Inorg. Nucl. Chem. Lett.*, 1972, **8**, 571.

34. W. G. Jackson and B. H. Dutton, *Inorg. Chem.*, 1989, **28**, 525.
35. W. L. Reynolds, S. Hafezi, A. Kessler, and S. Holly, *Inorg. Chem.*, 1979, **18**, 2860.
36. See, for example, D. A. Buckingham, C. R. Clark, and W. S. Webley, *J. Chem. Soc., Dalton Trans.*, 1980, 2255; *Aust. J. Chem.*, 1980, **33**, 263; W. G. Jackson, G. A. Lawrance, and A. M. Sargeson, *Inorg. Chem.*, 1980, **19**, 1001.
37. For example, D. A. Buckingham, C.R. Clark, and W. S. Webley, *Inorg. Chem.*, 1982, **21**, 3353.
38. A. A. Holder and T. P. Dasgupta, *J. Chem. Soc., Dalton Trans.*, 1996, 2637.
39. W. G. Jackson, *Inorg. Chem.*, 1988, **27**, 777.
40. R. C. Elder, M. J. Heeg, M. D. Payne, M. Trkula, and E. Deutsch, *Inorg. Chem.*, 1978, **17**, 431.
41. F. Basolo, J. G. Bergmann, R. E. Meeker, and R. G. Pearson, *J. Am. Chem. Soc.*, 1956, **78**, 2676.
42. D. A. Buckingham, B. M. Foxman, and A. M. Sargeson, *Inorg. Chem.*, 1970, **9**, 1790; N. J. Curtis and G. A. Lawrance, *Inorg. Chem.*, 1986, **25**, 1033.
43. Y. Kitamura, S. Taneda, and K. Kuroda, *Inorg. Chim. Acta*, 1989, **156**, 31.
44. P. O. Whimp, M. F. Bailey, and N. F. Curtis, *J. Chem. Soc. (A)*, 1970, 1956.
45. C. K. Poon and M. L. Tobe, *J. Chem. Soc. (A)*, 1967, 2069; C. J. Cooksey and M. L. Tobe, *Inorg. Chem.*, 1978, **17**, 1558.
46. D. A. House and M. L. Tobe, *J. Chem. Soc., Dalton Trans.*, 1989, 853.
47. R. Niththyananthan and M. L. Tobe, *Inorg. Chem.*, 1969, **8**, 1589.
48. M. L. Tobe, in *Studies on Chemical Structure and Reactivity*, ed. J. H. Ridd, Methuen, London, 1966, Chapter 11.
49. N. E. Brasch, D. A. Buckingham, C. R. Clark, and K. S. Finnie, *Inorg. Chem.*, 1989, **28**, 3386.
50. W. G. Jackson, *Inorg. Chim. Acta*, 1974, **10**, 51.
51. J. H. Worrell, *Inorg. Chem.*, 1975, **14**, 1699.
52. M. L. Tobe, *Adv. Inorg. Bioinorg. Chem.*, 1983, **2**, 1.
53. F. J. Garrick, *Nature*, 1937, **139**, 507.
54. R. G. Pearson and F. Basolo, *J. Am. Chem. Soc.*, 1956, **78**, 4878; C. K. Ingold, R. S. Nyholm, and M. L. Tobe, *J. Chem. Soc.*, 1956, 1691; R. S. Nyholm and M. L. Tobe, *J. Chem. Soc.*, 1956, 1707; R. G. Pearson, P. M. Henry, and F. Basolo, *J. Am. Chem. Soc.*, 1957, **79**, 5379.
55. M. L. Tobe, *Inorg. Chem.*, 1968, **7**, 1260.
56. S. Singh and R. Shankar, *Inorg. Chem.*, 1989, **28**, 2695.
57. B. Brønnum, H. S. Johansen, and L. H. Skibsted, *Inorg. Chem.*, 1988, **27**, 1859.
58. D. D. Brown and C. K. Ingold, *J. Chem. Soc.*, 1953, 2674.
59. S. Balt, H. J. Gamelkoorn, H. J. A. M. Kuipers, and W. Renkema, *Inorg. Chem.*, 1983, **22**, 3072; S. Balt, H. J. Gamelkoorn, and W. Renkema, *J. Chem. Soc., Dalton Trans.*, 1983, 2415.
60. S. Balt, W. E. Renkema, and P. C. M. van Zijl, *Inorg. Chim. Acta*, 1980, **45**, L241.
61. E. Ahmed, M. L. Tobe, and M. Humanes, *J. Chem. Soc., Dalton Trans.*, 1990, 2175.
62. N. E. Dixon, W. G. Jackson, W. Marty and A. M. Sargeson, *Inorg. Chem.*, 1982, **21**, 688.
63. Review: D. A. House, *Coord. Chem. Rev.*, 1992, **114**, 249.
64. W. G. Jackson, M. L. Randall, A. M. Sargeson, and W. Marty, *Inorg. Chem.*, 1983, **22**, 1013; W. G. Jackson, D. P. Fairlie, and M. L. Randall, *Inorg. Chim.*

Acta, 1983, **70**, 197; W. G. Jackson and C. N. Hookey, *Inorg. Chem.*, 1984, **23**, 668; W. G. Jackson, C. N. Hookey, M. L. Randall, P. Comba, and A. M. Sargeson, *Inorg. Chem.*, 1984, **23**, 2473.
65. E. Ahmed and M. L. Tobe, *Inorg. Chem.*, 1974, **13**, 2956.
66. C. K. Poon and M. L. Tobe, *J. Chem. Soc., Chem. Commun.*, 1968, 156; J. Lichtig, M. E. Sosa, and M. L. Tobe, *J. Chem. Soc., Dalton Trans.*, 1984, 581.
67. R. A. Henderson and M. L. Tobe, *Inorg. Chem.*, 1977, **16**, 2576.
68. D. A. House and H. K. J. Powell, *J. Chem. Soc., Chem. Commun.*, 1969, 382.
69. For example, K. Ishihara and Y. Kitamura, *Polyhedron*, 1996, **15**, 2969.
70. Y. Kitamura, G. A. Lawrance, and R. van Eldik, *Inorg. Chem.*, 1989, **28**, 333.
71. F. P. Rotzinger, *Inorg. Chem.*, 1988, **27**, 768, 772.
72. E. Ahmed, C. Chatterjee, C. J. Cooksey, M. Humanes, M. L. Tobe, and G. Williams, *J. Chem. Soc., Dalton Trans.*, 1989, 645; E. Ahmed, M. L. Tobe, and M. Humanes, *J. Chem. Soc., Dalton Trans.*, 1990, 2175; and see p. 7 of Ref. 52.
73. See Section VIII.A of Ref. 52.
74. R. W. Hay, *Inorg. Chim. Acta*, 1980, **45**, L83; R. W. Hay and P. R. Norman, *J. Chem. Soc., Chem. Commun.*, 1980, 734.
75. D. A. House, P. R. Norman, and R. W. Hay, *Inorg. Chim. Acta*, 1980, **45**, L117.
76. P. Comba and A. M. Sargeson, *J. Chem. Soc., Chem. Commun.*, 1988, 51; and see Section 7.1.5.4 of Ref. 24.
77. C. Kutal and A. W. Adamson, in *Comprehensive Coordination Chemistry*, Vol. 1, eds G. Wilkinson, R. D. Gillard, and J. A. McCleverty, Pergamon, Oxford, 1987, Chapter 7.3; P. C. Ford, *Coord. Chem. Rev.*, 1982, **44**, 61.
78. Review: P. A. Lay, *Comments Inorg. Chem.*, 1991, **11**, 235.
79. N. E. Dixon, G. A. Lawrance, P. A. Lay, and A. M. Sargeson, *Inorg. Chem.*, 1983, **22**, 846; N. E. Dixon, G. A. Lawrance, P. A. Lay, and A. M. Sargeson, *Inog. Chem.*, 1984, **23**, 2940; N. J. Curtis, G. A. Lawrance, P. A. Lay, and A. M. Sargeson, *Inorg. Chem.*, 1986, **25**, 484.
80. See p. 292 of Ref. 24 and p. 164 of Ref. 7.
81. R. Duval and C. Duval, in *Nouveau Traité de Chimie Minérale*, Vol. 14, ed. P. Pascal, Masson, Paris, 1959, p. 415.
82. C. S. Garner and D. A. House, in *Transition Metal Chemistry*, Vol. 6, ed. R. L. Carlin, Marcel Dekker, New York, 1970, p. 59.
83. T. W. Swaddle and W. E. Jones, *Can. J. Chem.*, 1970, **48**, 1054.
84. G. A. Lawrance and R. van Eldik, *J. Chem. Soc., Chem. Commun.*, 1987, 1105; N. J. Curtis, G. A. Lawrance and R. van Eldik, *Inorg. Chem.*, 1989, **28**, 333.
85. R. van Eldik, in *High Pressure Chemistry and Biochemistry*, eds R. van Eldik and J. Jonas, Reidel, Dordrecht, 1987, p. 344.
86. T. W. Swaddle and G. Guastalla, *Inorg. Chem.*, 1968, **7**, 1915; T. W. Swaddle and W. E. Jones, *Can. J. Chem.*, 1970, **48**, 1054.
87. H. Bruce, D. Reinhard, M. J. Saliby, and P. S. Sheridan, *Inorg. Chem.*, 1987, **26**, 4024.
88. R. G. Pearson, R. A. Munson, and F. Basolo, *J. Am. Chem. Soc.*, 1958, **80**, 504.
89. P. Kita, *Pol. J. Chem.*, 1988, **62**, 653.
90. D. A. House, *Inorg. Chem.*, 1988, **27**, 2587.
91. D. A. House, K. B. Reddy, and R. van Eldik, *Inorg. Chim. Acta*, 1991, **186**, 5.
92. G. A. Lawrance, K. Schneider, and R. van Eldik, *Inorg. Chem.*, 1984, **23**, 3922.
93. P. Guardado, G. A. Lawrance, and R. van Eldik, *Inorg. Chem.*, 1989, **28**, 976.
94. R. D. Gillard, *J. Chem. Soc. (A)*, 1967, 917.

95. D. P. Rillema, J. F. Endicott, and J. R. Barber, *J. Am. Chem. Soc.*, 1973, **95**, 6987.
96. J. F. Bunnett, *Accounts Chem. Res.*, 1978, **11**, 413.
97. A concise account and key references will be found in C. Kutal and A. W. Adamson, in *Comprehensive Coordination Chemistry*, Vol. 1, eds G. Wilkinson, R. D. Gillard, and J. A. McCleverty, Pergamon, Oxford, 1987, Chapter 7.3.3.1.1.
98. A. W. Adamson, *Coord. Chem. Rev.*, 1968, **3**, 169; A. D. Kirk, *Coord. Chem. Rev.*, 1981, **39**, 225; J. F. Endicott, R. B. Lessard, D. Lynch, M. W. Perkovic, and C. K. Ryn, *Coord. Chem. Rev.*, 1990, **97**, 65; F. Scandola, C. A. Bignozzi, C. Chiorboli, M. T. Indelli, and M. A. Rampi, *Coord. Chem. Rev.*, 1990, **97**, 229.
99. P. C. Ford, D. Wink, and J. DiBenedetto, *Progr. Inorg. Chem.*, 1983, **30**, 213 L.
100. S. Forster, *Chem. Rev.*, 1990, **90**, 331.
101. A. W. Adamson, *J. Phys. Chem.*, 1967, **71**, 798.
102. B. R. Hollebone, C. H. Langford, and N. Serpone, *Coord. Chem. Rev.*, 1981, **39**, 181.
103. L. Mønsted and O. Mønsted, *Pure Appl. Chem.*, 1990, **62**, 1095.
104. D. A. House, *Mech. Inorg. Organomet. React.*, 1991, **7**, 142–145.
105. L. G. Vanquickenborne, B. Coussens, D. Postelmans, A. Ceulemans, and K. Pierloot, *Inorg. Chem.*, 1992, **31**, 539.
106. L. Mønsted and O. Mønsted, *Acta Chem. Scand.*, 1984, **38A**, 679; 1986, **40A**, 637.
107. K. Angermann, R. van Eldik, H. Kelm, and F. Wasgestian, *Inorg. Chem.*, 1981, **20**, 955; K. Angermann, R. van Eldik, H. Kelm, and F. Wasgestian, *Inorg. Chim. Acta*, 1981, **49**, 247; J. F. Endicott and C. K. Ryu, *Comments Inorg. Chem*, 1987, **6**, 91.
108. M. L. Tobe, in *Comprehensive Coordination Chemistry*, Vol. 1, eds G. Wilkinson, R. D. Gillard, and J. A. McCleverty, Pergamon, Oxford, 1987, Chapter 7.1.
109. R. A. Bauer and F. Basolo, *Inorg. Chem.*, 1969, **8**, 2231, 2237; K. W. Bowker, E. R. Gardner, and J. Burgess, *Trans. Faraday Soc.*, 1970, **66**, 2065; K. W. Bowker, E. R. Gardner, and J. Burgess, 1971, **67**, 3076; J. Burgess, K. W. Bowker, E. R. Gardner, and F. M. Mekhail, *J. Inorg. Nucl. Chem.*, 1979, **41**, 1215.
110. S. Suvachittanont and R. van Eldik, *Inorg. Chem.*, 1989, **28**, 3660.
111. D. A. Palmer and H. Kelm, *J. Inorg. Nucl. Chem.*, 1978, **40**, 1095.
112. H. K. J. Powell, *Inorg. Nucl. Chem. Lett.*, 1972, **8**, 891.
113. T. W. Swaddle, *Can. J. Chem.*, 1977, **55**, 3166.
114. L. Randaccio, S. Geremia, R. Dreos-Garlatti, G. Tauzher, F.Asaro, and G. Pellizer, *Inorg. Chim. Acta*, 1992, **194**, 1; L. Seibles and E. Deutsch, *Inorg. Chem.*, 1977, **16**, 2273.
115. See, for example, A. J. Thirst and D. H. Vaughan, *J. Inorg. Nucl. Chem.*, 1981, **43**, 2889; M. M. Muir, and L. M. Torres, *Inorg. Chim. Acta*, 1989, **164**, 33.
116. A. Poë and C. Vuik, *J. Chem. Soc., Dalton Trans.*, 1976, 661.
117. S. Balt and A. Jelsma, *Inorg. Chem.*, 1981, **20**, 733.
118. P. C. Ford, *Coord. Chem. Rev.*, 1982, **44**, 61; P. C. Ford, *J. Chem. Ed.*, 1983, **60**, 829.

119. C. Kutal and A. W. Adamson, in *Comprehensive Coordination Chemistry*, Vol. 1, eds G. Wilkinson, R. D. Gillard, and J. A. McCleverty, Pergamon, Oxford, 1987, Chapter 7.3.3.1.2.
120. R. E. Mahnken, M. A. Billadeau, E. P. Nikonowicz, and H. Morrison, *J. Am. Chem. Soc.*, 1992, **114**, 9253.
121. W. Weber, R. van Eldik, H. Kelm, J. DiBenedetto, Y. Ducommun, H. Offen, and P. C. Ford, *Inorg. Chem.*, 1983, **22**, 623.
122. T. L. Kelly and J. F. Endicott, *J. Am. Chem. Soc.*, 1972, **94**, 278; *J. Phys. Chem.*, 1972, **76**, 1937; values from Table 3 of Ref. 97.
123. See p. 357 of Ref. 85.
124. L. H. Skibsted, *Inorg. Chem.*, 1985, **24**, 3791; L. H. Skibsted, *Acta Chem. Scand.*, 1988, **42A**, 189.
125. L. G. Vanquickenborne and A. Ceulemans, *Coord. Chem. Rev.*, 1983, **48**, 157.
126. For example, S. F. Clark and J. D. Petersen, *Inorg. Chem.*, 1981, **20**, 280; J. D. Petersen, *Inorg. Chem.*, 1981, **20**, 3123.
127. R. M. Carlos, M. E. Frink, E. Tfouni, and P. C. Ford, *Inorg. Chim. Acta*, 1992, **193**, 159.
128. L. Mønsted, O. Mønsted, and L. H. Skibsted, *Acta Chem. Scand.*, 1989, **43**, 128.
129. H. Taube, *Pure Appl. Chem.*, 1991, **63**, 651.
130. W. D. Harman, D. P. Fairlie, and H. Taube, *J. Am. Chem. Soc.*, 1986, **108**, 8223.
131. W. D. Harman, M. Sekine, and H. Taube, *J. Am. Chem. Soc.*, 1988, **110**, 5725.
132. C.-K. Poon, C.-M. Che, and Y.-P. Kan, *J. Chem. Soc., Dalton Trans.*, 1980, 128; C.-K. Poon, T.-C. Lau, C.-L. Wong, and Y.-P. Kan, *J. Chem. Soc., Dalton Trans.*, 1983, 1641; C.-K. Poon and T.-C. Lau, *Inorg. Chem.*, 1983, **22**, 1664.
133. M. T. Fairhurst and T. W. Swaddle, *Inorg. Chem.*, 1979, **18**, 3241.
134. J. A. Marchant, T. Matsubara, and P. C. Ford, *Inorg. Chem.*, 1977, **16**, 2160.
135. For example, M. M. Doeff and D. A. Sweigart, *Inorg. Chem.*, 1981, **20**, 1683.
136. P. C. Ford and C. Sutton, *Inorg. Chem.*, 1969, **8**, 1544.
137. J. F. Endicott and H. Taube, *J. Am. Chem. Soc.*, 1962, **84**, 4984; P. C. Ford, J. R. Kuempel, and H. Taube, *Inorg. Chem.*, 1968, **7**, 1976; P. C. Ford, *Coord. Chem. Rev.*, 1970, **5**, 75.
138. M. L. Fetterolf and H. W. Offen, *Inorg. Chem.*, 1987, **26**, 1070.
139. J. A. Broomhead and L. Kane-Maguire, *Inorg. Chem.*, 1969, **8**, 2124.
140. J. A. Broomhead, F. Basolo, and R. G. Pearson, *Inorg. Chem.*, 1964, **3**, 826.
141. J. A. Broomhead and N. A. Pasha, *Transition Met. Chem.*, 1992, **17**, 209.
142. S. D. Pell and J. N. Armor, *J. Am. Chem. Soc.*, 1975, **97**, 5012.
143. V. V. Tatarchuk and A. V. Belyaev, *Koord. Khim.*, 1978, **4**, 1059.
144. P. A. Lay and W. D. Harman, *Adv. Inorg. Chem.*, 1991, **37**, 219.
145. K. Kalyanasundaram, *Coord. Chem. Rev.*, 1982, **46**, 159.
146. For example, B. Durham, J. V. Caspar, J. K. Nagle, and T. J. Meyer, *J. Am. Chem. Soc.*, 1982, **104**, 4803; T. K. Foreman, J. B. S. Bonilha, and D. G. Whitten, *J. Phys. Chem.*, 1982, **86**, 3436.
147. S. Tachiyashiki, H. Ikezawa, and K. Mizumachi, *Inorg. Chem.*, 1994, **33**, 623.
148. M. L. Fetterolf and H. W. Offen, *Inorg. Chem.*, 1987, **26**, 1070.
149. S. Tachiyashiki and K. Mizumachi, *Chem. Lett.*, 1989, 1153.
150. P. C. Ford, D. A. Chaisson, and D. H. Stuermer, *J. Chem. Soc., Chem. Commun.*, 1971, 530; D. S. Caswell and T. G. Spiro, *Inorg. Chem.*, 1987, **26**, 18.
151. R. Santana da Silva and E. Tfouni, *Inorg. Chem.*, 1992, **31**, 3313.

152. F. Blau, *Chem. Ber.*, 1888, **21**, 1077; *Monatsh.*, 1889, **10**, 375; *Monatsh.*, 1898, **19**, 647.
153. J. H. Baxendale and P. George, *Nature*, 1948, **162**, 777; P. Krumholz, *Nature*, 1949, **163**, 725; *Trans. Faraday Soc.*, 1950, **46**, 736; T. S. Lee, I. M. Kolthoff, and D. L. Leussing, *J. Am. Chem. Soc.*, 1948, **70**, 3596; J. Burgess and R. H. Prince, *J. Chem. Soc.*, 1963, 1097.
154. W. W. Brandt and D. K. Gullstrom, *J. Am. Chem. Soc.*, 1952, **74**, 3532.
155. J. Burgess and R. H. Prince, *J. Chem. Soc.*, 1963, 5752; *J. Chem. Soc.* (*A*), 1967, 431.
156. J.-M. Lucie, D. R. Stranks, and J. Burgess, *J. Chem. Soc., Dalton Trans.*, 1975, 245.
157. J. Burgess, S. A. Galema, and C. D. Hubbard, *Polyhedron*, 1991, **10**, 703; J. Burgess and C. D. Hubbard, *Comments Inorg. Chem.*, 1995, **17**, 283.
158. I. N. de Carvalho and M. Tubino, *J. Braz. Chem. Soc.*, 1991, **2**, 56.z.
159. F. Basolo, J. C. Hayes, and H. M. Neumann, *J. Am. Chem. Soc.*, 1954, **76**, 3807; and see pp. 218–220 of Ref. 7.
160. H. Ogino, Y. Orihara, and N. Tanaka, *Inorg. Chem.*, 1980, **19**, 3178.
161. G. F. Smith and F. P. Richter, *Phenanthroline and Substituted Phenanthroline Indicators—Their Preparation, Properties, and Applications to Analysis*, G. F. Smith Chemical Company, Columbus, Ohio, 1944; F. H. Case, *A Review of the Synthesis of Organic Compounds containing the Ferroin Group*, G. F. Smith Chemical Company, Columbus, Ohio, 1960; D. Blair and H. Diehl, *Analyt. Chem.*, 1961, **33**, 86; A. A. Schilt, *Analytical Applications of 1,10-Phenanthroline and Related Compounds*, Pergamon, Oxford, 1969; L. L. Stookey, *Analyt. Chem.*, 1970, **42**, 779.
162. P. Krumholz, *Inorg. Chem.*, 1965, **4**, 609; J. Burgess and R. H. Prince, *J. Chem. Soc. (A)*, 1967, 434; J. Burgess, *J. Chem. Soc. (A)*, 1968, 497; P. Krumholz, *Struct. Bonding*, 1971, **9**, 139.
163. F. P. Dwyer, N. S. Gill, E. C. Gyarfas, and F. Lions, *J. Am. Chem. Soc.*, 1957, **79**, 1269; J. Dekkers and H. A. Goodwin, *Austr. J. Chem.*, 1967, **20**, 69; E. R. Gardner, F. M. Mekhail, and J. Burgess, *Int. J. Chem. Kinet.*, 1974, **6**, 133; J. Burgess and G. M. Burton, *Revista Latinoamer. Quim.*, 1980, **11**, 107.
164. S. C. Jackels, J. Zektzer, and N. J. Rose, *Inorg. Synth.*, 1977, **17**, 139; V. L. Goedken, *Inorg. Synth.*, 1980, **22**, 87; A. Al-Alousy and J. Burgess, *Transition Met. Chem.*, 1987, **12**, 565; M. J. Blandamer, J. Burgess, J. Fawcett, S. Radulović, and D. R. Russell, *Transition Met. Chem.*, 1988, **13**, 120.
165. D. W. Margerum and L. P. Morgenthaler, *J. Am. Chem. Soc.*, 1962, **84**, 706.
166. G. Nord, *Comments Inorg. Chem.*, 1976, **4**, 15, 1921; O. Farver, O. Mønsted, and G. Nord, *J. Am. Chem. Soc.*, 1979, **101**, 6118; N. Serpone, G. Ponterini, M. A. Jamieson, F. Bolletta, and M. Maestri, *Coord. Chem. Rev.*, 1983, **50**, 209; R. D. Gillard, *Coord. Chem. Rev.*, 1983, **50**, 303; E. C. Constable, *Polyhedron*, 1983, **2**, 551.
167. R. D. Gillard, *Inorg. Chim. Acta*, 1974, **11**, L21; R. D. Gillard, *Coord. Chem. Rev.*, 1975, **16**, 67.
168. M. J. Blandamer, J. Burgess, P. P. Duce, K. S. Payne, R. Sherry, P. Wellings, and M. V. Twigg, *Transition Met. Chem.*, 1984, **9**, 163.
169. A. Gameiro, R. D. Gillard, M. M. Rashad Bakhsh, and N. H. Rees, *J. Chem. Soc., Chem. Commun.*, 1996, 2245, and references therein.
170. J. Burgess, A. J. Duffield, and R. Sherry, *J. Chem. Soc., Chem. Commun.*, 1980, 350.

171. J. Burgess and C. D. Hubbard, *J. Am. Chem. Soc.*, 1984, **106**, 1717; J. Burgess and C. D. Hubbard, *Inorg. Chem.*, 1988, **27**, 2548; M. J. Blandamer, J. Burgess, J. Fawcett, P. Guardado, C. D. Hubbard, S. Nuttall, L. J. S. Prouse, S. Radulović, and D. R. Russell, *Inorg. Chem.*, 1992, **31**, 1383.
172. Reviews: D. H. Macartney, *Rev. Inorg. Chem.*, 1988, **9**, 101; G. Stochel, *Coord. Chem. Rev.*, 1992, **114**, 269.
173. E. A. Abu-Gharib, Razak bin Ali, M. J. Blandamer, and J. Burgess, *Transition Met. Chem.*, 1987, **12**, 371.
174. There is a possibility of misinterpretation for a single feeble nucleophile reacting with a complex—a sloping straight line for the k_{obs} versus nucleophile concentration plot may result from a *D* mechanism or from associative attack.
175. J. M. Malin, H. E. Toma, and E. Giesbrecht, *J. Chem. Ed.*, 1977, **54**, 385.
176. E. A. Abu-Gharib, Razak bin Ali, M. J. Blandamer, and J. Burgess, *Transition Met. Chem.*, 1987, **12**, 371.
177. M. J. Blandamer, J. Burgess, and R. I. Haines, *J. Chem. Soc., Dalton Trans.*, 1978, 244.
178. N. E. Katz, P. J. Aymonino, M. A. Blesa, and J. A. Olabe, *Inorg. Chem.*, 1978, **17**, 556; D. H. Macartney and A. McAuley, *Can. J. Chem.*, 1981, **20**, 748; K. J. Pfenning, Liangshiu Lee, H. D. Wohlers, and J. D. Petersen, *Inorg. Chem.*, 1982, **21**, 2477; A. Yeh and A. Haim, *J. Am. Chem. Soc.*, 1985, **107**, 369; D. H. Macartney and L. J. Warrack, *Can. J. Chem.*, 1989, **67**, 1774; Gyu-Hwan Lee, L. Della Ciani, and A. Haim, *J. Am. Chem. Soc.*, 1989, **111**, 2535; K. B. Reddy and R. van Eldik, *Inorg. Chem.*, 1991, **30**, 596; S. Alshehri, J. Burgess, R. van Eldik, and C. D. Hubbard, *Inorg. Chim. Acta*, 1995, **240**, 305, and references therein.
179. G. Stochel and R. van Eldik, *Inorg. Chim. Acta*, 1991, **190**, 55.
180. I. de Sousa Moreira and D. W. Franco, *Inorg. Chem.*, 1994, **33**, 1607, and references therein.
181. A. Haim and W. K. Wilmarth, *Inorg. Chem.*, 1962, **1**, 573; A. Haim, R. J. Grassi, and W. K. Wilmarth, *Adv. Chem. Ser.*, 1965, **49**, 31.
182. A. Haim, *Inorg. Chem.*, 1982, **21**, 2887.
183. M. G. Burnett and W. M. Gilfillan, *J. Chem. Soc., Dalton Trans.*, 1981, 1578; M. H. M. Abou-El-Wafa, M. G. Burnett, and J. F. McCullagh, *J. Chem. Soc., Dalton Trans.*, 1986, 2083; K. H. Halawani and C. F. Wells, *Int. J. Chem. Kinet.*, 1992, **24**, 1043.
184. H.-T. Macholdt and H. Elias, *Inorg. Chem.*, 1984, **23**, 4315; H.-T. Macholdt and H. Elias, *Transition Met. Chem.*, 1985, **10**, 323; J. Burgess and A. E. Smith, *Transition Met. Chem.*, 1987, **12**, 140.
185. J. E. Byrd and W. K. Wilmarth, *Inorg. Chim. Acta Rev.*, 1971, **5**, 7; W. K. Wilmarth, J. E. Byrd, H. N. Po, H. K. Wilcox, and P. H. Tewari, *Coord. Chem. Rev.*, 1983, **51**, 181; W. K. Wilmarth, J. E. Byrd, and H. N. Po, *Coord. Chem. Rev.*, 1983, **51**, 209; M. G. Burnett, *Chem. Soc. Rev.*, 1983, **12**, 267; M. H. M. Abou-El-Wafa and M. G. Burnett, *Polyhedron*, 1984, **3**, 895.
186. J. G. Leipoldt, W. Purcell, and H. Meyer, *Polyhedron*, 1991, **10**, 1379.
187. A. D. Kirk and D. M. Kneeland, *Inorg. Chem.*, 1995, **34**, 1536.
188. N. Y. M. Iha and J. F. de Lima, *Inorg. Chem.*, 1991, **30**, 4576.
189. P. Ascenzi, C. Ercolani, and F. Monacelli, *Inorg. Chim. Acta*, 1994, **219**, 199.
190. Review: D. V. Stynes, *Pure Appl. Chem.*, 1988, **60**, 561.
191. For example, M. Landergren and L. Baltzer, *J. Chem. Soc., Perkin Trans. II*, 1992, 355; F. A. Walker, U. Simonis, H. Zhang, J. M. Walker, T. M. Ruscitti,

C. Kipp, M. A. Amputch, B. V. Castillo, S. H. Cody, D. L. Wilson, R. E. Graul, G. J. Yong, K. Tobin, J. T. West, and B. A. Barichievich, *New J. Chem.*, 1992, **16**, 609; G. Levey, D. A. Sweigart, J. G. Jones, and A. L. Prignano, *J. Chem. Soc., Dalton Trans.*, 1992, 605.

192. For example, M. Momenteau, *Pure Appl. Chem.*, 1986, **58**, 1493.
193. Z. Bradić, P. C. Wilkins, and R. G. Wilkins, *Recl. Trav. Chim. Pays-Bas*, 1987, **106**, 174; J. Springborg, P. C. Wilkins, and R. G. Wilkins, *Acta Chem. Scand.*, 1989, **43**, 967.
194. T. G. Traylor, D. Magde, Jikun Luo, K. N. Walda, D. Bandyopadhyay, Guo-Zhang Wu, and V. S. Sharma, *J. Am. Chem. Soc.*, 1992, **114**, 9011.
195. M. J. Blandamer, J. Burgess, S. J. Hamshere, R. D. Peacock, J. H. Rogers, and H. D. B. Jenkins, *J. Chem. Soc., Dalton Trans.*, 1981, 726; D. H. Devia and A. G. Sykes, *Inorg. Chem.*, 1981, **20**, 910; M. Åberg and J. Glaser, *Inorg. Chim. Acta*, 1993, **206**, 53.
196. J. Szalma, V. I. Kravtsov, L. Ya. Smirnova, and G. P. Tsayun, *Magy. Kem. Foly.*, 1980, **86**, 293 (*Chem. Abstr.*, 1980, **93**, 121 171*k*).
197. W. Preetz and H.-D. Zerbe, *Z. Anorg. Allg. Chem.*, 1981, **479**, 7.
198. H. Schulz and W. Preetz, *Z. Anorg. Allg. Chem.*, 1982, **490**, 55.
199. M. M. Taqui Khan, G. Ramachandraiah, and R. S. Shukla, *Polyhedron*, 1992, **11**, 3075.
200. J. Burgess, B. A. Goodman, and J. B. Raynor, *J. Chem. Soc. (A)*, 1968, 501.
201. G. Mestroni, E. Alessio, M. Calligaris, W. M. Attia, F. Quadrifoglio, S. Cauci, G. Sava, S. Zorzet, S. Pacor, C. Monti-Bragadin, M. Tamaro, and L. Dolzani, *Progr. Clin. Biochem. Med.*, 1989, **10**, 71; E. Alessio, G. Balducci, M. Calligaris, G. Costa, W. M. Attia, and G. Mestroni, *Inorg. Chem.*, 1991, **30**, 609.
202. Reviews: W. R. Mason, *Coord. Chem. Rev.*, 1972, **7**, 241; A. Peloso, *Coord. Chem. Rev.*, 1973, **10**, 123.
203. R. C. Johnson, F. Basolo, and R. G. Pearson, *J. Inorg. Nucl. Chem.*, 1962, **24**, 59; A. A. Grinberg and A. A. Korableva, *Russ. J. Inorg. Chem.*, 1966, **11**, 409.
204. C. E. Skinner and M. M. Jones, *J. Am. Chem. Soc.*, 1969, **91**, 1984.
205. T. G. Appleton, R. D. Berry, J. R. Hall, and J. A. Sinkinson, *Inorg. Chem.*, 1991, **30**, 3860; G. Frommer, H. Preut, and B. Lippert, *Inorg. Chim. Acta*, 1992, **193**, 111.
206. M. Schmidt and G. G. Hoffmann, *Z. Anorg. Allg. Chem.*, 1979, **452**, 112.
207. For example, H. Köpf and P. Köpf-Maier, *Structure Bond.*, 1988, **70**, 103.
208. J. H. Toney and T. J. Marks, *J. Am. Chem. Soc.*, 1985, **107**, 947.
209. Razak bin Ali, J. Burgess, and A. T. Casey, *J. Organomet. Chem.*, 1989, **362**, 305; J. Burgess and S. A. Parsons, *Polyhedron*, 1993, **12**, 1959; J. Burgess and S. A. Parsons, *Appl. Organomet. Chem.*, 1993, **7**, 343.
210. A. Samotus, A. Kanas, W. Glug, J. Szklarzewicz, and J. Burgess, *Transition Met. Chem.*, 1991, **16**, 614.
211. J. G. Leipoldt, S. S. Basson, A. Roodt, and W. Purcell, *Polyhedron*, 1992, **11**, 2277.
212. A. Roodt, J. G. Leipoldt, L. Helm, A. Abou-Hamdan, and A. E. Merbach, *Inorg. Chem.*, 1995, **34**, 560.
213. J. Lu and M. J. Clarke, *Inorg. Chem.*, 1989, **28**, 2315; L. Helm, K. Deutsch, E. A. Deutsch, and A. E. Merbach, *Helv. Chim. Acta*, 1992, **75**, 210.
214. L. Moggi, F. Bolletta, V. Balzani, and F. Scandola, *J. Inorg. Nucl. Chem.*, 1966, **28**, 2589.
215. B. Sieklucka, *Progr. React. Kinet.*, 1989, **15**, 175.

216. B. Sieklucka and A. Samotus, *J. Inorg. Nucl. Chem.*, 1980, **42**, 1003.
217. S. I. Ali and H. Kaur, *Transition Met. Chem.*, 1991, **16**, 450; 1992, **17**, 304; *J. Photochem. Photobiol. A*, 1991, **61**, 183; 1992, **68**, 147; J. Szklarzewicz and A. Samotus, *Transition Met. Chem.*, 1995, **20**, 174; S. I. Ali and K. Majid, *Transition Met. Chem.*, 1997, **22**, 53.
218. W. U. Malik, S. P. Srivastava, K. K. Thallam, and V. K. Gupta, *Acta Chim. Acad. Sci. Hung.*, 1982, **109**, 345; V. K. Gupta, A. Kumar, and P. Singh, *Proc. Indian Natl. Sci. Acad., Part A*, 1991, **57**, 485.
219. S. Raman, *J. Inorg. Nucl. Chem.*, 1981, **43**, 1855; K. M. Rao, T. S. Reddy, and S. B. Rao, *Analyst*, 1988, **113**, 983; D. Sicilia, S. Rubio, and D. Perez Bendito, *Talanta*, 1991, **38**, 1147, and references therein.

5 Stereochemical change

Chapter contents

5.1 Introduction

Until about forty years ago the observation and study of stereochemical change required systems that reacted slowly because it was necessary to start with a single stereoisomer and observe the changes that took place. Whether one used optical activity (rotation or circular dichroism) changes, as in the case of tetrahedral carbon compounds and dissymmetric octahedral complexes, or ultraviolet–visible spectrophotometry (a technique only generally available since 1950) when studying *cis* ↔ *trans* changes in octahedral or square-planar complexes, it was always necessary to start well away from equilibrium. All of these techniques require the presence of a stereochemical 'signpost' in the complex. With sufficiently elaborate signposting (generally requiring a more elaborate means of investigation) a large amount of stereochemical information could be gained about the changes being observed. In recent years we have seen the development of chromatographic techniques to separate isomers and thereby improve the analysis of the data, but this type of approach is strictly limited.

The study of stereochemical change has changed dramatically as a result of the development of the techniques of nuclear magnetic resonance (NMR). Not only does this provide a means of assigning structures to compounds in solution but it also allows systems to be examined at equilibrium. When the exchange of magnetic nuclei between non-equivalent sites takes place at a rate that is within the time-scale of NMR, i.e. the frequency with which the resonating nucleus exchanges its sites is of a comparable magnitude to the difference (in Hz) of the chemical shifts of the nucleus in the different environments, the kinetics of the process can be studied. This technique is encountered in several chapters in connection with various fast reactions. It is also possible to tell, from the presence or absence of spin–spin coupling, whether or not the bonds between the coupled nuclei remain intact during the site-exchange process. Rearrangements that accompany ligand exchange therefore, in favourable circumstances, can be distinguished from a truly intramolecular process in which all bonds remain intact.

5.2 Classification of stereochemical change

The ways in which ligands can change their relative positions in the coordination shell can be listed in order of increasing complexity, as is done in the following sections.

5.2.1 Pseudorotation[1]

This process involves no change in the coordination number, and all the metal–ligand bonds remain intact. The transition state is generally another geometry of the same coordination number and the process can be looked upon as a vibration that has sufficient energy to carry the molecule over this energy barrier into another minimum. It is found in its simplest form in homoleptic molecules (molecules in which all ligands are chemically identical) as, for example, in the case of PF_5. Here the ^{31}P and ^{19}F NMR spectra are consistent with all five fluorines being in magnetically equivalent positions, even though the ground-state structure is trigonal bipyramidal with two axial and three equatorial fluorines. Discussion of these processes will form a major part of this chapter and will be subdivided according to the coordination number.

5.2.2 Pseudorotation in an intermediate of higher or lower coordination number

To some extent, this category overlaps with Section 5.2.3, but it is possible to make a distinction between stereochemical changes that fall into this category and those that are the direct consequence of the geometric requirements of substitution.

As will be seen below, the barrier to pseudorotation in four- and six-coordinate complexes tends to be particularly high, whereas in three- and five-coordinate complexes this barrier is frequently low. Many cases have now been documented in which the barriers to a change in coordination number (loss or gain of a ligand) are small enough to allow pseudorotation of an intermediate to be a favoured pathway for stereochemical change. There are many well-documented examples in the isomerization reactions of four-coordinate planar d^8 metal-ion complexes.

5.2.3 Isomerization due to ligand substitution

When the geometry of the transition state or intermediate in the act of ligand substitution is such that the relative positions of the ligand with respect to a signpost can change there will be some ligand rearrangement. This is clearly an integral part of the act of substitution, since it is a direct consequence of the geometrical properties of the act of substitution and it has been considered in that part of this book devoted to the process. If ligands are exchanged with others of the same type then this will appear as an isomerization or a site-exchange process.

Although it is not always possible to devise the definitive experiment, reactions of the above two categories can frequently be distinguished. For example, the racemization of *sec*-octyl iodide in the presence of iodide ions has been shown to be the consequence of passage through the trigonal bipyramidal transition state with the entering and leaving groups occupying axial positions, and the racemization of aliphatic carbon compounds undergoing dissociative substitution is due to the planar nature of the three-coordinate intermediate rather than pseudorotation of a pyramidal carbenium ion.

Sometimes it is far less easy to decide. Aquation of *trans*-$[Co(en)_2LCl]^{n+}$, where Cl is replaced by water and L is the stereochemical signpost, results in stereochemical change when L = OH, Cl, Br, NCS, RCO_2, or py, but complete retention when L = NH_3, NO_2, or CN. The arguments in favour of a trigonal bipyramidal intermediate when L is a potential π-donor have been discussed in Chapter 4. Nevertheless, it has been shown, by calculation that those ligands that lead to stereochemical change are also those that lead to the smallest barrier to pseudorotation in the five-coordinate intermediate,[2] the implication being that stereochemical change is determined after bond fission. The *cis* ↔ *trans* isomerization and racemization of aqua complexes of the specific type $[Co(en)_2L(H_2O)]^{(n+1)+}$ produce an interesting problem. Clearly exchange of H_2O with H_2O, should have a similar steric course to the replacement of Cl by H_2O, but how do the complexes where L = NH_3, NO_2, CN, or H_2O isomerize? Water exchange in the amine complex is about one hundred times faster than *trans* ↔ *cis* and *D* ↔ *L* changes;[3] the behaviour of $[Co(en)_2(H_2O)_2]^{3+}$ is very similar. In the latter case it was shown that the activation parameters for water exchange are completely different from those for isomerization. For the exchange

$$trans\text{-}[Co(en)_2(H_2O)_2]^{3+} + H_2{}^*O \longrightarrow trans\text{-}[Co(en)_2(H_2O)(H_2{}^*O)]^{3+} + H_2O$$

$k_{25} = 1.1 \times 10^{-5}\,s^{-1}$, $\Delta H^\ddagger = 109\,kJ\,mol^{-1}$, $\Delta S^\ddagger = +17\,J\,K^{-1}\,mol^{-1}$, and $\Delta V^\ddagger = +5.9\,cm^3\,mol^{-1}$,[4,5] whereas for the isomerization

$$trans\text{-}[Co(en)_2(H_2O)_2]^{3+} \longrightarrow cis\text{-}[Co(en)_2(H_2O)_2]^{3+}$$

$k_{25} = 7 \times 10^{-6}\,s^{-1}$, $\Delta H^{\ddagger} = 130\,kJ\,mol^{-1}$, $\Delta S^{\ddagger} = +105\,J\,K^{-1}\,mol^{-1}$, and $\Delta V^{\ddagger} = +14.3\,cm^3\,mol^{-1}$.[3,6]

The inference here is that the main pathway for substitution is different from that of isomerization. Water exchange takes place through a transition state in which the five donors that are not involved retain the same positions relative to one another throughout the act of reaction (the positive $\Delta V^{\ddagger}$ suggesting that the entering water is not involved), while the isomerization requires a subsequent rearrangement, presumably through a trigonal bipyramidal entity. There is more than one way to achieve this. Thus, for example, the rearrangement may be the result of an alternative pathway for water substitution involving a trigonal bipyramidal intermediate in a *D* process, or it may be the result of a relatively low probability intramolecular rearrangement of the square pyramidal intermediate involved in water exchange (through a trigonal bipyramidal transition state) that takes place before it is trapped by the capture of water. Experiments to distinguish between these possibilities have yet to be conceived. It is unlikely, in this instance, that the isomerization occurs while all six donors are attached to the cobalt and yet both $[Co(en)_2(H_2O)(OH)]^{2+}$ and $[Co(en)_2(NH_3)(OH)]^{2+}$ isomerize much faster than the corresponding aqua complexes.[2,3] In the former case, the water ligand is labile and exchange with stereochemical change is likely, but with *trans*-$[Co(en)_2(NH_3)(OH)]^{2+}$ there is no exchange between coordinated ^{18}O and $H_2{}^{18}O$.[2] The rate of isomerization of *trans*-$[Co(en)_2(NO_2)(H_2O)]^{2+}$ is independent of pH, even when the complex is almost entirely in the form of the hydroxo species, i.e. the aqua-nitro and hydroxo-nitro complexes isomerize at exactly the same rate.[7] The suggestion that isomerization involves ring opening is supported by the volumes of activation determined for these reactions.[8] That the *cis* ⟶ *trans* isomerization of *cis*-$[Co(en)_2(SO_3)(H_2O)]^{+}$ is suppressed in basic solution[9] (where the hydroxo complex is formed) does not contradict such a mechanism because, although the very strong *trans* effect of SO_3 might be expected to facilitate ring opening *trans* to $-SO_3$, such reactions are highly stereospecific, and ring opening and closing would not lead to stereochemical change.

A number of ternary metal(IV) complexes MX_4L_2, where X is a halide, L a Lewis base, exist in solution as a mixture of *cis* and *trans* isomers. There is clearly a question as to whether isomerization in such systems is intra- or intermolecular, especially in view of kinetic studies of solvent exchange at this type of system where L = donor solvent (cf. Chapter 7), suggesting the possibility of a facile intermolecular pathway for isomerization. However, comparisons of kinetic parameters for exchange and isomerization favour a difference in activation modes. Thus both for $SnCl_4(SMe_2)_2$ and for $TiCl_4(tmp)_2$ (tmp = trimethylphosphate, $OP(OMe)_3$), both $\Delta S^{\ddagger}$ and $\Delta V^{\ddagger}$ are large and positive for exchange of SMe_2 or tmp, but have much smaller, though still positive, values for isomerization.[10] Lewis-base exchange is dissociative, probably *D*, while isomerization takes place by an intramolecular twist in a transition state somewhat larger than the initial state. In the case of $ZrCl_4(tmp)_2$, solvent exchange is now associative, but isomerization is again believed to involve an intramolecular twist. As the rate constant for isomerization increases slightly with increase in pressure, the transition state must be slightly more compact than the initial state. These mechanistic differences between the zirconium complex and the others can be rationalized in terms of the larger size of zirconium.[11]

5.3 Inversion, pseudorotation, and coordination number

5.3.1 A general consideration of pseudorotation and consequent topological changes

The act of pseudorotation is conveniently discussed in terms of what we might call *complementary geometries*. Certain bond angles in the parent compound change until the coordination shell achieves its complementary geometry, thereby forming the transition state or possibly an intermediate for the act of rearrangement. Further motion regains the original geometry but may well lead to a change in the relative positions of the donor atoms. The concept of complementary geometry has been introduced because, in appropriate circumstances, the geometries can be reversed. For example, a tetrahedral species may invert through a square-planar transition state or intermediate, while a square-planar species may isomerize through a tetrahedral one.

The ease with which these transformations can occur will depend upon the energy difference between the complementary geometries (of the ground state and the transition state), which constitutes the barrier to the process.[12] When the energies of the complementary geometries are close, either because of the nature of the ligands or because of the inherent properties of the geometries (this is particularly true for coordination number five and for coordination numbers greater than six), the barrier to pseudorotation will be low and the process will be fast at all but the lowest temperatures.

In systems where the geometry provides non-equivalent sites it is possible to gain considerable information from very simple homoleptic species. For example, the axial and equatorial positions in trigonal bipyramidal PF_5 afford different magnetic environments and, in principle, this should be reflected in both the ^{31}P and ^{19}F NMR spectra were it not for the fact that site exchange is too fast. It is therefore necessary to use heteroleptic systems to slow down the processes and to provide more information about the detailed mechanism. In geometries where all sites are symmetrically equivalent, e.g. the octahedron, it is always necessary to label sites with different ligands. In the extreme case we can label the N sites with N different ligands but, while it is a standard undergraduate exercise to work out the number of stereoisomers for compounds of the type $[ML_1L_2L_3 \ldots L_N]$, the actual synthesis, separation, and identification of a complete set of isomers for any system with $N > 4$ is a vanishingly rare achievement. Anyone approaching a funding agency with a proposal to carry out such a research would rightly be given short shrift. As will be seen below, the way to tackle the problem is to choose a system so as to maximize the amount of stereochemical information that is gained from the minimum stereochemical complexity. The use of bi- and multidentate ligands is of great value in this area.

When such mixed-ligand complexes are used, account must be taken of the differences in the energies of the various stereoisomers. This may be due to electronic or steric reasons and, in systems where the barrier to pseudorotation is small, may be the major factor determining the rate of pseudorotation. This is particularly true in five-coordinate systems, where the eventual site exchange might require the passage through isomeric intermediates of the same geometry, but with unfavourable ligand distributions. This difference in energy of the isomers must be added to the pseudorotation barrier in order to get the barrier for site exchange. Obviously for pseudorotation to be observed, the barrier must be smaller than the barrier for bond fission, otherwise processes resulting from such bond fission will dominate the behaviour.

5.3.2 Inversion in three-coordinate systems

Three different ligands can be arranged about a pyramidal three-coordinate centre in two enantiomeric ways. The intramolecular inversion of these enantiomers is probably the simplest pseudorotation to visualize and is the inversion of the pyramid through flattening to a trigonal-planar transition state (the complementary geometry, Fig. 5.1). Such a process is an important feature in the stereochemistry of three-coordinate pyramidal nitrogen. The part of the vibrational spectrum of NH_3 assigned to the symmetric bend (ν_2 mode) is consistent with a doubling of energy levels, because the degeneracy produced by the two equivalent potential wells is removed by the ability of the molecule to tunnel from one side of the barrier to the other (Fig. 5.2).[13] The estimation of barriers from vibrational frequencies and bond parameters was described three decades ago for forty-six pyramidal XY_3 molecules,[14] with a slightly later extensive review and tabulation of data[15] giving a detailed account of inversions of three-coordinate pyramidal main-group elements (C, N, P, As, S, Se, etc.). Even when the ligands are much more massive than hydrogen, inversion can be very rapid. This is why nobody has yet been able to resolve an amine with three different substituents, the separation of the isomers being a much slower process than their rearrangement. ^{1}H NMR line broadening of the spectrum of dibenzylmethylamine is observed at −135 °C, where the inversion starts to become slow on the NMR time-scale. Inversion at nitrogen can be slowed further by making it part of a small ring, as for example in substituted aziridines, **1**,[16] where rate constants are often on the

NMR time-scale at around room temperature. The aziridine derivative *N*-chloroazabicycloheptane, **2**, can even be resolved into isomers (invertomers).[17]

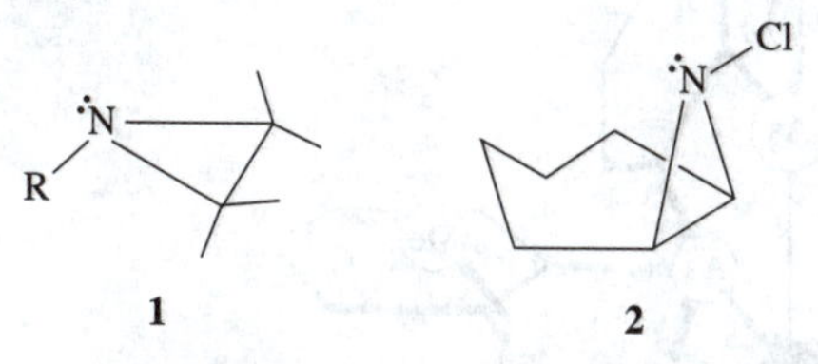

Pyramidal phosphorus[18] and arsenic compounds possess much higher barriers to pseudorotation. Not only can these pyramidal molecules be resolved, but they remain optically stable towards racemization at room temperature. Barriers to inversion of 90 kJ mol^{-1} or more are common—for example, the inversion barriers for PMe^{n}PrPh[19] and for AsMeEtPh[20] are about 120 and 180 kJ mol^{-1}, respectively. Resolution is usually carried out when the central atom is tetrahedral with four different ligands, one of which is then removed stereospecifically. Often the fourth group is a metal

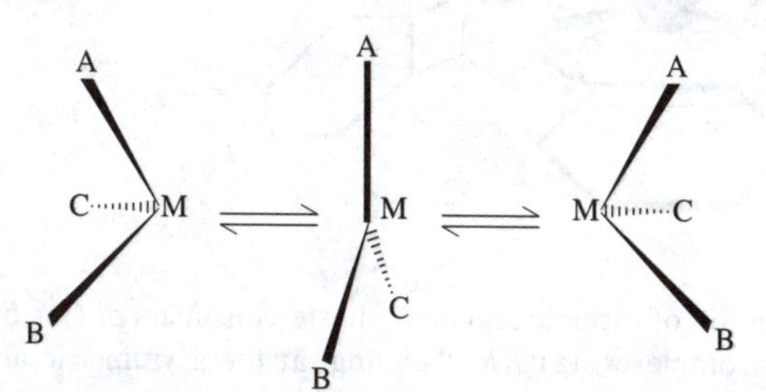

Fig. 5.1 Inversion of a trigonal pyramid by way of a planar complementary geometry.

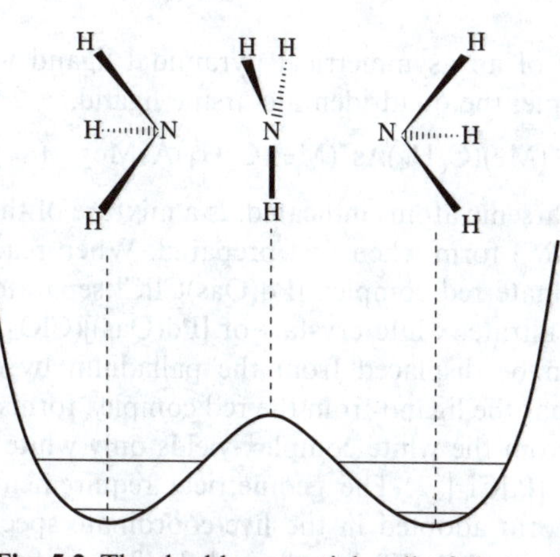

Fig. 5.2 The double potential well of the symmetrical bend of NH_3: rate constants of the magnitude 10^5 s^{-1} at 25 °C.

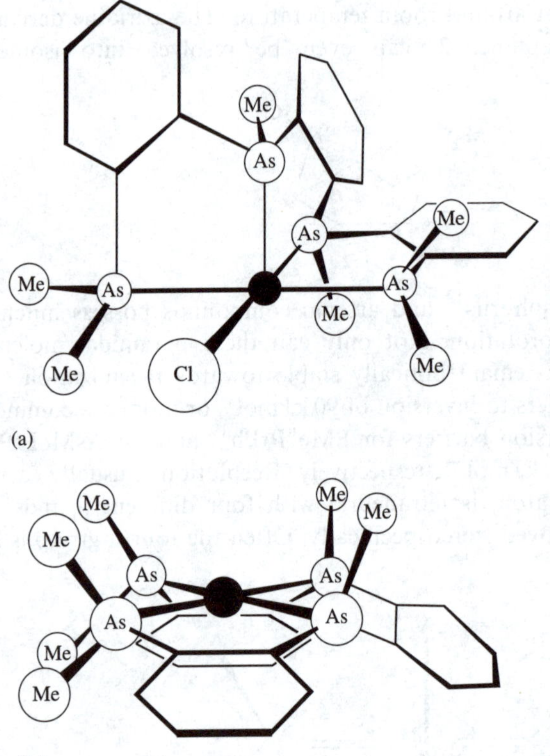

Fig. 5.3 Separation of racemic and *meso* diastereoisomers of Qas by means of their Pd(II) complexes. (a) *RR*: 'bending' at the asymmetric arsenic to accommodate the methyl groups does not favour a coplanar arrangement of four arsenics. (b) *RS*: Coplanar arrangement of four arsenics readily achieved.

ion, i.e. the resolution is of an asymmetrical pyramidal ligand when it is coordinated.[21] For example, the quadridentate arsine ligand,

$$Me_2As(C_6H_4)\mathbf{As}^*(Me)(C_6H_4)\mathbf{As}^*(Me)(C_6H_4)AsMe_2 \quad (= \mathit{Qas})$$

with the two asymmetric arsenic atoms indicated, is a mixture of the *RR*, *SS* racemate and the *meso* (*RS*) form when first prepared. When reacted with $[PdCl_4]^{2-}$, the five-coordinate red complex [Pd(Qas)Cl]Cl separates and, if $NaClO_4$ is added to the filtrate, white crystals of $[Pd(Qas)](ClO_4)_2$ can be obtained. The ligand can be displaced from the palladium by excess of cyanide, and it is found that the ligand from the red complex forms only red complex and the ligand from the white complex yields only white product, when reacted with fresh $[PdCl_4]^{2-}$. The geometrical requirements of the ligand make the folded form adopted in the five-coordinate species much more stable for the racemate (*RR*, *SS*) ligand, and the flat configuration is much more stable for the *meso* (*RS*) ligand (Fig. 5.3).[22] This is not a resolution but the separation of two diastereoisomers. The resolution of the *RR*, *SS* racemate should be achievable readily if an optically active ion is used in place of chloride. The experiment, however, demonstrates the optical stability of three-coordinate arsenic.

Racemization at pyramidal sulphur could occur by intramolecular inversion or by intermolecular S_N1 or S_N2 processes. Studies of inversion at the 1-adamantylethylmethylsulphonium cation, where neither S_N1 nor S_N2 processes are at all likely, and at the t-butylethylmethylsulphonium cation indicated the operation of simple intramolecular inversion.[23] Inversion in pyramidal sulphoxides, with barriers in the range 140–200 $kJ\,mol^{-1}$, is more difficult than in sulphonium cations. Separations of *meso* and racemic forms of chelating disulphoxides, e.g. $(C_6H_5)S(O)CH_2CH_2S(O)(C_6H_5)$, can be achieved by fractional crystallization of the mixture of the Pt(II) complexes $[Pt(S-S)Cl_2]$ (where the ligand is *S*, *S*-bonded), and here too the separated sulphoxides retain their chirality over very long periods of time. Much effort has been devoted to studying kinetics of inversion at sulphur coordinated to metals such as platinum—a simple and readable review of the main features of these studies is available.[24]

5.3.3 *Inversion and pseudorotation in four-coordinate systems*

The two complementary geometries of four-coordination are the tetrahedron and the plane and it is easy to visualize the inversion of a tetrahedral molecule by way of a planar transition state (Fig. 5.4). Here the angles between the pairs of ligands increases from 109.5° to 180°, at which point all four ligands are coplanar and further movement in the same direction will lead to the inverted tetrahedron. This vibration can be conveniently represented by making use of the fact that alternate corners of a cube occupy the same positions as the vertices of a tetrahedron and it is convenient to represent this inversion as a synchronous movement of the ligands along the appropriate cube edges. There are three ways of doing this, which give the three possible isomers of the four-coordinate planar [M(A)(B)(C)(D)] species. It is interesting to note that the inversion of a tetrahedron can be achieved by a single vibrational mode, whereas the isomerization of the planar species requires a change in the direction of the ligands and therefore the mediation of a tetrahedral intermediate.

In most areas of the Periodic Table the energy difference between the tetrahedral and planar forms is large. In a typical tetrahedral carbon compound the barrier to this intramolecular inversion is estimated to lie between 400 and 800 $kJ\,mol^{-1}$, so that it is much easier to break a bond than to invert the molecule in this way. It is possible to choose ligands to minimize this energy difference (to reverse it and synthesize a stable planar carbon complex would be quite an achievement) and thus obtain an

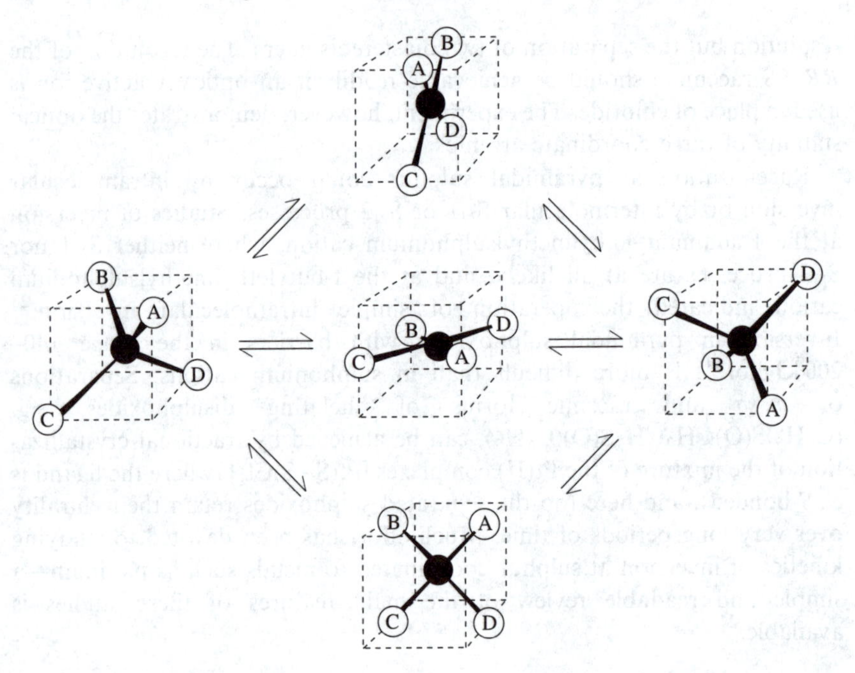

Fig. 5.4 Pathways for pseudorotation of tetrahedral and four-coordinate planar complexes.

asymmetric carbon compound that racemizes intramolecularly, but no one has yet succeeded.

In transition metal chemistry, especially of the elements of the first row, tetrahedral and square-planar complexes can approach each other in energy. There are many examples of low-spin d^8 four-coordinate complexes of Ni^{II} which are planar and of high-spin complexes which are tetrahedral.[25] In the case of $[Ni(PR_3)_2X_2]$ the balance point is reached when X = Br and the phosphine has both alkyl and aryl substituents. With $[Ni(PPh_2Me)_2Br_2]$ the isomeric form obtained depends upon the solvent from which it is crystallized. From n-butanol the green tetrahedral high-spin ($\mu = 3.20$ B.M.) allogon (allogons are substances that are isomeric through differences in their coordination geometry) is obtained, but from benzene it is the red planar diamagnetic allogon that separates on cooling. The existence, at equilibrium, of such a pair of complexes with mutually complementary geometries, whose rate of interconversion is not too fast to measure, indicates that in this case the transition state has a geometry that lies between that of the two complementary geometries. It is quite clear that the question of whether a process is synchronous (with a single transition state separating reagent and product) or non-synchronous (with two transition states and an intermediate) is as important in intramolecular rearrangements as it is in ligand substitution. In the case of the allogons of $[Ni(PR_3)_2X_2]$ the geometric change is also accompanied by a change in spin multiplicity, and the transition state may correspond to the spin crossover point.

The energy difference between favoured low-spin planar and unstable high-spin tetrahedral geometry becomes very much larger for the complexes of the second- and third-row elements, namely Pd(II) and Pt(II), and thermal activation of intramolecular rearrangement is unknown. However, photochemically catalysed isomerization is well known and it is not uncommon for the position of isomeric equilibrium in the excited state to be quite different from that of the ground state. Thus, irradiation of *cis*-$[Pt(Me_2SO)_2Cl_2]$ in dichloromethane gives significant amounts of the *trans* isomer, and the kinetics of the dark reaction back to the stable *cis* isomer can be studied.[26] In a similar way, *cis*-$[Pt(PEt_3)_2(C_6H_5)Cl]$ changes to the *trans* isomer in the dark but is re-formed upon irradiation.

The difference between first-row elements and second- and third-row elements is less marked for the copper–silver–gold triad. Stereochemical change in tetrahedral complexes of these metals in their 1+ oxidation state is very fast for monodentate ligands, but considerably slower for bis-ligand complexes containing bidentate phosphines such as $Ph_2PCH_2CH_2PEt_2$ (eppe), **3**. The copper(I), silver(I), and gold(I) complexes $[M(eppe)_2]^+$ undergo inversion by a mechanism involving intermediates containing monodentate eppe. The order of reactivity is of inversion rates increasing in the order $Cu^I < Au^I < Ag^I$. The various stereo forms of **3** and its copper(I) and silver(I) complexes provide further information on activation barriers and what controls them.[27]

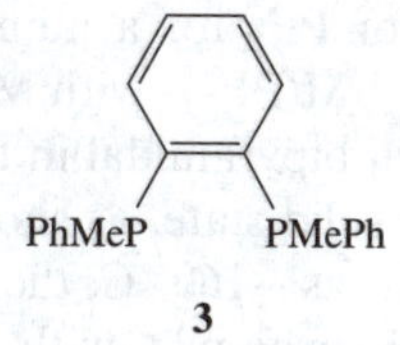

5.3.4 *Pseudorotation in five-coordinate systems*

The complementary geometries of five-coordinate species are usually energetically close to one another. Not only is this true for homoleptic trigonal bipyramidal (D_{3h}) and square pyramidal (C_{3v}) species but also for the distorted structures that lie between the two. It is interesting to look at Muetterties's compilation of the various possible distorted structures between the two extremes.[28] The stereochemical lability of five-coordinate species was noted and commented upon as soon as their NMR spectra were measured. In spite of the fact that electron diffraction in the gas phase indicates a trigonal bipyramidal geometry, the ^{19}F–^{31}P coupling pattern of the ^{31}P NMR spectrum of PF_5 and the ^{19}F spectrum itself indicate that all five fluorines are magnetically equivalent. The presence of the coupling tells us that the P–F bonds remain intact and so there is no rapid intermolecular fluorine exchange. On lowering the temperature there is no indication of line broadening, which would indicate that the site exchange is becoming slow enough to affect the spectrum, even at the point where all common solvents have started to freeze.[29] It is necessary to look at the solid-state NMR of a matrix at much lower temperature to see the resolution. To slow pseudo-rotation at trigonal bipyramidal phosphorus sufficiently to observe separate signals corresponding to axial and equatorial positions, it is necessary to utilize such factors as ligand bulk, electronic effects, and constraints imposed by chelate-ring geometry, as for instance in **4**,[30] to make inversion considerably more difficult (see below).

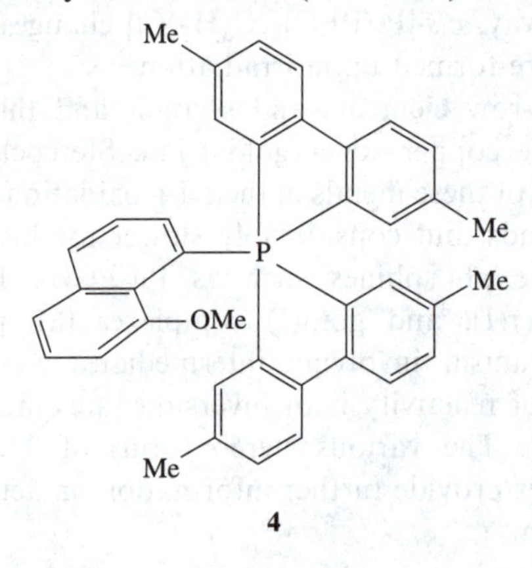

4

The situation is similar to that just described for PF_5 for a number of other simple D_{3h} five-coordinate species, e.g. AsF_5,[31] $M(PF_3)_5$ with M = Fe, Ru, or Os,[32] and $Fe(CO)_5$. Thus $Fe(CO)_5$ is trigonal bipyramidal in the gas phase, as shown by electron diffraction, and in the solid state, as shown by NMR.[33] But in solution, even at temperatures as low as −160 °C, the NMR spectrum indicates equivalent axial and equatorial carbonyls, with persistence of $^{57}Fe-^{13}C$ coupling confirming intramolecular inversion. The barrier to this inversion has been estimated at 2–10 kJ mol^{-1}, which is extremely low. The only homoleptic species for which it has proved possible to observe multi-line NMR spectra corresponding to slow pseudorotation is the $[Rh\{P(OMe)_3\}_5]^+$ cation, and then only below −100 °C.[34]

The simplest rearrangement that can be drawn to account for the rapid exchange of ligands between axial and equatorial sites requires that one of the three ligands in the trigonal plane acts as a reference 'pivot' while the angle between the other two opens up from 120 to 180°. At the same time, the two axial ligands bend away from the pivot, decreasing their angle from 180 to 120°. At this point a pair of equatorial ligands have become axial and a pair of axial ligands have become equatorial. This is the *Berry twist*.[35] Between the two limits there is a point where the four moving ligands are coplanar and the geometry of the species (transition state or intermediate) is a square-based pyramid with the pivot ligand apical (Fig. 5.5). Any other

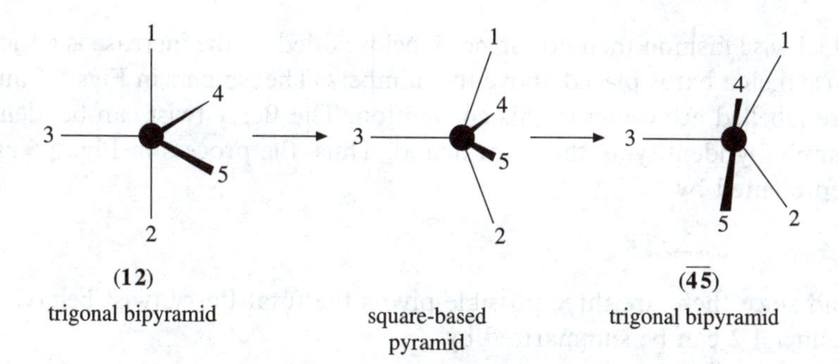

Fig. 5.5 The Berry twist to exchange the axial (1,2) ligands with a pair of equatorial ligands (4,5) using the third (3) as pivot.

process that leads to simultaneous exchange of the axial ligands with a pair of equatorial ligands is said to be permutationally equivalent and would not be experimentally distinguishable from the Berry twist.

A second permutational change involves the simultaneous exchange of a single axial ligand with a single equatorial ligand. One mechanism that has been postulated for this requires that an axial and an equatorial ligand remain fixed while the other three rotate in the manner of a turnstile.[36] In addition to exchanging an axial with an equatorial ligand, this rearrangement also inverts the sequence of ligands in the trigonal plane. This has been referred to as the *turnstile twist* and is depicted in Fig. 5.6. The Berry and turnstile twists have been reviewed and compared several times.[37]

Although there are no examples of trigonal bipyramidal species with five different ligands attached, it is instructive to examine the problem of rearrangements in trigonal bipyramidal substrates from this starting point. There are ten enantiomeric pairs of stereoisomers of the species $[M(L_1)(L_2)(L_3)(L_4)(L_5)]$. A particular diasteroisomer can be specified by listing the two axial ligands and the enantiomers can be distinguished by assigning a priority in much the same way as the Cahn–Ingold–Prelog rules[38] for the tetrahedron, according to the sequence in the trigonal plane. Thus, if we look down the axis in the direction of lower to higher priority and see the priority of the ligands in the trigonal plane increasing in a

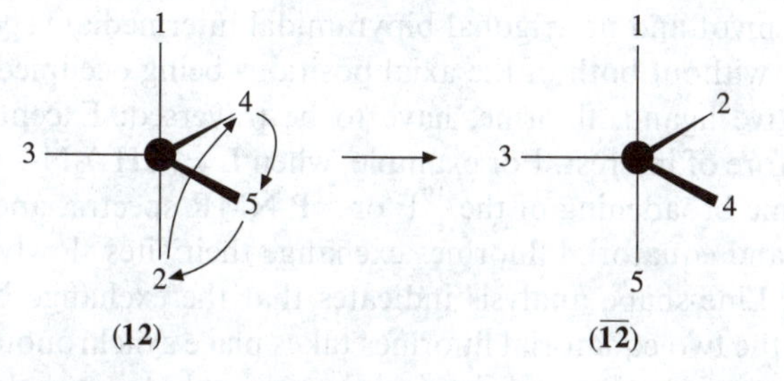

Fig. 5.6 The 'turnstile twist' to exchange one axial ligand (2) with one equatorial ligand (5) while reversing the sequence in the trigonal plane.

clockwise fashion then no further label is added. If the increase is anticlockwise then a bar is placed above the numbers. The isomers in Figs 5.5 and 5.6 are labelled according to this convention. The Berry twist can be identified simply by identifying the pivot ligand. Thus, the process in Fig. 5.5 can be represented by

$$\mathbf{1\,2} \overset{\mathbf{3}}{\longleftrightarrow} \overline{\mathbf{4\,5}}$$

and since there are three possible pivots the total Berry twist behaviour of isomer **1 2** can be summarized by

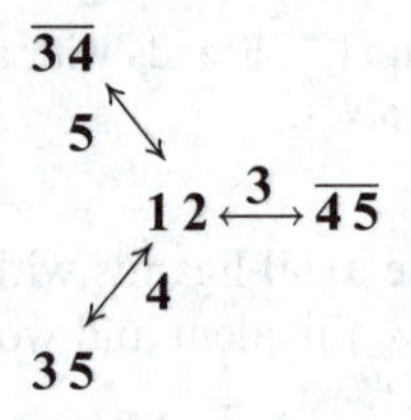

The whole system can be represented by a 20-vertex figure with a connectivity of three. Muetterties points out that in using such representations '. . . stereochemical problems can be solved at a glance, whereas paper and pencil solutions are laborious and errors are readily made.'[39] There are many ways of doing this but the figures are always interpenetrating. The Desargues–Levi map shown in Fig. 5.7 as used by Mislow[40] is quite useful. Enantiomers are located at opposite ends of diagonals (i.e. they are related by inversion through the centre). It can be seen by inspection that, since the figure is closed, all isomers can be interconverted by a series of Berry twists. An inversion, for example, requires a sequence of at least five successive twists and can follow a number of different pathways; the shortest closed cycle requires a sequence of six twists (a racemization cycle requires a sequence of at least ten twists). Readers familiar with the use of group theory will find their experience useful in using a diagram of this sort.

In homoleptic complexes (i.e. when all five ligands are chemically identical, e.g. PF_5 or $Fe(CO)_5$) the barrier to this twist is low and the examples studied are non-rigid even at the lowest temperatures studied. As a rule, compounds of the type $[P(L)F_4]$ are also non-rigid down to the lowest temperatures. This is because axial ↔ equatorial exchange can be achieved by using L as the pivot and no trigonal bipyramidal intermediate species of higher energy, i.e. without both of the axial positions being occupied by the most electronegative ligand, fluorine, have to be traversed. Exceptions to this rule are therefore of interest. For example, when $L = (CH_3)_2N-$ or RS– cooling leads to line broadening of the ^{19}F or ^{31}P NMR spectra, and below −70 °C the axial and equatorial fluorines exchange their sites slowly on the NMR time-scale. Line-shape analysis indicates that the exchange between the two axial and the two equatorial fluorines takes place synchronously and constitutes a proof of the Berry twist (or a process that is topologically equivalent to it).[41] It has been suggested that the higher barrier to twisting in

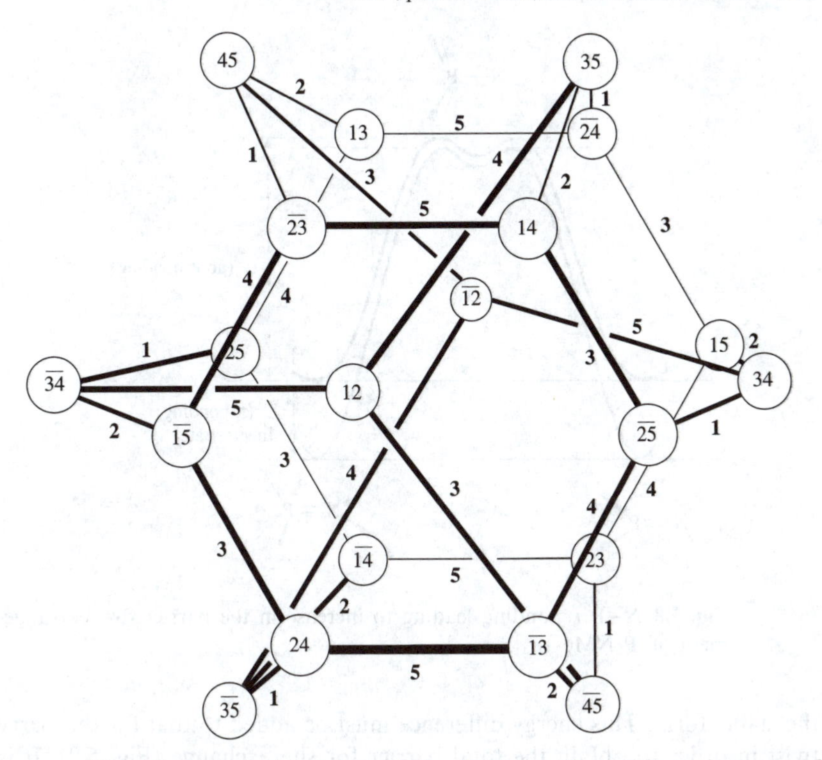

Fig. 5.7 Berry twist relationships between the various isomers of $[M(L_1)(L_2)(L_3)(L_4)(L_5)]$.

these compounds is the result of π-bonding between the lone pair on N (or S) and an empty *d*-orbital on P and that, since this is most effective when the donor atom is in the trigonal plane, then when it is apical in the square pyramid, the loss of some of this π-bonding in the transition state (intermediate) is a further contribution to the energy barrier (Fig. 5.8). In contrast to Berry pseudorotation for PF_4NMe_2, site exchange of fluorine atoms in PF_3Me_2 and PF_2Me_3 is intermolecular, involving P–F bond breaking.[42]

Another, much more common, cause of non-fluxionality in five-coordinate phosphorus compounds occurs when the system has to pass through isomeric forms of low stability in the process of site exchange through pseudorotation. PF_3Me_2 has a ^{19}F NMR spectrum with two peaks with areas in the ratio 2:1, each coupled to ^{31}P. There is no indication of line broadening, even at the highest temperatures studied, indicating that there is no rapid axial ↔ equatorial exchange. In order to make such an exchange through a sequence of Berry twists it is necessary to traverse trigonal bipyramidal intermediates where one or both of the methyl groups are in axial positions. Since the most stable trigonal bipyramidal phosphorus complexes are the ones with the most electronegative groups in the axial positions, these intermediate isomers will be at a much higher energy than

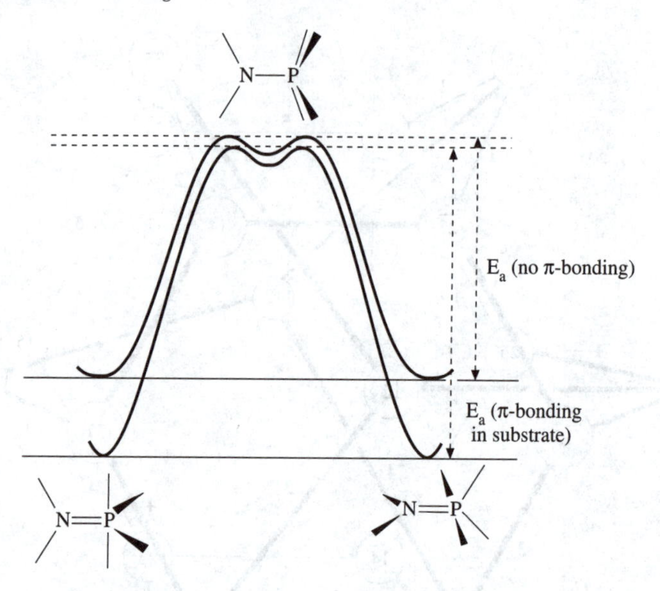

Fig. 5.8 N–P π-bonding leading to increase in the barrier for rearrangement of $[P(NMe_2)F_4]$.

the stable form. This energy difference must be added to that for the Berry twist in order to obtain the total barrier for site exchange (Fig. 5.9). It is convenient to use the network representation (Fig. 5.7) to see how the axial ↔ equatorial exchange might occur. If CH_3 takes positions 1 and 2 and F occupies 3, 4, and 5, then the most stable isomers (fluorines in both axial positions) will occupy the six corners ($\mathbf{34}, \overline{\mathbf{34}}, \mathbf{35}, \overline{\mathbf{35}}, \mathbf{45}, \overline{\mathbf{45}}$), the less stable axial–equatorial isomers occupy the hexagon vertices ($\mathbf{13}, \overline{\mathbf{13}}, \mathbf{14}, \overline{\mathbf{14}}, \mathbf{15}, \overline{\mathbf{15}}, \mathbf{23}, \overline{\mathbf{23}}, \mathbf{24}, \overline{\mathbf{24}}, \mathbf{25}, \overline{\mathbf{25}}$), and the least stable species (both methyl groups axial) lie at the centres ($\mathbf{12}, \overline{\mathbf{12}}$). The exchange of one axial with one equatorial ligand can be achieved by three twists, $\overline{\mathbf{34}} \xrightarrow{2} \overline{\mathbf{15}} \xrightarrow{3} \mathbf{24} \xrightarrow{1} \overline{\mathbf{35}}$, passing through the less stable axial–equatorial isomer. Site exchange can be achieved with two twists but this goes through the even less stable bis-axial isomer and yields an enantiomer, $\overline{\mathbf{34}} \xrightarrow{5} \mathbf{12} \xrightarrow{4} \mathbf{35}$. A similar exchange can be achieved by passing through the less unstable axial–equatorial isomer, but a sequence of four consecutive twists is required:

$$\overline{\mathbf{34}} \xrightarrow{\ 1\ } \mathbf{25} \xrightarrow{\ 4\ } \mathbf{13} \xrightarrow{\ 5\ } \overline{\mathbf{24}} \xrightarrow{\ 1\ } \mathbf{35}$$

Of course, in this system the three fluorines are chemically equivalent, and $\overline{\mathbf{35}}$ is indistinguishable from $\mathbf{35}$.

When the electronegativity difference between the two ligands is considerably less than that between CH_3 and F, the isomeric instability contribution to the barrier is less. The temperature-dependent NMR spectra

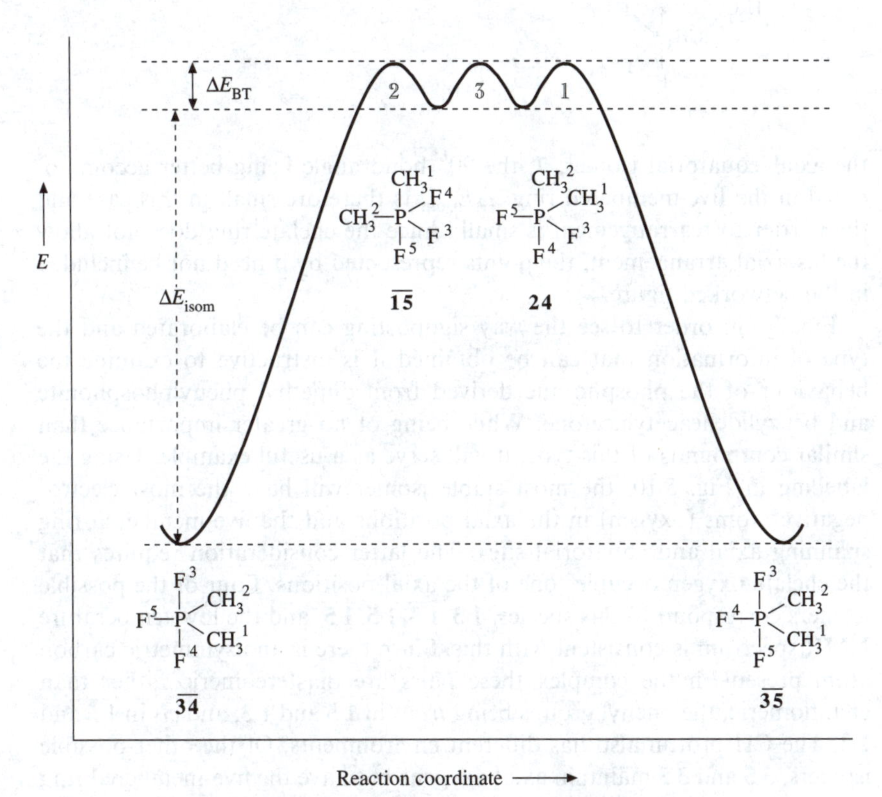

Fig. 5.9 Barriers (ΔE_{BT} for Berry twist + ΔE_{isom} for difference in isomer stability) for axial–equatorial fluorine exchange in $[P(CH_3)_2F_3]$.

of PF_3Cl_2 and PF_3Br_2 indicate rapid axial ↔ equatorial fluorine exchange at room temperature and only broaden and resolve below −60 °C.[43]

When one or more chelate is present in the coordination shell, the scope for signposting is improved and the control of the reaction rate is greatly increased. For example, **5** behaves just like the dimethyltrifluorophosphorane and there is no evidence for axial ↔ equatorial exchange even at the highest temperatures studied, but the five-membered ring analogue, **6**, is

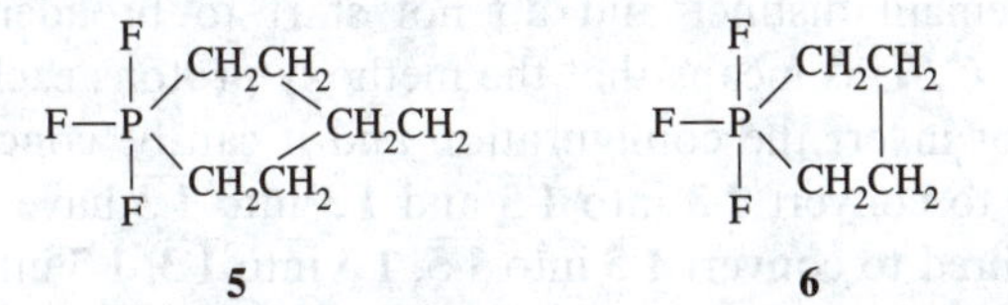

only non-fluxional at low temperatures.[43] Here it is thought that the strain induced in the five-membered ring when it is in the trigonal plane raises the energy of the ground state to the point where it is not much more stable than

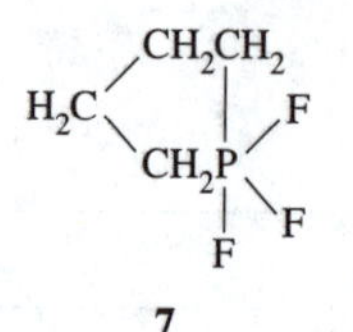

7

the axial–equatorial isomer, **7**, the 90° bond angle being better accommodated in the five-membered ring. ΔE_{isom} is therefore small in this case and the barrier to rearrangement is small. Since the chelate ring does not allow the bis-axial arrangement, the points represented by it need not be included in the networked figure.

Finally, in order to see the way signposting can be elaborated and the type of information that can be obtained it is instructive to examine the behaviour of the phosphorane derived from dimethyl phenylphosphonite and benzylideneacetylacetone. While being of no greater importance than similar compounds of this type, it will serve as a useful example. Using the labelling in Fig. 5.10, the most stable isomer will have the most electronegative atoms (oxygen) in the axial positions and the five-membered ring spanning axial and equatorial sites. The latter consideration requires that the chelate oxygen occupies one of the axial positions. Four of the possible isomers correspond to this species, **1 3**, $\overline{\mathbf{1\,3}}$, **1 5**, $\overline{\mathbf{1\,5}}$, and the low-temperature NMR spectrum is consistent with this. Since there is an asymmetric carbon atom present in the complex these pairs are diastereomeric rather than enantiomeric, the phenyl groups being *trans* in **1 5** and $\overline{\mathbf{1\,3}}$, and *cis* in $\overline{\mathbf{1\,5}}$ and **1 3**. The CH proton also has different environments. Of the other possible isomers, **3 5** and $\overline{\mathbf{3\,5}}$ maintain axial oxygens but have the five-membered ring spanning equatorial positions with consequent ring strain; **1 4**, $\overline{\mathbf{1\,4}}$, **2 3**, $\overline{\mathbf{2\,3}}$, **2 5**, and $\overline{\mathbf{2\,5}}$ place one carbon atom in an axial position, although they retain the axial–equatorial ring spanning and should be less stable; **3 4**, $\overline{\mathbf{3\,4}}$, **4 5** and $\overline{\mathbf{4\,5}}$ contain both the axial carbon and the strained bis-equatorial spanning by the ring and should be even less stable. The isomers with both axial sites occupied by C, **2 4** and $\overline{\mathbf{2\,4}}$, are likely to be the least stable. An examination of the network diagram shows that none of the stable forms is adjacent and so interconversion by way of Berry twists requires passing through less stable intermediate isomers. The ^{1}H NMR spectrum indicates fluxionality above 0 °C, where the protons of the axial and equatorial methoxy groups coalesce. However, the protons of the two different CH groups remain distinct and do not start to broaden and coalesce until above 50 °C. This means that the methoxy protons exchange by pathways that do not invert the configuration and it can be concluded that the pathways used to convert **1 3** into $\overline{\mathbf{1\,5}}$ and **1 5** into $\overline{\mathbf{1\,3}}$ have lower barriers than those required to convert **1 5** into $\overline{\mathbf{1\,5}}$, **1 3** into $\overline{\mathbf{1\,3}}$, **1 5** into **1 3**, and $\overline{\mathbf{1\,5}}$ into $\overline{\mathbf{1\,3}}$. The shortest pathway for the conversion, by way of **2 4** or $\overline{\mathbf{2\,4}}$, is ruled out because this would require synchronous site exchange of the methoxy protons and the CH protons. The coalescence at 50 °C might indicate that this barrier is, at last, being crossed, but the sequence,

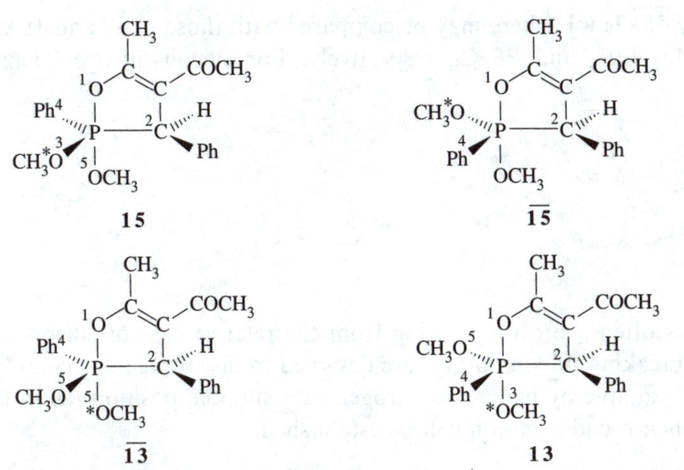

Fig. 5.10 The phosphorane produced by the reaction of dimethyl phenylphosphonite with benzylidene acetylacetone—the labelling scheme is for the stable isomer.

$$\overline{\mathbf{13}} \xrightarrow{4} \mathbf{25} \xrightarrow{3} \overline{\mathbf{14}} \xrightarrow{5} \mathbf{23} \xrightarrow{4} \mathbf{15}$$

would achieve the same effect. Experiment tells us that the pathway

$$\mathbf{13} \xrightarrow{4} \mathbf{25} \xrightarrow{1} \mathbf{34} \xrightarrow{2} \overline{\mathbf{15}}$$

has the lower barrier. It will be seen by inspection of Fig. 5.7 that passage through the other electronegativity-allowed isomer, **35** or $\overline{\mathbf{35}}$, is of no use because any rearrangements require a subsequent passage through the very unstable **24** or $\overline{\mathbf{24}}$. The activation energies of the slower rearrangements are in the region of 90 kJ mol^{-1} and so approach those related to bond-breaking processes. In this system, ring opening can be observed but only when the temperature is raised still further.

We have dwelt at length on pseudorotation at five-coordinate phosphorus since not only is this process of considerable intrinsic interest but it is also of importance in mechanisms of associative reactions of four-coordinate phosphorus compounds. The possibility of pseudorotation in the transition state or intermediate of such reactions is clearly of central importance in controlling product stereochemistry, as has been long realized.[44]

The relative rarity of five-coordination at other *sp*-block elements, apart from arsenic and antimony, rather restricts examination of pseudorotation at other centres. There have, however, been some studies of five-coordinate silicon species. Barriers for intramolecular ligand exchange within trigonal bipyramidal $RSiF_4^-$ are lower than within $R_2SiF_3^-$, where the operation of the Berry pseudorotation mechanism has been established.[45] Exchange within the $RSiF_4^-$ anions is generally fast on the NMR time-scale, but rates depend strongly on the bulk of the alkyl group R. Barriers of up to nearly 55 kJ mol^{-1} here may be compared with those of 37 and 18 kJ mol^{-1} for Me_2NPF_4 and PF_4Cl, respectively. For silicon–nitrogen compounds

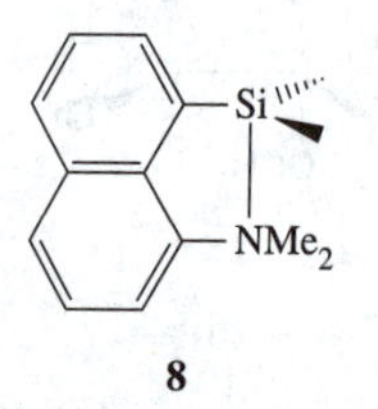

there is often a problem arising from the relative ease of silicon–nitrogen bond breaking. If compounds are designed to discourage this complication, as for example by fixing the nitrogen in a suitable position (as in the unit **8**),[46] then pseudorotation can be established.

5.3.5 Pseudorotation in six-coordinate systems[47]

Six-coordination is dominated by the octahedral geometry and, with few exceptions, the rates of isomerization of octahedral complexes are usually of the same magnitude or slower than their rates of ligand substitution. Alternative geometries being far less stable, the barriers to intramolecular rearrangement are large. The most likely complementary geometry is the trigonal prism. An octahedron (which is a regular trigonal antiprism) can be converted into a trigonal prism by rotating one face through an angle of 60 ° with respect to the opposite face on the common threefold axis. If this rotation is continued for a further 60° a new octahedron is generated. This serves to exchange all three *trans* pairs of ligands (Fig. 5.11). If the six octahedral sites are separately labelled there will be fifteen enantiomeric pairs of isomers. In spite of the fact that octahedral complexes of Co(III) and Pt(IV) are substitutionally inert and, in the latter case at least, it is possible to formulate feasible synthetic pathways, very few examples exist of octahedral complexes with six different ligands and in none of these cases have all thirty isomers been isolated and identified. As in the much less challenging case of four-coordinate square-planar complexes, there is little

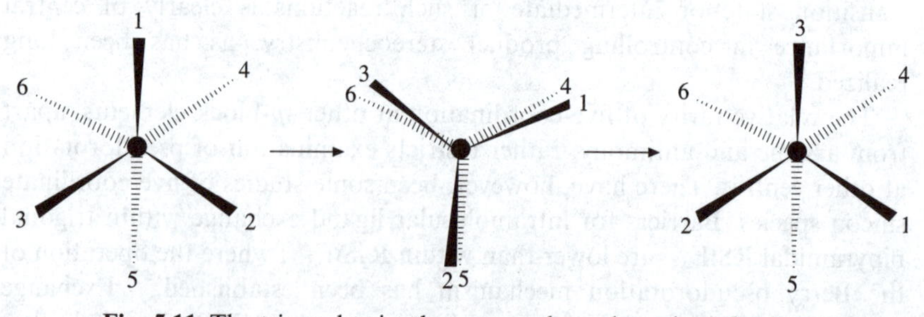

Fig. 5.11 The trigonal twist that passes through a trigonal-prismatic transition state (intermediate) and changes all *trans* partners.

purpose to serve in carrying out such an exercise. Following the approach used in the previous section, it should be possible to represent interconversions of octahedral species by using a 30-vertex figure with the appropriate connectivity (the trigonal twist has a connectivity of six since there are three threefold axes and the rotation can be clockwise or anticlockwise). The reader will be saved the chore of having to follow this through, but anyone with sufficient interest might like to try. On the whole it remains a pointless exercise, because octahedral systems usually contain sufficient bidentate or multidentate ligands to restrict the number of site permutations considerably.

The simplest studies can be made with tris(chelate) complexes, which, when the bidentate ligands are symmetrical, can exist as Δ and Λ enantiomers. Some of the earliest studies of reactivity of inorganic complexes involved rates of racemization. Unsymmetrical chelates give *fac* and *mer* diastereomers, each of which exists as enantiomeric pairs (Fig. 5.12). A twist about the unique threefold axis of $[M(AA)_3]$ can go in only one direction and will result in a $\Delta \leftrightarrow \Lambda$ change. This was mooted by several people in the mid-1950s, though Bailar normally receives the credit—it is usually called the *Bailar twist*. Only the *fac*-$[M(AB)_3]$ isomer has a true threefold axis and a Bailar twist will also lead to a $\Delta \leftrightarrow \Lambda$ change without any diastereomeric rearrangement. The equivalent twist on the pseudo-threefold axis of the *mer*-$[M(AB)_3]$ isomer also leads to a $\Delta \leftrightarrow \Lambda$ change but does not interconvert *fac* and *mer*.

A second twist, topologically equivalent to one proposed by Rây and Dutt many years ago[49] for the racemization of tris(chelate) complexes, involves a trigonal twist about a pseudo-threefold axis of the octahedron where the rotating face contains both ends of one chelate and is connected to its opposite partner by only one of the three chelates. This also leads to $\Delta \leftrightarrow \Lambda$ changes in symmetrical $[M(AA)_3]$ complexes, but when the ligands are not symmetrical, the three possible twist axes are not equivalent. As a result, the Rây and Dutt twist can lead not only to $\Delta \leftrightarrow \Lambda$ changes but also to *fac* $\leftrightarrow$ *mer* rearrangements. All three twists will lead to enantiomerization, but the occurrence of diastereoisomerization will depend upon the axis used. A meridional isomer can therefore undergo enantiomerization without necessarily changing to the facial isomer (Fig. 5.13). Springer and Sievers[50] have provided a useful analysis and comparison of the Bailar and the Rây and Dutt twists, while Johnson[12,51], has considered factors, energetic and geometric (chelate bite), that might determine which mechanism operates in specific cases.

In most cases the energy barriers for these trigonal twists are large and the main pathway for stereochemical change in octahedral complexes involves bond breaking. Until NMR was applied to the problem it was particularly difficult to ascertain whether or not bond breaking was involved when the coordination shell was occupied by bidentate or multidentate ligands, though the fate of a monodentate ligand can be monitored by isotopic labelling. Nuclear magnetic resonance spectroscopy not only

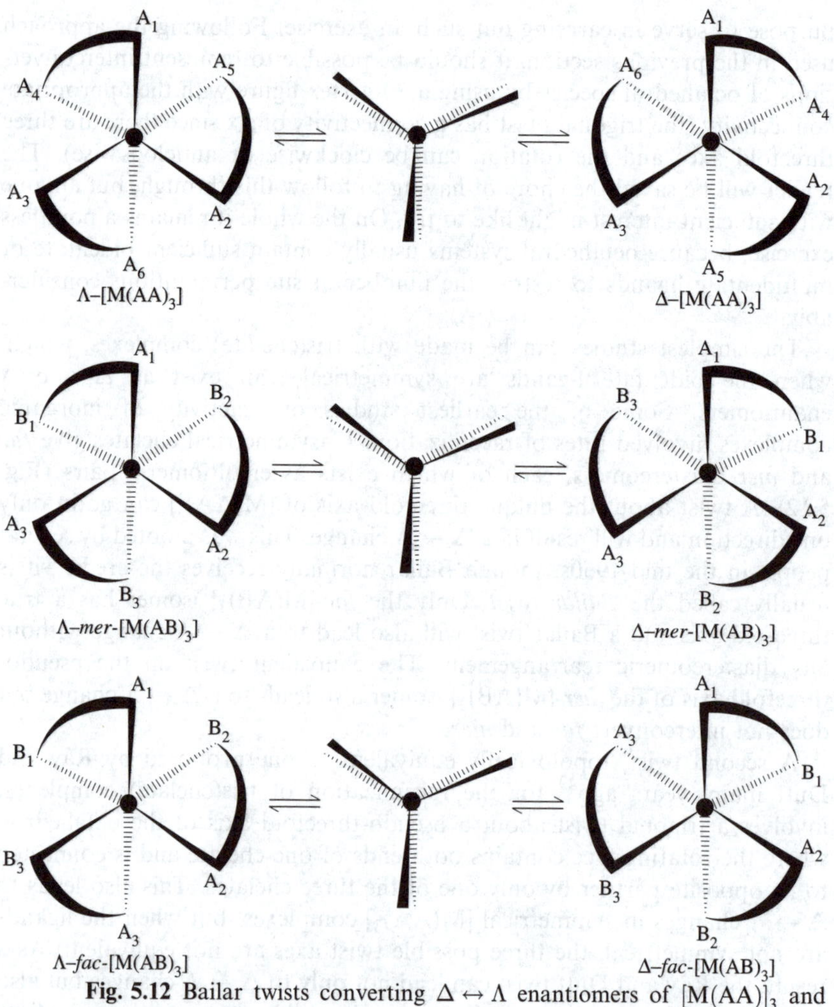

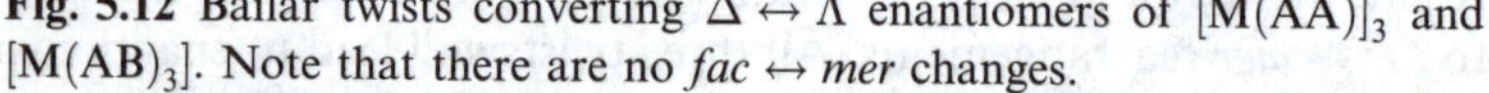

Fig. 5.12 Bailar twists converting Δ ↔ Λ enantiomers of $[M(AA)]_3$ and $[M(AB)_3]$. Note that there are no *fac* ↔ *mer* changes.

provides a means of identifying the sites exchanging and measuring the rates, but it can also provide, through the persistence of spin–spin coupling between appropriate nuclei, proof of bonds remaining intact during site exchange. Rearrangements with bond breaking are discussed below.

Using the argument that any modification that lowers the energy of the complementary geometry with respect to that of the substrate will lower the barrier to intramolecular rearrangement, one should examine the factors that lead to trigonal distortions in six-coordinate complexes and, indeed, examine the compounds where trigonal-prismatic geometry is found. The Mo atom in MoS_2 is in a trigonal-prismatic environment, as is Re^{VI} in $[Re(S_2C_2Ph_2)_3]$.[52] In many other tris-(bis-sulphur)-donor complexes, the extent of trigonal distortion away from regular octahedral geometry can be

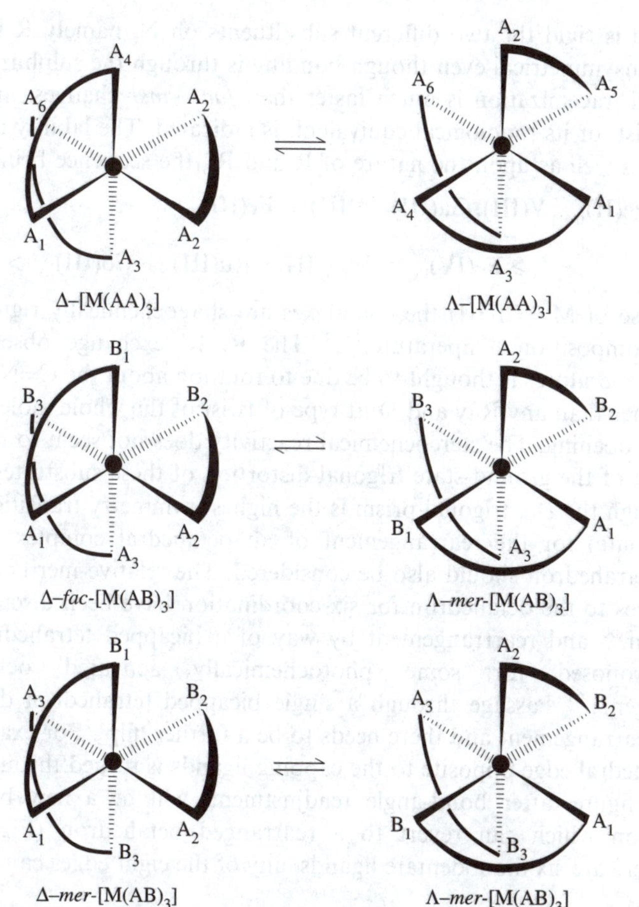

Fig. 5.13 Rây and Dutt twists leading to Δ ↔ Λ changes in $[M(AA)_3]$ and also to *fac* ↔ *mer* and *mer* ↔ *mer* changes in $[M(AB)_3]$.

large. Complexes of this sort have been, and remain, of considerable interest to the inorganic chemist. This stereochemistry has been explained in terms of steric effects (a small ligand bite favours the D_{3h} form)[53] or electronic effects (for example, the possibility of inter-ligand S- - -S interactions).[54] There is an excellent review of the earlier work.[55] It is of interest to note that the S_6 complexes of the saturated trithia-macrocycle, 1,4,7-trithianonane (=9-ane-S_3), $[M(9\text{-ane-}S_3)_2]^{2+}$, have a regular octahedral coordination shell with each ligand occupying opposite trigonal faces when M = Fe(II), Co(II), Ni(II).[56] Here there is no strain arising from the ligand bite angle, and the electron distribution in the saturated thioether is unambiguous. There is no information about the fluxionality of these complexes and it would be easy enough, with minor modifications in the ligand, to test for this.

It has been known for a long time that tris(dithiocarbamate) complexes, $[M(RR'NCS_2)_3]$, are frequently stereochemically non-rigid in solution on the NMR time-scale, and frequently trigonally distorted in the crystal. Since the ligand is rigid the two different substituents on N, namely R and R′, make it unsymmetrical even though bonding is through the sulphur atoms. In general, racemization is much faster than *fac* ↔ *mer* changes, and so a Bailar twist, or its topological equivalent, is indicated. The lability depends upon M as well as upon the nature of R and R′, the sequence being:

$$\mathrm{Fe(II)_{h.s.}, V(III), Ga(III), In(III) > Fe(III)_{h.s.}}$$

$$\geqslant \mathrm{Fe(IV)_{l.s.} > Mn(III) > Ru(III) \gg Co(III)_{l.s.} > Rh(III)}$$

In the case of M = Cr(III) the complexes are stereochemically rigid up to their decomposition temperature.[57,58] The R, R′ exchange observed at higher temperatures is thought to be due to rotation about the C=N double bond rather than any Rây and Dutt type of twist of the whole molecule, or even ring opening. The stereochemical reactivity does not seem to relate to the extent of the ground-state trigonal distortion of these substrates.[59]

Although the D_{3h} trigonal prism is the highest-symmetry transition state (intermediate) for the rearrangement of an octahedral complex, the bicapped tetrahedron should also be considered. The relative merits of these alternatives to the octahedron for six-coordination have been discussed by Hoffmann,[60] and rearrrangement by way of a bicapped tetrahedron has been proposed for some photochemically activated octahedral arrangements.[61] Passage through a single bicapped tetrahedron does not lead to rearrangement and there needs to be a further 'flip'. For example, if the tetrahedral edge opposite to the capping ligands is turned through 90 °, the new figure, after bond-angle readjustment, will be a new bicapped tetrahedron which can revert to a rearranged octahedron (Fig. 5.14). When there are six monodentate ligands, any of the eight edges can become

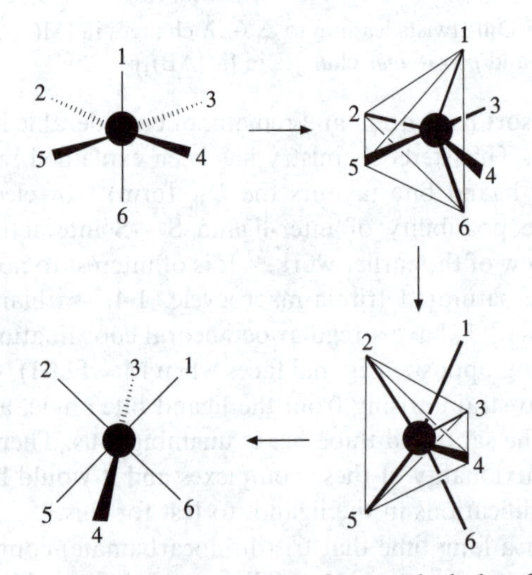

Fig. 5.14 Rearrangement of an octahedral molecule by way of a bicapped tetrahedron and a 90° 'edge flip'.

the tetrahedral edge that flips. Such a pathway could lead to the racemization of tris-bidentate octahedral complexes. Flipping the tetrahedral edge through 180° will serve to exchange a pair of *cis* ligands, while the other four maintain their relative positions. A mechanism of this sort has been considered, but later discarded in favour of the trigonal twist, for intramolecular rearrangement of *fac*-$[Cr\{P(OR)_3\}_3(CO)_2(CX)]$ (X = S, Se) to its *mer* isomer with CO *trans* to CO. This is relatively slow, with $k = ca\ 10^3\ s^{-1}$ at 25 °C and not very dependent upon the nature of X, including X = O.[62]

5.3.6 *Polytopal rearrangements in seven-coordinate systems*

There are a number of geometries encountered for seven-coordination,[63] with stereochemical non-rigidity a characteristic feature. The four most common regular geometries, the D_{5h} pentagonal bipyramid, the C_{3v} capped octa-hedron, the C_{2v} capped trigonal prism, and the C_s polyhedron with opposed tetragonal and trigonal faces can easily be interconverted by small changes in bond angle. There are detailed discussions in the literature.[37,64]

Dithiocarbamate complexes provide useful examples of these polytopal rearrangements. For example, complexes of the type $[(\eta^5\text{-}C_5H_5)M(Me_2NCS_2)_3]$, (M = Ti, Zr, Hf) may be considered (with considerable reluctance on the part of one of the authors) as seven-coordinate pentagonal bipyramidal molecules with the cyclopentadienyl ligand occupying one of the axial sites (Fig. 5.15). The titanium complex is rigid on the NMR time-scale below −50 °C but, on raising the temperature, the signals from the methyl groups above and below the pentagonal plane broaden and merge. At higher temperatures the methyl signals of the unique axial–equatorial ligand broaden and merge but even higher temperatures are required to exchange this ligand with the fully equatorial set. The first two processes are thought to involve rotation about the partly double C–N bond (with or without Ti–S bond fission) but the polytopal change, described as a double facial twist mechanism with a capped trigonal-prismatic intermediate, is thought to be responsible for the axial–equatorial site exchange.[65] The low-temperature site exchange of the Ti complex is much faster than that of the analogous Zr and Hf species, suggesting that M–S bond breaking might be important. The reference just cited is well worth examining for its permutational analysis of rearrangements in pentagonal bipyramidal $[M(LL')_3X]$ complexes.

A digonal twist mechanism has been invoked to account for the equatorial exchanges in the analogous $[(\eta^5\text{-}C_5H_5)Zr(\beta\text{-diketonate})_3]$ complexes (included in Fig. 5.15 for comparison).[66] This, however, would have a rather crowded pentagonal bipyramidal intermediate with the bulky cyclopentadienyl ligand in the pentagonal plane.

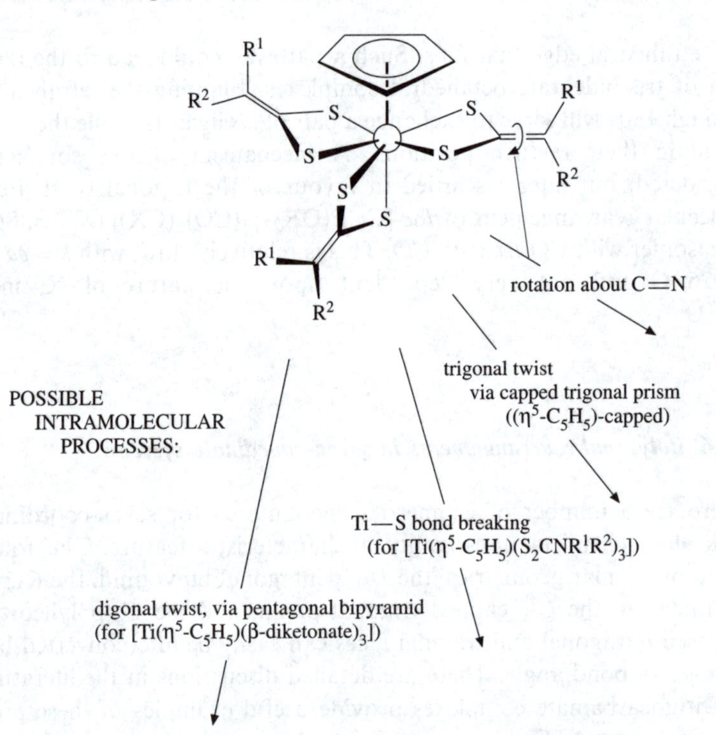

Fig. 5.15 Pentagonal bipyramidal $[M(\eta^5\text{-}C_5H_5)(S_2CNR^1R^2)_3]$; pathways for the exchange of R^1 and R^2 (with or without S site exchange).

5.4 The special case of hydrido complexes[67]

A large number of complexes exist that contain one or more hydride ligands and many of these are stereochemically non-rigid at room temperature. Frequently, the hydrogens seem to exert only a small influence on the geometry, which seems to be determined by the best arrangement of the other ligands, and it is possible to account for the observed site exchanges in terms of migration of the hydrogen, either by tunnelling (cf. the case of NH_3, Section 5.3.2) or by face–edge–face transfer, while the positions of the other ligands are barely altered.

The structures of compounds of the type $[HML_4]^{n-}$ ((i) $L = PF_3$, M = Fe, Ru, Os, $n = 1$; (ii) $L = PF_3$, M = Co, Rh, Ir, $n = 0$; (iii) $L = P(OPh)_3$, M = Rh, $n = 0$; (iv) $L = PPh(OEt)_2$, M = Co, $n = 0$) can be described as having a distorted tetrahedral arrangement of the four L ligands with the hydrogen capping one of the tetrahedral faces (this would be referred to as a capped tetrahedron and might be looked upon as a distorted trigonal bipyramid with the central atom displaced out of the trigonal plane away from H). At room temperature the NMR spectra indicate that all magnetic nuclei are in rapid site exchange and low temperatures are needed to obtain spectra consistent with the instantaneous structure, i.e. three ligands in

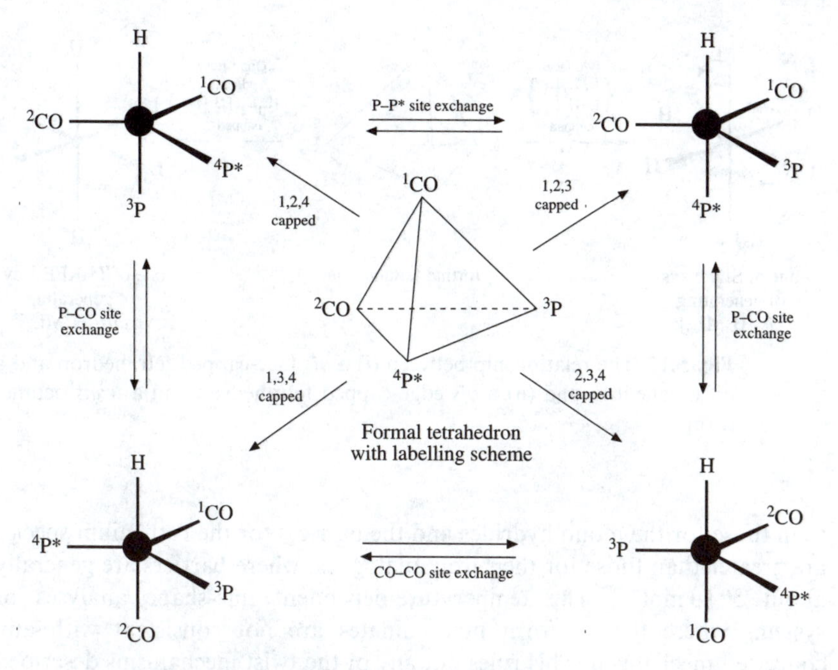

Fig. 5.16 Tetrahedral jump mechanism for site exchange in $[HIr(CO)_2(PR_3)_2]$.

equivalent environments that differ from that of the fourth. The change is consistent with the migration of H from one tetrahedral face to another (the *tetrahedral jump mechanism*) and the barrier is quite sensitive to the nature of M.[68] The $[HFe(CO)_4]^-$ anion, whose geometry is midway between a hydride-face-capped tetrahedron and a trigonal bipyramid, is fluxional even in the solid state (whereas $Fe(CO)_5$ is not), where magic-angle spinning ^{13}C NMR indicates a barrier of some 30 kJ mol^{-1}.[33]

In the complexes typified by $[HIr(CO)_2\{P(p\text{-tol})_3\}_2]$ the extra signposting leads to two stereoisomeric forms and phosphine–phosphine site exchange (no isomerization) has a lower barrier than phosphine–CO site exchange (isomerization). Here too the tetrahedral jump mechanism has been invoked, although the change is said to have some features of the Berry twist (Fig. 5.16).[67]

Six-coordinate H_2ML_4 species (M = Fe, Ru; L = a phosphorus-donor ligand) can be looked upon as distorted *cis*-octahedral species, but the solid-state structure of $[H_2Fe\{P(OEt)_2Ph\}_4]$ indicates a nearly tetrahedral arrangement of the four phosphorus-donor atoms about the central metal with the hydrogens capping two of the faces.[69] Occasionally these are in equilibrium with their *trans* isomers, which can be looked upon as a tetrahedron of phosphorus atoms with the hydrides above opposite edges (Fig. 5.17). Unlike most six-coordinate species, these compounds are stereochemically non-rigid, although the barriers are considerably higher

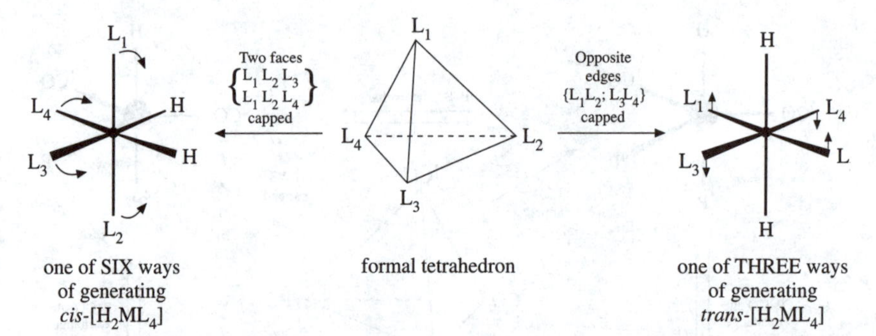

Fig. 5.17 The relationship between (i) a *bis* face-capped tetrahedron and a *cis* octahedron and (ii) a *bis* edge-capped tetrahedron and a *trans* octahedron.

than those for the mono hydrides and the barriers for the ruthenium species are greater than those for their iron analogues, where barriers are generally about 50 kJ mol^{-1}. The temperature-dependent line-shape analyses of systems where the *cis* form predominates are not consistent with any concerted mechanism. This rules out any of the twist mechanisms described above, and it has been suggested that, here too, the tetrahedral jump mechanism is operating with the hydrogen jumping to a site above one of the uncapped tetrahedral faces. Jesson has proposed that the hydrogen passes through a tetrahedral edge, but passage through the *trans* isomer would require a concerted move of both hydrogens and is thus ruled out.[67]

It is interesting to note that some eight-coordinate species $[H_4ML_4]$ are stereochemically rigid, which is unusual for eight-coordinate species, but fast- and slow-exchange NMR spectra have been obtained for a number of $[H_4ML_4]$ with M = Mo, W; L = PR_3. Activation barriers estimated from variable-temperature NMR are in the range 50–70 kJ mol^{-1}. Their trigonal dodecahedral structures, established for one compound by X-ray crystallography, can be looked upon as a distorted L_4 tetrahedron with each of the four faces capped by H. It is quite difficult to understand the low-temperature NMR fully, even harder to try to elucidate the mechanism for pseudorotation. This may occur through a D_{2d} transition state, or by tetrahedral tunnelling or a tetrahedral jump.[70] In these systems the tetrahedral jump mechanism would require a concerted motion of at least three hydrogens.

Other mixed-hydride complexes with coordination numbers greater than six, e.g., H_4OsL_3, H_5ReL_3 (L is a tertiary phosphine), and $H_6W(PR_3)_3$, as well as the binary $[ReH_9]^{2-}$, all have NMR spectra that indicate stereochemical non-rigidity. Rapid pseudorotation is to be expected for these high coordination numbers, both from qualitative considerations of the small changes in geometry required for interconversions,[63,71] and from the small values which have been calculated for the barriers involved.[72]

5.5 Rearrangement by way of pseudorotation of an intermediate of higher coordination number

The most likely place to find this mechanism is among the coordinatively unsaturated four-coordinate planar d^8 metal complexes, where substitution takes place with complete retention of configuration and where thermally activated rearrangements are unknown. Under conditions where the five-coordinate adduct could live long enough to undergo a polytopal rearrangement before dissociating this mechanism, which could take a great variety of detailed courses,[73] would be feasible and probably accompanied by ligand exchange. The kinetics of isomerization will be a simple second-order process involving the substrate and the extra nucleophile, which would function as a catalyst if there was no net substitution. It was long known that complexes of the type $[Pt(L)_2Cl_2]$ (originally for L = PEt_3, $AsEt_3$, and $SbEt_3$[74] and later for SR_2 and $S(O)R_2$) isomerize in the presence of catalytic amounts of L. A great deal of attention has been paid to this problem. However, the mechanism sought is by no means the only one consistent with the kinetic rate law (an unfortunate but not infrequent occurrence) and an alternative that involves simple substitution with no stereochemical change is a plausible alternative.

$$cis\text{-}[X{-}Pt(L)_2{-}X] + L \longleftrightarrow \left[X{-}Pt(L)_3\right]^+ \cdots X^- \longleftrightarrow trans\text{-}[X{-}Pt(L)_2{-}X] + L$$

This mechanism, first proposed by Drew,[75] has been well established for the isomerization of palladium(II) complexes where L is an amine in aqueous solution, where the rate depends upon the nucleophilicity of the amine.[76] Since this mechanism involves an ionic species as intermediate it was predicted that the addition mechanism would be favoured in non-polar solvents, whereas the two-stage substitution would be favoured in polar solvents. The observation that the isomerization of *cis*-$[Pt(PBu_3)_2Cl_2]$ in cyclohexane was catalysed by PBu_3 at a rate that was first-order in catalyst and was much reduced by the addition of polar solvents was taken as compelling evidence for the addition mechanism.[77] A report that the catalyst phosphine was not incorporated into the complex led to some speculation as to the nature of the rearrangement,[78] but a detailed NMR study of the $[Pt(PMe_3)_2Cl_2] + PMe_3$ system in non-polar solvents showed that the ionic species $[Pt(PMe_3)_3Cl]^+$ formed readily when an excess of phosphine was added to a CD_2Cl_2 solution of the complex, whereas there was no evidence for the uncharged five-coordinate $[Pt(PMe_3)_3Cl_2]$ species.[79] Of course, the amount of intermediate needed could easily be less than that required for detection by NMR, but the rapidity of the PMe_3 exchange makes it unlikely that the cationic species constitutes a dead-end reaction. The successive displacement mechanism has also been demonstrated for the Me_2S-catalysed isomerization of *cis*- and *trans*-$[Pt(Me_2S)_2Cl_2]$, where the rate of formation of $[Pt(Me_2S)_3Cl]^+$ in the presence of a chloride scavenger is the same as that for the isomerization in its absence.[80] There is no evidence for ionic intermediates in the Me_2SO-catalysed *trans* ↔ *cis* isomerization of $[Pt(Me_2SO)_2Cl_2]$,[81] and the observation that the product of the reaction between *cis*-$[Pt(Me_2SO)_2Cl_2]$ and α-amino acids (N–OH) is the unexpected *trans-O,S*-$[Pt(Me_2SO)(N{-}O)Cl]$ (it is reasoned that the more strongly nucleophilic end should have entered first and *trans* to the stronger *trans*-effect sulphoxide) has led to a detailed examination of isomerization and exchange in this system.[82] It was concluded that rearrangement of the five-coordinate $[Pt(Me_2SO)_2(N{-}O)Cl]$ provided a pathway for the isomerization to a more stable product. There is an excellent critical review on this sort of reaction, though only covering the literature up to 1980.[83]

5.6 Rearrangement by way of pseudorotation of an intermediate of lower coordination number

The best-documented examples of isomerization through the rearrangement of an intermediate of lower coordination number are to be found in the *cis* ↔ *trans* rearrangements of complexes of the type $[Pt(PEt_3)_2(R)X]$, where R is a σ-bonded alkyl or aryl group and X is usually a labile ligand such as Cl or Br (the cationic methanolo derivative rearranges much more rapidly). The system is usually discussed in the context of the dissociative pathway for substitution in four-coordinate planar d^8 metal complexes (see Section 3.11.2), which indeed it is, but in all these cases there is a much more facile stereoretentive associatively activated pathway for substitution.

The mechanism of these rearrangements has been shown to involve the formation of a three-coordinate, T-shaped intermediate,

$$\text{P—Pt(P)(R)—X} \underset{k_X}{\overset{k_i}{\rightleftharpoons}} \text{P—Pt}^+\text{(P)(R)} + \text{X}^- \xrightarrow{k_r} \text{R—Pt}^+\text{(P)(P)} + \text{X}^- \xrightarrow{\text{fast}} \text{R—Pt(P)(P)—X}$$

and gives rise to the rate law,

$$+\mathrm{d}[trans]/\mathrm{d}t = \frac{k_i k_r [cis]}{k_X[X^-] + k_r}$$

and, at constant $[X^-]$, the condition under which the study was usually carried out, the observed kinetics are indeed first-order, with

$$k_{obs} = \frac{k_i k_r}{k_X[X^-] + k_r}$$

The retardation of the isomerization by added X^- indicates quite clearly

Table 5.1 The effect of substituent, Y, upon the solvolytic (k_1) and isomerization (k_i) rate constants for complexes of the type *cis*-[Pt(PEt$_3$)$_2$(Y-C$_6$H$_4$)Br].[a]

Y	k_1 (s^{-1})	10^3k_i (s^{-1})
H	3.33	2.95
m-CH$_3$	3.20	4.43
p-CH$_3$	3.90	4.76
m-OCH$_3$	3.27	2.07
p-OCH$_3$	5.29	3.30
m-F	2.90	0.194
p-F	3.60	0.440
m-Cl	3.10	0.144
p-Cl	3.32	0.160
m-CF$_3$	3.05	0.070
p-CF$_3$	3.07	0.103

[a] In methanol at 30.0 °C.

that rearrangement takes place after the act of dissociation and that early trapping of this intermediate gives rise to a product of *cis* configuration. The k_i pathway can be readily distinguished from the associative solvolysis whose rate constant, k_1, is generally very much greater unless there is considerable steric hindrance. A selection of values for k_i for complexes of the type *cis*-[Pt(PEt$_3$)$_2$(Y-C$_6$H$_4$)Br] are compared with separately determined k_1 values in Table 5.1.[84] It has been seen already (Table 3.13) that the rate constants for the associative solvolysis of these complexes are very sensitive to steric hindrance from the alkyl or aryl group when it is *cis* to the leaving group. The data in Table 5.1 show that they are insensitive to the electron displacement properties of the substituted aryl group. On the other hand, the rate constants for dissociation, k_i, are insensitive to steric hindrance but increase markedly with increasing electron donation from the aryl group. Furthermore, whereas $\Delta S^\ddagger$ and $\Delta V^\ddagger$ for the associative solvolysis are negative, they are almost all positive for the k_i process. The sole exception is the 2,4,6-trimethylphenyl complex, where $k_1 = k_i$ but the entropies and volumes of activation are consistent with an associatively activated process. There was a great deal of argument about the interpretation of the mechanism but it turns out that, in the case of the mesityl complex (Fig. 5.18), dissociation of the methanolo complex offers a more facile pathway for the formation of the fluxional three-coordinate intermediate, and the rate-determining step is the associative solvolysis that generates it. A general method for the formation of the *cis*-[Pt(PEt$_3$)$_2$(aryl)(MeOH)]$^+$ involves addition of acid to a methanolic solution of *cis*-[Pt(PEt$_3$)$_2$(aryl)(CH$_3$)]. Methane is generated instantly and the subsequent isomerization of the methanolo complex can be studied kinetically. Volumes of activation of all the isomerizations (including that of the

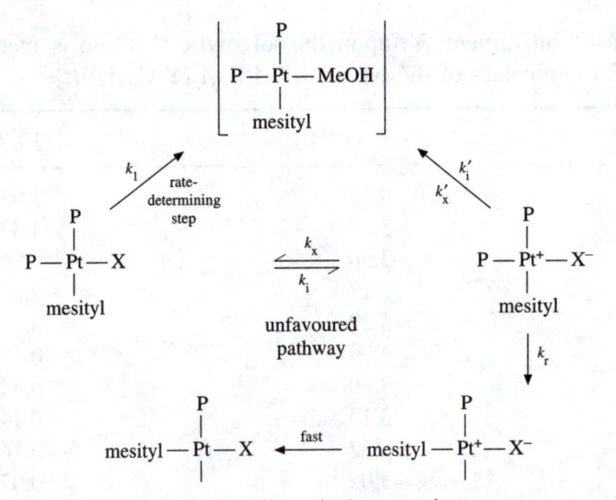

Fig. 5.18 Solvolysis, dissociation, and rearrangement in isomerization of the mesityl complex *cis*-$[Pt(PEt_3)_2(C_6H_2Me_3)X]$.

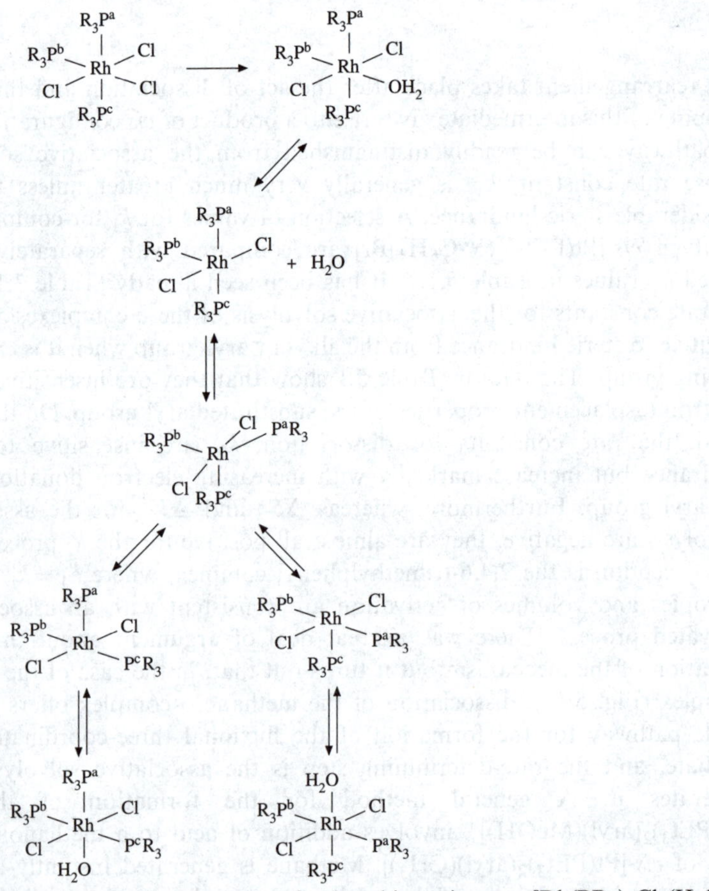

Fig. 5.19 Rearrangement of phosphines in *mer*-$[Rh(PR_3)_3Cl_2(H_2O)]^+$ through pseudorotation of five-coordinate intermediate.

aryl = mesityl complex) are positive and fully consistent with dissociative activation.[85]

Isomerization of a six-coordinate complex by way of a fluxional five-coordinate intermediate is less common, but the behaviour of *mer*-$[Rh(PMe_2Ph)_3X_2Y]$ serves as a good example (Fig 5.19). When X = Y = halide the complex is stereochemically rigid and the NMR spectra are consistent with two magnetically inequivalent environments for the phosphorus. However, when Y (*trans* to P) = H_2O all three phosphorus atoms are stereochemically equivalent at room temperature, although site exchange can be frozen out by reducing the temperature. The evidence points strongly to a reversible dissociation of water accompanied by a rapid intramolecular rearrangement of the five-coordinate intermediate (rate constants for isomerization and water exchange are similar).[86]

References

1. Review: E. L. Muetterties, *Accounts Chem. Res.*, 1970, **3**, 266.
2. L. G. Vanquickenborne and K. Pierloot, *Inorg. Chem.*, 1984, **23**, 1471.
3. D. F. Martin and M. L. Tobe, *J. Chem. Soc.*, 1962, 1388.
4. W. Kruse and H. Taube, *J. Am. Chem. Soc.*, 1961, **83**, 1280.
5. S. B. Tong, H. R. Krouse, and T. W. Swaddle, *Inorg. Chem.*, 1976, **15**, 2643.
6. D. R. Stranks and N. Vanderhoek, *Inorg. Chem.*, 1976, **15**, 2639.
7. M. N. Hughes, *J. Chem. Soc.*, 1967, 1284.
8. Y. Kitamura, T. Takamoto, and K. Yoshitani, *Inorg. Chem.*, 1988, **27**, 1382.
9. W. G. Jackson, *Inorg. Chem.*, 1988, **27**, 777.
10. C. T. G. Knight and A. E. Merbach, *Inorg. Chem.*, 1985, **24**, 576; E. Turin, R. M. Nielson and A. E. Merbach, *Inorg. Chim. Acta*, 1987, **134**, 79.
11. U. Frey, L. Helm, and A. E. Merbach, *Helv. Chim. Acta*, 1990, **73**, 199.
12. A. Rodger and B. F. G. Johnson, *Inorg. Chim. Acta*, 1988, **146**, 35; B. F. G. Johnson and A. Rodger, *Inorg. Chem.*, 1989, **28**, 1003; B. F. G. Johnson, J. Cook, D. Ellis, S. Hefford, and A Whittaker, *Polyhedron*, 1990, **9**, 2221.
13. P. W. Atkins, *Quanta, A Handbook of Concepts*, 2nd edn, Clarendon Press, Oxford, 1974, p. 194.
14. G. W. Koeppl, D. S. Sagatys, G. S. Krishnamurthy, and S. I. Miller, *J. Am. Chem. Soc.*, 1967, **89**, 3396.
15. J. B. Lambert, in *Topics in Stereochemistry*, Vol. 6, eds N. L. Allinger and E. L. Eliel, Wiley–Interscience, New York, 1971, p. 19.
16. S. J. Brois, *J. Am. Chem. Soc.*, 1967, **89**, 4242.
17. D. Felix and A. Eschenmoser, *Angew. Chem. Int. Edn. Engl.*, 1968, **7**, 224.
18. R. D. Baechler and K. Mislow, *J. Am. Chem. Soc.*, 1970, **92**, 3090, and references therein.
19. H. D. Munro and L. Horner, *Tetrahedron*, 1970, **26**, 4621.
20. G. H. Senkler and K. Mislow, *J. Am. Chem. Soc.*, 1972, **94**, 291.
21. B. Bosnich and S. B. Wild, *J. Am. Chem. Soc.*, 1970, **92**, 459.
22. B. Bosnich, S. T. D. Lo, and E. A. Sullivan, *Inorg. Chem.*, 1975, **14**, 2305; T. D. B. Morgan and M. L. Tobe, *Inorg. Chim. Acta*, 1971, **5**, 563.
23. R. Scartazzini and K. Mislow, *Tetrahedron Lett.*, 1967, 2719.

24. E. W. Abel, *Chem. Brit.*, 1990, 148.
25. J. K. Beattie, *Adv. Inorg. Chem.*, 1988, **32**, 2.
26. G. Annibale, M. Bonivento, L. Canovese, L. Cattalini, G. Michelon, and M. L. Tobe, *Inorg. Chem.*, 1985, **24**, 797.
27. S. J. Berners-Price, C. Brevard, A. Pagelot, and P. J. Sadler, *Inorg. Chem.*, 1986, **25**, 596; G. Salem, A. Schier, and S. B. Wild, *Inorg. Chem.*, 1988, **27**, 3029.
28. E. L. Muetterties, *J. Am. Chem. Soc.*, 1969, **91**, 1636, 4115.
29. H. S. Gutowsky, D. W. McCall, and C. P. Slichter, *J. Chem. Phys.*, 1953, **21**, 279.
30. D. Hellwinkel and H. J. Wilfinger, *Tetrahedron Lett.*, 1969, 3423.
31. E. L. Muetterties and W. D. Phillips, *J. Am. Chem. Soc.*, 1959, **81**, 1084.
32. P. Meakin, E. L. Muetterties, and J. P. Jesson, *J. Am. Chem. Soc.*, 1972, **94**, 5271.
33. B. E. Hanson and K. H. Whitmire, *J. Am. Chem. Soc.*, 1990, **112**, 974.
34. J. P. Jesson and P. Meakin, *J. Am. Chem. Soc.*, 1973, **95**, 1344.
35. R. S. Berry, *J. Chem. Phys.*, 1960, **32**, 933.
36. See pp. 725–6 of I. Ugi, F. Ramirez, D. Marquarding, H. Klusacek, G. Gokel, and P. Gillespie, *Angew. Chem. Int. Edn. Engl.*, 1970, **9**, 703.
37. For example, I. Ugi, D. Marquarding, H. Klusacek, P. Gillespie, and F. Ramirez, *Accounts Chem. Res.*, 1971, **4**, 288.
38. R. S. Cahn, C. K. Ingold, and V. Prelog, *Angew. Chem. Int. Edn. Engl.*, 1966, **5**, 385.
39. E. L. Muetterties, Stereochemical Non-Rigidity in *Reaction Mechanisms in Inorganic Chemistry*, ed. M. L. Tobe, M.T.P. Int. Rev. Sci., Inorg. Chem. Series 1, Vol. 9. Butterworths, London, 1972, p. 59.
40. K. E. DeBruin, K. Naumann, G. Zon, and K. Mislow, *J. Am. Chem. Soc.*, 1969, **91**, 7031.
41. G. M. Whitesides and H. L. Mitchell, *J. Am. Chem. Soc.*, 1969, **91**, 5384.
42. T. A. Furtsch, D. S. Dierdorf, and A. H. Cowley, *J. Am. Chem. Soc.*, 1970, **92**, 5759.
43. E. L. Muetterties, W. Mahler, and R. Schmuzler, *Inorg. Chem.*, 1963, **2**, 613.
44. Reviews: F. H. Westheimer, *Accounts Chem. Res.*, 1968, **1**, 70; K. Mislow, *Accounts Chem. Res.*, 1970, **3**, 321.
45. S. E. Johnson, R. O. Day, and R. R. Holmes, *Inorg. Chem.*, 1989, **28**, 3182; S. E. Johnson, J. S. Payne, R. O. Day, J. M. Holmes, and R. R. Holmes, *Inorg. Chem.*, 1989, **28**, 3190.
46. R. J. P. Corriu, M. Mazhar, M. Poirier, and G. Royo, *J. Organomet. Chem.*, 1986, **306**, C5.
47. Review: J. I. Musher, *Inorg. Chem.*, 1972, **11**, 2335.
48. J. C. Bailar, *J. Inorg. Nucl. Chem.*, 1958, **8**, 165.
49. P. Rây and N. K. Dutt, *J. Indian Chem. Soc.*, 1943, **20**, 81.
50. C. S. Springer and R. E. Sievers, *Inorg. Chem.*, 1967, **6**, 852.
51. A. Rodger and B. F. G. Johnson, *Inorg. Chem.*, 1988, **27**, 3061.
52. R. Eisenberg and J. A. Ibers, *J. Am. Chem. Soc.*, 1965, **87**, 3776.
53. D. L. Kepert, *Inorg. Chem.*, 1972, **11**, 1561.
54. E. I. Stiefel, R. Eisenberg, R. C. Rosenberg, and H. B. Gray, *J. Am. Chem. Soc.*, 1966, **88**, 1956.
55. J. A. McCleverty, *Prog. Inorg. Chem.*, 1968, **10**, 49.
56. K. Weighardt, H. J. Kuipers, and J. Weiss, *Inorg. Chem.*, 1985, **24**, 3067.
57. L. Que and L. H. Pignolet, *Inorg. Chem.*, 1974, **13**, 351.

58. D. J. Duffy and L. H. Pignolet, *Inorg. Chem.*, 1974, **13**, 2045.
59. R. Hoffmann, B. F. Beier, E. L. Muetterties, and A. R. Rossi, *Inorg. Chem.*, 1977, **16**, 511.
60. R. Hoffmann, J. M. Howell, and A. R. Rossi, *J. Am. Chem. Soc.*, 1976, **98**, 2484.
61. J. Burdett, *Inorg. Chem.*, 1976, **15**, 212.
62. A. A. Ismail, F. Sauriol and I. S. Butler, *Inorg. Chem.*, 1989, **28**, 1007.
63. D. L. Kepert, *Progr. Inorg. Chem.*, 1979, **25**, 41.
64. E. L. Muetterties, *Accounts Chem. Res.*, 1970, **3**, 266.
65. R. C. Fay, J. R. Weir, and A. H. Bruder, *Inorg. Chem.*, 1984, **23**, 1079.
66. T. J. Pinnavia, J. J. Howe, and R. E. Teets, *Inorg. Chem.*, 1974, **13**, 1074.
67. J. P. Jesson, in *M.T.P. Int. Rev. Sci., 2nd Series., Inorg. Chem.*, Vol. 9, ed. M. L. Tobe, Butterworths, London, 1974; G. G. Hlatky and R. H. Crabtree, *Coord. Chem. Rev.*, 1985, **65**, 1.
68. P. Meakin, E. L. Muetterties, and J. P. Jesson, *J. Am. Chem. Soc.*, 1972, **94**, 5271.
69. L. J. Guggenberger, D. D. Titus, M. T. Flood, R. E. Marsh, A. A. Orio, and H. B. Gray, *J. Am. Chem. Soc.*, 1972, **94**, 1135.
70. W. G. Kita, M. K. Lloyd, and J. A. McCleverty, *J. Chem. Soc., Chem. Commun.*, 1971, 420; P. Meakin, L. J. Guggenberger, W. G. Peet, E. L. Muetterties, and J. P. Jesson, *J. Am. Chem. Soc.*, 1973, **95**, 1467.
71. D. L. Kepert, *Progr. Inorg. Chem.*, 1978, **24**, 179; M. C. Favas and D. L. Kepert, *Progr. Inorg. Chem.*, 1981, **28**, 309.
72. L. J. Guggenberger and E. L. Muetterties, *J. Am. Chem. Soc.*, 1976, **98**, 7221.
73. See Scheme 8 on p. 319 of M. L. Tobe, Substitution Reactions, in *Comprehensive Coordination Chemistry*, Vol. 1, eds G. Wilkinson, R. D. Gillard, and J. A. McCleverty, Pergamon Press, Oxford, 1987, Chapter 7.1.
74. J. Chatt and R. G. Wilkins, *J. Chem. Soc.*, 1953, 70, and references therein.
75. H. D. K. Drew and G. H. Wyatt, *J. Chem. Soc.*, 1934, 56.
76. L. Cattalini and M. Martelli, *J. Am. Chem. Soc.*, 1969, **91**, 312.
77. P. Haake and R. M. Pfeiffer, *J. Am. Chem. Soc.*, 1970, **92**, 5243.
78. P. Haake and R. M. Pfeiffer, *J. Am. Chem. Soc.*, 1970, **92**, 4996.
79. R. Favez, R. Roulet, A. A. Pinkerton, and D. Schwarzenbach, *Inorg. Chem.*, 1980, **19**, 1356.
80. R. Roulet and C. Barbey, *Helv. Chim. Acta*, 1973, **56**, 2179.
81. G. Annibale, M. Bonivento, L. Canovese, L. Cattalini, G. Michelon, and M. L. Tobe, *Inorg. Chem.*, 1985, **24**, 797.
82. L. E. Erickson, T. A. Ferrett, and L. F. Buhse, *Inorg. Chem.*, 1983, **22**, 1461.
83. G. K. Anderson and R. J. Cross, *Chem. Soc. Rev.*, 1980, **9**, 185.
84. R. Romeo, D. Minniti, and S. Lanza, *Inorg. Chem.*, 1979, **18**, 2362.
85. W. J. Louw, R. van Eldik, and H. Kelm, *Inorg. Chem.*, 1980, **19**, 2878, and references therein.
86. A. J. Deeming and G. P. Proud, *Inorg. Chim. Acta*, 1985, **100**, 223.

6 Substitution reactions of carbonyl and related compounds

Chapter contents

6.1 Introduction

It is convenient, for a variety of reasons, to consider the substitution reactions of metal carbonyl complexes and their derivatives[1,2,3] separately from the reactions of the more classical coordination complexes, often referred to nowadays as *Werner complexes*, and it is instructive to see why this is so. Any model for bonding between carbon monoxide and a metal requires a measure of covalency considerably greater than that needed to account for the classical Werner complexes. Carbon monoxide is a poor σ-donor and needs to act as a π-acceptor as well. The synergism that results strengthens the bonding sufficiently to lead to stable compounds, even of metals in zero or negative oxidation states, under appropriate circumstances. Electrons which would otherwise be localized on the metal become strongly bonding if they are placed in orbitals of π-symmetry appropriate to the coordination number and geometry of the metal. It is therefore no longer possible to rely on models where the filled bonding orbitals are associated with the ligand, and the non-bonding and lowest energy antibonding orbitals are associated with the metal. Considerations of charge and size, and of ligand-field stabilization energy, give way to the stability of closed shells—and odd-electron species become very reactive. Consequently, a particular electron configuration formally assigned to the metal will be associated with a strongly preferred coordination number and geometry.

Eighteen electrons rule supreme in determining the composition of stable species and the rule, which must be broken in the act of substitution, breaks more readily in the direction of a 16-electron valence shell intermediate. Therefore, one might predict that a dissociative mechanism is preferred. Having said that, one should not lose sight of the fact that carbon, a stickler for obeying the analogous 8-electron rule for *p*-block elements, can still have associatively activated nucleophilic pathways which, at first sight, appear to require a 10-electron valence shell in the five-coordinate transition state derived from a tetrahedral (aliphatic) reaction centre. However, this I_a (S_N2) transition state has the two extra electrons in a molecular orbital that is mainly ligand-based. It is the *A* mechanism that requires the 10-electron valence shell about the reaction centre and this is not found for carbon. The same is probably true for the 18-electron carbonyl complexes, with the I_a mechanism still possible but the *A* mechanism unlikely. Associative attack at carbonyls has been reviewed.[4]

The preceding paragraph implies a stark contrast between carbonyl ligands, on the one hand, and classical ligands such as water, ammonia, and halides on the other. As so often, things are not so clear-cut in practice, with several types of ligand bridging the gap between carbonyls and Werner complexes. Thus cyanide forms many very stable complexes of the Werner type, but it is isoelectronic with carbon monoxide and, like CO, it is able to indulge in π-bonding to metal centres with appropriate orbitals. Cyanide's strong σ-donating ability and its negative charge help to stabilize metal ions in high oxidation states, but its π-bonding properties permit even zero oxidation state complexes, such as $[Ni(CN)_4]^{4-}$ and $[Cr(CN)_6]^{6-}$, to exist, at least in the solid state and in liquid ammonia if not in aqueous solution. The other important carbonyl-to-Werner linking ligands are phosphines and related compounds. These can stabilize metal(0) species, e.g. $Pt(PR_3)_n$, and also form many complexes of the $M(PR_3)_2X_2$ type on reaction with first-row transition metal halides.

The π-bonding in carbonyls, as in cyanide and diimine (e.g. 2,2′-bipyridyl; 1,10-phenanthroline) complexes, can have an important effect on substitution mechanisms. By reducing electron density in the metal t_{2g} orbitals, approach by a potential nucleophilic ligand becomes less difficult, and S_N2/I_a attack easier. The other mechanistic effect of the metal–CO interactions is that the metal polarizes the system such that the carbon has a marked δ+ charge, with the result that nucleophilic attack sometimes takes place at the carbonyl-carbon rather than at the metal centre. Both features appear in the following pages.

6.2 Substitution reactions of simple mononuclear complexes

Binary mononuclear carbonyl complexes are found for the even-electron valence states. They are summarized in Table 6.1. The d^8 metals also form

Table 6.1 Simple mononuclear carbonyl molecules and ions that obey the eighteen-electron rule.

Configuration			
d^{10}	$Ni(CO)_4$	$Pd(CO)_4$ [a]	$Pt(CO)_4$ [a]
	$[Co(CO)_4]^-$	$[Rh(CO)_4]^-$	
	$[Fe(CO)_4]^{2-}$	$[Ru(CO)_4]^{2-}$	$[Os(CO)_4]^{2-}$
		$[Tc(CO)_4]^{3-}$	
d^8	$Fe(CO)_5$	$Ru(CO)_5$	$Os(CO)_5$
	$[Mn(CO)_5]^-$	$[Tc(CO)_5]^-$	$[Re(CO)_5]^-$
	$[Cr(CO)_5]^{2-}$ [b]	$[Mo(CO)_5]^{2-}$ [b]	$[W(CO)_5]^{2-}$ [b]
d^6			$[Ir(CO)_6]^{3+}$ [c]
	$[Fe(CO)_6]^{2+}$ [c]		
	$[Mn(CO)_6]^+$ [d]	$[Tc(CO)_6]^+$	$[Re(CO)_6]^+$
	$Cr(CO)_6$	$Mo(CO)_6$	$W(CO)_6$
	$[V(CO)_6]^-$ [b]	$[Nb(CO)_6]^-$ [b]	$[Ta(CO)_6]^-$ [b]
	$[Ti(CO)_6]^{2-}$ [e]	$[Zr(CO)_6]^{2-}$ [e]	

[a] Unstable except at very low temperature.
[b] Isolable as R_4N^+ salts.
[c] Recently established candidates for possible kinetic studies.
[d] Isolable as $AlCl_4^-$ salt.
[e] Isolable as $K[222]^+$ salt.

polynuclear complexes with bridging CO and/or metal–metal bonds, to the extent that any kinetic study of $Ru(CO)_5$ or of $Os(CO)_5$ must be carried out in the presence of an excess of CO to prevent this oligomerization. The 17-electron $[V(CO)_6]$ will be discussed separately, as will the binuclear compounds of the other odd-electron reaction centres (e.g. of Mn, Re), which form metal–metal-bonded or carbonyl-bridged species to avoid the presence of unpaired electrons.

The uncharged mononuclear complexes possess the advantage that they are soluble in non-polar solvents and possess uncharged leaving groups. Their reactions with neutral nucleophiles can therefore be carried out in non-coordinating solvents—though it will be seen below that in some cases almost any molecule will coordinate, if only transiently, to a coordinatively unsaturated carbonyl fragment. The relationship between kinetic order and molecularity here becomes much more clear-cut than in the case of aqueous solutions of Werner complexes.

6.2.1 Carbon monoxide exchange

Historically, the logical and, hopefully, simple study of CO exchange between the complex and the free ligand proved to be a disaster. The observation that $[Ni(CO)_4]$ exchanged its CO with ^{14}CO at a rate that was independent of [CO] was consistent with a *D* mechanism, and the fact that the rate of the reaction

$$[Ni(CO)_4] + PPh_3 \longrightarrow [Ni(CO)_3(PPh_3)] + CO$$

was also independent of the concentration of triphenyl phosphine seemed to confirm this. Unfortunately, the two processes appeared to have significantly different enthalpies of activation (although the rate-determining step should have been the same in both cases) and some very imaginative mechanistic proposals were made until it was realized that the $[Ni(CO)_4]$/*CO exchange also took place in the gas phase, a reaction not available to the phosphine substitution.[5]

The $[Ni(CO)_4]$/*CO exchange is the only homoleptic exchange involving 18-electron carbonyl systems that can be studied adequately, most other such species being substitutionally inert except at very high temperatures. These are precisely the conditions under which the systems become very sensitive to catalysis by impurities or decomposition products. $Fe(CO)_5$ presents a well-documented case in which the substitutional lability decreases with increasing purification and where, even with the purest material, it is thought that the exchange observed above 90 °C is being catalysed by the products of the thermal decomposition. The exchange of *CO with $M(CO)_6$ has to be followed at such high temperatures that the results in the literature are nearly as unreliable as the long extrapolation needed to make reactivity comparisons at 25 °C. Carbon monoxide exchange can be monitored with $[Co(CO)_4]^-$, but here there is a problem with marked effects of ion-pairing on reactivity.[6]

There is a very great range of reactivities for carbon monoxide exchange with binary[1,2,7,8] and with ternary[8] carbonyls. In the first row of the *d*-block, the half-life for such exchange at ambient temperatures is of the order of minutes for $Ni(CO)_4$ and $V(CO)_6$, but a matter of years for $Fe(CO)_5$. The reactivity sequence is [2]

$$Co(CO)_4, Mn(CO)_5 > V(CO)_6 > Ni(CO)_4 > Cr(CO)_6 > Fe(CO)_5$$

This order reflects coordination number and *d*-electron configuration. The ionic binary carbonyls $V(CO)_6^-$, $Mn(CO)_5^-$, and $Mn(CO)_6^+$ all undergo carbon monoxide exchange extremely slowly. Relative reactivities on descending the Periodic Table all show a maximum for the second-row element:

$$Cr(CO)_6 < Mo(CO)_6 > W(CO)_6$$

$$Mn_2(CO)_{10} < Tc_2(CO)_{10} > Re_2(CO)_{10}$$

$$Fe(CO)_5 \ll Ru(CO)_5 \gg Os(CO)_5$$

It may be noted that the reactivity order has, for the Group VI hexacarbonyls, been shown to parallel force constants derived from solution infrared and Raman spectra rather than mean bond dissociation energies, though the theoreticians have managed to calculate dissociation energies for loss of the first CO which are in the above $Cr(CO)_6$ < $Mo(CO)_6$ > $W(CO)_6$

order.[9] These calculations suggested that σ-bonding is maximal for tungsten, π-bonding maximal for chromium, and that molybdenum has the worst combination.

In general, reactivity sequences are consistent with dissociative activation, reflecting the ease with which the 18-electron rule is broken and the coordinatively unsaturated transition state (intermediate) formed. Thus, whereas stable five- or four-coordination is a very rare occurrence when there is a formal low-spin d^6 metal centre, the occurrence of stable four-coordination about a formal low-spin d^8 metal centre is common (Chapter 3) although not in the lowest oxidation states. On the other hand, d^{10} metals in low oxidation states often have coordination numbers of 2 or 3 (again see Chapter 3). Furthermore, the lower coordination numbers are more common with the elements of the second and third rows.

6.2.2 Substitution by other nucleophiles

Because of the technical difficulties involved in following the kinetics of CO exchange (even though ^{13}C NMR studies might nowadays be easier than using radioactive ^{14}C), the bulk of the published work has dealt with the replacement of CO by other nucleophiles. While introducing new problems such as the possibility of subsequent reactions which would interfere with the kinetic analysis, or ambiguities as to the actual site of the rate-determining attack, the advantages are overwhelming.

The generally dissociative nature of the process has been confirmed for the reactions of $Ni(CO)_4$ (Table 6.2)[10] and the pentacarbonyls $M(CO)_5$,[11] especially $Fe(CO)_5$[12] and $Ru(CO)_5$,[13] with nucleophiles. However, the hexacarbonyls frequently follow a mixed-order rate law,

$$-d[M(CO)_6]/dt = \{k_1 + k_2[L]\}[M(CO)_6]$$

corresponding to two parallel pathways. In general, the bulk of the reaction goes by the k_1 pathway—indeed the first investigators missed the k_2 term because they worked at insufficiently high concentrations of L—though the k_1:k_2 balance changes significantly on going from chromium to molybdenum and tungsten (Table 6.3).[14] For a given $M(CO)_6$, k_1 is essentially independent of the nature of the incoming nucleophile, as expected for a dissociative pathway. The values of k_2 are modestly sensitive to the nature

Table 6.2 Kinetic parameters indicating dissociative ligand substitution in $Ni(CO)_4$.

Entering ligand	$10^4 k$ (s^{-1})[a]	$\Delta H^‡$ (kJ mol^{-1})	$\Delta S^‡$ (J K^{-1} mol^{-1})
$C^{18}O$	52	102	59
PPh_3	50	101	54[b]

[a] In hexane, at 20 °C.
[b] The activation volume for reaction of $Ni(CO)_4$ with $P(OEt)_3$ in heptane is +8 cm^3 mol^{-1}.

Table 6.3 Relative reactivities for the k_1 and k_2 paths in substitution at Group VI hexacarbonyls, in decalin.

	Temperature (°C)	$10^4 k_1$ (s^{-1})	$10^4 k_2$ ($dm^3\ mol^{-1}\ s^{-1}$)	k_1/k_2
$Cr(CO)_6$	130.7	1.38	0.45	3.1
$Mo(CO)_6$	112.0	2.13	1.77	1.2
$W(CO)_6$	165.7	1.15	0.89	1.3

Table 6.4 Values of $10^4 k_2$ ($dm^3\ mol^{-1}\ s^{-1}$) for the reaction $[M(CO)_6] + L \rightarrow [M(CO)_5L] + CO$, in decalin.

Metal:	Cr	Mo[a]	W
Temperature:	131 °C	112 °C	116 °C
$AsPh_3$		1.03	
$P(OPh)_3$		1.43	
PPh_3	0.43	1.77	0.89
$P(OCH_2)_3CEt$		3.6	
$C_6H_{11}NH_2$		4.2	
$PhCH_2NH_2$	0.79	4.4	
$P(OEt)_3$	0.45	6.7	1.70
P^nBu_3	0.85	20.5	7.1

[a] Rate constants for reaction with MeCN and with PhCN (in $C_2H_2Cl_2$ at 65 °C) differ by only about 25%.

of L (Table 6.4),[15,16] but do not really cover as wide a range as one might expect for an S_N2/I_a mechanism (similar comments[14] apply to the carbonyl halide anions $[Mo(CO)_5X]^-$). Hence the original suggestion that the kinetics indicated parallel dissociatively and associatively activated pathways was in some quarters brushed aside in favour of what has been described as parallel D and I_d mechanisms.[16] However, if activation entropies[15] and volumes[17] for the k_1 and $k_2[L]$ terms (Table 6.5) are considered, then the I_a mechanism does seem, at least to the present authors, to be the more attractive. The lower activation enthalpies for the $k_2[L]$ terms than for the respective k_1 terms are also consistent with, if hardly diagnostic of, I_a substitution. It is interesting that the activation enthalpies (k_1 term) for ^{14}CO exchange with the hexacarbonyls in the gas phase are equal, within their uncertainties, to the corresponding activation enthalpies for nucleophilic substitution in solution. Activation parameters for the k_1 terms are, of course, independent of entering nucleophile for a given hexacarbonyl (see Table 6.5).

It is likely that the confusion between the I_a and I_d assignments for the $k_2[L]$ term arises from misuse of these labels. If the distinction between the D and I_d mechanism rests with the lifetime of the intermediate then it would appear unlikely that the two mechanisms could coexist. It is likely that in the present case the I_d label has been wrongly applied to a synchronous process

Table 6.5 Activation parameters for substitution reactions of hexacarbonyls.

Compound	k_1 term			k_2[L] term		
Nucleophile	$\Delta H^{\ddagger}$ (kJ mol^{-1})	$\Delta S^{\ddagger}$ (J K^{-1} mol^{-1})	$\Delta V^{\ddagger}$ (cm^3 mol^{-1})	$\Delta H^{\ddagger}$ (kJ mol^{-1})	$\Delta S^{\ddagger}$ (J K^{-1} mol^{-1})	$\Delta V^{\ddagger}$ (cm^3 mol^{-1})
$Cr(CO)_6$						
P^nBu_3	168	+95		107	−61	
PPh_3			+15			
$Mo(CO)_6$						
P^nBu_3	133	+28		91	−62	
PPh_3			+10			
^{14}CO	129					
$W(CO)_6$						
P^nBu_3	167	+58		122	−29	
PPh_3						−10

with a fairly open transition state. It seems that the main reason for not wanting associative activation is that it would require a 20-electron valence shell in the transition state but, as has been pointed out elsewhere (Section 6.1 above), it is important to know where the extra two electrons are located.

In recent years, it has become possible to generate coordinatively unsaturated species by laser flash photolysis and to examine their subsequent reactions directly. The kinetics can be monitored by, for example, time-resolved picosecond absorption spectroscopy,[18] and dynamical factors probed through solvent and pressure effects.[19] The hexacarbonyls [$M(CO)_6$] give [$M(CO)_5$] species which react with their environment without any discrimination at diffusion-controlled rates, coordinating such unlikely ligands as nitrogen,[20] cyclohexane, sulphur hexafluoride, or even xenon,[21] with a facility equal to that with which they grab chlorobenzene, piperidine, or phosphines. Indeed liquid xenon and other super-critical fluids are useful solvents for the characterization of these unstable intermediates.[22] The stabilities of these adducts are reflected in their rates of dissociation[23]—an estimate of 35 kJ mol^{-1} has been made from kinetic data for the strength of tungsten–xenon bonding in [$W(CO)_5Xe$].[24] The [$M(CO)_5$] species are generally believed to be square pyramidal (SP), but evidence has been presented that [$Cr(CO)_5$] can also exist in a trigonal bipyramidal (TBP) form. Both the extreme transience of these [$M(CO)_5$] intermediates, and the expected higher reactivity of the trigonal bipyramidal geometry, can be illustrated by the rate constants estimated for reaction of [$Cr(CO)_5$] with cyclohexane, where the values are 9×10^9 s^{-1} for the SP form, 2×10^{10} s^{-1} for the TBP form.[25] The overall process of substitution leading to the thermodynamically preferred product can be looked upon as a cascade where the solvent (solv) and the nucleophile (L) compete according to their relative concentrations for the intermediate generated by desolvation:

$$[M(CO)_6] \xrightarrow{h\nu} [M(CO)_5] + CO$$

$$[M(CO)_5] + solv \longrightarrow [M(CO)_5(solv)]$$

$$\downarrow L$$

$$[M(CO)_5] + L \longrightarrow [M(CO)_5L]$$

The dissociative nature of thermal substitution at $[M(CO)_5(solv)]$ is supported by the determination of activation volumes for reactions of a number of such species with such incoming groups (L) as pyridine, piperidine, or 1-hexene.[26] Values of up to $+12\,cm^3\,mol^{-1}$, dependent on solvent and entering group, together with exhaustive but irritatingly dispersed evidence from other kinetic parameters,[27] suggest the operation of a *D* mechanism in most, if not all, cases. Unsurprisingly, activation volumes for overall photosubstitution

$$[M(CO)_6] + L \longrightarrow [M(CO)_5L] + CO$$

are also positive, being $+5$ to $+10\,cm^3\,mol^{-1}$ for M = Cr (depending on L and solvent), $+10\,cm^3\,mol^{-1}$ for M = Mo, and $+9\,cm^3\,mol^{-1}$ for M = W.[28]

Similar chemistry applies to analogous tetrahedral systems. The kinetics of the thermal reaction of $Ni(CO)_3(N_2)$ with carbon monoxide, to reform the $Ni(CO)_4$ from which the $Ni(CO)_3(N_2)$ was photolytically generated, follow a two-term rate law, suggesting parallel dissociative and associative pathways in a four-coordinate analogue of the $[M(CO)_5(solv)]$ systems.[29] Activation energies of about 40 and $20\,kJ\,mol^{-1}$ for the k_1 and k_2 pathways, respectively, give some idea of the energetics, the former leading to an estimate of the Ni–N bond strength in $Ni(CO)_3(N_2)$.

A detailed study of reactions

$$cis\text{-}[W(CO)_4(L)(pip)] + L' \longrightarrow [W(CO)_4(L)(L')] + pip$$

(pip = piperidine; L, L′ are a variety of phosphine and phosphite ligands) involving both thermal and photochemical activation has shown that the initial reaction is a stereoretentive displacement of piperidine and that any *trans* product is the result of subsequent intramolecular rearrangements. The 'bare' intermediate produced by laser flash photolysis is square pyramidal, with L in a basal position, and behaves in its subsequent reactions in exactly the same way as that produced thermally. When the reaction is carried out in cyclohexane it can be shown that a molecule of solvent jumps into the vacant site at a rate that is diffusion-controlled and that its subsequent replacement by other nucleophiles takes place at a rate that is determined by the desolvation of the 'solvento' species. The concept of a non-interfering solvent becomes very shaky in these systems.[30] When the reaction of *cis*-$[W(CO)_4(L)(pip)]$ is carried out by pulsed laser flash photolysis in chlorobenzene it is then possible to determine the activation volume for replacement of the chlorobenzene by pyridine. The value obtained, $+11.3\ \mathrm{cm^3\ mol^{-1}}$, is consistent with simple dissociative loss of the coordinated chlorobenzene.[31]

The presence of a charge, as in $[Mn(CO)_6]^+$ or $[V(CO)_6]^-$, does not have a profound effect upon the ligand-substitution lability, which remains very low in these octahedral $t_{2g}{}^6$ species. Indeed, photochemical activation is needed to substitute into $[V(CO)_6]^-$ within any reasonable length of time. The effect of the charge is much more important in relation to nucleophilic attack at the coordinated carbonyl—see Section 6.4 below. Substitution into the charged d^{10} carbonyl $[Co(CO)_4]^-$ by phosphines or phosphites, is, not unexpectedly, much easier.[32]

6.3 Reactions of substituted carbonyl complexes that obey the 18-electron rule

It is possible to extend the discussion from the substitution reactions of homoleptic carbonyl complexes to those involving the displacement of CO from derivatives of these species and even to the displacement of ligands other than CO. In the limit, it is possible to extend the discussion to the reactions of species that contain no CO ligands, e.g. $[M(PR_3)_4]$, $[M\{P(OR)_3\}_4]$, and $[M(PF_3)_4]$ (M = Ni, Pd, Pt; R = alkyl or aryl). Such complexes permit the establishment of steric, e.g. cone angle, and electronic effects on reactivity[33] in a way that is clearly impossible for homoleptic carbonyls. Provided that the highly covalent requirements remain, the 18-electron rule still dominate the mechanistic behaviour. This *George Washington's axe* approach brings us round to the point where organometallic chemistry meets Werner complex chemistry, a meeting briefly documented at the end of this chapter. In general, there are two consequences of replacing CO by other ligands, and these will be considered separately.

6.3.1 Ligands that do not change their electron donicity during the act of substitution

General mechanistic features do not change on going from $M(CO)_n$ to $M(CO)_{n-1}L$, $M(CO)_{n-2}L_2$, and so on. But lability, depending as one might expect upon the nature of the leaving group, now often L rather than CO, is much affected by the nature of the complex. The available kinetic data and mechanistic conclusions and speculations would, if discussed in detail, produce a very substantial chapter, since there are no succinct general guidelines. Electronic factors, σ- and π-bonding, geometric (*cis* and *trans* effects) and steric factors (cone angles), and various other parameters, all make varying contributions. Photochemical substitution also, as for Werner complexes, often gives different products, or ratios of products, from thermal substitution. Thus compounds $[M(CO)_5(amine)]$ lose only the amine under thermal conditions, but photochemically lose amine and carbon monoxide, in a ratio dependent on the wavelength of irradiation. There are several comprehensive and detailed reviews of this area,[1,2,34] which may be consulted for details of kinetic parameters and of explanations and rationalizations in terms of a variety of ligand, structural, and bonding effects.

Returning to the $[M(CO)_5(amine)]$ compounds referred to in the preceding paragraph, we should mention the question of their thermal substitution. Kinetic evidence from rate-constant dependences on the nature and concentration of the incoming ligand, and on added leaving ligand (the so-called mass-law retardation effect), all indicate a limiting dissociative (*D*) mechanism, as for the analogous pentacyanoferrates (Section 4.7.5.2). However, volumes of activation are surprisingly small, though they are independent of entering group in so far as this has been checked. Substitution at the (substituted) pyridine compounds $[M(CO)_5(Xpy)]$ for a number of groups X (H, 4-Me, 4-CN) and of entering groups have $\Delta V^{\ddagger}$ between 0 and $+3\ cm^3\ mol^{-1}$.[35] Similar reactions for *cis*-$[Mo(CO)_4(py)_2]$ have $\Delta V^{\ddagger}$ between +3 and $+5\ cm^3\ mol^{-1}$; with only substitution at *cis*-$[W(CO)_4(4Mepy)_2]$, at $+8\ cm^3\ mol^{-1}$, being characterized by an activation volume on the edge of the expected range.[36] Moving from nitrogen donors to the oxygen donor tetrahydrofuran, the mechanistic pattern for substitution changes markedly. Reactions of $M(CO)_5(thf)$ with piperidine, triphenyl phosphine, or triethyl phosphite are characterized by activation volumes of between −2 and −4 for M = Cr, between −4 and −8 for M = Mo, and between −4 and −15 for M = W.[37] Both the increasingly negative values and the greater range of values, on descending the group from chromium to tungsten, suggest a gradual change in mechanism from essentially dissociative to associative. The respective activation entropies are consistent with this interpretation. A similar pattern, contrasting with the $[M(CO)_5(amine)]$ pattern discussed above, has been established for replacement of a number of bidentate nitrogen and sulphur-donor ligands from compounds $[M(CO)_4(LL)]$ by phosphites. Here $\Delta V^{\ddagger}$ values for the chromium compounds are between +9 and $+14\ cm^3\ mol^{-1}$, firmly indicating dissociative activation, but phosphite substitution at analogous molybdenum complexes is associative, with negative $\Delta V^{\ddagger}$ values between −9 and $-11\ cm^3\ mol^{-1}$ (with one exception, involving the particularly bulky and rigid ligand $Me_3CSCH{=}CHSCMe_3$).[38]

Irradiation of $[M(CO)_4(phen)]$ plus triethyl phosphine in the ligand field band of the former leads to dissociative substitution ($\Delta V^{\ddagger} = +9.6, +5.7, +8.1\ cm^3\ mol^{-1}$ for M = Cr, Mo, W); similarly for $[W(CO)_5(Xpy)]$ plus triethyl phosphite, $\Delta V^{\ddagger} = +6$ to $+10\ cm^3\ mol^{-1}$. On the other hand, irradiation of $[M(CO)_4(phen)]$ plus triethyl phosphine in the charge-transfer band of the former leads to associative substitution for M = Mo, W ($\Delta V^{\ddagger} = -13, -12\ cm^3\ mol^{-1}$), though still dissociative for M = Cr. The

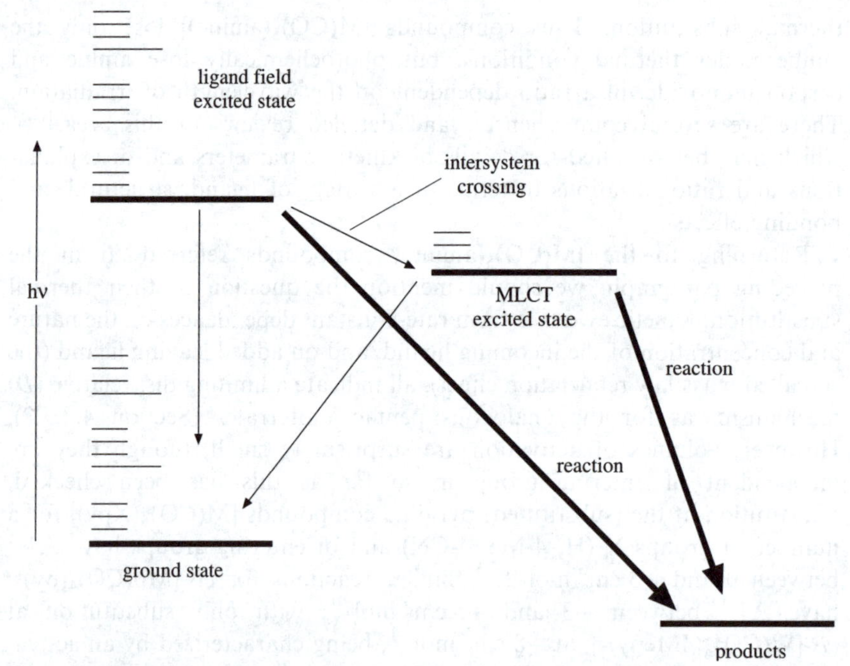

Fig. 6.1 Excited states of $[W(CO)_4(phen)]$ relevant to photosubstitution (MLCT = metal-to-ligand charge transfer).

relevant photochemistry for the tungsten compound is summarized in Fig. 6.1.[39]

Photolysis of the hexacarbonyls $M(CO)_6$ in the presence of bidentate ligands LL initially gives complexes $M(CO)_5(LL)$, containing monodentate LL. Ring closure takes place thermally to give $M(CO)_4(LL)$, at rates which are very dependent on the nature, particularly the flexibility or rigidity, of LL. Thus, for example, ring closure when LL = 1,2-ethanediamine is 10^8 times slower than for LL = 1,10-phenanthroline. Activation parameters ($\Delta H^\ddagger, \Delta S^\ddagger$, and $\Delta V^\ddagger$) have been determined for a number of these reactions. Activation enthalpies as a function of central metal M show a not unexpected minimum at molybdenum; activation entropies give ambiguous indications in some cases, especially when one inspects their variation as a function of substituent for LL = various alkyl derivatives of diazabutadiene or pyridine 2-carbaldehyde Schiff-base derivatives. However, activation volumes, in many cases supported by activation entropies, give a self-consistent picture of mechanistic trends (Table 6.6). There is a clear difference between ring closure controlled by dissociation of carbon monoxide for the chromium-diimine compounds and associative ring closure for their molybdenum and tungsten analogues. There is also a clear trend from associative to dissociative activation for the molybdenum $-2,2'$-bipyridyl series as ligand bulk increases; for the tungsten analogues the trend is merely from fully associative to less associative.[40] The terpyridyl

Table 6.6 Kinetic parameters for chelate ring-closure reactions of Group VI pentacarbonyl diimine complexes, $M(CO)_5(LL) \rightarrow M(CO)_4(LL) + CO$.

	Chromium		Molybdenum		Tungsten	
	$\Delta S^{\ddagger}$ ($J\,K^{-1}\,mol^{-1}$)	$\Delta V^{\ddagger}$ ($cm^3\,mol^{-1}$)	$\Delta S^{\ddagger}$ ($J\,K^{-1}\,mol^{-1}$)	$\Delta V^{\ddagger}$ ($cm^3\,mol^{-1}$)	$\Delta S^{\ddagger}$ ($J\,K^{-1}\,mol^{-1}$)	$\Delta V^{\ddagger}$ ($cm^3\,mol^{-1}$)
$M(CO)_5(en)$	−145	−11.9	−92	−5.4	−172	−12.3
$M(CO)_5(bipy)$			−26	−3.9	−70	−10.9
$M(CO)_5(4,4'\text{-}Me_2bipy)$			−20	−5.6		−8.4
$M(CO)_5(4,4'\text{-}Ph_2bipy)$			−4	+5.4		−6.4
$M(CO)_5(4,4'\text{-}{}^tBu_2bipy)$				+6.2		−4.5
$M(CO)_5(phen)$	−42	+6.2	−8	−2.9	−6	−8.2
$M(CO)_5({}^iPr_2dab)$	−47	+17.2	−40	−9.5	−53	−13.7

analogue $Mo(CO)_4$(terpy) contains bidentate terpy with a pendant, uncoordinated pyridyl group. It, and related rhenium(I) and platinum(IV) complexes, are fluxional in solution, undergoing what is picturesequely termed 'tick-tock twists'. These involve exchanging coordinated and uncoordinated pyridines intramolecularly, probably by an associative transition state with all three pyridine nitrogens interacting with the metal in the transition state.[41]

The ternary complexes $M(CO)_5X$ with M = Mn, Tc, Re, deserve mention as they are effectively the simplest, and certainly the most studied, mononuclear carbonyls for these elements—the binary carbonyls $M_2(CO)_{10}$ appear later in this chapter (Section 6.6.1). ^{13}CO exchange is considerably faster at $Mn(CO)_5X$ than at the binary $Mn(CO)_6{}^+$ cation. Here, and in other substitution reactions, kinetic characterization is complicated by the possibility of exchange at positions *cis* and *trans* to the halide, and the further possibility of pseudorotation within any transient five-coordinate intermediate.[42] These reactions have been used to illustrate the application of the principle of microscopic reversibility, and to expose some published misconceptions as to its use.[43] Rate constants for dissociation of carbon monoxide from *cis*-$[Mn(CO)_4LBr]$ provide good illustrations of very marked *cis* effects, in fact of a similar order of magnitude to *trans* effects in square-planar platinum(II) complexes. Both *cis* and *trans* effects in this type of system can be evaluated from kinetic data for similar chromium(0) and molybdenum(0) complexes, *cis* effects from $[Cr(CO)_5L]$ and $[Cr(CO)_5X]^-$, and from *cis*-$[Mo(CO)_4L_2]$, *trans* effects from *trans*-$[Cr(CO)_4LL']$.[44]

Some rate constants and activation parameters for substitution at $Mn(CO)_5X$ and $Re(CO)_5X$ are given in Table 6.7. The reactivity sequence $Mn(CO)_5Cl > Mn(CO)_5Br > Mn(CO)_5I$ parallels carbonyl stretching frequencies, consistent with strongest π-interactions in the iodide. It is noteworthy that it is possible to correlate activation energies with carbonyl stretching frequencies not only for these three $Mn(CO)_5X$ compounds, but also to include in this same correlation points for $Cr(CO)_6$ and for $Rh(CO)(PPh_3)_2Cl_3$. The resultant correlation extends over an activation energy range of more than $100\,kJ\,mol^{-1}$.[1] Positive activation entropies (Table 6.7), large and positive activation volumes (e.g. +21 to $+22\,cm^3\,mol^{-1}$ for three substitutions at $Mn(CO)_5X$),[17,45] and rate constants which do not depend on the nature and concentration of the incoming group (Table 6.7) all signal dissociative mechanisms.

$[(\eta^5\text{-}C_5Me_5)Mn(CO)_3]$ and $[HMn(CO)_5]$ are analogues of the compounds discussed in the preceding paragraphs. Photolysis of the former produces transient $[(\eta^5\text{-}C_5Me_5)Mn(CO)_2(\text{solvent})]$ (cf. Section 6.2.2 above), providing a dissociative route to substitution products.[46] However, thermal substitution at $[HMn(CO)_5]$ follows a second-order rate law. Such reactions provide a good illustration of relative reactivities for a number of nucleophiles, with triphenyl phosphine reacting nearly a hundred times faster than triphenyl arsine, tri-n-butyl phosphine reacting faster still.[47]

Table 6.7 Kinetic parameters for substitution at Group VII carbonyl halides $M(CO)_5X$.

	X	Solvent	T(K)	$10^5 k(s^{-1})$	$\Delta H^{\ddagger}(kJ\ mol^{-1})$	$\Delta S^{\ddagger}(J\ K^{-1}\ mol^{-1})$
$Mn(CO)_5X + AsPh_3$	Cl	$CHCl_3$	313.2	260	115	+66
	Br			33	125	+79
	I			2	135	+87
$Re(CO)_5X + bipy$	NCO	$C_6H_5CH_3$	333.2	105	98	+8
	Cl	CCl_4		42[a]	114	+26
	Br			7.2[b]	122	+41
	I			0.9	133	+54
$Re(CO)_5X + {}^*CO$	Cl	$C_6H_5CH_3$		20[a]	122	+44
	Br			6.8[b]	132	+59
	I			1.1	138	+64

[a] Cf. 36×10^{-5}, 42×10^{-5} for analogous reactions with triphenyl phosphine, pyridine.
[b] Cf. 7×10^{-5} for analogous reactions both with triphenyl phosphine and with pyridine.

However, it is not clear whether these reactions are simple S_N2 substitution or involve ligand migration and insertion—formyl analogues of the well-established acetyl formation $MnMe(CO)_5 + L \rightarrow Mn(COMe)(CO)_4L$ (Chapter 10).

6.3.2 Ligands that change their electron donicity during the act of substitution

In the search for ways in which to extend the original studies of the substitution reactions of simple mononuclear carbonyl complexes it was reasoned that if a two-electron donor CO molecule was replaced by the three-electron donor NO, analogous sets of complexes with the same coordination numbers, geometries and 18-electron valence shells could be produced using the element one place to the left in the Periodic Table (Table 6.8).

In order to extend their knowledge of tetrahedral reaction centres and to find more examples to compare with $Ni(CO)_4$, Thorsteinson and Basolo[48] examined the substitution reactions of $[Co(CO)_3(NO)]$. In all cases it is the CO that is replaced and the rates, no longer independent of the concentration of the entering ligand, are also very sensitive to its nature. The rate law, under first-order conditions, took the form $k_{obs} = k_1 + k_2[L]$. The value of k_1 was very much less than that for $[Ni(CO)_4]$ under comparable conditions, the k_2 pathway dominating the rate of substitution. Clearly, the introduction of one NO ligand changed the mode of activation from dissociative to associative. This has proved to be a common mechanistic feature of the mixed carbonyl–nitrosyl complexes, with the solitary exception of *trans*-$[V(CO)_4(NO)(PMe_3)]$, where the very strongly electron-donating trimethylphosphine overrides the effect of the nitrosyl ligand to give a dissociative mechanism. In general, reactivities of carbonyl–nitrosyls are very much greater than those of the homoleptic carbonyls.

$[Co(CO)_3(NO)]$ and $[Fe(CO)_2(NO)_2]$ are useful substrates in that their associative mode of substitution permits the assessment of nucleophilicities of a range of incoming groups towards low oxidation state metal centres.

Table 6.8 Comparison of known mixed mononuclear nitrosyl–carbonyl complexes with the corresponding homoleptic carbonyls.

FOUR BONDING PAIRS; TETRAHEDRAL				
$[Ni(CO)_4]$	$[Co(CO)_3(NO)]$	$[Fe(CO)_2(NO)_2]$	$[Mn(CO)(NO)_3]$	$[Cr(NO)_4]$
FIVE BONDING PAIRS; TRIGONAL BIPYRAMIDAL				
$[Fe(CO)_5]$	$[Mn(CO)_4(NO)]$			
SIX BONDING PAIRS; OCTAHEDRAL				
$[Cr(CO)_6]$	$[V(CO)_5(NO)]$			

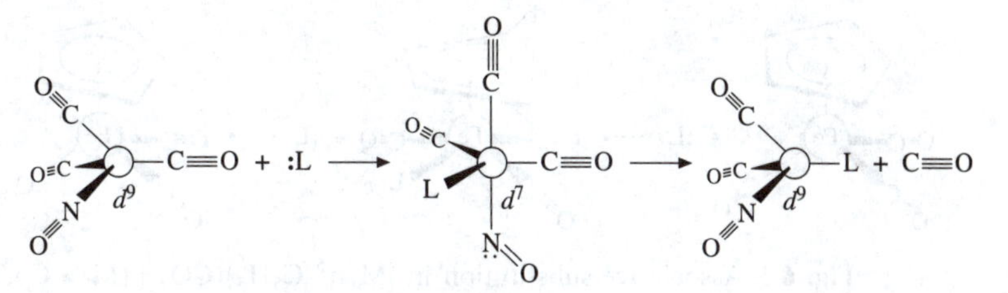

Fig. 6.2 Associative substitution in $[Co(NO)(CO)_3]$, showing the electron migration to the nitrosyl ligand.

There is a reasonably convincing linear correlation of logarithms of rate constants with the proton basicities of the entering ligands for a given ligand type—the correlation lines for nitrogen donors and phosphorus donors are widely separated, but each covers a range of around 10^4 in rate constants.[1]

The role of the nitrosyl group in promoting the associative mechanism arises from its ability to act either as a three-electron donor when it functions as $-N{\equiv}O^+$, isoelectronic with $-C{\equiv}O$, or a one-electron donor when it functions as $-N{\equiv}O^-$. Examples of this type of behaviour are well known in stable complexes, and can be distinguished, not only from the N–O bond length but also from the geometry (M–N–O being linear and bent, respectively). The 18-electron valence shell of the central metal atom can be maintained in the intermediate of higher coordination number by the transfer of two electrons to the nitrogen of the nitrosyl ligand, which changes from a three-electron donor to a one-electron donor. This is formally a two-electron oxidation of the metal (Fig. 6.2).

This type of process is not restricted to nitrosyl complexes but is to be found in a range of complexes where one of the spectator ligands has the ability to reduce its electron donicity by two. Open-sandwich complexes fall into this category, one of the first examples being the displacement of CO from $[M(\eta^5\text{-}C_5H_5)(CO)_2]$ (M = Co, Rh, Ir), which, unlike the analogous reaction of the pseudo-isoelectronic $Fe(CO)_5$, took place readily and with an associative mechanism.[49] In the intermediate of higher coordination number the cyclopentadienyl ligand changes from a five- to a three-electron donor (Fig. 6.3). This type of reaction is well known in organometallic chemistry, where it goes under the name of *ring slippage*, but it is conceptually no different from the process found in the nitrosyl complexes.[50]

The presence of a ligand that can provide this ring slippage does not necessarily ensure an associative process. The reaction

$$[(\eta^5\text{-}C_5H_5)M(CO)_3L] + R_3P \longrightarrow [(\eta^5\text{-}C_5H_5)M(CO)_3(R_3P)] + L$$

(L = CO, C_4H_8E (E = O, S, Se, Te)) is dissociative for M = V, but goes by parallel dissociative and associative (with ring slippage) pathways for M = Nb and Ta. It is suggested that the associatively activated process must involve a large enough reaction centre to accommodate the increase in coordination number.[51]

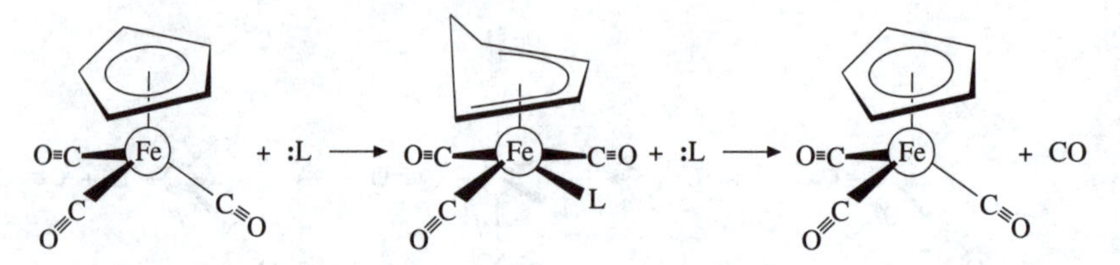

Fig. 6.3 Associative substitution in $[M(\eta^5\text{-}C_5H_5)(CO)_2]$ (M = Co, Rh, Ir) showing the electron migration to the cyclopentadienyl ligand.

Table 6.9 Rate constants for the replacement of pyridine in $[M(CO)_3(PCy_3)_2(py)]$ by $P(OMe)_3$.[a]

M	k_1 s^{-1}	$10^{-3}k_{-1}$ ($dm^3\ mol^{-1}\ s^{-1}$)	$10^{-3}k_2$ ($dm^3\ mol^{-1}\ s^{-1}$)
Cr	606	49	1.5
Mo	37.3	2000	120
W	0.125	860	55

[a] In toluene at 25.0 °C. Reference 52 also gives activation parameters $\Delta H^{\ddagger}$ and $\Delta S^{\ddagger}$ and reaction parameters ΔH° and ΔS° for several of the reactions in these systems.

Facilitation of what is, at first sight at least, a dissociative mode of activation by increasing the bulk of some of the ligands about the metal can be seen in the substitution reactions of the type

$$[M(CO)_3(PCy_3)_2(py)] \underset{k_{-1}}{\overset{k_1}{\rightleftharpoons}} [M(CO)_3(PCy_3)_2] + py$$

$$[M(CO)_3(PCy_3)_2] + P(OMe)_3 \xrightarrow{k_2} [M(CO)_3(PCy_3)_2\{P(OMe)_3\}]$$

where M = Cr, Mo, or W, and PCy_3 is the bulky tricyclohexyl phosphine. Two of the five-coordinate intermediates (Cr, W) have been prepared independently in crystalline form—and their structure determined—and the three rate constants for each system unscrambled (Table 6.9).[52] There is a strong agostic interaction in solid $[M(CO)_3(PCy_3)_2]$ between one of the cyclohexyl hydrogens and the vacant site on the metal (Cr····H, W····H = 2.24, 2.27 Å). It is suggested that the common 17 kJ mol^{-1} activation energy for the addition of pyridine to the three five-coordinate intermediates is evidence for concerted attack at the agostic bond.

6.4 Reaction at coordinated carbonyl

Although nucleophilic substitution at metal carbonyls normally takes place at the metal, there is an alternative possibility of a nucleophile attacking at the carbonyl-carbon, which has a δ+ charge. The best established example is

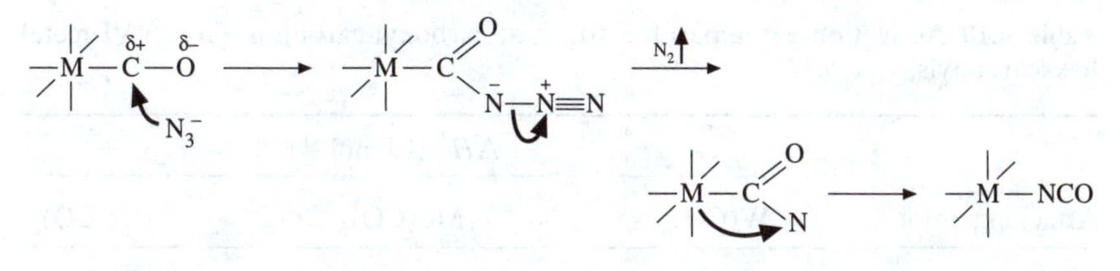

Fig. 6.4 Mechanism of azide attack at Group VI hexacarbonyls, showing initial attack at carbonyl-carbon, transient azide and nitrene intermediates, and final isocyanate product.

that of the reaction of azide with hexacarbonyls, where the key piece of evidence is provided by the product:

$$[M(CO)_6] + N_3^- \longrightarrow [M(CO)_5(NCO)]^- + N_2$$

It is difficult to imagine that CO would be displaced from the metal by azide, but would then return to form cyanate; easier to propose nucleophilic attack by the azide at carbonyl-carbon followed by a Curtius rearrangement, as outlined in Fig. 6.4. All the kinetic information—second-order rate law, much faster reaction than with phosphine and phosphite ligands, and negative activation entropies—is (with the arguable exception of an activation volume of around zero for $[Cr(CO)_6]$ plus azide) consistent with this mechanism. There are several other nucleophiles which attack by this route, for example alkyl and aryl anions as in the reactions:

$$[M(CO)_6] + LiR \longrightarrow [M(CO)_5(COR)]^- + Li^+$$

and, probably, thiocyanate, cyanate, and cyanide. While it is easy to distinguish between attack at C and attack at M when the products do not rearrange, it is necessary to summon up extra evidence when the answer is not obvious from the study of the products. For the reactions with (pseudo)halides:

$$[M(CO)_6] + X^- \longrightarrow [M(CO)_5X]^- + CO$$

(M = Mo, W; X = Br, NCO, NCS, CN) it has been proposed that the rate-determining step is again the attack of X^- on the carbonyl-carbon, to give a much more labile leaving group that dissociates rapidly to $[M(CO)_5]$. This transient fragment then picks up a second X^- to give the observed product. Table 6.10 shows how activation enthalpies for these reactions vary both with the nature of the attacking anion and with the nature of the central metal. In respect to the latter, there is a regular trend Cr > Mo > W, indicating a dominant role for metal radius in determining ease of attack at carbonyl-carbon.

The cationic species $[Mn(CO)_6]^+$ and $[Re(CO)_6]^+$ are much more susceptible to this sort of process. They give the expected products, e.g. $[Re(CO)_5(NCO)]$ on reaction with azide or hydrazine,[54] and may yield stable addition products with, for example, amines or alkoxides:

Table 6.10 Activation enthalpies for attack at carbonyl-carbon in Group VI metal hexacarbonyls.

	$\Delta H^{\ddagger}$ (kJ mol^{-1})		
Attacking anion	$W(CO)_6$	$Mo(CO)_6$	$Cr(CO)_6$
Azide	53.5	63.9	76.1
Thiocyanate	66.9		
Cyanate	86.5		
Cyanide	91.1		

$$[Re(CO)_6]^+ + 2RNH_2 \longrightarrow [Re(CO)_5(CONHR)] + RNH_3^+$$

$$[Mn(CO)_4(phen)]^+ + OR^- \longrightarrow [Mn(CO)_4(phen)(CO_2R)]$$

The addition products with water and with ammonia are less stable:

$$\text{e.g. } [Re(CO)_6]^+ + H_2{}^{18}O \longrightarrow [Re(CO)_5(C^{18}O)]^+ + H_2O$$

though $Mn_2(CO)_{10}$ disproportionates in liquid ammonia to give the reasonably stable $[Mn(CO)_3(NH_3)_3]^+$ cation.[55] The base-catalysed exchange of the carbonyl oxygens of $[Re(CO)_6]^+$ with labelled water ($H_2{}^{18}O$) in alkaline aqueous solution has been attributed to attack by OH^- on the carbonyl carbon:[56]

$$[Re(CO)_6]^+ + {}^{18}OH^- \longrightarrow [Re(CO)_5(CO^{18}OH)]$$

$$[Re(CO)_5(CO^{18}OH)] \longrightarrow [Re(CO)_5(C^{18}O)]^+ + OH^-$$

This chemistry of cationic $[M(CO)_6]^+$, the first suggestion of this type of nucleophilic attack at carbon of coordinated carbonyl, should be contrasted with that of anionic $[M(CO)_5]^-$. Thus, for example, $[Mn(CO)_5]^-$ reacts with water to give $[Mn(H)(CO)_5]$ plus hydroxide.[57] Another situation where the mechanism involves nucleophilic attack by hydroxide at coordinated carbonyl is that of reaction with iron pentacarbonyl. But here further reaction with hydroxide leads to liberation of CO_2 from the coordinated carboxylate in the initial product, and final formation of the tetracarbonylferrate anion:

$$(OC)_4Fe-C{\equiv}O \longrightarrow (OC)_4Fe-C\begin{matrix}{\nearrow}O \\ \searrow O-H\end{matrix} \longrightarrow [Fe(CO)]_4{}^{2-} + CO_3{}^{2-} + H_2O$$

Sometimes a reaction can be even more complicated than indicated by analogy arguments and unsophisticated kinetic studies. For example, the reaction

$$[(\eta^7\text{-}C_7H_7)Mo(CO)_3]^+ + I^- \longrightarrow [(\eta^7\text{-}C_7H_7)Mo(CO)_2I] + CO$$

in acetone proceeds at a rate that is first-order in complex and first-order in $[I^-]$. Since the C_7H_7 moiety can potentially change from a seven-electron to

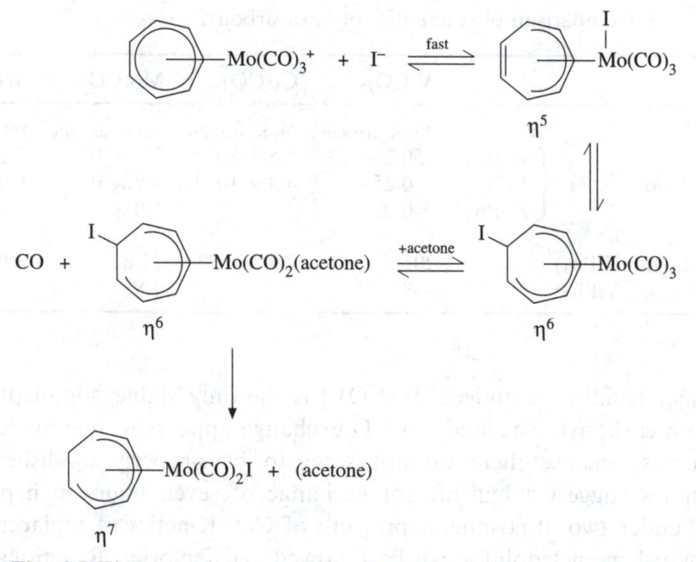

Fig. 6.5 The mechanism of the displacement of CO from $[Mo(\eta^7\text{-}C_7H_7)(CO)_3]^+$ (and its W equivalent) by I^-.

a five-electron donor, a simple associatively activated process might be predicted. However, a detailed stopped-flow NMR study reveals what simple spectrophotometry misses. This is not a simple substitution but a multistage process in which the iodide does indeed add to the molybdenum to give the pentahapto intermediate of higher coordination number, but in a reversible fashion. Instead of losing CO and forming the final product, there is a rate-determining transfer of I^- from the metal to the ligand, which is followed by a solvolytic displacement of CO. The reaction is completed by the transfer of the I^- from the ligand to the metal with the expulsion of the acetone (Fig. 6.5).[58] Although Nature is usually kinder to mechanists than she is in this case, such examples of the invalidity of Occam's Razor when it comes to assigning mechanism are by no means uncommon. It proves just how important it is to investigate a reaction by as many independent techniques as possible.

Nucleophilic attack at coordinated nitric oxide provides a parallel to that at coordinated carbonyl. Such reactions of nitrosyls have been much less studied, but the various factors which affect such reactions, particularly through their effects on N–O bond cleavage, have been rehearsed.[59]

6.5 Substrates with a 17-electron valence shell

In contrast to the enormous range of substrates containing metal centres with 18-electron valence shells that have been examined, $[V(CO)_6]$ offers one of the few 17-electron systems whose reactions can be studied under non-

Table 6.11 Comparison of reactivities of hexacarbonyls.

		$V(CO)_6$	$Cr(CO)_6$	$Mo(CO)_6$	$W(CO)_6$
		298 K; hexane	404 K; decalin	385 K; decalin	439 K; decalin
$k_2(dm^3\ mol^{-1}\ s^{-1})$	P^nBu_3	50.2	8.5×10^{-5}	2.1×10^{-3}	7.1×10^{-4}
	PPh_3	0.25	4.3×10^{-5}	1.8×10^{-4}	8.9×10^{-5}
	$AsPh_3$	~0.03		1.0×10^{-4}	
$k_2(P^nBu_3)/k_2(PPh_3)$		201	2.0	11.6	8.0
$k_2(PPh_3)/k_2(AsPh_3)$		~8		1.8	

transient conditions. Indeed, $[V(CO)_6]$ is the only stable homoleptic odd-electron carbonyl. The study of CO exchange appears to present technical difficulties because there do not seem to be properly published data. Footnotes suggest a half-life for exchange of seven hours in heptane at 10 °C under two atmospheres pressure of CO. Kinetics of replacement of CO by other nucleophiles can be followed satisfactorily. Reactions of the type

$$[V(CO)_6] + L \longrightarrow [V(CO)_5L] + CO$$

follow a second-order rate law (first-order in incoming nucleophile), with second-order rate constants ranging from 0.02 $dm^3\ mol^{-1}\ s^{-1}$ for $L = AsPh_3$ through 0.2, 50 for PPh_3 and P^nBu_3 to 132 for PMe_3, at 25.0 °C in hexane.[60] This rate law and the sensitivity of rate to nature of the incoming nucleophile, together with relatively low activation enthalpies in the range 30–45 $kJ\ mol^{-1}$ and strongly negative activation entropies in the range −100 to −120 $J\ K^{-1}\ mol^{-1}$, are all indicative of an associative mechanism. It is remarkable how very much more reactive $[V(CO)_6]$ is than the 18-electron $t_{2g}^{\ 6}$ $[Cr(CO)_6]$—a factor of about 10^{10} for P^nBu_3 as nucleophile. It is also noteworthy that steric factors are very much less important than electronic factors in determining reactivity in these $[V(CO)_6]$ substitutions—there must be a rather long bond to, i.e. rather a weak interaction with, the incoming nucleophile in the transition state.

One must assume that, with one less electron than other carbonyls, the reluctant dissociative mode of activation is replaced by a facile associative pathway. The reaction centre can apparently tolerate a 19-electron valence shell in the transition state. It has been suggested that this is stabilized by a two-centre three-electron bond.[61] $[V(CO)_6]$ exerts considerably more discrimination than the Group VI hexacarbonyls (Table 6.11)—the former is thus a considerably more convincing candidate for associative substitution than the latter.

The 17-electron $[Mn(CO)_5]$ and $[Re(CO)_5]$, which can be generated as transients from $[Mn_2(CO)_{10}]$ and $[Re_2(CO)_{10}]$, respectively, also react associatively with phosphines, but with second-order rate constants of the order of 10^7–10^9 $dm^3\ mol^{-1}\ s^{-1}$.[62] The time-scales for these reactions are similar to those for dimerization of these pentacarbonyl radicals, characterized by rate constants of 9×10^8 and 3×10^9 $dm^3\ mol^{-1}\ s^{-1}$ for $[Mn(CO)_5]$ and $[Re(CO)_5]$, respectively.

Displacement of a second CO, from $[V(CO)_6]$ or from $[Mn(CO)_5]$ or $[Re(CO)_5]$, is much slower. Reactions of $[V(CO)_5L]$ ($L = PR_3$, $P(OR)_3$, AsR_3) have been studied but in general both L and CO are replaceable. However, qualitative comparisons show that reactivities of $[V(CO)_5(PR_3)]$ and of $[Mn(CO)_4(PR_3)]$ are 10^3–10^6 times less than those of the parent homoleptic carbonyls. Mononuclear manganese pentacarbonyl derivatives $[Mn(CO)_3(PR_3)_2]$ are often much more stable than $[Mn(CO)_5]$, as the bulkiness of phosphine ligands L can prevent Mn–Mn bond formation. Reactions of the type

$$2[Mn(CO)_3(PR_3)_2] + 2CO \longrightarrow [Mn_2(CO)_8(PR_3)_2] + 2L$$

follow rate laws which are approximately first-order in carbon monoxide, with faster reaction for compounds with less bulky phosphine ligands. Rate law and steric effects both suggest predominantly associative activation.[63]

When the entering nucleophile is 'hard', e.g. a nitrogen or oxygen donor, disproportionation is observed, e.g.

$$3[V(CO)_6] + 6py \longrightarrow [V(py)_6]^{2+} + 2[V(CO)_6]^- + 6CO$$

The kinetics also follow a simple second-order rate law and it is suggested that the rate-determining step is indeed the replacement of the first CO by pyridine. However, the rate constant for the reaction with pyridine ($k_2 = 0.70\ dm^3\ mol^{-1}\ s^{-1}$), leading to disproportionation, is not much smaller than that for the entry of PMe_3 ($k_2 = 1.22\ dm^3\ mol^{-1}\ s^{-1}$) at the same temperature (25.0 °C).[64] Rate constants for reaction with other 'hard' donors cover a fairly wide range, down to approximately $10^{-4}\ dm^3\ mol^{-1}\ s^{-1}$ for reaction with diethyl ether. Mechanisms that involve a preliminary disproportionation of the 17-electron species

$$2[V(CO)_6] \longrightarrow [V(CO)_6]^+ + [V(CO)_6]^-$$

followed by rapid substitution can be ruled out because they are inconsistent with the experimentally established rate law.

The relative importance of the associative and dissociative pathways has been compared for *CO exchange with substitution-inert $[(\eta^5\text{-}C_5Me)_2V(CO)]$ (the reaction of $[(\eta^5\text{-}C_5Me_5)_2Fe(CO)_2\cdot]$ with $P(OMe)_3$ is very fast and associative[65]). It is possible to examine the kinetics and equilibrium of addition of CO to the fifteen-electron $[(\eta^5\text{-}C_5Me_5)_2V]$ complex and from this calculate that the first-order rate constant for the dissociation of CO from the seventeen-electron $[(\eta^5\text{-}C_5Me_5)_2V(CO)]$ is $1.4 \times 10^{-5}\ s^{-1}$. This can be compared with the directly measured rate constant for the associative exchange of CO with this substrate, $k_2 = 48\ dm^3\ mol^{-1}\ s^{-1}$.[66] Associative attack at substrates of this type is subject to electronic effects—thus associative substitution in the η^5-6,6-dimethylcyclohexadienyl analogue **1** is considerably slower than at $[(\eta^5\text{-}C_5Me_5)_2V(CO)]$. Whereas associative

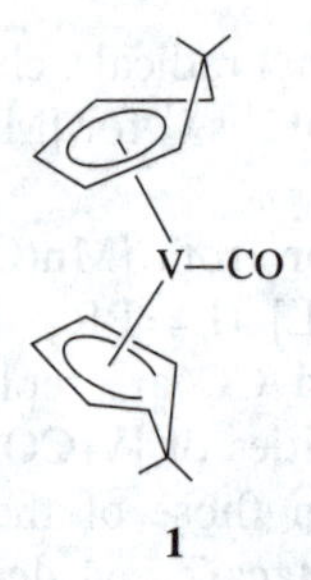

attack at seventeen-electron carbonyls is many orders of magnitude faster than at eighteen-electron carbonyls, dissociative substitution occurs at similar rates for 17- and 18-electron analogues. This difference can be understood in terms of the relative (in)stabilities of 19- and 20-electron and of 15- and 16-electron intermediates/transition states.

6.6 Binuclear and cluster carbonyls

This is an area of substantial and growing interest. It extends from the simple binuclear species such as $[M_2(CO)_{10}]$ (M = Mn, Tc, Re), where there is a single metal–metal bond and no bridging carbon monoxides, to small-cluster carbonyl complexes such as $[Os_3(CO)_{12}]$, and on to the very large clusters containing several dozen metal atoms, which may serve as models for linking homogeneous catalysis (Chapter 10) to heterogeneous catalysis. The most detailed mechanistic work has been done on the Mn, Tc, and Re systems, where our discussion starts. There is also a respectable body of kinetic information on substitution at tri- and tetranuclear carbonyls (Sections 6.6.2 and 6.6.3).

6.6.1 Manganese and rhenium carbonyls

The kinetics of the first two stages in phosphine or phosphite substitution in the Group VII decacarbonyls $M_2(CO)_{10}$, M = Mn, Tc, or Re

$$M_2(CO)_{10} \longrightarrow M_2(CO)_9(PR_3) \longrightarrow M_2(CO)_8(PR_3)_2$$

follow simple first-order rate laws. At least four mechanisms have been put forward for these substitution reactions:

(a) direct substitution in the dimer with the metal–metal bond acting as a spectator ligand;
(b) homolytic fission of the M–M bond followed by rapid substitution in the 17-electron $M(CO)_5$ fragments;
(c) heterolytic fission of the M–M bond followed by rapid substitution in the 16- and 18-electron $M(CO)_5$ fragments;[67] and

(d) an isomerization pathway involving insertion of a terminal carbonyl into the M–M bond.[68]

The key experiment in distinguishing between (a) and (b), the main contenders, was that involving $^{185}Re_2(CO)_{10}$ and $^{187}Re_2(CO)_{10}$. There was no evidence for the formation of $^{185}Re^{187}Re(CO)_{10}$ after many substitution half-lives, showing that there could be no M–M homolysis in thermal substitution at $Re_2(CO)_{10}$. An equivalent experiment cannot be carried out for manganese, as there are no suitable isotopes of this metal, but ^{13}CO-labelling experiments give indirect support for the hypothesis that Mn–CO rather than Mn–Mn bond breaking is also rate-determining in thermal substitution at $Mn_2(CO)_{10}$. Photolysis of $Mn_2(CO)_{10}$[69] results in parallel Mn–Mn and Mn–CO cleavage; the 17-electron $[Mn(CO)_5]$ radical (cf. above) resulting from the former can undergo a variety of substitution and redox reactions.

The behaviour of carbonyls such as $Co_2(CO)_8$ differs considerably from that of the carbonyls $M_2(CO)_{10}$, M = Mn, Tc, or Re. $Co_2(CO)_8$ reacts with strong nucleophiles—very rapidly with P^nBu_3, less quickly with PPh_3—to form $[Co(CO)_3(PR_3)_2]^+$, as the $[Co(CO)_4]^-$ salt, with liberation of carbon monoxide. The kinetics of these reactions are complicated, with the rate law apparently changing with temperature, suggesting a radical mechanism. Reactions with weak nucleophiles, such as $AsPh_3$, are very much slower—reaction with $AsPh_3$ is about a thousand times slower than with P^nBu_3 at an initial incoming ligand concentration of 0.05 mol dm^{-3}. Now the products are binuclear, successively $[Co_2(CO)_7(AsPh_3)]$ and $[Co(CO)_6(AsPh_3)_2]$, with a rate law zero-order in $AsPh_3$ and marked inhibition by added carbon monoxide, indicating a dissociative mechanism.[70] In the case of the binuclear cobalt carbonyl, it is possible to insert GeI_2 or SnI_2

$$[Co_2(CO)_8] + GeI_2 \longrightarrow (OC)_4Co\text{–}GeI_2\text{–}Co(CO)_4$$

a reaction not achievable for the Group VII compounds. Similar reactions occur with anionic species such as $[Cr_2(CO)_8]^{2-}$ and $[W_2(CO)_8]^{2-}$.[71]

6.6.2 Trinuclear clusters

Kinetic and product studies on substitution at the trinuclear carbonyls $M_3(CO)_{12}$, M = Fe, Ru, or Os, gave the simplest results for the ruthenium compound. For reaction of $Ru_3(CO)_{12}$ with phosphines and phosphites (L) products $Ru_3(CO)_9L_3$ were obtained, but the simple two-term rate law

$$-d[Ru_3(CO)_{12}]/dt = \{k_1 + k_2[L]\}[Ru_3(CO)_{12}]$$

was followed, for replacement of the first CO ligand, as the second and third substitution steps are much faster than the first. For the k_1 term, which is dominant for the weaker ligands, $\Delta S_1^{\ddagger}$ is large and positive, as expected for a dissociative path. Values of k_2 vary as expected for associative attack,[72] though whether this is simple S_N2 or a more complicated bimolecular fragmentation process (F_N2) has been a matter for discussion. This discussion entailed detailed consideration of correlations of k_2 values for several cluster carbonyls with steric and electronic effects.[73] The kinetic characteristics[74] and activation volumes[75] for substitution reactions (k_1 path) of the $Ru_3(CO)_{12}$ derivatives $[Ru_3H(CO)_{11}]^-$ (with PPh_3) and $Ru_3(CO)_{11}(OAc)$ (with $P(OMe)_3$) are $+21\ cm^3\ mol^{-1}$ and $+16\ cm^3\ mol^{-1}$ (in 90% THF/10% MeOH—see Chapter 8 for medium effects on this value), respectively, consistent with dissociative activation. Under photochemical conditions, substitution may well be accompanied by break-up of the cluster (cf. thermal F_N2 above), as in the production of $Ru(CO)_5$ from the reaction of $Ru_3(CO)_{12}$ with carbon monoxide in cyclohexane.[13]

Kinetics of thermal substitution at $Os_3(CO)_{12}$ are complicated by the smaller differences between reactivities for the three steps. The intermediacy both of $Os_3(CO)_{11}L$ and of $Os_3(CO)_{10}L_2$ has to be allowed for in a full consecutive-reactions kinetic analysis. When L in the latter = MeCN, attempts to substitute the good leaving group MeCN by pyridine gave not the expected $Os_3(CO)_{10}(py)_2$ but $HOs_3(CO)_{10}(C_5H_4N)$—a reaction of C–H bond activation (Chapter 10) rather than simple substitution. In the case of $Fe_3(CO)_{12}$, analysis of the kinetics is complicated by the dependence of products on temperature, with $Fe_3(CO)_{11}L$ and $Fe_3(CO)_{10}L_2$ being the major products at 50 °C, but mononuclear $Fe(CO)_4L$ and $Fe(CO)_3L_2$ at 80 °C. Substitution at mixed-metal trinuclear carbonyls has also been studied, with the main point of interest being that CO is displaced from $Fe_2Ru(CO)_{12}$ and $FeRu_2(CO)_{12}$ considerably faster than from the parent $Fe_3(CO)_{12}$. This greater lability is reflected in lower values for $\Delta H^{\ddagger}$.[76]

6.6.3 Tetranuclear clusters

Kinetics of substitution at tetrairidium dodecacarbonyl have been quite extensively studied,[77] with results available for replacement of the first three carbonyls, namely

$$[Ir_4(CO)_{12}] \longrightarrow [Ir_4(CO)_{11}(PR_3)] \rightarrow [Ir_4(CO)_{10}(PR_3)_2] \longrightarrow [Ir_4(CO)_9(PR_3)_3]$$

The rate law for the first step is the common two-term type[78]

$$-d[Ir_4(CO)_{12}]/dt = \{k_1 + k_2[PR_3]\}[Ir_4(CO)_{12}]$$

with k_2 ranging from $\gg 130 \times 10^{-4}\ dm^3\ mol^{-1}\ s^{-1}$ for P^nBu_3 down to $2.7 \times 10^{-4}\ dm^3\ mol^{-1}\ s^{-1}$ for $AsPh_3$, and corresponding activation entropies $\Delta S_2^{\ddagger}$ of -70 and $-110\ J\ K^{-1}\ mol^{-1}$ (all in chlorobenzene solution) being compatible with associative substitution for this k_2 term. For this first step in the substitution sequence, the k_2 term is dominant for incoming phosphorus-donor ligands, contributing >95% of the reaction under normal laboratory conditions. Only for $AsPh_3$ does the k_1 term represent an important pathway, with $\Delta S_1^{\ddagger}$ of +27 J K^{-1} mol^{-1} being consistent with dissociative activation.

For the second step in the substitution sequence, the k_2 term is only significant for the strong nucleophiles P^nBu_3 and CN^tBu, while for the third step only the k_1 term can be detected under normal laboratory conditions. As a consequence of this variation in rate law it is impossible to make quantitative reactivity comparisons for the three steps, but in a qualitative sense it can be said that substitution gets appreciably faster going from first to second to third step in the sequence. As mentioned above, there is a similar speeding up on going from the first to the second step in substitution at the binuclear carbonyl $Mn_2(CO)_{10}$. Working in the opposite direction, steric effects can be more important than electronic effects in determining reactivity in reactions

$$Ir_4(CO)_8(PR_3)_4 + CO \longrightarrow Ir_4(CO)_9(PR_3)_3 + PR_3$$

where correlation of rate constants with cone angles has been established, and in carbon monoxide exchange at $Ir_4(CO)_9(PR_3)_3$.[79]

The operation of the two-term rate set out above, and the importance of cone angles, also apply to substitution at rhodium analogues. This has been established, for instance, in an investigation of such complexes as $Rh_4(CO)_{10}(PCy_3)_2$, $Rh_4(CO)_9(HC\{PPh_2\}_3)$, and **2**. However, here there is

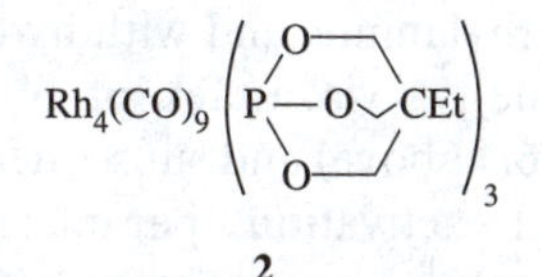

2

the complication that the associative path may lead to substitution or to fragmentation, depending on the ligand.[80] Kinetics of substitution at $Co_4(CO)_{12}$ are complicated by autocatalysis.[81]

Substitution at mixed-metal polynuclear carbonyls introduces the possibility of isomers in which a ligand is bonded to one metal or the other. Reaction of $Ir_2Rh_2(CO)_{12}$ with triphenyl phosphine gives a mixture of the kinetically favoured isomer containing PPh_3 bonded to Rh and the thermodynamically favoured isomer containing PPh_3 bonded to Ir $\{K(Ir–P/Rh–P) \sim 3\}$. In the first high-pressure ^{31}P NMR kinetic study, activation volumes of +13 and +10 cm^3 mol^{-1} were established for PPh_3 moving from Ir to Rh and Rh to Ir, respectively, indicating dissociative intermolecular pathways—substitution rather than pseudorotation (cf. Section 5.2.3) in both directions.[82]

6.7 Substitution at related compounds

As mentioned earlier in this chapter, ligands such as phosphines, nitriles, and alkyls can behave very similarly to carbon monoxide—certainly their complexes show the same overriding deference to the eighteen-electron rule. We are therefore concluding this chapter with a few examples of kinetic and mechanistic studies of such complexes.

Phosphines have appeared in earlier sections as replacing ligands for carbon monoxide, with CO loss becoming successively more difficult and slower as more CO ligands are replaced by PR_3, or $P(OR)_3$, Nonetheless, it is possible to prepare homoleptic phosphine complexes, such as $W(PMe_3)_6$ and $Os(PMe_3)_5$, and study their substitution kinetics. $W(PMe_3)_6$ reacts dissociatively in C_6D_6, with a transient $W(PMe_3)_5$ intermediate leading to **3** as product. Activation parameters $\Delta S^{\ddagger} = +58$ J K^{-1} mol^{-1} and $\Delta H^{\ddagger} = 108$ kJ mol^{-1} are consistent, in sign and magnitude, respectively, with dissociative activation here.[83]

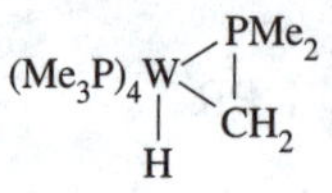

$(Me_3P)_4W$, PMe_2, CH_2, H

3

Substitution kinetics have been studied for nitrile and for isonitrile replacements. $[Fe(CNMe)_6]^{2+}$, with its six metal–carbon bonds, is truly organometallic. It, and derivatives such as $[Fe(CNMe)_4(CN)_2]$, undergo stepwise photoaquation or photosolvolysis in coordinating solvents such as acetonitrile.[84] $[Fe(CNMe)_6]^{2+}$ reacts with methylamine, and with hydrazine or hydrazinium, to give $[Fe(CNMe)_4(carbene)]^{2+}$ via attack of N_2H_4 or $N_2H_5^+$ at the coordinated ligand (cf. Section 6.4 above) and subsequent ring closure. The second-order rate law and activation parameters of $\Delta H^{\ddagger} = 59$ kJ mol^{-1} and $\Delta S^{\ddagger} = -58$ J K^{-1} mol^{-1} are consistent with this mechanism.[85]

Substitution at $[Fe(MeCN)_6]^{2+}$, in the present context better written as $[Fe(NCMe)_6]^{2+}$, the non-organometallic isomer of $[Fe(CNMe)_6]^{2+}$, has also been studied, albeit rather qualitatively. The first two steps in substitution by $P(OMe)_3$ are complete within a few minutes, even at temperatures as low as 243 K. Further substitution stages are succcessively slower, with several weeks elapsing before there is any sign of the ultimate product, $[Fe\{P(OMe)_3\}_6]^{2+}$.[86] $[Fe(MeCN)_6]^{2+}$, a much more kineticist-friendly species than $[Fe(CO)_6]^{2+}$, together with $[Ru(MeCN)_6]^{2+}$ and $[Ru(MeCN)_6]^{3+}$, provides a link to Werner complexes and solvento cations—see Chapter 7 for their solvent-exchange kinetic parameters. Another link is provided by the ternary complex $[Ru(C_5H_5)(MeCN)_3]^+$, where activation parameters of $\Delta V^{\ddagger} = +11$ cm^3 mol^{-1} and $\Delta S^{\ddagger} = 60$ J K^{-1} mol^{-1} provide strong indications of a dissociative mechanism for acetonitrile exchange. Of particular relevance to this chapter is the comparison with $\Delta V^{\ddagger} = 0$ cm^3 mol^{-1} and $\Delta S^{\ddagger} = +33$ J K^{-1} mol^{-1} for acetonitrile exchange at $[Ru(MeCN)_6]^{2+}$. The cyclopentadienyl ligand has a marked effect on kinetic parameters, suggesting a significant mechanistic change from I_d to D on going from the binary solvento cation to the ternary $[Ru(C_5H_5)(MeCN)_3]^+$ cation.[87]

References

1. There are many reviews, but that of F. Basolo, *Polyhedron*, 1990, **9**, 1503 is extremely readable; the early studies are recalled briefly in F. Basolo, *J. Organomet. Chem.*, 1990, **383**, 579 and linked with more recent work in F. Basolo, *Pure Appl. Chem.*, 1988, **60**, 1193.
2. Review: J. D. Atwood, in *Encyclopaedia of Inorganic Chemistry*, Vol. 4, ed. R. B. King, Wiley, Chichester, 1994, p. 2119.
3. For a fuller treatment of several aspects of this chapter the reader is referred to J. D. Atwood, *Inorganic and Organometallic Reaction Mechanisms*, Brooks/Cole, Monterey, 1985, Chapter 4.
4. A. J. Poë, *Pure Appl. Chem.*, 1989, **60**, 1209.
5. F. Basolo and A. Wojcicki, *J. Am. Chem. Soc.*, 1961, **83**, 520.
6. F. Ungváry and A. Wojcicki, *J. Am. Chem. Soc.*, 1987, **109**, 6848.
7. J. D. Atwood, *J. Organomet. Chem.*, 1990, **383**, 59.
8. F. Basolo and R. G. Pearson, *Mechanisms of Inorganic Reactions*, 2nd edn, Wiley, New York, 1967, p. 541.
9. L. H. Jones, R. S. McDowell, and M. Goldblatt, *Inorg. Chem.*, 1969, **8**, 2349; J. A. Connor, H. A. Skinner, and Y. Virmani, *J. Chem. Soc., Faraday Trans. I*, 1972, **68**, 1754; T. Ziegler, V. Tschinke, and C. Ursenbach, *J. Am. Chem. Soc.*, 1987, **109**, 4825.
10. J. P. Day, F. Basolo, R. G. Pearson, L. F. Kangas, and P. M. Henry, *J. Am. Chem. Soc.*, 1968, **90**, 1925.
11. B. J. Huber and A. J. Poë, *Inorg. Chim. Acta*, 1994, **227**, 215.
12. M. Poliakoff and E. Weitz, *Accounts Chem. Res.*, 1987, **20**, 408.
13. R. Huq, A. J. Poë, and S. Chawla, *Inorg. Chim. Acta*, 1980, **38**, 121; W. R. Hastings, M. R. Roussel, and M. C. Baird, *J. Chem. Soc., Dalton Trans.*, 1990, 203.
14. G. R. Dobson, *Accounts Chem. Res.*, 1976, **9**, 300.
15. R. J. Angelici and J. R. Graham, *J. Am. Chem. Soc.*, 1966, **88**, 3658; J. R. Graham and R. Angelici, *Inorg. Chem.*, 1967, **6**, 2082.
16. J. E. Pardue and G. R. Dobson, *Inorg. Chim. Acta*, 1967, **20**, 207.
17. R. van Eldik, in *High Pressure Chemistry and Biochemistry*, eds R. van Eldik and J. Jonas, Reidel, Dordrecht, 1987, p. 333.
18. J. D. Simon and Xiaoliang Xie, *J. Phys. Chem.*, 1987, **91**, 5538.
19. C. H. Langford and L. E. Shaw, *Coord. Chem. Rev.*, 1997, **159**, 221.
20. Yo-ichi Ishikawa, P. A. Hackett, and D. M. Rayner, *J. Phys. Chem.*, 1989, **93**, 652.
21. R. N. Perutz and J. J. Turner, *J. Am. Chem. Soc.*, 1975, **97**, 4791.
22. G. E. Gadd, M. Poliakoff, and J. J. Turner, *Organometallics*, 1987, **6**, 391; S. M. Howdle, M. A. Healy, and M. Poliakoff, *J. Am. Chem. Soc.*, 1990, **112**, 4804.
23. Xue-Zhong Sun, M. W. George, S. G. Kazarian, S. M. Nikiforov, and M. Poliakoff, *J. Am. Chem. Soc.*, 1996, **118**, 10525.
24. B. H. Weiller, *J. Am. Chem. Soc.*, 1992, **114**, 10910.

25. Liang Wang, Xinming Zhu, and K. G. Spears, *J. Am. Chem. Soc.*, 1988, **110**, 8695; Liang Wang, Xinming Zhu, and K. G. Spears, *J. Phys. Chem.*, 1989, **93**, 2.
26. S. Zhang, V. Zang, H. C. Bajaj, G. R. Dobson, and R. van Eldik, *J. Organomet. Chem.*, 1990, **397**, 279; S. Wieland and R. van Eldik, *Organometallics*, 1991, **10**, 3110.
27. S. Zhang, G. R. Dobson, V. Zang, H. C. Bajaj, and R. van Eldik, *Inorg. Chem.*, 1990, **29**, 3477; G. R. Dobson and S. Zhang, *J. Coord. Chem.*, 1990, **21**, 155; S. Zhang and G. R. Dobson, *Polyhedron*, 1990, **9**, 2511; S. Zhang and G. R. Dobson, *Inorg. Chim. Acta*, 1991, **181**, 103.
28. S. Wieland and R. van Eldik, *J. Phys. Chem.*, 1990, **94**, 5865.
29. J. J. Turner, M. B. Simpson, M. Poliakoff, and W. B. Maier, *J. Am. Chem. Soc.*, 1983, **105**, 3898.
30. K. J. Asali, S. S. Basson, J. S. Tucker, B. C. Hester, J. E. Cortes, H. H. Awad, and G. R. Dobson, *J. Am. Chem. Soc.*, 1987, **109**, 5386.
31. H. H. Awad, G. R. Dobson, and R. van Eldik, *J. Chem. Soc., Chem. Commun.*, 1987, 1839.
32. F. Ungváry and A. Wojcicki, *J. Am. Chem. Soc.*, 1987, **109**, 6849.
33. R. H. Crabtree, *The Organometallic Chemistry of the Transition Elements*, Wiley, New York, 1988, pp. 72–73.
34. J. A. S. Howell and P. M. Burkinshaw, *Chem. Rev.*, 1983, **83**, 557.
35. H.-T. Macholdt and R. van Eldik, *Transition Met. Chem.*, 1985, **10**, 323.
36. J. Burgess and A. E. Smith, *Transition Met. Chem.*, 1987, **12**, 140.
37. S. Wieland and R. van Eldik, *Organometallics*, 1991, **10**, 3110.
38. H. H. Awad, C. B. Dobson, G. R. Dobson, J. G. Leipoldt, K. Schneider, R. van Eldik, and H. Wood, *Inorg. Chem.*, 1989, **28**, 1654; K. J. Schneider and R. van Eldik, *Organometallics*, 1990, **9**, 1235.
39. S. Wieland and R. van Eldik, *J. Chem. Soc., Chem. Commun.*, 1989, 367; S. Wieland, K. B. Reddy, and R. van Eldik, *Organometallics*, 1990, **9**, 1802.
40. K. B. Reddy and R. van Eldik, *Inorg. Chim. Acta*, 1990, **169**, 13; K. B. Reddy and R. van Eldik, *Organometallics*, 1990, **9**, 1418; S. Zhang, V. Zang, G. R. Dobson, and R. van Eldik, *Inorg. Chem.*, 1991, **30**, 355; S. Cao, K. B. Reddy, E. M. Eyring, and R. van Eldik, *Organometallics*, 1994, **13**, 91.
41. E. W. Abel, K. G. Orrell, A. G. Osborne, H. M. Paine, and V. Sik, *J. Chem. Soc., Dalton Trans.*, 1991, 111.
42. J. D. Atwood and T. L. Brown, *J. Am. Chem. Soc.*, 1975, **97**, 3380.
43. W. G. Jackson, *Inorg. Chem.*, 1987, **26**, 3004.
44. Details and references, for the Mn, Cr, and Mo series of complexes, can be traced through pp. 115 to 122 of Atwood's book (see Ref. 3).
45. Razak bin Ali, J. Burgess, and P. Guardado, *Transition Met. Chem.*, 1988, **13**, 126.
46. B. S. Creaven, A. J. Dixon, J. M. Kelly, C. Long, and M. Poliakoff, *Organometallics*, 1987, **6**, 2600.
47. B. H. Byers and T. L. Brown, *J. Organomet. Chem.*, 1977, **127**, 181.
48. E. M. Thorsteinson and F. Basolo, *J. Am. Chem. Soc.*, 1966, **88**, 3929.
49. H. G. Schuster-Woldan and F. Basolo, *J. Am. Chem. Soc.*, 1966, **88**, 1657.
50. Reviews: F. Basolo, *Inorg. Chim. Acta*, 1985, **100**, 33; J. M. O'Connor and C. P. Casey, *Chem. Rev.*, 1987, **87**, 307; F. Basolo, *New J. Chem.*, 1994, **18**, 19.
51. J. W. Freeman and F. Basolo, *Organometallics*, 1991, **10**, 256.

52. K. Zhang, A. A. Gonzalez, S. L. Mukerjee, Shou-Jiau Chou, C. D. Hoff, K. A. Kubat-Martin, D. Barnhart and G. J. Kubas, *J. Am. Chem. Soc.*, 1991, **113**, 9170.
53. H. Werner, W. Beck, and H. Engelman, *Inorg. Chim. Acta*, 1969, **3**, 331.
54. R. J. Angelici and G. C. Faber, *Inorg. Chem.*, 1971, **10**, 514.
55. D. J. Darensbourg and J. A. Froelich, *J. Am. Chem. Soc.*, 1977, **99**, 4726; H. Behrens, E. Ruyter, and H. Wakamatsu, *Z. Anorg. Allg. Chem.*, 1967, **349**, 241.
56. E. L. Muetterties, *Inorg. Chem.*, 1965, **4**, 1841.
57. T. Kruck and M. Noack, *Chem. Ber.*, 1964, **97**, 1693.
58. P. Powell, L. J. Russell, E. Styles, A. J. Brown, O. W. Howarth, and P. Moore, *J. Organomet. Chem.*, 1978, **149**, C41.
59. P. Legzdins and M. A. Young, *Comments Inorg. Chem.*, 1995, **17**, 239.
60. Qi-Zhen Shi, T. G. Richmond, W. C. Trogler, and F. Basolo, *J. Am. Chem. Soc.*, 1984, **106**, 71; W. C. Trogler, *Int. J. Chem. Kinetics*, 1987, **19**, 1025.
61. M. J. Therien and W. C. Trogler, *J. Am. Chem. Soc.*, 1988, **110**, 4942; Zhenyang Lin and M. B. Hall, *J. Am. Chem. Soc.*, 1992, **114**, 6574.
62. D. R. Kidd and T. L. Brown, *J. Am. Chem. Soc.*, 1978, **100**, 4095; T. L. Brown, *Ann. N. Y. Acad. Sci.*, 1980, **333**, 80; T. R. Herrinton and T. L. Brown, *J. Am. Chem. Soc.*, 1985, **107**, 5700.
63. S. B. McCullen, H. W. Walker, and T. L. Brown, *J. Am. Chem. Soc.*, 1982, **104**, 4007.
64. Qi-Zhen Shi, T. G. Richmond, W. C. Trogler, and F. Basolo, *J. Am. Chem. Soc.*, 1982, **104**, 4032; T. G. Richmond, Qi-Zhen Shi, W. C. Trogler, and F. Basolo, *J. Am. Chem. Soc.*, 1984, **106**, 76; W. C. Trogler, *Int. J. Chem. Kinetics*, 1987, **19**, 1025.
65. A. J. Dixon, S. J. Gravelle, L. J. van de Burgt, M. Poliakoff, J. J. Turner, and E. Weitz, *J. Chem. Soc., Chem. Commun.*, 1987, 1023.
66. R. M. Kowaleski, F. Basolo, W. C. Trogler, K. W. Gedridge, T. D. Newbound, and R. D. Ernst, *J. Am. Chem. Soc.*, 1987, **109**, 4860, N. C. Hallinen, G. Morelli, and F. Basolo, *J. Am. Chem. Soc.*, 1988, **110**, 6585.
67. B. F. G. Johnson, *J. Organomet. Chem.*, 1991, **415**, 109.
68. A. J. Poë, *Inorg. Chim. Acta*, 1987, **129**, L17.
69. T. J. Meyer and J. V. Caspar, *Chem. Rev.*, 1985, **85**, 187.
70. M. Absi-Halabi, J. D. Atwood, N. P. Forbes, and T. L. Brown, *J. Am. Chem. Soc.*, 1980, **102**, 6248.
71. J. K. Ruff, *Inorg. Chem.*, 1967, **6**, 2081.
72. J. P. Candlin and A. C. Shortland, *J. Organomet. Chem.*, 1969, **16**, 289.
73. N. M. J. Brodie, L. Chen, and A. J. Poë, *Int. J. Chem. Kinet.*, 1988, **20**, 467.
74. D. J. Taube and P. C. Ford, *Organometallics*, 1986, **5**, 99.
75. D. J. Taube, R. van Eldik, and P. C. Ford, *Organometallics*, 1987, **6**, 125; J. Anhaus, H. C. Bajaj, R. van Eldik, L. R. Nevinger, and J. B. Keister, *Organometallics*, 1989, **8**, 2903.
76. R. Shojaie and J. D. Atwood, *Inorg. Chem.*, 1987, **26**, 2199.
77. K. J. Karel and J. R. Norton, *J. Am. Chem. Soc.*, 1974, **96**, 6812; D. C. Sonnenberger and J. D. Atwood, *Inorg. Chem.*, 1981, **20**, 3243; D. C. Sonnenberger and J. D. Atwood, *J. Am. Chem. Soc.*, 1982, **104**, 2113.
78. See Fig. 1.6 on p. 12 of Ref. 3.
79. D. J. Darensbourg and B. J. Baldwin-Zuschke, *J. Am. Chem. Soc.*, 1981, **104**, 3906; *Inorg. Chem.*, 1981, **20**, 3846.

80. N. M. J. Brodie and A. J. Poë, *J. Organomet. Chem.*, 1990, **383**, 531.
81. J. R. Kennedy, F. Basolo, and W. C. Trogler, *Inorg. Chim. Acta*, 1988, **146**, 75.
82. G. Laurenczy, G. Bondietti, A. E. Merbach, B. Moullet, and R. Roulet, *Helv. Chim. Acta*, 1994, **77**, 547.
83. D. Rabinovich and G. Parkin, *J. Am. Chem. Soc.*, 1990, **112**, 5381.
84. L. L. Costanzo, S. Giuffrida, G. de Guidi, and G. Condorelli, *Inorg. Chim. Acta*, 1985, **101**, 71; L. L. Costanzo, S. Giuffrida, G. de Guidi, and G. Condorelli, *J. Organomet. Chem.*, 1986, **315**, 73.
85. D. H. Cuatecontzi, Miller, S. and J. D. Miller, *Inorg. Chim. Acta*, 1980, **38**, 157.
86. C. J. Barbour, J. H. Cameron, and J. M. Winfield, *J. Chem. Soc., Dalton Trans.*, 1980, 2001.
87. W. Luginbühl, P. Zbinden, P. A. Pittet, T. Armbruster, H.-B. Bürgi, A. E. Merbach, and A. Ludi, *Inorg. Chem.*, 1991, **30**, 2350.

7 Solvent exchange and complex formation

Chapter contents

7.1 Introduction

It is convenient and helpful to consider these two special cases of substitution together,[1,2] since they share the key step of loss of a coordinated solvent molecule from a solvento cation:

$$\text{e.g.}\quad [M(H_2O)_6]^{n+} + H_2O \longrightarrow [M(H_2O)_5(H_2{}^*O)]^{n+} + H_2O$$

$$[M(H_2O)_6]^{n+} + L \longrightarrow [M(H_2O)_5L]^{n+} + H_2O$$

Indeed, in many cases where activation is dissociative, the rate-limiting steps are essentially identical, while even for associative processes stretching of the M^{n+} to leaving water (solvent) bond is a major component of the transformation of the initial state into the transition state. It is also convenient to deal with this group of substitutions in a separate chapter, since it is useful to discuss patterns of mechanism and of reactivity in relation to the Periodic Table for these reactions without the division according to coordination number and stereochemistry which controlled the arrangement of Chapters 3 and 4 (similar comments apply to substitution at carbonyls, Chapter 6). This is not to deny the importance of these two factors, but rather to recognize the comparable importance of others. One also needs to deal with coordination (solvation) numbers greater than six, which make only the briefest of appearances in Chapters 3 and 4.

This organization of material also avoids problems with solvated cations of uncertain or unknown solvation number. A number of these still exist, despite the great advances in recent years in knowledge of solvation (hydration) numbers and geometry of solvated cations, especially of aqua-ions, from NMR and from X-ray diffraction and neutron diffraction studies of salt solutions.[3] It is clearly highly desirable to know the solvation number of a cation—otherwise one does not actually know the nature of the key reactant in its solvent-exchange and complex-formation reactions. NMR spectroscopy has proved invaluable in establishing solvation numbers, but it can only be used over a limited range of lifetimes for solvent molecules in cation primary solvation shells. Solvent exchange with solvento cations is too fast for NMR techniques for roughly half the cations of interest to inorganic chemists. X-ray diffraction, and related techniques such as EXAFS and neutron diffraction, operate on a very much shorter time-scale, and can give cation–solvent distances for solvento cations in solution with considerable accuracy. These techniques are, unfortunately, less useful for establishing solvation numbers.[4] Currently they provide estimates with uncertainties of about one solvent molecule, which means that they are of little help in deciding whether a cation has six or seven or eight solvent molecules in its primary solvation sphere. X-ray and neutron methods are also of restricted value in determining the geometry of solvation shells, though in some cases it is possible to establish this from, say, metal–oxygen and oxygen–oxygen distances.[5] Of course the situation is very different for solid crystalline materials, and the detailed geometry of solid hydrates and other solvates obtained from classical X-ray diffraction crystal structure determinations is very useful in providing models for the analogous solvento cations in solution. In many cases, particularly for the earlier *sp*-block elements and for transition metals, it is reasonably certain that octahedral, tetrahedral, and square-planar solvento cations have essentially identical geometries in the solid state and in solution. However, there remain difficulties over a number of cations. Thus, for instance, the very fast solvent exchange rates for the alkali metal cations means that their hydration numbers in aqueous solution are not well defined, and for, for instance, K^+aq, indirect evidence for a tetrahedral aqua-cation in solution contrasts with the six-coordination found in the very few structurally characterized salts, such as $KF.2H_2O$, $KF.4H_2O$, and alums, which contain hydrated K^+ ions. There have been long-standing problems with the lanthanide cations in solution, though now it is becoming clearer that nine-coordination is normal for the larger ions towards the left of the 4*f* series, eight-coordination towards the end of the series. The nature of $Y^{3+}aq$ in solution must be considered uncertain, in view of the existence of octa- and ennea-hydrated forms in the solid state, in the hydrated yttrium salts of octachloroditechnetate(3−)[6] and ethylsulphate,[7] respectively. The former contains the approximately square-antiprismatic $[Y(H_2O)_8]^{3+}$ cation, despite the fact that the salt is an enneahydrate, $Y[Tc_2Cl_8]9H_2O$, and that early thermal analysis[8] suggested three weakly bound waters, six firmly bound, i.e. adumbrated octahedral $[Y(H_2O)_6]^{3+}$! The case of $Sc^{3+}aq$ is also uncertain. The $[Sc(H_2O)_9]^{3+}$ unit has recently been established in X-ray diffraction studies of crystals of hydrated scandium triflate,[9] but several hydrated salts contain octahedral ScO_6 units, albeit with the oxygens coming from coordinated oxoanions as well as from coordinated waters.[10] The binuclear di-μ-hydroxo anion $[(H_2O)_5Sc(OH)_2Sc(H_2O)_5]^{4+}$ contains seven-coordinate Sc^{3+}.[11] It may well be that Sc^{3+} has a hydration number of six, or perhaps seven or eight, in aqueous solution. An instant increase from a hydration number of six for Al^{3+} to nine for Sc^{3+} seems distinctly implausible—specific hydrogen-bonding effects in the solid state stabilizing $[Sc(H_2O)_9]^{3+}$ in the triflate seem less implausible. It should be added that there is good NMR evidence for the existence of six-coordinate Sc^{3+} in a number of organic-donor solvents. We have dealt with the rather obscure scandium and yttrium as they illustrate actual and potential problems rather well. In particular, we must emphasize that whereas in most inorganic reaction mechanisms it is the nature of the transition state that is debatable, in solvent exchange and complex formation the nature of the initial state may be equally open to debate. However, we should end this section by reminding readers that solvation numbers and coordination geometry are known, or may be assumed with confidence, for the majority of ions. There are many solvates which are undisputably octahedral, two cations (Pd^{2+} and Pt^{2+}) whose solvates are square-planar, and a few solvates which are known to be tetrahedral (Be^{2+} solvates; $[Cu(MeCN)_4]^+$).

7.2 Solvent exchange—determination of kinetic parameters

The exchange of coordinated solvent with bulk solvent is probably the simplest act of substitution that can be considered. All ligands, the entering group, and the leaving group are the same and thus the reaction comes into the category of a 'no reaction' reaction. In such a process there is no change in the free energy of the system as a result of the exchange and the changes that take place during activation must be symmetrical about the transition state in a synchronous process, or about the intermediate in a two-stage sequence. Any theoretical treatment is thereby simplified—as also in zero-free-energy-change outer-sphere electron-transfer processes discussed in Chapter 9.

The problem of studying these 'no reaction' reactions is how to measure their kinetics, required both to establish the effect of cation variation and of solvent variation on reactivity, and to establish mechanisms of solvent exchange. The early work was done with isotopic labelling, e.g. $H_2{}^{18}O$ exchange, and involved stopping or quenching the reaction, separating free and coordinated water and then measuring isotopic ratios by mass spectrometry. The solvent could be sampled rapidly by drawing off and condensing some of the vapour. $[Cr(H_2O)_6]^{3+}$ was the first and most convenient aqua-cation to be studied in this way.[12] The coordinated solvent itself could be monitored if the solvento complex cation could be precipitated by adding a suitable counterion. Classical isotopic label exchange studies require a slow reaction (or the ability to slow down the reaction) and therefore can only be applied to solvento ions of Cr(III), Rh(III) and, if one has a human lifetime or two to spare, Ir(III). In principle, the fast mixing and sampling technique that Taube used to assign a solvation number to the Al^{3+} ion in water[13] could be used to determine the half-life for its exchange of water, but there are now easier ways to do this.

The availability of nuclear magnetic resonance has changed the picture completely. Using line-broadening techniques, ^{1}H NMR can be used to study the water-exchange kinetics of diamagnetic aqua-ions, and ^{17}O NMR can be used for some paramagnetic species—the ^{17}O line width is much greater than the difference in the chemical shift of bulk solvent water and bound water unless paramagnetic species are present to Knight shift the latter. By these means the range of measurable rate constants has been extended to $10^{6}\ s^{-1}$. Much of the early NMR work, published in the 1960s, has been repeated using improved techniques and instrumentation, and temperature and pressure dependences investigated. Precise activation parameters ($\Delta H^{\ddagger}, \Delta S^{\ddagger}, \Delta V^{\ddagger}$) are now available for many aqua-cations and other solvento cations.[14] Technical problems in obtaining NMR spectra at pressures up to about 2 kbar were solved several years ago (cf., e.g., Fig. 7.1), so that key $\Delta V^{\ddagger}$ information is now available. Activation parameters for further systems continue to be published. Systems of higher lability have been investigated by other relaxation techniques and, when all else fails, approximate rate constants can be evaluated from the kinetics of complex formation.

There have been several attempts to characterize the secondary hydration shells around cations, and to obtain estimates for kinetic parameters for exchange processes involving such water molecules. The most promising candidate for such experiments is Cr^{3+}aq. Here the cation has a reasonably large charge : radius ratio, a long half-life for exchange involving the primary hydration shell, and favourable properties for NMR monitoring. A combination of real experiments and molecular modelling has provided good evidence for a reasonably well-defined secondary hydration shell containing about a dozen water molecules, whose mean lifetime in this

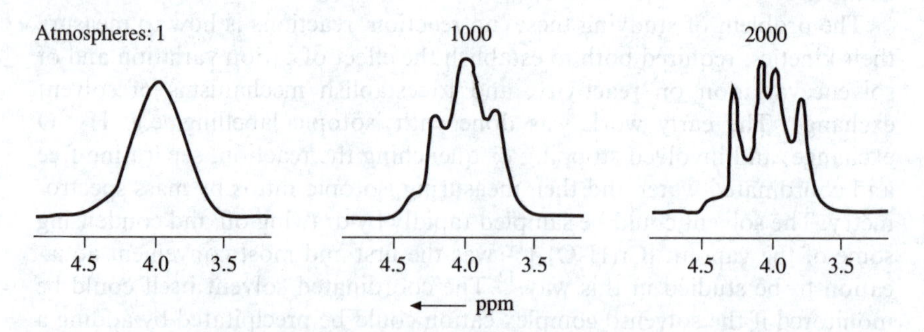

Fig. 7.1 The effect of increasing pressure on the 60 Mz ^{1}H NMR spectrum of a trimethyl phosphate–nitromethane solution containing Al^{3+}, at 341.3 K.

environment is 130–140 ps. This lifetime corresponds to a rate constant for exchange with bulk solvent of $8 \times 10^9\ s^{-1}$.[15]

7.3 Solvent exchange—reactivities

7.3.1 Cation variation

A general picture of reactivity patterns for water exchange at aqua-cations is given in Fig. 7.2. This emphasizes the very great range of reactivities, and gives some idea of the time-scales involved and appropriate experimental techniques. A clearer picture of underlying factors can be built up by dealing with selected groups of aqua-cations. The effects of size and charge are, as ever, best illustrated by concentrating on cations of the *sp*-block. Table 7.1 shows just the variation expected for spherical ions with a closed-shell configuration such as s^2p^6(or d^0), d^5 (high-spin) and d^{10}—the reactivity decreases with increasing charge and increases with increasing size. Figure 7.3 gives further illustration of the importance of ion size, showing remarkably smooth dependences of reactivity on ionic radius for 1+, 2+, and 3+ aqua-cations and for fringe species such as U^{3+}aq and d^5 Mn^{2+}aq, as well as for the *sp*-block cations. It is interesting that hydration number and stereochemistry do not seem to have a major disruptive effect on the trends. Indeed, aqua-ions of high hydration number have high exchange rate constants despite steric crowding, which might inhibit the formation of associative transition states. Thus rate constants increase steadily on descending Group II, despite increasing hydration numbers from four to six to (probably) eight on going from Be^{2+}aq down to Ba^{2+}aq. Moreover, the mechanism of water exchange seems not to interfere with the trends, as will be seen in detail later when the 3*d* sequences of transition metal 2+ and 3+ cations are documented—crystal (ligand) field effects control reactivity, regardless of the change in mechanism traversing the series from one end to the other.

Kinetic parameters for water exchange at first-row transition metal 2+ and 3+ aqua-cations are given in Table 7.2. Rate constants for water exchange at the aqua-ions of the *d*-block (d^1 to d^9) are very sensitive to the electron configuration, oxidation state, spin multiplicity, and principal quantum number of the metal ion, varying from $> 10^8\ s^{-1}$ for the $3d^4$ and $3d^9$ $[Cr(H_2O)_6]^{2+}$ and $[Cu(H_2O)_6]^{2+}$ ions to $< 10^{-9}\ s^{-1}$ for the low-spin $5d^6$ $[Ir(H_2O)_6]^{3+}$ ion at 298 K.[16] This shows up clearly in Fig. 7.4, where $\log_{10} k_{exch}$ and activation enthalpies for water exchange at $[M(H_2O)_6]^{n+}$ ($n = 2$ or 3) are shown as a function of d^n, the electron configuration of M^{n+}. Logarithms of rate constants are used in Fig. 7.4 as they are proportional to $\Delta G^\ddagger$ and hence to the activation barrier to reaction. Data are not available for divalent d^1 and d^2 ions (Sc^{2+} and Ti^{2+} are not yet known in aqueous solution), while exchange kinetics data for tervalent d^4 $[Mn(H_2O)_6]^{3+}$ have not yet been reported—the ion is well known, but inconveniently strongly

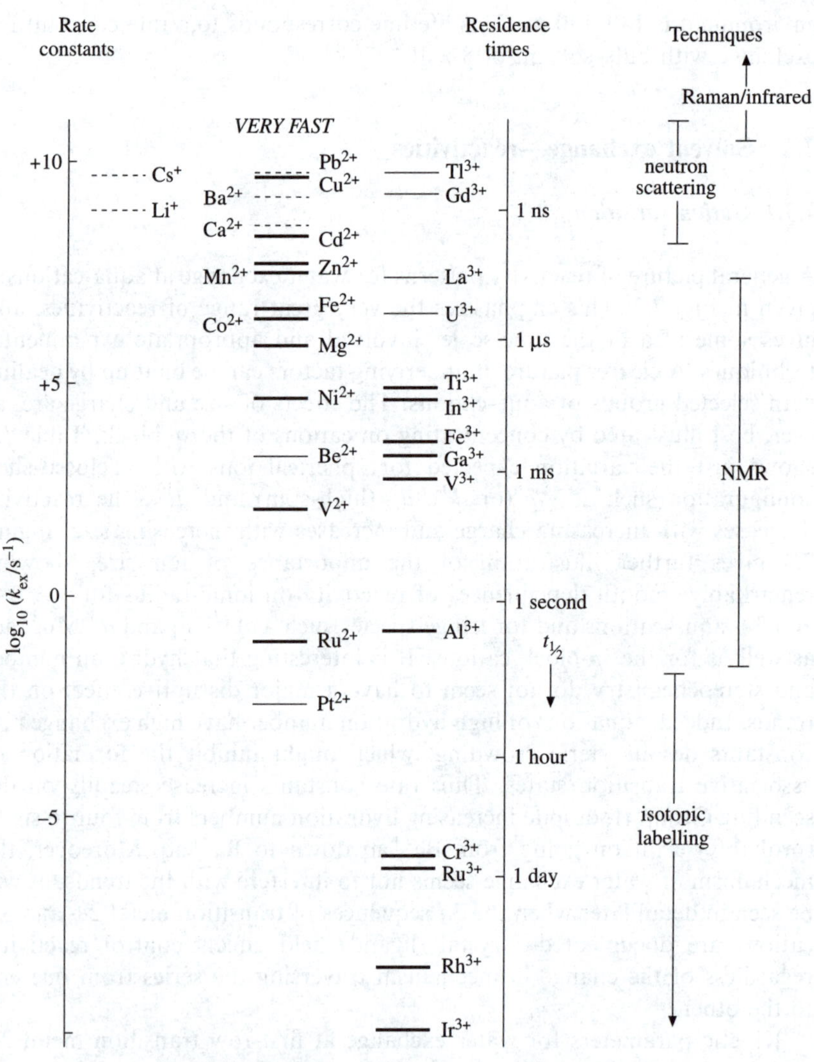

Fig. 7.2 Rate constants for water exchange, and mean residence times for primary hydration shell water molecules, for aquametal ions, at 298.2 K. The right-hand side of this figure indicates techniques appropriate for measuring water-exchange rate constants over various parts of this range of reactivities.

oxidizing. The simple aqua-ions of d^7 Ni(III) and d^8 Cu(III) have not yet been prepared and are also likely to be highly oxidizing; the preparation of the d^9 Zn^{3+} aqua-ion would present a considerable challenge. Apart from $[Co(H_2O)_6]^{3+}$, which sits near a high-spin/low-spin crossover, all the $3d^n$ aqua-ions are high-spin when there is a choice.

Extremely high lability is observed in the high-spin d^4 ($[Cr(H_2O)_6]^{2+}$) and

Table 7.1 Values for $\log k_{exch}$ for solvent (water) exchange at spherical s^2 and s^2p^6 metal aqua-cations in water at 298 K.

Effect of cation charge	Na^+	Mg^{2+}	Al^{3+}
$\log(k_{exch}/s^{-1})$	9.0	5.7	−0.8
Effect of cation size	Be^{2+}	Mg^{2+}	Ca^{2+}
Radius (Å)[a]	0.41	0.86	1.26
$\log(k_{exch}/s^{-1})$	2.5	5.7	7–8

[a] These are Shannon and Prewitt radii, for the appropriate coordination number in each case.

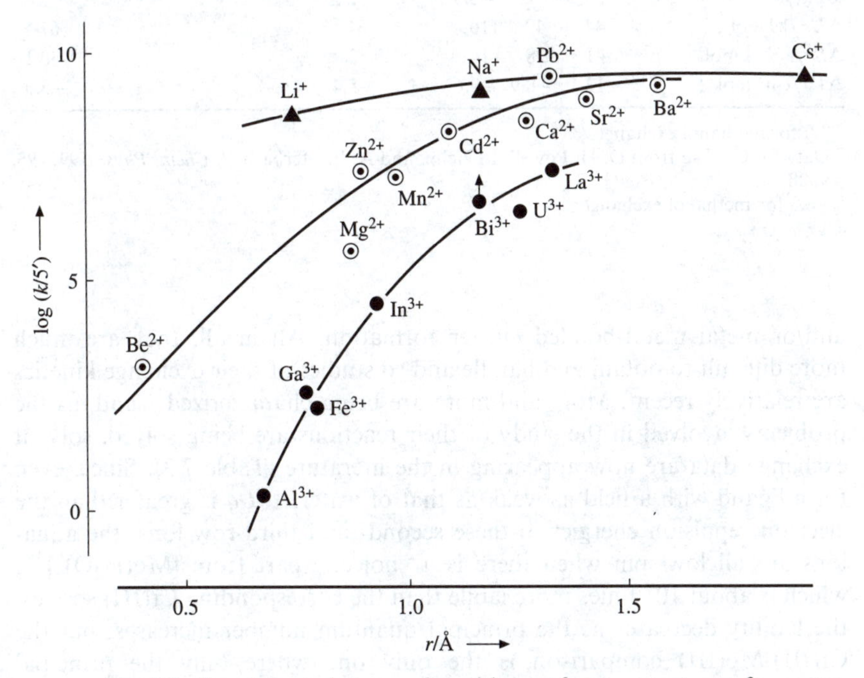

Fig. 7.3 Correlations between logarithms of rate constants for water exchange and ionic radius for a selection of 1+, 2+, and 3+ cations.

d^9 ($[Cu(H_2O)_6]^{2+}$) ions. Here, although the metal ions are formally six-coordinate and, time-averaged, all six ligands are equivalently bound, the Jahn–Teller effect leads to a tetragonal distortion in which two of the waters are bound more loosely than the other four. Combined with facile vibrations that exchange the distortion between the three pairs of mutually *trans* ligands and which might even be the rate-limiting process, this is responsible for the high lability (Fig. 7.5).

Water, as a ligand, is less favoured by the $4d^n$ and $5d^n$ ions—the aqua complexes of the +II and +III oxidation states are frequently strong reducing agents and/or readily hydrolysed and given to polymerization

Table 7.2 Rate constants ($\log_{10} k$ at 298 K) and activation parameters for the exchange of solvent water with aqua-cations $[M(H_2O)_6]^{n+}$ of the 3*d* series.

Configuration	$3d^0$	$3d^1$	$3d^2$	$3d^3$	$3d^4$	$3d^5$	$3d^6$	$3d^7$	$3d^8$	$3d^9$	$3d^{10}$
M	Ca^{2+}			V^{2+}	Cr^{2+}	Mn^{2+}	Fe^{2+}	Co^{2+}	Ni^{2+}	Cu^{2+}	Zn^{2+}
$\log (k/s^{-1})$	8.4			1.9	>8.5^a	7.3	6.6	6.5	4.5	9.7^b	7.5
$\Delta H^‡/kJ\ mol^{-1}$				62		33	41	47	57	12	
$\Delta S^‡/J\ K^{-1}\ mol^{-1}$				0		6	21	37	32	−22	
$\Delta V^‡/cm^3\ mol^{-1}$				−4.1		−5.4	+3.8	+6.1	+7.2	^{c}	
M		Ti^{3+}	V^{3+}	Cr^{3+}		Fe^{3+}					Ga^{3+}
$\log (k/s^{-1})$		5.26	2.7	−5.5		2.2					2.6
$\Delta H^‡/kJ\ mol^{-1}$		44	49	110		64					67
$\Delta S^‡/J\ K^{-1}\ mol^{-1}$		+1	−28	+16		+12					+30.1
$\Delta V^‡/cm^3\ mol^{-1}$		−12.1	−8.9	−9.6		−5.4					+5.0

^{a} 8.0 for methanol exchange.
^{b} Data for Cu^{2+}aq from D. H. Powell, L. Helm, and A. E. Merbach, *J. Chem. Phys.*, 1991, **95**, 9528.
^{c} +8.3 for methanol exchange.

and/or metal–metal-bonded cluster formation. All in all, they are much more difficult to obtain and handle and so studies of their exchange kinetics are relatively recent. More and more are being characterized[17] and, as the problems involved in the study of their reactions are being solved, solvent exchange data are now appearing in the literature (Table 7.3). Since, even for a ligand with a field as weak as that of water, $10Dq$ is greater than the electron repulsion energies in these second- and third-row ions, the aqua-ions are all low-spin when there is a choice. Apart from $[Mo(H_2O)_6]^{3+}$, which is about 10^4 times more labile than the corresponding Cr(III) species, the lability decreases as the principal quantum number increases; but the Cr(III)/Mo(III) comparison is the only one where only the principal quantum number is changed. The 3*d* Fe(II), Fe(III), and Co(III) are all high-spin whereas their 4*d* and 5*d* congeners Ru(II), Ru(III), Rh(III), and Ir(III) are low-spin. The decrease in lability with an increase in principal quantum number is better seen in the general substitutional lability of the more formal complexes of these metals, where it may be separated from the effect of the decrease in the affinity for the 'hard' ligand water as the reaction centre becomes softer. Nevertheless, there is a rule in organometallic chemistry that the 4*d* species is more labile than its 3*d* congener, and both are much more labile than the 5*d* analogue.

The lanthanide ($4f^n$) aqua-ions are all very labile, with k_{298} increasing from $4 \times 10^7\ s^{-1}$ at Yb^{3+}aq to $1.1 \times 10^9\ s^{-1}$ at Gd^{3+}aq.[18,19] A molecular dynamics simulation study of Ln^{3+}aq ions has as a useful by-product produced good estimates for k_{exch} values.[20] Solvent exchange in

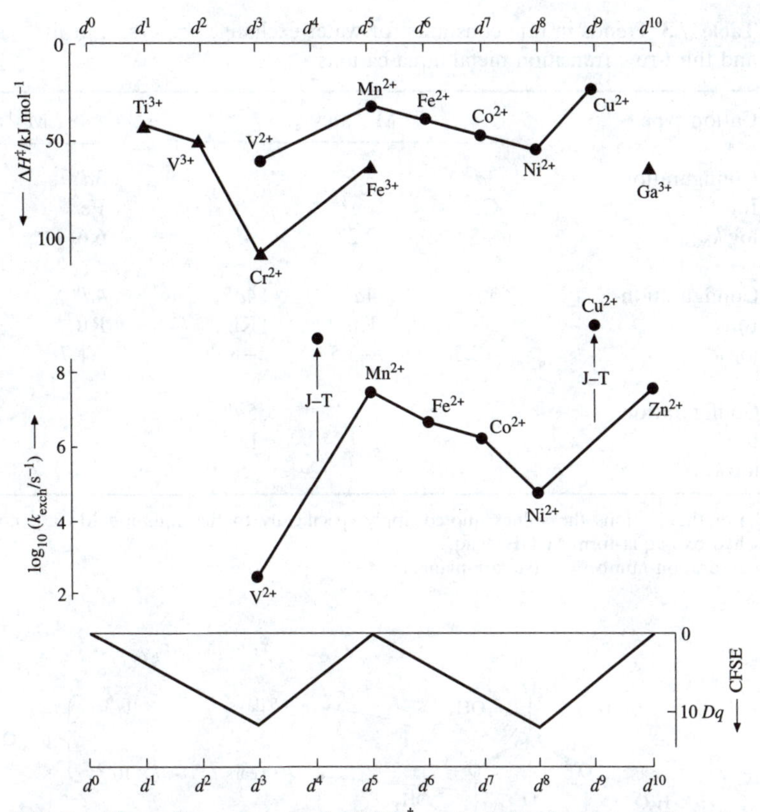

Fig. 7.4 Plots of $\log_{10} k_{exch}$ and of $\Delta H^{\ddagger}$ against the electron configuration of M^{n+} for water exchange at high-spin $[M(H_2O)_6]^{n+}$ ions. (J–T indicates the effects of Jahn–Teller distortion).

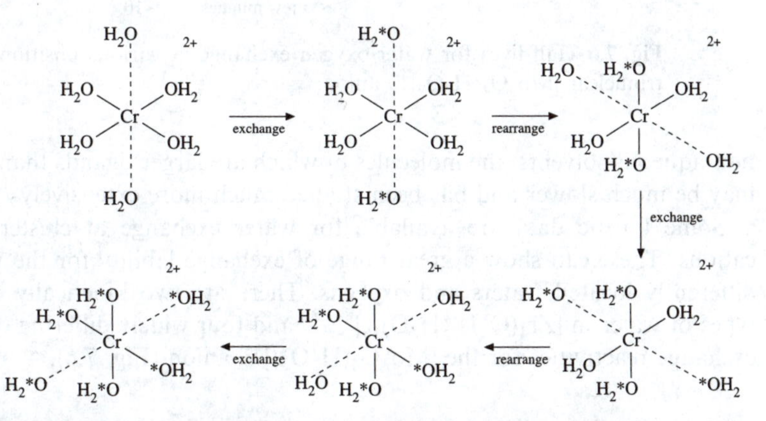

Fig. 7.5 The effects of Jahn–Teller distortion on water exchange. The longer H_2O-Cr^{2+} bonds are indicated by dashed lines—very rapid exchange of such waters must be followed by rapid rearrangement to lead to the very high rate constants established for water exchange at Cr^{2+}aq and at Cu^{2+}aq.

Table 7.3 Trends in rate constants for water exchange, k_{exch}/s^{-1}, at first-, second-, and third-row transition metal aqua-cations.

Cation type	$M^{3+}aq$[a]			$M^{2+}aq$	
Configuration	$3d^3$	$3d^5$		$3d^6$	$3d^8$
Ion	Cr^{3+}	Fe^{3+}		Fe^{2+}	Ni^{2+}
$\log k_{exch}$	−5.5	2.2		6.6	4.5
Configuration	$4d^3$	$4d^5$	$4d^6$	$4d^6$	$4d^8$
Ion	Mo^{3+}	Ru^{3+}	Rh^{3+}	Ru^{2+}	Pd^{2+}[b]
$\log k_{exch}$	−2.3	−5.5	−8.7	−1.7	2.7
Configuration			$5d^6$		$5d^8$
Ion			Ir^{3+}		Pt^{2+}[b]
$\log k_{exch}$			< -9		−2.6

[a] For the 3+ ions the values quoted apply specifically to the aqua-ion $M^{3+}aq$, not to the hydroxo-aqua-form $M(OH)^{2+}aq$.
[b] Hydration number = 4 (square-planar).

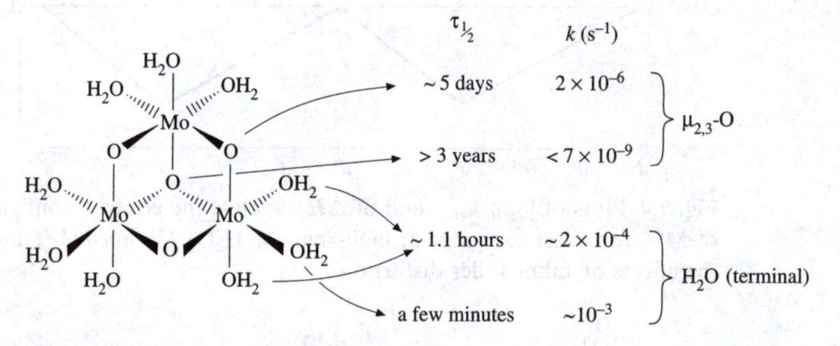

Fig. 7.6 Half-lives for water/oxygen exchange at various positions in the trinuclear $[Mo_3O_4(H_2O_9]^{4+}$ anion.

non-aqueous solvents, the molecules of which are larger ligands than water, may be much slower and has been studied much more extensively.

Some kinetic data are available for water exchange at cluster aqua-cations. These can show a great range of exchange lability for the various differently located waters and oxygens. There are two kinetically distinct types of water in $[Zr_4(OH)_8(H_2O)_{16}]^{8+}$,[21] and four widely differing oxygen-exchange reactivities for the $[Mo_3O_4(H_2O)_9]^{4+}$ anion (Fig. 7.6).[22]

7.3.2 *Solvent variation*

Although water is a solvent of great importance and ubiquity, the ability to generate other $[M(solvent)_x]^{n+}$ cations has long been with us and data for

Table 7.4 Dependence of rate constant[a] for solvent exchange, in logarithmic form, on solvent nature for a selection of metal cations.

	$\log_{10}(k_{exch}/s^{-1})$					
	Ni^{2+}	Co^{2+}	Mn^{2+}	Al^{3+}	Ga^{3+}	Cr^{3+}
Hexamethylphosphoramide				3.7[b]		
Liquid ammonia	5.0	6.9				−5.1
Water	4.5	6.3	7.3	0.1	2.6	−6.3
Ethanol	4.0					
Trimethyl phosphate				−1.0	0.8	
Dimethylformanide	3.8	5.5	6.3	−0.8	0.2	−7.3
Acetonitrile	3.7	5.3	7.2			
Dimethyl sulphoxide	3.7	5.2		−0.5	0.3	−7.5
Methanol	2.7	4.0	5.6			

[a] At 298.2 K—these values should be considered approximate guides, as there are significant uncertainties in many of them, due to differences in experimental conditions, technical difficulties, and the frequent need for extrapolation.
[b] Second-order rate constant for exchange at tetrahedral $[Al(hmpa)_4]^{3+}$.

solvent-exchange kinetics are very extensive. Frequently these solvents are more convenient to study from the point of view of the availability of magnetic nuclei, e.g. ^{19}F, ^{13}C, and ^{31}P, in addition to ^{1}H (which need not be potentially acidic, as it is in aqua complexes) and ^{17}O, and the exchanges are frequently slower. Although the breadth of cover achieved by aqua-ions has yet to be reached it is only a matter of time before comparably full data become available. It is already quite clear that, especially in cases where orbital symmetry and directed valence do not require a fixed coordination number and geometry, the mechanistically controlling feature seems to be the relative sizes of the metal ion and the ligand. The dominant coordination number can change with temperature, for example $[Ga(MeOH)_6]^{3+}$ goes to $[Ga(MeOH)_7]^{3+}$ below −45 °C,[23] and it can also change with pressure, as in the case of $[Nd(dmf)_8]^{3+}$, which is converted into $[Nd(dmf)_9]^{3+}$ when the pressure is high enough.[24]

The extent to which rate constants for solvent exchange vary with the nature of the solvent is shown, for a selection of cations, in Table 7.4.[25] The main comment to be made is that the range of reactivity for a given ion is really rather small, certainly in dramatic contrast to the enormous range of reactivity for various cations with a given solvent (e.g. water—see Fig. 7.1 above). There is also no clear correlation with solvent donor numbers, even if one ignores water and the alcohols and restricts attention to solvents for which donor numbers have been properly (i.e. calorimetrically) determined.

The consequences of going from a standard monodentate solvent to solvents whose molecules may act as bi- or even tridentate ligands have been little probed. However, it has been claimed that solvent exchange of ethane-

1,2-diol (glycol) with $[Ni(glycol)_3]^{2+}$ and of propane-1,2,3-triol (glycerol) with $[Ni(glycerol)_2]^{2+}$ have very similar rate constants, about $3 \times 10^3\ s^{-1}$ at 298 K,[26] which value is remarkably similar to that for water exchange at $[Ni(H_2O)_6]^{2+}$ ($\sim 3 \times 10^4\ s^{-1}$ at 298 K). There is good evidence[27] that glycol and glycerol are bidentate and terdentate (*fac* geometry) in their respective solvento-nickel(II) cations in solution, so the ease of exchange is perhaps a little surprising. Exchange of ethane-1,2-diamine at $[Ni(en)_3]^{2+}$ or at $[Cu(en)_3]^{2+}$ in ethane-1,2-diamine as solvent is, in contrast, very much slower than ammonia exchange (albeit in aqueous media) at the respective cations. The activation volume for exchange of ethane-1,2-diamine with $[Ni(en)_3]^{2+}$, at $+11.4\ cm^3\ mol^{-1}$,[28] is similar to values for exchange of other solvent molecules at this cation, and has been interpreted in terms of a dissociative mechanism involving an intermediate with two monodentate en molecules bonded to the Ni^{2+}. The same mechanism has been proposed for en exchange at $[Cu(en)_3]^{2+}$ in ethane-1,2-diamine, despite a significantly negative activation entropy.[29] The latter system is particularly interesting in view of the demonstrated (X-ray diffraction; EXAFS) Jahn–Teller distortion in the $[Cu(en)_3]^{2+}$ cation (Cu–N = 1.92 or 1.97; 2.22 or 2.27 Å),[30] which may well be the reason for the very low activation enthalpy of $9.2\ kJ\ mol^{-1}$; axial/equatorial exchange is believed to be very facile, with a rate constant in the region of $10^{11}\ s^{-1}$.

7.4 Solvent exchange—mechanisms

7.4.1 Determination of mechanism—the molecularity of solvent exchange

When studying solvent exchange it is not possible to study the effects of varying the nature or the concentration of the entering group and so the major classical tool for the assignment of mechanism is unavailable. To diagnose mechanisms of solvent exchange, the most common approach is the interpretation of activation parameters, specifically $\Delta S^\ddagger$ and $\Delta V^\ddagger$.[31] However, the assignment of mechanism from a knowledge of the activation parameters of a reaction is fraught with pitfalls, especially in interpreting entropies of activation. Large experimental errors are inherent in the long extrapolation needed to estimate $\Delta S^\ddagger$ values from the intercepts of plots of ln (k/T) against T^{-1}.[32] There is also the problem that the range of temperatures over which kinetic data can be obtained from series of temperature-dependent NMR spectra is often rather short.[33] The consequences of these problems are well exemplified by the range of activation parameters reported for dimethyl sulphoxide[34] and for acetonitrile[35] exchange at Ni^{2+} (Table 7.5—where the chronological order demonstrates that there is not a systematic time-shift, but conceals a very close correlation of $\Delta H^\ddagger$ with $\Delta S^\ddagger$). The great sensitivity of $\Delta H^\ddagger$ and $\Delta S^\ddagger$ to solvation changes on going to the transition state and the absence of any simple model to estimate entropy makes any attempt to assign the sign and the magnitude

Table 7.5 Activation parameters determined by NMR spectroscopy from 1967 to 1978 for dimethyl sulphoxide exchange and for acetonitrile exchange at Ni^{2+}.

Dimethyl sulphoxide			Acetonitrile			
Year	$\Delta H^{\ddagger}$ kJ mol^{-1}	$\Delta S^{\ddagger}$ J K^{-1} mol^{-1}	Year	$\Delta H^{\ddagger}$ kJ mol^{-1}	$\Delta S^{\ddagger}$ J K^{-1} mol^{-1}	NMR nucleus
1967	33.5	−67	1967	49	−15	^1H
1968	36.4	−46	1967	46	−37	^1H
1969	49.0	−4	1967	49	−16	^1H
1969	50.6	−4	1971	67	+43	^1H
1969/1971	54.4	+13	1971	63.2	+42	^{14}N
1971	33.9	−59	1973	68	+50	^{14}N
1972	25.9	−84	1973	60	+23	^1H
1973	50.6	+4	1973	39.5	−33	^{14}N
			1978	64.6	+38	^1H

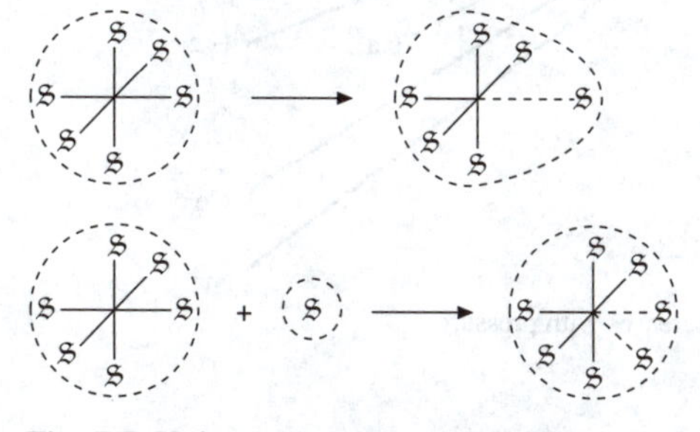

Fig. 7.7 Volume changes on transition state formation for (a) dissociative and (b) associative solvent exchange.

of $\Delta S^{\ddagger}$ to the molecularity of the process rather risky. It has been claimed that volumes of activation can be determined with greater precision than entropies of activation, though a closer examination shows that what is gained by avoiding the long extrapolation needed to evaluate $\Delta S^{\ddagger}$ is often largely lost by the smaller variation of the magnitude of the rate constant over the range of pressure (up to about 2 kbar) that can usually be achieved. Nevertheless, it does not require tremendous precision to establish whether the rate constant decreases or increases when the pressure is increased, i.e. whether $\Delta V^{\ddagger}$ is positive or negative (cf. Figs 7.7 and 7.8). The mechanistic significance of an activation volume of zero can be deemed, as we have already seen, to be debatable. Though volumes of activation are just as sensitive to solvation changes as the other activation parameters (especially when the charge distribution changes on going from the ground state to the transition state), they can be estimated from relatively simple molecular models.

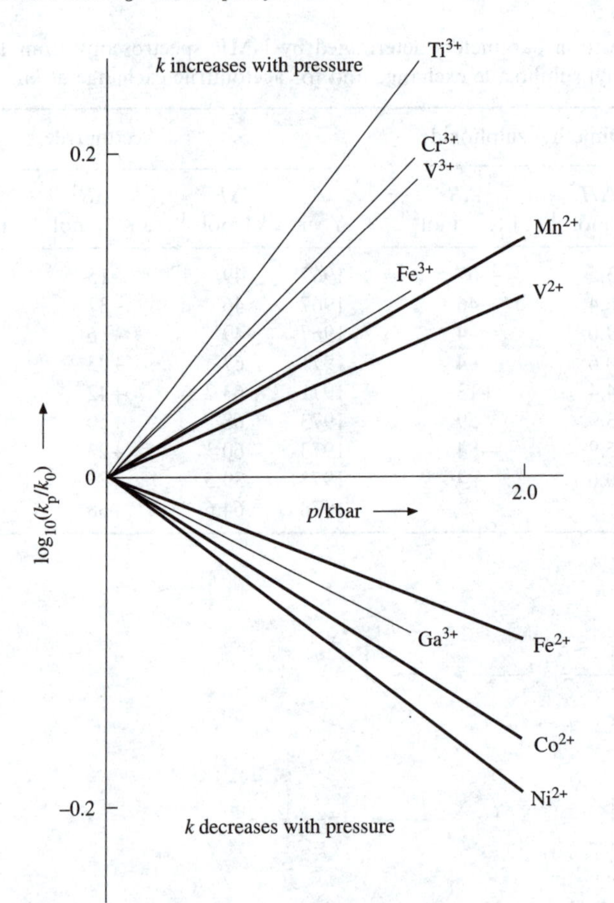

Fig. 7.8 Pressure effects on rate constants for water exchange at 2+ and 3+ transition metal ions (first row of *d*-block).

In terms of the ease with which $\Delta V^{\ddagger}$ data can be used to assign molecularity, these solvent-exchange reactions come second only to reactions involving the exchange of uncharged nucleophiles with uncharged substrates in non-polar solvents (cf. Section 6.2.2). The problems of measuring NMR spectra at high pressure have now been largely overcome, with $\Delta V^{\ddagger}$ values for solvent exchange now available for many systems. Two books devoted to kinetics at high pressure[36] provide an extremely useful collection of germane reviews and are compulsory reading for anyone who is interested in the technique and its applications.

Based on the assumption that the coordinated solvent molecules are closely packed and rigidly held, while the bulk solvent retains a much more open structure, a limiting *D* mechanism should have a positive $\Delta V^{\ddagger}$ value equal in magnitude to the difference in the partial molar volumes of the coordinated and the free solvent molecules, while the limiting *A* mechanism should have a negative $\Delta V^{\ddagger}$ value of similar magnitude. The partial molar volume of the free solvent molecule is readily determined, while that of the coordinated solvent can be determined from crystallographic data. Alternatively, the volume changes on complete dissociation of a solvent molecule from, or incorporation of an incoming solvent molecule into, an initial-state solvento cation may be calculated by a range of methods. For water exchange, an activation volume of between -13 and $-15\,cm^3\,mol^{-1}$ (cf. the partial molar volume of water, which is 18 $cm^3\,mol^{-1}$) can be forecast for a limiting associative, *A*, process. The reported activation volume of $-13.6\ cm^3\,mol^{-1}$ for water exchange at Be^{2+}aq thus indicates an *A* mechanism.[37] Interchange mechanisms ought to have smaller volumes of activation, between the limits of approximately $\pm 14\,cm^3\,mol^{-1}$. The so-called More O'Ferrall diagram[38] for the smooth change in volume characteristics from associative to limiting dissociative via interchange mechanisms of continuously changing balance between bond making and bond breaking, shown for the general case of symmetrical ligand exchange as Fig. 2.4 (p.42), is particularly relevant here. It has been claimed that, once corrections have been made for electrostriction and other solvation phenomena, the change in the sign of $\Delta V^{\ddagger}$ from positive to negative represents a change from an I_d to an I_a mechanism. However, this seems to introduce a new definition of I_d and I_a which does not necessarily coincide with that introduced by Langford and Gray and which was discussed at length in Chapter 2. As we saw there, a synchronous process with a single transition state in which the entering and leaving groups are equivalently bound (surely this must be an I_a mechanism) can have $\Delta V^{\ddagger}$ values lying between the two limits of a *D* mechanism and an *A* mechanism, either positive or negative depending upon the openness or tightness of the transition state. Furthermore, it is wrong to ignore what happens to the other ligands in the substrate during the act of substitution. A decrease in coordination number will be accompanied by some expansion of the spectator ligands, thereby increasing $\Delta V^{\ddagger}$ still further, while an increase in the coordination number might cause the spectator ligands to 'pull in their stomachs' to make more room.[39] It seems likely that dissociative/associative activation differences actually determine the observed activation volume through initial-state rather than transition-state effects.[40] Clearly, volumes of activation, while providing information which is extremely useful in the clarification of mechanism, are not the universal panacea that some enthusiasts might suggest. Since this is not an advanced text and some of our best friends (not to mention one of the present authors!) are very keen on volumes of activation, the discussion must close at this point and those wishing to know more should consult the specialist reviews and textbooks.[36,41] Perhaps a good test of the reader's aptitude for research into reaction mechanisms is whether this statement is greeted with relief or with disappointment.

When these activation volume arguments are applied to the solvent-exchange reactions of the divalent and tervalent $3d^n$ metal ions (Table 7.6), the conclusions are clear. For the divalent ions, the values of $\Delta V^{\ddagger}$ change from negative to positive as we cross the *d*-block. Since the ionic radii also

Table 7.6 Activation volumes ($cm^3\ mol^{-1}$) for solvent exchange at metal ions.

	d^0	d^1	d^2	d^3	d^4	d^5	d^6	d^7	d^8	d^9	d^{10}
				V^{2+}		Mn^{2+}	Fe^{2+}	Co^{2+}	Ni^{2+}	Cu^{2+}	
Water				−4.1		−5.4	−3.8	+6.1	+7.2		
Methanol						−5.0	+0.4	+8.9	+11.4	+8.3	
Acetonitrile						−7.0	+3.0	+7.7	+9.6		
Dimethylformamide						+2.4	+8.5	+6.7	+9.1		
	Sc^{3+}	Ti^{3+}	V^{3+}	Cr^{3+}		Fe^{3+}					Ga^{3+}
Water		−12.1	−8.9	−9.3		−5.4					+5.0
Dimethylformamide		−5.7		−6.3		−0.9					+7.9
Dimethyl sulphoxide			−10.1	−11.3		−3.1					+13.1
Trimethyl phosphate	−20.7										+20.7
		← ASSOCIATIVE							DISSOCIATIVE →		

decrease as the number of non-bonding electrons increases, it has been concluded in some quarters that size determines molecularity, in the same way as it controls molecularity in relatively electrovalent complexes (cf. Al ↔ In), larger ions being able to increase their coordination numbers in forming the transition states, while the smaller ones cannot. The tervalent cations Ti^{3+} ($3d^1$), V^{3+} ($3d^2$), Cr^{3+} ($3d^3$), and Fe^{3+} ($3d^5$; $t_{2g}{}^3e_g{}^2$) all exchange their water molecules with negative volumes of activation, and the associative mode is indicated. It is, however, quite clear that ionic radius alone is not controlling the mechanism, because Cr^{3+} (radius 0.62 Å) is smaller than Ni^{2+} (0.69 Å) and yet it is the latter that has the dissociative mechanism. If we take the view that all of these exchanges occur by an I_a mechanism but vary in the extent to which the transition state is compacted, then not only can we account for the dependence of the volume of activation upon ionic radius but we can also explain the solvated ions of higher charge exchange with a negative $\Delta V^\ddagger$ in terms of the higher charge tightening up the transition state.

The size of the solvent molecules can also be significant. As shown in Table 7.6, the changeover from negative to positive values for activation volumes occurs between Mn and Fe for water, methanol, and acetonitrile, but before Mn for dimethylformamide. The larger solvent makes associative activation more difficult, and thus DMF exchange occurs dissociatively.

Despite the clear importance of size factors, we must not lose sight of the fact that both the charge and the electron configuration of the reaction centre almost certainly have an important part to play. Electrons in the t_{2g} orbitals, pointing between the six solvent molecules, will discourage the entry of the seventh ligand so that the dissociative mode of activation becomes less favoured as the t_{2g} shell fills up. The tervalent ions studied only extend to Fe^{3+} (high-spin, $t_{2g}{}^3e_g{}^2$, for the hexaaqua-ion), whereas the postulated mechanistic changeover in the divalent series was found at Fe^{2+} ($t_{2g}{}^4e_g{}^2$). The fact that the exchange of solvent water with the coordinated water of Cr^{3+} ($t_{2g}{}^3$) has a negative $\Delta V^\ddagger$, while that of the corresponding reaction of the Ga^{3+} ion ($t_{2g}{}^6e_g{}^4$) is positive, even though their ionic radii are similar (0.62 Å), tells us that we must not ignore the various electron distribution models, whether they be repulsive, as in the electrostatic (crystal field) model, or based on the bonding/non-bonding (π)/antibonding (σ^*) occupancy in the various molecular orbitals. The fact that $\Delta V^\ddagger$ for methanol exchange at Cu^{2+} is slightly less positive than at Ni^{2+} encourages invocation of the Jahn–Teller effect. Since at this point there are more parameters than facts, it is possible to account for everything.

A consideration of steric and electronic factors and their likely effects on $\Delta V^\ddagger$ has in fact proved very successful in establishing an associative ↔ dissociative trend in mechanism on crossing the first row of the *d*-block transition metals. It is much more difficult to establish mechanistic trends down triads in the transition metals, due to the paucity of simple aqua-ions in the second and third rows of the *d*-block. The one triad where the members of these two rows exist, namely Rh^{3+}aq and Ir^{3+}aq, is ruled

Table 7.7 Activation volumes for water exchange at complexes $[M(NH_3)_5(solv)]^{3+}$

d^3		d^6	
Complex	$\Delta V^‡/cm^3\ mol^{-1}$	Complex	$\Delta V^‡/cm^3\ mol^{-1}$
Non-aqueous			
$[Cr(NH_3)_5(DMSO)]^{3+}$	−3.2	$[Co(NH_3)_5(DMSO)]^{3+}$	+2.0
$[Cr(NH_3)_5(DMF)]^{3+}$	−7.4	$[Co[NH_3)_5(DMF)]^{3+}$	+2.6
Aqueous			
$[Cr(NH_3)_5(H_2O)]^{3+}$	−5.8	$[Co(NH_3)_5(H_2O)]^{3+}$	+1.2
		$[Rh(NH_3)_5(H_2O)]^{3+}$	−4.1(35 °C)
		$[Ir(NH_3)_5(H_2O)]^{3+}$	−3.2(71 °C)

out by the instability of the first-row member, Co^{3+}aq. Comparison of Fe^{3+}aq with Ru^{3+}aq is complicated by the high-spin → low-spin change. For a complete and comparable sequence it is necessary to turn to the series of aqua-ammine complexes, $[M(NH_3)_5(H_2O)]^{3+}$, where the $\Delta V^‡$ values (Table 7.7) indicate a changeover from dissociative to associative activation on going down the triad. As is also shown in Table 7.7, there is an associative → dissociative trend from chromium(III) to cobalt(III), consistent with the overall trend shown in Table 7.6 for simple solvento ions.

Just as it is necessary to turn to ternary complexes to establish mechanistic trends for water exchange going down the Periodic Table for the *d*-block, so it is also necessary to consider ternary aqua complexes (or non-aqueous systems, *v.i.*) to establish the mechanistic trend across the 4*f* series of elements. Activation volumes for water exchange for the series Tb^{3+} to Tm^{3+} are all equal within experimental uncertainty, at $-6\,cm^3\ mol^{-1}$, though the somewhat erratic trend in activation entropies does suggest rather more dissociative character to an I_a mechanism as the number of *f* electrons increases.[19] However, if one examines the sequence of activation volumes for water exchange at $[Ln(pdta)(H_2O)_2]^-$, pdta = **1**, where $\Delta V^‡ = -7.6, -5.5, -6.5, -1.2$, and $+7.4\ cm^3\ mol^{-1}$ for Tb^{3+}, Dy^{3+}, Er^{3+}, Tm^{3+}, and Yb^{3+}, respectively, then a trend from associative interchange to dissociative is apparent.[42]

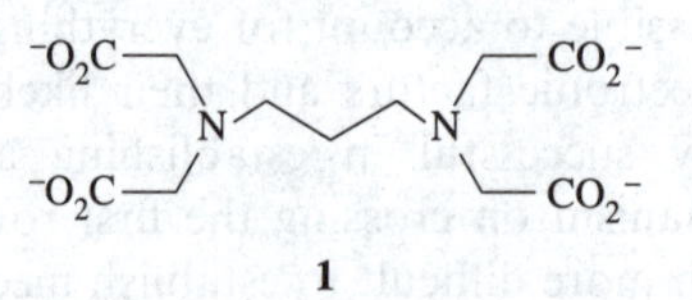

1

One has to be wary of comparisons between binary and ternary complexes—the presence of the other ligands L may make a large difference

Table 7.8 The steric effects of spectator ligands—methylamine versus ammonia—on activation volumes ($\Delta V^{\ddagger}/cm^3\ mol^{-1}$) for water exchange.

Complex	Cr^{III}	Rh^{III}	Co^{III}
$[M(NH_3)_5(H_2O)]^{3+}$	−5.8	−4.1	+1.2
$[M(NH_2Me)_5(H_2O)]^{3+}$	−3.8	+1.2	+5.7

in the reactivity, and possibly a significant change in the mechanism, for water exchange at, say, $[ML_5(solv)]^{n+}$ as compared with $[M(solv)_6]^{n+}$. Relevant reactivity trends have been documented for, *inter alia*, series of chromium(III)-ammine-water,[43] of nickel(II)-ammine or amine-water,[44] and of ternary aqua-manganese(II) and -cobalt(II)[45] complexes. Replacement of several aqua-ligands by ammonia can change rate constants for water exchange by several powers of ten. A change in ligand bulk, from NH_3 to RNH_2, even just to $MeNH_2$, affects reactivity and can also have a significant effect on mechanism. Steric crowding causes a shift towards a greater degree of dissociative character, best seen in activation volumes (Table 7.8).[46]

Major changes in 'spectator' ligands can have dramatic consequences, with enormous labilization at normally extremely inert centres such as the low-spin inert rhodium(III), ruthenium(II), and osmium(II). Taube's work on osmium complexes containing π-acid ligands is seminal.[47] The replacement of some coordinated solvent molecules by benzene or by η^5-cyclopentadienyl makes an enormous difference to the lability of the remaining coordinated solvent molecules[48]—e.g. the 10^{14}-fold increase in water exchange rates at rhodium(III) on going from $[Rh(H_2O)_6]^{3+}$ to $[Rh(\eta^5\text{-}C_5Me_5)(H_2O)_3]^{2+}$. Acetonitrile exchange at $[Ru(\eta^6\text{-}C_6H_6)(MeCN)_3]^{2+}$ and at $[Ru(\eta^5\text{-}C_5H_5)(MeCN)_3]^{+}$ is approximately 10^5 and 10^{10} times, respectively, faster than at $[Ru(MeCN)_6]^{2+}$. These huge differences in reactivity are accompanied by changes in the character of the mechanism of solvent exchange. This is illustrated in Table 7.9, for water exchange at ruthenium(II) and rhodium(III) and for acetonitrile exchange at ruthenium(II). The discussion of these kinetic and mechanistic differences makes much of the differences in metal–ligand bond strengths, particularly those ascribable to π-bonding, established by X-ray diffraction determinations of the crystal structures of salts of the various cations.[49]

A rather different case of spectator ligands affecting mechanism, this time through steric rather than electronic effects, is afforded by water exchange at pharmacologically relevant gadolinium complexes such as $[Gd(dtpa)(H_2O)]^{2-}$ and $[Gd(dota)(H_2O)]^{-}$, whose dynamic behaviour has been thoroughly characterized by a combination of ESR, ^{17}O NMR, and NMRD.[50] The last-named technique, nuclear magnetic resonance dispersion, involves measuring the excess longitudinal proton relaxation time caused by the gadolinium contrast agent. The bulky $dtpa^{5-}$ (**2**) and

Table 7.9 The effects of spectator ligands on kinetic parameters and mechanisms for solvent exchange at rhodium(III) and at ruthenium(II).

	k_{298} (s^{-1})	$\Delta H^{\ddagger}$ (kJ mol^{-1})	$\Delta S^{\ddagger}$ (J K^{-1} mol^{-1})	$\Delta V^{\ddagger}$ (cm^3 mol^{-1})	Substitution mechanism
$[Rh(H_2O)_6]^{3+}$	2.2×10^{-9}	131	+29	−4.2	I_a
$[Rh(\eta^5\text{-}C_5Me_5)(H_2O)_3]^{2+}$	1.6×10^{5}	66	+75	+0.6	$I_{(d)}$
$[Ru(H_2O)_6]^{2+}$	1.8×10^{-2}	88	+16	−0.4	I
$[Ru(\eta^6\text{-}C_6H_6)(H_2O)_3]^{2+}$	11.5	76	+30	+1.5	I_d
$[Ru(MeCN)_6]^{2+}$	8.9×10^{-11}	140	+33	+0.4	I
$[Ru(\eta^6\text{-}C_6H_6)(MeCN)_3]^{2+}$	4.1×10^{-5}	103	+15	+2.4	I
$[Ru(\eta^5\text{-}C_5H_5)(MeCN)_3]^{+}$	5.6	87	+60	+11.1	D

$dota^{4-}$ (**3**) ligands crowd around the sole coordinated water molecule, obstruct approach of incoming water to the gadolinium, and force the mechanism to limiting dissociative. This is evidenced by the activation volumes, +12.5 and +10.5 cm^3 mol^{-1}, respectively, which are markedly more positive than those for unhindered analogues.[51] In similar vein, activation volumes for water exchange at the bis(methylamide)dtpa-aqua-lanthanide complexes $[Ln(bma\text{-}dtpa)(H_2O)]$ are significantly more positive than for water exchange at the less hindered pdta ($pdta^{4-}$ = **1**) complexes $[Ln(pdta)(H_2O)_2]^-$ mentioned above, again suggesting the favouring of dissociative activation by steric crowding of the coordinated water.[52] The effect of the replacement of waters in Cu^{2+}aq by tren (**4**) might be deemed an extreme example of spectator ligand effects. Coordination of tren to Cu^{2+} causes a change of stereochemistry, to give five-coordinate $[Cu(tren)(H_2O)]^{2+}$, at which water exchange is about 10^3 times slower than at Cu^{2+}aq, and takes place by the I_a mechanism ($\Delta V^{\ddagger} = -4.7$ cm^3 mol^{-1}).[53]

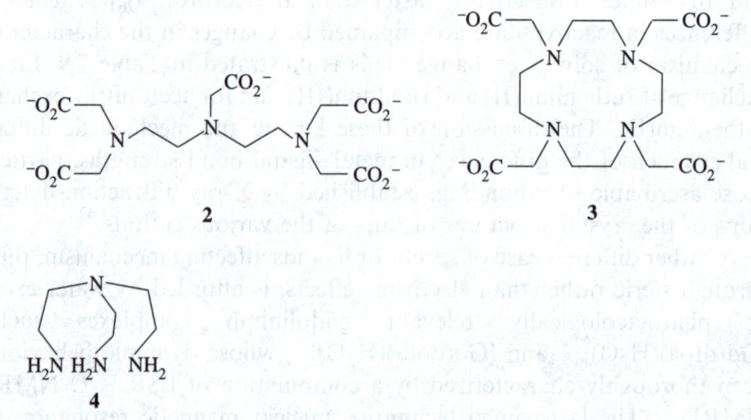

We have thus far concentrated very much on $\Delta V^{\ddagger}$ to the almost complete exclusion of $\Delta S^{\ddagger}$. Table 7.2 showed that there was some measure of consistency between the indications from activation entropies and volumes for water exchange at transition metal aqua-cations, though the latter gave the clearer indication of the changeover from dissociative to associative activation on crossing the Periodic Table. Activation entropies and volumes for solvent exchange at square-planar Pd^{2+} and Pt^{2+} also agree in indicating associative activation.[54] The selection of values of activation entropies and volumes for solvent exchange at some *sp*-, *d*-, and *f*-block 3+ cations given in Table 7.10[55] again generally shows good agreement between the mechanistic indications from these two parameters—the uncertainties on several of the activation entropies are such that some of the small positive values could well turn out to be small and negative when more precise values are obtained. Figure 7.9 shows that the correlation between activation volumes and entropies is qualitative rather than quantitative, for solvent exchange at the 3+ cations included in Table 7.10, at the ruthenium(II) complexes included in Table 7.9, and at some square-planar solvento complexes. Some of the scatter in Fig. 7.9 may be ascribed to the rather large uncertainties in some of the activation parameters, but even so the correlation between $\Delta S^‡$ and $\Delta V^‡$ here seems less close than that demonstrated for substitution at a number of d^3 and d^6 complexes of the ammine-halide type.[56]

Table 7.10 indicates a changeover from dissociative to associative

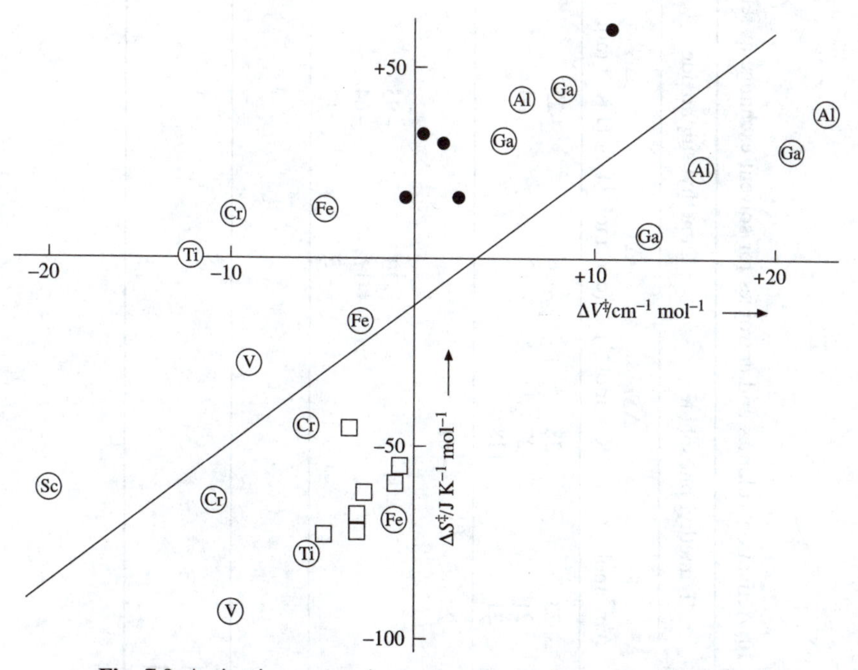

Fig. 7.9 Activation entropies and volumes for solvent exchange: ○ at $[M^{III}(solvent)_6]^{3+}$; ● at binary and ternary Ru^{2+} solvento ions; □ at square-planar $[M^{II}(solvent)_4]^{3+}$.

Table 7.10 Activation volumes and entropies for solvent exchange at selected 3+ cations.

	Trimethyl phosphate		Dimethyl sulphoxide		Dimethylformamide		Water	
Cation	$\Delta V^{\ddagger}$ ($cm^3\ mol^{-1}$)	$\Delta S^{\ddagger}$ ($J\ K^{-1}\ mol^{-1}$)	$\Delta V^{\ddagger}$ ($cm^3\ mol^{-1}$)	$\Delta S^{\ddagger}$ ($J\ K^{-1}\ mol^{-1}$)	$\Delta V^{\ddagger}$ ($cm^3\ mol^{-1}$)	$\Delta S^{\ddagger}$ ($J\ K^{-1}\ mol^{-1}$)	$\Delta V^{\ddagger}$ ($cm^3\ mol^{-1}$)	$\Delta S^{\ddagger}$ ($J\ K^{-1}\ mol^{-1}$)
Al^{3+}	+23	+38	+16	+22			+6	+42
Ga^{3+}	+21	+27	+13	+4	+8	+45	+5	+30
In^{3+}	−21	−118						
Sc^{3+}	−20	−61						
Ti^{3+}					−6	−74	−12	+1
V^{3+}			−10	−95			−9	−28
Cr^{3+}			−11	−64	−6	−44	−10	+12
Fe^{3+}			−3	−17	−1	−69	−5	+12
Ru^{3+}						+39	−8	−48
Rh^{3+}							−4	+29
Ln^{3+}					+5 to +12		−6 to −7	−17 to −38

Table 7.11 Kinetic parameters for water exchange at the 2+ and 3+ aqua-ions of iron and ruthenium.

	$\log_{10}(k_{298}/s^{-1})$	$\Delta H^{\ddagger}$ (kJ mol^{-1})	$\Delta S^{\ddagger}$ (J K^{-1} mol^{-1})	$\Delta V^{\ddagger}$ (cm^3 mol^{-1})	Mechanism
Fe^{2+}aq	6.6	41	+21	+3.8	I_d
Fe^{3+}aq	2.2	65	+12	−5.4	I_a
Ru^{2+}aq	−1.8	88	+16	−0.4	I
Ru^{3+}aq	−5.4	90	−48	−8.3	I_a

activation as cation size increases on descending Group 13 of the Periodic Table, and the decreasingly associative nature of solvent exchange on going across the transition metals from left to right. We should perhaps add here that the observed variation in rate constants for solvent exchange as one traverses, for instance, the first row of transition metal 2+ cations (Table 7.2) represents the resultant of a fairly steady change in $\Delta S^{\ddagger}$ stemming from this steady change in mechanism and a very bumpy change in $\Delta H^{\ddagger}$ arising from ligand field ('crystal field stabilization/activation energies', CFSE/CFAE) effects (Fig. 7.4).

It is difficult to compare reactivity and mechanism of solvent exchange at different oxidation states of a given metal in a given solvent, as data are rarely available for, e.g., $[M(solvent)_6]^{2+}$ and $[M(solvent)_6]^{3+}$. The strongly reducing properties of Cr^{2+}aq and the strongly oxidizing properties of Mn^{3+}aq preclude such comparisons for chromium and manganese in aqueous media. However, comparisons can be made for iron and for ruthenium in water,[57,58] as detailed in Table 7.11, and for ruthenium in dimethylformamide.[59]

An important feature of the chemistry of most 3+ aqua-cations is the hydrolysis of coordinated water, $M^{n+}aq \rightleftharpoons MOH^{(n-1)+}aq + H^{+}aq$. Kinetics of water exchange at the hydroxo species $MOH^{(n-1)+}$aq have been determined for a number of metal(III) cations; activation parameters for water exchange at M^{n+}aq and at $MOH^{(n-1)+}$aq are collected in Table 7.12.[57,60] There is a wide range of activation enthalpies, whose variation is tolerably consistent with expectations from ionic radii and crystal field effects. Activation entropies and volumes indicate likely mechanisms. Thus for, e.g., gallium, water exchange at both the hexaaqua and the hydroxo-aqua forms is dissociative in character, but for iron there is a clear indication that whereas exchange of water at Fe^{3+}aq is associatively activated, at $FeOH^{2+}$aq the mechanism is dissociative (see also below under complex formation, Section 7.5). At both chromium and ruthenium, water exchange involves associative interchange, but at $CrOH^{2+}$aq and at $RuOH^{2+}$aq the character of the interchange step is intermediate between associative and dissociative—in other words there is only weak interaction between the metal ion and the incoming water in the transition state.

Table 7.12 Activation parameters for water exchange at 3+ aqua-cations.

	$\Delta H^{\ddagger}$ (kJ mol^{-1})	$\Delta S^{\ddagger}$ (J K^{-1} mol^{-1})	$\Delta V^{\ddagger}$ (cm^3 mol^{-1})	Assigned mechanism
Ga^{3+}	67	+30	+5	I_d
$GaOH^{2+}$			+6	I_d
Ti^{3+}	43	−12	+1	I_a, (A)
Cr^{3+}	109	+12	−10	I_a
$CrOH^{2+}$			+3	I
Fe^{3+}	65	+12	−5	I_a
$FeOH^{2+}$			+7	I_d
Ru^{3+}	90	−48	−8	I_a
$RuOH^{2+}$			+1	I
Rh^{3+}	131	+29	−4	I_a
$RhOH^{2+}$			+2	I
Ir^{3+}	131	+2	−6	I_a
$IrOH^{2+}$			+1	I

When we come to 4+ cations of *d*-block elements, then the important species in aqueous media is the oxo-cation $[MO(H_2O)_5]^{2+}$, with analogous species in non-aqueous solvents. Such complexes contain two different types of solvent molecule, *cis* and *trans* to the oxide ligand, with very different reactivities. Kinetic parameters are available for a number of solvent exchange reactions at VO^{2+}, but it is difficult to obtain good values for exchange of equatorial molecules and usually impossible to obtain values for exchange of axial molecules by NMR methods at this paramagnetic centre. The determination of kinetic parameters for solvent exchange at diamagnetic TiO^{2+} is much more straightforward; values have been obtained for exchange of both axial and equatorial (about 400 times slower at 298 K) dimethyl sulphoxide. Activation enthalpies are virtually the same for the two positions, and activation volumes are both positive (+1.6 and +4.8 cm^3 mol^{-1} for axial and equatorial DMSO exchange), suggesting dissociative activation.[61] This last reaction was in fact carried out in a dimethyl sulphoxide–nitromethane solvent mixture, presumably to avoid the DMSO freezing at the low temperatures required for the NMR experiments. This use of mixed solvents leads us into the next section, where another use of mixed solvents, to provide information of reaction orders with respect to exchanging solvent, is described.

7.4.2 Exchange reactions of solvento cations in non-coordinating solvents

The reaction of solvent exchange is very simple, with reactant ≡ product and $\Delta G^{\circ} = 0$. However, as stated earlier, this very simplicity brings the major problem when one tries to establish mechanism that it is clearly impossible to determine the order of reaction in exchanging solvent since this reagent is present in large and unvariable excess, by definition. An obvious way round this problem is to use as working solvent a dilute solution of the exchanging solvent in an inert second solvent. Subject to occasional solubility difficulties (a dilute solution of water in heptane would be invaluable for this and many other problems of solution chemistry!), this approach can work quite well. Although it can be argued that there is little justification for treating this type of reaction strictly as solvent exchange, it is possible to obtain valuable kinetic and mechanistic information from such systems to supplement that from genuine solvent exchange. We shall start by dealing with the negative argument, then focus on the ways in which information from this type of reaction usefully supplements studies of 'pure' solvent exchange.

In genuine solvent exchange, and indeed in any solvolytic reaction, the reagent is also the solvent and is present at all times within the solvation shell of the substrate and any intermediate that might be generated by it. Its concentration is invariant and so the dependence of rate on the concentration of nucleophile is a meaningless concept. Nowadays the term 'solvato (or solvento) complex' has been given to any homoleptic complex (usually cationic) whose ligands are uncharged and, in the liquid state, have solvent properties. Such complexes can be isolated as salts of non-coordinating anions and are conceptually no different from any other type of coordination complex. The fact that liquid ammonia is a good solvent is no reason for calling the cation of $[Co(NH_3)_6](ClO_4)_3$ a solvato complex (if liquid CO was a suitable solvent would the metal carbonyls be called solvato complexes?). It is possible to dissolve such complexes in a non-coordinating solvent and to examine the rates of exchange of ligand with free ligand added in stoichiometrically controllable amounts. The process is therefore a classical zero-free-energy substitution reaction equivalent to genuine solvent exchange, *v.s.*:

$$[MS_N]^{n+} + {}^*S \rightleftharpoons [MS_{(N-1)}{}^*S]^{n+} + S$$

Studies of such reactions reveal a wide range of kinetic behaviour. There is now a considerable amount of information in the literature, including rate laws and activation parameters, in particular volumes of activation (the exchange of neutral ligands keeps electrostriction effects to a minimum even though the substrate is always charged). There are examples of reactions in which the rate is independent of the concentration or nature of the entering group and where $\Delta V^\ddagger$ is large and positive (indicating a *D* mechanism). There are others where the kinetics are clearly second-order and $\Delta V^\ddagger$ is large and negative (indicating an *A*, or possibly an I_a, mechanism). There are also cases where the plots of k_{obs} against solvent concentration, $[S]$, obey the relationship $k_{obs} = kK[S]/(1 + K[S])$, indicating the importance of pre-aggregation and suggesting a dissociatively activated interchange. The mechanism adopted by a particular ion often depends on the nature of

Table 7.13 Kinetic rate laws, volumes of activation, and assigned mechanisms for exchange of free and coordinated solvent molecules at solvento cations in the non-coordinating solvent CD_3NO_2.

Cation	Coordinating solvent[a]	SN[b]	Dependence on [coord. solv.] independent (D, I_d)	curved (I_d)	linear (A, I_a)	$\Delta V^{\ddagger}/cm^3\ mol^{-1}$ dissoc.	assoc.
Be^{2+}	TMU	4	✓				
	TMP	4			✓		
	DMSO	4			✓		−2.5
	DMF	4			✓		−3.1
Mg^{2+}	DMF	6	✓			+8.5	
Zn^{2+}	TMTU[c]	4	✓				
	HMPA[c]	4			✓		
Al^{3+}	TMP	6	✓			+22.5	
	DMSO	6	✓			+15.6	
	DMF	6	✓			+13.7	
	HMPA	4			✓		
Ga^{2+}	DMSO	6	✓			+13.1	
	DMF	6	✓			+7.9	
	TMP	6	✓			+20.7	
In^{3+}	TMP	6			✓		−22.9
Sc^{3+}	TMP	6			✓		−18.7
	DMA,DEA	6	✓[d]		✓[d]		
	TMU	6	✓				
Tb^{3+}	DMF	8		✓		+5.2	
Ho^{3+}	DMF	8		✓		+5.2	
Er^{3+}	DMF	8	✓[d]	✓[d]		+5.4	
Tm^{3+}	DMF	8	✓	✓		+7.4	
Yb^{3+}	DMF	8	✓			+11.8	

[a] Abbrevations: DEA = diethylacetamide, $MeCONEt_2$; DMA = dimethylacetamide, $MeCONMe_2$; DMF = dimethylformamide, $HCONMe_2$; DMSO = dimethyl sulphoxide, Me_2SO; HMPA = hexamethylphosphoramide, $(Me_2N)_3PO$; TMTU = tetramethylthiourea, $SC(NMe_2)_2$; TMU = tetramethylurea, $OC(NMe_2)_2$; TMP = trimethyl phosphate, $(MeO)_3PO$.
[b] SN = solvation number.
[c] Non-coordinating solvent CD_2Cl_2.
[d] Parallel pathways indicated by rate law.

the neutral ligand *S* and, once again, steric factors often determine whether a higher or lower coordination number is favoured in the transition state. A far from comprehensive selection of data, including some for tetrahedral four-coordinate and for eight-coordinate species, is collected in Table 7.13. This table compares rate-law information with that from activation parameters, and indicates the likely mechanisms operating for solvent exchange in genuine solvent exchange and in non-coordinating solvent media.

7.5 Complex formation[62]

7.5.1 The Eigen–Wilkins mechanism[63]

The replacement of one or more solvent molecules in a solvento cation by a ligand or ligands is a very common and important reaction. Such reactions were not, however, much studied in the early days of inorganic studies since they are generally too fast to monitor by the long-established conventional techniques. Only ligand substitution at Cr^{3+}aq was slow enough for kinetic studies by early kineticists. The situation changed dramatically with the advent of stopped-flow techniques, which permitted the study of substitution kinetics at such centres as Mg^{2+}, Al^{3+}, Co^{2+}, and, above all, Ni^{2+}. The fact that displacement of solvent from hexasolventonickel(II) cations took place on a particularly appropriate time-scale for stopped-flow kinetics resulted in an enormous body of kinetic data on such reactions. It soon became clear that there was an apparent inconsistency between the experimentally derived second-order rate law for complex formation

$$+\mathrm{d[complex]}/\mathrm{d}t = k\mathrm{[metal\ ion][ligand]}$$

and the established wisdom that substitution at centres such as Co^{2+}, Ni^{2+}, Cu^{2+}, Zn^{2+}, and Mg^{2+} was dissociative in character. For proposing a simple and elegant solution to this problem, Manfred Eigen and Ralph Wilkins have had the distinction, rare in inorganic chemistry, of having their names attached to the general mechanism for complex formation. The name of the man who probably actually did the kinetics seems to be remembered only in Germany, where the mechanism is often referred to as the Eigen–Tamm–Wilkins mechanism. Although originally developed to take care of the apparent inconsistency for the M^{2+} plus ligand reactions just mentioned, the basic Eigen–Wilkins mechanism can be extended to cover all solvento-metal ions and a wide range of ligands, as will become apparent in the following pages.

The basic Eigen–Wilkins mechanism is shown in Fig. 7.10. It comprises pre-association of the reactants in a rapidly established equilibrium followed by rate-limiting replacement (or interchange) of solvent ligand by incoming ligand. The way in which a second-order rate law and a marked dependence of rate constant on the nature of the incoming ligand can be consistent with dissociative activation is set out in Table 7.14,[64] for the most studied solvento ion, Ni^{2+}aq, reacting with a wide range of ligands. There are also kinetic data available for environmental ligands such as fulvic acid,[65] but their accuracy is insufficient for inclusion in Table 7.14. The left-hand column of rate constants, obtained from the dependence of observed first-order rate constants on incoming ligand concentration—under conditions where $[Ni^{2+}aq] \gg [L]$ both to give first-order kinetics and to avoid formation of ML_2, ML_3, etc.—shows a 10^9-fold range of values. It is immediately apparent that the charge on the incoming group is of paramount importance and thus that pre-association is likely to play a key role in determining the

MECHANISM:

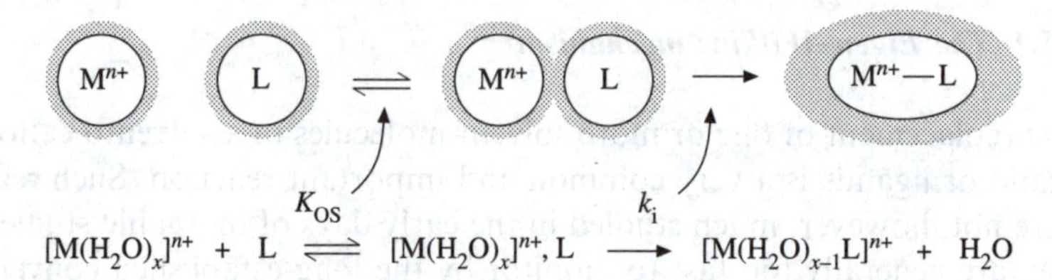

KINETICS:

General rate law:

$$+\frac{d[M^{n+}]}{dt} = \frac{K_{os}k_i[M^{n+}]\,[L]}{1 + K_{os}[L]}$$

Under usual experimental conditions, of $[M^{n+}] \gg [L]$:

$$+\,d[M^{n+}]\,/\,dt = K_{os}k_i[M^{n+}]\,[L]$$

Whence: $k_i = K_{os}k_i$

Fig. 7.10 The Eigen–Wilkins complex formation sequence: mechanism, equations, and kinetics.

ligand effects on reactivity. In order to estimate interchange rate constants, these k_f values have to be divided by the respective outer-sphere association constants K_{OS}. Unfortunately, these are almost never directly available—as complete reaction takes only a fraction of a second there is hardly time to measure an association constant! Just occasionally, as for instance in the reaction between Ni^{2+}aq and the methyl phosphate dianion, a curved plot for k_f versus [L] permits the estimation of K_{OS}.[66] Also, for the exceptional cases of substitution-inert Cr^{3+}aq and aqua-cobalt(III) complexes such as $[Co(NH_3)_5(H_2O)]^{3+}$, there may well be time to measure K_{OS} directly (cf. Section 7.5.4.1 below). Otherwise, values of K_{OS} can be estimated from measurements of slowly or non-reacting analogues or, more usually, calculated. Such calculations for association between ions can be made using the electrostatically based Fuoss equation,[67,68] and between an ion and a dipolar but uncharged ligand by using Eigen's extension of this equation.[69] Such estimates as are available are shown in the K_{OS} column in Table 7.14. Division of k_f by K_{OS} gives an estimate of the interchange rate constant k_i for each reaction. In a few systems it is possible to obtain k_i experimentally by ultrasonics. For Ni^{2+}aq such a value, of *ca* $10^4\ s^{-1}$, has been obtained for interchange within the Ni^{2+}, SO_4^{2-} ion-pair (at 20 °C).[68] The values for k_i in the right-hand column of Table 7.14 cover a very much smaller range than the k_f values; indeed, the k_i values may be considered equal within experimental uncertainty and the uncertainties in calculating K_{OS} values. Moreover, they are similar to the water exchange rate for Ni^{2+}aq under analogous conditions (cf. Section 7.3.1 above). The k_i values thus indicate a dissociative mechanism for the kinetic, ligand interchange, step as hoped for at the start of this analysis.

Table 7.14 Complex formation at Ni^{2+}aq : rate constants and outer-sphere association constants.

Incoming ligand	Measured $10^{-3}k_f$ ($dm^3\ mol^{-1}\ s^{-1}$)	Estimated K_{OS} (molar scale)	Derived $10^{-3}k_i$ (s^{-1})
$TetrenH_3^{3+}$	0.0035		
$TetrenH_2^{2+}$	0.032		
N,N,N-Me_3en^+	0.5	0.02	25
N-MethylimidazoleH$^+$	0.23	0.02	12
ImidazoleH$^+$	0.5	0.02	20
$[Co(NH_3)_5(oxalate)]^+$	3.9		
Ammonia	5	0.15	33
Hydrogen fluoride	3	0.15	20
Imidazole	2.8–7	0.15	19–43
1,10-Phenanthroline	4.1	0.15	26
Diglycine	21	0.17	12
Fluoride$^-$	8	1	8
Acetate$^-$	100	3	30
Glycinate$^-$	21	2	10
Cytidine monophosphateH$^-$	23		
Oxalate^{2-}	75	13	6
Malonate^{2-}	450	95	5
Methylphosphate^{2-}	290	40	7
Ribose monophosphate^{2-}	23		
Flavine adenine dinucleotide^{2-}	200		
Pyrophosphate^{3-}	2100	88	24
Tripolyphosphate^{4-}	6800	570	12
Adenosine 5′-triphosphate^{4-}	4×10^6		

An analogous analysis can be carried out for complex formation from Co^{2+}aq. Here there are many fewer kinetic data, but information is available[70] for nearly a dozen cobalt(III) complexes of the $[Co(NH_3)_5L]^{n+}$ type as incoming ligands (contrast the solitary $[Co(NH_3)_5(oxalate)]^{n+}$ entry in Table 7.14). These complexes provide more interesting examples of positively charged ($n =$ up to 3) incoming groups than the protonated polyamines in Table 7.14. These formation reactions between M^{2+}aq and $[Co(NH_3)_5L]^{n+}$ species provide a model for, and a link to, the initial stage in the inner-sphere mechanism for electron transfer that plays such a large part in Chapter 9.

Table 7.15 shows that a similar analysis for complex-formation reactions of the *sp*-block ion Mg^{2+}aq[71] is also consistent with dissociative activation for the rate-determining ligand interchange step. Rate constants specifically for the interchange step have been measured by ultrasonics for incoming ligands sulphate, thiosulphate, chromate, and acetate.[72] Relaxation times in solutions of magnesium sulphate and of magnesium chromate indicate k_i

Table 7.15 Complex formation at Mg^{2+}aq : rate constants and outer-sphere association constants.

Incoming ligand	Measured $10^{-5}k_f$ ($dm^3\ mol^{-1}\ s^{-1}$)	Estimated K_{OS} (molar scale)	Derived $10^{-5}k_i$ (s^{-1})
8-Hydroxyquinoline	0.13	0.2	0.7
CalcimycinH (antibiotic A 23187)	1.2		
8-Hydroxyquinolinate$^-$	6.0	2.2	2.9
Fluoride$^-$	0.55	1.6	0.4
5-Nitrosalicylate$^-$	7.1	2.0	3.6
Hydrogen carbonate$^-$	5.0	0.9	5.6
Carbonate^{2-}	0.15	3.5	0.4
PyrophosphateH$_2^{2-}$	5.4	13	0.4
Adenosine diphosphateH^{2-}	10	9	1.1
Adenosine diphosphate^{3-}	30		
Adenosine triphosphateH^{3-}	30	30	1.0
Adenosine triphosphate^{4-}	125	120	1.0

values of $1 \times 10^5\ s^{-1}$ for both, comfortingly close to $k = 7 \times 10^5\ s^{-1}$ for water exchange (especially bearing in mind the larger number of water molecules available in the vicinity of the $Mg(H_2O)_5^{2+}$ transition state). Some ligands of biological and of geochemical relevance have been included for Mg^{2+}aq, as several bio-ligands have been included in the preceding Ni^{2+}aq tabulation, to emphasize the generality of the Eigen–Wilkins treatment. Similar analyses of kinetic data applied to the somewhat less extensive results for Co^{2+}aq, Cu^{2+}aq, and Zn^{2+}aq, and for hexasolvento derivatives of these 2+ transition metal ions, are also consistent with dissociative activation at these centres. As in the case of Ni^{2+}, there are some direct estimates of interchange rate constants from ultrasonics experiments.[73]

Supporting evidence is available from activation parameters. $\Delta H_i^\ddagger$ and $\Delta S_i^\ddagger$ values for the interchange step in complex-formation reactions of Ni^{2+}aq both cover only very short ranges. Volume profiles for complex formation reactions of Ni^{2+}aq are particularly revealing.[74] Figure 7.11 shows that the charge on the incoming ligand has a significant effect on the volume change on going from the reactants to the ion-pair or ion-association complex. However, the volume change for the next step to the interchange transition state is essentially equal for all incoming groups. It is only in the final stage, the breakdown of interchange transition state to products, that large differences between ligands are found. Incidentally, the lack of any correlation between activation volume and overall volume change (initial reactants to products) is another piece of evidence in favour of a dissociative mechanism for these reactions. In this connection we may note that activation volumes for exchange at Ni^{2+} of a series of nitriles

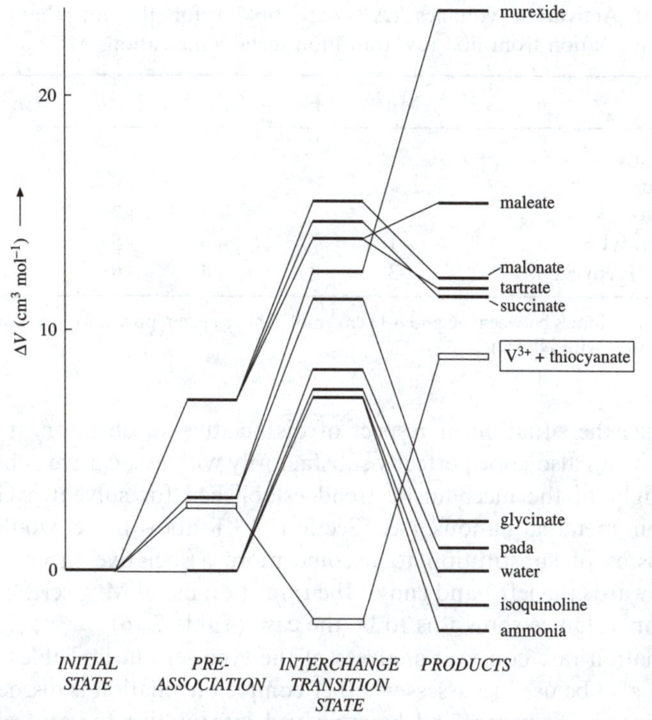

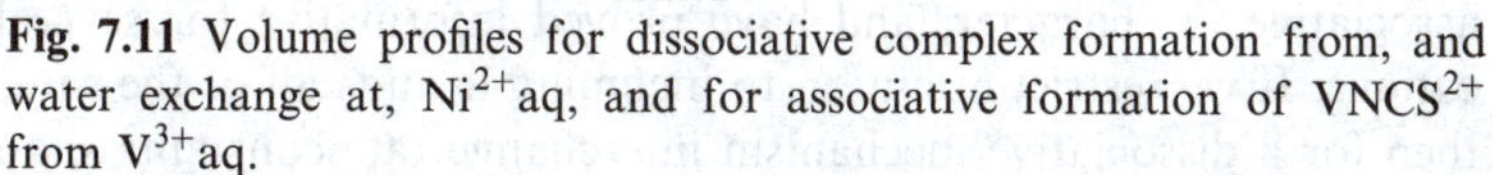

Fig. 7.11 Volume profiles for dissociative complex formation from, and water exchange at, Ni^{2+}aq, and for associative formation of $VNCS^{2+}$ from V^{3+}aq.

covering a wide range of partial molar volumes vary over only a very small range.[75]

We have, a little prematurely, but to provide an illuminating comparison, included the volume profile for associative formation of thiocyanato-vanadium(III)[76] in Fig. 7.11. This shows the same volume increase on formation of the outer-sphere associated species, but then the associative interchange transition state is formed with a marked *decrease* in volume. Interestingly, this provides another example of the lack of correlation between activation volumes and overall volume changes, for the overall change in volume on going from V^{3+}aq plus NCS^- to $VNCS^{2+}$aq is markedly positive, despite the negative activation volume for the associative interchange step.

7.5.2 Dissociative and associative activation

Thus far we have concentrated on ions M^{2+}aq at the right-hand end of the first row of the transition metals, and Mg^{2+}aq, where dissociative activation is expected. Although the Eigen–Wilkins mechanism was developed to

Table 7.16 Activation volumes ($\Delta V^{\ddagger}/\mathrm{cm^3\,mol^{-1}}$) for the interchange step in complex formation from first-row transition metal aqua-cations M^{2+}aq.

Ligand	V^{2+}	Mn^{2+}	Fe^{2+}	Co^{2+}	Ni^{2+} [a]	Cu^{2+}	Zn^{2+}
Thiocyanate	−5						
Glycinate					+8	+9	+4
Ammonia				+5	+7		
2,2′-Bipyridyl		−1		+6	+5		+4
2,2′,6′,2″-Terpyridyl		−3	+4	+4	+6		

[a] Activation volumes between +6 and +11 $\mathrm{cm^3\,mol^{-1}}$ have been reported for reaction of several other ligands with Ni^{2+}aq.

rationalize the situation in respect of dissociative substitution at such 2+ cations, it can also cope perfectly satisfactorily with associative substitution. In the light of the mechanistic trend established for solvent exchange at transition metal aqua-ions (see Section 7.3.1 above), we would expect mechanisms of substitution to become more associative in character on going towards the left-hand end of the transition metal M^{2+} series; reported activation volumes show this to be the case (Table 7.16).

Formation rate constant analyses of the type set out in Tables 7.14 and 7.15 can also be used to assess whether complex formation is dissociative or associative in character, and have proved informative for several M^{3+}aq cations. If we restrict attention to incoming ligands all of the same charge, then for a dissociative mechanism interchange rate constants should cover only a very small range. The big ranges shown for Ti^{3+}aq and V^{3+}aq in Table 7.17, and the difference of 50 times in rate constants for reaction of Mo^{3+}aq with thiocyanate or chloride, all suggest markedly associative character to accommodate the marked dependence of rate constant on the nature of the incoming group. The fact that water exchange occurs considerably more rapidly than the slowest substitutions given in Table 7.17 furnishes another piece of evidence in favour of associative activation in substitution at these 3+ centres. The contrast between the volume profiles

Table 7.17 Rate constants for complex formation from Ti^{3+}aq (left) and V^{3+}aq (right) and uni-negative ligands, in aqueous soluton at 298.2 K.

Ligand	$10^{-3}k_f/\mathrm{dm^3\,mol^{-1}\,s^{-1}}$	Ligand	$k_f/\mathrm{dm^3\,mol^{-1}\,s^{-1}}$
Acetate	1800	4-Aminosalicylate	7000
OxalateH	390	Salicylate	1400
3-CNacac	160	OxalateH	1300
Chloroacetate	110	Azide	(900)
Thiocyanate	8	Thiocyanate	110
		Bromide	10
		Chloride	3

for complex formation proceeding by dissociative interchange at Ni^{2+} and by associative interchange at V^{3+} has already been illustrated, in Fig. 7.11 above.

As in the case of solvent exchange, it is easy to show a changeover from associative to dissociative substitution on traversing the first row of the *d*-block elements from left to right, but it is more difficult to establish trends descending the Periodic Table. Substitution at Cr^{3+}aq is associative in character, so it is hardly surprising that substitution at Mo^{3+}aq is also associative. There are no other direct comparisons for M^{3+}aq. For M^{2+}aq, it follows from activation volumes of $+7.1\ cm^3\ mol^{-1}$ and $-5.5\ cm^3\ mol^{-1}$ for reaction of Zn^{2+}aq and of Cd^{2+}aq, respectively, with 2,2′-bipyridyl that there is a change from dissociative to associative activation on descending this group.[77] Conflicting mechanistic indications from various formation reactions of Ru^{2+}aq[78] rule out any meaningful comparisons involving this cation.

7.5.3 Metal(III) hydroxo-aqua- and aqua-cations

The hydrolysis behaviour of the majority of M^{3+}aq ions should not be forgotten—in other than acid media the hydroxo form $M(OH)^{2+}$aq will be an important or the dominant form. In most cases, $M(OH)^{2+}$aq is more reactive than M^{3+}aq; the exceptions are vanadium(III) and titanium(III). Kinetic parameters have been published for reaction of the two forms for a number of metals M, especially iron, chromium, and aluminium, with a variety of ligands. Grant and Jordan documented reactivity trends for Fe^{3+}aq and $Fe(OH)^{2+}$aq some years ago,[79] kinetics of the environmentally significant iron(III)–fulvic acid system have been established,[80] and a recent review gives overall cover of kinetics of complexation of iron(III) in aqueous solution.[81] The standard compilations of kinetic data for Cr^{3+}aq and $Cr(OH)^{2+}$aq are those by Espenson[82] and Thusius,[83] with more recent data traceable through reports on kinetics of reaction of chromium(III) with L-histidine[84] and with octacyanomolybdate(IV).[85] Rate constants for complex formation from aluminium(III), from $Al(OH)^{2+}$aq as well as Al^{3+}aq and $Al(OH)^{2+}$aq, were collated in 1992;[71] more recent kinetic studies[86] have dealt with more exotic ligands without disturbing the established pattern. Rate constant patterns for aluminium, chromium, and iron are analysed in Table 7.18. It is instructive to compare reactivity trends for the two forms for the three metals. In the case of aluminium the data are consistent with dissociative activation for both Al^{3+}aq and $Al(OH)^{2+}$aq, whereas in the case of chromium associative activation is suggested for both Cr^{3+}aq and $Cr(OH)^{2+}$aq. However, for iron, complex formation seems to be associative from Fe^{3+}aq, but dissociative from $Fe(OH)^{2+}$aq. This mechanistic difference is confirmed by available activation volumes[87] (Table 7.19). Not only is there a clear sign difference between Fe^{3+}aq and $Fe(OH)^{2+}$aq, there is also a wide range of values for the former, again consistent with an important role

Table 7.18 Rate constants for complex formation from the hexaaqua and hydroxo-aqua forms, $M^{3+}aq$ and $M(OH)^{2+}aq$, of aluminium(III), chromium(III), and iron(III), in aqueous solution at 298.2 K.

	$k_f/dm^3\ mol^{-1}\ s^{-1}$			$10^8 k_f/dm^3\ mol^{-1}\ s^{-1}$			$k_f/dm^3\ mol^{-1}\ s^{-1}$	
Ligand	$Al^{3+}aq$	$AlOH^{2+}aq$	Ligand	$Cr^{3+}aq$	$CrOH^{2+}aq$	Ligand	$Fe^{3+}aq$	$FeOH^{2+}aq$
Kojic acid	0.83	825	$Iodide^-$	0.008	260	Mandelic acid	slow	7500
Ferron	~1.4	2000	*S*-$Thiocyanate^-$	0.41	490	Acethydroxamic acid	1.2	2000
Acethydroxamic acid	1.7	2300	$Chloride^-$	3.0	2300	Hydroxybenzoic acid	<10	<3500
Desferrioxamine	0.13	2700	*N*-$Thiocyanate^-$	73	2200	Desferrioxamine	282	4100
$Salicylate^-$	~1	~2000	$edtaH_2^{2-}$	150	2800	Hydrazoic acid	4000	6800
$Citrate^{2-}$	80		$edtaH^{3-}$		60 000	$Chloride^-$	4.8	5500
$Sulphate^{2-}$	~1000					$Thiocyanate^-$	122	23 200
$Hexacyanoferrate^{3-}$	4900					$Trichloroacetate^-$	63	7800
						$Dichloroacetate^-$	118	1900
						$Monochloroacetate^-$	1500	4100
			$CrOH^{2+}$	~600	20 000	$FeOH^{2+}$	6.8	670
						$[Co(NH_3)_5(salH)]^{2+}$	5.6	770

Table 7.19 Activation volumes for complex formation at iron(III) in aqueous solution.

	$\Delta V^{\ddagger}/cm^3\ mol^{-1}$	
Ligand	$Fe^{3+}aq$	$Fe(OH)^{2+}aq$
Thiocyanate$^-$	−1.2	+6.6
Chloride$^-$	−9.9	+7.8
Bromide$^-$	−19	
Acethydroxamic acid	−6	+5
Desferrioxamine	−5	+4

for the (hydrated) entering group in the transition state. Similar analyses indicate dissociative mechanisms for formation from both $Ga^{3+}aq$ and $Ga(OH)^{2+}aq$,[88] associative—on the basis of rather few data—for $In^{3+}aq$ and $In(OH)^{2+}aq$.[89] In the case of rhodium both associative and dissociative formation have been claimed from both $Rh^{3+}aq$ and $Rh(OH)^{2+}aq$.[90]

Included in Table 7.18 are rate constants for $M(OH)^{2+}aq$ as entering ligand, in other words for the special case of binuclear complex formation. Kinetics of formation of polynuclear hydroxo-bridged aqua-cations have been briefly discussed elsewhere[71]—the case of chromium(III) has been particularly well characterized, thanks to the slowness of reactions and consequent possibilities of characterizing intermediates and products with some confidence.[91]

We should close this section with a word of warning. For reaction of a 3+ aqua-cation with a ligand which may react as HL or L^-, there is a potential mechanistic ambiguity, as it is not possible to distinguish kinetically between the two reactions $M^{3+}aq$ plus L^- and $M(OH)^{2+}aq$ plus HL. For a ligand such as chloride there is no problem, and for several organic ligands the rate constant calculated for the $M^{3+}aq$ plus L^- path is higher than the diffusion-controlled limit, thus firmly indicating that $M(OH)^{2+}aq$ reacts with HL. However, for many ligands it can be very difficult to tell which path operates.[92]

7.5.4 Anation at ternary aqua complexes

7.5.4.1 Cobalt(III), chromium(III), and ruthenium(II) aquapentaammines

Reactions of cobalt(III) and chromium(III) complexes of the type

$$[M(NH_3)_5(H_2O)]^{3+} + L \longrightarrow [M(NH_3)_5L]^{3+} + H_2O$$

are a special case of complex formation, entirely analogous to the reactions discussed in the preceding section, but proceeding at very much slower rates. It is therefore much easier to obtain spectroscopic and kinetic evidence for

Table 7.20 Kinetic parameters for anation at $[Ru(NH_3)_5(H_2O)]^{2+}$, in aqueous solution at 298 K.

	$10^3k_f(dm^3\ mol^{-1}\ s^{-1})$	$\Delta H^‡/(kJ\ mol^{-1})$	$\Delta S^‡(J\ K^{-1}\ mol^{-1})$
pyH^+	3.1	68	−54
$isonicH^+$	3.6		
$imidH^+$	2.7		
pyridine (py)	92	67	−38
pyrazine	56	73	−24
isonicotinamide (isonic)	105		
imidazole (imid)	200	69	−27
3-picoline	91	68	−28
dinitrogen	71	77	−10
acetonitrile	280	67	−28
benzyl isonitrile	100		
acetic acid	36		
$cyanoacetate^-$	1200	63	−34
$thiocyanate^-$	*ca* 1000		

the pre-association equilibrium. Werner recognized that ion-pairing might play a key role, and over half a century ago comparison of the rate laws for formation and aquation of $[Co(NH_3)_5Cl]^{2+}$ demonstrated the inclusion in the former of a term involving chloride concentration.[93] Since then several investigators have furnished spectroscopic evidence for the intermediacy of ion-pairs, for example in the reactions of $[Co(NH_3)_5(H_2O)]^{3+}$ with sulphate[94] and with chloride.[95] Kinetic evidence for the intermediacy of ion-pairs can sometimes be obtained from plots of rate constant versus incoming ligand concentration. The curvature of such plots shows significant ion-pairing; conversion into reciprocal rate constant versus reciprocal concentration format gives straight lines, from whose intercept and slope ion-pairing constants and interchange rate constants may be estimated. Such a treatment of the $[Co(NH_3)_5(H_2O)]^{3+}$ plus chloride, thiocyanate, or sulphate systems with a subsequent 'Eigen–Wilkins' analysis gave a roughly constant value for the interchange rate constant of approximately $2 \times 10^{-5}\ dm^3\ mol^{-1}$, indicating the expected dissociative activation. Kinetic parameters for anation at $[Ru(NH_3)_5(H_2O)]^{2+}$ (Table 7.20)[96] follow a pattern very similar to that shown by $Ni^{2+}aq$ and $Mg^{2+}aq$ (cf. Tables 7.14 and 7.15 above), suggesting dissociative activation in these reactions too. But an 'Eigen–Wilkins' kinetic analysis of a number of anation reactions of $[Cr(NH_3)_5(H_2O)]^{3+}$ (Table 7.21) leads to a less clear-cut result, suggesting that here, as in a number of other substitutions at this complex,[97] the mode of activation may well be on the dissociative/associative borderline.[98] Kinetic and spectroscopic results on a number of complex formation reactions of $[Cr(H_2O)_6]^{3+}$ were also strongly indicative of the operation of

Table 7.21 Eigen–Wilkins analysis of kinetics of anation of the $[Cr(NH_3)_5(H_2O)]^{3+}$ cation.

Incoming ligand	$10^{-3}k_f$ ($dm^3\,mol^{-1}\,s^{-1}$)	K_{OS} (molar scale)	$10^{-3}k_i$ (s^{-1})
H_3PO_4	0.46	0.32	1.4
$H_2PO_4^-$	2.6	1.79	1.5
NCS^-	4.2	<0.7	>6
$C_2O_4H^-$	6.5	1.16	5.6
$H_3N^+CH_2CO_2^-$	7.8	0.55	14.2
$C_2O_4^{2-}$	29	4.5	6.4

the two-stage mechanism of ion-pairing followed by rate-limiting interchange, though here the interchange step appears to have somewhat more associative character (cf. water exchange, Section 7.4.1).[99] Kinetic parameters for ternary ammine- or amine-aqua complexes of metal(III) cations have been conveniently collected together.[100]

7.5.4.2 Pentacyanometallates

Dissociation kinetics of pentacyanometallates $[M(CN)_5L]^{n-}$ were discussed in Chapter 4 (Section 4.7.5.2), and a few formation and exchange reactions of such complexes were mentioned then in connection with arguments as to whether substitution at pentacyanometallates is limiting dissociative, *D*, or dissociative interchange, I_d. They will appear again in Chapter 9 in connection with intramolecular electron transfer in binuclear species (Section 9.4.4). There is a bountiful supply of data (not always as consistent as one might hope between different groups of investigators) on kinetic parameters for formation of complexes $[M(CN)_5L]^{n-}$ from $[M(CN)_5(H_2O)]^{n-}$, particularly for $M = Fe^{II}$. Selected values, both for simple organic ligands L and for some ruthenium complexes as incoming L, are listed in Table 7.22.[101,102,103] The dependence of k_f values on the charge of the incoming ligand is very reminiscent of formation reactions from such cations as $Ni^{2+}aq$ (Section 7.5.1) and thus readily interpreted in terms of dissociative interchange. In this connection it may be remarked that rate constants for the sequence of reactions of tris-bipyrazine-ruthenium(II), $[Ru(bipz)_3]^{2+}$, where bipz = **5**, with $[Fe(CN)_5(H_2O)]^{3-}$, decrease steadily from 3500 $dm^3\,mol^{-1}\,s^{-1}$ for attachment of the first $[Fe(CN)_5]$ to a peripheral ligand nitrogen to 0.3 $dm^3\,mol^{-1}\,s^{-1}$ for attachment of the final $[Fe(CN)_5]$ to the sixth peripheral ligand nitrogen,[104] as the

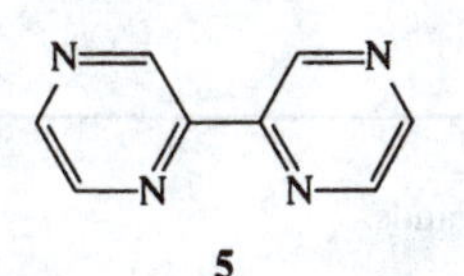

Table 7.22 Kinetic parameters for formation reactions from aquapentacyanoferrates(II).

	k_f ($dm^3\ mol^{-1}\ s^{-1}$)	$\Delta H^\ddagger$ ($kJ\ mol^{-1}$)	$\Delta S^\ddagger$ ($J\ K^{-1}\ mol^{-1}$)	$\Delta V^\ddagger$ ($cm^3\ mol^{-1}$)
L^{3+}				
pz–$M^{III}(NH_3)_5^{3+}$ [a]	5500–8800			
L^{2+}				
NC–py–$Ru^{II}(NH_3)_5^{2+}$	4600			
pz–$Ru^{II}(NH_3)_5^{2+}$	3700			
L^{+}				
N-methyl-4,4′-bipyridylium	2820	68	+50	
N-methylpyrazinium	2380	70	+42	
4,4′-bipyridylium	1610			
L				
4,4′-bipyridyl	586	64	+21	
methionine	535	70	+42	+18
pyrazine	380	64	+21	
pyridine	365	67	+29	
histidine	315	64	+21	+17
carbon monoxide	310	63	+15	
dimethyl sulphoxide	240	64	+17	
imidazole	240	64	+13	+16
thiourea	194	60	+1	
L^{-}				
glutathionate	219	76	+54	+14
β-alaninate	57	60	−10	+17
4-nitroimidazolate	45			
cyanide	30	77	+42	+14
glycinate	28 [b]	61	−13	+16
Ru(edta)L (L = pz, 4,4′-bipy, etc.)	100–200			
L^{2-}				
glutamate, tyrosinate	9			
4,4′-bipy–$Co(CN)_5^{2-}$	3			
NC–py–$Co(CN)_5^{2-}$	2			

[a] M = Co, Rh, Ru.
[b] Between 18 and 30 for other singly charged amino acid ligands.

charge on the ruthenium-centred species goes from 2+ to (formally) 13−, i.e. $z_A z_B$ goes from favourable 6− to very unfavourable 39+. In contrast to the I_d mechanism adumbrated by these formation rate constant trends, the large positive activation volumes indicate something close to a *D* mechanism. Indeed the fact that activation volumes for anation at the iron(III) analogue are considerably smaller, between +3 and +6 $cm^3\ mol^{-1}$, certainly suggests a greater interchange character for substitution at iron(III) than at iron(II) in these pentacyanometallates.[102,105] High-pressure effects on photochemical as well as on thermal kinetic behaviour of pentacyanometallates have been reviewed.[106]

7.5.5 Tetrahedral and square-planar solvento cations

Solvent exchange at Be^{2+} proceeds by dissociative or associative pathways depending on the bulk and donor properties (nucleophilicity) of the solvent, as mentioned in Section 7.4.2 above. There is very little information on mechanisms of complex formation at Be^{2+}aq. An early analysis[107] of a very limited number of k_f values into their (estimated) K_{OS} and k_i components failed to give a clear indication of mechanism, with a subsequent discussion rather cleverly side-stepping a direct associative/dissociative choice.[108] A more recent report[109] of $\Delta V^\ddagger = -7.1\ cm^3\ mol^{-1}$ and $\Delta S^\ddagger = -84\ J\ K^{-1}\ mol^{-1}$ for reaction with 4-isopropyl-tropolone (hinokitiol or β-thujaplicin) suggests a predominantly associative mechanism, at least in this case. A subsequent examination of the kinetics of complex formation with nitrosalicaldehydopentaamminecobalt(III) as a rather exotic entering group failed to give clear additional information due to uncertainty as to the importance of chelate ring closure in the overall kinetic scheme (cf. following section).[110]

The presumed-tetrahedral copper(I) aqua-cation Cu^+aq survives just long enough for some electron transfer kinetic studies to have been carried out, but the mechanism of substitution at this transient aqua-ion seems not to have been established. However, kinetic patterns for substitution at the much more stable $[Cu(MeCN)_4]^+$ indicate associative (I_a) activation.[111]

For square-planar complexes associative activation is the normal mode, and has been demonstrated from rate law (Fig. 7.12), reactivity trend (Table 7.23),[112] and activation volume[113] (Table 7.24) determinations. Figure 7.12 shows the expected resemblance to analogous plots for substitution at platinum(II) complexes in general, as discussed in Chapter 3—see, for example Fig. 3.10. Figure 7.12 and Table 7.23 show the expected increase in k_2 values with increasing affinity of the incoming ligand for Pt^{2+}. Both entropies and volumes of activation (Table 7.24) are also consistent with associative substitution at the two square-planar aqua-ions Pd^{2+}aq and Pt^{2+}aq. The volume profile for reaction of Pd^{2+}aq with dimethyl sulphoxide to form $[Pd(dmso)(H_2O)_3]^{2+}$, shown in Fig. 7.13, quantifies the compact nature of the transition state for this formation reaction, and also shows the similarity with associative water exchange at Pd^{2+}aq.[114]

Table 7.23 Rate constants for reaction of Pt^{2+}aq with a range of incoming ligands, in water at 298.2 K.

Entering ligand	k_2 ($dm^3\ mol^{-1}\ s^{-1}$)	Entering ligand	k_2 ($dm^3\ mol^{-1}\ s^{-1}$)	Entering ligand	k_2 ($dm^3\ mol^{-1}\ s^{-1}$)
H_2O	2.8×10^{-5}	Cl^-	0.027	Me_2S	3.6
Me_2SO	8.4×10^{-5}	Br^-	0.31	I^-	7.7
$HgCl^+$	9.2×10^{-5}	SCN^-	1.3	$SC(NH_2)_2$	13.9
$H_2C{=}CH_2$	1.0×10^{-2}	Et_2S	1.9		

Table 7.24 Activation parameters for substitution at Pd^{2+}aq and at Pt^{2+}aq.

	$\Delta H^{\ddagger}$(kJ mol^{-1})		$\Delta S^{\ddagger}$(J K^{-1} mol^{-1})		$\Delta V^{\ddagger}$(cm^3 mol^{-1})	
	Pd^{2+}aq	Pt^{2+}aq	Pd^{2+}aq	Pt^{2+}aq	Pd^{2+}aq	Pt^{2+}aq
H_2O	50	90	−60	−43	−2.2	−4.6
Me_2S	31	61	−54	−40	−4.0	−15.3
Et_2S	43	57	−19	−60	−8.7	−17.0
$S(CH_2)_4O$ [a]	34	62	−54	−46	−6.6	−13.9
$S(CH_2)_4S$ [b]	45	50	−17	−84	−10.1	−20.1

[a] 1,4-thioxane.
[b] 1,4-dithiane.

There has been some interest in kinetics of anation at ternary complexes of the *cis*-$[Pt(LL)_2(H_2O)_2]^{2+}$ type, in view of their relevance to the mechanism of action of anticancer drugs of the cisplatin and carboplatin type. Systems studied have included *cis*-$[Pt(LL)_2(H_2O)_2]^{2+}$, with LL = $(NH_3)_2$ or related methyl-substituted derivatives, reacting with amino acids, *N*-acetylglycine, or nucleosides,[115] or indeed with DNA.[116]

7.5.6 Formation of chelate rings and macrocyclic complexes

The preceding discussion of complex formation mechanisms has focused on the formation of monoligand complexes with monodentate ligands

$$[M(H_2O)_6]^{n+} + L \longrightarrow [M(H_2O)_5L]^{n+} + H_2O$$

or the formation of the first metal ion–ligand-donor bond for multidentate ligands

$$[M(H_2O)_6]^{n+} + LL \longrightarrow [M(H_2O)_5(-L{-}L)]^{n+} + H_2O$$

$$[M(H_2O)_5(-L-L)]^{n+} \longrightarrow [M(H_2O)_4(\overset{L}{\underset{L}{\langle\ \rangle}})]^{n+} + H_2O$$

The detailed sequence for a hexasolvento cation is shown in Fig. 7.14. The

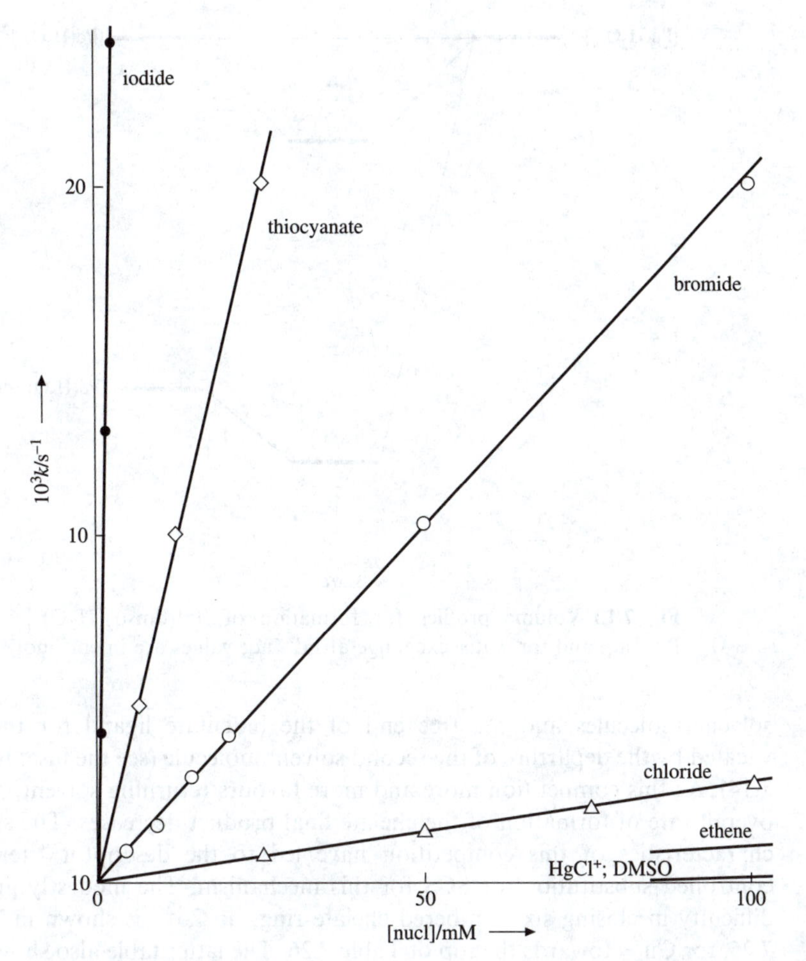

Fig. 7.12 Dependence of observed first-order rate constants on the nature and concentration of the incoming ligand for substitution at Pt^{2+}aq.

first stage is identical with the Eigen–Wilkins mechanism described above; the overall rate of formation of the chelate may be controlled by the first stage or by the rate of ring closure. The resultant depends primarily on the nature of the metal ion. For a cation such as Ni^{2+}, which undergoes relatively slow solvent exchange and solvent replacement, loss of the first solvent molecule and formation of the monodentate intermediate $[Ni(solv)_5(L–L)]^{n+}$ is normally followed by loss of the second solvent molecule at a similar rate and rapid ring closure, to give an overall rate of formation of a chelate similar to that of its monodentate analogue (see, e.g., Ni^{2+} plus glycinate or oxalate in Table 7.14). When the time-scale for solvent loss is shorter, as for Co^{2+} or, more dramatically, Mn^{2+} or Cu^{2+}, or when there is some difficulty in ring closure, as in the formation of a large chelate ring, then there may be competition between returning or replacing

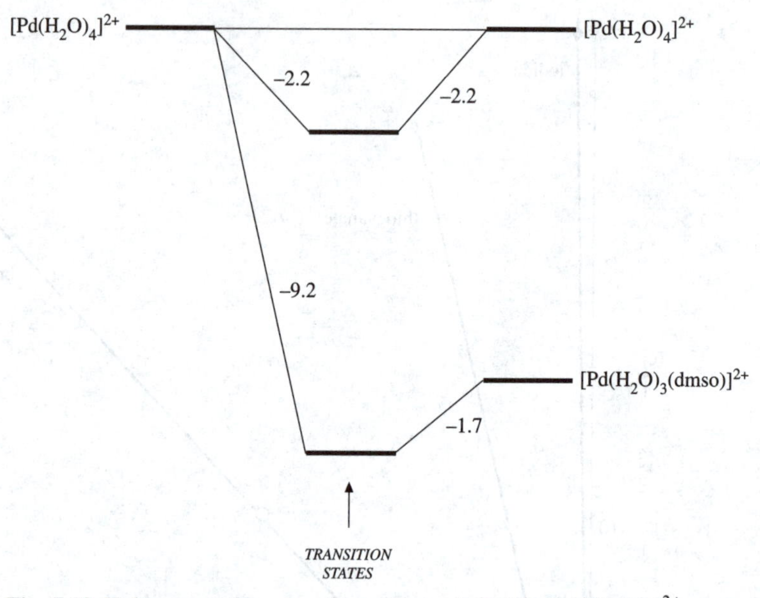

Fig. 7.13 Volume profiles for formation of $[Pd(dmso)(H_2O)_3]^{2+}$ from Pd^{2+}aq, and for water exchange at Pd^{2+}aq; values are in $cm^3\ mol^{-1}$.

solvent molecules and the free end of the bidentate ligand for the site vacated by the departure of the second solvent molecule (see the inset in Fig. 7.14). As this competition more and more favours returning solvent, so the overall rate of formation of the chelate final product decreases. The spatial characteristics of this competition have led to the descriptor 'sterically controlled substitution' or SCS for this mechanism. The modestly greater difficulty in closing six-membered chelate rings at Co^{2+} is shown in Table 7.25, for Cu^{2+} towards the top of Table 7.26. The latter table also shows the greater difficulty in closing a seven-membered chelate ring, and documents the increasing slowness of complex formation on going from open-chain (garland) multidentate ligands through flexible macrocycles to rigid macrocycles; Cu^{2+} has been chosen to show this sequence as the difference between fastest and slowest is the most dramatic. For the ultimate step in the type of sequence shown in Table 7.26 we have to switch from Cu^{2+} to earlier first-row M^{2+}. Rate constants have been determined for complex formation between Mn^{2+}, Fe^{2+}, and Co^{2+} and the Fe_3S_4 unit in aconitase. Very small and very similar rate constants for these three cations, 1.0, 1.9, and 5.9 $dm^3\ mol^{-1}\ s^{-1}$, respectively, reflect the great difficulty of approach of these aqua-cations through the narrow channel which forms the approach route to the Fe_3S_4 unit, which is buried deep inside the 754-amino-acid protein.[117] One would forecast with some confidence that the rate constant for reaction with Cu^{2+} would be very similar, and thus in turn very similar to that for the Cu^{2+} plus porphyrin reaction at the foot of Table 7.26. The use of a strongly reducing lower oxidation state, for example Cr^{2+}

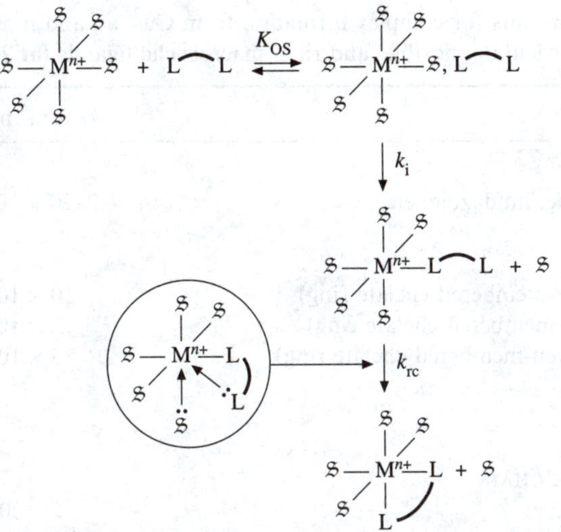

Fig. 7.14 Details of the SCS mechanism for complex formation from solvento-metal ions and chelating ligands. The inset relating to k_{rc} shows competition between ring closure and solvent return.

Table 7.25 Rate constants, $k_f(\mathrm{dm^3\ mol^{-1}})$ for complex formation from Co^{2+}aq and bidentate ligands forming five- and six-membered chelating rings (at 298.2 K).

Complex formation with monodentate ligands:		charge 1 –	$k_f = 1–3 \times 10^6$
		charge 2 –	$k_f = 1–3 \times 10^7$
Formation of five-membered ring:			
	α-alaninate$^-$	α-aminobutyrate$^-$	iminodiacetate^{2-}
$k_f =$	2×10^6	2.5×10^5	1×10^7
Formation of six-membered ring:			
	β-alaninate$^-$	β-aminobutyrate$^-$	iminodiacetate^{2-}
$k_f =$	2×10^6	2.5×10^5	1×10^7

or Cu^+,[118] is generally needed to insert a fourth metal atom into incomplete cuboidal molybdenum clusters $[Mo_3S_4(H_2O)_9]^{n+}$; substitution in the former product, namely $[Mo_3CrS_4(H_2O)_{12}]^{n+}$, is mentioned in the next section. Replacement of one metal ion M^{II} in $[Mo_3M^{II}S_4(H_2O)_{10}]^{4+}$ by another is, as expected, slow — replacement of Ni^{2+} by Cu^{2+} has a half-life of several hours at room temperature.[119]

Table 7.26 Rate constants for complex formation from Cu^{2+}aq and monodentate, bidentate, linear polydentate, flexible, and rigid macrocyclic ligands (at 298.2 K).

Entering ligand	k_f ($dm^3\ mol^{-1}\ s^{-1}$)
MONODENTATE	
Ammonia, pyridine, imidazole, etc.	$2–20 \times 10^8$
BIDENTATE	
α-Alaninate(→ five-membered chelate ring)	10×10^8
β-Alaninate(→ six-membered chelate ring)	2×10^8
L-Carnosine(→ seven-membered chelate ring)	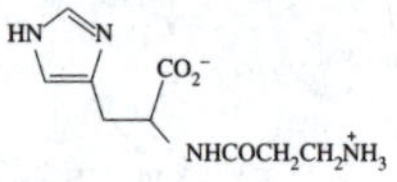 5×10^4
TETRADENTATE: OPEN-CHAIN	
2,3,2-tet	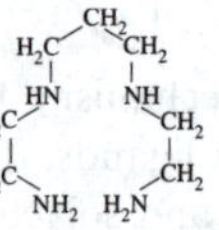 10^7
TETRADENTATE: FLEXIBLE	
tet-*a*	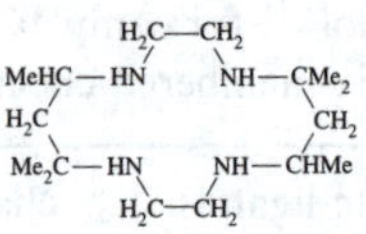 1.6×10^3
tet-*b*	3.6×10^3
trans-[14]-diene	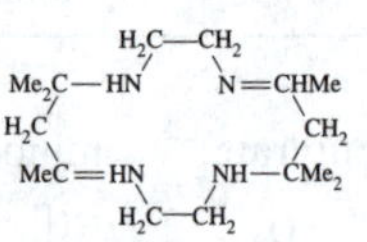 5.6×10^3
Cryptand [221]	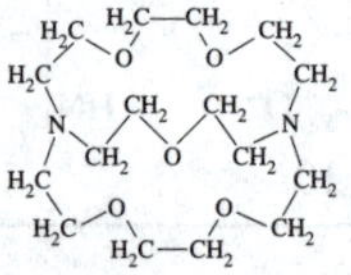
TETRADENTATE: RIGID	
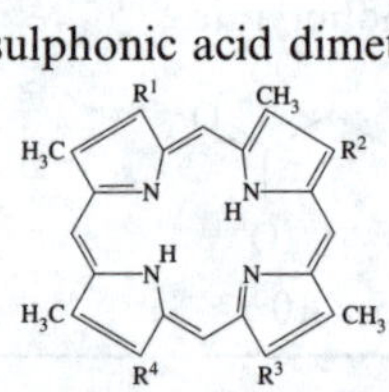 Deuteroporphyrin-2,4-disulphonic acid dimethyl ester ($R^1 = R^2 = SO_3H$; $R^3 = R^4 = CH_2CH_2CO_2CH_3$)	4.3
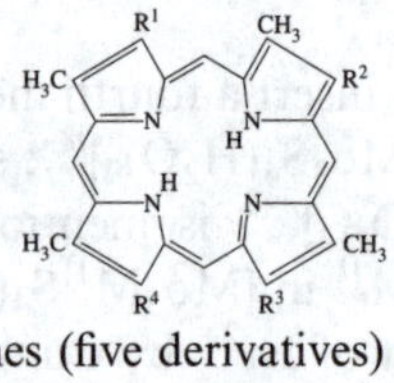 Haematoporphyrin IX ($R^1 = R^2 = CH(OH)CH_3$; $R^3 = R^4 = CH_2CH_2CO_2H$)	2.0×10^{-2}
meso-Tetraphenylporphines (five derivatives)	$1–20 \times 10^{-3}$

[a] In dimethyl sulphoxide.

Table 7.27 Rate constants, k_f ($dm^3\ mol^{-1}\ s^{-1}$), for formation of complexes from $[Cu(OH)_3]^-$ and tetraaza-macrocyclic ligands in alkaline aqueous solution (at 298.2 K).

Ligand (LLLL)	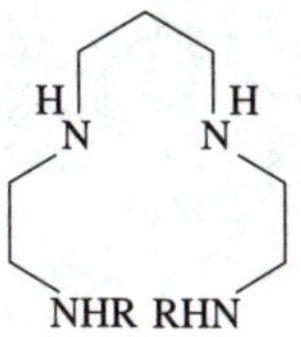		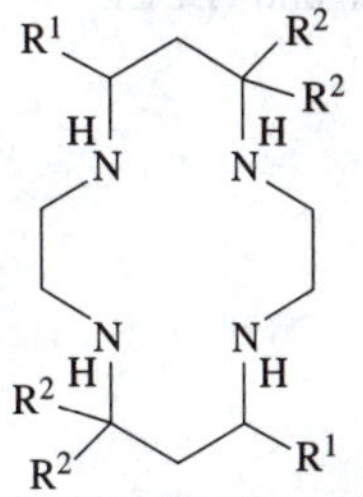		
	$R = H$	Et	$R^1 = H$	Me	Me
			$R^2 = H$	H	Me
k_f for $[Cu(OH)_3]^-$ plus LLLL	1.0×10^7	0.3×10^7	2.7×10^6	5.6×10^5	$\sim 10^4$

It is difficult to make a direct comparison between a macrocyle and its open-chain analogue. Thus, for example, for tetraaza ligands there are two primary $-NH_2$ groups in the latter which are absent from the former, making a profound difference to the ligand-donor properties. This problem can be largely compensated for by replacing $-NH_2$ by $-NHR$, preferably with R = for example Et rather than Me to avoid the marked inductive release of electron density by the methyl group directly affecting the donor nitrogens. Table 7.27 shows an open-chain versus macrocycle comparison which takes some heed of this difficulty. Unfortunately, the strongly basic properties of amines often prevent accurate determination of their rate constants for reaction with Cu^{2+} itself, hence the appearance of hydroxo-copper(II) in Table 7.27. The ligand basicity problem can be avoided by switching to sulphur-donor ligands.[120] There is very much less kinetic information on such reactions, but Table 7.28 does show that the number of five- and six-membered chelate rings generated by the macrocyclic ligands on forming the respective Cu^{2+} complexes has little effect on rate constants until the most restricting and sterically constraining ring system, that giving four five-membered rings, is reached.

The next stage involves moving from macrocyclic ligands to the even more sterically demanding and restricting encapsulating ligands, with cryptand and crown ether ligands forming the link between these two types of ligand. There is a comprehensive review (340 references),[121] with subsequent updates,[122] which covers not only rate constants and activation parameters but also equilibrium constants and enthalpies and entropies of reaction for overall formation, for reactions of metal ions with crowns, cryptands, and other macrocyclic and encapsulating ligands. The cryptand [221] appears in the ligand sequence in Table 7.26. Complex formation from

Table 7.28 Rate contants, k_f(dm^3 mol^{-1} s^{-1}), for formation of complexes from Cu^{2+}aq and tetrathia-macrocyclic ligands, in 80% methanol solution (at 298.2 K).

Ligand (LLLL)			
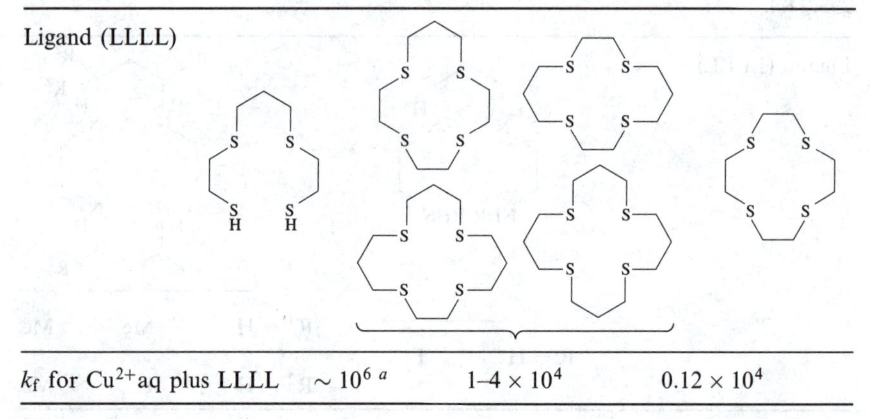			
k_f for Cu^{2+}aq plus LLLL	$\sim 10^6$ [a]	1–4×10^4	0.12×10^4

[a] Estimated.

cryptands, and from crown ethers, is, of course, much more important for alkali metal and alkaline earth cations than for transition metal cations. Such ligands are flexible, the crown ethers considerably more so than the cryptands. Values of k_f for reactions of alkali metal and alkaline earth cations, and also for Pb^{2+} and Hg^{2+}, with crown ethers are almost all within the range 1–10×10^8 dm^3 mol^{-1} s^{-1} (in water at 298.2 K); factors such as the rigidity and ring size of the crown ether have only a small effect. Formation rate constants cover a range of less than tenfold on going from Li^+ to Cs^+. Table 7.29 shows that Na^+ and K^+ react at similar rates with bioligands such as valinomycin, with cryptates and with simple crown ethers—the great range of stability constants for such complexes arises not from formation rate constants but from dissociation rate constants, which span 10^8-fold going from the cryptand [221] to valinomycin.[123] However, it should be added that alkaline earth cations

Table 7.29 Rate constants for reaction of Na^+ and of K^+ with crowns and crypts, in methanol at 298 K.

	k_f(dm^3 mol^{-1} s^{-1})	
Ligand	Na^+	K^+
Dicyclohexyl-18-crown-6	2.6×10^8	
Dibenzo-30-crown-10		6×10^8
Cryptand [221]	8.7×10^7	3.8×10^8
Cryptand [222]	2.7×10^8	4.7×10^8
Valinomycin	1.2×10^7	3.5×10^7
Monensin	2.2×10^8	
Monactin	3×10^8	

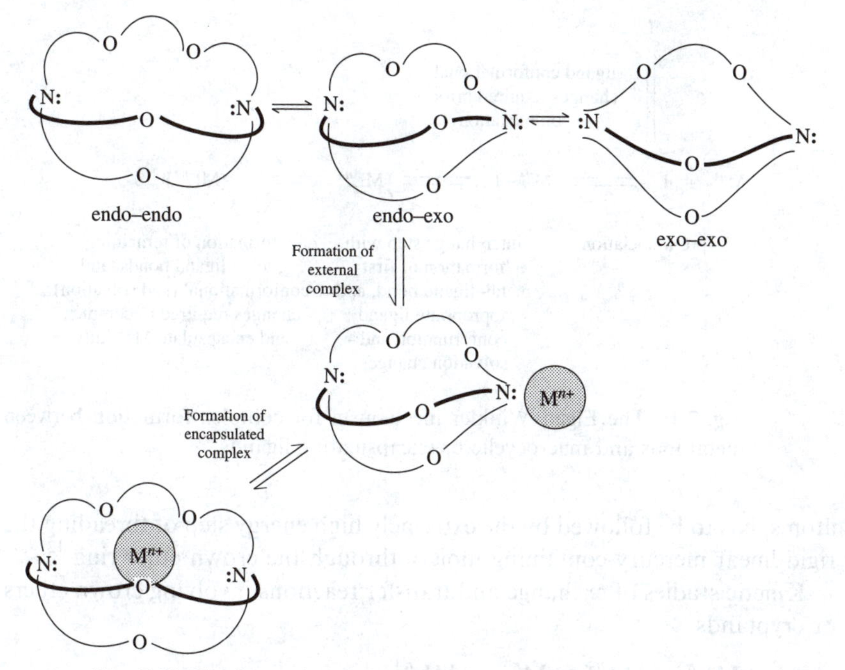

Fig. 7.15 Conformational changes for cryptands and for cryptate formation.

react with cryptates much more slowly, often of the order of thousands of times, than with crown ethers. The conformational changes required for complexation by cryptands (Fig. 7.15) must be having a significant effect here.[124] For these and similar systems it has proved convenient to add up to two stages of ligand conformational change[125] (accompanied by relevant solvation changes), before and after cation complexation, to the two-step Eigen–Wilkins mechanism, giving the (slightly modified) Eigen–Winkler mechanism[126] of Fig. 7.16. The three steps, and indeed any pre-complexation conformational change in the ligand, will have various relations between their respective kinetic and equilibrium parameters, resulting in either complicated kinetics or a kinetically simple pattern open to a variety of mechanistic interpretations.

In the preceding paragraphs we have been dealing with simple solvated cations where exchange rates for solvent molecules in the primary solvation shell are faster, often very much faster, than rates of complex formation. If the metal ion to be complexed already bears firmly bound ligands, then these may make the introduction of the metal-containing species into the cavity of the crown ether or cryptand considerably more difficult. Rate constants for reactions of $Hg(CN)_2$ or $Hg(CF_3)_2$ with crown ethers lie in the range 10^{-2} down to 10^{-7} dm^3 mol^{-1} s^{-1}, corresponding to half-lives of up to three months. It was proposed that in these reactions initial rapid binding of the crown ether to the mercury, through perhaps three of the oxygen-donor

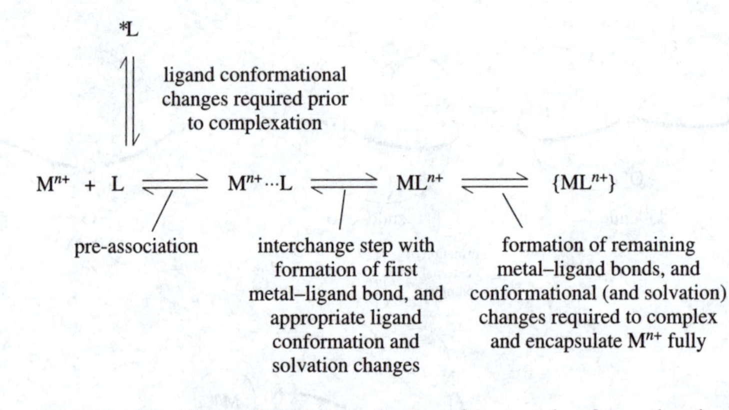

Fig. 7.16 The Eigen–Winkler mechanism for complex formation between metal ions and macrocyclic or encapsulating ligands.

atoms, has to be followed by the extremely high energy step of threading the rigid linear mercury-containing moiety through the crown ether ring.[127]

Kinetic studies of exchange and transfer reactions involving crown ethers or cryptands

$$ML^{n+} + {}^*M^{n+} = M^{n+} + {}^*ML^{n+}$$

$$ML^{n+} + {}^*L = M^*L^{n+} + L$$

are closely related to the simple formation reactions of the type

$$M^{n+} + L = ML^{n+}$$

just discussed. The two main points of interest are the mechanism of the exchange or transfer reaction, whether it is unimolecular (rate-limiting M---L bond breaking) or bimolecular (M---L---*M transition state), and the effects of solvent. At first it was assumed that the former mechanism was normal,[128] but even as early as 1970 the latter mechanism had been invoked in discussion of the selective transport, by a 'carrier relay' mechanism, of K^+ through synthetic membranes charged with valinomycin.[129] The bimolecular mechanism was first demonstrated for the reaction of $Pb([211])^{2+}$ with cryptand [221] or [222] and for K^+ exchange with 18-crown-6 in 1,3-dioxolane.[130] Solvent effects were subsequently shown often to be important in determining mechanism, with such reactions as alkali metal cation exchange with benzo-15-crown-5, 18-crown-6, dibenzo-18-crown-6, and dibenzo-24-crown-8 occuring by the bimolecular mechanism in solvents of low donicity, by rate-determining dissociation in high-donicity media.[131] Temperature may also affect the predominant mechanism, as in the exchange of dibenzo-30-crown-10 at Cs^+, where the dissociative mechanism operates in, e.g., propylene carbonate or methanol at −10 °C, the bimolecular mechanism at temperatures below this.[132] Photochemical activation[133] and solvation effects may also be very important in determining reactivity. The alkaline earth 2+ cations react with crown ethers and cryptands much more slowly than do the alkali metal 1+ cations—indeed reactions such as that of Sr^{2+} with dibenzo-18-crown-6 are sufficiently slow for stopped-flow monitoring,[134] as is the reaction of Cu^{2+} with the cryptand [221] in dimethyl sulphoxide. The transfer chemical potential of Cu^{2+} from water into dimethyl sulphoxide is $-41\,kJ\,mol^{-1}$, which gives some idea of the extra energy required to desolvate Cu^{2+} in dimethyl sulphoxide compared with water.[135] We have now slipped into the realms of solvent effects, the subject of the following chapter. However, before embarking on that chapter we shall deal with clusters, and digress to the topic of tailoring ligands to make metal centres, which are normally very labile, undergo substitution very slowly. In the course of this digression we shall be dealing with dissociation as well as formation of complexes. Though some of this material is not strictly within the remit of the present chapter it is convenient to discuss these topics here, as they are closely connected and in some cases involve coordination numbers not included in the earlier chapters devoted to substitution.

7.5.7 Substitution at clusters

There have been a number of kinetic studies of substitution at, or into, metal clusters. Some examples have already appeared in this chapter—water exchange at $Mo_3O_4^{4+}$aq (end of Section 7.3.1) and the incorporation of M^{2+}aq ions into incomplete cuboidal Fe_3S_4 clusters (Section 7.5.6, as an example of difficult and thus particularly slow substitution at M^{2+}aq). The series of Mo_3E_4 clusters $[Mo_3S_{4-x}O_x(H_2O)_9]^{4+}$ have been fully studied in relation to substitution of water by thiocyanate. Rate constants decrease steadily as sulphur in the core is replaced by oxygen, with a more than 200-fold decrease in rate constant going from the Mo_3S_4 to the Mo_3O_4 cluster.[136] The complete cubes $[Mo_4S_4(H_2O)_{12}]^{n+}$, where $n = 4$ or 5, react very much more slowly. The all-Mo^{III} cluster ($n = 4$) reacts with thiocyanate in two consecutive stages, substitution followed by slower linkage isomerization of the thiocyanate. The initial substitution step has, reassuringly, a similar rate constant to those for complex formation from Mo^{3+}aq. The $Mo^{III}Mo^{IV}$ cluster ($n = 5$) reacts by two concurrent paths, believed to correspond to substitution at the Mo^{III} and at the Mo^{IV} in parallel. Comparison of rate constants for thiocyanate and for chloride substitution at the 5+ cluster suggests a mechanism of dissociative interchange.[137] Another mixed-valence cluster for which kinetic data for substitution are available is the anion of Delépine's salt, $[Ir_3(\mu_3\text{-}N)(\mu\text{-}SO_4)_6(H_2O)_3]^{4-}$. Chloride, bromide, and azide substitute in a one-step process. This behaviour could result from an initial slow substitution step followed by two faster replacements of water by incoming ligand. Alternatively, the simple kinetic pattern could reflect equivalence of the three iridium atoms. Although Mössbauer spectra at 4 K indicate the presence of distinct Ir^{III}

and Ir^{IV} centres, these could easily have become equivalent by 298 K.[138] Mixed-valence μ_3-oxo trinuclear species of the type $[M_3O(RCO_2)_6(H_2O)_3]^{n\pm}$, where the three metal centres may be the same or different, are important in relation to intramolecular electron transfer (Section 9.4.4). They have been relatively little studied in respect to replacement of the coordinated waters by other ligands, though it seems that replacement of coordinated waters is often slow relative to the respective M^{n+}aq cations. The mechanism of substitution at the μ_3-oxo-tri-molybdenum and -tungsten species appears to be dissociative.[139]

Stopped-flow studies of the reaction of the mixed-metal cluster $[Mo_3CrS_4(H_2O)_{12}]^{4+}$, whose preparation by Cr^{2+} reduction of, and incorporation into, $[Mo_3S_4(H_2O)_9]^{4+}$ was mentioned in the previous section, with thiocyanate revealed a single substitution step on this time-scale. This can be attributed to replacement of one of the waters on chromium, presumed to be in oxidation state 3+, greatly labilized by the three μ_3-sulphides.[140] The relative lability of iron-bonded and cobalt-bonded ligands in $[CoFe_3S_4(smes)_4]^{2-}$ (smes = the mesitylthiolate anion) has also been probed, by competion rather than by direct kinetics, the iron sites being found to be the more reactive.[141] Currently, new mixed-metal clusters of the $Mo_3MS_4{}^{n+}$aq and $W_3MS_4{}^{n+}$aq type, (with M = for example, In, Ge, Sn) are being synthesized and characterized; it is to be hoped that substitution kinetics will shortly be investigated to allow appropriate inter-metal reactivity comparisons to be made.[142]

Some of the hydroxo-bridged polynuclear species, the first stage in whose formation was briefly mentioned as a special case of complex formation from hydroxo-metal(III) species $M(OH)^{2+}$aq, are sufficiently stable and inert to act as substrates for water replacement kinetic studies. Thus, for example, kinetics of reactions of the now well-characterized Al_{13}-cluster have been established for ligands such as fulvate[143] and ferron[144] (7-iodo-8-hydroxyquinoline-5-sulphonate). The latter system provides a possible kinetic mode of estimating the polynuclear/mononuclear ratio in samples.

7.5.8 Slow substitution at labile centres

Several exceptionally slow formation reactions have been encountered in earlier sections; in this section and the next we concentrate on an important special case. This involves polydentate pendent-arm, often macrocyclic, ligands of the polyaminocarboxylate type. We shall deal with formation and dissociation reactions in parallel. The sometimes very large decreases in reactivity consequent on going from complexes containing simple monodentate ligands to those containing macrocyclic ligands and, even more, to encapsulating ligands, is of great importance in relation to the use of various metals in medicine, especially in diagnosis and therapy involving radioisotopes, but also in the use of gadolinium in magnetic resonance imaging, MRI. If an aggressively radioactive isotope such as ^{90}Y is to be used to kill cancer cells, it must be targeted to the tumour, and free $^{90}Y^{3+}$ must not be allowed to circulate around the body, localize in, say, the skeleton, and thereby cause dangerous radiation damage. Similarly, for less aggressive isotopes used in diagnosis—such as ^{67}Ga, ^{111}In, ^{99m}Tc, and so on—it is desirable for these isotopes to remain firmly bound in the complex which is to localize in the diseased tissue. Again for the non-radioactive, but chemically toxic, gadolinium(III) used as contrast enhancer in MRI, it is desirable for administration to be of a very stable and very inert complex. The ions Y^{3+}, Ga^{3+}, In^{3+}, and Gd^{3+} are all labile, with half-lives with respect to substitution by simple ligands between 10^{-3} and 10^{-9} s, so an enormous decrease in reactivity is needed to increase these half-lives to the required time-scale of weeks or more. That this should be feasible is suggested, for instance, by the range of reactivities reported for substitution at Cu^{2+} in Table 7.26 above. In the case of the 3+ ions just mentioned, the most successful approach has been through the use of macrocycles with pendant arms to encapsulate the ions.[145] The tetraaza-macrocyclic ligand with pendant carboxylate arms dota^{4-}, **6**, with R = H, forms extremely stable and inert complexes with large 3+ ions such as Gd^{3+}, Y^{3+}, Ga^{3+}, and In^{3+}.[146] Thus, for example, [Gd(dota)]$^{-}$ is stable in acid solution for many months. Analogues with phosphorus oxoanion groups in place of carboxylate are nearly as inert,[147] while the Gd^{3+} complex of the macrocycle **7**, where phosphate pendant arms complement the six-nitrogen macrocyclic ring incorporating two 2,2′-bipyridyl units, is even more inert, even under conditions of photochemical activation.[148] Whereas ligands of type **6** are well matched to Gd^{3+}, which favours eight- or nine-coordination, the triaza-analogue nota^{3-}, **8**, is more appropriate for such ions as Ga^{3+} and In^{3+}, which favour octahedral stereochemistry. Nonetheless, a number of Gd^{3+} complexes of ligands of the very similar type **9** are also stable and inert.[149] The case of gadolinium is of particular kinetic interest, for the requirements for a pharmacologically satisfactory complex include both the presence of a very strongly binding polydentate ligand and the presence of one or more rapidly exchanging water molecules for efficient action as an MRI contrast agent.[150]

Very large effects on reactivity can best be illustrated by the case of Y^{3+}.[151] The half-life for water exchange at Y^{3+}aq is about 10^{-9} s at 37 °C, the half-life for dissociation of dtpa (**2**) from its Y^{3+} complex is over one day, while the half-life for dissociation of the complex of the 4-nitrobenzyl derivative of dota, **6**, with R = 4-$C_6H_4NO_2$, is more than 200 days at this temperature. By this stage, the half-life for dissociation is considerably longer than the half-life (64 hours) for radioactive decay of the ^{90}Y, so the risk of dispersal of this potentially very harmful isotope around the body is negligible. These dramatic changes in reactivity are illustrated in Fig. 7.17. There is one drawback to the use of M^{3+} complexes of the dota family of ligands, **6**, and that is that complexes [M^{III}(dota)]$^{-}$ are charged. Uncharged species are to be preferred, for more easily crossing body membranes; this can be arranged without change in (potential) coordination number by replacing one of the acetate pendant arms in dota by the alcohol function CH_2OH, to give the ligand **10**.

In a chapter ostensibly dedicated to formation rather than dissociation, it is fitting to close this section with the comment that reactions of aqua-cations with ligands of the nota and dota types, which are fairly rigid, are also generally very slow. A couple of examples will suffice. The half-life for reaction of Ce^{3+}aq with dota is in the range 10^2–10^4 s (pH dependent),[152] while the plan to deliver ^{111}In as In^{3+} complexed with a *C*-functionalized triazamacrocyle bound to a monoclonal antibody foundered on the extreme slowness of incorporation into the functionalized macrocyclic ring—this was so slow that the ^{111}In ($\tau_{1/2} = 67$ hours) had lost most of its radioactivity before incorporation had been achieved![153]

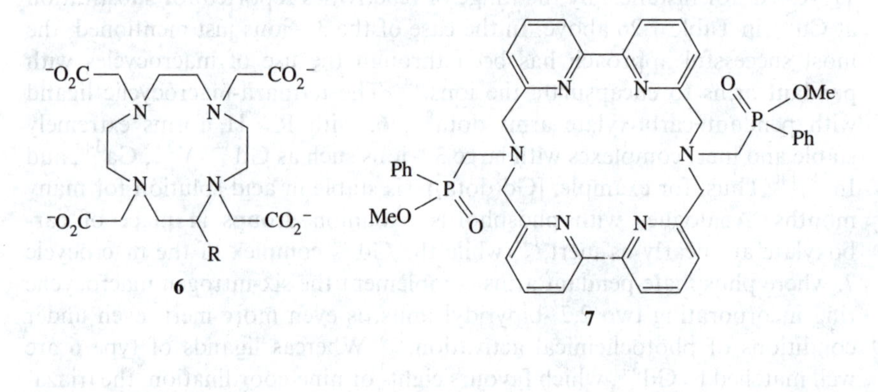

6 **7**

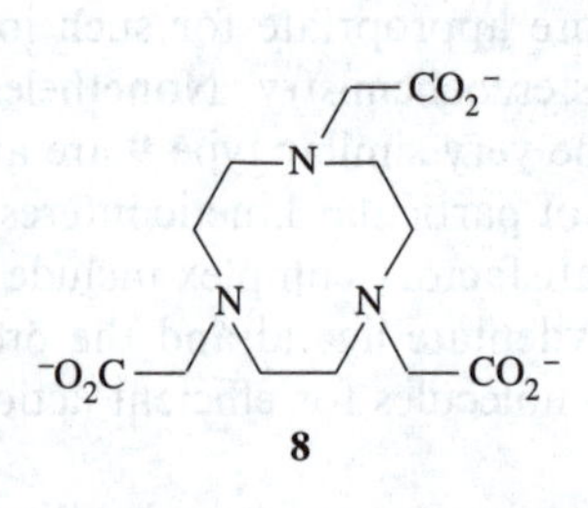

8

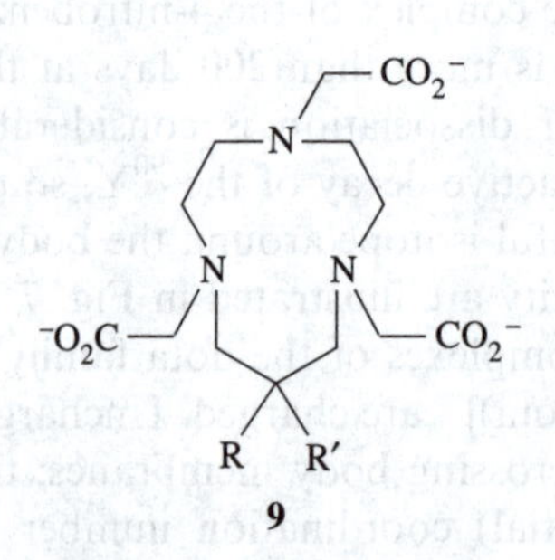

9

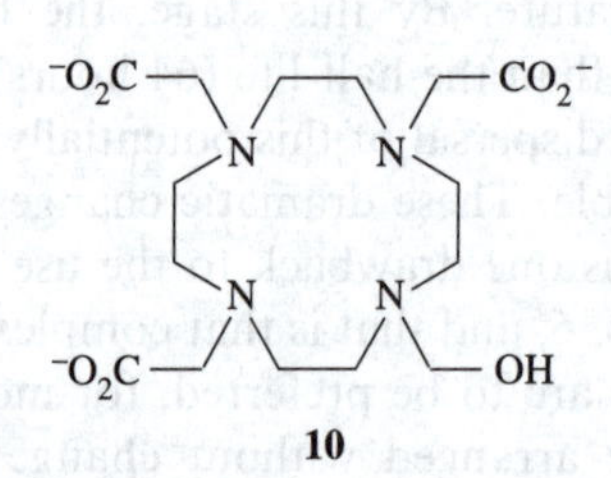

10

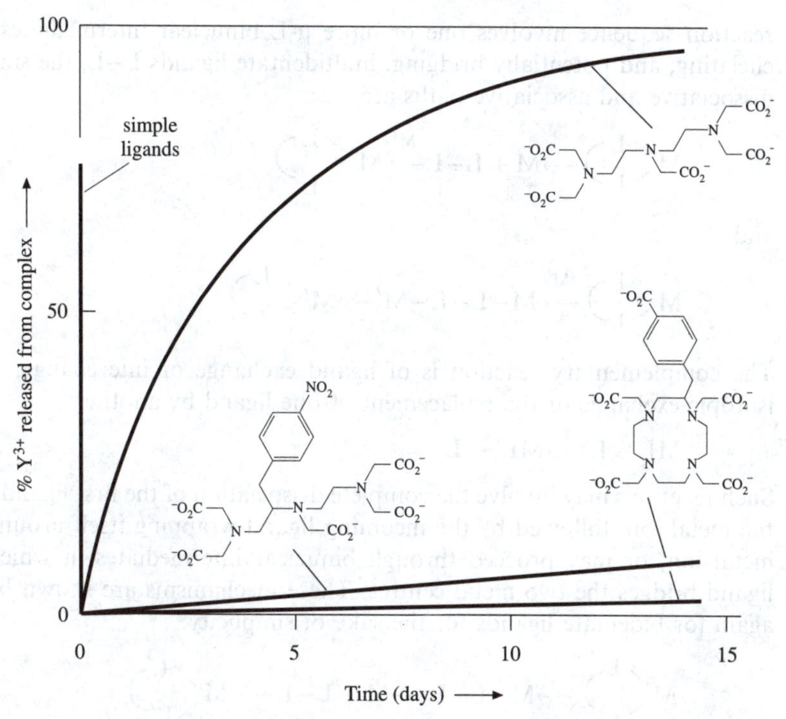

Fig. 7.17 Inertness and lability of yttrium(III) complexes.

7.5.9 Ligand transfer and exchange reactions

These are special cases of substitution, and in general follow the kinetic and mechanistic patterns already outlined. Indeed, a few examples of simple ligand exchange have already been mentioned, for instance for hexahalogenometallates (Table 4.21 in Section 4.7.6). Studies of kinetics of exchange of labelled cyanide with hexacyanometallates were important many years ago in demonstrating the non-correlation of kinetics with thermodynamics—rate constants did not relate inversely to stability constants, as one might intuitively have expected. However, in this section we are primarily concerned with ligand transfer and exchange reactions involving multidentate ligands, including cases where complexity produces particularly slow reaction (cf. preceding section).

Transfer of a ligand from one metal ion to another:

$$ML + M' \rightarrow M'L + M$$

is complex formation at M′, dissociation at M. If the ligand L is polydentate, reaction may involve stepwise dissociation of L from M, then stepwise attachment to M′. Alternatively, partial dissociation of the ligand from M may be followed by now-available donor atoms starting to complex M′ before the ligand has fully dissociated from M. In this case the reaction sequence involves one or more μ-L binuclear intermediates. For chelating, and potentially bridging, multidentate ligands L–L, the simplest dissociative and associative paths are

$$M\langle^{L}_{L}\rangle \longrightarrow M + L{-}L \xrightarrow{M'} M'\langle^{L}_{L}\rangle$$

and

$$M\langle^{L}_{L}\rangle \xrightarrow{M'} M{-}L{-}L{-}M' \longrightarrow M'\langle^{L}_{L}\rangle$$

The complementary reaction is of ligand exchange or interchange, either isotopic exchange or the replacement of one ligand by another:

$$ML + L' \longrightarrow ML' + L$$

Such reactions may involve the complete dissociation of the first ligand from the metal ion, followed by the incoming ligand wrapping itself around the metal ion, or may proceed through binuclear intermediates in which the ligand bridges the two metal centres. These mechanisms are shown below, again for bidentate ligands for the sake of simplicity:

$$M\langle^{L}_{L}\rangle \longrightarrow M{-}L{-}L \longrightarrow M + L{-}L \xrightarrow{L{-}L'} M\langle^{L}_{L'}\rangle$$

$$M\langle^{L}_{L}\rangle + L{-}L' \longrightarrow L{-}L{-}M{-}L{-}L' \longrightarrow M\langle^{L}_{L'}\rangle + L{-}L$$

As dissociation can occur stepwise

$$M\langle^{L}_{L}\rangle \longrightarrow M{-}L{-}L \longrightarrow M + L{-}L$$

as can the formation of the product complex, there are a number of candidates for the rate-determining step—or the kinetics may be bi- or multiphasic. Further kinetic complication may ensue if, say, the first step in dissociation is reversible:

$$M\langle^{L}_{L}\rangle \overset{L{-}L'}{\rightleftharpoons} L{-}L{-}M{-}L{-}L' \longrightarrow \langle^{L'}_{L}\rangle M + L{-}L$$

The kinetics and mechanisms of reactions of these types have been well documented, at least for the particularly active period from 1969 to 1980,[154] so we need only give a few examples for illustration. Transfer of *N*-(hydroxyethyl)-ethylenediamine triacetate, $hedta^{3-}$ (**11**), from Pb^{2+} to Ga^{3+} is dissociative, whereas analogous transfers of $edta^{4-}$ (**12**), $dtpa^{5-}$ (**2**), and the cyclohexyl derivative $cydta^{4-}$ (**13**) proceed through polyaminocarboxylate ligand-bridged intermediates.[155] Transfer of $edta^{4-}$ from Eu^{3+} to the actinide cations Cm^{3+}, Bk^{3+}, and Cf^{3+} involves acid-catalysed dissociation and binuclear intermediates.[156] In the case of cydta transfer from several lanthanide complexes $[Ln(cydta)]^-$ to Er^{3+}, activation volumes of between -4.7 and -0.6 cm^3 mol^{-1} suggest that the mechanism is of dissociative dechelation.[157]

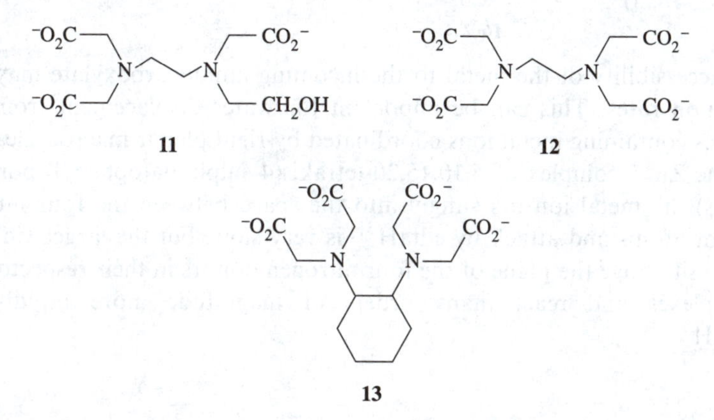

Reactions of this type can be rather slow, even when they involve normally labile metal centres. Examples of such slow ligand-transfer reactions include $[Eu(egta)]^-$ ($egta^-$ = **14**) plus Cu^{2+}, which has fairly recently received detailed kinetic investigation.[158] Even the relatively simple chelating ligand **15** is reluctant to transfer from very labile Y^{3+} to even more labile Cu^{2+}—the rate constant is 6.3 s^{-1} at 30 °C for the major rate-determining dissociation pathway.[159] Slow transfer of ligands from one metal to another can be important in environmental chemistry, where the terms *adjunctive* and *disjunctive* are used for mechanisms involving a binuclear intermediate or rate-limiting dissociation, respectively.[160] This terminology has been applied to the model system of transfer of edta between Cu^{2+} and Ca^{2+}, where the dominant pathway switches from disjunctive to adjunctive as the concentration of Ca^{2+} increases.[161] This behaviour is, of course, consistent with the rate laws for the two pathways. These studies were extended to various ligand- and metal-exchange systems containing, *inter alia*, Ca^{2+}, Pb^{2+}, Cd^{2+}, Ni^{2+}, and Cu^{2+}, with ligands such as edta and humic acid.[160,162] Other slow ligand replacements include reactions of $[Cu(edta)]^{2-}$ and related complexes with tetraaza-macrocycles,[163] of $[Ni(edta)]^{2-}$ with $pdta^{4-}$ ($pdta^{4-}$ = **1** above),[164] and for a recent bioinorganic example, the removal of iron from the two metal-binding sites of transferrin by $edta^{4-}$ (**12**) or its phosphate analogue $edtp^{4-}$ (**16**).[165]

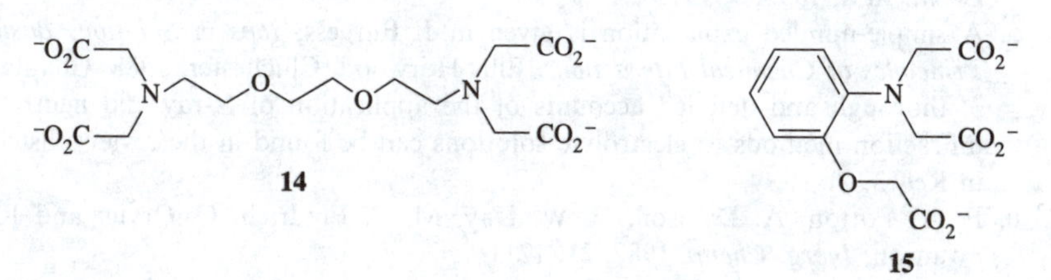

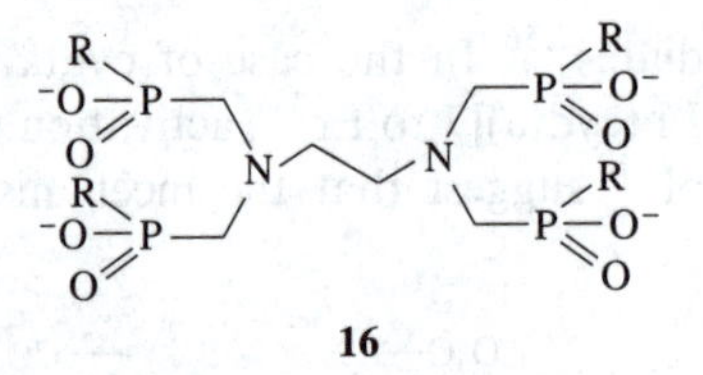

16

Accessibility of the metal to the incoming aminocarboxylate may affect reaction rates. This can be important for water displacement from complexes containing metal ions coordinated by rigid planar macrocycles. Thus in the Zn^{2+} complex of 5,10,15,20-tetrakis(4-sulphonatophenyl)-porphyrin (tpps) the metal ion fits snugly into the space between the four nitrogen-donor atoms and attack by $edtaH^{3-}$ is very slow, but the larger Cd^{2+} and Pb^{2+} sit above the plane of the four nitrogen donors in their respective tpps complexes and react many orders of magnitude more rapidly with $edtaH^{3-}$ [166]

References

1. Review: S. F. Lincoln and A. E. Merbach, *Adv. Inorg. Chem.*, 1995, **42**, 1. The introduction to this review cites a number of earlier reviews of this area.
2. Books which give good coverage of this class of reactions include E. F. Caldin, *Fast Reactions in Solution,* Blackwell, Oxford, 1964; D. N. Hague, *Fast Reactions,* Wiley–Interscience, New York,1971; R. G. Wilkins, *Kinetics and Mechanism of Reactions of Transition Metal Complexes*, 2nd edn, VCH, Weinheim, 1991.
3. See, e.g., J. E. Enderby and G. W. Nielson, *Rep. Progr. Phys.*, 1981, **44**, 593; *Water and Aqueous Solutions*, eds G. W. Nielson and J. E. Enderby, Adam Hilger, Bristol, 1986; J. E. Enderby, S. Cummings, G. J. Herdman, G. W. Nielson, P. S. Salmon, and N. Skipper, *J. Phys. Chem.*, 1987, **91**, 5851; S.-I. Ishiguro and H. Ohtaki, *J. Coord. Chem.*, 1987, **15**, 237; M. Magini (ed.), *X-Ray Diffraction of Ions in Aqueous Solution: Hydration and Complex Formation*, CRC Press, Florida, 1988; H. Ohtaki and T. Radnai, *Chem. Rev.*, 1993, **93**, 1157.
4. There are innumerable other ways of estimating solvation numbers, of varying degrees of unsatisfactoriness. They, and their results, are exhaustively (but uncritically) reviewed in J. F. Hinton and E. S. Amis, *Chem. Rev.*, 1971, **71**, 627; somewhat less comprehensively but more digestibly in Chapters 2 to 5 of J. Burgess, *Metal Ions in Solution*, Ellis Horwood, Chichester, 1978. The most recent and helpful review of ion solvation is that by H. Ohtaki and T. Radnai, *Chem. Rev.*, 1993, **93**, 1157.
5. A simple-minded explanation is given in J. Burgess, *Ions in Solution*: *Basic Principles of Chemical Interactions*, Ellis Horwood, Chichester, 1988, Chapter 3; thorough and detailed accounts of the application of X-ray and neutron diffraction methods to electrolyte solutions can be found in the reviews listed in Ref. 3.
6. F. A. Cotton, A. Davison, V. W. Day, M. F. Fredrich, C. Orvig, and R. Swanson, *Inorg. Chem.*, 1982, **21**, 1211.

7. R. W. Broach, J. M. Williams, G. P. Felcher, and D. G. Hinks, *Acta Cryst.*, 1979, **35B**, 2317.
8. J. D. Eakins, D. G. Humphreys, and C. E. Mellish, *J. Chem. Soc.*, 1963, 6012.
9. C. B. Castellani, O. Carugo, M. Giusti, and N. Sardone, *Eur. J. Solid State Inorg. Chem.*, 1995, **32**, 1089.
10. A caesium alum, stable only below 0 °C, presumably contains octahedral Sc^{3+}aq—see B. I. Bashkov, L. N. Komissarova, F. M. Spiridonov, and V. M. Shatskii, *Vestn. Mosk. Univ., Khim.*, 1972, **13**, 598 (*Chem. Abs.*, 1973, **78**, 35 015w). See also ref. 13 of Chapter 4 (p.196).
11. F. Matsumoto, Y. Ohki, Y. Susuki, and A. Ouchi, *Bull. Chem. Soc. Jpn*, 1989, **62**, 2089.
12. R. A. Plane and J. P. Hunt, *J. Am. Chem. Soc.*, 1954, **76**, 5960.
13. H. H. Baldwin and H. Taube, *J. Chem. Phys.*, 1960, **33**, 206.
14. For example, A. E. Merbach, *Pure Appl. Chem.*, 1987, **59**, 161.
15. A. Bleuzen, F. Foglia, E. Furet, L. Helm, A. E. Merbach, and J. Weber, *J. Am. Chem. Soc.*, 1996, **118**, 12 777.
16. J. Burgess, *Analyst*, 1992, **117**, 605, and references therein.
17. D. T. Richens, *The Chemistry of Aqua Ions—Synthesis, Structure, and Reactivity*, Wiley, Chichester, 1997.
18. R. V. Southwood-Jones, W. L. Earl, K. E. Newman, and A. E. Merbach, *J. Chem. Phys.*, 1980, **73**, 5909; D. H. Powell, A. E. Merbach, G. González, E. Brücher, K. Micskei, M. F. Ottaviani, K. Köhler, A. von Zelewsky, O. Ya. Grinberg, and Y. S. Lebedev, *Helv. Chim. Acta*, 1993, **76**, 2129.
19. E. N. Rizkalla, *Radiochim. Acta*, 1993, **61**, 181.
20. T. Kowall, F. Foglia, L. Helm, and A. E. Merbach, *J. Am. Chem. Soc.*, 1995, **117**, 3790.
21. M. Åberg and J. Glaser, *Inorg. Chim. Acta*, 1993, **206**, 53.
22. K. R. Rodgers, R. K. Murmann, E. O. Schlemper, and M. E. Shelton, *Inorg. Chem.*, 1985, **24**, 1313.
23. D. Richardson and T. L. Alger, *J. Phys. Chem.*, 1975, **79**, 1733.
24. D. Pisanello, L. Helm, P. Meier, and A. E. Merbach, *J. Am. Chem. Soc.*, 1983, **105**, 4528.
25. Data have been taken from J. Burgess, *Ions in Solution: Basic Principles of Chemical Interactions*, Ellis Horwood, Chichester, 1988; Y. Ducommun and A. E. Merbach, in *Inorganic High Pressure Chemistry: Kinetics and Mechanisms*, ed. R. van Eldik, Elsevier, Amsterdam, 1986, Chapter 6; A. E. Merbach, P. Moore, O. W. Howarth, and C. H. McAteer, *Inorg. Chim. Acta*, 1980, **39**, 129.
26. R. G. Pearson and R. D. Lanier, *J. Am. Chem. Soc.*, 1964, **86**, 765; T. Larsson and J. Kowalewski, *Magn. Reson. Chem.*, 1988, **26**, 1020.
27. S. E. Okan, P. S. Salmon, D. C. Champeney, and I. Petri, *Mol. Phys.*, 1995, **84**, 325.
28. S. Soyama, M. Ishii, S. Funahashi, and M. Tanaka, *Inorg. Chem.*, 1992, **31**, 536.
29. Y. Inada, K. Ozutsumi, S. Funahashi, S. Soyama, T. Kawashima, and M. Tanaka, *Inorg. Chem.*, 1993, **32**, 3010.
30. T. Fujita and H. Ohtaki, *Bull. Chem. Soc. Jpn*, 1983, **56**, 3276; A. Filipponi, P. D'Angelo, N. V. Pavel, and A. Di Cicco, *Chem. Phys. Lett.*, 1994, **225**, 150.

31. See, e.g., the review by A. E. Merbach, *Pure Appl. Chem.*, 1987, **59**, 161.
32. In addition to its usefulness as a source of kinetic data, the review by J. O. Edwards, F. Monacelli, and G. Ortaggi, *Inorg. Chim. Acta*, 1974, **11**, 47, is well worth reading for its treatment of this problem.
33. For a detailed explanation of these difficulties, see *Inorganic High Pressure Chemistry: Kinetics and Mechanisms*, ed. R. van Eldik, Elsevier, Amsterdam, 1986, pp. 44–45, 73–74, and 404–408; for a brief account see J. Burgess, *Metal Ions in Solution*, Ellis Horwood, Chichester, 1978, p. 313 (Fig. 11.2).
34. J. Burgess, *Metal Ions in Solution*, Ellis Horwood, Chichester, 1978, p. 313.
35. K. E. Newman, F. K. Meyer, and A. E. Merbach, *J. Am. Chem. Soc.*, 1979, **101**, 1470.
36. R. van Eldik (ed.), *Inorganic High Pressure Chemistry: Kinetics and Mechanisms*, Elsevier, Amsterdam, 1986; R. van Eldik and J. Jonas (eds), *High Pressure Chemistry and Biochemistry* (NATO ASI Series C, vol. 197), Reidel, Dordrecht, 1987.
37. M. A. Lee, N. W. Winter, and W. H. Casey, *J. Phys. Chem.*, 1994, **98**, 8641.
38. R. A. More O'Ferrall, *J. Chem. Soc. (B)*, 1970, 274; D. A. Jencks and W. P. Jencks, *J. Am. Chem. Soc.*, 1977, **99**, 7948.
39. T. W. Swaddle, *Adv. Inorg. Bioinorg. Mech.*, 1983, **2**, 95.
40. T. W. Swaddle, *J. Chem. Soc., Chem. Commun.*, 1982, 832.
41. N. S. Isaacs, *Liquid Phase High Pressure Chemistry*, Wiley, Chichester, 1981; R. van Eldik, T. Asano, and W. J. Le Noble, *Chem. Rev.*, 1989, **89**, 549.
42. N. Graeppi, D. H. Powell, G. Laurenczy, and A. E. Merbach, *Inorg. Chim. Acta*, 1995, **235**, 311.
43. D. F. Shriver, P. W. Atkins, and C. H. Langford, *Inorganic Chemistry*, 2nd edn, Oxford University Press, Oxford, 1994, p. 635.
44. See Table V of Ref. 1.
45. D. N. Hague and M. S. Zetter, *Trans. Faraday Soc.*, 1970, **66**, 1176; A. H. Zeltmann and L. O. Morgan, *Inorg. Chem.*, 1970, **9**, 2522; P. E. Hoggard, H. W. Dodgen, and J. P. Hunt, *Inorg. Chem.*, 1971, **10**, 71.
46. G. González, B. Moullet, M. Martinez, and A. E. Merbach, *Inorg. Chem.*, 1994, **33**, 2330.
47. H. Taube, *Pure Appl. Chem.*, 1979, **51**, 901; Yann Hung, Wei-Jen Kung, and H. Taube, *Inorg. Chem.*, 1981, **20**, 457; H. Taube, *Pure Appl. Chem.*, 1991, **63**, 651.
48. L. Dadci, H. Elias, U. Frey, A. Hörnig, U. Koelle, A. E. Merbach, H. Paulus, and J. S. Schneider, *Inorg. Chem.*, 1995, **34**, 306; R. Boča and H. Elias, *Polyhedron*, 1996, **15**, 2425.
49. W. Luginbühl, P. Zbinden, P. A. Pittet, T. Armbruster, H.-B. Bürgi, A. E. Merbach, and A. Ludi, *Inorg. Chem.*, 1991, **30**, 2350.
50. D. H. Powell, O. M. Ni Dhubhghaill, D. Pubanz, L. Helm, Y. S. Lebedev, W. Schlaepfer, and A. E. Merbach, *J. Am. Chem. Soc.*, 1996, **118**, 9333.
51. K. Micskei, L. Helm, E. Brücher, and A. E. Merbach, *Inorg. Chem.*, 1993, **32**, 3844.
52. D. Pubanz, G. Gonzalez, D. H. Powell, and A. E. Merbach, *Inorg. Chem.*, 1995, **34**, 4447.
53. D. H. Powell, A. E. Merbach, I. Fábián, S. Schindler, and R. van Eldik, *Inorg. Chem.*, 1994, **33**, 4468.
54. See Table XIV of Ref. 1.

55. Most of the data in this table have been taken from I. Dellavia, L. Helm, and A. E. Merbach, *Inorg. Chem.*, 1992, **31**, 2230; data for lanthanide cations from Ref. 19.
56. M. V. Twigg, *Inorg. Chim. Acta*, 1977, **24**, L84.
57. I. Rapaport, L. Helm, A. E. Merbach, P. Bernhard, and A. Ludi, *Inorg. Chem.*, 1988, **27**, 873.
58. G. Laurenczy, I. Rapaport, D. Zbinden, and A. E. Merbach, *Magn. Reson. Chem.*, 1991, **29**, S45.
59. R. J. Judd, R. Cao, M. Biner, T. Armbruster, H.-B. Bürgi, A. E. Merbach, and A. Ludi, *Inorg. Chem.*, 1995, **34**, 5080.
60. A. Cusanelli, U. Frey, D. T. Richens, and A. E. Merbach, *J. Am. Chem. Soc.*, 1996, **118**, 5265.
61. I. Dellavia, L. Helm, and A. E. Merbach, *Inorg. Chem.*, 1992, **31**, 4151.
62. R. G. Wilkins, *Kinetics and Mechanism of Reactions of Transition Metal Complexes*, 2nd edn, VCH, Weinheim, 1991, Chapter 4.
63. R. G. Wilkins and M. Eigen, *Adv. Chem. Ser.*, 1965, **49**, 55.
64. Almost all the data in this table have been taken from Table 12.1 on p. 353 of J. Burgess, *Metal Ions in Solution*, Ellis Horwood, Chichester, 1978.
65. J. A. Lavigne, C. H. Langford, and M. K. S. Mak, *Analyt. Chem.*, 1987, **59**, 2616.
66. H. Brintzinger and G. G. Hammes, *Inorg. Chem.*, 1966, **5**, 1286.
67. R. M. Fuoss, *J. Am. Chem. Soc.*, 1958, **80**, 5059.
68. M. Eigen, *Z. Electrochem.*, 1960, **64**, 115.
69. M. Eigen, W. Kruse, G. Maass, and L. de Maeyer, *Progr. React. Kinet.*, 1964, **2**, 287.
70. N. Das and R. Das, *Transition Met. Chem.*, 1996, **21**, 239.
71. Data are taken from Table 3 of J. Burgess, *Analyst*, 1992, **117**, 605 and Table 12.8 on p. 365 of J. Burgess, *Metal Ions in Solution*, Ellis Horwood, Chichester, 1978.
72. M. Eigen and K. Tamm, *Z. Elektrochem.*, 1962, **66**, 93, 107; H. Diebler, M. Eigen, G. Ilgenfritz, G. Maass, and R. Winkler, *Pure Appl. Chem.*, 1969, **20**, 93: G. Atkinson, M. M. Emara, and R. Fernandez-Prini, *J. Phys. Chem.*, 1974, **78**, 1913.
73. M. Eigen and K. Tamm, *Z. Elektrochem.*, 1962, **66**, 93, 107.
74. K. Kojima, T. Inoue, M. Izaki, and R. Shimozaura, *Bull. Chem. Soc. Jpn*, 1986, **59**, 139; T. Inoue, K. Kojima, and R. Shimozaura, *Bull. Chem. Soc. Jpn*, 1986, **59**, 1683.
75. M. Ishii, S. Funahashi, K. Ishihara, and M. Tanaka, *Bull. Chem. Soc. Jpn*, 1989, **62**, 1852.
76. P. J. Nichols, Y. Ducommun, and A. E. Merbach, *Inorg. Chem.*, 1983, **22**, 3993; R. van Eldik, in *Inorganic High Pressure Chemistry: Kinetics and Mechanisms*, ed. R. van Eldik, Elsevier, Amsterdam, 1986, Chapter 3.
77. Y. Ducommun, G. Laurenczy, and A. E. Merbach, *Inorg. Chem.*, 1988, **27**, 1148.
78. A. Patel, P. Leitch, and D. T. Richens, *J. Chem. Soc., Dalton Trans.*, 1991, 1029; N. Aebischer, G. Laurenczy, A. Ludi, and A. E. Merbach, *Inorg. Chem.*, 1993, **32**, 2810.
79. M. Grant and R. B. Jordan, *Inorg. Chem.*, 1981, **20**, 55.
80. C. H. Langford and T. R. Khan, *Can. J. Chem.*, 1975, **53**, 2979.

81. M. Biruš, N. Kujundzić, and M. Pribanić, *Progr. React. Kinet.*, 1993, **18**, 171.
82. J. H. Espenson, *Inorg. Chem.*, 1969, **8**, 1554.
83. D. Thusius, *Inorg. Chem.*, 1971, **10**, 1106.
84. Kabur-ud-Din, G. J. Khan, Z. Khan, and M. Z. A. Rafiquee, *J. Chem. Res.*, 1996, (*S*) 178; (*M*) 0926.
85. S. I. Ali, S. Sharma, and Z. Khan, *Transition Met. Chem.*, 1996, **21**, 222.
86. M. Sado, T. Ozawa, K. Jitsukawa, and H. Einaga, *Polyhedron*, 1995, **14**, 2985; M. Sado, T. Ozawa, K. Jitsukawa, and H. Einaga, *Polyhedron*, 1996, **15**, 103.
87. M. Birus and R. van Eldik, *Inorg. Chem.*, 1991, **30**, 4559; M. R. Grace and T. W. Swaddle, *Inorg. Chem.*, 1992, **31**, 4674.
88. B. Perlmutter-Hayman, F. Secco, E. Tapuhi, and M. Venturini, *J. Chem. Soc., Dalton Trans.*, 1977, 2220; E. Mentasti, C. Baiocchi, and L. J. Kirschenbaum, *J. Chem. Soc., Dalton Trans.*, 1985, 2615; A. Campisi and P. Tregloan, *Inorg. Chim. Acta*, 1985, **100**, 251.
89. B. Perlmutter-Hayman, F. Secco, and M. Venturini, *Inorg. Chem.*, 1985, **24**, 3828.
90. G. Laurenczy, I. Rapaport, D. Zbinden, and A. E. Merbach, *Magn. Reson. Chem.*, 1991, **29**, S45; A. K. Ghosh, S. Ghosh, and G. S. De, *Transition Met. Chem.*, 1996, **21**, 358.
91. H. Stünzi and W. Marty, *Inorg. Chem.*, 1983, **22**, 2145; A. Drljaca and L. Spiccia, *Polyhedron*, 1996, **15**, 4373.
92. See, e.g., S. Gouger and J. Stuehr, *Inorg. Chem.*, 1974, **13**, 379.
93. B. Adell, *Z. Anorg. Allg. Chem.*, 1941, **246**, 303.
94. H. Taube and F. A. Posey, *J. Am. Chem. Soc.*, 1953, **75**, 1463; F. A. Posey and H. Taube, *J. Am. Chem. Soc.*, 1956, **78**, 15.
95. C. H. Langford and W. R. Muir, *J. Am. Chem. Soc.*, 1967, **89**, 3141.
96. R. J. Allen and P. C. Ford, *Inorg. Chem.*, 1972, **11**, 679; R. E. Shepherd and H. Taube, *Inorg. Chem.*, 1973, **12**, 1392; J. F. Ojo, O. Olubuyide, and O. Oyetunji, *J. Chem. Soc., Dalton Trans.*, 1987, 957.
97. C. Postmus and E. L. King, *J. Phys. Chem.*, 1955, **59**, 1217; F. Basolo and R. G. Pearson, *Mechanisms of Inorganic Reactions*, 2nd edn., Wiley, New York, 1967, p. 197.
98. M. Ferrer and A. G. Sykes, *Inorg. Chem.*, 1979, **18**, 3345.
99. D. Banerjea and S. D. Chaudhuri, *J. Inorg. Nucl. Chem.*, 1968, **30**, 871; 1970, 32, 1617; P. De and G. S. De, *Transition Met. Chem.*, 1993, **18**, 353.
100. J. O. Edwards, F. Monacelli, and G. Ortaggi, *Inorg. Chim. Acta*, 1974, **11**, 47.
101. D. H. Macartney and A. McAuley, *Inorg. Chem.*, 1981, **20**, 748; K. J. Pfenning, Liangshiu Lee, H. D. Wohlers, and J. D. Petersen, *Inorg. Chem.*, 1982, **21**, 2477; H. E. Toma, A. A. Batista, and H. B. Gray, *J. Am. Chem. Soc.*, 1982, **104**, 7509; D. H. Macartney and A. McAuley, *Can. J. Chem.*, 1989, **67**, 1774; Gyu-Hwan Lee, L. Della Ciana, and A. Haim, *J. Am. Chem. Soc.*, 1989, **111**, 2535; Hung-Yi Huang, Wen-Jang Chen, Chang-Chan Yang, and A. Yeh, *Inorg. Chem.*, 1991, **30**, 1862
102. G. Stochel, J. Chatlas, P. Martinez, and R. van Eldik, *Inorg. Chem.*, 1992, **31**, 5480.
103. A. Das and H. C. Bajaj, *Polyhedron*, 1997, **16**, 1023.
104. H. E. Toma and A. B. P. Lever, *Inorg. Chem.*, 1986, **25**, 176.
105. G. Stochel and R. van Eldik, *Inorg. Chim. Acta*, 1991, **190**, 55.

106. G. Stochel, *Coord. Chem. Rev.*, 1992, **114**, 269.
107. W. G. Baldwin and D. R. Stranks, *Aust. J. Chem.*, 1968, **21**, 2161.
108. H. Strehlow and W. Knoche, *Ber. Bunsenges. Phys. Chem.*, 1969, **73**, 427; B. Gruenewald, W. Knoche, and N. H. Rees, *J. Chem. Soc., Dalton Trans.*, 1976, 2338.
109. M. Inamo, K. Ishihara, S. Funahashi, Y. Ducommun, A. E. Merbach, and M. Tanaka, *Inorg. Chem.*, 1991, **30**, 1580.
110. A. C. Dash, P. Mohanty, and A. N. Acharya, *Transition Met. Chem.*, 1995, **20**, 406.
111. U. M. Frei and G. Geier, *Inorg. Chem.*, 1992, **31**, 187, 3132.
112. L. I. Elding, *Inorg. Chim. Acta*, 1978, **28**, 255; Ö. Gröning, T. Drakenberg, and L. I. Elding, *Inorg. Chem.*, 1982, **21**, 1820.
113. S. Elmroth, Z. Bugarcic, and L. I. Elding, *Inorg. Chem.*, 1992, **31**, 3551.
114. B. Hellquist, L. I. Elding, and Y. Ducommun, *Inorg. Chem.*, 1988, **27**, 3620.
115. T. G. Appleton, J. R. Hall, and P. D. Prenzler, *Inorg. Chem.*, 1989, **28**, 815; J. Arpalahti and B. Lippert, *Inorg. Chem.*, 1990, **29**, 104; J. Arpalahti, M. Mikola, and S. Mauristo, *Inorg. Chem.*, 1993, **32**, 3327.
116. W. Schaller, H. Reisner, and E. Holler, *Biochemistry*, 1987, **26**, 943.
117. K. Y. Faridoon, H.-Y. Zhuang, and A. G. Sykes, *Inorg. Chem.*, 1994, **33**, 2209.
118. M. Nasreldin, Yue-Jin Li, F. E. Mabbs, and A. G. Sykes, *Inorg. Chem.*, 1994, **33**, 4283.
119. T. Shibahara, T. Asano, and G. Sakane, *Polyhedron*, 1991, **10**, 2351.
120. B. C. Westerby, K. L. Juntunen, G. H. Leggett, V. B. Pett, M. J. Koenigsbauer, M. D. Purgett, M. J. Taschner, L. A. Ochrymowycz, and D. B. Rorabacher, *Inorg. Chem.*, 1991, **30**, 2109.
121. R. M. Izatt, J. S. Bradshaw, S. A. Nielsen, J. D. Lamb, J. J. Christensen, and D. Sen, *Chem. Rev.*, 1985, **85**, 271.
122. R. M. Izatt, K. Pawlak, J. S. Bradshaw, and R. L. Bruening, *Chem. Rev.*, 1991, **91**, 1721; R. M. Izatt, K. Pawlak, J. S. Bradshaw, and R. L. Bruening, *Chem. Rev.*, 1995, **95**, 2529.
123. B. G. Cox and H. Schneider, *Pure Appl. Chem.*, 1990, **62**, 2259.
124. L. Echegoyen, A. Kaifer, H. Durst, R. A. Schultz, D. M. Dishong, D. M. Goli, and G. W. Gokel, *J. Am. Chem. Soc.*, 1984, **106**, 5100; F. Eggers, T. Funck, K. H. Richmann, H. Schneider, E. M. Eyring, and S. Petrucci, *J. Phys. Chem.*, 1987, **91**, 1961.
125. P. B. Chock, *Proc. Natl. Acad. Sci., U.S.A.*, 1972, **69**, 1939.
126. M. Eigen and R. Winkler, in *The Neurosciences: Second Study Program*, ed. F. O. Schmitt, Rockefeller Press, New York, 1970, pp. 685–696.
127. J. Rebek, S. V. Luis, and L. R. Marshall, *J. Am. Chem. Soc.*, 1986, **108**, 5011.
128. See Ref. 1–4 in H. D. H. Stöver and C. Detellier, *J. Phys. Chem.*, 1989, **93**, 3174.
129. H. K. Wipf, A. Olivier, and W. Simon, *Helv. Chim. Acta*, 1970, **53**, 1605.
130. B. G. Cox, J. Garcia-Rosas, and H. Schneider, *J. Am. Chem. Soc.*, 1982, **104**, 2434; E. Schmidt and A. I. Popov, *J. Am. Chem. Soc.*, 1983, **105**, 1873.
131. For example, E. Schmidt and A. I. Popov, *J. Am. Chem. Soc.*, 1983, **105**, 1873; K. M. Brière and C. Detellier, *J. Phys. Chem.*, 1987, **91**, 6097.
132. M. Shamsipur and A. I. Popov, *J. Phys. Chem.*, 1988, **92**, 147.
133. E. Grell and R. Warmuth, *Pure Appl. Chem.*, 1993, **65**, 373.
134. B. G. Cox, P. Firman, and H. Schneider, *Inorg. Chim. Acta*, 1982, **64**, L263.

135. B. G. Cox, P. Firman, and H. Schneider, *Inorg. Chem.*, 1982, **21**, 2320.
136. B.-L. Ooi, M. Martinez, and A. G. Sykes, *J. Chem. Soc., Chem. Commun.*, 1988, 1324; B.-L. Ooi and A. G. Sykes, *Inorg. Chem.*, 1988, **27**, 310; B.-L. Ooi and A. G. Sykes, *Inorg. Chem.*, 1989, **28**, 3799.
137. Yue-Jin Li, M. Nasreldin, M. Humanes, and A. G. Sykes, *Inorg. Chem.*, 1992, **31**, 3011.
138. E. F. Hills, D. T. Richens, and A. G. Sykes, *Inorg. Chem.*, 1986, **25**, 3144.
139. Y. Sasaki, A. Toliwa, and T. Ito, *J. Am. Chem. Soc.*, 1987, **109**, 6341; K. Nakata, A. Nagasawa, N. Soyama, Y. Sasaki, and T. Ito, *Inorg. Chem.*, 1991, **30**, 1575.
140. C. A. Routledge, M. Humanes, Yue-Jin Li, and A. G. Sykes, *J. Chem. Soc., Dalton Trans.*, 1994, 1275.
141. Jian Zhou, M. J. Scott, Zhengguo Hu, Gang Peng, E. Münck, and R. H. Holm, *J. Am. Chem. Soc.*, 1992, **114**, 10843.
142. V. P. Fedin, M. N. Sokolov, and A. G. Sykes, *J. Chem. Soc., Dalton Trans.*, 1996, 4089.
143. B. J. Plankey and H. H. Patterson, *Environ. Sci. Technol.*, 1987, **21**, 595.
144. D. R. Parker and P. M. Bertsch, *Environ. Sci. Technol.*, 1992, **26**, 908.
145. See R. C. Hider and A. D. Hall, *Progr. Med. Chem.*, 1991, **28**, 41, and S. Jurisson, D. Berning, Wei Jia, and Dangshe Ma, *Chem. Rev.*, 1993, **93**, 1137 for full treatments; J. Burgess, *Chem. Soc. Rev.*, 1996, **25**, 85 for a superficial overview; and J. Burgess, *Mech. Inorg. Organomet. React.*, 1991, **7**, 169; 1994, **8**, 145 for specifically kinetic aspects.
146. C. J. Broan, J. P. L. Cox, A. S. Craig, R. Kataky, D. Parker, A. Harrison, A. M. Randall, and G. Ferguson, *J. Chem. Soc., Perkin Trans. 2*, 1991, 87.
147. D. Parker, K. Pulukkody, T. J. Norman, A. Harrison, L. Royle, and C. Walker, *J. Chem. Soc., Chem. Commun.*, 1992, 1441.
148. N. Sabbatini, M. Guardigli, F. Bolletta, I. Manet, and R. Ziessel, *Angew. Chem. Int. Edn. Engl.*, 1994, **33**, 1501.
149. E. Brucher, S. Cortes, F. Chavez, and A. D. Sherry, *Inorg. Chem.*, 1991, **30**, 2092.
150. K. Kumar and M. F. Tweedle, *Pure Appl. Chem.*, 1993, **65**, 515.
151. M. K. Moi, C. F. Meares, and S. J. DeNardo, *J. Am. Chem. Soc.*, 1988, **110**, 6266; J. P. L. Cox, K. J. Jankowski, R. Kataky, D. Parker, N. R. A. Beeley, B. A. Boyce, M. A. W. Eaton, K. Millar, A. T. Millican, A. Harrison, and C. Walker, *J. Chem. Soc., Chem. Commun.*, 1989, 797.
152. E. Brücher, G. Laurenczy, and Z. Makra, *Inorg. Chim. Acta*, 1987, **139**, 141.
153. A. S. Craig, I. M. Helps, K. J. Jankowski, D. Parker, N. R. A. Beeley, B. A. Boyce, M. A. W. Eaton, A. T. Millican, K. Millar, A. Phipps, S. K. Rhind, A. Harrison, and C. Walker, *J. Chem. Soc., Chem. Commun.*, 1989, 794.
154. See the appropriate section in the Chemical Society/Royal Society of Chemistry Specialist Periodical Report, *Inorganic Reaction Mechanisms*, Vols 1–7 (1971–1981).
155. T. Nozaki, K. Kasuga, and N. Suemitsu, *Nippon Kagaku Kaishi*, 1974, 485.
156. K. R. Williams and G. R. Choppin, *J. Inorg. Nucl. Chem.*, 1974, **36**, 1849.
157. E. Brücher and H. Kelm, *J. Coord. Chem.*, 1974, **4**, 133.
158. E. R. Souaya, *Z. Phys. Chem. (Leipzig)*, 1988, **269**, 1017.
159. Zhang Hualin and L. Peiyi, *Transition Met. Chem.*, 1991, **16**, 352.
160. J. G. Hering and F. M. M. Morel, in *Aquatic Chemical Kinetics*, ed. W. Stumm, Wiley, New York, 1990, Chapter 5.

161. J. G. Hering and F. M. M. Morel, *Environ. Sci. Technol.*, 1988, **22**, 1469.
162. J. G. Hering and F. M. M. Morel, *Environ. Sci. Technol.*, 1988, **22**, 1234; J. G. Hering and F. M. M. Morel, *Geochim. Cosmochim. Acta*, 1989, **53**, 611; J. G. Hering and F. M. M. Morel, *Environ. Sci. Technol. 1*, 1990, **24**, 242.
163. M. Kodama and E. Kimwa, *J. Chem. Soc., Dalton Trans.*, 1978, 247.
164. J. D. Carr and N. D. Danielson, *Inorg. Chim. Acta*, 1977, **25**, L27.
165. W. R. Harris and Gang Bao, *Polyhedron*, 1997, **16**, 1069.
166. M. Tabata and K. Suenaga, *Bull. Chem. Soc. Jpn*, 1991, **64**, 469.

8 Medium effects

Chapter contents

8.1 Introduction

Medium effects on reactivity[1] can be very large. Thus, for example, solvolysis of t-butyl chloride is about a million times faster in water than in methanol.[2] Accelerations of similar magnitude have been reported for, for instance, mercury(II)-catalysed aquation of $[Co(NH_3)_5Br]^{2+}$ in micellar media compared with water.[3] At the other extreme, dissociation of the $[Rh\{P(OMe)_3\}_5]^+$ cation has exactly the same rate constant in five different solvents.[4] The aims of this chapter are twofold. Firstly, we shall try to give an overall idea of medium effects on reactivity, to give some idea of how solvation affects kinetic parameters. To this end we shall cover ranges of solvents, series of binary aqueous media, aqueous solutions of electrolytes, and 'organized media' such as micelles and microemulsions. Secondly, we shall endeavour to outline how solvent and salt effects can be useful in establishing mechanisms.

It should be borne in mind that a sufficiently drastic change in medium may affect not only kinetic parameters but also rate laws and mechanism—and, in some cases, reaction products. Thus the solvolysis of t-butyl chloride gives t-butyl alcohol in media where hydroxyl groups are available, but butene in non-hydroxylic media. Dihydropyrazoles (**1**) also give alkenes as predominant products in non-polar media, but give mainly substituted cyclopropanes in polar solvents.[5] Tetrabutoxytin reacts with trimethylsilylacetate in benzene with ester elimination

$$Sn(O^tBu)_4 + Me_3SiOAc \longrightarrow Sn(O^tBu)_3(OSiMe)_3 + {}^tBuOAc$$

but in the donor solvent pyridine the t-butoxy group leaves in the form of its trimethylsilyl derivative[6]

$$Sn(O^tBu)_4 + Me_3SiOAc \longrightarrow Sn(O^tBu)_3(OAc)(Py) + Me_3SiO^tBu$$

The phospha-alkyne $^tBuC{\equiv}P$ reacts with triethylaluminium to give a polycyclic oligomer (Al:P = 1:4) in diethyl ether, but a cage species (Al:P = 2:3; one Al in an AlP_3 cage) in n-hexane.[7] Moving on to classical Werner complexes, the product of solvolysis of, for instance, $[Co(NH_3)_5Cl]^{2+}$ will change from the aqua complex $[Co(NH_3)_5(H_2O)]^{3+}$ to $[Co(NH_3)_5(MeOH)]^{3+}$ to $[Co(NH_3)_5(DMSO)]^{3+}$ as the solvent changes. Further, the product ratio may well not reflect the solvent composition for solvolyses in binary mixtures of coordinating solvents. Further examples appear in the following sections, and in Fig. 8.1. Special cases include solvent effects on the steric course of a reaction (cf. the last entry in Fig. 8.1) and on selectivity (e.g. in Diels–Alder reactions[8]). There is also the possibility of a major solvent effect in systems where there is a balance between kinetic and thermodynamic control.[9]

We have sandwiched the topic of medium effects between the chapters on substitution and that on oxidation–reduction reactions because medium

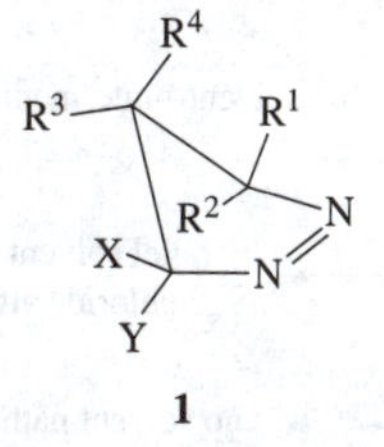

effects on reactivity have played an important role in establishing mechanisms for electron transfer, whereas medium effects on substitution kinetics have tended to provide information on the role of solvation and solvent structure in reactions of established mechanism. Solvent and salt effects on reactivity have played only a modest part in the elucidation of substitution mechanisms at inorganic centres—such approaches as that of Grunwald and Winstein[2] (detailed in Section 8.2.1 below) are less helpful in the diagnosis of mechanisms across the wide range of compounds and complexes of inorganic chemistry than they are in the limited world of substitution at carbon. We therefore need to know something about medium effects before dealing in detail with electron transfer (Chapter 9), but on the other hand need to know about substitution mechanisms (Chapters 3–7) before discussing medium effects on substitution processes.

8.2 Solvent effects

8.2.1 History and background

Magnitudes of solvent effects on reactivity can be very large, as intimated in the preceding section. Kinetic data for a number of solvent-sensitive reactions were tabulated in Table 1.1 of Chapter 1. Figure 8.2 puts the magnitude of solvent effects on rate constants into the general context of

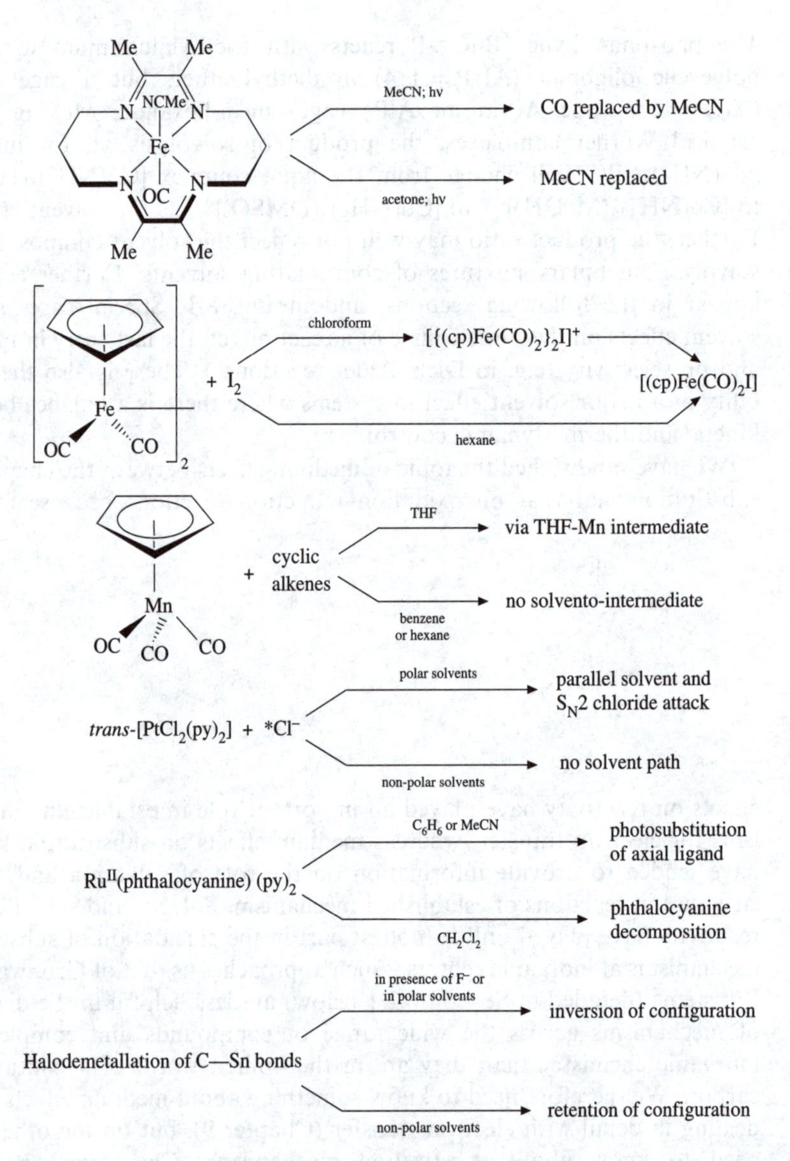

Fig. 8.1 Solvent control of mechanism and of products.

reactivity ranges. The realization of the important part that the solvent may play in determining reactivity goes back a long way. Berthelot, in 1862, was probably the first to discuss solvent effects, in his case in comparing esterification of ethanol by acetic acid in benzene and in ethanol.[10] The first systematic study (1890) was carried out by Menschutkin, of the reaction which bears his name (i.e. $Et_3N + EtI \rightarrow [Et_4N]I$).[11] The next major development came in the mid-1930s. Hughes and Ingold started to systematize, using then current views on ion solvation,[12] solvent effects on

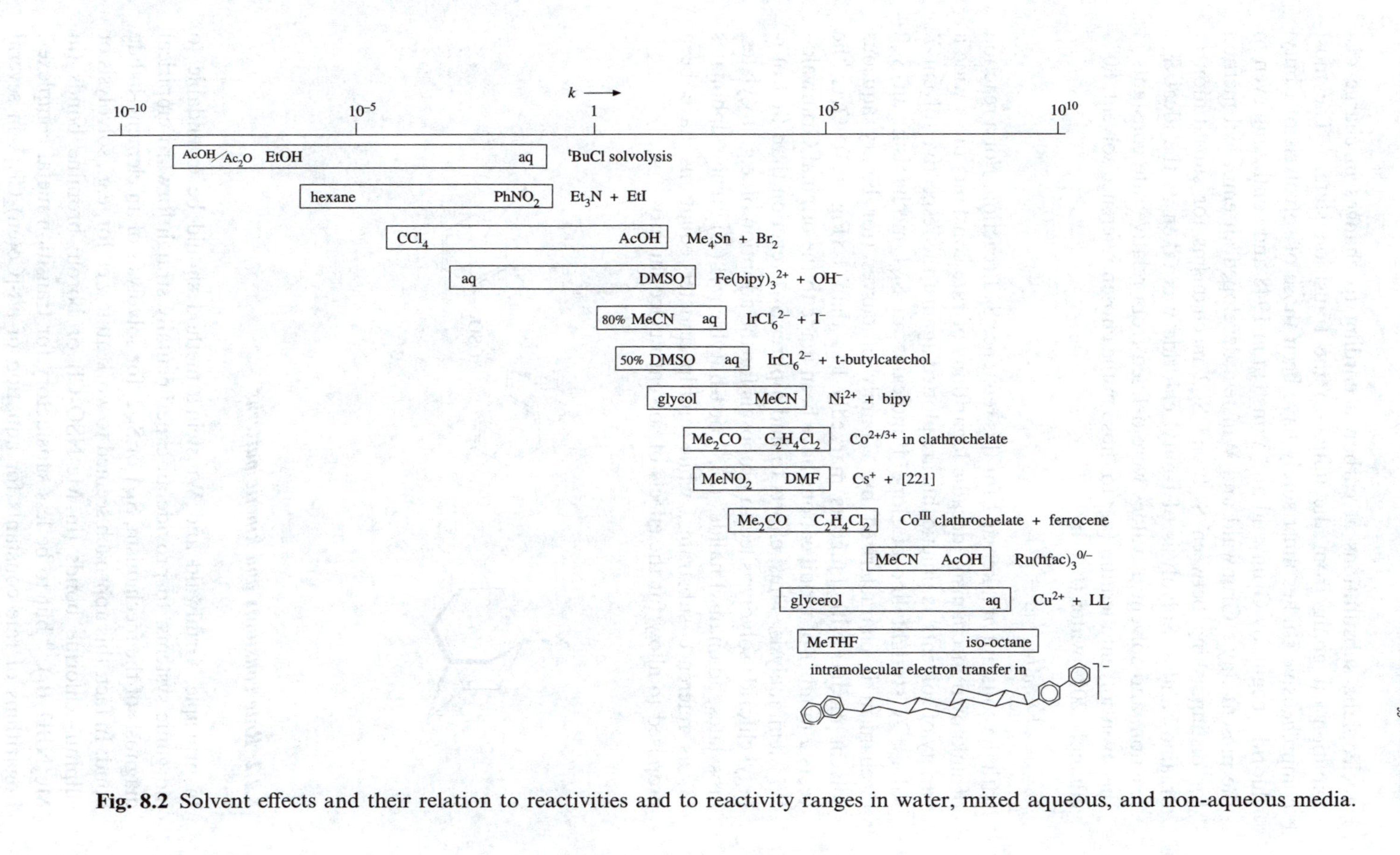

Fig. 8.2 Solvent effects and their relation to reactivities and to reactivity ranges in water, mixed aqueous, and non-aqueous media.

nucleophilic substitutions at carbon according to changes in charge distribution on going from the initial to the transition state.[13] The most comprehensive studies must surely have been those of solvolysis of t-butyl chloride, begun by Grunwald and Winstein in 1948 and continuing even to the present day.[2] Grunwald and Winstein developed an empirical method for distinguishing between S_N1 and S_N2 mechanisms for substitution at organic halides. Solvolysis of t-butyl chloride was taken as the ideal S_N1 reaction, and solvent Y values were defined from relative rate constants in the given solvent medium (k_S) to those in the chosen reference solvent, 80% ethanol– 20% water (k_O), by

$$\log_{10}(k_S/k_O) = mY$$

with the constant m taken as 1 for this reference S_N1 reaction. For a reaction of unknown mechanism, decadic logarithms of rate constants in as wide a variety of solvents as practicable are plotted against the respective solvent Y values. A straight line of slope 1 indicates an S_N1 mechanism; an S_N2 mechanism is characterized by a family of curves (for binary aqueous solvent media) with tangents markedly less than 1 (Fig. 8.3). Over the years a number of modifications have been made to the original Grunwald–Winstein analysis. Thus, efforts to find a purer S_N1 reaction than that of t-butyl chloride solvolysis led to proposals for the adoption of, e.g., tosylates, brosylates, adamantyl halides (**2**),[14] or t-butylbenzyl (Z)-arene sulphonates (**3**) as reference substrates, while multiparameter equations have been proposed to allow for the effects of solvent nucleophilicity.[15]

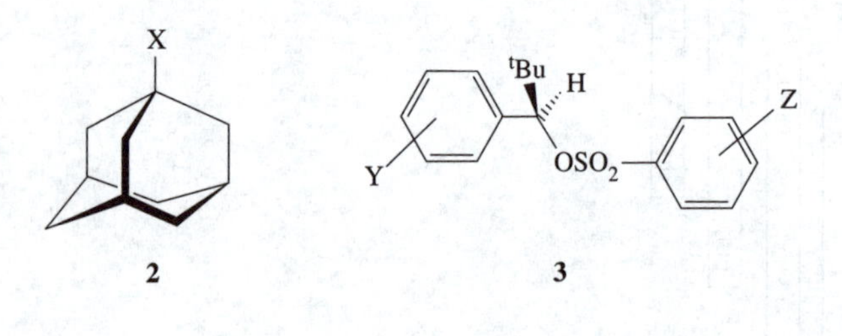

8.2.2 *Rate constants and kinetic patterns*

In principle, Grunwald and Winstein's method should be extendable to inorganic systems, to provide an experimentally straightforward empirical diagnosis of the mechanism, S_N1 or S_N2, for solvolysis of an element–halide bond. In fact, this approach seemed to work quite well for, e.g., solvolysis of sulphur–chlorine bonds (in Me_2NSO_2Cl) or boron–bromine bonds (in Me_3NBH_2Br),[16] but to be less satisfactory for transition metal complexes. Logarithms of rate constants for aquation of *cis*-$[Co(en)_2Cl_2]^+$ in several

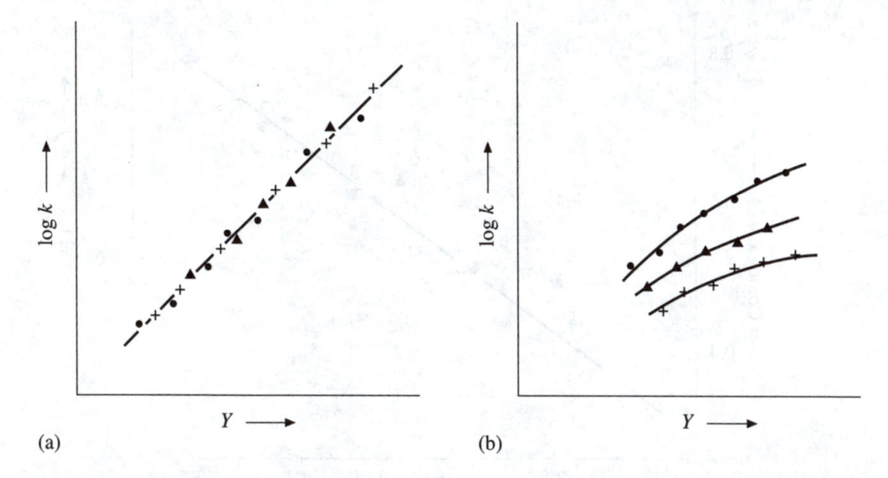

Fig. 8.3 Grunwald–Winstein treatment of solvent effects on reactivities for (a) S_N1 (I_d) and (b) S_N2 (I_a) organic reactions. The various different symbols signify different cosolvents for series of binary aqueous solvent media.

series of binary aqueous solvent mixtures (water-rich to ensure that the aqua complex $[Co(en)_2(OH_2)Cl]^+$ was the sole product) against solvent m values clustered close to a common straight line (Fig. 8.4), which indicated an S_N1 process, but the slope of the line was rather less than 0.3, which in organic applications would signal an S_N2 process.[17] Interestingly, an analogous treatment of similar chromium(III) complexes gave a line of slope *ca* 0.1, which at least is consistent with more associative attack at Cr^{III} than at Co^{III}.[18] Presumably the solvation of inorganic moieties of the $[M(NH_3)_5]^{3+}$ and $[M(en)_2]^{3+}$ type changes less with solvent composition than that of Me_3C^+. With appropriate modifications for solvent nucleophilicity the Grunwald–Winstein treatment was applied with some success to ruthenium(III)–chloride complexes, showing hydrolysis at ruthenium(III) to be slightly more associative in character than at cobalt(III).[19] More successful, because much closer to the original organic systems, were the attempts to diagnose mechanisms of isomerization of halogeno-alkene and -alkyne complexes of the type

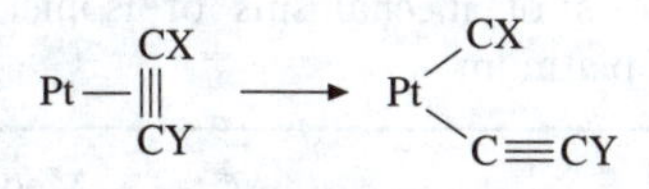

As shown in Fig. 8.5(a), slopes of plots of logarithms of rate constants against solvent Y values give straight lines, whose slopes m (Table 8.1) range from 0.16 to 0.86. The latter value is close to 1.0, suggesting that the isomerization of the tetrachloroethene complex approximates to an intermolecular, dissociative (S_N1) process (Fig. 8.5(b)), with rate-limiting breaking of a carbon–chlorine bond followed by rapid scavenging of the (almost)

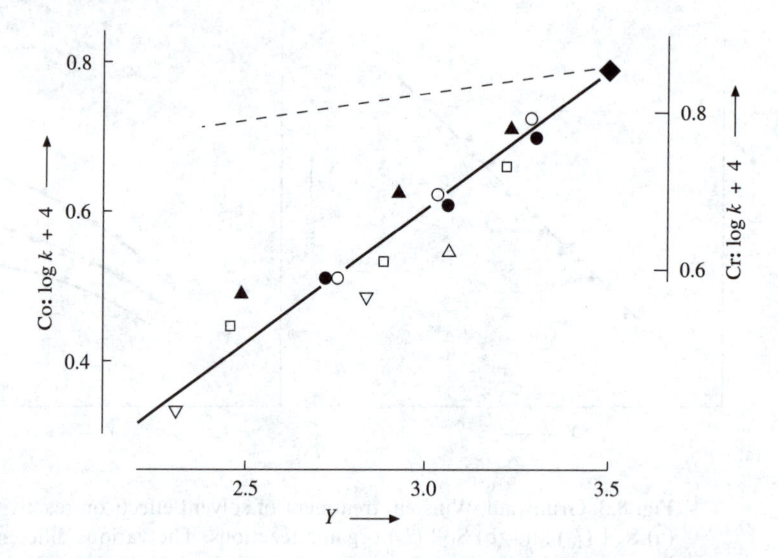

Fig. 8.4 Grunwald–Winstein analysis of solvent effects on reactivity for metal(III)-chloride aquation. The main plot is for aquation of *cis*-$[Co(en)_2Cl_2]^+$ in water-rich binary aqueous solvent mixtures with cosolvents methanol (○), ethanol (●), acetone (▲), dioxan (□), formic acid (△), and acetic acid (▽). The dashed line corresponds to aquation of *cis*-$[Cr(en)_2Cl_2]^+$ in a similar series of media. The values for aquation in water are shown as ◆.

free chloride by the platinum. On the other hand, the small solvent effects observed for the PhC≡Cl and $F_2C{=}CFBr$ complexes suggest that these isomerizations take place intramolecularly (Fig. 8.5(c)), with the halide forming a bond to the platinum as it leaves from carbon, with very little negative charge generation (δ in $X^{\delta-}$ close to zero).[20] A number of other empirical solvent parameters have been pressed into service in the search for methods of distinguishing between dissociative and associative substitution at inorganic centres. We might mention the correlations with Reichardt's spectroscopically based E_T parameter of rate constants for the (oxidative)

Table 8.1 Solvent effects in the diagnosis of mechanisms of isomerization of haloalkene and haloalkyne complexes of platinum.

Reaction[a]	m[b]	Mechanism
$(Ph_3P)_2Pt(Cl_2C{=}CCl_2) \rightarrow (Ph_3P)_2PtCl(CCl{=}CCl_2)$	0.86	intermolecular
$(Ph_3P)_2Pt(Cl_2C{=}CHCl) \rightarrow (Ph_3P)_2PtCl(CH{=}CCl_2)$	0.54	↓
$(Ph_3As)_2Pt(F_2C{=}CFBr) \rightarrow (Ph_3As)_2PtBr(CF{=}CF_2)$	0.32	↓
$(Ph_3P)_2Pt(PhC{\equiv}CCl) \rightarrow (Ph_3P)_2PtCl(C{\equiv}CPh)$	0.16	intramolecular

[a] In each case from a $\pi(\eta^2)$ to a $\sigma(\eta^1)$ complex.
[b] Slope of Grunwald–Winstein plot [cf. Fig. 8.5(a)].

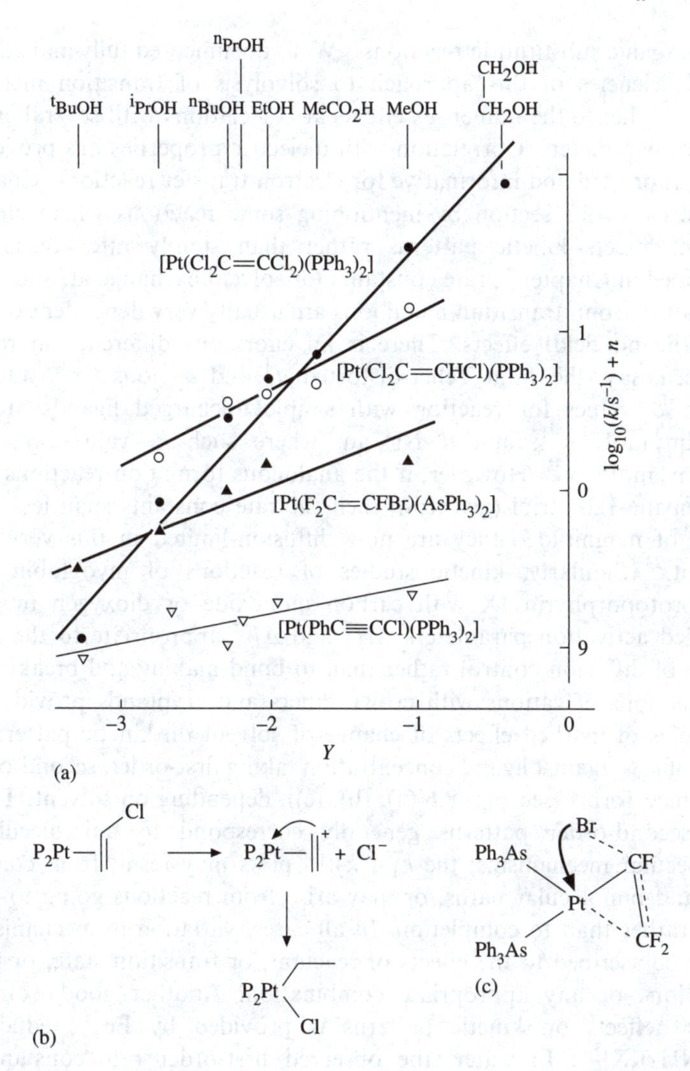

Fig. 8.5 (a) Grunwald–Winstein analysis of solvent effects on isomerization reactions of haloalkene and haloalkyne complexes of platinum. Intermolecular and intramolecular mechanisms are outlined in (b) and (c), respectively.

addition of methyl iodide, dihydrogen, and dioxygen to Vaska's compound, *trans*-[IrCl(CO)(PPh$_3$)$_2$].[21] Here the slopes of the correlation lines for the two diatomic molecules are markedly different from that for methyl iodide, suggesting a considerable difference in transition states (see Chapter 10 for mechanistic details of these reactions). There is also a reasonably good correlation of logarithms of rate constants with dielectric properties, specifically with the function $(\varepsilon - 1)(2\varepsilon + 1)$. It is quite rare to find simple and informative correlations of reactivities with solvent dielectric properties for inorganic substitution reactions—Wells documented fully and admirably the deficiencies of this approach to solvolysis of transition metal complexes[22]—hence the numerous efforts at correlations with several empirical solvent parameters. Correlation with dielectric properties has proved more straightforward and informative for electron transfer reactions (Chapter 9).

We close this section by mentioning some reactions where change of solvent affects kinetic patterns rather than simply rate constants. As described in Chapter 7, rate constants for solvent exchange at, and complex formation from, transition metal ions are usually very dependent on crystal field (ligand field) effects. There is an enormous difference in reactivity between, say, the Jahn–Teller-distorted d^4 and d^9 ions Cr^{2+} and Cu^{2+}, where k_f values for reaction with simple uncharged ligands are about 10^9 dm^3 mol^{-1} s^{-1}, and d^8 Ni^{2+}aq, where such k_f values are around 10^3 dm^3 mol^{-1} s^{-1}. However, if the analogous formation reactions are run in propane-1,2,3-triol (glycerol), then the rate constants span less than an order of magnitude—they are now diffusion-limited in this very viscous solvent.[23] Similarly, kinetic studies of reactions of myoglobin and of ferroprotoporphyrin IX with carbon monoxide or dioxygen in glycerol revealed activation parameters $\Delta H^\ddagger$ and $\Delta V^\ddagger$ appropriate to the requirements of diffusion control rather than to bond making and breaking.[24]

Reactions of cations with crown ethers and cryptands provide several examples of marked effects of change of solvent on kinetic patterns, with plots of k_{obs} against ligand concentration taking first-order, second-order, or combined forms (see Fig. 8.6 (a), (b), (c)), depending on solvent. The first- and second-order patterns generally correspond to unimolecular and bimolecular mechanisms; the $k_1 + k_2[L]$ plots may result from concurrent uni- and bimolecular paths, or may arise from reactions going to equilibrium rather than to completion. In all cases, variation in mechanism may safely be ascribed to the effects of reactant, or transition state, or product solvation—or any appropriate combination! Another good example of solvent effects on kinetic patterns is provided by Fe^{2+} reduction of $[Co(NH_3)_5X]^{2+}$. In water, the observed first-order rate constant (k_{obs})

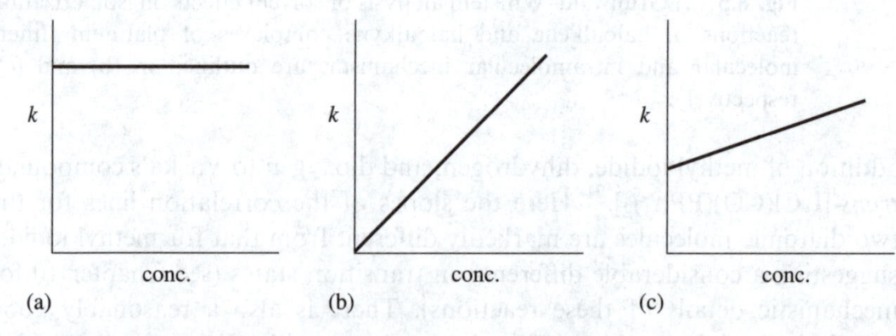

Fig. 8.6 Outline kinetic patterns for exchange reactions of crown ether and cryptand complexes: (a) first-order form; (b) second-order form; (c) combined form.

depends linearly on Fe^{2+}aq concentration (Fe^{2+} in large excess), but in dimethylformamide there is a curved dependence of k_{obs} on Fe^{2+} concentration, from which the intermediacy of a transient binuclear species may be inferred (and its equilibrium and electron transfer rate constant deduced, as detailed for Cr^{2+}aq reduction of $[Co(NH_3)_5(NCacac)]^{2+}$ in Section 9.4.1).[25]

8.2.3 Activation parameters

The potential pitfalls in using activation entropies and volumes as indicators of mechanism for reactions carried out in polar solvents can in fact be turned to advantage in the assessment of solvation of reactants and transition states. In principle, $\Delta S^{\ddagger}$ and $\Delta V^{\ddagger}$ should be good indicators of mechanism. For a dissociative process the transition state is looser and larger than the initial state, giving positive values for $\Delta S^{\ddagger}$ and $\Delta V^{\ddagger}$. On the other hand, the transition state for an associative process has less freedom and is more compact than the separate reagents in the initial state, indicating negative $\Delta S^{\ddagger}$ and $\Delta V^{\ddagger}$ values. However, the use of these activation parameters in the diagnosis of mechanism is reliable only in the absence of strong substrate–solvent interactions, i.e. of strong solvation, for solvation changes consequent on transition-state formation may make a significant contribution to the experimentally obtained $\Delta S^{\ddagger}$ and $\Delta V^{\ddagger}$ values:

$$\Delta X^{\ddagger} = \Delta X^{\ddagger}_{\text{intrinsic}} + \Delta X^{\ddagger}_{\text{solvation}}$$

Indeed, the solvation contribution may even dominate over the intrinsic $\Delta S^{\ddagger}$ and $\Delta V^{\ddagger}$ values, and $\Delta X^{\ddagger}$ may therefore change its sign if solvation and intrinsic factors are of opposite sign. Thus for the classic dissociative reaction of organic chemistry, t-butyl chloride solvolysis, activation volumes are actually *negative*, ranging from -2 to $-32\,\text{cm}^3\,\text{mol}^{-1}$ as the solvation contribution becomes ever more important than the intrinsic volume of about $+10\,\text{cm}^3\,\text{mol}^{-1}$ expected for simple unimolecular dissociation.[26] Conversely, activation volumes for bimolecular attack of hydroxide or methoxide at sulphonium ions R_3S^+ are in the region of $+5$ to $+15\,\text{cm}^3\,\text{mol}^{-1}$, in contrast to the $-10\,\text{cm}^3\,\text{mol}^{-1}$ expected for the formation of a bimolecular transition state. Sometimes the solvation contribution is not sufficient to result in a change of sign, but reveals itself in a marked solvent variation of $\Delta S^{\ddagger}$ or $\Delta V^{\ddagger}$ values (cf. the range for t-butyl chloride solvolysis just mentioned), as in the example shown in Table 8.2. Volume profiles may be informative (see Chapter 1, Figure 1.3).

Selections of $\Delta S^{\ddagger}$ or $\Delta V^{\ddagger}$ values for several reactions involving inorganic and organometallic species are given in Tables 8.3 and 8.4 to illustrate the importance of solvation, especially for reactions in water. The reactions included in Table 8.3 have been selected to show how $\Delta S^{\ddagger}$ varies with the charges on the reactants—there is a clear pattern despite the fact that at least three of the reactions are two-stage, and thus the $\Delta S^{\ddagger}$ values shown in Table 8.3 are composite quantities. However, these $\Delta S^{\ddagger}$ values are often

Table 8.2 Solvent effects on activation volumes for bimolecular reactions.

Solvation effects	Reaction and reactants	Transition state	$\Delta V^{\ddagger}$ ($cm^3\ mol^{-1}$)
NEGLIGIBLE	Diels–Alder: no charge separation →		
	isoprene + maleic anhydride	(cyclic transition state)	−32 to −39[a]
SIGNIFICANT	Menschutkin: charge separation →		
	$Et_3N + EtI$	$Et_4\overset{\delta+}{N}\cdots\overset{\delta-}{I}$	− 12 to − 58

[a] These negative values are particularly large as the transition state is unusually compact—both new carbon–carbon bonds are almost completely formed in what is a 'late' transition state.

Table 8.3 Comparison of observed and calculated activation entropies for bimolecular reactions involving various charge products.

Reaction	Charge product z_Az_B	$\Delta S^{\ddagger}$ ($J\ K^{-1}\ mol^{-1}$) measured	calculated[b]
$S_2O_3^{2-} + SO_3^{2-}$	4+	−126	−170
$[ML_5X]^{2+} + Hg^{2+}$ [a]	4+	−60 to −104	−170
$BrCH_2CO_2^- + S_2O_3^{2-}$	2+	−71	−85
$ClCH_2CO_2^- + OH^-$	1+	−50	−40
$BrCH_2CO_2Me + S_2O_3^{2-}$	0	+25	0
$[Co(NH_3)_5Br]^{2+} + OH^-$	2−	+92	+85
$[Co(NH_3)_5(H_2O)]^{3+} + Cl^-$	3−	+142	+125
$ReCl_6^{2-} + Hg^{2+}$	4−	+142	+170

[a] For various combinations of $ML_5 = Cr(OH_2)_5$, $Cr(NH_3)_5$, $Co(NH_3)_5$, $Rh(NH_3)_5$ with X = Cl, Br.
[b] According to Laidler's simple solvation model.

considerably smaller in magnitude than forecast on a simple solvation model, particularly for reactions involving the larger ions or complexes.[27] Values of $\Delta S^{\ddagger}$ show relatively small solvation contributions (i.e. are markedly negative for these bimolecular reactions) when there is charge augmentation. For a reaction between two 2+ ions the increasing field due to the 4+ in the transition state is at least partially compensated by the increase in size of the transition state. But towards the bottom of Table 8.3, $\Delta S^{\ddagger}$ changes sign, to become markedly positive when there is charge diminution, and thus considerable de-electrostriction of solvating water,

Table 8.4 The influence of solvation on activation volumes, $\Delta V^{\ddagger}$ ($cm^3\ mol^{-1}$), for some inorganic and organometallic reactions.

Negligible solvent effects		Significant solvent effects [a]	
DISSOCIATION			
$Cr(CO)_6 + PPh_3$ (k_1 term) [b]	+15	$[Co(NH_3)_5Cl]^{2+}$ aquation	−9
$[Co(NH_3)_5(SO_3)]^+$ (NH_3 loss) [c]	+6	$[Co(NH_3)_5(SO_4)]^+$ aquation	−19 [d]
$[Fe(phen)_3]^{2+}$ aquation	+15		
$[Fe(CN)_5(4CN\,py)]^{3-}$ substitution	+19		
BIMOLECULAR REACTIONS			
$W(CO)_6 + P^nBu_3$ (k_2 term) [b]	−10	$[Co(NH_3)_5(N_3)]^{2+} + Fe^{2+}aq$	+14
trans-$[IrCl(CO)(PPh_3)_2] + H_2$	−19	$[RhCl_6]^{3-} + Hg^{2+}aq$	+22
		$[Fe(gmi)_3]^{2+}$ [e] $+ OH^-$	+17

[a] All values quoted here refer to aqueous solution.
[b] See Chapter 6 (p244) for the rate law for $M(CO)_6 + PR_3$ reactions.
[c] Due to the strong *trans* labilizing effect of sulphite.
[d] The higher charge on sulphate than on chloride explains the more negative $\Delta V^{\ddagger}$ for the former complex.
[e] gmi is MeN=CHCH=NMe.

on transition state formation. Table 8.4 compares and contrasts $\Delta V^{\ddagger}$ values for reactions with and without major solvational contributions.

Solubility limitations make it difficult to establish solvent effects on $\Delta V^{\ddagger}$ values for reactions of classical coordination complexes over a wide range of solvents, in contrast to the range of solvents for which $\Delta V^{\ddagger}$ values for, e.g., the Menschutkin reaction $Et_3N + EtI$ have been determined. Several redox reactions involving cobalt(III) complexes show significant solvent effects on $\Delta V^{\ddagger}$, and $\Delta S^{\ddagger}$, in water and in dipolar aprotic solvents such as acetonitrile, dimethyl sulphoxide, and dimethylformamide.[28] Hydroxide or cyanide attack at low-spin iron(II)-diimine complexes in water is characterized by $\Delta V^{\ddagger}$ values between +10 and +22 $cm^3\ mol^{-1}$,[29] far from the −10 $cm^3\ mol^{-1}$ expected for bimolecular attack. These values may be contrasted with $\Delta V^{\ddagger} = +4$ and $-9\ cm^3\ mol^{-1}$ for cyanide attack at the closely related complex $[Mo(CO)_4(bipy)]$ in methanol and in dimethyl sulphoxide, respectively.[30] It is tempting to interpret these results in terms of extensive desolvation of the attacking OH^- or CN^- on entering the transition state in aqueous solution, somewhat less desolvation in methanol, and negligible desolvation in dimethyl sulphoxide, in which CN^- is only very lightly solvated. There are some fascinating ligand effects on the variation of $\Delta V^{\ddagger}$ with solvent composition for base hydrolysis of low-spin iron(II)-diimine complexes in binary aqueous mixtures, but these take us rather beyond the scope of this book.[31]

While in the area of $t_{2g}^{\,6}$ complexes, we might illustrate the obvious fact that if intrinsic and solvation effects on $\Delta V^{\ddagger}$ or $\Delta S^{\ddagger}$ are equally balanced, then these activation parameters may have zero values. Thus, for example, change of pressure produces negligible changes in rate constants for base hydrolysis of $[Fe(fz)_3]^{4-}$ and of $[Mo(CO)_4(fz)]^{2-}$ and for peroxodisulphate oxidation of the $[Fe(CN)_4(bipy)]^{2-}$ anion.[32] The substrates here are all markedly hydrophilic, from their negative charges and from the presence of either several cyanide ligands or the sulphonated diimine ligand ferrozine (fz, or sometimes ppsa from its systematic name, **4**).

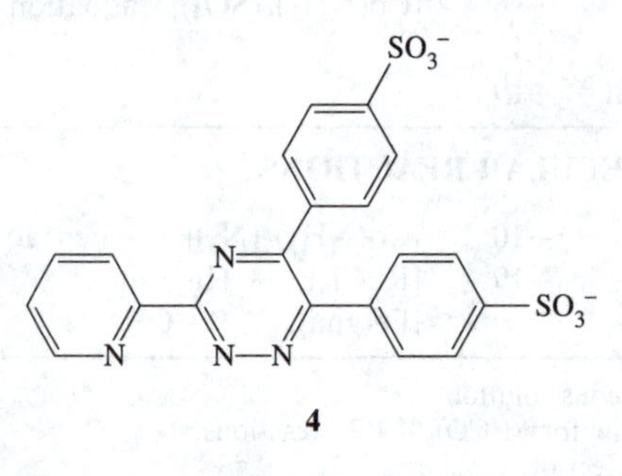

8.2.4 Initial-state and transition-state solvation

Rate constants and activation parameters are composite quantities, representing the *difference* between the initial state and the transition state (Fig. 8.7(a)). To understand solvent effects on kinetic parameters one needs to consider solvation of both the initial state[34] and the transition state.[35] A marked increase in rate constant on going from one solvent medium to another may be due to stabilization (decrease in chemical potential) of the transition state, destabilization of the initial state, or, as shown in Fig. 8.7(b), both. Alternatively, both initial and transition states may be considerably, but almost equally, affected by change of solvent (Fig. 8.7(b)). In such a situation the rate constant changes very little despite the solvation changes. This last case may be exemplified by dissociation of pentacyanoferrates, where, for instance, the rate constant for dissociation of 4-cyanopyridine from $[Fe(CN)_5(4\text{-}CNpy)]^{3-}$ increases by only about 15% on going from water to 80% methanol.[36] However, both the initial and transition states are strongly destabilized (the cyanide ligands are very hydrophilic), by 31 and 30 kJ mol^{-1}, respectively, on transfer from water into 80% methanol. These large transfer chemical potentials should be compared with a total activation barrier, $\Delta G^{\ddagger}$, of 90 kJ mol^{-1} in water.

Kinetic results give information on the activation barrier, and how this changes on transfer from one medium to another. To obtain information on solvent effects on initial and transition states separately it is necessary to have information on the initial state, i.e. the reactant(s)—clearly there is no way of obtaining information on transition-state solvation by direct experiment! The dissection of solvent effects on reactivity into their initial- and transition-state components will now be illustrated for the straightforward bimolecular (S_E2) reaction between tetraethyltin and mercury(II)

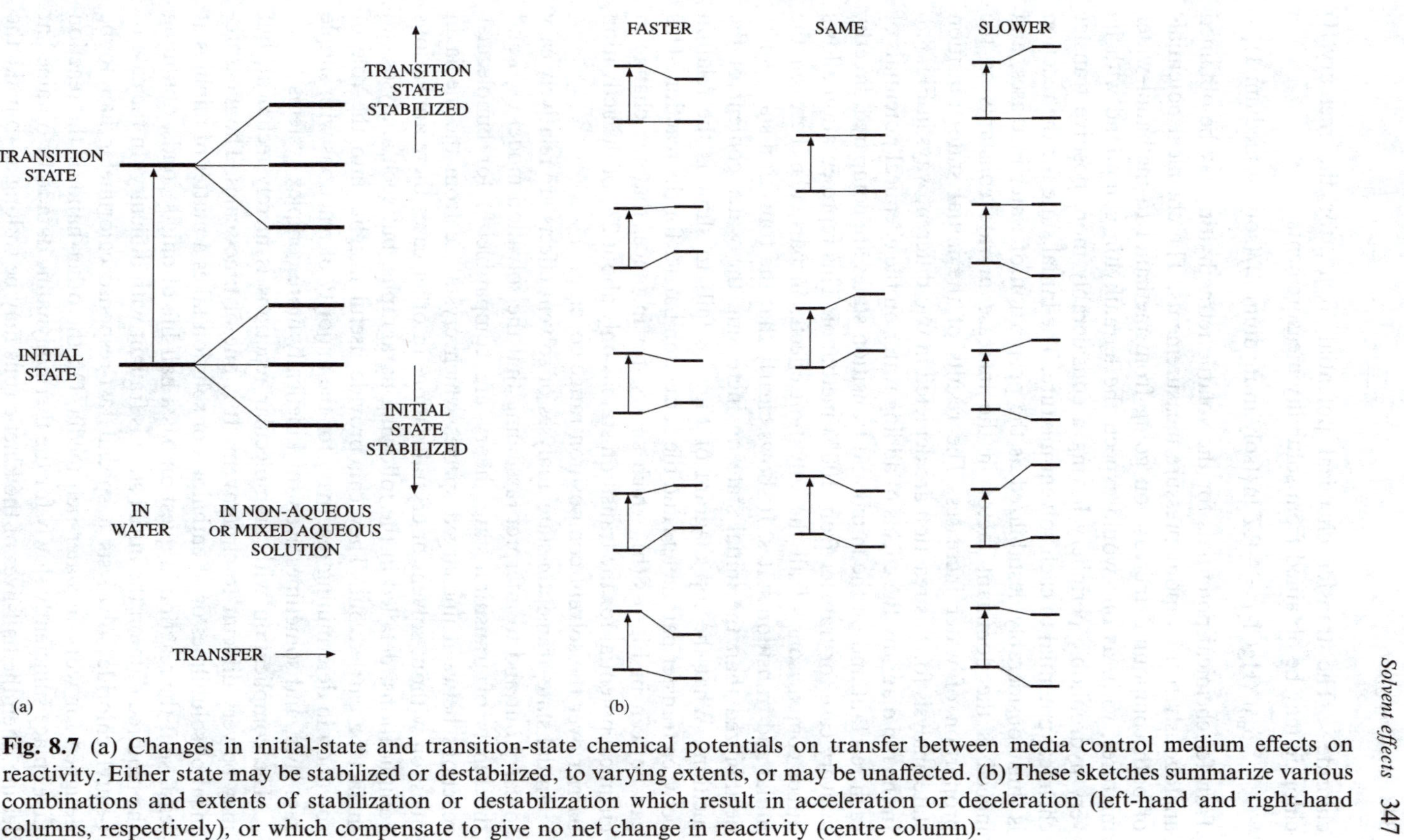

Fig. 8.7 (a) Changes in initial-state and transition-state chemical potentials on transfer between media control medium effects on reactivity. Either state may be stabilized or destabilized, to varying extents, or may be unaffected. (b) These sketches summarize various combinations and extents of stabilization or destabilization which result in acceleration or deceleration (left-hand and right-hand columns, respectively), or which compensate to give no net change in reactivity (centre column).

chloride.[37] The transfer chemical potential, $\delta_m\mu^\circ$, for the mercury(II) chloride may be obtained from solubility measurements:

$$\delta_m\mu^\circ(\mathrm{HgCl_2}) = -RT\ln\{(\text{soly in medium 2})/(\text{soly in medium 1})\}$$

Transfer chemical potentials for the volatile tetraethyltin may be obtained analogously from vapour pressure measurements. The chemical potentials of both compounds increase on going from methanol to methanol–water mixtures, with, as one would expect, the tetraalkyltin compound with its very hydrophobic periphery having a considerably more positive transfer chemical potential to each solvent mixture. The initial state for the reaction is thus considerably destabilized as the proportion of water increases, and indeed is the dominant factor in the increase in rate constants as the proportion of water increases. The results of the initial state–transition state analysis for this reaction are depicted in two different ways in Fig. 8.8. In the top diagram, the effects of adding water on the chemical potentials of each reactant and on the initial and transition states are compared directly, in every case for transfer from 100% methanol. This representation allows direct comparisons of all the solvation effects, in particular between the initial and transition states. It shows clearly that the transition state is less destabilized than the initial state on increasing the water content of the medium. What the top diagram of Fig. 8.8 fails to show is the relation between transfer parameters and the activation barrier for the reaction. This can be seen in Fig. 8.8(b), which shows, for instance, that the change in chemical potential for the transition state is only about 2% of the activation barrier over the solvent composition range covered.

Initial state–transition state analyses of solvent effects are relatively easy to carry out and to interpret (assuming that the solvation model used and the tenets of transition-state theory are compatible![38]) for bimolecular reactions between uncharged species which crystallize from the media in question without solvent of crystallization. In other cases there are various complications, detailed in the following paragraphs, but such analyses can in fact be carried out. They can provide useful insights into the role of solvation in determining reactivity for the majority of reactions with simple rate laws, and sometimes even for kinetically more complex systems.

The complication with unimolecular reactions is merely technical, but may be very difficult to circumvent. It is one of time-scales. The measurement of solubilities or of enthalpies of solution takes a matter of minutes at least, so if the reaction in question has a half-life of only seconds, obtaining the required data on the initial state is fraught with difficulty. In the case of t-butyl chloride solvolysis a special fast-response calorimeter had to be designed for work in water-rich media. On the other hand, if the reaction takes place sufficiently slowly for the thermodynamic data to be obtained at leisure, then the half-lives of the kinetic runs may be so long as to make the obtaining of kinetic parameters as a function of medium composition an impossibly extended process. All this is avoided in bimolecular processes, of course, since the initial-state thermodynamic parameters can be determined

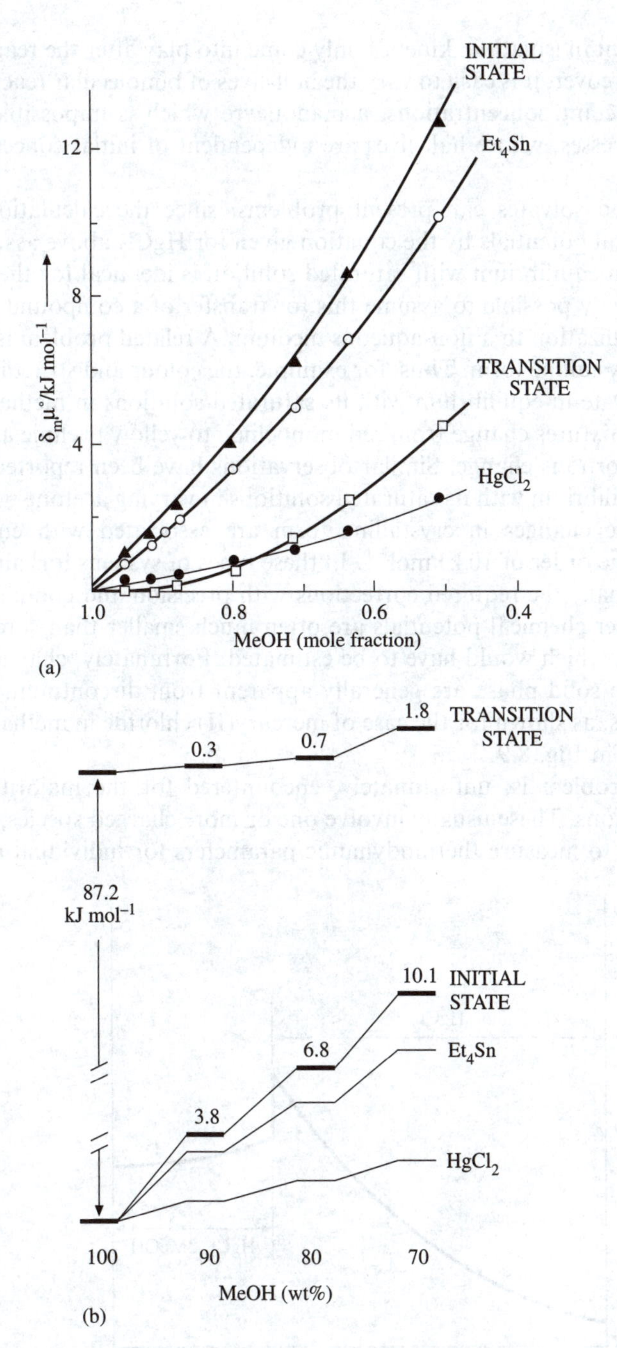

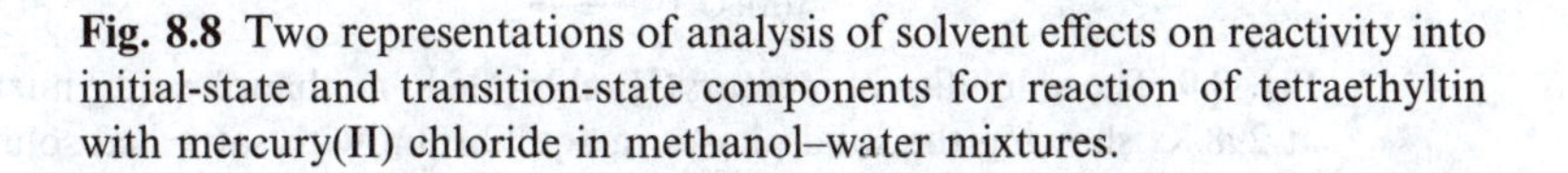
Fig. 8.8 Two representations of analysis of solvent effects on reactivity into initial-state and transition-state components for reaction of tetraethyltin with mercury(II) chloride in methanol–water mixtures.

for each reactant in isolation; kinetics only come into play after the reagents are mixed! Moreover, it is easy to vary the half-lives of bimolecular reactions by varying reactant concentrations, a manouevre which is impossible for first-order processes, whose half-lives are independent of initial concentration.

Hydrates and solvates can present problems, since the calculation of transfer chemical potentials by the equation given for $HgCl_2$ above assumes that the solid in equilibrium with saturated solution is identical for the two media. It is hardly possible to assume this for transfer of a compound with water of crystallization to a non-aqueous medium. A related problem is one of change of crystalline form. Thus, for example, the colour and structure of thallium(I) picrate in equilibrium with its saturated solutions in methanol–water solvent mixtures change from red monoclinic to yellow triclinic as the cosolvent proportions change. Similar observations have been reported for K_2PtBr_6 in equilibrium with its saturated solutions in varying acetone–water mixtures. These changes in crystalline form are associated with energy differences of the order of 10 kJ mol^{-1}. In these types of systems it is almost impossible to make the required corrections with precision and confidence, since the transfer chemical potentials are often much smaller than some of the corrections, which would have to be estimated. Fortunately, changes in solvation in the solid phase are generally apparent from discontinuities in solubility trends, as shown for the case of mercury(II) chloride in methanol–water mixtures in Fig. 8.9.

The third problem is, unfortunately, encountered for the majority of inorganic reactions. These usually involve one or more charged species, and it is impossible to measure thermodynamic parameters for individual ions.

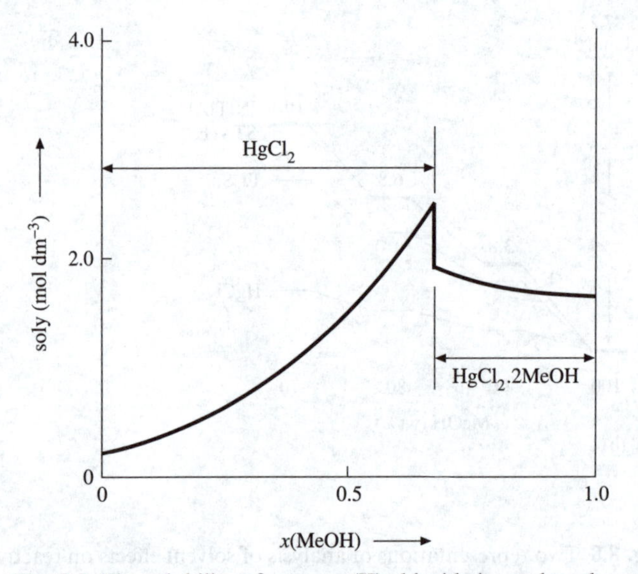

Fig. 8.9 The solubility of mercury(II) chloride in methanol–water mixtures at 298 K, showing the solid phases in equilibrium with saturated solution.

This is not the place to become involved in either the philosophy or the assumptions involved in estimating transfer parameters for ionic species.[39] Suffice it to say that by making extra-thermodynamic assumptions of the type that transfer parameters for K^+ and Cl^- or, better, $AsPh_4{}^+$ and $BPh_4{}^-$ are equal, one can estimate single-ion transfer parameters from measurements on appropriate salts. Though there are often significant differences in numerical values between analyses based on different single-ion extra-thermodynamic assumptions, often similar *qualitative* patterns of solvation effects on reactivity are obtained. The reduction of hexachloroiridate(IV) by iodide provides an example of a successful initial state–transition state analysis of an ion–ion reaction.[40] Here there is the relatively unusual feature that transfer chemical potentials for the transition state can be estimated, since the transition state can be modelled by hexachloroiridate(III). Transfer chemical potentials for $[IrCl_6]^{3-}$ and $[IrCl_6]^{2-}$ can be deduced from solubility measurements on their respective potassium salts. The results of this analysis illustrate the importance of desolvation of hydrophilic anions as the water content of mixed solvents decreases, and of the importance of solvation changes on bringing the two reactants together prior to electron transfer (Chapter 9—Sections 9.5.2.1 and 9.6.1). The large increases in the chemical potential of hydrophilic anions on going from water into mixed and non-aqueous solvents—especially dipolar aprotic media such as dimethyl sulphoxide—cause large increases in rate constants for nucleophilic substitutions. Such effects are particularly marked for fluoride and for hydroxide, but are also significant for, e.g., other halides and for cyanide. An early example involving halide ions was the reduction of dimethyl sulphoxide by iodide in DMSO–water mixtures of varying composition.[41] More recently, there have been striking demonstrations of rapidly increasing rates of base hydrolysis of low-spin iron(II) complexes as the DMSO content of DMSO–water mixtures increases.[42] Several reactions of diimine complexes which are slow in water become too fast to follow by conventional techniques at high DMSO contents, while extremely inert encapsulated analogues (e.g. **5**), which are apparently totally unaffected by hydroxide in aqueous solution undergo base hydrolysis quite rapidly when the DMSO content approaches 100%. This type of behaviour is often put to good effect in preparative chemistry, particularly by organic chemists (e.g.

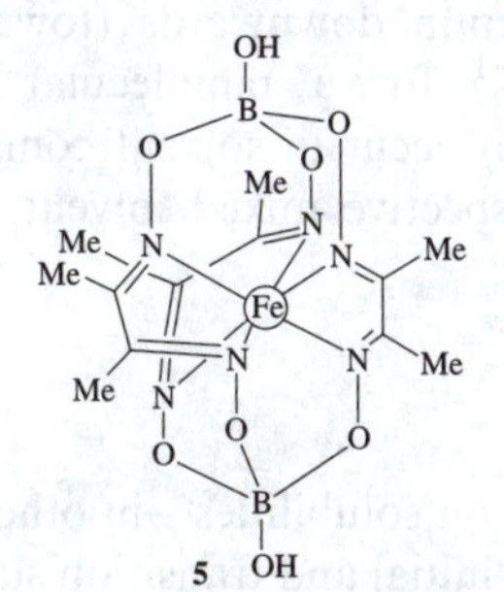

5

'naked' fluoride), and can be useful in the preparation of derivatized ligands. Thus reactions of 3-hydroxy-4-pyranones, such as maltol, with primary amines to give more strongly complexing 3-hydroxypyridinones normally proceed at a satisfactory rate in aqueous alcohol. However, such reactions do not take place at a perceptible rate for amines such as trifluoroethylamine—the N-CF_3CH_2-pyridinone can only be generated in dry acetonitrile or dry dimethyl sulphoxide.

8.2.5 Solvent structure

The exceptional case of glycerol as a very viscous and structured solvent has already been alluded to in Section 8.2.2 above. Solvent structural effects on kinetics can also be important in less extreme media. Water and alcohols are strongly hydrogen-bonded liquids; addition of successive amounts of an alcohol to water generally leads first to an increase in hydrogen-bonding and then a decrease. One might expect that these structural effects would be reflected in trends of kinetic parameters in series of solvents and of solvent mixtures. Such is indeed the case, though the reflection is in the activation parameters $\Delta H^{\ddagger}$, $\Delta S^{\ddagger}$, and $\Delta V^{\ddagger}$, rather than in rate constants (k, i.e. $\Delta G^{\ddagger}$). The best-documented area of study has been that of formation of complexes, particularly from Ni^{2+}.[43] In single solvents the importance of structural effects can be seen on correlating activation enthalpies with enthalpies of vaporization for the respective solvents, while in binary aqueous solvent mixtures kinks in $\Delta H^{\ddagger}$ versus $\Delta S^{\ddagger}$ plots (Fig. 8.10) pinpoint extrema in structural properties. The dependence of activation parameters on solvent composition, represented in the form shown in Fig. 8.11, highlights the relation to solvent structure, in that the extrema of the Fig. 8.11 plots correspond with the compositions of maximum structuredness established from a variety of spectroscopic and thermodynamic data.[44]

A reflection of solvent structural effects in $\Delta V^{\ddagger}$ trends is shown in Fig. 8.12, for base hydrolysis of the simplest and smallest iron(II)-tris-diimine complex, the $[Fe(gmi)_3]^{2+}$ (gmi is MeN=CHCH=NMe) cation.[45] Here the positive activation volumes in water for a bimolecular process betoken a large contribution from hydroxide desolvation on entering the transition state, while the points of inflection downwards (towards the intrinsic $\Delta V^{\ddagger}$ value of *ca* $-10\ cm^3\ mol^{-1}$ for a bimolecular process unencumbered by solvational complications) occur in solvent composition regions where the structuredness of the respective mixed-solvent series is maximal.

8.3 Salt effects[46]

Added electrolytes can have a marked effect on solubilities—in other words on solvation of solutes. If such effects on the initial and transition states of a

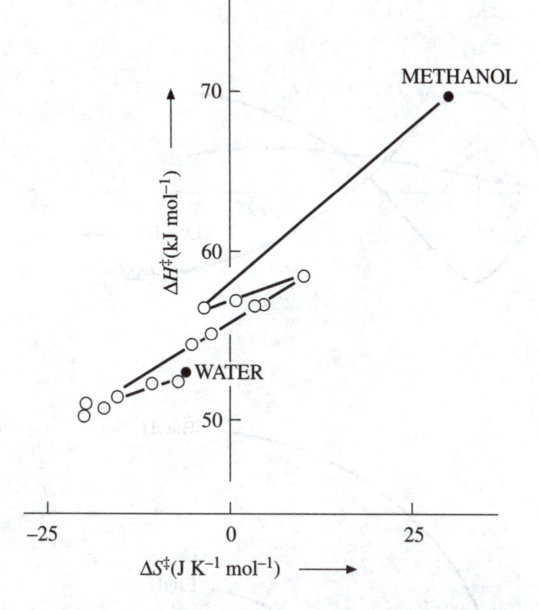

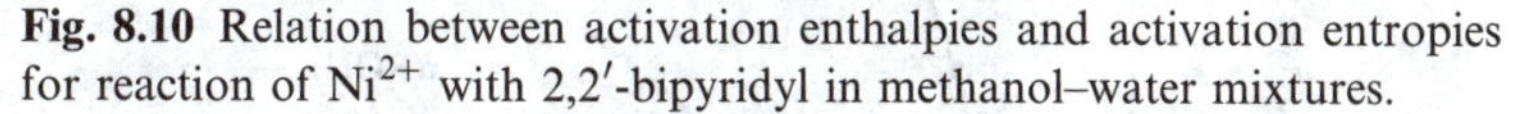

Fig. 8.10 Relation between activation enthalpies and activation entropies for reaction of Ni^{2+} with 2,2′-bipyridyl in methanol–water mixtures.

reaction differ, then this will be reflected in kinetic parameters. The study of effects of added electrolytes on rate constants for reactions in solution goes back more than a century, as does the study of the effects of added electrolytes on solubilities. Setchenow was the first to systematize salt effects on solubilities in aqueous solution;[47] subsequently a fairly extensive and clear pattern was built up for effects of inorganic salts on solubilities of organic solutes.[48] Information on inorganic solutes is very sparse—some data have been available for some time on argon, dioxygen, and on a solitary complex, $[Co(NH_3)_3(NO_2)_3]$.[49] More recently, data have been obtained for a few more inorganic complexes, and for one or two ligands,[50] so that a fuller picture of salt effects on solubilities of such species is beginning to emerge.

The available information is summarized in Fig. 8.13, where it should be noted that the function plotted along the y axis is the negative of that normally used in Setchenow plots—we have plotted log (S/S_0) rather than log (S_0/S) for consistency with related graphs and treatments elsewhere in the present book. Figure 8.13 shows comparisons of both effects of different salts on a given solute and of the given salts KBr and Et_4NBr on a range of solutes. In the former comparisons, it is interesting to note that the (anti)clockwise order of the plots is always the same, regardless of whether the alkali metal salts are salting-in or salting-out the solute; the usually relatively large effects of the tetraalkylammonium salts are clear. In the latter comparisons, added KBr and added Et_4NBr both have larger effects

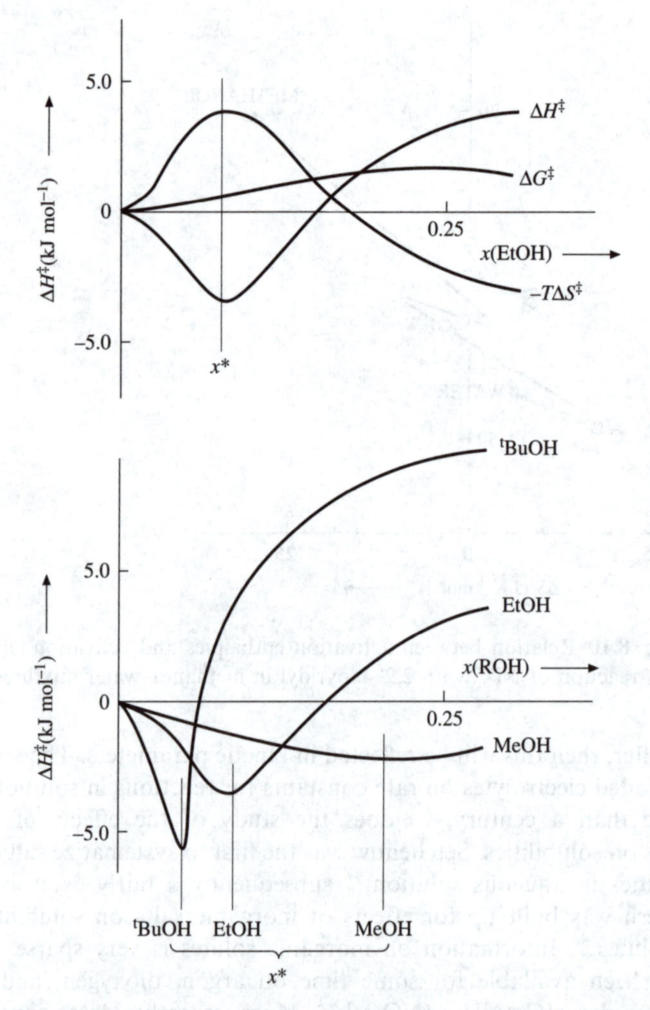

Fig. 8.11 Variation of activation parameters with solvent composition for reaction of Ni^{2+} with 2,2′-bipyridyl in alcohol–water mixtures, mole fraction x(alcohol). The thin vertical lines indicate solvent compositions (mole fraction x^*) at which the structural effects of the added alcohol are maximal.

on the inorganic complexes than on the organic and simple inorganic (Me_3NSO_3) solutes. It is clear that added salts will affect solvation (hydration) of inorganic complexes, are likely to affect transition states for their reactions, and may be expected to have significant effects on kinetic parameters. In the following discussion the effects of simple inorganic salts will be considered first (Section 8.3.1), followed by a consideration of the often bigger tetraalkylammonium salts (Section 8.3.2), which in turn lead to micelles and microemulsions (Section 8.4).

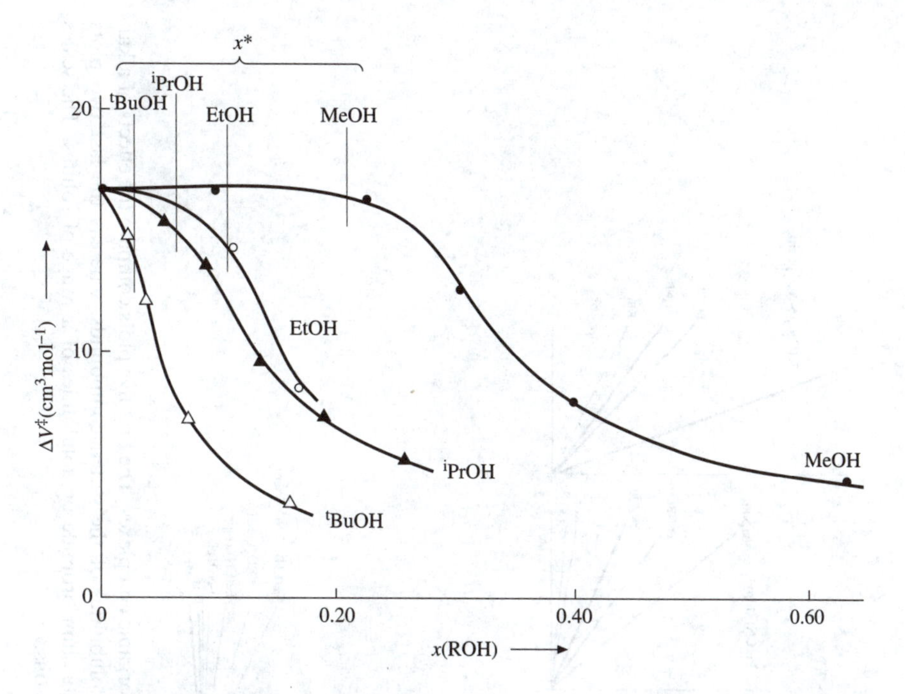

Fig. 8.12 Dependence of activation volumes on solvent composition for base hydrolysis of $[Fe(gmi)_3]^{2+}$ in binary aqueous mixtures, mole fraction x(ROH). The thin vertical lines indicate solvent compositions (mole fraction x^*) at which cosolvent structural effects are maximal.

8.3.1 Classical salt effects

8.3.1.1 History

As mentioned above, kinetic salt effects have been studied for over a century, with much investigation of so-called 'cation catalysis' of reactions between organic substrates and such inorganic nucleophiles as hydroxide and thiocyanate in the opening decades of the present century. It was Brønsted who, in 1922, recognized in his treatment of primary salt effects that the observed kinetic trends could be rationalized in terms of activity coefficients of the initial state and of the transition state.[51]

8.3.1.2 Debye–Hückel and Brønsted–Bjerrum

In 1925, Brønsted,[52] following hints by Bjerrum,[53] incorporated the Debye–Hückel theory on activity coefficients for solutes into his kinetic analysis.[54] The Debye–Hückel limiting law applies only to very dilute solutions, but the extended Debye–Hückel theory and equation can be used for analysis of kinetic data for reactions carried out under concentration conditions more normal in a kinetics laboratory.

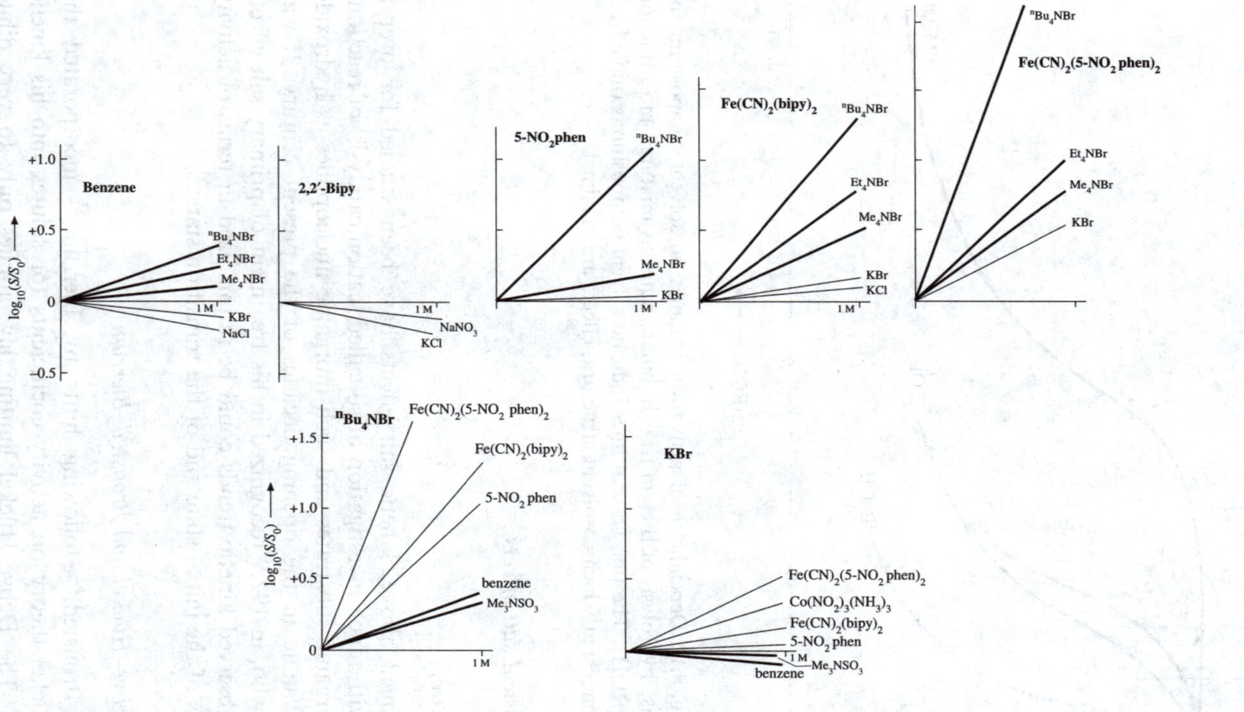

Fig. 8.13 Salt effects on solubilities of uncharged compounds and inorganic complexes. The first five plots compare the effects of alkali metal (thin lines) and tetraalkylammonium (thick lines) salts on solubilities of the stated compounds, whereas the last two plots compare the effects of tetra-n-butylammonium bromide and of potassium bromide on solubilities of a range of solutes. The scales for the x and y axes are the same for all plots to facilitate comparisons.

8.3.1.3 Livingston diagrams

In 1927, Brønsted and Livingston applied the former's analysis to base hydrolysis and mercury(II)-catalysed aquation of $[Co(NH_3)_5Cl]^{2+}$. Luckily for them the two-stage nature of both of these reactions caused no complications—they found good agreement between experiment and prediction, and Livingston[55] went on to elaborate the diagram that bears his name. The classic version of the Livingston diagram is shown as Fig. 8.14;[27] a more up-to-date version (based on extended rather than simple Debye–Hückel theory) is shown as Fig 8.15.[56] In the early days such diagrams were of most interest as rationalizations of observed salt effects. Later their use was often recommended for determining the charge product for reactants forming the transition state of a bimolecular reaction, though examples of situations where such determinations were carried out successfully and found truly informative are rather rare. In recent years Livingston diagrams have come into their own, proving very useful in estimating local charges at electron transfer sites on the surfaces of metalloproteins. This type of application will be described in detail in Chapter 9 (see Section 9.5.2.2).

8.3.2 Specific ion effects

8.3.2.1 General features

The salt effects discussed in the preceding section are non-specific; Livingston diagrams (ancient or modern) are based simply on ionic strength effects. Consideration of the facts presented at the start of Section 8.3 suggests that if one is dealing with a wide range of ions, particularly if ions of the tetraalkylammonium type are included, then there may well be specific ion effects in addition to the general Debye–Hückel effects just discussed. Olson and Simonson[57] are usually credited with pointing this out, in relation to the iodide–peroxodisulphate reaction, but that specific ion effects might well be superimposed on classical electrostatic effects had actually been adumbrated rather earlier.

To start with a simple, all-inorganic system, effects of added cations M^{n+}aq on hydrolysis of pyrosulphate, $S_2O_7^{2-}$, have been documented for $n = 1$ (M = Li → Cs), $n = 2$ (Mg, Mn, Cd), $n = 3$ (M = La, $[Co(NH_3)_6]$), and $n = 4$ (M = Th).[58] The trends established, for the rather small effects observed, reflect the effects of charge and size in the expected manner. Better illustrations of specific ion effects tend to be provided by comparisons of the effects of added inorganic and organic cations on reactions involving organic substrates. Remarkably different effects of the normally well-behaved alkali metal cations are well illustrated by base hydrolysis of methyl acetate (Fig. 8.16(a)). Effects of a much wider range of cations are shown in Fig. 8.16(b), this time for base hydrolysis (depolymerization) of so-called diacetone alcohol, $Me_2C(OH)CH_2COCH_3$. The alkali metal cations have small and similar effects this time, but Tl^+ has a much bigger

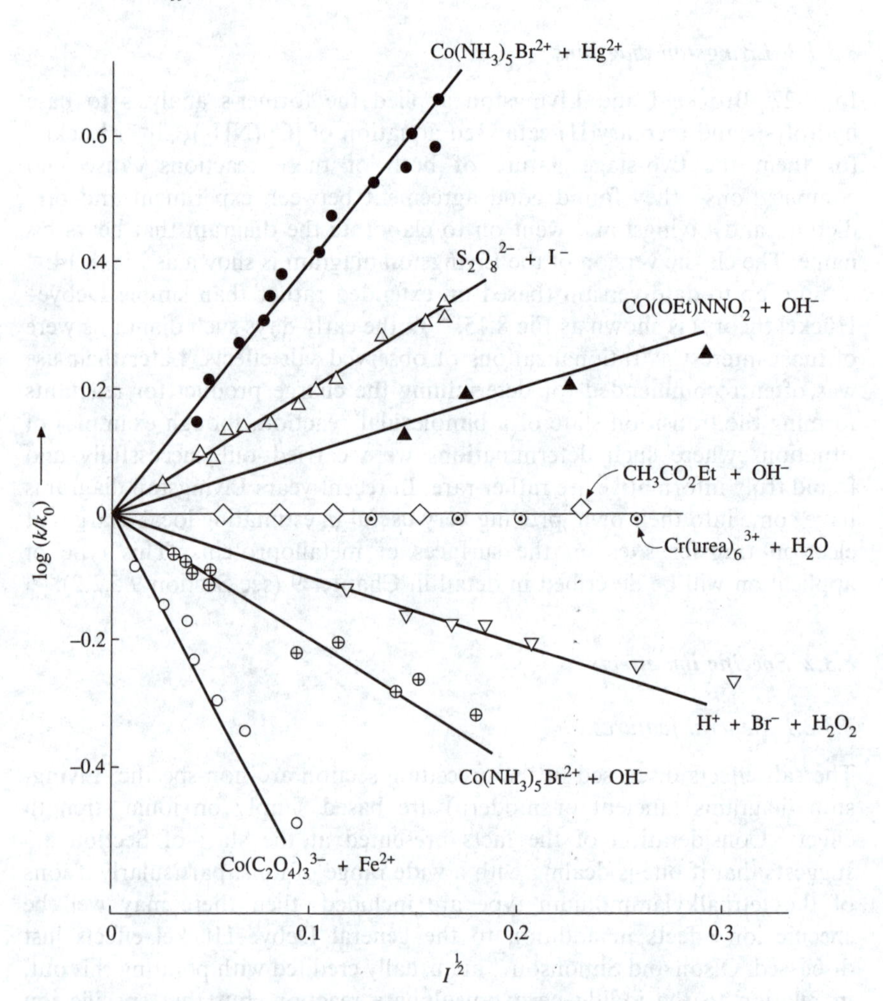

Fig. 8.14 Livingston diagram: original format.

effect, as do the 2+ cations of calcium and barium. The complex cation $[Co(NH_3)_6]^{3+}$ has an effect midway between Ca^{2+} and Ba^{2+}, which is perhaps where one might expect it to come on simple electrostatic grounds. The tetraalkylammonium cations, on the other hand, have large accelerating effects, in marked contrast to the simple cations.

Returning to specific salt effects in inorganic systems, reactions such as oxidation of iodide by hexacyanoferrate(III) or by hydrogen peroxide,[59] and isomerization of the diaquabis(oxalato)chromium(III) anion (Fig. 8.17), have been extensively studied. Large effects of cation charge, and generally opposite effects of organic (tetraalkylammonium) and inorganic (alkali metal and alkaline earth) cations, are apparent for all these reactions. Figure 8.18 shows further examples of markedly differing effects of these two types of added salts on rate constants for a varied selection of inorganic

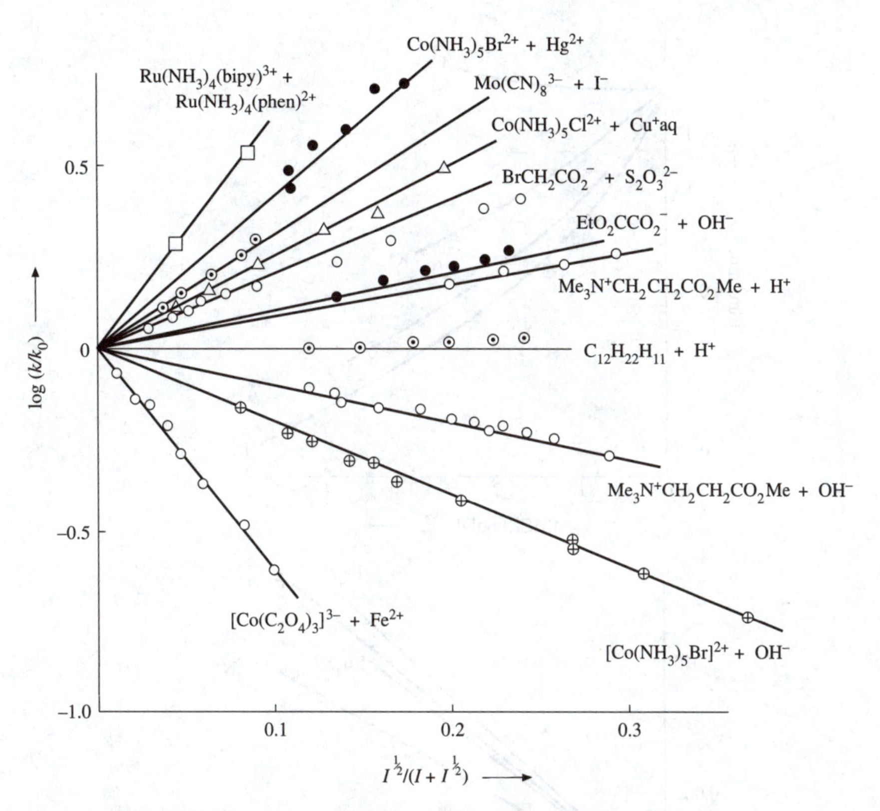

Fig. 8.15 Livingston diagram: Pethybridge and Prue format.

reactions. In contrast to the pattern common to the reactions mentioned so far, for some reactions effects of added R_4N^+ are very similar to those of alkali metal cations.[60] Thus, for example, 0.015 mol dm^3 Me_4N^+, Et_4N^+, and K^+ all cause an acceleration of about 50% in the monobromoacetate plus thiosulphate reaction, long popular as a model bimolecular substitution essentially free from side reactions[61]

$$BrCH_2CO_2{}^- + S_2O_3{}^{2-} = S_2O_3CH_2CO_2{}^{2-} + Br^-$$

The effect of K^+ is intermediate between those of the two R_4N^+ cations.

It should be borne in mind that the salt effects under consideration here may be the result of one or more of a number of factors, involving various ion–ion and ion–solvent interactions, for the reactant(s) and the added electrolyte(s). At one extreme, an added electrolyte may affect reactivity indirectly, by the effect of its constituent ions on water activity and thence on hydration of the reactant(s) and/or transition state. At the other extreme, one or other of the ions of the added electrolyte may participate in extensive ion-pairing with one or other of the reactants. This has been known for many years[62] and extensively studied for a range of anion–anion reactions, including manganate(VI)/(VII)[63] and hexacyanoferrate(II)/(III)[64] exchange.

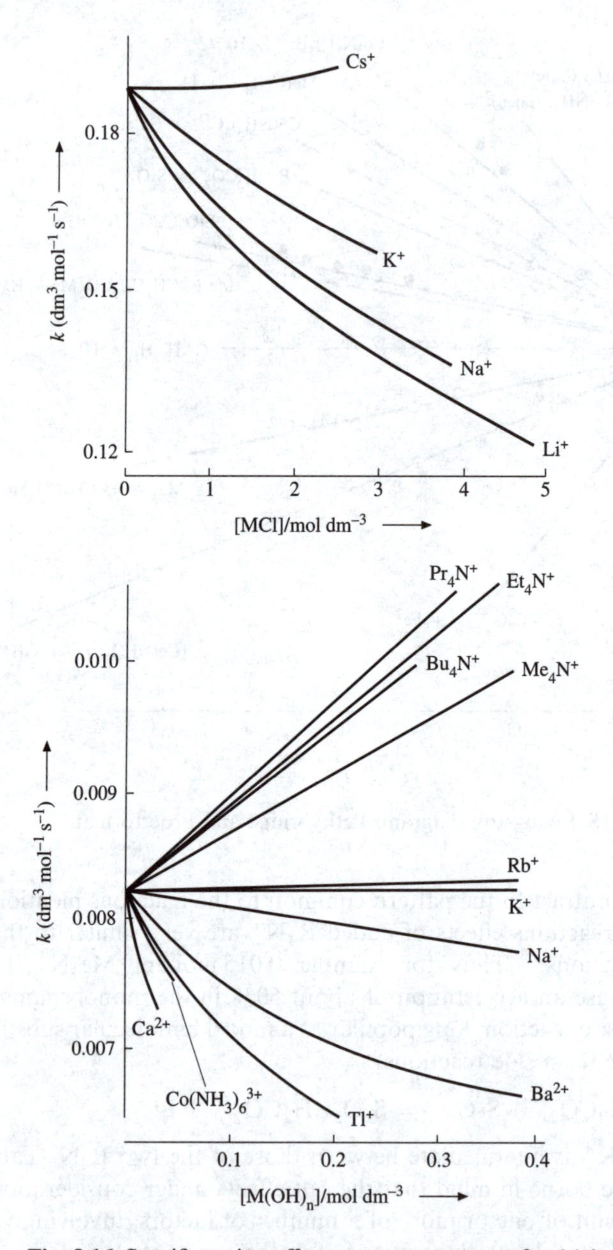

Fig. 8.16 Specific cation effects on rate constants for (a) base hydrolysis of methyl acetate and (b) depolymerization of diacetone alcohol, in both cases at 298 K.

Thus peroxodisulphate oxidation of hexacyanoferrate(II) is markedly affected by the nature and concentration of cations present, with reaction between the respective potassium salts involving $KS_2O_8^-$ oxidation of $KFe(CN)_6^{3-}$ as a major pathway.[65] The peroxodisulphate–

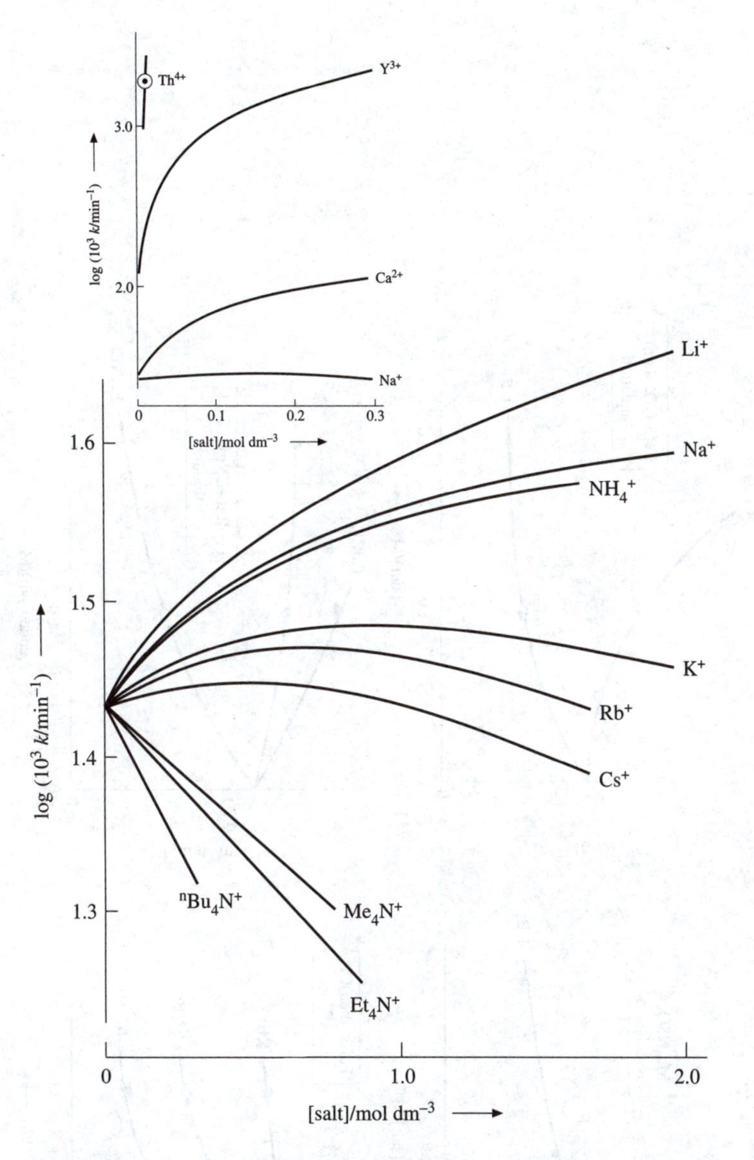

Fig. 8.17 Salt effects on rate constants for isomerization of the $[Cr(C_2O_4)_2(H_2O)_2]^-$ anion, at 298 K. In all cases the added salt was the nitrate.

hexacyanoferrate(II),[66] peroxodisulphate–iodide,[67] and above-mentioned thiosulphate–bromoacetate[68] reactions have probably been the most studied in terms of the range of potentially associating cations added. In the thiosulphate–bromoacetate case, as in a number of other reactions, Brønsted behaviour is observed at low ionic strengths for added 1:1 electrolytes, but progressively larger deviations are observed on increasing

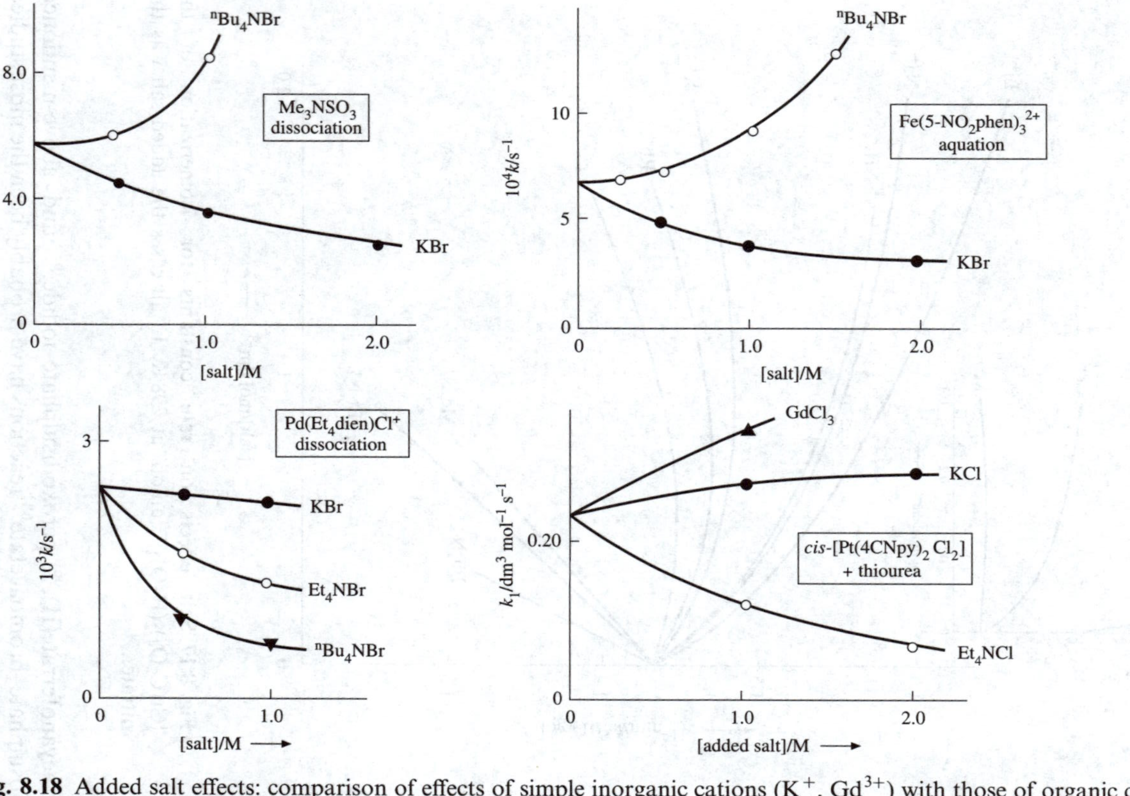

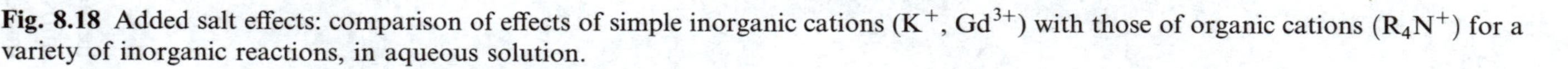
Fig. 8.18 Added salt effects: comparison of effects of simple inorganic cations (K^+, Gd^{3+}) with those of organic cations (R_4N^+) for a variety of inorganic reactions, in aqueous solution.

cation charge ($M^+ \rightarrow Mg^{2+}, Ca^{2+}, Ba^{2+} \rightarrow La^{3+}$) or electrolyte concentration. Ion association in this system has been encouraged by carrying out the reaction in the presence of added electrolytes in propanol–water solvent mixtures. More recently, the importance of ion-pairing has been demonstrated for peroxodisulphate oxidation of octacyanomolybdate(IV)—the 4- charge on the latter and the use of particularly strong electrolyte solutions both encouraged extensive ion-pairing here.[69]

8.3.2.2 Initial-state–transition-state analysis

As in the case of solvent and mixed-solvent effects on kinetics (Section 8.2), so also in the case of salt effects it is in principle possible to analyse reactivity trends into initial state and transition state contributions. Unfortunately, this can at the moment only be done with much confidence for unimolecular and bimolecular reactions involving, respectively, one and two uncharged species, which is rather constricting to the inorganic chemist. Transition-state solvation is undoubtedly important in determining salt effects on reactivity for a number of reactions involving charged complexes. This has been established qualitatively for reactions such as linkage isomerization of $[Co(NH_3)_5(ONO)]^{2+}$ and redox reactions such as iodide plus bromate and peroxodisulphate oxidation of $[Fe(phen)_3]^{2+}$.[70] On the other hand, salt effects on intramolecular electron transfer within the μ-pyrazine carboxylato complex $[(H_3N)_4Co(pzc)Fe(CN)_5]^-$ seem to operate primarily on the initial state, reflecting mainly the variation in redox power of the cobalt portion of the binuclear complex.[71] The problem in relation to carrying out a quantitative initial state–transition state analysis lies in the scarcity of data for separating transfer chemical potentials for electrolytes into ionic components. We shall here merely show results for two inorganic reactions[72] (Fig. 8.19 (a) and (b)) to show that such analyses are possible, and for hydroxide attack at t-butyl chloride (Fig. 8.19 (c)) to show the similarity of patterns for chloride leaving from an inorganic centre (Pt) and from carbon.

8.4 Organized media[73]

It is only a small step from $[^nBu_4N]Br$ to $[(C_{16}H_{33})Me_3N]Br$—from $C_{16}H_{36}N^+$ to $C_{19}H_{42}N^+$—but the effect on rate constants can be enormous. Whereas the R_4N^+ cations have a marked but modest effect on reactivities through their effects on the structure of solvent water, ions such as $[(C_{16}H_{33})Me_3N]^+$ and anionic equivalents such as $[C_{12}H_{25}OSO_3]^-$ can have very large effects on reactivity through the formation of micelles, on whose surface appropriate reactions can undergo dramatic rate enhancements. Such effects are considered in the next section, while reactivities in another type of 'organized aqueous media', namely microemulsions, whose function is often that of solubilization, are considered in the following section.

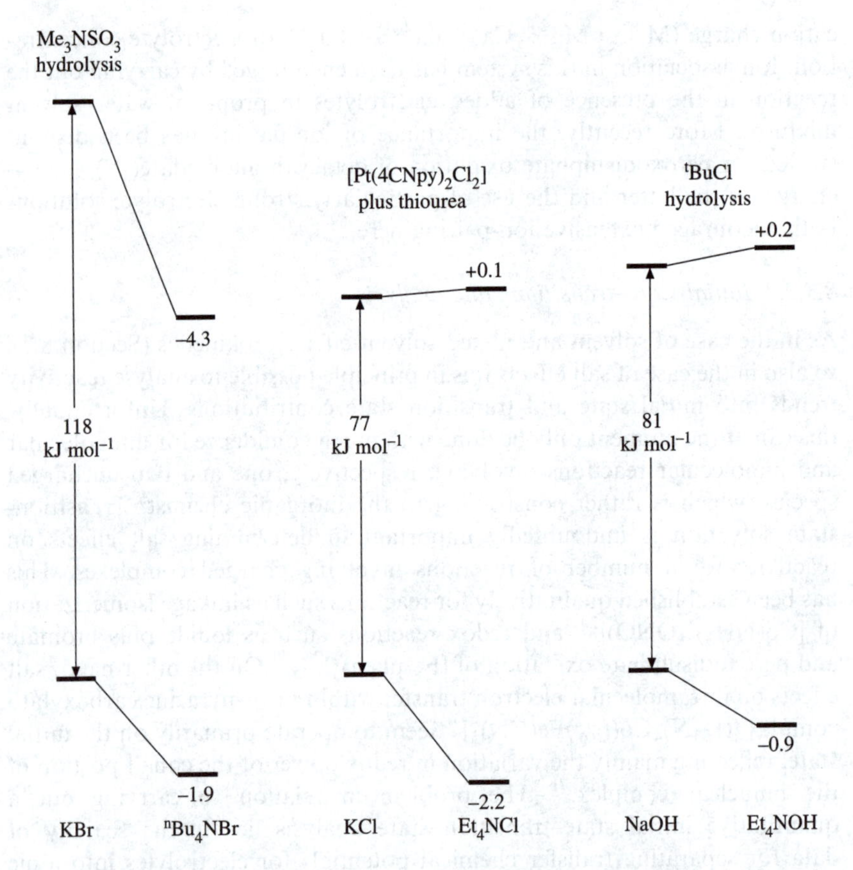

Fig. 8.19 Initial state–transition state dissections of salt effects on reactivity. In each case the left → right comparison involves transfer from an alkali salt or hydroxide to its tetraalkylammonium analogue; in all cases the salt concentration is $1.0\,\text{mol dm}^{-3}$.

8.4.1 Micelles

Some examples of surfactants which form micelles in aqueous media are given in Table 8.5, while some types of micelle structure are shown in Fig. 8.20. Diameters of micelles are generally in the range 30–60 Å, though the recently examined micellar 'clumps' (hexadecanoate/MeCN/phosphate buffer), in which a 10^8-fold acceleration of hydrolysis of the amide linkage in $C_{12}H_{25}N^+Me_2CH_2CONHC_6H_4NO_2$ is observed in the presence of $10^{-5}\,\text{mol dm}^{-3}$ catalyst,[74] must be larger. Figure 8.20 also gives an idea of the relation of micelles on the one hand to monolayers on surfaces and on the other hand to biological membranes. The dependence of rate constant on surfactant concentration may be of the form shown in Fig. 8.21, with an abrupt change in rate constant trend in the region of the critical micelle concentration (c.m.c.) of the surfactant. The rate enhancement for the

Table 8.5 Common surfactants, with their critical micelle concentrations (c.m.c.) and aggregation numbers (N) at 298.2 K.

			10^3 c.m.c.	N
ANIONIC				
Sodium hexadecylsulphate	$Na[Me(CH_2)_{15}(OSO_3)]$	SHDS, NaHDS	0.52	100
Sodium dodecylsulphate	$Na[Me(CH_2)_{11}(OSO_3)]$	SDS, NaLS [a]	8.1	62
Sodium decanoate (laurate)	$Na[Me(CH_2)_{10}(CO_2)]$		24	56
Sodium perfluorooctanoate	$Na[F_3C(CF_2)_6(CO_2)]$	NaPFO	36	23
Sodium dodecylbenzenesulphonate	$Na[Me(CH_2)_{11}-C_6H_4-OSO_2]$	NaDBS, SDBS	1.6	
CATIONIC				
Hexadecyltrimethylammonium bromide	$[Me(CH_2)_{15}NMe_3]Br$	CTAB [b]	0.9	61
Dodecyltrimethylammonium bromide	$[Me(CH_2)_{11}NMe_3]Br$	DTAB	65	47
Hexadecylpyridinium chloride	$[Me(CH_2)_{15}-NC_5H_5]Cl$	CPC	0.9	95
ZWITTERIONIC				
Dodecyldimethylammonium acetate (n-dodecyl betaine)	$Me(CH_2)_{11}NMe_2CH_2CO_2H$	DoDAA	*ca* 2	73
UNCHARGED				
1-(1,1,3,3-Tetramethylbutyl)-4-polyoxyethylene(9.5)benzene (Triton X-100)	$Me_3CCH_2CMe_2-C_6H_4-(OCH_2CH_2)_{9-10}OH$	TX-100	0.24	143

[a] Dodecyl = lauryl.
[b] Hexadecyl = cetyl.

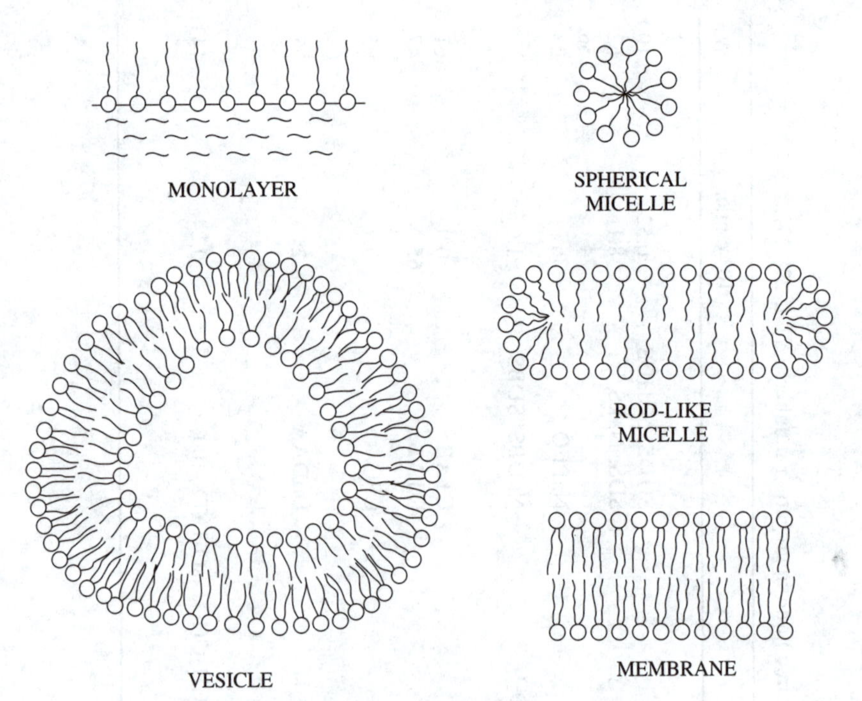

Fig. 8.20 Organized media—from monolayers to membranes.

Ni^{2+}aq plus pada reaction of Fig. 8.21 is up to about 1000 times.[75] Micellar effects on the Pd^{2+}aq plus glutathione reaction, in the presence of SDS, are much less, but exhibit the interesting feature of a crossover from acceleration to retardation in the pH region 3–4. This may be attributed to the change in the charge of the glutathione ($pK_2 = 3.5$).[76]

Reactions between pairs of cations or pairs of anions may be dramatically accelerated by appropriately charged surfactants or polyelectrolytes. At the simplest level such very large rate enhancements can be attributed to the large increase in local concentrations of ions at the micelle–water interface when the charges on the two reactants are of the same sign, opposite to that on the surfactant head groups (Fig. 8.22). Such an explanation could hold for, e.g., Hg^{2+} catalysis of dissociation (aquation) of the $[Co(NH_3)_5Br]^{2+}$ cation (approaching 10^6-fold enhancement),[77] or of Fe^{2+}aq reduction of $Fe(diimine)_3{}^{3+}$ cations in the presence of SDS (up to 10^4-fold enhancement).[78] Rate enhancements for cerium(IV) oxidation of $Fe(diimine)_3{}^{3+}$ cations depend markedly on the hydrophobicity of the coordinated diimine.[79] In reactions involving lipophilic complexes, micellar effects on rates may be attributable largely to favourable interactions between lipophilic complex and lipophilic micelle centre. In general, reactivities can be rationalized with the aid of the so-called pseudophase model, due to Menger and Portnoy,[80] and thereby related to Brønsted's

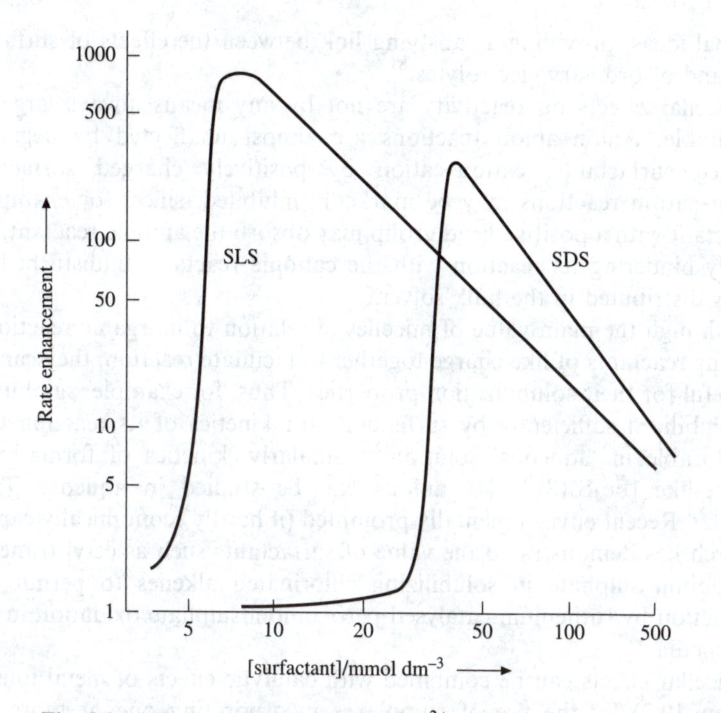

Fig. 8.21 Micellar effects on the Ni^{2+} plus pada reaction. The rate enhancement maxima correspond to the critical micelle concentrations of the respective surfactants; SDS = sodium decyl sulphate and SLS = sodium dodecyl (lauryl) sulphate.

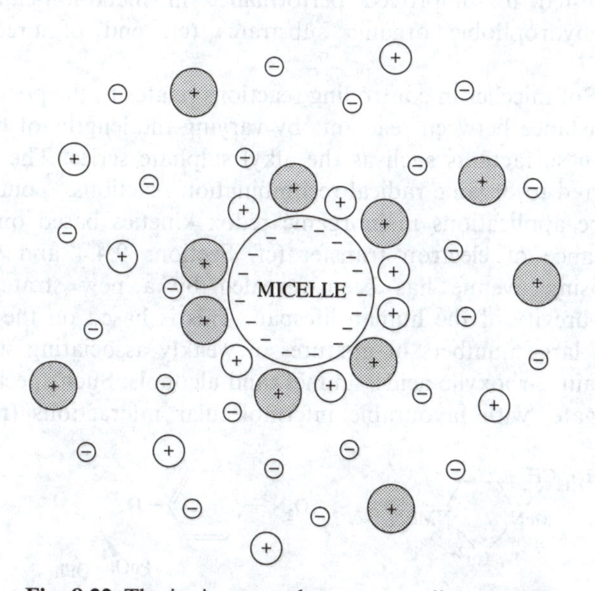

Fig. 8.22 The ionic atmosphere surrounding a micelle.

original ideas, providing a satisfying link between the effects of surfactant salts and of ordinary electrolytes.[81]

Micellar effects on reactivity are not by any means always large and favourable. Anion–anion reactions are almost unaffected by negatively charged surfactants, cation–cation by positively charged surfactants. Anion–cation reactions may be markedly inhibited, since, for example, a surfactant with a positive head group may absorb the anionic reactant, thus greatly hindering its reaction with the cationic reactant, undisturbed and evenly distributed in the bulk solvent.

Although the main value of micelles in relation to inorganic reactions is to bring reactants of like charge together to facilitate reaction, they can also be useful for their solubilization properties. Thus, for example, sulphur can be solubilized sufficiently by surfactants for kinetics of its reaction to be monitorable in 'aqueous' solution.[82] Similarly, kinetics of formation of cubane-like $[Fe_4S_4(SPh)_4]^{2-}$ anions can be studied in aqueous Triton X-100.[83] Recent environmentally prompted (if hardly economically capped) research has demonstrated the value of surfactants such as cetyl trimethylammonium sulphate in solubilizing chlorinated alkenes to permit their destruction by ruthenium-catalysed peroxomonosulphate oxidation in aqueous media.[84]

Micellar effects can be combined with catalytic effects of metal ions (cf. Section 10.2) by the use of complexes incorporating one or more long hydrocarbon chains. Thus the complex **6** produces a 10^5-fold enhancement of hydrolysis of the 4-nitrophenyl phosphate derivative **7**. The catalytic effect was suggested to be apportionable in a ratio of approximately 10 to 1 for the micelle compared with the Cu^{2+}.[85] The use of surfactant-like ligands has been suggested for improved performance in metal-ion-catalysed oxidations of hydrophobic organic substrates (cf. end of preceding paragraph).[86]

A further use of micelles in controlling reactions relates to the possibility of tuning the distance between reactants by varying the lengths of hydrocarbon chains in surfactants such as the alkyl sulphate series. The initial suggestion referred to organic radical recombination reactions,[87] but there should surely be applications in inorganic redox kinetics based on such control of distance of electron transfer (cf. Sections 9.4.4 and 9.6.5). Another promising avenue has been revealed by a new strategy to circumvent the brevity of the human lifespan. This is based on the rapid investigation of large numbers of mixtures of weakly associating species, such as long-chain carboxylic acids, amines, and alcohols. Such species are likely to aggregate, with favourable intermolecular interactions (mixed-

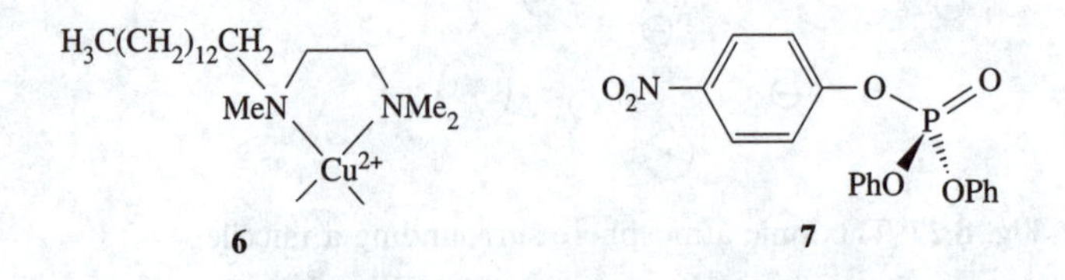

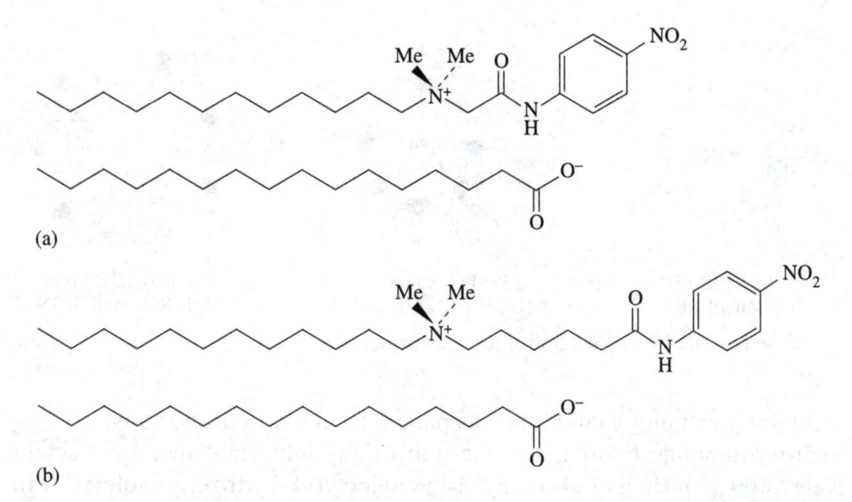

Fig. 8.23 (a) Favourable interaction between a long-chain alkylammonium amide and hexadecanoate; (b) introduction of $-(CH_2)_n-$ spacer leading to loss of hexadecanoate catalysis.

micelle formation) possibly leading to large increases in reactivity. Hexadecanoate reduces the time-scale for hydrolysis of the amide $C_{12}H_{25}N^+Me_2CH_2CONHC_6H_4NO_2$ from over a year to a few minutes. The favourable interaction responsible is sketched in Fig. 8.23(a); replacement of –NH– {or O} by $-(CH_2)_nNH-$ {or by $-(CH_2)_nO-$} separates the influence from the reaction centre [Fig. 8.23(b)] and nullifies the catalytic effect of the hexadecanoate. Replacement of hexadecanoate by the shorter chain dodecanoate increases the time-scale to a matter of hours.[74] Again the initial relevance is to organic chemistry, especially to enzyme catalysis,[88] but the inorganic potential is clear.

8.4.2 Microemulsions

The value of microemulsions usually resides in their solubilization properties—in their ability to dissolve recalcitrant compounds to give a microheterogeneous optically transparent system. Indeed, microemulsions combine the advantages of homogeneity with those of phase transfer conditions, as stated in the report of yield optimization for palladium dichloride catalysis of the reaction between styrene and hydrogen peroxide.[89] The relation of microemulsions to micelles is shown in Fig. 8.24, which shows both oil-in-water and water-in-oil microemulsions. Although the microemulsion appears uniform to the eye, in practice there is a real possibility of potential reactants being partitioned between different microregions. Reactants may be thus solubilized but separated, and hence discouraged from reacting.

Formation reactions from Ni^{2+}aq have been fairly extensively studied;

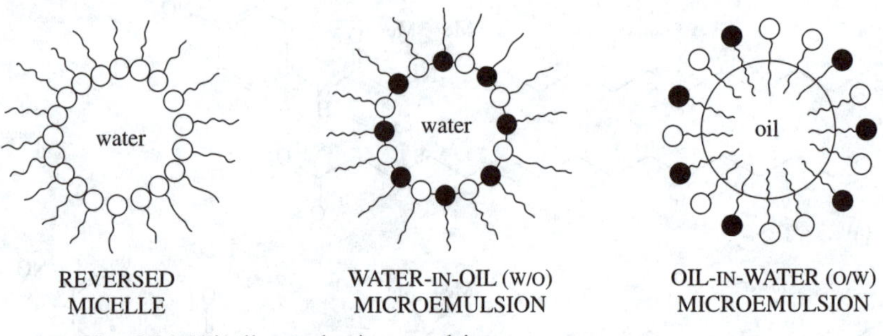

Fig. 8.24 Micelles and microemulsions.

including incoming ligands such as pada (cf. micellar effects, above),[75,90] 8-hydroxyquinoline (oxine), and the industrially important metal extractant Kelex 100 (which is 7-(4-methyl-1-methyloctyl)-8-hydroxyquinoline).[91] In all cases the mechanism remains as in water, dissociative interchange within the pre-association complex. In some cases there are large rate enhancements due to locally high concentrations of the reactants, but for the reaction with Kelex 100 there was disappointingly little increase in rate. Incorporation of M^{2+} ions into porphyrins can be greatly accelerated in microemulsions, as in the case of Cu^{2+} in alcohol-rich water–toluene–2-propanol media.[92]

Substitution at square-planar centres can be exemplified by the study of reactions of $[Pd(LL)(en)]^{2+}$, where LL = 2,2′-bipyridyl or its 4,4′-dimethyl derivative, with 1,2-ethanediamine or its *N*,*N*-dimethyl derivative in heptane–AOT–water microemulsions (AOT = the surfactant sodium bis(2-ethylhexyl)sulphosuccinate). Reaction takes place at the AOT/water interface, with rate enhancements much affected by steric factors.[93]

Base hydrolysis of iron(II)-diimine complexes, as of organic substrates such as 2,4-dichloronitrobenzene, is generally accelerated in microemulsion media, often by factors of more than a thousand-fold. Moreover, some of the reactions of the iron(II) complexes exhibit a change of rate law on going from water or binary aqueous solvent mixtures into microemulsions. There is the additional complication that some of the complexes of this type are large, and may have volumes comparable with the usual droplet size in microemulsions.[94] Kinetic studies of related $t_{2g}{}^6$ molybdenum(0)-diimine-carbonyl complexes, such as $Mo(CO)_4(bipy)$, are facilitated by their solubility in microemulsions—in contrast to their extreme insolubility in water.[95]

Whereas there have been many investigations of micellar effects on electron transfer reactions (cf. preceding section), there have been rather few studies of redox reactions in microemulsions. The interested reader can find a discussion of the factors operating in the reports of, for example, hexachloroiridate(IV) oxidations of benzene diols and of the polynuclear cobaltitungstate $[CoW_{12}O_{40}]^{5-}$.[96] Again there can be large rate enhancements, as in the 10^4-fold increase in the rate of oxygenation of Fe^{2+} aq.[97]

In certain cases, reactions in microemulsion media take place at the interfaces between the microdomains. It is therefore relevant to mention a couple of recent investigations of kinetics at macro (i.e. normal) interfaces. That of kinetics of complexation of Cu^{2+} by the ligand **8** at the water/1,2-dichloroethane interface is obviously connected to some of the systems just mentioned.[98] That of reaction of $[Co(NH_3)_5Cl]^{2+}$ with *N*,*N*-dimethyl-*p*-phenylenediamine at silver or gold surfaces marks the connection to liquid/solid interfaces.[99]

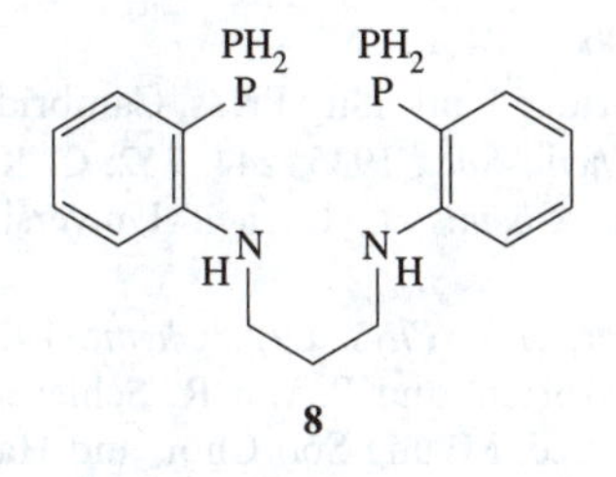

8.4.3 Gels

In a sense, gels are one of the most organized media, approximating to 'solid liquids'. However, the water which is enclosed in the multitudinous cavities in the apparently solid gel actually has properties very similar to ordinary bulk water.[100] This is reflected in the really remarkably small differences between rate constants for reactions in gels and in bulk water. Thus both for aquation of $[Fe(5NO_2phen)_3]^{2+}$ and for cyanide attack at $[Fe(5Mephen)_3]^{2+}$ there is less than 20% difference in rate constants on going from water to aqueous gels. There seems to have been very little interest in recent years in kinetics of inorganic reactions in gel media, but the formation reaction of Ni^{2+}aq with dithio-oxamide and related ligands has been studied in a gelatin matrix with a view to establishing medium effects on products and their distribution.[102]

References

1. J. Burgess and E. Pelizzetti, *Progr. React. Kinet.*, 1992, **17**, 1, provides an extensive and exhaustively referenced (860 references) review of the ground covered in this chapter; a much abbreviated version (15 pages) may be found in J. Burgess and E. Pelizzetti, *Gazz. Chim. Ital.*, 1988, **118**, 803.
2. E. Grunwald and S. Winstein, *J. Am. Chem. Soc.*, 1948, **70**, 846; A. H. Fainberg and S. Winstein, *J. Am. Chem. Soc.*, 1956, **78**, 2770; P. R. Wells, *Chem. Rev.*, 1963, **63**, 171.
3. N. Ise and F. Matsui, *J. Am. Chem. Soc.*, 1968, **90**, 4242; J.-R. Cho and H. Morawetz, *J. Am. Chem. Soc.*, 1972, **94**, 375.
4. A. D. English, P. Meakin, and J. P. Jesson, *J. Am. Chem. Soc.*, 1976, **98**, 7590.
5. M. Hamaguchi and T. Nagai, *J. Chem. Soc., Chem. Commun.*, 1987, 1299.

6. J. Caruso, C. Roger, F. Schwertfeger, M. J. Hampden-Smith, A. L. Rheingold, and G. Yap, *Inorg. Chem.*, 1995, **34**, 449.
7. B. Breit, A. Hoffmann, U. Bergsträsser, L. Ricard, F. Mathey, and M. Regitz, *Angew. Chem. Int. Edn. Engl.*, 1994, **33**, 1491.
8. Chao-Jun Li, *Chem. Rev.*, 1993, **93**, 2023.
9. M. E. Brown, K. J. Buchanan, and A. Goosen, *J. Chem. Ed.*, 1985, **62**, 575; R. B. Snadden, *J. Chem. Ed.*, 1985, **62**, 653.
10. M. Berthelot and L. Péan de Saint-Gilles, *Ann. Chim. Phys., Ser. III*, 1862, **66**, 5 (see pp. 61–67).
11. N. Menschutkin, *Z. Phys. Chem.*, 1890, **6**, 41.
12. R. W. Gurney, *Ions in Solution*, Cambridge University Press, Cambridge, 1936.
13. E. D. Hughes and C. K. Ingold, *J. Chem. Soc.*, 1935, 244, 252; C. K. Ingold, *Structure and Mechanism in Organic Chemistry*, Cornell University Press, Ithaca, NY, 1953.
14. T. W. Bentley and P. van R. Schleyer, *Adv. Phys. Org. Chem.*, 1977, **14**, 1; T. W. Bentley, C. T. Bowen, D. H. Morten, and P. van R. Schleyer, *J. Am. Chem. Soc.*, 1981, **103**, 5466; Ikchoon Lee, Myung Soo Choi, and Hai Whang Lee, *J. Chem. Res.* 1994, (*S*) 92, (*M*) 0568.
15. P. R. Wells, *Chem. Rev.*, 1963, **63**, 171; D. N. Kevill and M. J. D'Souza, *J. Chem. Res.*, 1994, (*S*) 190.
16. O. Rogne, *J. Chem. Soc. (B)*, 1969, 663; J. R. Lowe, S. S. Uppal, C. Weidig, and H. C. Kelly, *Inorg. Chem.*, 1970, **9**, 1423.
17. J. Burgess, *J. Chem. Soc. (A)*, 1970, 2703; J. Burgess and M. G. Price, *J. Chem. Soc. (A)*, 1971, 3108.
18. J. Burgess, *J. Chem. Soc., Dalton Trans.*, 1973, 825.
19. L. A. P. Kane-Maguire and G. Thomas, *J. Chem. Soc., Dalton Trans.*, 1975, 1324.
20. W. J. Bland, J. Burgess, and R. D. W. Kemmitt, *J. Organomet. Chem.*, 1968, **15**, 217; 1969, **18**, 199; J. Burgess, M. M. Hunt, and R. D. W. Kemmitt, *J. Organomet. Chem.*, 1977, **134**, 131; J. Burgess, M. E. Howden, R. D. W. Kemmitt, and N. S. Sridhara, *J. Chem. Soc., Dalton Trans.*, 1978, 1577.
21. H. Stieger and H. Kelm, *J. Phys. Chem.*, 1973, **77**, 290; R. Schmidt, M. Geis, and H. Kelm, *Z. Phys. Chem. (Frankfurt)*, 1974, **92**, 223.
22. C. F. Wells, *J. Chem. Soc., Faraday Trans. I*, 1977, **73**, 1851.
23. E. F. Caldin and M. W. Grant, *J. Chem. Soc., Faraday Trans. I*, 1973, **69**, 1648.
24. E. F. Caldin and B. B. Hasinoff, *J. Chem. Soc., Faraday Trans. I*, 1975, **71**, 515; B. Hasinoff, *J. Phys. Chem.*, 1978, **82**, 2630.
25. B. A. Matthews and D. W. Watts, *Aust. J. Chem.*, 1976, **29**, 97; B. A. Matthews, J. V. Turner, and D. W. Watts, *Aust. J. Chem.*, 1976, **29**, 551; K. R. Beckham and D. W. Watts, *Aust. J. Chem.*, 1979, **32**, 1425; L. Spiccia and D. W. Watts, *Aust. J. Chem.*, 1979, **32**, 2275.
26. T. Asano and W. J. Le Noble, *Chem. Rev.*, 1978, **78**, 407.
27. K. J. Laidler, *Reaction Kinetics: Reactions in Solution*, Vol. 2, Pergamon, Oxford, 1963, pp. 11–15.
28. See pp. 103–104 of J. Burgess and E. Pelizzetti, *Progr. React. Kinet.*, 1992, **17**, 1; T. W. Swaddle, *J. Mol. Liq.*, 1995, **65/66**, 237.
29. J. Burgess, S. A. Galema, and C. D. Hubbard, *Polyhedron*, 1991, **10**, 703.
30. J. Burgess, A. J. Duffield, and R. Sherry, *J. Chem. Soc., Chem. Commun.*, 1980, 350.

31. The interested reader can follow this topic through J. Burgess and C. D. Hubbard, *J. Am. Chem. Soc.*, 1984, **106**, 1717; J. Burgess and C. D. Hubbard, *Inorg. Chem.*, 1988, **27**, 2548; M. J. Blandamer, J. Burgess, J. Fawcett, P. Guardado, C. D. Hubbard, S. Nuttall, L. J. S. Prouse, S. Radulović, and D. R. Russell, *Inorg. Chem.*, 1992, **31**, 1383.
32. J. Burgess and C. D. Hubbard, *Inorg. Chim. Acta*, 1982, **64**, L71; M. J. Blandamer, J. Burgess, P. P. Duce, K. S. Payne, R. Sherry, P. Wellings, and M. V. Twigg, *Transition Met. Chem.*, 1984, **9**, 163.
33. Reviews: A. J. Parker, *Chem. Rev.*, 1969, **61**, 1; M. J. Blandamer and J. Burgess, *Coord. Chem. Rev.*, 1980, **31**, 93; *Pure Appl. Chem.*, 1979, **51**, 2087; *Pure Appl. Chem.*, 1982, **54**, 2285; *Pure Appl. Chem.*, 1983, **55**, 55; J. Burgess, *Pure Appl. Chem.*, 1991, **63**, 1677.
34. There are innumerable papers and reviews dealing with the solvation of simple ions and compounds in water, and non-aqueous and mixed solvents; thermochemical approaches to characterizing solvation of transition metal complexes can be traced through M. J. Blandamer and J. Burgess, *Transition Met. Chem.*, 1988, **13**, 1.
35. The direct observation of transition states has been described as one of the 'Holy Grails' of chemistry: J. C. Polanyi and A. H. Zewail, *Accounts Chem. Res.*, 1995, **28**, 119.
36. M. J. Blandamer, J. Burgess, K. W. Morcom, and R. Sherry, *Transition Met. Chem.*, 1983, **8**, 354.
37. M. H. Abraham, *J. Chem. Soc.(A)*, 1971, 1061.
38. J. Bertrán and F. Sánchez Burgos, *J. Chem. Ed.*, 1984, **61**, 416.
39. A detailed, and fully referenced, discussion of the calculation of transfer chemical potentials for ions and of such related matters as standard states will be found in Appendices A and C of M. J. Blandamer, J. Burgess, B. Clark, P. P. Duce, A. W. Hakin, N. Gosal, S. Radulović, P. Guardado, F. Sánchez, C. D. Hubbard, and E. A. Abu-Gharib, *J. Chem. Soc., Faraday Trans.*, 1986, **82**, 1471. The main body of this paper contains examples both of the derivation of transfer chemical potentials for inorganic ions and complexes and of their use in initial state–transition state analysis of reactivity trends, in methanol–water mixtures.
40. M. J. Blandamer, J. Burgess, S. J. Hamshere, C. White, R. I. Haines, and A. McAuley, *Can. J. Chem.*, 1983, **61**, 1361.
41. J. H. Krueger, *Inorg. Chem.*, 1966, **88**, 132.
42. A. Al-Alousy, S. Alshehri, M. J. Blandamer, N. J. Blundell, J. Burgess, H. J. Cowles, S. Radulović, P. Guardado, and C. D. Hubbard, *J. Chem. Soc., Faraday Trans.*, 1993, **89**, 1041.
43. H. P. Bennetto and E. F. Caldin, *J. Chem. Soc. A*, 1971, 2190, 2198, 2207; H. P. Bennetto, *J. Chem. Soc. A*, 1971, 2211; E. F. Caldin and P. Godfrey, *J. Chem. Soc., Faraday Trans. I*, 1974, **70**, 2260.
44. F. Franks and D. J. G. Ives, *Quart. Rev. Chem. Soc.*, 1966, **20**, 1.
45. J. Burgess and C. D. Hubbard, *J. Am. Chem. Soc.*, 1984, **106**, 1717.
46. A. D. Pethybridge and J. E. Prue, *Progr. Inorg. Chem.*, 1972, **17**, 327 provides a thorough treatment of salt effects on kinetics of inorganic reactions, starting at the very beginning (S. Arrhenius, *Z. Phys. Chem.*, 1887, **1**, 110).
47. J. Setchenow, *Z. Phys. Chem.*, 1889, **4**, 117; *Ann. Chim.*, 1892, **25**, 226.
48. F. A. Long and W. F. McDevit, *Chem. Rev.*, 1952, **51**, 119.

49. See the preceding reference, and V. F. Sergeeva, *Russ. Chem. Rev.*, 1965, **34**, 309.
50. M. J. Blandamer, J. Burgess, and J. C. McGowan, *J. Chem. Soc., Dalton Trans.*, 1980, 616.
51. J. N. Brønsted, *Z. Phys. Chem.*, 1922, **102**, 169, 195.
52. J. N. Brønsted, *Z. Phys. Chem.*, 1925, **115**, 337.
53. N. Bjerrum, *Z. Phys. Chem.*, 1924, **108**, 82; 1925, **118**, 251.
54. V. K. La Mer, *Chem. Rev.*, 1932, **10**, 179.
55. R. S. Livingston, *J. Chem. Ed.*, 1930, **7**, 2899.
56. From Fig. 1 on p. 332 of Ref. 46.
57. A. R. Olson and T. R. Simonson, *J. Chem. Phys.*, 1949, **17**, 1167.
58. E. Thilo and F. V. Lampe, *Z. Anorg. Allg. Chem.*, 1963, **319**, 387; E. Thilo and F. V. Lampe, *Chem. Ber.*, 1964, **97**, 1775.
59. B. Surfleet and P. A. H. Wyatt, *J. Chem. Soc. (A)*, 1967, 1564.
60. A. Indelli, *J. Phys. Chem.*, 1961, **65**, 973.
61. V. K. La Mer, *J. Am. Chem. Soc.*, 1929, **51**, 3341.
62. C. W. Davies, *Progr. React. Kinet.*, 1961, **1**, 161; C. W. Davies, *Ion Association*, Butterworths, London, 1962, Chapter 13.
63. J. C. Sheppard and A. C. Wahl, *J. Am. Chem. Soc.*, 1957, **79**, 1020; L. Gjertsen and A. C. Wahl, *J. Am. Chem. Soc.*, 1959, **81**, 1572.
64. A. C. Wahl, *Z. Elektrochem.*, 1960, **64**, 90; R. J. Campion, C. F. Deck, P. King, and A. C. Wahl, *Inorg. Chem.*, 1967, **6**, 672; A. Lowenstein and G. Ron, *Inorg. Chem.*, 1967, **6**, 1604.
65. R. W. Chlebek and M. W. Lister, *Can. J. Chem.*, 1966, **44**, 437.
66. J. Holluta and W. Herrmann, *Z. Phys. Chem.*, 1933, **A166**, 453; M. R. Kershaw and J. E. Prue, *Trans. Faraday Soc.*, 1967, **63**, 1198.
67. A. Indelli and J. E. Prue, *J. Chem. Soc.*, 1959, 107.
68. V. K. La Mer and R. W. Fessenden, *J. Am. Chem. Soc.*, 1932, **54**, 2351; P. A. H. Wyatt and C. W. Davies, *Trans. Faraday Soc.*, 1949, **45**, 774; G. Corsaro and M. C. Morris, *J. Electrochem. Soc.*, 1961, **108**, 689.
69. F. Sánchez, M. J. Nasarre, M. M. Graciani, R. Jimenez, M. L. Moyá, J. Burgess, and M. J. Blandamer, *Transition Met. Chem.*, 1988, **13**, 150.
70. J. Burgess, F. Sánchez, E. Morillo, A. Gil, J. I. Tejera, A. Galán, and J. M. García, *Transition Met. Chem.*, 1986, **11**, 166.
71. M. Galán, M. Domínguez, R. Andreu, M. L. Moyá, F. Sánchez, and J. Burgess, *J. Chem. Soc., Faraday Trans.*, 1990, **86**, 937.
72. M. J. Blandamer, J. Burgess, and P. P. Duce, *J. Inorg. Nucl. Chem.*, 1981, **43**, 3103; *Transition Met. Chem.*, 1983, **8**, 308.
73. For a brief treatment of inorganic reactions and their reactivity trends in organized aqueous media, see J. Burgess, *Coll. Surfaces*, 1990, **48**, 185. Much fuller documentation can be found in Ref. 1 of this chapter.
74. F. M. Menger and Z. X. Fei, *Angew. Chem. Int. Edn. Engl.*, 1994, **33**, 346.
75. P. D. I. Fletcher and B. H. Robinson, *J. Chem. Soc., Faraday Trans. I*, 1983, **79**, 1959.
76. M. S. Tosić, V. M. Vasić, J. M. Nedeljković, and L. A. Ilić, *Polyhedron*, 1997, **16**, 1157.
77. N. Ise and F. Matsui, *J. Am. Chem. Soc.*, 1968, **90**, 4242; J.-R. Cho and H. Morawetz, *J. Am. Chem. Soc.*, 1972, **94**, 375.
78. E. Pelizzetti and E. Pramauro, *Inorg. Chem.*, 1980, **19**, 1407.

79. M. Vincenti, E. Pramauro, E. Pelizzetti, S. Diekmann, and J. Frahm, *Inorg. Chem.*, 1985, **24**, 4533.
80. F. M. Menger and C. E. Portnoy, *J. Am. Chem. Soc.*, 1967, **89**, 4698.
81. P. L. Cornejo, R. Jiménez, M. L. Moyá, F. Sánchez, and J. Burgess, *Langmuir*, 1996, **12**, 4981.
82. R. Steudel and G. Holdt, *Angew. Chem. Int. Edn. Engl.*, 1988, **27**, 1358.
83. D. Diaz, A. S. Erokhin, I. V. Berezin, and A. K. Yatsimirsky, *Transition Met. Chem.*, 1987, **12**, 87.
84. M. Bressan, L. Forti, and A. Morvillo, *J. Chem. Soc., Chem. Commun.*, 1994, 253.
85. F. M. Menger, *Angew. Chem. Int. Edn. Engl.*, 1991, **30**, 1086.
86. A. Gupta and J. Skarzewski, *J. Chem. Res.*, 1994, (*S*) 206, (*M*) 1117.
87. V. F. Tarasov, N. D. Ghatlia, N. I. Avdievich, I. A. Shkrob, A. L. Buchachenko, and N. J. Turro, *J. Am. Chem. Soc.*, 1994, **116**, 2281.
88. A. J. Kirby, *Angew. Chem. Int. Edn. Engl.*, 1994, **33**, 551.
89. N. Alandis, I. Rico-Lattes, and A. Lattes, *New J. Chem.*, 1994, **18**, 1147.
90. P. D. I. Fletcher, J. R. Hicks, and V. C. Reinsborough, *Can. J. Chem.*, 1983, **61**, 1594.
91. C. Tondre and M. Boumezioud, *J. Phys. Chem.*, 1989, **93**, 846.
92. B. A. Keiser and S. L. Holt, *Inorg. Chem.*, 1982, **21**, 2323.
93. F. P. Cavasino, C. Sbrizioli, M. L. Turco Liveri, and V. Turco Liveri, *J. Chem. Soc., Faraday Trans.*, 1994, **90**, 311.
94. M. J. Blandamer, J. Burgess, and B. Clark, *J. Chem. Soc., Chem. Commun.*, 1983, 659; M. J. Blandamer, J. Burgess, B. Clark, P. P. Duce, and J. M. W. Scott, *J. Chem. Soc., Faraday Trans. I*, 1984, **80**, 739; M. J. Blandamer, J. Burgess, B. Clark, A. W. Hakin, M. W. Hyett, S. Spencer, and N. Taylor, *J. Chem. Soc., Faraday Trans. I*, 1985, **81**, 2357.
95. H. Elias, H.-T. Macholdt, K. J. Wannowius, M. J. Blandamer, J. Burgess, and B. Clark, *Inorg. Chem.*, 1986, **25**, 3048.
96. C. Minero, E. Pramauro, and E. Pelizzetti, *Langmuir*, 1988, **4**, 101; *Coll. Surfaces*, 1989, **35**, 237.
97. S. Ozeki and K. Inouye, *J. Phys. Chem.*, 1984, **88**, 5360.
98. Yufei Cheng, D. J. Schiffrin, P. Guerriero, and P. A. Vigato, *Inorg. Chem.*, 1994, **33**, 765.
99. Yao-Hong Chen, U. Nickel, and M. Spiro, *J. Chem. Soc., Faraday Trans.*, 1994, **90**, 617.
100. See several contributions in *Gels and Gelling Processes*, Faraday Society Discussion No. 57, The Chemical Society, London, 1974.
101. M. J. Blandamer, J. Burgess, and J. R. Membrey, *J. Chem. Soc., Faraday Trans. I*, 1975, **71**, 145.
102. O. V. Mikhailov, *Transition Met. Chem.*, 1996, **21**, 363.

9 Oxidation–reduction reactions

Chapter contents

9.1 Introduction

Oxidation–reduction, or *redox*, reactions involve changes of oxidation state. They involve two species, the *oxidant*, which takes the electron(s), and the *reductant*, which gives the electron(s), and often follow simple second-order rate laws. However, this kinetic simplicity cloaks a range of mechanisms, involving electron transfer, atom transfer or group transfer, ligand substitution, addition or dissociation. There are numerous instances of two-stage mechanisms with transient intermediates. The study and the explanation of this great variety of systems can range in sophistication from preparative inorganic chemistry at one extreme to solid-state physics at the other. We shall try to cover this ground in one chapter. Fortunately, there are two excellent books devoted to the kinetics and mechanisms of inorganic redox reactions,[1,2] which are both comprehensive and lucid, and are thus the place to turn for amplification of topics that are covered too sketchily in this chapter. There is also a review volume relating electron transfer in inorganic systems to those in organic and biological systems,[3] and an extensive bibliography of reviews on electron transfer.[4]

It is possible to assign mechanisms to certain redox reactions between pairs of transition metal complexes with an unusually high degree of confidence, as was demonstrated by Taube more than three decades ago. The area of electron transfer reactions has been well studied ever since Harcourt and Esson's reports on hydrogen peroxide/iodide and manganate(VII) (permanganate)/oxalate kinetics in 1865–7. This area, or at least that major part which covers electron transfers involving transition metal complexes, is also fairly well understood.[5–10] We shall therefore start the main body of this chapter with this type of reaction. Reactions involving *s*- and *p*-block oxidants and reductants provide more variety and, at the same time, more complications. Thus, for example, stable oxidation states of the *s*- and *p*-block elements are generally separated by two units, e.g. tin(II)/(IV), phosphorus(III)/(V), sulphur(II)/(IV)/(VI) or iodine(−I)/(I)/(III)/(V)/(VII). As most transition metal complexes usually act as one-electron oxidants or reductants, reactions between them and the *sp*-block elements must either involve a large number of highly implausible termolecular mechanisms or transient intermediate species containing the elements in unusual oxidation states. This topic is discussed in some detail in Section 9.7.2.

Although reactions between a two-electron oxidant and a two-electron reductant may be straightforward, as in oxygen-atom transfer reactions such as

$$SO_3^{2-} + ClO^- \longrightarrow SO_4^{2-} + Cl^-$$

there are many cases where the mechanism is less simple. Thus oxidations by the peroxodisulphate anion, which is stoichiometrically a two-electron oxidant,

$$S_2O_8^{2-} + 2e^- \longrightarrow 2SO_4^{2-}$$

often proceed by way of an initial homolysis to $[SO_4]^{-\cdot}$ radicals, which are one-electron oxidants and may well initiate radical pathways.

Certain species can act as one-electron and two-electron reagents (Table 9.1); in some cases the one- and two-electron paths are concurrent. Evidence comes from product studies, including the results of labelling experiments, as well as from kinetics. In the case of sulphite reductions, one has to be wary of assigning mechanism from products, for the sulphite radical which is the initial product of one electron transfer can dimerize to give dithionate or, sometimes, undergo a second one-electron reaction to give sulphate. The observation of dithionate and sulphate as products of permanganate oxidation of sulphite arises from such a mechanism—this redox reaction consists solely of one-electron transfers, and not of parallel one- and two-electron transfers.[11] In this context it is worth mentioning the possibility of favouring one-electron or two-electron paths by employing appropriate mono- or binuclear complexes of the type L_5ML' and $L_5M{-}L'M{-}L_5$, respectively.

One- and two-electron transfers are the norm, but there are a few instances of the simultaneous transfer of more than two electrons. Chromium(VI) oxidations usually involve a sequence of one- or two-electron transfers, but there is at least one well-documented example of a three-electron transfer to give the stable oxidation state chromium(III) in one step,[12] and even a claim that chromium(V) oxidation of hydroxylamine

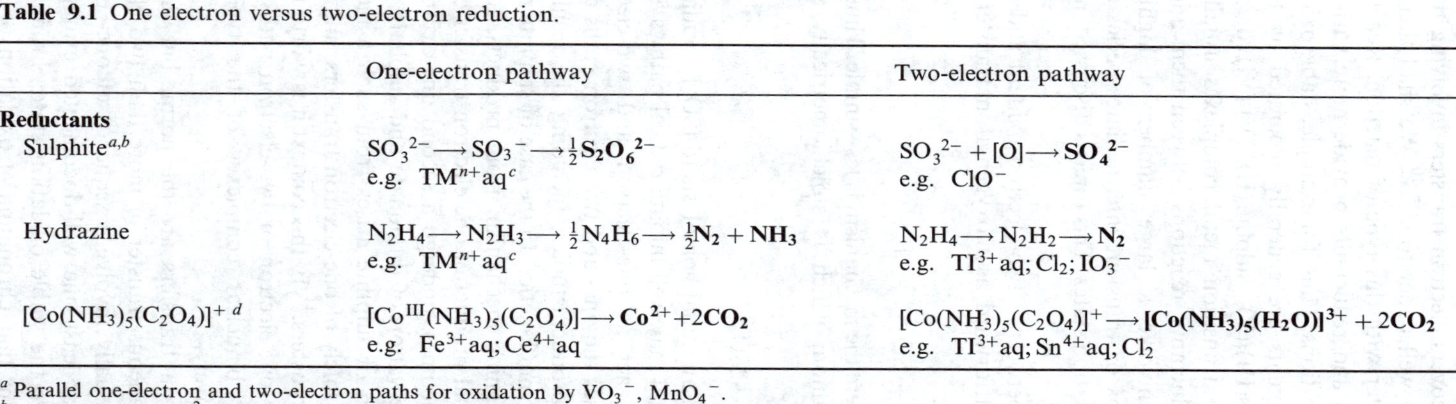

Table 9.1 One electron versus two-electron reduction.

	One-electron pathway	Two-electron pathway
Reductants		
Sulphite[a,b]	$SO_3^{2-} \longrightarrow SO_3^{-} \longrightarrow \frac{1}{2}\mathbf{S_2O_6^{2-}}$ e.g. $TM^{n+}aq$[c]	$SO_3^{2-} + [O] \longrightarrow \mathbf{SO_4^{2-}}$ e.g. ClO^-
Hydrazine	$N_2H_4 \longrightarrow N_2H_3 \longrightarrow \frac{1}{2}N_4H_6 \longrightarrow \frac{1}{2}\mathbf{N_2} + \mathbf{NH_3}$ e.g. $TM^{n+}aq$[c]	$N_2H_4 \longrightarrow N_2H_2 \longrightarrow \mathbf{N_2}$ e.g. $Tl^{3+}aq$; Cl_2; IO_3^-
$[Co(NH_3)_5(C_2O_4)]^+$[d]	$[Co^{III}(NH_3)_5(C_2O_4^{\cdot})] \longrightarrow \mathbf{Co^{2+}} + \mathbf{2CO_2}$ e.g. $Fe^{3+}aq$; $Ce^{4+}aq$	$[Co(NH_3)_5(C_2O_4)]^+ \longrightarrow \mathbf{[Co(NH_3)_5(H_2O)]^{3+}} + \mathbf{2CO_2}$ e.g. $Tl^{3+}aq$; $Sn^{4+}aq$; Cl_2

[a] Parallel one-electron and two-electron paths for oxidation by VO_3^-, MnO_4^-.
[b] Production of $S_4O_6^{2-}$ from silver(III) oxidation of thiosulphate suggests two consecutive one-electron transfers.
[c] TM^{n+} = e.g. Mn^{3+}, Fe^{3+}, Co^{3+}, Ce^{4+}—i.e. one-electron-preferred oxidants.
[d] $C_2O_4^{2-}$ = oxalate.

produces a nitrosyl chromium(I) species with simultaneous transfer of four electrons.[13]

9.2 Inner-sphere and outer-sphere mechanisms

Two distinct mechanisms are recognized for redox reactions. These are the inner-sphere and the outer-sphere mechanisms (Fig. 9.1), which will be outlined in this section and then described in detail in the sections that follow. The essential difference between these mechanisms lies in the key role played by substitution prior to electron transfer in the inner-sphere route. Outer-sphere redox reactions simply involve electron transfer.

Although the differences between the two mechanisms are clear, it is not always an easy matter to establish which mechanism operates in specific cases. There are two reasons for this. In favourable cases it is possible, by ways set out in Section 9.3, to establish beyond any reasonable doubt that the inner-sphere mechanism operates, but in many systems where an inner-sphere electron transfer mechanism is suspected, either the reactants or the products, or possibly both, are substitutionally labile. Then it is not possible to establish unequivocally the precise nature of the actual species reacting or the products that are initially formed. The other source of uncertainty applies to the outer-sphere mechanism, which, since it involves nothing more than the transfer of an electron from the reductant to the oxidant, is singularly difficult to demonstrate in an unambiguous manner. The methodology of proving this mechanism usually relies on providing convincing evidence that the alternative inner-sphere mechanism is not available. Thus, there are many reactions that indubitably go by the inner-sphere mechanism, many that are almost certainly outer-sphere processes, and an

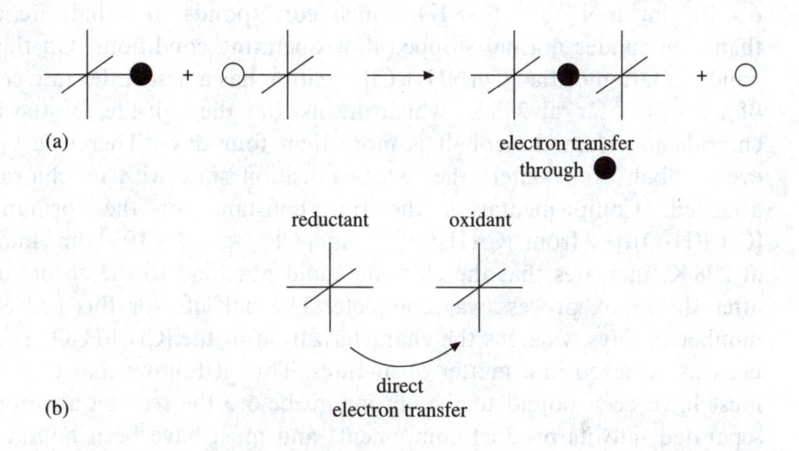

Fig. 9.1 The two mechanisms for the transfer of electrons from reductant to oxidant. (a) The inner-sphere mechanism. (b) The outer-sphere mechanism.

uncomfortably large number of reactions in the grey area in between. The question of deciding which is the more likely mechanism in particular cases is taken up in Section 9.6.2.

9.3 The inner-sphere mechanism: general

The inner-sphere mechanism for electron transfer between transition metal complexes is perhaps the most unequivocally characterized mechanism in inorganic chemistry. The demonstration of its operation in the particular case of the reduction of $[Co(NH_3)_5Cl]^{2+}$ by $[Cr(H_2O)_6]^{2+}$ is described in Section 9.3.1 below. This was the key experiment of Henry Taube, who was able to select the ideal system as a result of the extensive review of substitution kinetics that he had then recently completed.[14]

9.3.1 Taube's experiment[2,15,16]

The nature of the products of chromium(II) reduction of the cobalt(III) cation $[Co(NH_3)_5Cl]^{2+}$ in acidic aqueous solution shows unequivocally that chloride is transferred from cobalt to chromium:

$$[Co(NH_3)_5Cl]^{2+} + [Cr(H_2O)_6]^{2+} + 5H^+ \longrightarrow [CrCl(H_2O)_5]^{2+} + [Co(H_2O)_6]^{2+} + 5NH_4{}^+$$

This transfer is an *essential* feature of the electron transfer process, as is demonstrated by the following considerations, summarized in Fig. 9.2, of the substitutional lability and inertness of the metal centres involved.

The redox reaction has a second-order rate constant of 6×10^5 dm^3 mol^{-1} s^{-1} (298 K), which corresponds to a half-life of less than 1 ms under normal stopped-flow operating conditions. On the other hand, aquation of the $[Co(NH_3)_5Cl]^{2+}$ cation has a first-order rate constant of 1.7×10^{-6} s^{-1} at 298 K, which means that the half-life for the loss of chloride ion from the cobalt is more than four days. Therefore virtually every cobalt atom enters the redox transition state with its chloride still attached. Complementarily, the rate constant for the formation of $[CrCl(H_2O)_5]^{2+}$ from $[Cr(H_2O)_6]^{3+}$ and Cl^-, $k = 3 \times 10^{-8}$ dm^3 mol^{-1} s^{-1} at 298 K, indicates that the chloride could not bind to the chromium(III) after the redox process was complete. The half-life for this process is a number of days, whereas the characterization of the $[CrCl(H_2O)_5]^{2+}$ product was achieved in a matter of minutes. Thus it follows that the chloride must have been bound to the chromium before the redox transition state separated into its product components and must have been bound to the cobalt as the transition state was formed. In other words, the chloride must be bound to both metals in the electron transfer transition state (the Franck–Condon principle that electron transfer takes place without

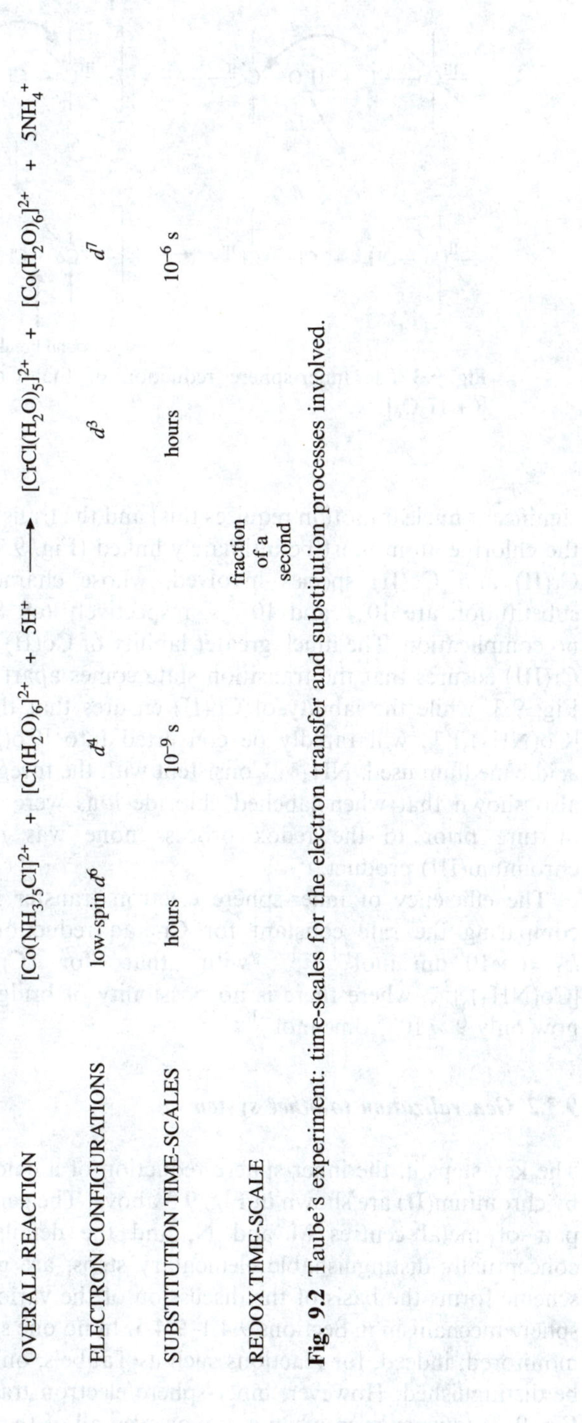

Fig. 9.2 Taube's experiment: time-scales for the electron transfer and substitution processes involved.

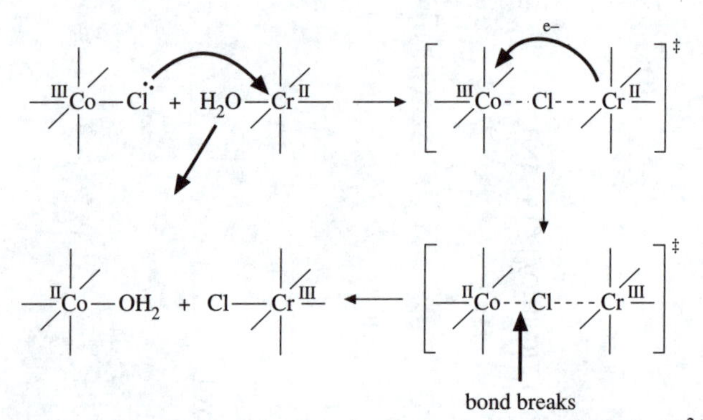

Fig. 9.3 The inner-sphere reduction of the $[Co(NH_3)_5Cl]^{2+}$ cation by $[Cr(H_2O_6]^{2+}$.

significant nuclear motion requires this) and the transfers of the electron and the chlorine atom must be intimately linked (Fig. 9.3). The labilities of the Cr(II) and Co(II) species involved, whose characteristic half-lives for substitution are 10^{-9} and 10^{-5} s, respectively (cf. Section 7.3.1), present no complication. The much greater lability of Co(II) with respect to that of Cr(III) ensures that the transition state comes apart at the bond shown in Fig. 9.3, while the lability of Co(II) ensures that the immediate product, $[Co(NH_3)_5]^{2+}$, will rapidly be converted into $[Co(H_2O)_6]^{2+}$ plus, in the acidic medium used, NH_4^+. Consistent with the foregoing discussion, it was also shown that when labelled chloride ions were added to the reaction mixture prior to the redox process none was incorporated into the chromium(III) product.

The efficiency of inner-sphere electron transfer may be illustrated by comparing the rate constant for Cr^{2+}aq reduction of $[Co(NH_3)_5Cl]^{2+}$, $k_2 = 6 \times 10^5$ dm^3 mol^{-1} s^{-1}, with that for Cr^{2+}aq reduction of $[Co(NH_3)_6]^{3+}$, where there is no possibility of bridge formation and k_2 is now only 9×10^{-5} dm^3 mol^{-1} s^{-1}.

9.3.2 Generalization to other systems

The key steps in the inner-sphere reduction of a chlorocobalt(III) complex by chromium(II) are shown in Fig. 9.3 above. The generalization of this to a pair of metal centres M and N, and the details of the sequence of conceptually distinguishable elementary steps, are given in Fig. 9.4. This scheme forms the basis of the discussion of the various facets of the inner-sphere mechanism in Sections 9.4.1–9.4.5. In no one system can each step be monitored; indeed, for reactions such as Taube's, only one kinetic step can be distinguished. However, inner-sphere electron transfer does involve the Fig. 9.4 sequence even when many or even all of the transients may escape detection by kinetic, spectroscopic, or other means. Some may, indeed, be

$L_5M^M(LL) + N^NL'_5(OH_2)$	reactants
⇅ *outer-sphere association*	
$L_5M^M(LL), N^NL'_5(OH_2)$	precursor complex
⇅ *ligand-bridge formation*	
$L_5M^M—LL—N^NL'_5$	bridged intermediate
⇅ *electron transfer to bridging ligand*	
$L_5M^M—LL^{\bullet}—N^{N+1}L'_5$	radical–ligand bridge
⇅ *electron transfer to M*	
$L_5M^{M-1}—LL—N^{N+1}L'_5$	post-electron transfer intermediate
⇅ *rupture of bridging-ligand bond*	
$L_5M^{M-1}(OH_2), (LL)N^{N+1}L'_5$	successor complex
⇅ *separation of successor complex*	
$L_5M^{M-1}(OH_2) + (LL)N^{N+1}L'_5$	products

Fig. 9.4 Constituent steps in the general inner-sphere mechanism for electron transfer.

better considered as transition states. In particular, the electron generally moves 'instantly' across the bridging ligand, with no meaningful radical–ligand bridge. This is discussed in Section 9.4.4, after consideration of reactions in which there *is* some indication of transient radical–ligand-bridged species.

The one essential requirement for the inner-sphere electron transfer is a suitable bridging ligand. Group 16 and 17 donor atoms (halogens, oxygen, sulphur) usually have electron pairs available to form a bridge to a second metal ion, but Group 15 atoms (nitrogen, phosphorus, arsenic) have a single lone pair which is fully occupied in bonding to M. Nevertheless, an available lone pair may not be an *essential* requirement for inner-sphere electron transfer. Thus, just as CH_3 can act as a bridge between I^- and OH^- in the S_N2 base hydrolysis of methyl iodide, so it may act as a bridge between Cr(II) and Co(III) in the reduction of methyl-cobalt(III) species by $[Cr(H_2O)_6]^{2+}$. Of course, bridge formation by a polyatomic ligand is easy when it contains more donor sites than are bound to M in the initial complex. Thus, in $[Co(NH_3)_5L]^{3+}$, pyridine, **1**, is not a potential bridging ligand, but ligands such as 4-cyanopyridine, **2**, pyrazine, **3**, and 4,4′-bipyridyl, **4**, are effective bridging ligands for the inner-sphere electron transfer.

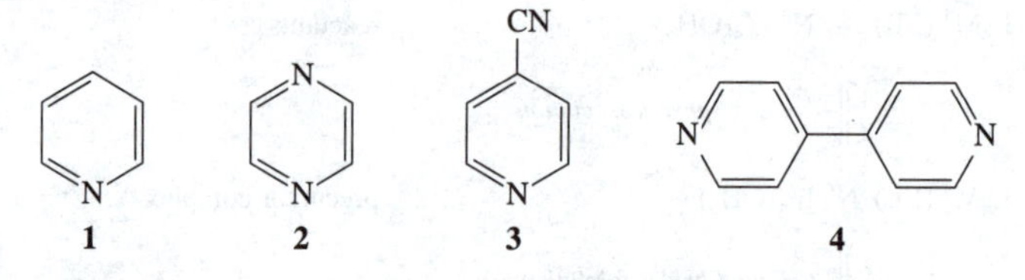

It is, obviously, necessary for the electronic structure of the bridging ligand to be such that electron transmission from one donor site to the other is possible. When there is no available bridging ligand, or no reasonable through-ligand electron transfer path, then electron transfer must take place by the outer-sphere mechanism (Section 9.5). When the oxidant (or reductant) contains two potential bridging ligands with suitable geometry, then double bridging between oxidant and reductant is possible, at least in principle. Double bridging has been established in Cr^{2+}aq reduction of *cis*-$[Co(en)_2(N_3)_2]^+$ and also probably occurs in analogous reactions involving certain *cis*-carboxylatocobalt(III) complexes.

A special case of the inner-sphere mechanism is provided by Cr^{2+}aq catalysis of aquation of chromium(III) complexes $[Cr(H_2O)_5X]^{2+}$ (X = for example, SCN, CN, or halide). Kinetic parameters are available for ten such reactions.[17] Activation entropies of between -60 and $-140\,J\,K^{-1}\,mol^{-1}$ are consistent with a bimolecular rate-determining step; hydration contributions should be small for reaction between like-charged complexes. Electron transfer from the substitution-inert Cr^{III} gives the extremely substitution-labile Cr^{II}. The ligand X^- is then immediately replaced by water:

$$[^*Cr(H_2O)_5X]^{2+} + Cr^{2+}aq \longrightarrow {}^*Cr^{2+}aq + Cr^{3+}aq + X^-$$

It was in Cr^{2+}aq catalysis of aquation of *cis*-$[Co(H_2O)_4(N_3)_2]^+$ that the double-bridged path for electron transfer was first demonstrated. On the basis of the species found among the reaction products, parallel single- and double-bridged paths were postulated.[18]

9.4 The inner-sphere mechanism: details

9.4.1 Formation of precursor complexes

In the Taube reaction, as indeed in the great majority of inner-sphere redox reactions, the formation of the association and precursor complexes (Fig. 9.3) are rapid, and usually reversible, steps preceding rate-limiting electron transfer. It was not until sixteen years after Taube's work that convincing evidence for the intermediacy of a precursor complex was presented, although indirect evidence for the existence of very transient precursor complexes had been obtained previously.

The first kinetic demonstration of the intermediacy of a precursor complex, with a lifetime of about ten seconds, was in the reduction of $[Co(NH_3)_5(nta)]$ (nta = nitrilotriacetate, $N(CH_2CO_2)_3{}^{3-}$) by $[Fe(H_2O)_6]^{2+}$.

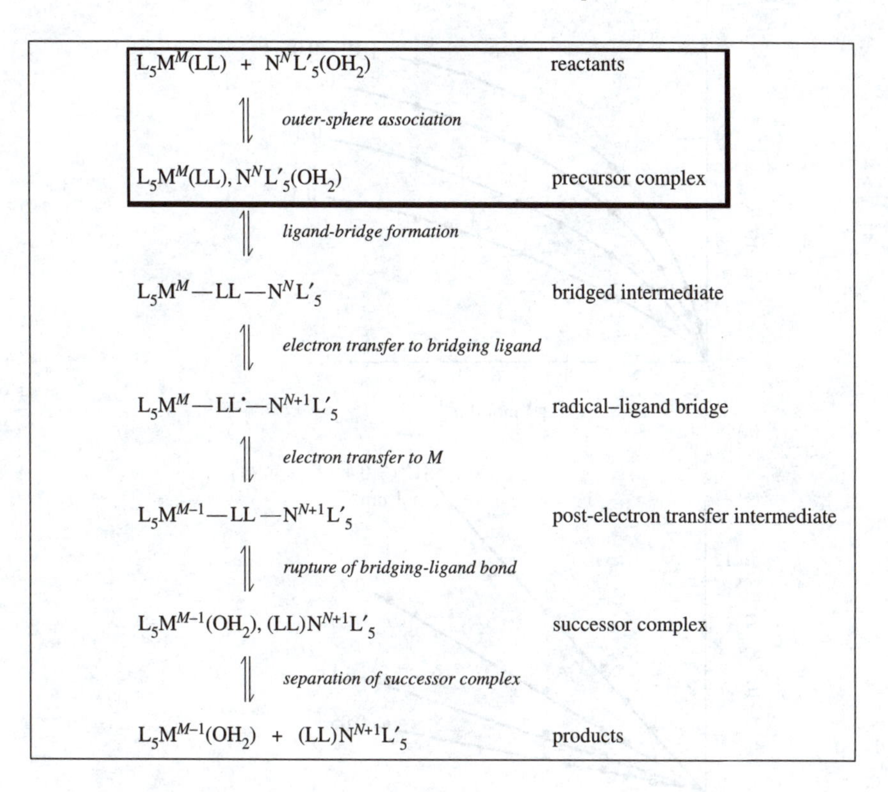

This, and later examples,[19] for instance the reduction of $[Co(NH_3)_5(3\text{-CN-acac})]^{2+}$ by $[Cr(H_2O)_6]^{2+}$ and the Fe^{2+} reduction of $[Co(NH_3)_5X]^{2+}$ (X = Cl, Br, I, N_3) in dimethylformamide mentioned in Chapter 8, exhibited saturation kinetics. At constant pH, and using a large excess of reducing agent, the experimental rate law for the former is

$$\text{rate} = k_{\text{obs}}[Co(NH_3)_5(3\text{-CN-acac})^{2+}]$$

and variation of the concentration of Cr(II) then shows that

$$k_{\text{obs}} = k_{\text{et}}K[Cr^{2+}\text{aq}]/(1+K[Cr^{2+}\text{aq}])$$

An astute reader might point out that, while such a rate law strongly suggests that there is a reversible pre-equilibrium between the oxidant and the reductant, it does not rule out the possibility that the main adduct is a 'dead end' and that the reaction involves an alternative pathway in which the two species come together and react. Systems which involve such complications are dealt with in Section 9.4.6 below. Figure 9.5 shows the curved plot corresponding to the effects of precursor formation plus the reciprocal plot whence the equilibrium constant for precursor formation, K, and the rate constant for the subsequent electron transfer, k_{et}, may be obtained. Equilibrium constants and the associated standard enthalpy and entropy changes for precursor complex formation are given in Table 9.2 and

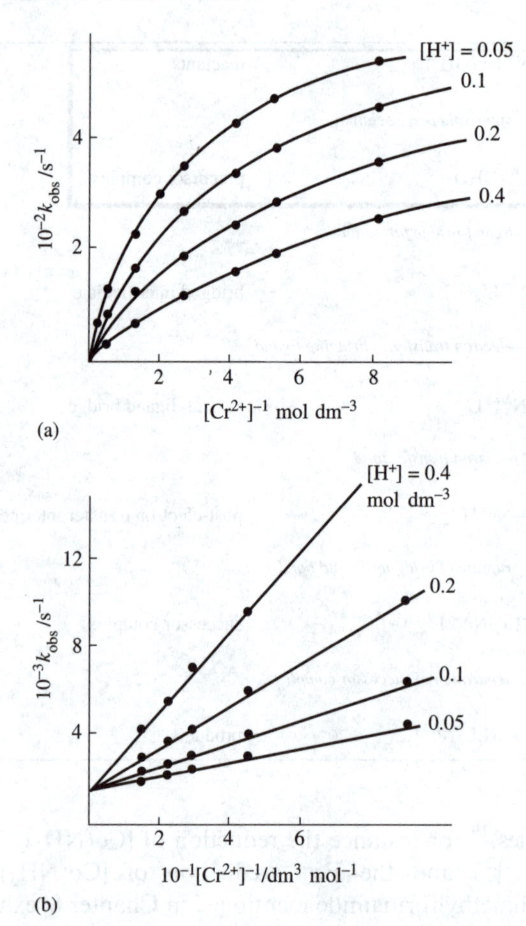

Fig. 9.5 $[Cr(H_2O)_6]^{2+}$ reduction of $[(NH_3)_5Co(NCacac)]^{2+}$ (I = 1.00 mol dm^{-3}; 25 °C). Each point represents an average of at least six runs. In (a), k_{obs} is plotted against $[Cr^{2+}]$ and the solid lines are drawn by using calculated values of k_{obs} from the above equation and the constants from Table 9.2. In (b), k_{obs}^{-1} is plotted against $[Cr^{2+}]^{-1}$ and the solid lines are from a linear regression of the data points.

compared with the thermodynamic constants for the formation of some non-redox analogues. Note the similarity of the magnitudes of the equilibrium constants for the transient redox species and the stable non-redox analogues. Entropy changes are negative because the increase in the charge on complex formation between ions of like sign presumably results in increased electrostriction, which increases the already negative entropy change for the combination of two species to give just one.

Another example of a system for which observed kinetic parameters can be dissected into pre-association equilibrium parameters and kinetic parameters for the actual electron transfer step is documented in Fig. 9.6. There

Table 9.2 Thermodynamic parameters for formation of precursor complexes and for analogous non-redox complexes.

	K (molar scale)	ΔH° (kJ mol^{-1})	ΔS° (J K^{-1} ml^{-1})
$[Co(NH)_5(ntaH)]^+ + Fe^{2+}$	1.1×10^6 [a]	+6	−96
$[Co(NH_3)_5(ntaH)]^+ + Co^{2+}$	1.8×10^7		
$[Co(NH_3)_5(ntaH)]^+ + Ti^{3+}$	40		
$[Co(NH_3)_5(3\text{-CNacac})]^{2+} + Cr^{2+}$	58	−30	−59
$[Co(NH_3)_5(NCS)]^{2+} + Hg^{2+}$		−22	−71
$[Co(NH_3)_5(fumarate)]^{2+} + Cu^+$	4.4×10^3	−48	−92
$[Ru(NH_3)_5(4\text{-vinylpy})]^{3+} + Cu^+$ [b]		−41	−75

[a] This is similar to K_1 for Fe^{2+}aq + *N*-methyliminodiacetate.
[b] The precursor complex is:

$(H_3N)_5Ru^{III}$—N(pyridine)–CH=CH$_2$ → Cu^I

is a notably low enthalpy barrier to electron transfer once the intermediate has been formed. The negative activation entropy has been attributed to the difficulty of transferring the electron to the cobalt, as there is a π/σ mismatch of orbitals between the ligand orbital through which the electron is transferred and the $d_z{}^2$ orbital on the cobalt which receives the electron. This electron transfer may therefore be non-adiabatic, involving the electron dwelling on the bridging ligand for longer than the time corresponding to a simple 'straight-through' transition state.[20]

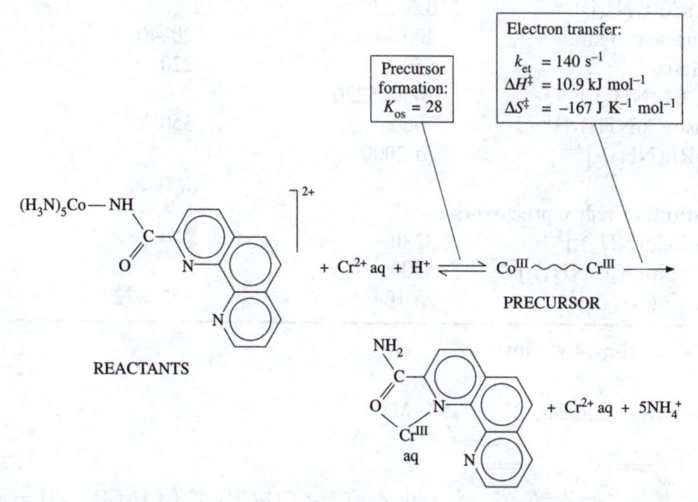

Fig. 9.6 An example of precursor formation and subsequent electron transfer.

The characterization of **5** and several of the other precursors cited in Table 9.2 is based on kinetic evidence and not entirely unequivocal, though rapid scanning spectrophotometry revealed the spectrum of the precursor in the $[Co(NH_3)_5(ntaH)]^+ + Ti^{3+}aq$ reaction. Somewhat firmer characterizations of precursor complexes can be obtained in the special case of the reduction by $[Fe(CN)_5(H_2O)]^{3-}$ of cobalt(III) complexes containing potentially bridging ligands, such as 4,4′-bipyridyl, **6**, 3- or 4-cyanopyridine, or

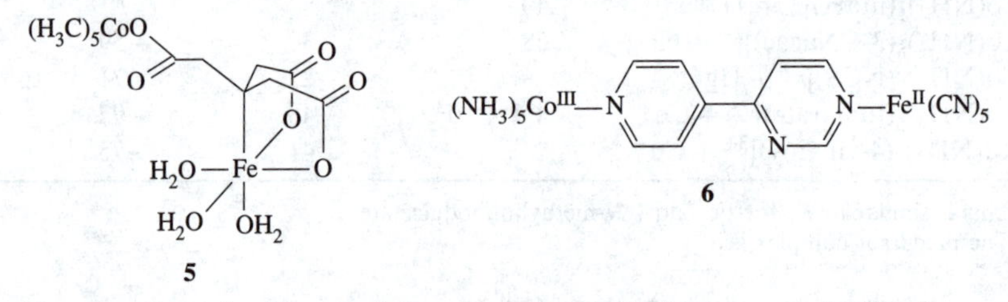

imidazole. Here, k_f, K_{OS}, and k_i values (Table 9.3), are very similar for each system, and, moreover, are very similar to values for non-redox substitution

Table 9.3 Formation rate constants (k_f) and derived pre-association equilibrium constants (K_{os}) and interchange rate constants (k_i) for reactions of $[Fe(CN)_5(H_2O)]^{3-}$ with oxidizing and non-oxidizing complexes.

		Eigen–Wilkins	
	$k_f(dm^3\ mol^{-1}\ s^{-1})$	K_{OS}(molar scale)	$k_i(s^{-1})$
Substitution (non-redox)			
$[(LL)Co(CN)_5]^{2-}$ [a]	0.5–3.0		
Uncharged ligands	200–400	20–40	*ca* 10
N-Me-pz$^+$ [b]	550	220	2.5
$[(LL)Co(NH_3)_5]^{3+}$ [c]	2050–5230		
$[(dmso)CoNH_3)_5]^{3+}$	7000	350	20
$[(pz)Rh(NH_3)_5]^{3+}$	*ca* 2000		
Substitution redox precursors			
$[(imid)Co(NH_3)_5]^{3+}$	3280	670	4.9
$[(4,4'\text{-bipy})Co(NH_3)_5]^{3+}$	5500		
$[(LL)Co(NH_3)_4]^{3+}$ [d]	*ca* 104	477–872	13–42

[a] LL = 4-CN-py; 4, 4′-bipy; pz.

[b] *N*-methylpyrazinium, N⟨pyrazine ring⟩N$^+$—Me.

[c] LL = N⟨pyridine⟩—Z—⟨pyridine⟩N with Z = CH_2; CH_2CH_2; $CH_2CH_2CH_2$; CH=CH; CO.

[d] LL = 3- or 4-CN-py (bonded through CN or py nitrogen).

at $[Fe(CN)_5(H_2O)]^{3-}$ by simple ligands and by inorganic complexes. Thus, for these redox reactions, we have Eigen–Wilkins formation of precursor complexes followed by electron transfer which is markedly slower and kinetically distinct (Section 9.4.4.5).[21]

9.4.2 The site of attack by the reductant

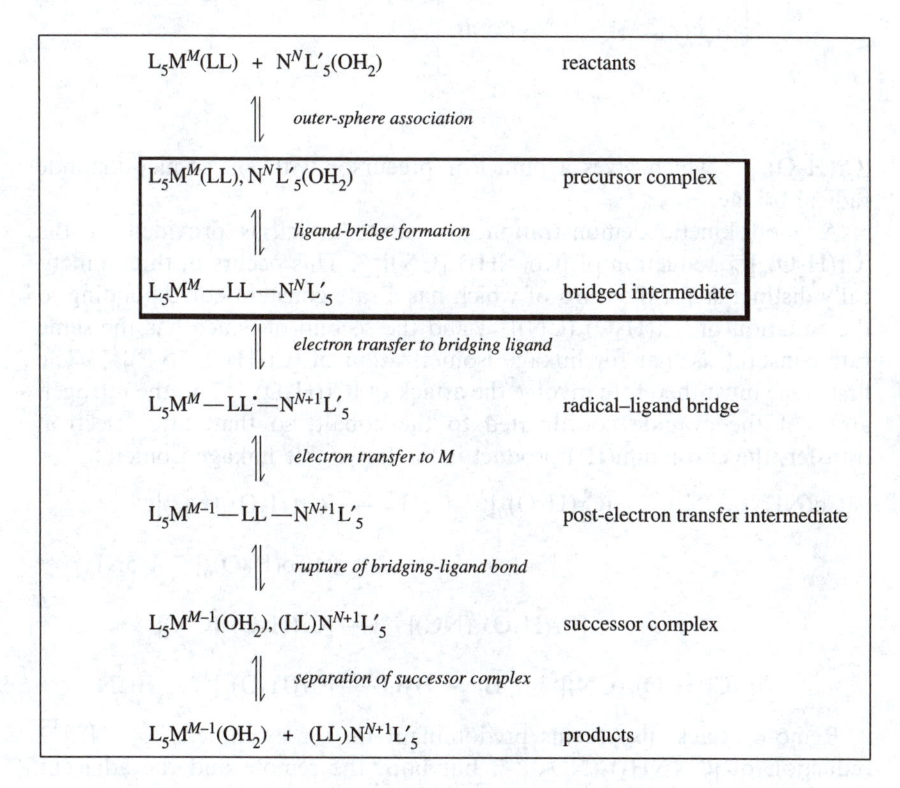

Many potential bridging ligands have alternative sites available for bridging to the second metal centre in the generation of precursor complexes (Fig. 9.7). Often the site of attack is determined by the balance between steric control, namely the ease of approach to a remote site, and the ease of electron transfer from an adjacent site. This simple choice may be

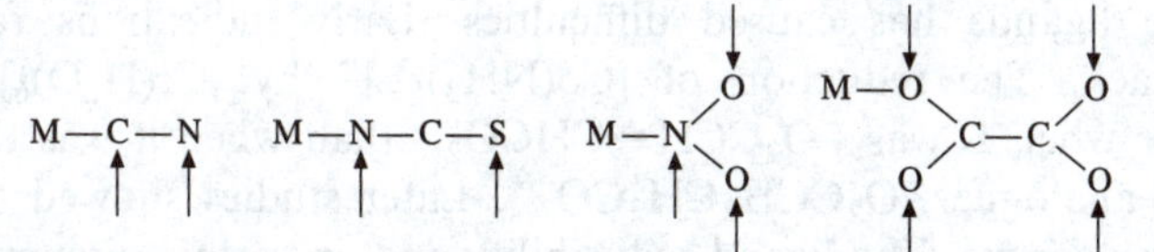

Fig. 9.7 Possible alternative sites of attack on some common bridging ligands.

complicated by such matters as the possibility of facile electron transfer from a remote site through a conjugated ligand and the possibility of the transfer of the electron to an easily reducible ligand to give a radical-bridged intermediate (cf. Section 9.4.3). Indeed, one of the early demonstrations of remote attack was that of the reduction of $[Co(NH_3)_5(isonic)]^{3+}$, **7**, by

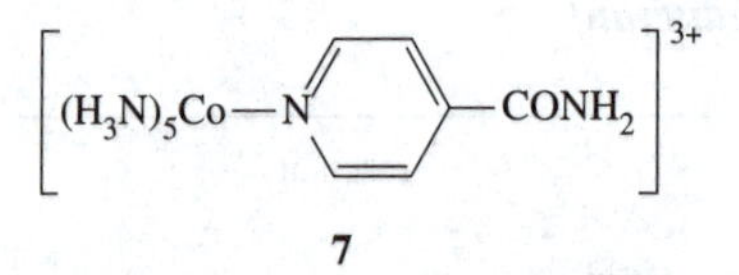

7

$[Cr(H_2O)_6]^{2+}$ which gives a binuclear precursor with an iso-nicotinamide radical bridge.

A good kinetic demonstration of remote attack is provided by the $[Cr(H_2O)_6]^{2+}$ reduction of $[Co(NH_3)_5(CN)]^{2+}$. This occurs in three kinetically distinct steps, the third of which has a rate constant corresponding to the equation of $[Cr(H_2O)_5(CN)]^{2+}$, and the second of which has the same rate constant as that for linkage isomerization of $[Cr(H_2O)_5(NC)]^{2+}$. The first stage must therefore involve the attack of $[Cr(H_2O)_6]^{2+}$ on the nitrogen atom of the cyanide coordinated to the cobalt, so that, after electron transfer, the chromium(III) product is the less stable linkage isomer:[22]

$$[Co(NH_3)_5(CN)]^{2+} + [Cr(H_2O)_6]^{2+} + 5H^+ \longrightarrow [Cr(H_2O)_5(NC)]^{2+} + [Co(H_2O)_6]^{2+} + 5NH_4^+$$

$$[Cr(H_2O)_5(NC)]^{2+} \longrightarrow [Cr(H_2O)_5(CN)]^{2+}$$

$$[Cr(H_2O)_5(CN)]^{2+} + H^+ + H_2O \longrightarrow [Cr(H_2O)_6]^{3+} + HCN$$

Remote attack also seems predominant in, for example, $[Co(CN)_5]^{3-}$ reduction of $[Co(NH_3)_5(NO_2)]^{2+}$, but both the remote and the adjacent attack pathways are significant in the $[Cr(H_2O)_6]^{2+}$ reduction of $[Co(NH_3)_5(SCN)]^{2+}$. Reduction of this cation by $[Co(CN)_5]^{3-}$ involves predominantly adjacent attack; presumably the 'soft' reductant prefers the 'soft' sulphur of the coordinated thiocyanate. This type of redox reaction sometimes proves useful in the preparation of unstable linkage isomers, for example *N*-bonded cyanide (as above) or thiocyanate bound through sulphur to 'hard' metal centres.

The identification of the site of attack by the reductant at coordinated carboxylate ligands has caused difficulties. Early indications favoured remote attack. The reduction of $[Co(NH_3)_5L]^+$ by $[Cr(H_2O)_6]^{2+}$ was much faster when L was $^-O_2CCH{=}CHCO_2^-$ than when it was the non-conjugated analogue $^-O_2CCH_2CH_2CO_2^-$. Later studies showed that the situation was confused by ligand reducibility and, in certain circumstances, chelation. Recent studies of a series of specifically designed complexes, **8**, have indicated that remote attack can give rise to very rapid reduction, at least by $[Cr(H_2O)_6]^{2+}$, $[V(H_2O)_6]^{2+}$, Eu^{2+}aq, and U^{3+}aq in such systems. The binuclear complexes, **8**, have the 'adjacent' carboxylate group incorpo-

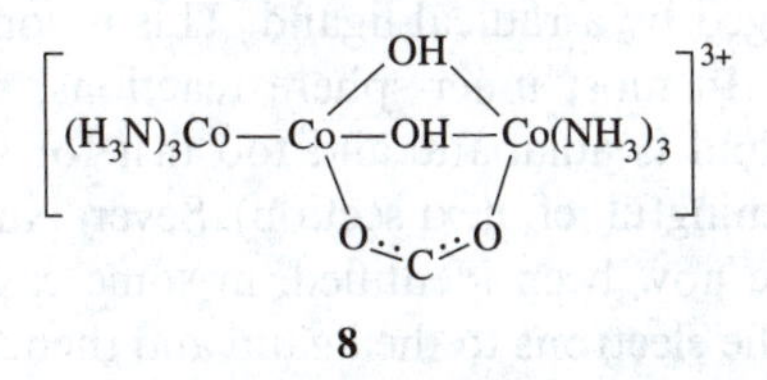

8

rated as the third bridge, but a series of groups, R, contain 'remote' pendant carboxylate groups. When these are conjugated through to the cobalt, reduction is 10^2–10^7 times faster than when there is no direct delocalized path from the 'remote' carboxylate to the cobalt.[23]

9.4.3 *Electron transfer to the bridging ligand*

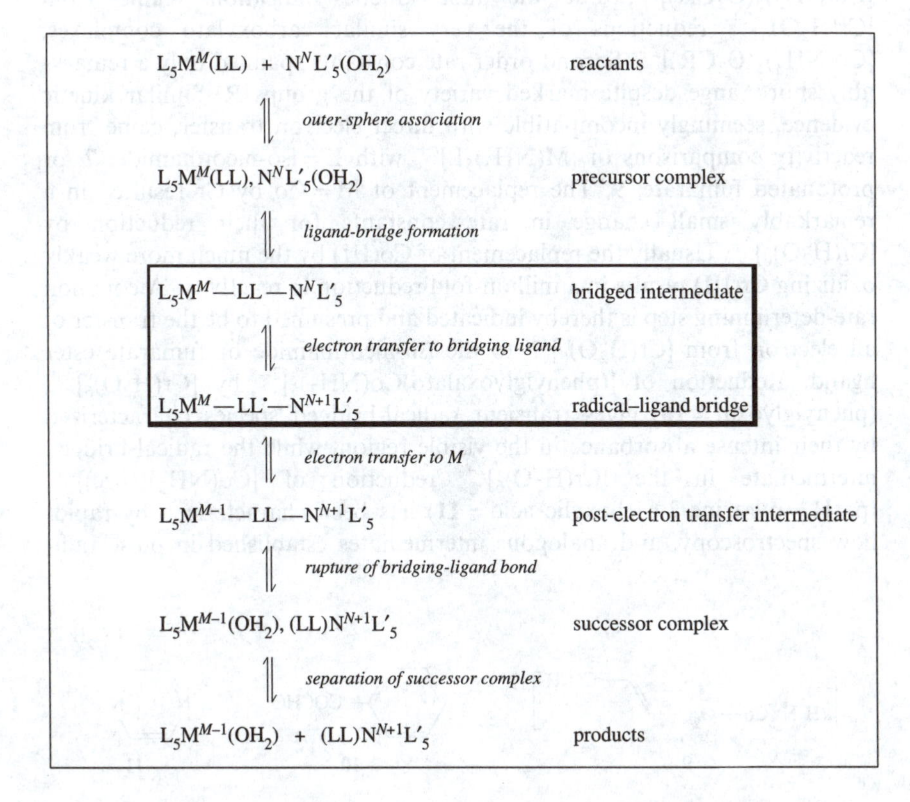

The inner-sphere mechanism involves linking the two metal centres through the atoms and the orbitals of a bridging ligand. The electron therefore traverses the bridge in the course of the net redox process. If it dwells on the ligand for an appreciable time, then there is an intermediate in which the metal centres are transiently bridged by a radical ligand. This is sometimes called the *chemical mechanism.*[24] In most inner-sphere reactions, electron transfer through the bridging ligand is adiabatic and too fast for such an intermediate to be chemically meaningful (cf. next section). Several instances of the 'chemical mechanism' have now been identified, in some cases with rate constants for the transfer of the electrons to the ligand and thence to the oxidizing metal centre either measured or estimated (Section 9.4.3.1). In connection with the latter step, we also mention some kinetic data for analogous electron transfer in mononuclear radical ion complexes.

9.4.3.1 Radical-bridged intermediates

The earliest suggestion of a radical-bridged intermediate in a redox reaction may have been made in connection with the oxidation of $[Co(NH_3)_5(O_2CH)]^{2+}$, but the first kinetic indications came from $[Cr(H_2O)_6]^{2+}$ reductions of the very similar carboxylato complexes, $[Co(NH_3)_5(O_2CR)]^{2+}$. Second-order rate constants spanned only a remarkably short range despite marked variety of the groups R. Similar kinetic evidence, seemingly incompatible with direct electron transfer, came from reactivity comparisons of $[M(NH_3)_5L]^{n+}$ with L = iso-nicotinamide, **7**, or protonated fumarate, **9**. The replacement of M = Co by Cr resulted in a remarkably small change in rate constants for their reduction by $[Cr(H_2O)_6]^{2+}$. Usually the replacement of Co(III) by the much more weakly oxidizing Cr(III) results in a million-fold reduction in reactivity. A common rate-determining step is thereby indicated and presumed to be the transfer of an electron from $[Cr(H_2O)_6]^{2+}$ to the *iso*-nicotinamide or fumarate ester ligand. Reduction of $[(phenylglyoxalato)Co(NH_3)_5]^{2+}$ by $[Cr(H_2O)_6]^{2+}$, (phenylglyoxal = **10**) gives transient radical-bridged species characterized by their intense absorbance in the visible region, while the radical-bridged intermediate in the $[Cr(H_2O)_6]^{2+}$ reduction of $[Co(NH_3)_5(pzc)]^{2+}$, (pzcH = pyrazine-2-carboxylic acid = **11**) has been characterized by rapid-flow spectroscopy, and analogous intermediates established in pulse radi-

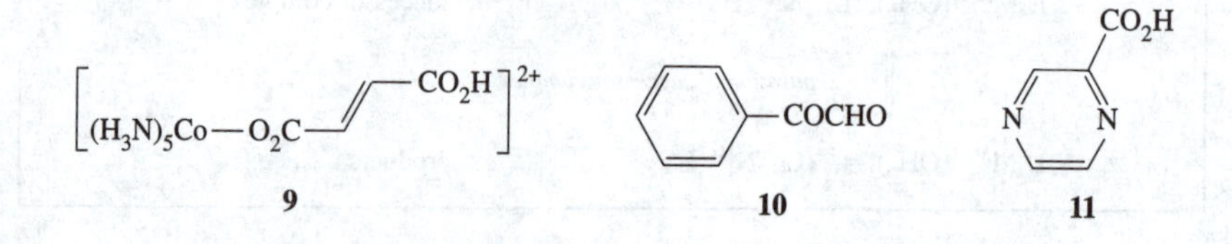

Table 9.4 Rate constants for intramolecular electron transfer from radical ligand to cobalt(III): $Co^{III}-L-L^{\cdot}-M \longrightarrow Co^{II}-L-L-M$.

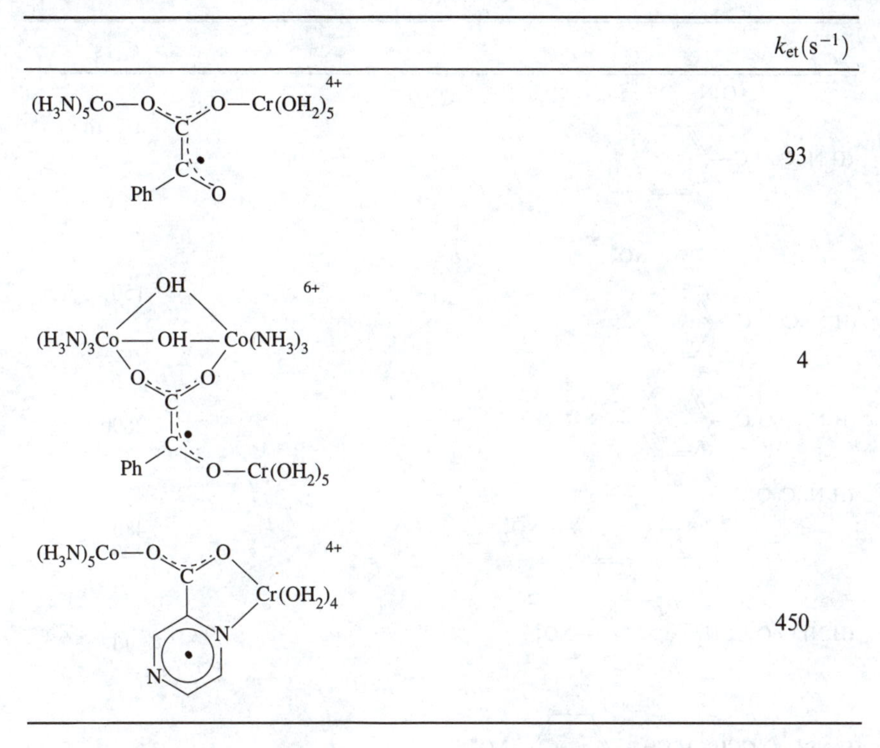

olysis studies of $[Co(NH_3)_5(imidH)]^{3+}$.[25] Structures of these intermediates, and rate constants for electron transfer from bridging ligand to the cobalt(III), are given in Table 9.4. These rate constants may be compared with those for electron transfer from radical ligand to metal in mononuclear analogues in the following section (see Table 9.5).

It will be noted that all the radical bridges mentioned in this section involve cobalt(III) as the oxidizing centre. Transfer of the electron from the bridging radical to cobalt(III) is slow, permitting the detection of the intermediate, because the electron has to move from an orbital of π-symmetry on the bridging ligand to the e_g orbital of σ-symmetry on the metal in order to reduce the $t_{2g}{}^6$Co(III) to $t_{2g}{}^6 e_g{}^1$Co(II). In contrast, reduction of Ru(III) to $t_{2g}{}^5$Ru(II) is rapid since the t_{2g} orbital on the Ru(III) to which the electron passes has the same π-symmetry as that on the bridging ligand. Hence no evidence has yet been obtained for the operation of the 'chemical mechanism' in ruthenium(III) oxidations.

Table 9.5 Rate constants for intramolecular electron transfer from coordinated nitrobenzoate and related radical ligands to cobalt(III): $Co^{III}-L^{\cdot} \longrightarrow Co^{II}-L$.

	k_{et}/s^{-1}
$[(H_3N)_5CoO_2C-C_6H_4-(o\text{-}NO_2^{\bullet})]^{2+}$	4×10^5
$[(H_3N)_5CoO_2C-C_6H_4-(m\text{-}NO_2^{\bullet})]^{2+}$	150
$[(H_3N)_5CoO_2C-C_6H_4-(p\text{-}NO_2^{\bullet})]^{2+}$	2600
$[(H_3N)_5CoO_2C-CH{=}CH-C_6H_4-NO_2^{\bullet}]^{2+}$	480
$[(H_3N)_5CoO_2CCH_2-C_6H_4-NO_2^{\bullet}]^{2+}$	390
$[(H_3N)_5CoO_2CCH_2CH_2CH_2-C_6H_4-NO_2^{\bullet}]^{2+}$	150
$[(H_3N)_5CoO_2CCH_2NHCO-C_6H_4-NO_2^{\bullet}]^{2+}$	5.8
$[(H_3N)_5Co-N(\text{imidazole radical})NH]^{3+}$	~3000
$(bipy)_2Co(bipy^{\bullet})^{2+}$	3.5^a

[a] The rate constant for intramolecular electron transfer within $(bipy)_2Cr(bipy^{\cdot})^{2+}$ is the same.

9.4.3.2 Electron transfer: radical ligand to metal

Mononuclear complexes containing radical ligands have been prepared in a variety of circumstances and, more or less, well characterized.[26] Treatment of $[Co(NH_3)_5L]^{2+}$ with excited-state $*[Ru(bipy)_3]^{2+}$ gives, for L = nitrobenzoate, a complex containing the nitrobenzoate radical anion coordinated to the $[Co(NH_3)_5]^{3+}$ moiety. When L = benzoate or acetate, no analogous complexes are formed, since these ligands are not sufficiently

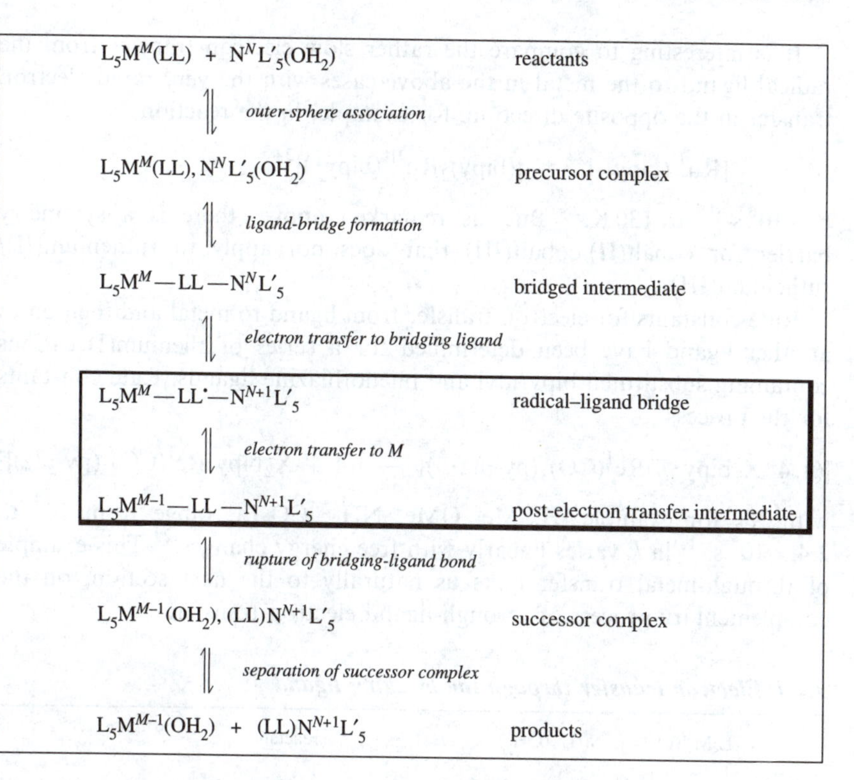

reducible. Rate constants for intramolecular electron transfer from radical anion ligands to cobalt(III) are shown in Table 9.5.[27]

The relative reactivities of the 2-, 3-, and 4-substituted series in Table 9.5 emphasize the importance of conjugation, which is still effective when $-CH{=}CH-$ is inserted between the aromatic ring and the carboxylate. If conjugation between $-NO_2$ and the cobalt is blocked, as when $-CH_2NHCO-$ is inserted between the ring and the carboxylate, then the rate constant is markedly reduced. If the geometry is suitable, the electron can bypass a non-transmitting $-(CH_2)_n-$ bridge by through-space transfer, as shown in **12**. Such through-space transfers may be nearly as rapid as through a delocalized ligand. Entropies of activation for these electron

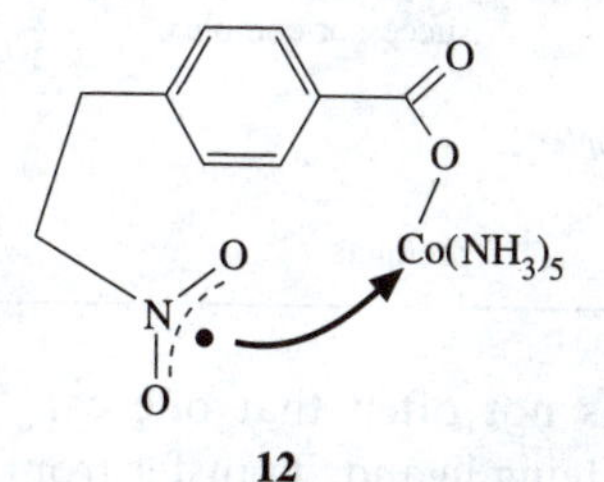

12

transfers are, as expected, close to zero; $\Delta S^{\ddagger} = +11\,J\,K^{-1}\,mol^{-1}$ for the parent 4-nitrobenzene complex.

It is interesting to compare the rather slow electron transfer from the radical ligand to the metal in the above cases with the very rapid electron transfer in the opposite direction, for example, in the reaction

$$^{*}[\mathrm{Ru^{II}(bipy)_3}]^{2+} \longrightarrow [(\mathrm{bipy})_2\mathrm{Ru^{III}(bipy\cdot)}]^{2+}$$

$k \sim 10^6\ \mathrm{s^{-1}}$ at 130 K.[28] But, as remarked above, there is a symmetry barrier for cobalt(II)/cobalt(III) that does not apply to ruthenium(II)/ruthenium(III).

Rate constants for electron transfer from ligand to metal and then on to another ligand have been determined for a series of rhenium(I) cations containing substituted bipyridyl and phenothiazine ligands. Rate constants for the process

$$[(4,4'\text{-}\mathrm{X_2bipy}\cdot^{-})\mathrm{Re^I(CO)_3}(\text{py-ptz}\cdot^{+})]^{+} \longrightarrow [(4,4'\text{-}\mathrm{X_2bipy})\mathrm{Re^I(CO)_3}(\text{py-ptz})]^{+}$$

with X = for example, H, Me, OMe, NH_2, CO_2Et, range from 10^7 to $2.4 \times 10^8\ \mathrm{s^{-1}}$; ln k varies linearly with free energy changes.[29] This example of through-metal transfer links us naturally to the next section, on the complementary process of through-ligand electron transfer.

9.4.4 Electron transfer through the bridging ligand

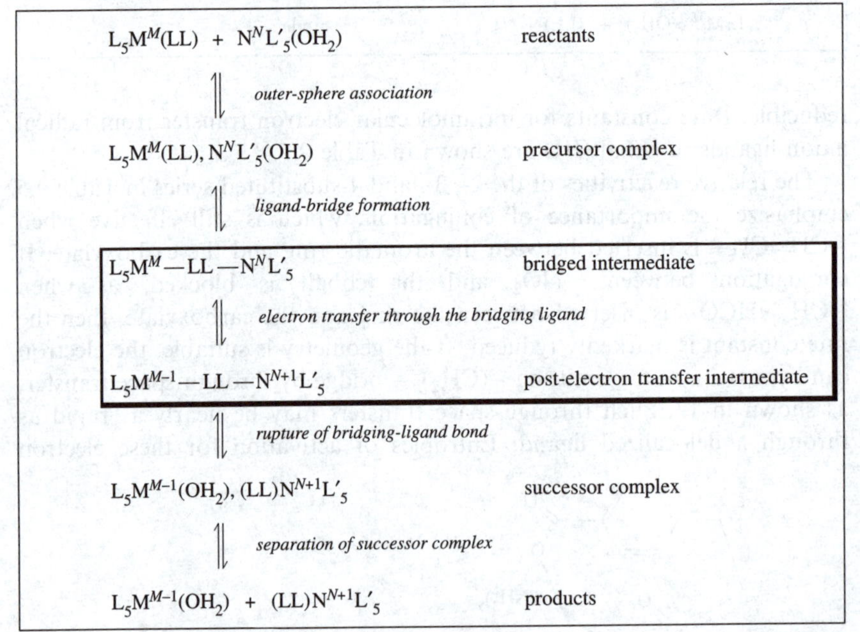

As stated in the preceding section, it is not often that one can 'see' the transferring electron dwelling on the bridging ligand. Transfer from N to M is usually fast on the time-scale of the formation of the precursor complex. For those precursor complexes whose formation was discussed in Section

9.4.1, electron transfer within the precursor leads directly in one kinetic step to the successor complex, which then usually dissociates rapidly into products. Monitoring d[precursor]/d*t* (and often +d[products]/d*t*) gives k_{et}, the rate constant for electron transfer through the bridging ligand from N to M. Such data can be supplemented by electron transfer information derived from model precursor complexes containing identical metal atoms in different oxidation states—the so-called *mixed-valence* (*MV*) complexes.[30,31] These may contain the two metals in identical coordination environments, as in $[(H_3N)_5Ru(pz)Ru(NH_3)_5]^{5+}$, or different, as in $[(H_3N)_4Ru(bipym)Ru(bipy)_2]^{5+}$. Electron transfer in this type of compound is evidenced by the appearance of charge-transfer bands. That such bands can arise from metal → metal (rather than, or as well as, M ↔ L charge (electron) transfer) is shown by the following pair of experiments. When solutions of *trans*-$[Co(en)_2(NH_3)(H_2O)]^{3+}$ and $[Fe(CN)_6]^{4-}$ are mixed, an intense red colour develops at a rate that corresponds to the replacement of the water coordinated to cobalt(III). However, when solutions containing $[Cr(NH_3)_5(H_2O)]^{3+}$ and $[Co(CN)_6]^{3-}$ are mixed no new absorption bands are seen. Metal ↔ ligand charge transfer bands are feasible for both, but whereas iron(II) → cobalt(III) charge transfer is possible for the former binuclear complex, chromium(III) → cobalt(III) charge transfer is wildly improbable in the latter complex. The extraction of kinetic data (k_{et} values) from charge-transfer spectra is not simple and will be discussed in Section 9.4.4.5 after consideration of mixed-valence complexes and their properties.

9.4.4.1 Generation of binuclear intermediates and mixed-valence complexes

There are quite severe restrictions on the types of binuclear systems, precursors or mixed-valence models which can be studied with a view to obtaining intramolecular electron transfer rate constants. These restrictions are imposed by the need for a fair degree of kinetic inertness for both metal centres, and an appropriate balance between substitution and redox rates; and there is, of course, the need for the sign of ΔG° for electron transfer in the required direction to be negative. Much of the work on mixed-valence complexes has been concerned with ruthenium(II)/ruthenium(III) species, as both metal centres are low-spin and substitutionally inert. The analogous systems of the congener osmium(II)/osmium(III) are less accessible because the extreme inertness to substitution causes preparative problems, and the tendency of Os(II) species to be oxidized by atmospheric oxygen is also a disincentive.

It is much more difficult to find suitable iron(II)/iron(III) pairs due to the prevalence of substitutional lability, the majority of iron(II) complexes being high-spin. However, $[Fe(CN)_5L]^{3-}$ derivatives are low-spin, due to the large ligand field of cyanide, and a series of binuclear Fe^{II}–LL–Co^{III} complexes have been prepared and the electron transfer rate constants, k_{et}, measured successfully. In fact, such complexes can be prepared very easily (Fig. 9.8). Substitution of the bridging ligand at the iron atom is fairly rapid (studiable

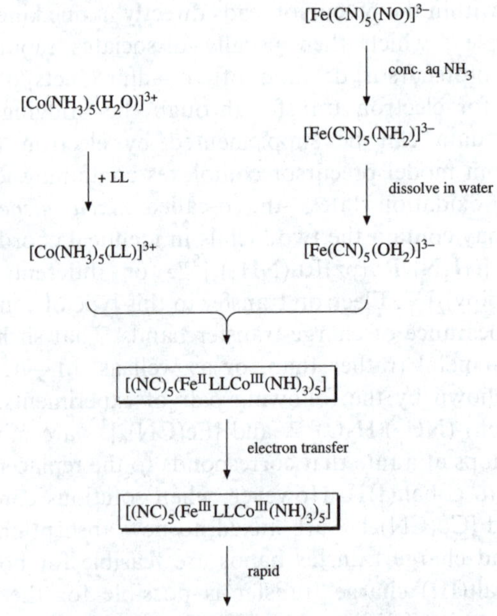

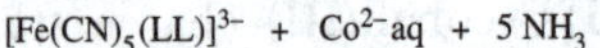

Fig. 9.8 Generation of $Co^{III}-LL-Fe^{II}$ binuclear species for the monitoring of intramolecular electron transfer.

by the stopped-flow technique) and so the slower subsequent electron transfer from iron to cobalt can be monitored easily in several systems of this type. There may be a significant difference between the detailed mechanism of the latter step carried out under thermal and photochemical conditions. Intramolecular thermal electron transfer in the pyrazine carboxylate-bridged species $[L_4Co(pzc)Fe(CN)_5]^-$, where $L=(NH_3)_4$ or $(en)_2$, is characterized by large positive activation volumes, +24 and +27 $cm^3\ mol^{-1}$ respectively (in aqueous solution), but under photochemical activation the observed activation volume is only +3 $cm^3\ mol^{-1}$. Under thermal activation the electron transfers in a single step from iron to cobalt, with the activation volume reflecting large hydration changes around the $Fe(CN)_5$ moiety as the charge on the metal increases from 2+ to 3+. Large hydration changes are to be expected in view of the big difference in partial molar volumes between the hexacyanoferrate(II) and (III) anions in aqueous solution (75 and 120 $cm^3\ mol^{-1}$, respectively). When the intramolecular electron transfer is photochemically excited, the electron transfers from the iron to the pzc ligand extremely quickly, and the rate-limiting step is now the orbitally forbidden transfer to the cobalt. As there is no change in hydration around the $Fe(CN)_5$ moiety this time, the activation volume is close to zero, as expected.[32]

An alternative approach involved making a binuclear complex containing both metal atoms in the reduced, or both in the oxidized, form and then carrying out a one-equivalent oxidation or reduction with a rapid oxidant ((Ce(IV)) or reductant (Eu(II); ascorbate; dithionite). This approach is satisfactory so long as the one-equivalent oxidant or reductant reacts quickly enough and gives the thermodynamically unstable isomer of the two possible redox species. In other words, fast reduction of M^{III}–LL–N^{III} should yield M^{III}–LL–N^{II}, which should then undergo relatively slow electron transfer to give the stable M^{II}–LL–N^{III}. Fortunately, there are several cases that conform to these stringent requirements, the most studied of which have been binuclear cobalt–ruthenium complexes. If one starts with a complex Co^{III}–LL–Ru^{III}, the ruthenium is more rapidly reduced, $d^5 \rightarrow d^6$ being more facile than $d^6 \rightarrow d^7$, cf. above, and one can then follow the electron transfer to give the thermodynamically favoured Co^{II}–LL–Ru^{III} product. Mixed-valence Ru^{III}–LL–Ru^{II} complexes can be made in an analogous way. For example, the mixed-valence dinuclear Schiff-base complex, **13**, can be made from the diruthenium(III) analogue using one equivalent of the enforcedly substitution-inert $[Co(sep)]^{2+}$ (sep = the encapsulating sepulchrate ligand, **14**) as reductant. Here there is no problem over which metal atom is reduced, but the requirements regarding substitution inertness still hold. The converse approach, using one-electron oxidation by Ce^{IV}, has been used to generate Os^{III}–LL–Co^{II} species from Os^{II}–LL–Co^{II}; spontaneous electron transfer takes place in the former to give Os^{II}–LL–Co^{III}. The choice of which cobalt(II) complex can be used as starting material is severely limited by the necessity of substitution inertness, though in recent years the number and variety of encapsulated cobalt complexes has increased considerably.[33]

Electrochemical oxidation or reduction can also be used to generate mixed-valence complexes from M^{II}–LL–N^{II} or M^{III}–LL–N^{III} species. For M = N = Ru, this can be useful in the final stage of the synthesis of mixed-valence complexes from $[Ru(NH_3)_5(H_2O)]^{3+}$ (Fig. 9.9). Sometimes, when M = N, it is possible to generate a mixed-valence complex by the simple expedient of mixing equimolar amounts of the 'both reduced' and the 'both oxidised' forms, e.g.

$$[(Cl(phen)_2Ru^{II}\text{-pz-}Ru^{II}(phen)_2Cl]^{4+} + [(Cl(phen)_2Ru^{II}\text{-pz-}Ru^{II}(phen)_2Cl]^{6+} = 2[(Cl(phen)_2Ru^{II}\text{-pz-}Ru^{II}(phen)_2Cl]^{5+}$$

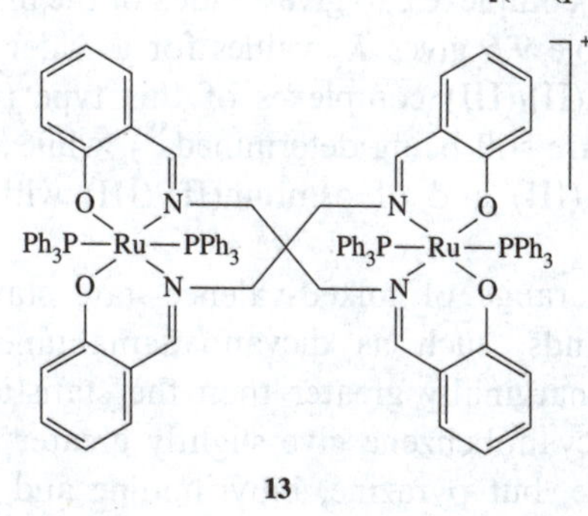

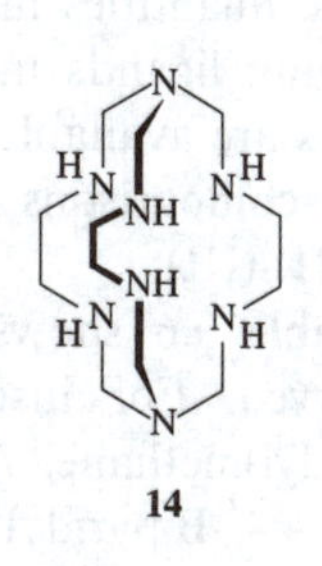

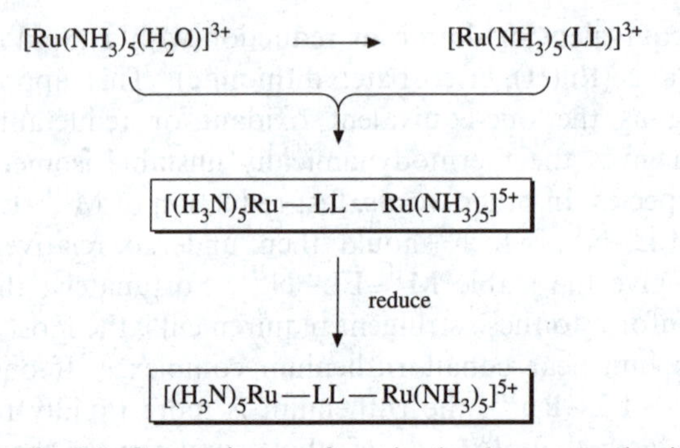

Fig. 9.9 Generation of mixed-valence ruthenium(II)–ruthenium(III) species.

9.4.4.2 Stability and comproportionation constants

As hinted in the previous section, mixed-valence complexes tend to be stable with respect to disproportionation into their non-mixed-valence binuclear parents. This stability is usually quantified in terms of K_c, the constant for the comproportionation equilibrium, which for an equilibrium of the type

$$\mathrm{Ru^{II}-LL-Ru^{II} + Ru^{III}-LL-Ru^{III} \rightleftharpoons 2Ru^{II}-LL-Ru^{III}}$$

takes the form

$$K_c = \frac{2[\mathrm{Ru^{II}-LL-Ru^{III}}]}{[\mathrm{Ru^{II}-LL-Ru^{II}}][\mathrm{Ru^{III}-LL-Ru^{III}}]}$$

For 50% formation of the mixed-valence complex, the value for $K_c = 4$. K_c can be measured by spectrophotometric titration, using an appropriate oxidant or reductant, but more usually it is obtained electrochemically.[34,35] Very good agreement between spectrophotometric and electrochemical estimates for K_c was demonstrated by values for the 4,4′-bipyridyl-bridged $[(H_3N)_5Ru(4,4'\text{-bipy})Ru(NH_3)_5]^{5+}$ cation.[36] If redox potentials for $n, n \Leftrightarrow n, n+1 \Leftrightarrow n+1, n+1$ differ by ΔE, then $\log K_c = 16.9(\Delta E/V)$. For $K_c = 4$, $\log K_c = 0.60$ and $\Delta E = 0.035$ V. Relationships between K_c, ΔE, and ΔG are depicted in Fig. 9.10 for a limited range of $[(H_3N)_5Ru{-}LL{-}Ru(NH_3)_5]^{5+}$ complexes, to give an idea of the magnitudes of the quantities involved. Table 9.6 gives K_c values for a wider range of bridging ligands in ruthenium(II)/(III) complexes of this type (yet more values are available,[37] indeed are still being determined[38]), while Table 9.7 gives comparisons of iron(II)/(III) and of osmium(II)/(III) with ruthenium(II)/(III).

Table 9.6 shows the great range of mixed-valence-state stabilization observed. For insulating ligands, such as dicyanoadamantane or bis(pyridyl)methane, K_c is only marginally greater than the statistical value of 4. 4,4′-Bipyridyl and 1,4-dicyanobenzene give slightly greater stabilization of the mixed-valence state, but pyrazine, bipyrimidine and 4-cyano-

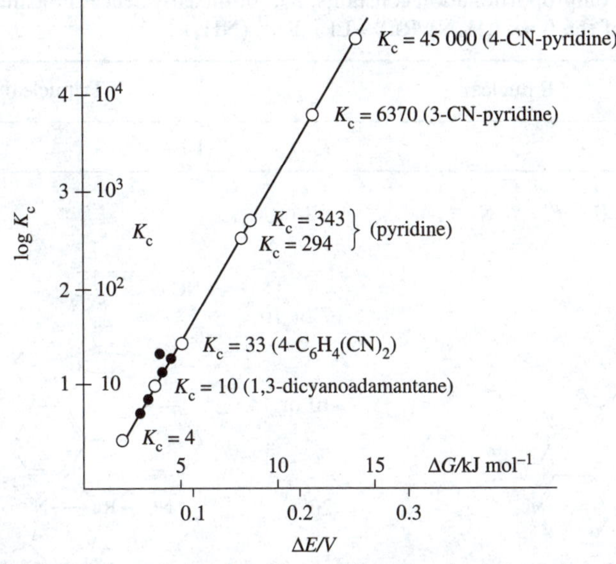

Fig. 9.10 Plot relating log K_c to differences in potential and Gibbs free energy, for $[(H_3N)_5Ru-LL-Ru(NH_3)_5]^{6+/5+/4+}$ complexes.

pyridine give much greater stabilization, as does NCNCN. The tiny and delocalized NCCN bridge results in more than a million-fold increase in stabilization. There are three more ruthenium(II)/(III) mixed-valence complexes with extremely high K_c values, the oxidized cyclam derivative[39] at the foot of Table 9.6, the tris-chloride-bridged complex, **15**, with $K_c = 6 \times 10^{13}$, and the pyrazine-2,5-dicarboxylate (pzdc) bridged complex, **16**, with $K_c = 5 \times 10^{13}$. Both NCCN and Cl appear to be intrinsically good bridges for stabilizing ruthenium(II)/(III) complexes, but in relation to pzdc as a bridge it is important to note that the value of K_c for the mixed non-bridging ligand species, **16**, is enormously larger, by a factor of perhaps 10^8, than that of K_c for the symmetrical complexes $[L_4Ru\text{-pzdc-}RuL_4]^{3+}$ with $L_4 = (NH_3)_4$ or $(bipy)_2$. In general, $(bipy)_2$ is less effective than $(NH_3)_4$ in the stabilization of mixed-valence complexes, probably because the bipyridyl ligands strongly favour the d^6 ruthenium(II) state. These non-bridging ligand effects may depend on the nature of the bridging ligand, for with pyrazine as bridging ligand, substitution of 5-nitro or 4,7-dimethyl groups in the terminal

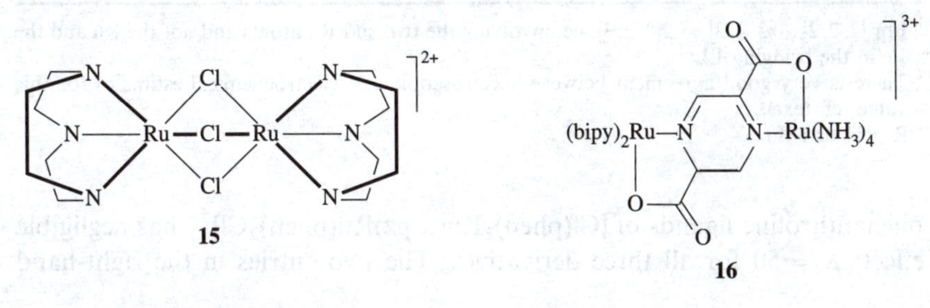

Table 9.6 Comproportionation constants, K_c, for mixed-valence ruthenium(II)/(III) complexes of the type, $[(H_3N)_5Ru^{II}-LL-Ru^{III}(NH_3)_5]^{5+}$.

Binuclear		Trinuclear	
LL	K_c	LL	K_c [a]
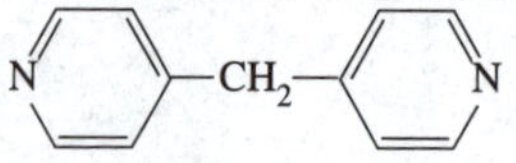 N—CH_2—N	7		
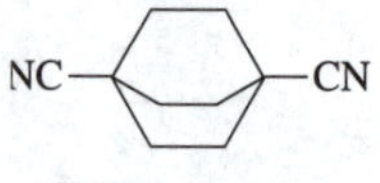 NC—CN	7 or 10	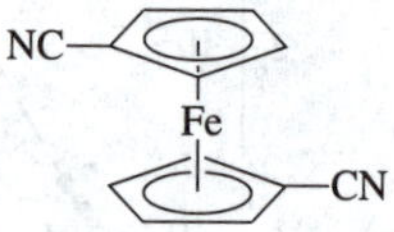 NC— Fe —CN	10
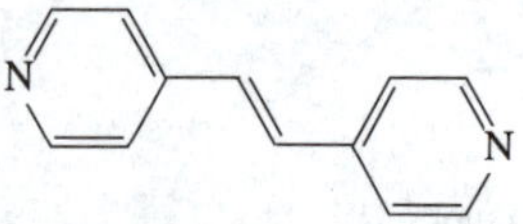 N N	10 or 14		
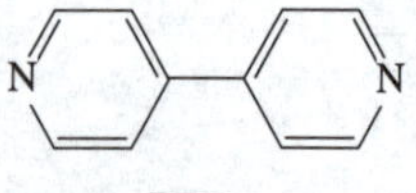 N N	20 [b]	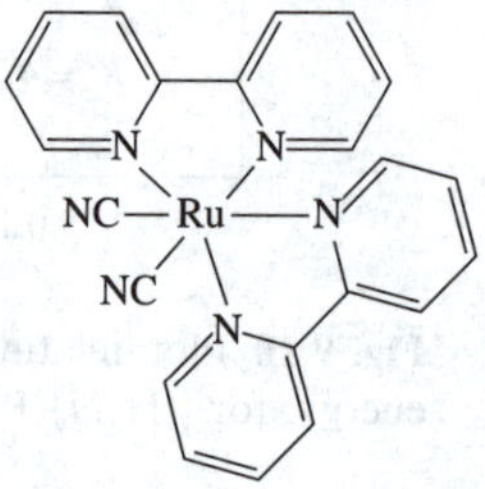 N N NC—Ru—N NC N	15
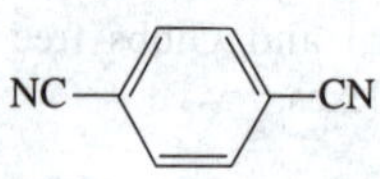 NC—CN	33		
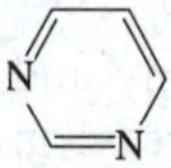 N N	230 or 340		
NCNCN	340		
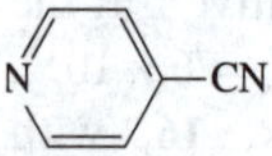 N —CN	4.5×10^5		
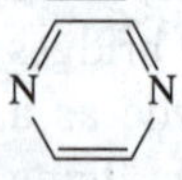 N N	1.3×10^6		
NC∸CR∸CN [c]	$> 10^{10}$		
N≡C–C≡N	$> 10^{13}$		
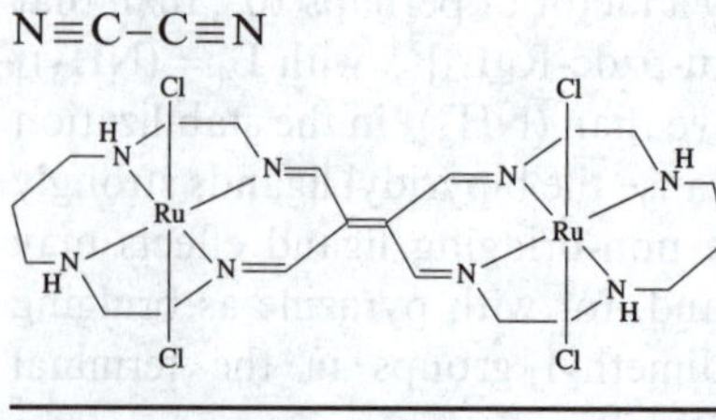 Cl Cl H N N= Ru =N N H H N N= Ru =N N H Cl Cl	3.5×10^{15}		

[a] For $[2,2,2] + [3,2,3] \Leftrightarrow 2[2,2,3]$, i.e. involving the two end Ru atoms and not the Ru and the Fe in the bridging LL.
[b] There is very good agreement between spectroscopic and electrochemical estimates for this value (cf. text).
[c] R = H or t-Bu.

phenanthroline ligands of $[Cl(phen)_2Ru(\mu\text{-pz})Ru(phen)_2Cl]^{5+}$ has negligible effect; $K_c = 50$ for all three derivatives. The two entries in the right-hand

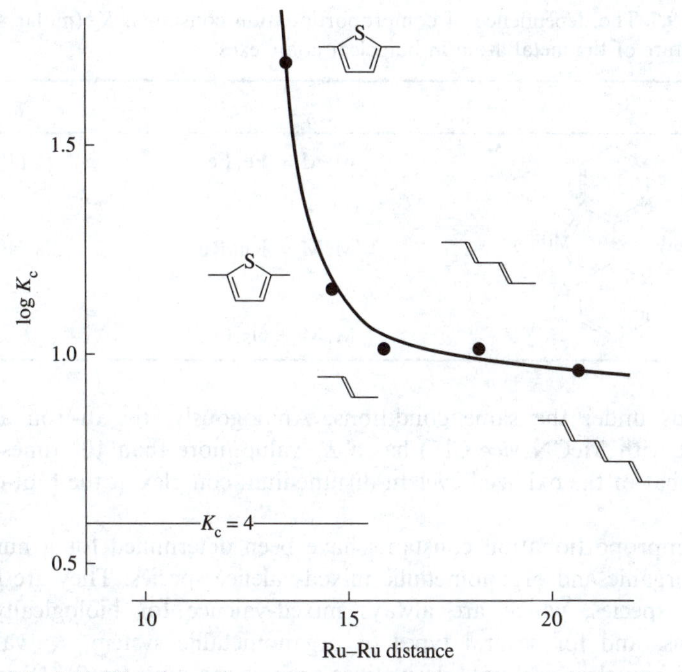

Fig. 9.11 The dependence of comproportionation constants ($\log K_c$) on inter-ruthenium distance for binuclear Ru^{II}/Ru^{III} complexes

$[(H_3N)_5Ru{-}N$ ⟨pyridine⟩ $-Y-$ ⟨pyridine⟩ $N{-}Ru(NH_3)_5]^{5+}$

with linking groups –Y– as shown on the plot.

column of Table 9.6 show that metal complexes acting as bridges are rather poor at stabilizing mixed-valence states even when delocalized.

One might expect communication between the two ruthenium atoms in a mixed-valence Ru^{II}/Ru^{III} species to decrease with increasing distance between the two atoms. It is not easy to test this theory, since it is impossible to vary distance without varying the nature of the communicating bridge. Moreover, it is an arduous and time-consuming task to prepare each of a series of binuclear complexes in crystals suitable for X-ray diffraction and solve its structure. Nonetheless, there are data available to support some sort of inverse correlation between $\log K_c$ and inter-Ru distance. Figure 9.11 shows one such set of data.[40]

Table 9.7[41] compares binuclear complexes as one descends the Periodic Table from iron through ruthenium to osmium. In these and other complexes, mixed-valence osmium species are generally about a million times more stable towards disproportionation than their ruthenium analogues. $[(H_3N)_5Os(N_2)Os(NH_3)_5]^{5+}$ has a particularly high K_c (10^{20} to 10^{21}). This very high stability is reflected in its kinetic behaviour, for this mixed-valence osmium complex is completely inert towards solvolysis in water at room temperature, whereas $[Os(NH_3)_5(N_2)]^{3+}$ has a half-life of only a few

Table 9.7 The dependence of comproportionation constants, K_c (molar scale), on the nature of the metal atom in binuclear complexes.

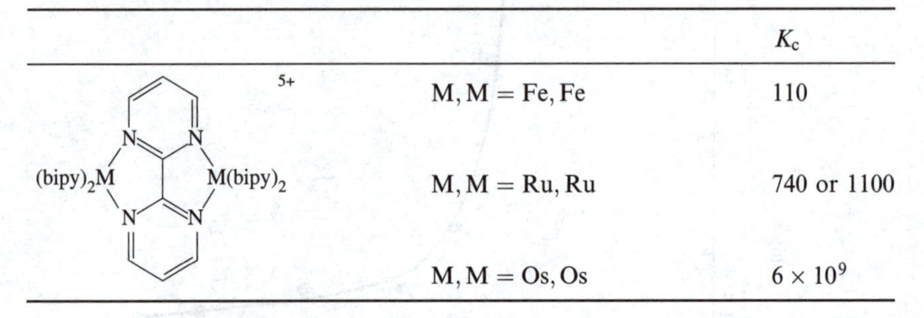

		K_c
	M, M = Fe, Fe	110
	M, M = Ru, Ru	740 or 1100
	M, M = Os, Os	6×10^9

seconds under the same conditions. Analogously, the di-iron analogue (albeit with MeCN *vice* Cl^-) has a K_c value more than 10^4 times smaller than that of the oxidized cyclam-diruthenium complex at the foot of Table 9.6.[42]

Comproportionation constants have been determined for a number of bioinorganic and organometallic mixed-valence species. They are high for Fe_4S_4 species, which are always mixed-valence for biologically active systems, and for several types of organometallic system. K_c values for simple binuclear carbonyl derivatives such as the tungsten(0)/(I) species **17** ($K_c = 3.6 \times 10^4$) reflect the expected good communication across the bridging diimine ligand. Biferrocene, **18**, has K_c values (somewhat solvent-dependent) approaching 10^6.[43] Arene-cyclopentadienyl di-iron complexes of the two complementary types shown as **19** and **20** also have high comproportiona-

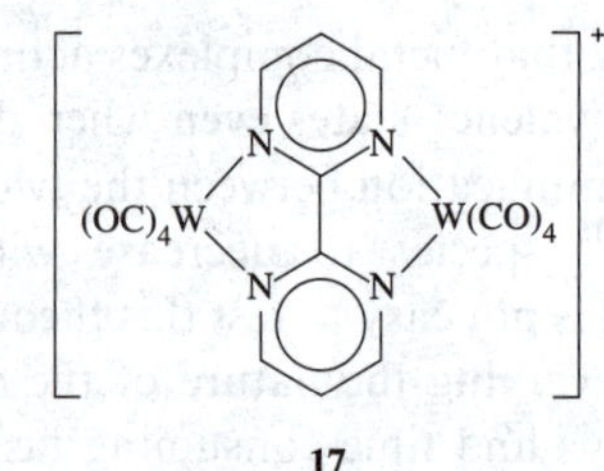

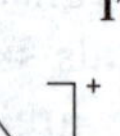

17

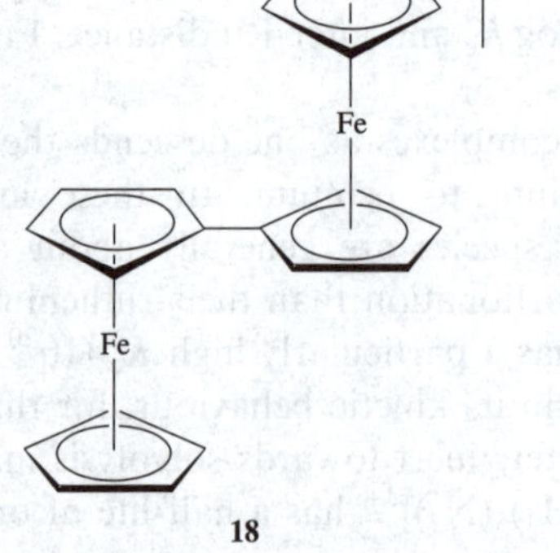

18

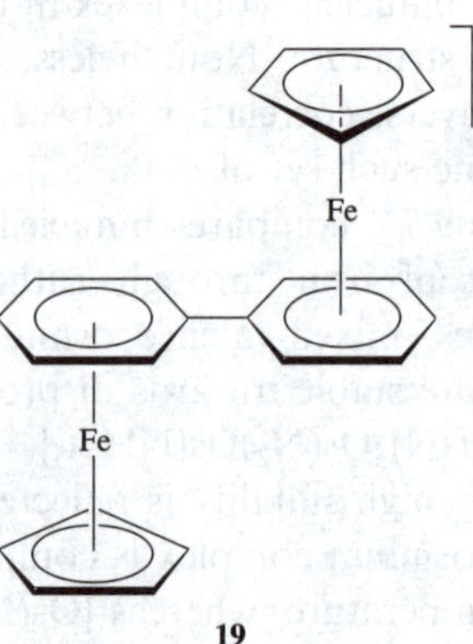

19

tion constants. Varying the number and position of alkyl groups on the benzene and cyclopentadiene rings affords an opportunity to tune K_c values, which, for instance, range from 10^6 to over 10^{12} for Fe^0/Fe^I species of type **20**.[44] Metal–metal communication is also possible across cyclophane and related bridges in units such as **21** and **22**, some of whose mixed-valence

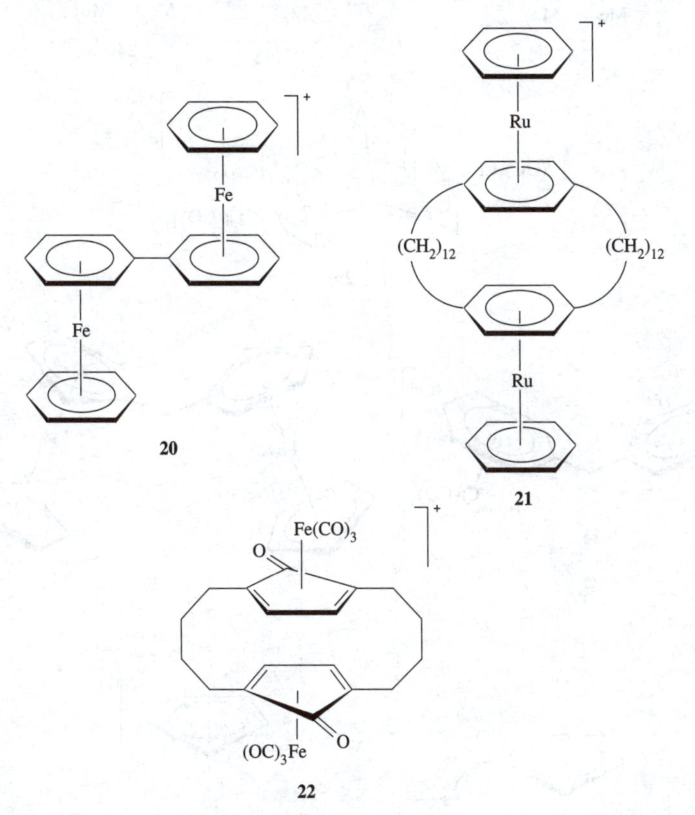

forms have had their electrochemical characteristics established.[45] Relatively small barriers to metal–metal communication of short non-conjugated units are apparent on comparing published values of ΔE for **23** and **24**, which are 370 and 290 mV ($K_c \sim 10^6$ and 10^5), respectively.[46] Another illustration of this is provided by $\Delta E = 140$ mV for **25**, where there is relatively efficient coupling through the alicyclic spacer—the chromium atoms are 6.5 Å apart. ΔE for the very similar compound **26** is only 45 mV ($K_c \sim 6$), almost down to the value of 36 mV ($K_c = 4$) for no stabilization. The $-CH_2-$ link proves a very efficient insulator here, as for diruthenium mixed-valence complexes (cf. the top left entry in Table 9.5 above), though its insulating properties are less evident in the biferrocene derivative **27** ($\Delta E = 170$ mV). Metal–metal communication also appears to be difficult in the bis-ferrocenyl species **28**, with $K_c = 21$ indicating that the delocalized nature of the bridge does not compensate for its length. The trinuclear rhodium compound **29** provides an interesting variant, in that

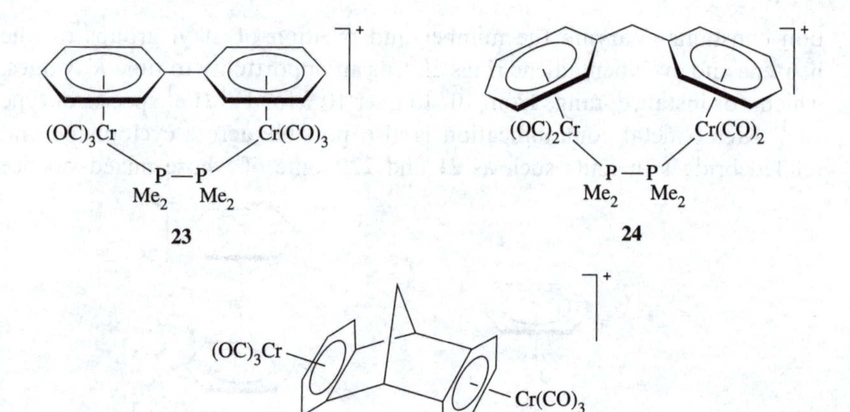
+
(OC)3Cr
Cr(CO)3
P—P
Me2 Me2
23
+
(OC)2Cr
Cr(CO)2
P—P
Me2 Me2
24
+
(OC)3Cr
Cr(CO)3
25

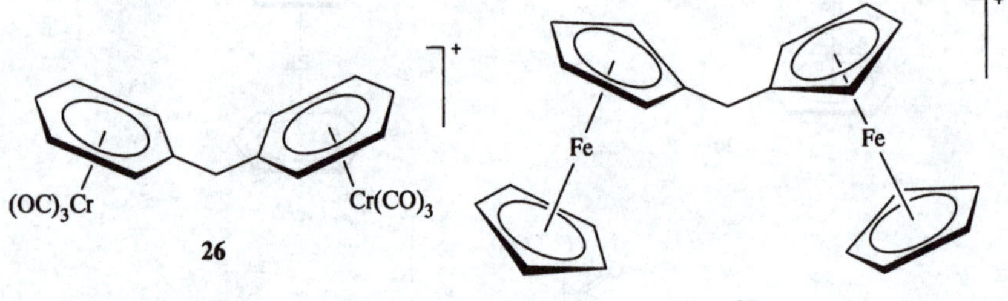
+
(OC)3Cr
Cr(CO)3
26
+
Fe
Fe

27

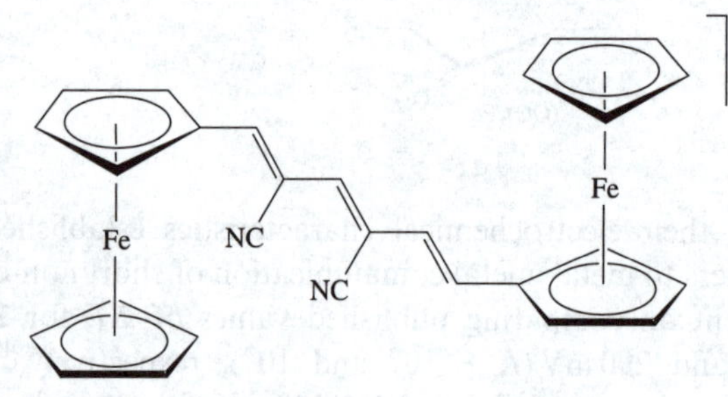
+
Fe
Fe
NC
NC
28

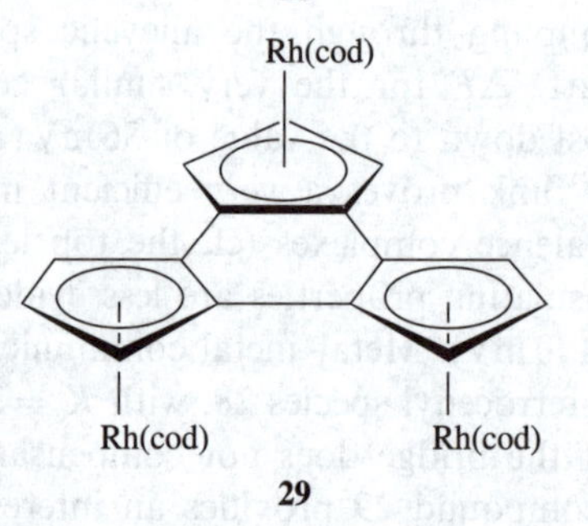
Rh(cod)
Rh(cod)
Rh(cod)
29

there is a much larger ΔE between the $[Rh]_3^+/[Rh]_3^{2+}$ couple (−160 mV) and the $[Rh]_3^{2+}/[Rh]_3^{3+}$ couple (+430 mV) than there is between the $[Rh]_3/[Rh]_3^+$ couple (−330 mV) and the $[Rh]_3^+/[Rh]_3^{2+}$ couple (−160 mV), suggesting respectively Class III and Class II behaviour in the two cases.[48]

Finally, we should mention some mixed-valence binuclear complexes in which K_c is greater than 4, not because of delocalization or charge transfer but because of steric constraints imposed by ligand and structure. Thus the cobalt(II)/(III) complex containing dtne, **30**, as bridging ligand with a saturated link, has $K_c = 33$, reflecting the fact that this ligand permits the cobalt atoms to be in close proximity.[49] A similar phenomenon has been established for bis-fulvalene-bis-iron, **31**, where the Fe–Fe distance is

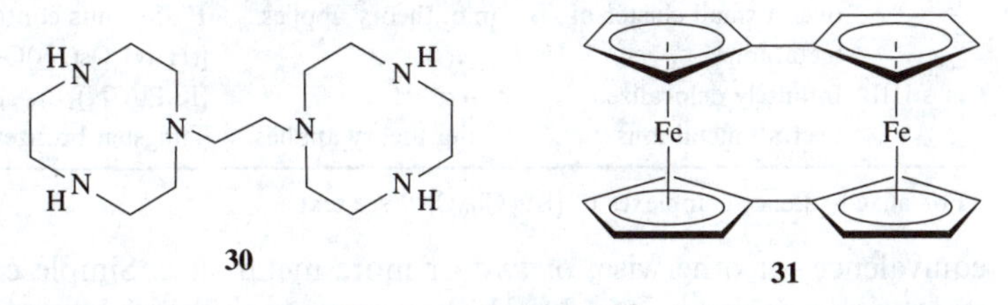

3.99 Å in the iron(II)/iron(II) form but only 3.64 Å in the mixed-valence iron(II)/iron(III) form. This short metal–metal distance gives a strong interaction, reflected in an intense intervalence charge transfer band in the visible–ultraviolet spectrum and Class IIIA behaviour.[50]

9.4.4.3 Classification of mixed-valence complexes

Comproportionation constants depend upon metal–ligand and metal–metal interactions with and through the bridge and they give a good thermodynamic indication of delocalization. To gain further insight and, in particular to get estimates for k_{et} for fast intramolecular electron transfer between metal centres, it is necessary to turn to spectroscopy. ESR, Mössbauer, and, especially, visible–ultraviolet spectroscopy are most relevant.

Before examining the role of spectroscopy in kinetics, we need to discuss briefly the classification of mixed-valence species, set out in Table 9.8.[51] Only certain types of mixed-valence complex, specifically Class II, are suitable for our present purposes. Class I compounds have such high barriers to electron transfer that spectroscopic manifestations can rarely be detected, while Class IIIA and Class IIIB species are fully delocalized. For Class III species there are therefore difficulties in relating observed absorptions to the precursor → successor process in the inner-sphere reaction sequence (Fig. 9.4); the main characteristic of these Class III species is that their metal centres all have the same, often fractional, oxidation state. The essentially complete delocalization which characterizes all Class III species can sometimes be established from Mössbauer or ESR spectroscopy, or from X-ray crystallography. Each of these techniques can show the

Table 9.8 Classification of mixed-valence complexes.

	Valence characteristics	Spectroscopic consequences	Examples
Class I	Localized or trapped	Any M → N charge transfer at very high energy (ν)	$[Co(NH_3)_6]_2[CoCl_4]_3$ '$GaCl_2$'=$Ga[GaCl_4]$
Class II	Significant delocalization but metal ions crystallographically distinguishable	M → N charge transfer in visible, near UV, or near IR	$Fe_4[Fe(CN)_6]_3 4H_2O$ Some Pt(II)/Pt(IV) chain compounds
Class IIIA	Completely delocalized over a small cluster of metal ions	$\pi \rightarrow \pi^*$ etc.; m.o. theory applies	$Nb_6Cl_{14} = [Nb_6Cl_{12}]Cl_2$ Fe_4S_4 units containing $Fe^{2.5+}$ $[(H_3N)_5Os(N_2)Os(NH_3)_5]^{5+}$ [a]
Class IIIB	Infinitely delocalized over all metal ions	$\pi \rightarrow \pi^*$ etc.; band theory applies	$[K_2Pt(CN)_4Br_{0.30}].3H_2O$ Tungsten bronzes

[a] For mixed-valence complexes of $[Ru_2Cl_{10}O]^{n-}$ see text.

equivalence, or otherwise, of two or more metal sites. Simple examples of structural indications of Class I character are afforded by red lead, gold dichloride, and the salt $[Co(NH_3)_6][CoCl_4]$. The first contains octahedral $Pb^{IV}O_6$ and pyramidal $Pb^{II}O_3$ units; the second contains linear Au^I and square-planar Au^{III}; while the cobalt salt contains a cation which is manifestly purely octahedral cobalt(III) and a tetrahedral cobalt(II) anion. Figure 9.12 shows a mixed-valence platinum compound where both coordination geometry and bond distances demonstrate completely localized Pt^{II} and Pt^{IV} sites. On the other hand, the centrosymmetric structures of such mixed valence binuclear species as $[Tc_2Cl_8]^{3-}$, $[(H_3N)_5Ru(\mu\text{-pz})Ru(NH_3)_5]^{5+}$ (the Creutz–Taube complex, in the form of its chloride),[53] and one form of the triangular oxo-trimanganese complex $[Mn_3O(OAc)_6(py)_3]$[54] indicate a high level of electron delocalization, i.e. Class III or borderline Class III/II character. Variable-temperature Mössbauer or ESR spectroscopy sometimes indicates equivalence at room temperature and non-equivalence at low temperatures, as for some oxo-trimanganese and oxo-tri-iron complexes.[55] Such temperature dependence[56] allows k_{et} to be estimated from the time constant for the signal at the coalescence temperature. Arrhenius parameters can then be determined from full line shape analysis around the coalescence temperature. However, it should be borne in mind that the temperature effects may actually be due to a changeover from Class II to Class III behaviour. Examples of mixed-valence species which show coalescence of signals on raising the temperature include several iron complexes of type **32** (Mössbauer) and dicopper species, such as **33** (ESR). Values for k_{et} are given in Section 9.4.4.5 below. Copper species such as **33** have K_c values within the range 4×10^6 to 7×10^8; K_c values of this order of magnitude or larger indicate Class III behaviour (cf. Tables 9.6 and 9.7 above). These Class III species with a high degree of

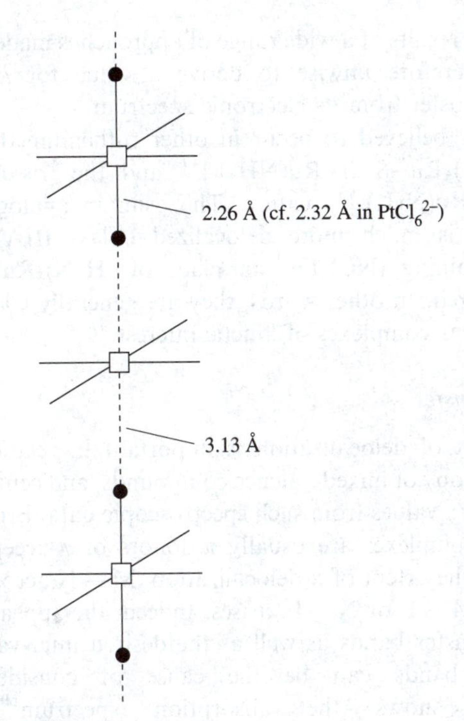

Fig. 9.12 Structural indications of Class I character: platinum–chloride distances in Wolffram's Red Salt, *trans*-$[Pt(EtNH_2)_4Cl_2][Pt(EtNH_2)_4]Cl_4.4H_2O$.

delocalization and high K_c tend to have very narrow charge transfer bands, whereas Class II species tend to have broader charge transfer bands showing marked solvatochromism.

Sometimes a relatively minor change in structure may cause a changeover between Class II and Class III behaviour. Thus bis-ferrocenes are Class II, but the bis-fulvalene iron complexes, **31**, are Class IIIA. The best-known mixed-valence complex, the Creutz–Taube ion, $[(NH_3)_5Ru(pz)Ru(NH_3)_5]^{5+}$, is close to the Class II/Class IIIA borderline—probably just on the Class III

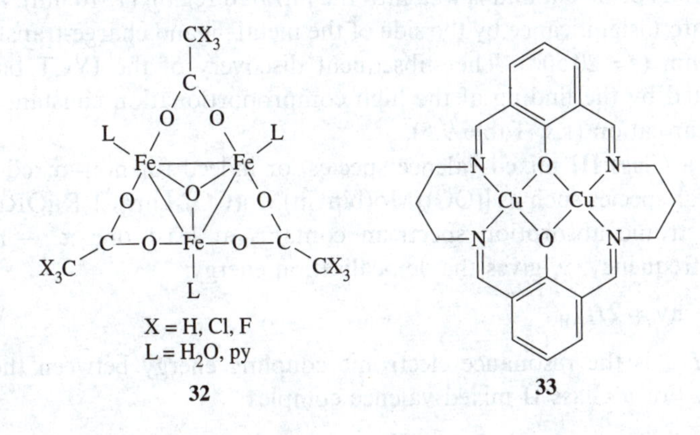

side—as is indicated by the results of a wide range of approaches made to its classification.[57,58] It is therefore unwise to derive a value for k_{et} for intramolecular electron transfer from its electronic spectrum.

Similar delocalization is believed to occur in other ruthenium(II)/(III) complexes such as $[(NH_3)_5Ru{-}S{-}S{-}Ru(NH_3)_5]^{4+}$ and the tris-chloro-bridged $[(NH_3)_3Ru(\mu\text{-}Cl)_3Ru(NH_3)_3]^{3+}$ cation. The osmium analogue of the Creutz–Taube cation is much more delocalized (Class IIIA), but binuclear complexes containing $(NC)_5Fe^{II}$ in place of $(H_3N)_5Ru^{II}$ are thought to be less delocalized. In other words, they are generally Class II, like the majority of binuclear complexes of kinetic interest.[59]

9.4.4.4 *Optical electron transfer*

The question of the extent of delocalization is important in relation to ultraviolet–visible spectroscopy of mixed-valence compounds, and central to the question of evaluating k_{et} values from such spectroscopic data. Bridging ligands in mixed-valence complexes are usually π-donors or π-acceptors. Therefore, the question of the extent of π-delocalization, $M \rightarrow N$ across the bridging ligand or simply $M \rightarrow L$ or $N \rightarrow L$, arises. Indeed, the appearance of metal–ligand charge transfer bands as well as the desired inter-valence charge transfer (IVCT) bands can be the cause of considerable difficulty. Figure 9.13 shows the absorption spectrum[60] of $[(H_3N)_5Ru{-}LL{-}Ru(NH_3)_5]^{5+}$ with LL = **34**. This complex, almost Class

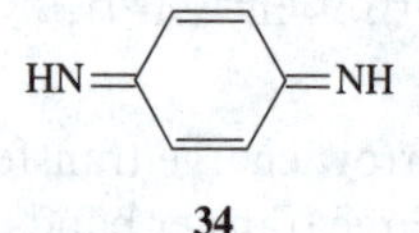

34

IIIA in character, has a very intense absorption at 634 nm, ($\varepsilon = 108\,000$, surely the highest extinction coefficient recorded for a ruthenium(II)/(III) complex) which dwarfs the near infrared IVCT bands ($\varepsilon \sim 1000$) (see Fig. 9.13). In the case of the Creutz–Taube cation the IVCT band was missed by Ford and Taube when they first prepared it. This band has a relatively small extinction coefficient and is well into the infrared region (1570 nm), and thus pales into insignificance by the side of the metal–ligand charge-transfer band at 565 nm ($\varepsilon = 20\,500$). The subsequent discovery of the IVCT band was prompted by the finding of the high comproportionation constant for this binuclear cation (*v.s.* Table 9.5).

For a Class III mixed-valence species, or indeed for non-mixed valence binuclear species such as $[(OC)_4Mo(bipym)Mo(CO)_4]$ or $[Cl_5RuORuCl_5]^{n-}$, the electronic absorption spectrum contains at least one $\pi^* \leftarrow \pi$ band, whose frequency, ν, gives the delocalisation energy:

$$h\nu = 2H_{AB}$$

Here H_{AB} is the resonance electronic coupling energy between the metal centres. For a Class II mixed-valence complex:

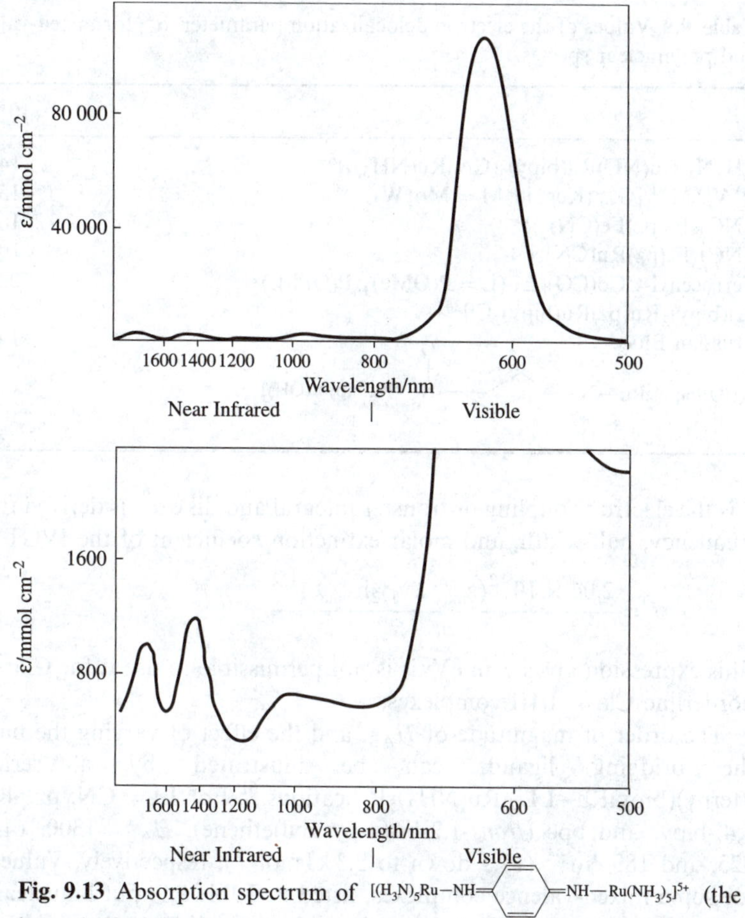

Fig. 9.13 Absorption spectrum of $[(H_3N)_5Ru{-}NH{=}C_6H_4{=}NH{-}Ru(NH_3)_5]^{5+}$ (the lower half of this figure shows the weaker absorptions expanded).

$$h\nu_{IVCT} = E_{op} = 4E_{th}$$

where E_{th} is the barrier to thermal electron transfer. If there is significant electron delocalization, especially in complexes near the Class II/Class III borderline, then some allowance has to be made for this:

$$E_{op} = 4E_{th} - H_{AB}$$

The issue is complicated by the extraction of parameters such as α^2 or J from spectra. α^2 is a delocalization parameter, which gives an estimate of the extent of delocalization of the extra electron in the ground state. It is obtained from the frequency, ν_{max}, half-width, $\Delta\nu_{1/2}$, and molar extinction coefficient, ε, of the IVCT absorption band:

$$\alpha^2 = \frac{4.2 \times 10^{-4}\varepsilon_{max}\Delta\nu_{1/2}}{\nu_{max}d^2}$$

Table 9.9 Values of the electron delocalization parameter, α^2, for mixed-valence bi- and polynuclear species.

	$10^3\alpha^2$
$[(H_3N)_5Ru(NC)Ru(bipy)_2(CN)Ru(NH_3)_5]^{6+}$	140
$P^VV^{IV}M^{VI}{}_{11}O_{40}$ (Keggin; M = Mo; W)	15; 34
$[(NC)_5Fe(pz)Fe(CN)_5)]^{5-}$	10 or less
$[(NC)_5Fe(pz)Ru(CN)_5]^{5-}$	10 or less
Ferrocenyl-$CCo(CO)_6L_3$ (L = $P(OMe)_3$; $P(OPh)_3$)	2.3; 3.5
$[Cl(bipy)_2Ru(pz)Ru(bipy)_2Cl]^{3+}$	2
Prussian Blue	1.6
$[(HO)(bipy)_2Ru—N(O^-)$–C_6H_{10}–$N(O^-)—Ru(bipy)_2(OH)]$	1.3

J is the electron coupling or transfer integral and, like α^2, is derived from the frequency, half-width, and molar extinction coefficient of the IVCT band:

$$J = \frac{2.06 \times 10^{-2}(\varepsilon_{max}\Delta\nu_{1/2}h\nu_{max})^{1/2}}{ar}$$

This expression gives J in eV; it is not permissible to use it for Class III or borderline Class II/III complexes.

The order of magnitude of H_{AB}, and the effect of varying the nature of the bridging ligand, can be illustrated by a series of $[(terpy)(bipy)Ru{-}LL{-}Ru(NH_3)_5]^{5+}$ cations.[61] For LL = CN, pz, 4CN py, 4,4′-bipy, and bpe (*trans*-1,2-bis(4-pyridyl)ethene), H_{AB} = 1300, 614, 421, 325, and 189 cm^{-1} (15.5 down to 2.3 kJ mol^{-1}), respectively. Values of α^2 for some mixed-valence complexes, listed in Table 9.9, put binuclear redox species into context with such species as heteropolytungstates and Prussian Blue. Delocalization energies range from 20 to 25 kJ mol^{-1} at the top of the table to 1 kJ mol^{-1} or less, in other words less than 1% delocalization, at the foot. This corresponds to a gradation from essentially Class III to essentially Class II. Values of J for three series of ruthenium(II)/(III) binuclear complexes are shown in Table 9.10. The top section of this table shows how J drops from 13.5 to 0.3 kJ mol^{-1} with the distance, d, between the metal centres, in a series of complexes in which the bridging ligands hold the metals apart at clearly defined distances. The lower portion of this table documents how J correlates inversely (and exponentially) with metal-centre separation for a given structure type, and how J is affected by conjugation or otherwise of the metal centres.

9.4.4.5 Kinetic parameters for electron transfer

The derivation of E_{th}, the barrier to thermal electron transfer, from optical spectra has been outlined in the previous section. Further assumptions are needed to obtain an estimate of the rate constant for intramolecular electron

Table 9.10 Values of the transfer integral, J, for mixed-valence diruthenium complexes, $[(H_3N)_5Ru(LL)Ru(NH_3)_5]^{5+}$ and their relation to the internuclear distance, d.

LL	J/eV	d/Å
	0.14	9.0
	0.0056	11.2
	0.0027	13.4

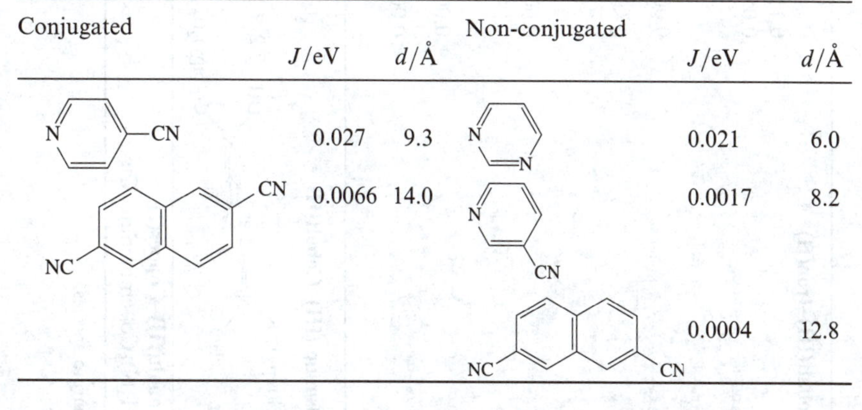

Conjugated	J/eV	d/Å	Non-conjugated	J/eV	d/Å
	0.027	9.3		0.021	6.0
	0.0066	14.0		0.0017	8.2
				0.0004	12.8

transfer, k_{et}, from these values.[62] These assumptions are that electron transfer is adiabatic, with no lingering of the electron on the bridging ligand, and has zero activation energy. It is also necessary to assume a value for the transmission coefficient. We can then use the relationship

$$k_{et} = (k_B T/h) \exp(-E_{th}/RT)$$

The number and magnitude of the assumptions used in the overall derivation of k_{et} from spectroscopic observation are somewhat disquieting. Indeed, such estimations may be several orders of magnitude in error; Haim believes that all k_{et} values obtained in this way may be much too large. The very rare cases where k_{et} can be estimated from both optical spectra and conventional kinetic techniques support this view, with differences of three orders of magnitude or more. Nevertheless, at the time of writing such estimates of k_{et} from optical spectra provide the only information about a variety of important systems. The reader should bear this in mind for the remainder of this section.

Kinetic data, i.e. k_{et} and some associated activation parameters, for a variety of intramolecular electron transfers are collected in Tables 9.11–9.14. The data in Tables 9.11 and 9.12 come from 'conventional' kinetic measure-

Table 9.11 Kinetic parameters, from conventional measurements, for intramolecular electron transfer.

	$k(s^{-1})$
Cobalt(III)—Ruthenium(II)	
$(H_3N)_5Co—O_2C$ — N—$Ru(NH_3)_4(SO_4)^{2+}$	~100
$(H_3N)_5Co—N$ N—$Ru(edta)^{+}$	21
$(H_3N)_5Co—N$ N—$Ru(NH_3)_4(H_2O)^{5+}$	0.30
$(H_3N)_5Co—N$ N—$Ru(NH_3)_4(H_2O)^{5+}$	0.044
$(H_3N)_5Co—N$ N—$Ru(NH_3)_4(H_2O)^{5+}$	0.0187
$(H_3N)_5Co—N$ N—$Ru(NH_3)_4(H_2O)^{5+}$	0.055
$(H_3N)_5Co—N$ —S— N—$Ru(NH_3)_4(H_2O)^{5+}$	0.0049
$(H_3N)_5Co—N$ N—$Ru(NH_3)_4(H_2O)^{5+}$	0.0012
(dien)Co O—C O N N—$Ru(NH_3)_5^{3+}$ O—C O	0.059
(dien)Co—NC— —CN—$Ru(NH_3)_5^{3+}$	$< 3 \times 10^{-6}$

	$k(s^{-1})$	$\Delta H^{\ddagger}(kJ\ mol^{-1})$	$\Delta S^{\ddagger}(J\ K^{-1}\ mol^{-1})$
Cobalt(III)–Iron(II)			
$(H_3N)_5Co$—nta—Fe aq	0.115	78	0
$(H_3N)_5Co—N$ N—$Fe(CN)_5$	0.055	103	+70[b]
$(H_3N)_5Co—NC$— N—$Fe(CN)_5$		98	+80
$(H_3N)_5Co—N$ N—$Fe(CN)_5$	0.0026		
$(H_3N)_5Co—N$ N—$Fe(CN)_5$	0.0014	102	+44
$(H_3N)_5Co—N$ N—$Fe(CN)_5$	0.0021		
$(H_3N)_5Co—O_2C$— N—$Fe(CN)_5$	0.00017		
$(H_3N)_5Co—O_2C$— N—$Fe(CN)_5$	< 0.00003		
Osmium(III)–Cobalt(II)			
$Cl(bipy)_2Os$ N= HO =N OH $Co(bipy)_2(H_2O)$	7.2×10^{-4}		
Cobalt(III)–Copper(I)			
$(H_3N)_5Co$–cinnamate–Cu^{I} [a]	$< 2 \times 10^{-7}$		

[a] Electron transfer is even slower in analogous μ-acrylato or μ-fumarato species.
[b] $\Delta V^{\ddagger} = +35$ or $+38\ cm^3\ mol^{-1}$.

Table 9.12 Dependence of electron transfer rate constant $k_{et}(s^{-1})$ on metal–metal separation, d_{M-M} (Å), for binuclear complexes $[(H_3N)_5Os^{II}–LL–Ru^{III}(NH_3)_5]^{4+}$, where LL = isonicotinyl(proline)$_n$.

n	d_{M-M}	k_{et}
0	9	$> 5 \times 10^9$
1	12	3.1×10^6
2	15	3.7×10^4
3	18	3.2×10^2
4	21	~ 50

Table 9.13 Rate constants and activation barriers, estimated by indirect methods (cf. text), for rapid intramolecular electron transfer in binuclear complexes.

	$k(s^{-1})$	Barrier[a] ($kJ\,mol^{-1}$)
Ruthenium(III)–Ruthenium(II)		
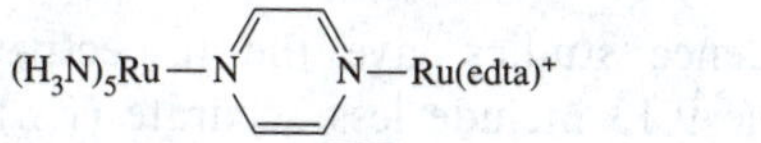 $(H_3N)_5Ru$—N N—Ru(edta)$^+$	10^{10}–10^{11}	$\sim$ 10–15
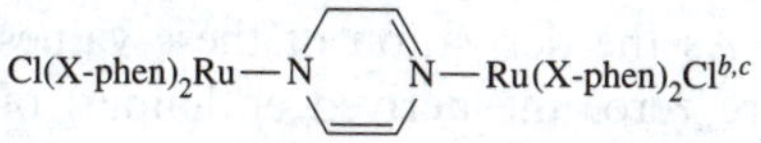 Cl(X-phen)$_2$Ru—N N—Ru(X-phen)$_2$Cl[b,c]	$\sim 6 \times 10^8$	~ 23
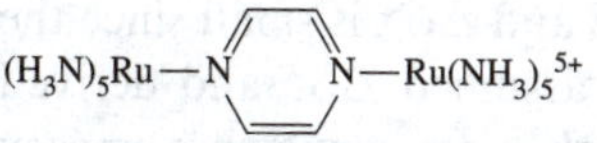 $(H_3N)_5Ru$—N N—$Ru(NH_3)_5^{5+}$	3×10^8	24
Ruthenium(III)–Iron(II)		
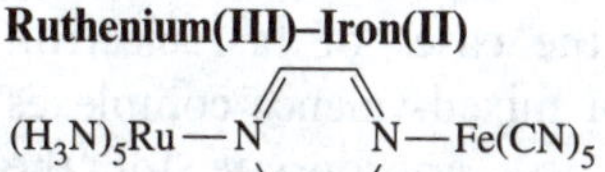 $(H_3N)_5Ru$—N N—$Fe(CN)_5$	2.7×10^{10}	13
Copper(II)–Copper(I)		
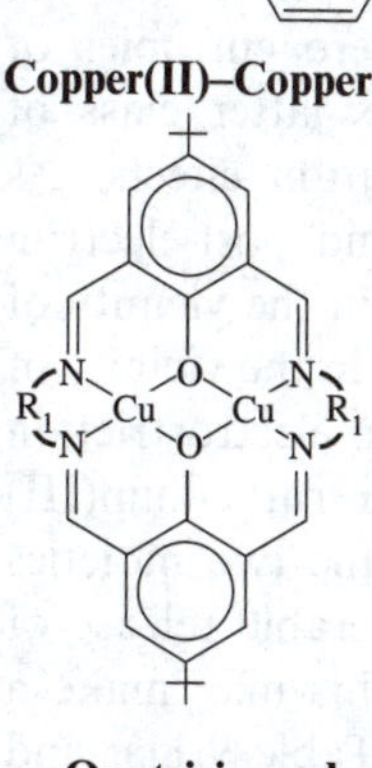 N O N R$_1$ Cu Cu R$_1$ N O N	$< 2 \times 10^7$ to 1.7×10^{10}	$\sim 15^c$
μ_3-Oxotrisiron clusters		
$Fe_3O(O_2CCH_3)_6L_3$ (L = H_2O or py)	$\sim 3 \times 10^8$	~ 24
$Fe_3O(O_2CCMe_3)_6(C_5D_5N)_3$	$\sim 10^{11}$	~ 10
Organometallic: Iron–Cobalt		
Ferrocenyl C $Co_3(CO)_6L_3$: L = $P(OMe)_3$	4×10^{10}	~ 13
L = $P(OPh)_3$	5×10^9	~ 17

[a] $\Delta G^\ddagger \approx \Delta H^\ddagger$ (see text).
[b] X = H, 5-NO_2, 4,7-Me_2, or 3,4,7,8-Me_4.
[c] E_a.

Table 9.14 Effect of spectator ligands on rate constants for intramolecular electron transfer.

		$k(s^{-1})$
Cobalt(III)–Ruthenium(II)		
$L_3Co(\mu\text{-}OH)_2(\mu\text{-}O_2C\text{-}C_5H_4N\text{-}Ru(NH_3)_5)CoL_3$	$L_3 = (NH_3)_3$	0.065
	L_3 = dien	0.0016
	L_3 = [9]aneN$_3$ (cyclic triamine)	0.000 052
Cobalt(III)–Iron(II)		
$L_5Co\text{-}N(pyrazine\text{-}carboxylate, O)\text{-}N\text{-}Fe(CN)_5^-$	$L_4 = (NH_3)_4$	0.013
	$L_4 = (en)_2$	$< 10^{-5}$

ments, and so the temperature-dependence studies give the respective Arrhenius parameters. The data in Table 9.13 include less accurate (*v.s.*) rate constants for fast electron transfer. As the derivation of these values assumes that the activation entropies are zero, the derived enthalpies of activation must therefore be small ($\Delta H^\ddagger \approx \Delta G^\ddagger$, and $\Delta G^\ddagger$ is small since the activation barrier is low). Comparison of such assumed $\Delta S^\ddagger$ and derived $\Delta H^\ddagger$ values with actual values in Table 9.11 makes one somewhat uneasy about the assumption that $\Delta S^\ddagger = 0$ used in the cases of fast electron transfer. This assumption may be acceptable for mixed-valence complexes of the $[L_5M{-}LL{-}ML'_5]^{5+}$ type, but much less appropriate for the $[(H_3N)_5M{-}LL{-}Fe(CN)_5]$ (M = Co or Ru) series. Non-zero entropies of activation for electron transfer within complexes of this latter class of binuclear complex can be understood in terms of solvation effects. As suggested by Fig. 9.14(a), the solvation shells for pre- and post-electron transfer entities are practically the same, with desolvation in the vicinity of the oxidizing centre closely matched by increased solvation in the vicinity of the reducing centre. The net solvation change attendant on electrostriction changes is thus close to zero. But for the cobalt(III) or ruthenium(III) pentacyanoferrate(II) series, Fig. 9.14(b), the charges on the two moieties change from 3+,3− to 2+,2− and there will be a considerable release of electrostricted solvent molecules into the bulk solvent. This may make a significant positive contribution to entropy changes (cf. Table 9.11), and thus to the work terms for such processes. Of course, the Franck–Condon principle requires that, if electron transfer is very fast, the solvent movement will lag well behind and the contribution to the entropy change will again be zero.

A major factor influencing k_{et} is the nature of the two metal centres.[63] The importance of this is immediately apparent on perusal of Tables 9.11–

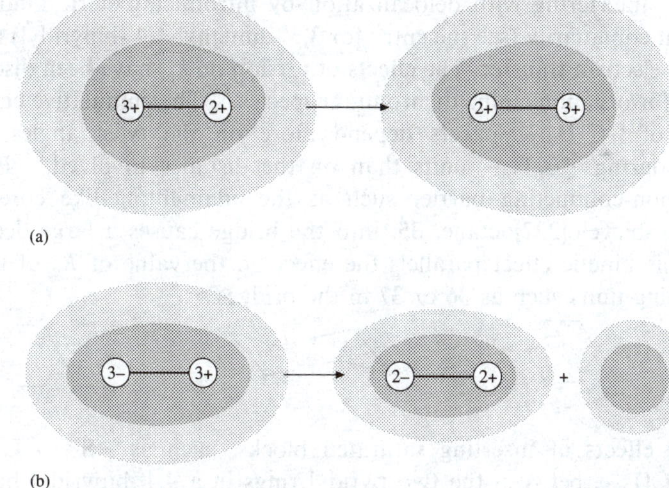

Fig. 9.14 Diagrammatic representation of solvation changes accompanying electron transfer in (a) an $[L_5M{-}LL{-}ML_5]^{5+}$ mixed-valence complex and (b) one of the $[(NC)_5Co{-}LL{-}Fe(CN)_5]$ type.

9.13. Dramatic comparisons can be made, for example between the binuclear cations $[(H_3N)_5M(pz)Ru(NH_3)_5]^{5+}$ with M = Co or Ru, where $k_{et} = 0.055\ s^{-1}$ and $2.7 \times 10^{10}\ s^{-1}$, respectively. Table 9.11 also indicates that the range of k_{et} values for a given pair of metal centres may cover several powers of ten, being strongly affected by the nature of the bridging ligand and the distance between the metal centres. Rate constants may be markedly affected by solvent—for details the reader will have to look elsewhere.[64] Finally, rate constants for electron transfer may be strongly influenced by the nature of the other, non-bridging, ligands (Table 9.14). A similar large effect of non-bridging ligands on K_c values was mentioned in Section 9.4.4.2. These non-bridging ligand effects arise from their influence on the driving force for electron transfer—the redox potentials for the hexaammine and bis-triazacyclononane cobalt(III)/(II) couples differ by 0.5 V. The effects of bridging ligands are more complicated, and important questions are still unanswered at the present time. It is clear from Tables 9.11 and 9.13 that pyrazine is a particularly good ligand for mediating electron transfer, that 4,4′-bipyridyl is much less effective, and that 1,2-bis-(4-pyridyl)-ethene is less effective still. Although full delocalization across the bridge is maintained in this series, the actual distance to be traversed by the electron increases steadily. Indeed, there is a linear correlation between $\log k_{et}$ and the reciprocals of the distances between the two metal atoms. Table 9.12 shows another series of complexes, this time with bridging ligands of constant type, where rate constants decrease dramatically with increasing distance between the metal centres. However, it should be noted that, with the driving force of 0.25 eV (24 kJ mol^{-1}) operative here, electron transfer still takes place reasonably rapidly over a separation of more than

20 Å.[65] Interfering with delocalization by introducing steric hindrance to prevent coplanarity (see the entry for 3,3′-dimethyl-4,4′-bipyridyl) results in slower electron transfer. The effects of torsion on k_{et} have been discussed in detail for organometallic dichromium species.[66] The cumulative effects of a series of $-C_6H_4-$ spacers depend more on the twist angles between neighbouring $-C_6H_4-$ units than on the distance involved.[67] Putting a rigid non-conducting barrier such as the adamantane-like core of 1,4-dicyano-bicyclo[222]octane, **35**, into the bridge causes a large decrease in k_{et}. This kinetic effect parallels the effect on the value of K_c of including insulating units such as **36** or **37** in the bridges.

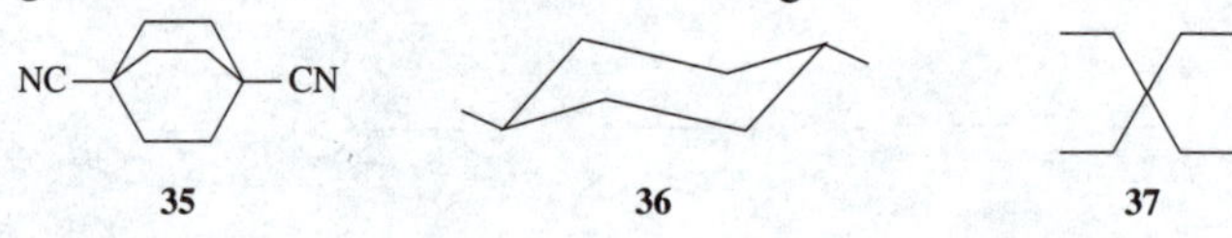

The effects of inserting saturated blocks, such as –S–, $-CH_2-$, or $-CH_2CH_2-$, between the two pyridyl rings in a 4,4′-bipyridyl bridge are less clear-cut, especially when one compares a $-CH_2CH_2-$ link with a –CH=CH– link. The rather small slowing produced by saturated linkages may indicate that there are contributions from 'through-space' electron transfer. Figure 9.15 shows relative metal–metal distances, with, for instance, M–M at the closest approach permitted in the $-CH_2CH_2-$ derivative only slightly longer than in the parent pyrazine-bridged complex. The M–M distance in the –CH=CH– derivative is considerably (~80%) longer. The importance of these M–M distances, and of the ability of flexible ligands to 'circle round', is neatly shown by the values of k_{et} for electron transfer from ruthenium to cobalt in the series of cations, **38**, with $n = 0$–4. From $n = 0$ to $n = 2$, k_{et} decreases as the ruthenium–cobalt distance increases, but as n increases further to 3 and to 4, k_{et} starts to increase again as direct, through-space, electron transfer, **39**, begins to make an important contribution.[68] Similar through-space electron transfer seems to be responsible for the IVCT band of the analogous compound, **40**.

The observations in the preceding paragraph parallel patterns established earlier for the way in which, for example, k_{et} varies for electron transfer within organic radicals such as **41**. A sufficiently large $-\Delta G$ can drive

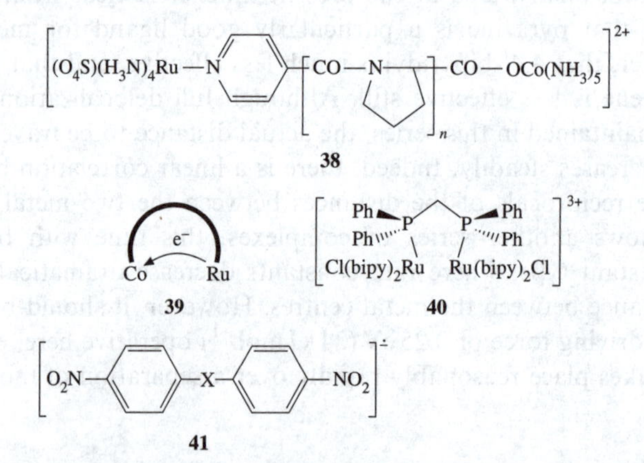

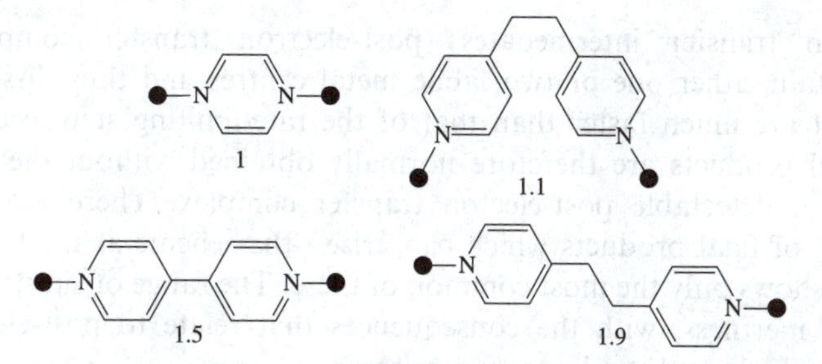

Fig. 9.15 Relative metal–metal distances at shortest approach for pyrazine and 4,4′-bipyridyl derivatives as bridging ligands. The symbol ● here stands for $Ru(NH_3)_5$.

electron transfer very quickly across predominantly saturated organic systems such as cyclohexane rings, spirocyclic butanes, polynorbornyl bridges,[69] steroids, and proteins for 10 or even 20 Å.[70] We shall return to long-range electron transfer in Section 9.6, as it is equally important in inner- and outer-sphere electron transfer processes.

9.4.5 Post-electron transfer intermediates

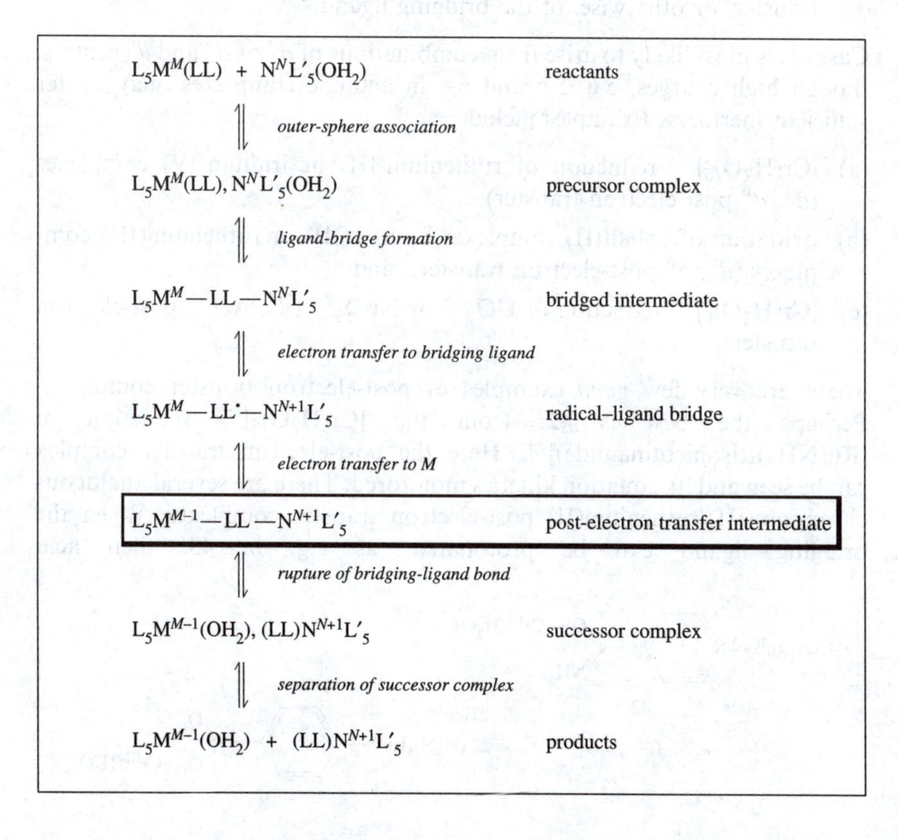

Post-electron transfer intermedates (post-electron transfer complexes) usually contain either one or two labile metal centres and thus dissociate at rates that are much faster than that of the rate-limiting step discussed above. Final products are therefore normally obtained without the intermediacy of a detectable post-electon transfer complex. There are three possible sets of final products which can arise—the scheme at the head of this section shows only the most common of these. The range of substitution lability and inertness, with the consequences that relate to post-electron transfer complexes and products, comprises:

(a) If both metal centres are labile after electron transfer, then no post-electron transfer complex will be observable and it will not be possible to prove whether or not ligand transfer occurs.
(b) If N^{n-1} is inert, then ligand transfer can be convincingly demonstrated (as in Taube's experiment; Section 9.3.1).
(c) If M^{m+1} is inert, then there will be no ligand transfer.
(d) If both M^{m+1} and N^{n-1} are inert, then the binuclear post-electron transfer complex will be slow to dissociate to the ultimate mononuclear products. Now not only should the post-electron transfer complex be detectable, but it should be possible to measure k_{diss} and to demonstrate transfer, or otherwise, of the bridging ligand.

Case (d) is most likely to arise from combinations of d^6 or d^3 and d^6 centres, though high charges, e.g. 5+ and 6+ in actinide complexes, may confer sufficient inertness. Examples include:

(a) $[Cr(H_2O)_6]^{2+}$ reduction of ruthenium(III) or iridium(IV) complexes (d^3, d^6 post-electron transfer);
(b) oxidation of cobalt(II) complexes by iron(III) or ruthenium(III) complexes (d^6, d^6 post-electron transfer); and
(c) $[Cr(H_2O)_6]^{2+}$ reduction of $UO_2{}^{2+}$ or $NpO_2{}^{2+}$ (d^3, Act^{5+} post-electron transfer).

There are very few *good* examples of post-electron transfer complexes. Perhaps the best is **42**, from the $[Cr(H_2O)_6]^{2+}$ reduction of $[Ru(NH_3)_5(\text{isonicotinamide})]^{3+}$. Here the post-electron transfer complex can be seen and its aquation kinetics monitored. There are several analogous chromium(III)/ruthenium(II) post-electron transfer complexes. When the bridging ligand can be protonated, as e.g. in **43**, then acid

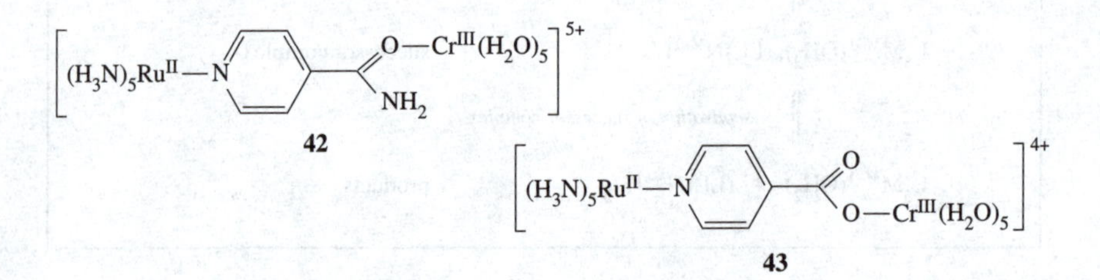

catalysis of dissociation to final product is important:

$$\text{d[products]}/\text{d}t = (k_1 + k_2[\text{H}^+])\ [\text{post-electron transfer complex}]$$

Reaction of the copper(III)-imineoxime complexes, **44** and **45**, with

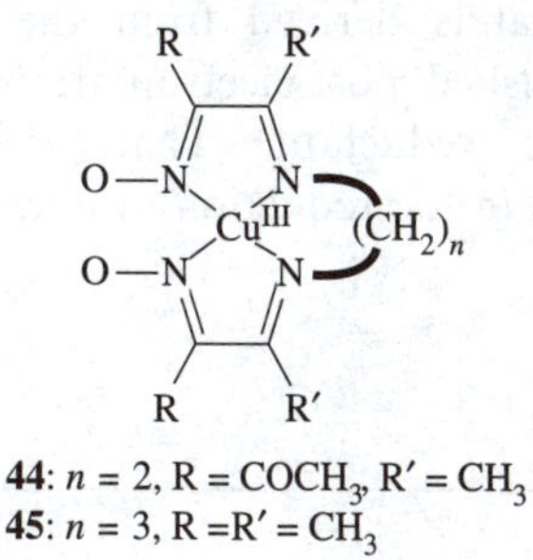

44: $n = 2$, R = $COCH_3$, R′ = CH_3
45: $n = 3$, R = R′ = CH_3

cobalt(II)-edta gives post-electron transfer complexes containing cobalt(III) bonded through oximate-O,N to copper(I), whose dissociation to mononuclear final products can be followed from the rate of formation of $[Co^{III}(edta)]^-$.[71]

Hexacyanoferrate(III) and pentacyanocobaltate(II) react to give the d^6, d^6 complex $[(NC)_5Fe(CN)Co(CN)_5]^{6-}$, which can be isolated as its potassium or barium salt and is robust enough to be oxidized to $[(NC)_5Fe{-}CN{-}Co(CN)_5]^{5-}$. The high formal charge on the uranium in $[CrOUO]^{4+}$ confers stability and permits the assignment of an oxidation state of 3+ to the chromium in the dark-green post-electron transfer complex from its ultraviolet spectrum. It has a half-life of minutes at 0 °C, decomposing to $[Cr(H_2O)_6]^{3+}$, U^{4+}aq, and UO_2^{2+} (U^V being, of course, unstable with respect to disproportionation).

There is sometimes a fine balance between alternative mechanistic pathways. This can be well illustrated by the reaction of the hexachloroiridate(IV) ion with $[Cr(H_2O)_6]^{2+}$. This was the original example (1954) of an inner-sphere electron transfer without ligand transfer. The transient deep-blue colour that was observed was attributed to the post-electron transfer complex, $[Cl_5Ir^{III}{-}Cl{-}Cr^{III}(H_2O)_5]$. This contains two substitution-inert centres and so is reluctant to dissociate. When it does come apart, the greater inertness of the third-row, d^6, iridium(III) ensures that the bridging chloride is not transferred to the chromium, which is merely a first-row d^3 ion. Subsequently, this reaction has been shown to be more complicated, with parallel chloride-transfer inner-sphere and simple outer-sphere pathways also contributing to the overall redox reaction. The post-electron transfer complex in the hexabromoiridate(IV) oxidation of pentacyanocobaltate(II), $[Br_5Ir^{III}{-}Br{-}Co^{III}{-}(CN)_5]^{5-}$, dissociates by Ir–Br and Co–Br fission in almost equal proportion.[72] The corresponding reaction with $[IrCl_6]^{2-}$ in place of $[IrBr_6]^{2-}$ goes by a single pathway, with no transfer of chloride to the cobalt. Another good example of an inner-sphere redox reaction with no transfer of the bridging ligand is the $[Cr(H_2O)_6]^{2+}$ reduction of the isonicotinamide of pentammine-ruthenium(III) that was mentioned above.

All the cases of post-electron transfer complexes and ligand transfer detailed above have their bridging ligands derived from the oxidant. There seems to be only one well-established post-electron transfer complex whose bridge comes from the reductant. That complex is $[(NC)_5Fe{-}CN{-}Cr(OH_2)_5]$, which occurs in the reduction of chromate(VI) by hexacyanoferrate(II).

9.4.6 Dead-end complexes

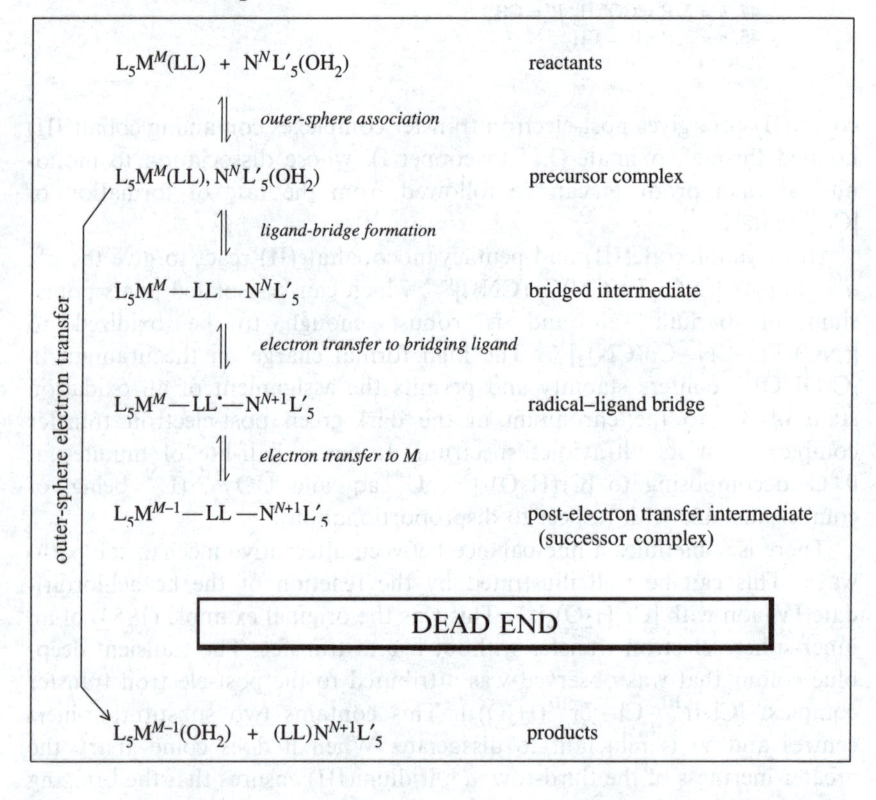

The post-electron transfer complex $[(NC)_5Fe^{II}{-}CN{-}Co^{III}(CN)_5]^{6-}$ referred to in the previous section is very inert and reluctant to dissociate to mononuclear products. In some closely related reactions the post-electron transfer complex is similarly inert to dissociation and is, in fact, bypassed by an alternative *outer-sphere* electron transfer path (cf. Section 9.5 below) which leads directly to mononuclear iron(II) and cobalt(III) products. This is shown in the section heading above. Thus, for example, $[Fe(CN)_6]^{3-}$ and $[Co(edta)]^{2-}$ react within milliseconds to give a cyano-bridged species,

$[(NC)_5Fe{-}CN{-}Co(edta)]^{5-}$ (Co^{II} has a d^7 electron configuration and is therefore labile). This, presumably, is an iron(II)–cobalt(III) post-electron transfer complex, with both metal ions in the low-spin d^6 configuration, and is very reluctant to dissociate. The final products, $[Fe(CN)_6]^{4-}$ and $[Co(edta)]^-$, are actually produced by slow outer-sphere electron transfer from $[Co(edta)]^{2-}$ to $[Fe(CN)_6]^{3-}$. The binuclear post-electron transfer complex that is formed initially eventually goes to the final product by way of the sequence:

$$\text{'successor'} \rightleftharpoons \text{'precursor'} \rightleftharpoons \text{reactants} \xrightarrow{\text{outer sphere}} \text{products!}$$

This is rather tricky to prove, as rate laws for this mechanism and for a simple inner-sphere mechanism are the same, and also the non-transfer of the bridging ligand does not rule out the direct inner-sphere route (cf. Section 9.4.5). However, a number of comparisons of rate constants and activation parameters for this and other related redox and substitution reactions makes the 'dead-end' mechanism almost certain.[73] There is further support from the close kinetic similarities between the two reactions $[Fe(CN)_5L]^{2+} + [Co(edta)]^{2-}$ with L = 4,4′-bipyridyl (when a direct inner-sphere reaction is possible) and L = pyridine (when it is impossible). It was possible to measure the rate constant for the return reaction, i.e. 'successor' → reactants, in the reduction of $[Ru(NH_3)_5(pz)]^{3+}$ with $[Co(edta)]^{2-}$, since this step is considerably slower ($k = 15\,s^{-1}$) than the outer-sphere electron transfer between the reactants. The operation of the 'dead-end' mechanism requires a very inert d^6, d^6 post-electron transfer complex and the choice is very limited, mainly because of the scarcity of suitable cobalt(II) reactants. In principle, there should be an analogous 'dead-end' mechanism in which the precursor rather than the post-electron transfer complex provides the barrier, but so far nobody has reported an example.

9.5 The outer-sphere mechanism

9.5.1 General features[74]

In this mechanism the redox step postulated is simply and solely the transfer of an electron—the coordination spheres of the two complexes are unchanged. The great simplicity of this mechanism (Fig. 9.16) makes positive proof very hard to obtain. There are a number of inorganic redox reactions where it is extremely difficult to imagine an inner-sphere mechanism operating, and which are therefore firmly believed to go by way of an outer-sphere mechanism. Such reactions are those in which both the oxidant and reductant are substitutionally inert (some examples are given in Table 9.15) and yet electron transfer is very rapid. A good example is the electron exchange reaction between ferrocene and the ferrocinium cation, where no plausible ligand bridging path can exist. $[Fe(Me_2phen)_3]^{n+}/IrX_6^{n-}$

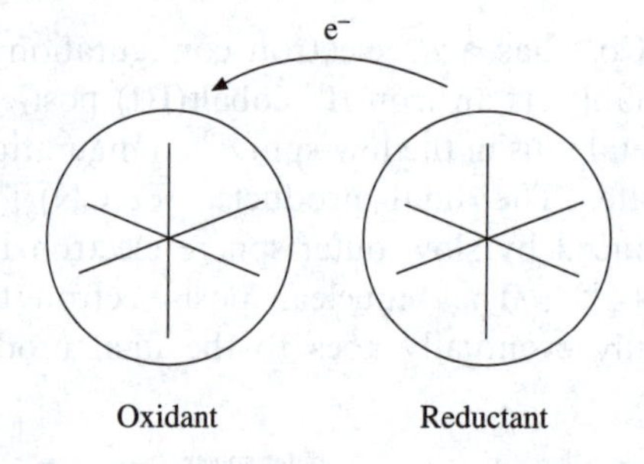

Fig. 9.16 The outer-sphere mechanism.

reactions of the type included in Table 9.16 occur at rates approaching the diffusion-controlled limit,[75] yet the individual reactants have half-lives with respect to substitution measured in hours. Table 9.16 gives a small selection of rate constants for outer-sphere redox reactions, to give some idea of the great range of rate constants for outer-sphere electron transfer, ranging from quite slow to nearly diffusion-controlled. More rate data, covering outer-sphere reactions involving bioinorganic species as well as simple inorganic complexes, are included in the rate constant correlation figures in Section 9.5.3 below. Activation enthalpies for reactions included in Table 9.16 range from 0 to $60\,\text{kJ}\,\text{mol}^{-1}$, and activation entropies from 0 to $-160\,\text{J}\,\text{K}^{-1}\,\text{mol}^{-1}$. The examples included in this table concentrate on substitutionally inert species such as permanganate, cyanide, and diimine complexes and the sepulcrate complexes of cobalt, **46**, where both the 2+ and 3+ species have enforced high inertness. The time-scales for the rate constants in Table 9.16 can be compared with their equivalents for substitution at the respective metal centres (cf. Table 9.15 and Chapter 4) to show how much faster electron transfer takes place in comparison with

Table 9.15 The substitution inertness of some outer-sphere redox partners.[a]

Complex		$k(\text{s}^{-1})$	Approximate $\tau_{1/2}$
$Fe(CN)_6^{3-}$	for cyanide exchange	$<10^{-6}$	>1 week
$Fe(CN)_6^{4-}$	for cyanide exchange	$<10^{-6}$	>1 week
$Fe(phen)_3^{2+}$	for aquation	7×10^{-5}	3 hours
$Fe(4,7\text{-}Me_2phen)_3^{2+}$	for aquation	2×10^{-5}	10 hours
$IrCl_6^{2-}$	for aquation	$\sim 10^{-9}$	~1 year
$IrCl_6^{3-}$	for aquation	9×10^{-6}	1 day
$Co(sep)^{2+}$	for racemization	no loss of optical activity after 2 hours[b]	
$Co(sep)^{2+}$	for aquation	no exchange with $^{60}Co^{2+}aq$ in 24 hours	

[a] In aqueous solution at 298.2 K.
[b] In neutral solution—the cage opens quite quickly in acid solution.

Table 9.16 Rate constants for outer-sphere electron transfer in aqueous solution at 298.2 K.[a]

	k_2 ($dm^3\ mol^{-1}\ s^{-1}$)
$Cr^{2+}aq + Co(NH_3)_6^{3+}$	9×10^{-5} to 7×10^{-3} [b]
$Cr^{2+}aq + Co(en)_3^{3+}$	3.4×10^{-4}
$Ru(NH_3)_6^{2+} + Co(NH_3)_6^{3+}$	0.011
$V^{2+}aq + Co(NH_3)_6^{3+}$	0.010
$Ru(CN)_6^{4-} + MnO_4^-$	0.938
$U^{3+}aq + Co(sep)^{3+}$	15.8
$Cr(bipy)_3^{2+} + Co(NH_3)_6^{3+}$	187
$Co(sep)^{2+} + Co^{3+}aq$	700
$MnO_4^{2-} + W(CN)_8^{3-}$	3×10^3
$V^{2+} + Co(phen)_3^{3+}$	3.8×10^3
$Co(sep)^{2+} + Co(bipy)_3^{3+}$	1.0×10^4
$Os(CN)_8^{4-} + MnO_4^-$	3.3×10^4
$Fe(CN)_6^{4-} + IrCl_6^{2-}$	3.8×10^5
cytochrome *c* + $Fe(C_5H_5)_2^+$	6.2×10^6
$^*Ru(bipy)_3^{2+} + Co(sep)^{3+}$	3×10^8
$Fe(4,7\text{-}Me_2phen)_3^{2+} + IrCl_6^{2-}$	1×10^9
$Cr(bipy)_3^{2+} + Fe^{3+}aq$	1.4×10^9
$Ru(NH_3)_6^{2+} + Ru(bipy)_3^{3+}$	3.7×10^9
$Fe(5,6\text{-}Me_2phen)_3^{3+} + IrBr_6^{3-}$	$\sim 10^{10}$ [c]
$Mo_4S_4(edta)_2^{4-} + Mo_4S_4(edta)_2^{2-}$	2.4×10^{10} [d]

[a] See Table 2.1 of Ref. 2 for more rate constants, and some activation parameters.
[b] Depending on pH, ionic strength, and author.
[c] Estimated from the reported value at 283 K.
[d] This is an estimated value (P. W. Dimmock, J. McGinnis, Bee-Lean Ooi, and A. G. Sykes, *Inorg. Chem.*, 1990, **29**, 1085).

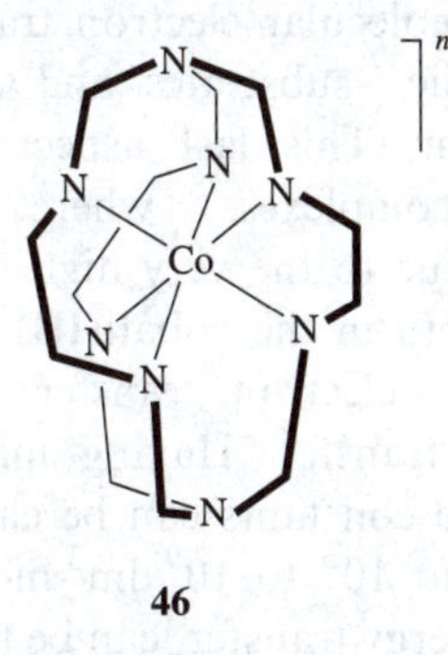

46

the substitution processes which would be required for the operation of the inner-sphere mechanism.

The excited-state species $^*Ru(bipy)_3^{2+}$ [76] appears in Table 9.16 and elsewhere in this book. It may be generated photochemically from ground-state $Ru(bipy)_3^{2+}$; such excitation requires an energy input of 202 kJ mol^{-1} (2.1 eV). Triplet-state $^*Ru(bipy)_3^{2+}$ has a lifetime of a little less than a microsecond, sufficiently long in relation to the time-scale for

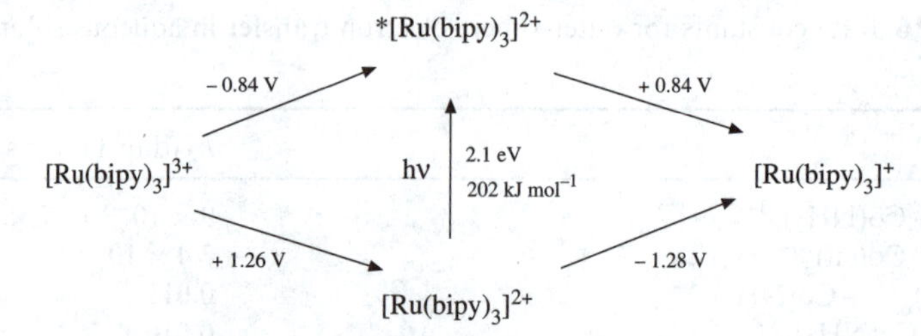

Fig. 9.17 Redox potentials for couples involving ground-state $Ru(bipy)_3^{2+}$ and excited-state $^*Ru(bipy)_3^{2+}$.

diffusion for it to have a rich chemistry. The excited state is both a better oxidant and a better reductant than the ground state, with disproportionation of $*Ru(bipy)_3^{2+}$ into $Ru(bipy)_3^+$ plus $Ru(bipy)_3^{3+}$, a thermodynamically favourable process. Relevant redox potentials are shown in Fig. 9.17. The main interest in $*Ru(bipy)_3^{2+}$ over the years has been in relation to the attractive but elusive possibility of harnessing its particular properties to the splitting of water. Its redox potentials indicate that it can be used both to reduce water to dihydrogen and to oxidize water to dioxygen, overall:

$$H_2O \xrightarrow{^*Ru(bipy)_3^{2+}} H_2 + \tfrac{1}{2}O_2$$

though there are major chemical and technical problems, way beyond the scope of the present book, involved in achieving this desirable result efficiently.

Excited-state inorganic complexes can interact by electron transfer, energy transfer, or, in appropriate cases, by atom transfer. Our main interest in $*Ru(bipy)_3^{2+}$ is as an activator for intramolecular electron transfer in bioinorganic (Section 9.6.5) and organometallic[77] substrates, and as a very reactive outer-sphere electron transfer agent. This last aspect can be illustrated in relation to cobalt(III) cage complexes,[78] where electron transfer is not complicated by dissociation, due to the very high substitution-inertness of this family of complexes, both in the cobalt(III) and the cobalt(II) form (see Table 9.15). However, electron transfer may be accompanied, or even eclipsed, by energy transfer. Having made due allowance for the latter, electron transfer rate constants can be calculated and are found to cover the range from about 10^6 to $10^9\ dm^3\ mol^{-1}\ s^{-1}$. These rate constants, and rate constants for energy transfer, can be tuned by changing the nature of the cage, including its geometry, the nature of the donor atoms, and the nature of apical substituents. There is, as one would expect, a correlation between these electron transfer rates and rate constants for the respective Co^{II}/Co^{III} self-exchange reactions (Section 9.5.4.2). It seems likely that electron transfer reactions of $*Ru(bipy)_3^{2+}$ and of other photoexcited complexes will figure much more prominently in discussions of outer-sphere redox reactions in the future.[79]

The chemistry presented so far in this chapter has been of great fundamental importance and interest, but in the main of rather little direct applicability to 'real life'. However, the $*Ru(bipy)_3^{2+}$ chemistry alluded to in the preceding paragraphs may prove to be of great value in dealing with environmental pollution problems and in the harnessing of solar energy. Another energy-related application has been the attempt to utilize outer-sphere redox reactions in the cleaning of pipework in nuclear reactors. The insoluble nickel ferrite which is deposited is particularly difficult to remove. It is resistant to all attempts to dissolve it in solutions of polyaminocarboxylates such as edta, but it can be dissolved through outer-sphere electron transfer reactions with picolinate-complexed vanadium(II).[80] Further examples of useful redox reactions will be alluded to in Section 9.7, especially in relation to oxoanion and peroxoanion oxidants (Section 9.7.4.2).

9.5.2 Pre-association and reactant orientation

The outer-sphere mechanism is much simpler than its inner-sphere counterpart, since there is not the extensive sequence of elementary steps involved in the latter (Fig. 9.3). Nevertheless, two steps are still necessary for the outer-sphere mechanism: (i) a fast pre-equilibrium association of the reagents, followed by (ii) the act of electron transfer.

$$M^M + N^N \xrightleftharpoons{\text{fast}} M^M, N^N, \qquad K_{ip} \text{ or } K_{OS}$$

$$M^M, N^N \xrightarrow{k_{et}} M^{M-1} + N^{N+1}$$

In principle, the second step might be looked upon as composite, with the electron transfer, $M^M, N^N \rightarrow M^{M-1}, N^{N+1}$ followed by the separation of the two components of the successor complex. In practice, this separation to give the final products is always fast, as there is no bond breaking involved as in the case of the inner-sphere successor complex. We shall therefore not be concerned with this separation step here, but it should be pointed out that its reverse is the initial step in the reverse redox reaction, $M^{M-1} + N^{N+1} \rightarrow M^M + N^N$, if the redox potentials of the M and N couples involved are almost equal and the reaction is reversible.

The pre-association of reactants prior to electron transfer brings them sufficiently close together for the transfer of the electron to occur with reasonable probability and speed. For large complexes, particularly bioinorganic molecules, it may be necessary for the oxidant and reductant to associate with a fairly specific orientation in order to make the electron transfer from one metal centre to the other reasonably facile. These matters of orientation and distance of electron transfer will be considered after we have dealt with the general features of pre-association equilibria in simpler systems. It is also possible that rapid transfer of an electron to the surface of an oxidant may be followed by relatively slow transfer through the ligand to the metal centre, effectively giving a radical–ligand complex, $M^{n+}-L\cdot$ en route to $M^{(n-1)+}-L$. This seems to be the case for $*[Ru(bipy)_3]^{2+}$ reductions of $[Co(NH_3)_5L]^{3+}$ with L = **47** or **48**, for these complexes with easily reducible ligands are reduced much more rapidly than their analogues with L = pyridine or **49**, where there are no reducible ligands. There are also cases where catalysis by readily reducible ligands, e.g. pyridine carboxylates (pcx), suggests that electron transfer takes place from $N^{(n+1)}–L\cdot$ rather than from N^{n+} or $N^{n+}-L$. This can be illustrated by the role suggested for $Eu^{3+}(pcx\cdot)$ in pcx catalysis of the reduction of $[Co(NH_3)_5(py)]^{3+}$ and of $[Co(en)_3]^{3+}$ by Eu^{2+}aq.[81]

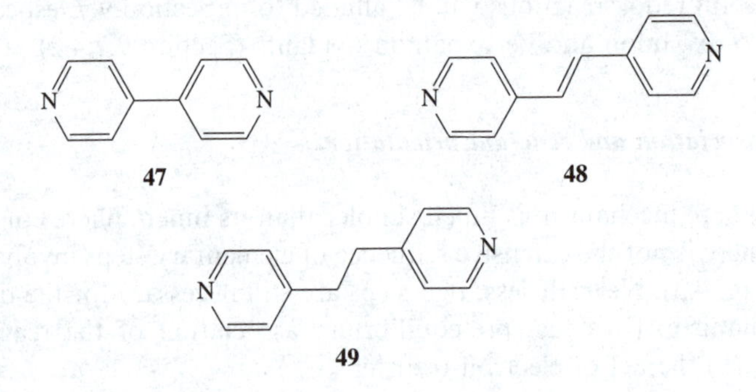

9.5.2.1 *Inorganic complexes*

The first convincing demonstrations of pre-association (ion-pairing) before electron transfer came from a series of cobalt(III)–iron(II) reactions.[82] Here both metals are in the low-spin d^6 configuration, the iron(II) being in the form of a pentacyano derivative. This ensures substitution-inertness prior to electron transfer. A large charge product (z_+z_-) encourages ion-pairing, and electron transfer is slow enough to permit observation of the pre-association of the reactants. The kinetic pattern for reaction of $[Co(NH_3)_5(H_2O)]^{3+}$ with $[Fe(CN)_6]^{4-}$ shows the saturation limit expected (Fig. 9.18). The pre-association equilibrium constant, K_{ip} or K_{OS}, and the rate constant for electron transfer, k_{et}, can be derived from the customary reciprocal plot. For the reaction cited, $k_{et} = 0.19\ s^{-1}$, so the half-life of the ion-pair is only about 4 s. However, the ion-pair has a high formation constant, $K_{ip} = 1500$ (molar scale). In the analogous reaction with $[Ru(NH_3)_5(py)]^{3+}$ the ion-pair can be crystallized as the salt $[Ru(NH_3)_5(py)]_4[Fe(CN)_6]_3.7H_2O$. This has a violet colour, attributable to the ruthenium–iron IVCT.

The importance of specific orientation in appropriate circumstances may be illustrated by the small but significant differences, with enantiomeric excesses in the 10–40 % range, between rate constants for D and L forms in, for example, the oxidation of $*[Ru(bipy)_3]^{2+}$ by $[Co(edta)]^-$,[83] a number of cobalt(II)/(III) reactions involving such complexes as $[Co(en)_3]^{3+}$,

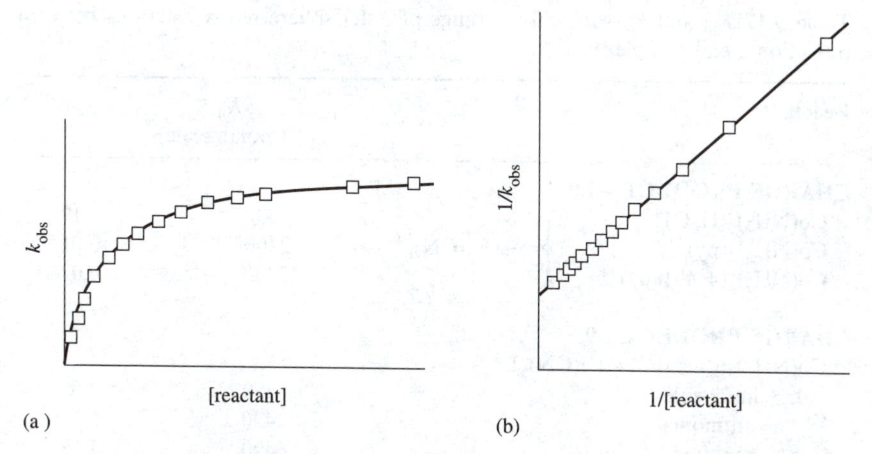

Fig. 9.18 (a) Typical saturation kinetics plot for the pre-equilibrium formation of a reactive intermediate and (b) the corresponding linear reciprocal plot.

$[Co(sep)]^{2+}$, $[Co(edta)]^-$, and $[Co(ox)_3]^{3-}$,[84] and some $[Co(edta)]^{2-}$ reductions of nickel(III) and nickel(IV) complexes.[85] The nickel(IV)/$[Co(edta)]^{2-}$ reactions demonstrate the possibilities of maximizing the enantiomeric excess through tuning by means of ligand substituents, here variously methyl, phenyl, and benzyl groups substituted into the hexadentate bis-oximate ligand. In general, differences in observed rate constants are ascribed to differences in the ion-association pre-equilibrium rather than to the electron transfer step. That association constants for ion-pairs of this type do differ for the two enantiomeric forms has been established, for example, for association between non-redox pairs such as $[Co(en)_3]^{3+}$ plus antimonyl dimethyltartrate, where K_{ip} values for the two enantiomers of the cation are 38 and 58.[86] The detailed orientation and stereochemistry for the $[Co(en)_3]^{2+}$,$[Co(edta)]^-$ ion-pair has been established by X-ray diffraction.[87]

Table 9.17 shows K_{ip} and k_{et} values[88] for a range of outer-sphere redox reactions between transition metal complexes. K_{ip}, of course, depends upon the charge product, z_+z_-, and the internuclear distance; values of this distance can be estimated from the magnitude of K_{ip}. For $*[Ru(bipy)_3]^{2+}$, $[Co(edta)]^-$, the cobalt–ruthenium distance is estimated to be about 11 Å, which corresponds closely to the value for a contact ion-pair. The internuclear distance is important in relation to k_{et}, which is principally controlled by the driving force for the reaction, $\Delta G(\Delta E)$. The variation of ΔG with the nature of ligand L determines the range of k_{et} values found for the oxidation of $[Fe(CN)_5L]^{3-}$ by $[Co(NH_3)_5(dmso)]^{3+}$ (Table 9.17); K_{ip} values vary very little in this series. The iron-to-cobalt distance varies only slightly, and there is a linear correlation between $\log k_{et}$ and the redox potentials of the $[Fe(CN)_5L]^{3-/2-}$ couples (Fig. 9.19), which are themselves

Table 9.17 K_{ip} and k_{et} values for a range of outer-sphere redox reactions between transition metal complexes.

Reactants	K_{ip} [a] (molar scale)	$k_{et}(s^{-1})$
CHARGE PRODUCT −12		
$Co(NH_3)_5(H_2O)^{3-}$ + $Fe(CN)_6{}^{4-}$	1500[b]	0.19
$Co(NH_3)_5(py)^{3+}$ + $Fe(CN)_6{}^{4-}$	2400[b]	0.015
$Co(NH_3)_5(4,4'\text{-bipy})^{3+}$ + $Fe(CN)_6{}^{4-}$	2300	0.024
CHARGE PRODUCT −9		
$Co(NH_3)_5(dmso)^{3+} + Fe(CN)_5L^{3-}$		
L = imidazole	450	2.6
ammonia	420	1.3
pyridine	490	0.15
isonicotinamide	600	0.05
pyrazine	360	0.0089
pyrazine-2-carboxamide	560	0.0047
$Co(NH_3)_5(py)^{3+} + Fe(CN)_5(py)^{3-}$	860	0.0068
$Co(phen)_3{}^{3+} + Co(ox)_3{}^{3-}$	650	0.24
CHARGE PRODUCT −8		
$Co(NH_3)_5(acetate)^{2+}$ + $Fe(CN)_6{}^{4-}$	300	0.000 37
$Co(NH_3)_5(benzoate)^{2+}$ + $Fe(CN)_6{}^{4-}$	240	0.000 62
$Co(NH_3)_5Cl^{2+}$ + $Fe(CN)_6{}^{4-}$	38	0.027
$Co(NH_3)_5(N_3)^{2+}$ + $Fe(CN)_6{}^{4-}$	49	0.000 62

[a] These values should be compared with values in the range 0.1–1 for association between a 2+/2+ pair of complexes.
[b] Significantly lower estimates have also been published—see I. Krack and R. van Eldik, *Inorg. Chem.*, 1986, **25**, 1743.

enthalpy-determined.[89] The importance of solvation changes consequent on initial ion association is well shown by the volume changes given in Table 9.18.

An alternative way to study the effect of distance is to examine rates of heterogeneous electron transfer at electrodes.[90] There is a 10^4-fold decrease in k_{et} for electron transfer from a gold or a mercury surface to the cobalt(III) in $[Co(NH_3)_5(\text{thioalkylcarboxylate}]^{2+}$ as the polymethylene chain lengthens from $-CH_2-$ to $-(CH_2)_5-$. Close packing at the electrode surface both gives an effectively constant 'K_{ip}' and prevents the polymethylene chain from buckling and thereby reducing the cobalt–sulphur (electrode surface) distance. It is interesting that rate constants for intermolecular electron transfer are sometimes remarkably similar to those for intramolecular electron transfer within an analogous binuclear inner-sphere precursor. This similarity will be documented in the comparison between inner- and outer-sphere mechanisms later in this chapter, in Section 9.6.3.

Table 9.18 Ion-association and electron-transfer components of overall volumes of activation for redox reactions between oppositely charged complexes.

Reaction	$z_A z_B$	$\Delta V^{\ddagger}$	=	$\Delta V_{os}^{\ddagger}$	+	$\Delta V_{et}^{\ddagger}$
$[Co(NH_3)_5Cl]^{2+} + [Fe(CN)_6]^{4-}$	−8	+23		−3		+26
$[Co(NH_3)_5(N_3)]^{2+} + [Fe(CN)_6]^{4-}$	−8	+2		−17		+19
$[Co(NH_3)_5(py)]^{3+} + [Fe(CN)_6]^{4-}$	−12	+47		+23		+24
$[(H_3N)_5CoO_2Co(NH_3)_5]^{5+} + [Mo_2O_4(edta)]^{4-}$	−20	+36		+24		+12

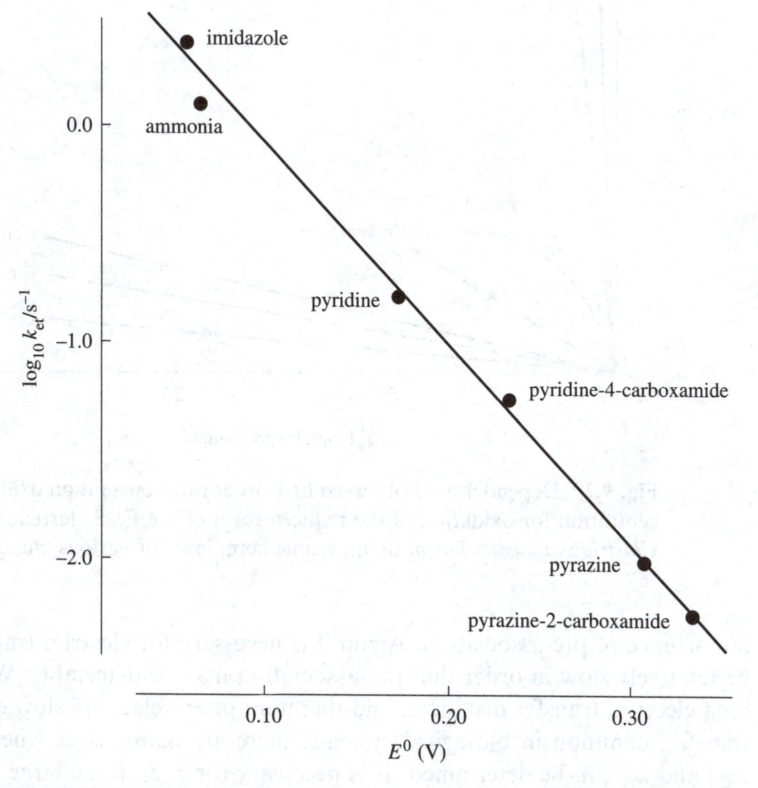

Fig. 9.19 Electron transfer within the precursor: a plot of $\log_{10} k_{et}$ against E^0 (versus SCE) for the $[Fe(CN)_5L]^{3-/2-}$ couples, for the ligands L shown.

9.5.2.2 Biochemical relevance

A similar picture applies, of course, to outer-sphere redox reactions involving bioinorganic substrates, reacting either with simple inorganic species or in biologically relevant pairs such as plastocyanin plus cytochrome *f*. This was recognized many years ago, when kinetic experiments on the cytochrome *c*/hexacyanoferrate equilibrium showed saturation kinetics and the

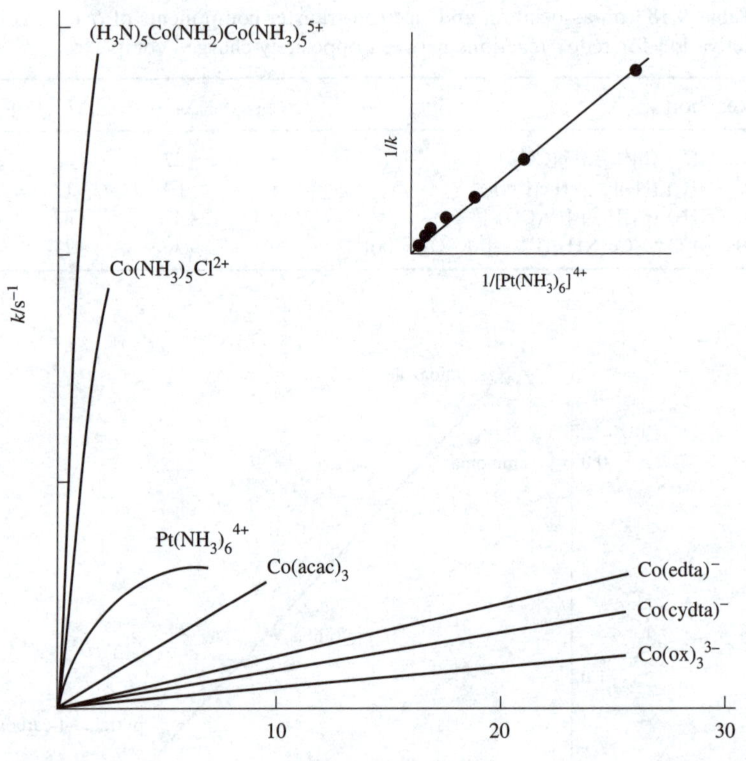

Fig. 9.20 Dependence of observed first-order rate constant on oxidant concentration for oxidation of the reduced form of the Fe_8S_8 ferredoxin from *Clostridium pasteurianum* by inorganic complexes of various charges.

importance of pre-association. Again it is necessary for electron transfer to be relatively slow in order that pre-association may be detectable. With the long electron transfer distances, and therefore often relatively slow electron transfer, common in biological systems, there are many cases where both K_{OS} and k_{et} can be determined. It is necessary for z_+z_- to be large enough for saturation kinetics to be observed (cf. Fig. 9.18 above). Thus, for example, curvature is not apparent in the k_{obs} versus [oxidant] plot for reaction of cytochrome *b* with $[Co(edta)]^-$, but it is apparent when the oxidant bears a sizeable positive charge, e.g. $[Co(NH_3)_6]^{3+}$, $[Pt(NH_3)_6]^{4+}$, or $[(H_3N)_5CoNH_2Co(NH_3)_5]^{5+}$. Similarly, these three 3+, 4+, and 5+ oxidants give saturation kinetics in ferredoxin oxidation, whereas $[Co(NH_3)_5Cl]^{2+}$ gives barely detectable curvature of the k_{obs} versus [oxidant] plot and $[Co(NH_3)_5(ox)]^+$, $[Co(acac)_3]$, $[Co(edta)]^-$, and $[Co(ox)_3]^{3-}$ all give no indication of equilibrium pre-association, the magnitude of K_{OS} being too small for K_{OS}[oxidant] to be significant compared with 1 (Fig. 9.20).[91]

It is often not at all clear what value to assign to the effective charge on a bioinorganic species. Sometimes it is estimated from the effect of the variation of ionic strength on the rate constants, an application of the classical Livingston diagram (Brønsted–Bjerrum out of Debye–Hückel—see Chapter 8.3.1.2). This approach is based on the use of inorganic complexes of known charge as oxidants or reductants, establishing charge products from ionic-strength dependences, and thence deducing the effective charge on the electron transfer site of the bioinorganic species. Oxidation of *Clostridium pasteurianum* flavodoxin semiquinone illustrates this approach well.[92] This appears to have an effective or 'interaction domain' charge of approximately 4−. This result was cross-checked by assumption of this value leading to apparent charges of (2.6)− and (1.0)− on $[Fe(CN)_6]^{3-}$ and $[Fe(edta)]^-$ from the effect of ionic strength variation on the relevant rate constants. Having checked the consistency of the approach for oxidants of known charge, such ionic-strength effects on the rate constants for the oxidation of flavodoxin semiquinone by various cytochromes *c*(III) can be used to assign effective charges of between 2− and 7+ for these cytochromes. Later work,[93] in which $[Fe(CN)_6]^{3-}$ and $[Co(phen)_3]^{3+}$ gave essentially equal estimates for each of three cytochromes *c*, and a series of five inorganic reductants all indicated a local charge of between 6+ and 7+ for the electron transfer site of horse heart cytochrome *c*(III), confirmed the acceptability of this analysis. These effective charges, essentially localized in the region to which the electron is transferred, may be very different from the overall charge on the compound. This is strikingly illustrated by *Pseudomonas denitrificans* cytochrome c_2, whose net charge is 7−,[94] but which behaves in electron transfer reactions as though its reaction centre or 'interaction domain' has a charge between 3+ and 4+. Azurin from *P. aeruginosa* has an overall charge between 1− and 2−, but its reaction centre behaves as though it bears a positive charge. Oxidation of cytochrome *f*(II), which also has an effective positive charge despite an overall negative charge, by $[(NC)_5Fe(CN)Co(CN)_5]^{5-}$ shows saturation kinetics. In this case it has been possible to separate the ionic-strength effects on the pre-association equilibrium constant, K_{OS}, from those on the actual electron transfer, k_{et}. As is often observed, the changes in medium affect K_{OS} much more than k_{et}—an increase of ionic strength from 0.025 to 0.20 mol dm^{-3} results in a 25-fold decrease in K_{OS} but only a three-fold increase in k_{et} for this reaction.[95]

There is evidence that active sites for electron transfer are often exposed edges of haem units. This is consistent with the effective charges established by the ionic-strength approach, and also with the effects of chemical modification of, and in the vicinity of, these sites. Such modification may be achieved by pH variation, by coordination of chromium(III) complexes,[95,96] or by reaction of such reagents as CDNP (4-carboxy-2,6-dinitrophenyl) with lysine residues.[97] Such tailoring of the surface and its effect on redox rate constants for, e.g., $[Co(phen)_3]^{3+}$ and $[Fe(CN)_6]^{3-}$ oxidations and $[Co(sep)]^{2+}$ and $[Fe(edta)]^{2-}$ reductions, can provide particularly detailed insight into electron-transfer processes in this type of system.[98]

Modification of the vicinity of the haem centre may reduce reactivity by impeding the approach of the oxidant, but a more effective way of inhibiting these redox reactions is through competition between the oxidant and a redox-inactive analogue. $[Cr(NH_3)_6]^{3+}$ and $[Cr(en)_3]^{3+}$ are effective inhibitors for $[Co(NH_3)_6]^{3+}$, whereas $[Zr(ox)_4]^{4-}$ and $[Mo(CN)_8]^{4-}$ are effective inhibitors for anionic oxidants. Oxidation of cytochrome *f*(II) from *Brassica oleracea* (i.e. cabbage leaves) by $[(NC)_5Fe(CN)Co(CN)_5]^{5-}$ shows saturation kinetics with $K_{OS} = 4100\,dm^3\,mol^{-1}$; this oxidation is inhibited by $[Zr(ox)_4]^{4-}$, with $K_{OS} = 530\,dm^3\,mol^{-1}$. Complementarily, oxidation of ferredoxin from *Spirulina platensis* (algae) by $[Co(NH_3)_6]^{3+}$ ($K_{OS} = 2070$ $dm^3\,mol^{-1}$) is inhibited by $[Cr(NH_3)_6]^{3+}$ ($K_{OS} = 1020\,dm^3\,mol^{-1}$). The redox reaction between plastocyanin and cytochrome *f* is inhibited by both cations such as $[Pt(NH_3)_6]^{4+}$, which blocks the negative copper(II) site in the plastocyanin, and anions such as $[Zr(ox)_4]^{4-}$, which blocks the positive iron(II) site of the cytochrome. Site blocking can sometimes be probed by observing the effects of chromium complexes on ^1H NMR spectra of the bioinorganic substrates. The paramagnetic chromium(III) causes line broadening of amino acid residues in the vicinity of the interaction, i.e. potential electron transfer, site. $[Cr(NH_3)_6]^{3+}$ and $[Cr(CN)_6]^{3-}$ can be used to probe sites with effective negative and positive charges, respectively. The association–kinetics–inhibition pattern developed in this and the preceding paragraphs is well and succinctly exemplified by the results for Fe^{II}/Fe^{III} purple acid phosphatase shown in Table 9.19.[99] Association constants determined from electron transfer kinetics may also

Table 9.19 Association, electron transfer, and inhibition in the chemistry of Fe^{II}/Fe^{III} purple acid phosphatase.

Complex	Kinetic behaviour	K_{OS}	Rate constants
$[Co(phen)_3]^{3+}$	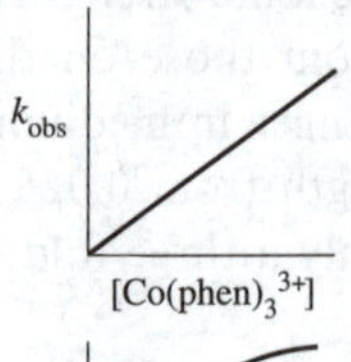 k_{obs} vs $[Co(phen)_3^{3+}]$ (linear)		$k_2 = 1.26\,dm^3\,mol^{-1}\,s^{-1}$
$[Fe(CN)_6]^{3-}$	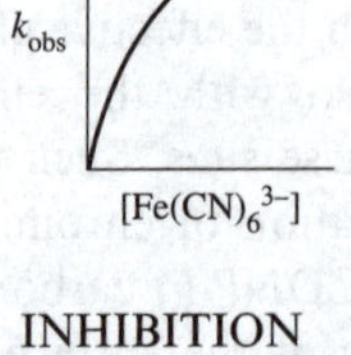 k_{obs} vs $[Fe(CN)_6^{3-}]$ (saturation curve)	540	$k_{et} = 1.0\,s^{-1}$
$[Cr(CN)_6]^{3-}$	INHIBITION	550	
$[Mo(CN)_8]^{4-}$	INHIBITION	1580	

be useful in relation to other reactions of metalloproteins. Thus the K_{OS} values of Table 9.19 have been used in analysing the kinetics of substitution reactions, especially of phosphates and related oxoanions, in purple acid phosphatase catalysis.[100] Protons can also inhibit electron transfer by interacting with negatively charged sites, but this is a more complicated matter as protonation can take place at a variety of sites in these metalloproteins.

We shall return to bioinorganic redox reactions again in Section 9.6.5, where the emphasis will be on electron transfer rates rather than on the initial pre-equilibrium association or docking of the two reactants.

9.5.3 Marcus theory

Outer-sphere electron transfer is of great interest and importance in that its relative simplicity permits theoretical analysis and calculation with a workably small number of assumptions. Moreover, it provides many examples of direct kinetic–thermodynamic correlations, where reactivities parallel thermodynamic driving forces for reaction. An early example was provided by cerium(IV) oxidation of a series of (substituted) tris-(1,10-phenanthroline) iron(II) complexes (Fig. 9.21),[101] a somewhat later example by iodide reduction of (substituted) tris-(1,10-phenanthroline)iron(III) complexes and of hexachloroiridate(IV).[102]

The main theoretical treatments have been those of Marcus[103] and Hush,[104] neatly compared by Cannon,[105] with support, and some inorganic applications, provided by Sutin.[106,107] Figure 9.22 shows the relationship between reactant(s), product(s), free-energy terms, and electron transfer routes. For his treatment of electron transfer under normal thermal conditions, Marcus started from the standard transition-state theory equation

$$k_{et} = \text{constant.exp}\,(-\Delta G^{\ddagger}/RT)$$

with $\Delta G^{\ddagger}$ given by

$$\Delta G^{\ddagger} = \tfrac{1}{4}\lambda(1 + \Delta G^{\circ}/\lambda)^2$$

Here λ is the nuclear reorganization energy or intrinsic free-energy barrier. This barrier has two components, λ_{int} arising from geometric changes within the reactant(s) and λ_{ext}, arising from solvation changes; $\lambda = \lambda_{int} + \lambda_{ext}$. The first contribution can be estimated from changes in bond lengths and force constants. The second contribution is more troublesome, with estimates based on simple electrostatic solvation models often less than satisfactory. For a bimolecular redox reaction, the work term, the work required to bring the two reactants together from infinity, can be a major contributor to k_2 through K_{OS} in the composite $k_2 = K_{OS}k_{et}$. In addition to these various components of the so-called 'nuclear factor', there is also an 'electronic factor'[108] to be incorporated in the analysis. This electronic factor expresses the probability of electron transfer taking place once the reagents are in their

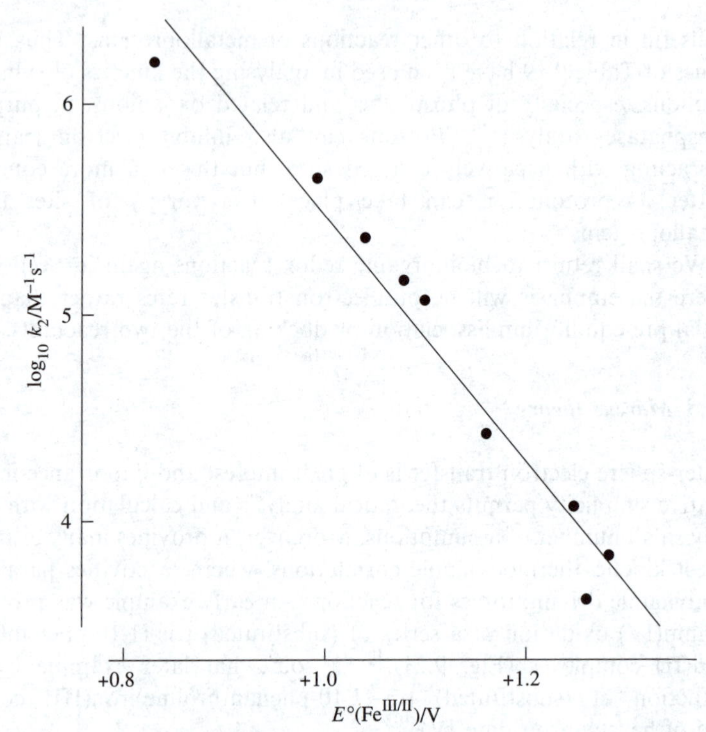

Fig. 9.21 Correlation between logarithms of rate constants for cerium(IV) oxidation of $[Fe(X\text{-}phen)_3]^{2+}$ cations (X = Me_4, Me_2, Me, H, Ph, Cl, SO_3^-, NO_2) in aqueous sulphuric acid and redox potentials for the respective iron(II)/(III) couples.

optimal geometric state. A low probability implies that the system passes through the transition state several times before electron transfer actually takes place, i.e. electron transfer is non-adiabatic. Basic Marcus theory and its applications assume adiabatic behaviour.

For systems conforming to Marcus's assumptions it can be deduced that the slope of a linear free-energy plot of $\Delta G^{\ddagger}$ against ΔG° should be 0.5. This was long ago shown to be the case for such reactions as reduction of $[Fe(phen)_3]^{3+}$ by Fe^{2+}aq, oxidation of $[Fe(phen)_3]^{2+}$ by manganese(III), cobalt(III), and cerium(IV), and reduction of a range of cobalt(III) complexes by $[Ru(NH_3)_6]^{2+}$,[109] and indeed the reactions included in Fig. 9.21. Oxidations of organic species by inorganic complexes can also give linear free-energy plots of this type with slopes equal to, or close to, 0.5.[110]

The full theoretical Marcus treatment[111] requires a number of assumptions and estimates, but several of the complications disappear if one assumes the operation of the Frank–Condon principle and one compares rate constants for a given reaction with rate constants for self-exchange for the component redox couples, for example if one compares k_{12} for the cross-reaction (with equilibrium constant K_{12})

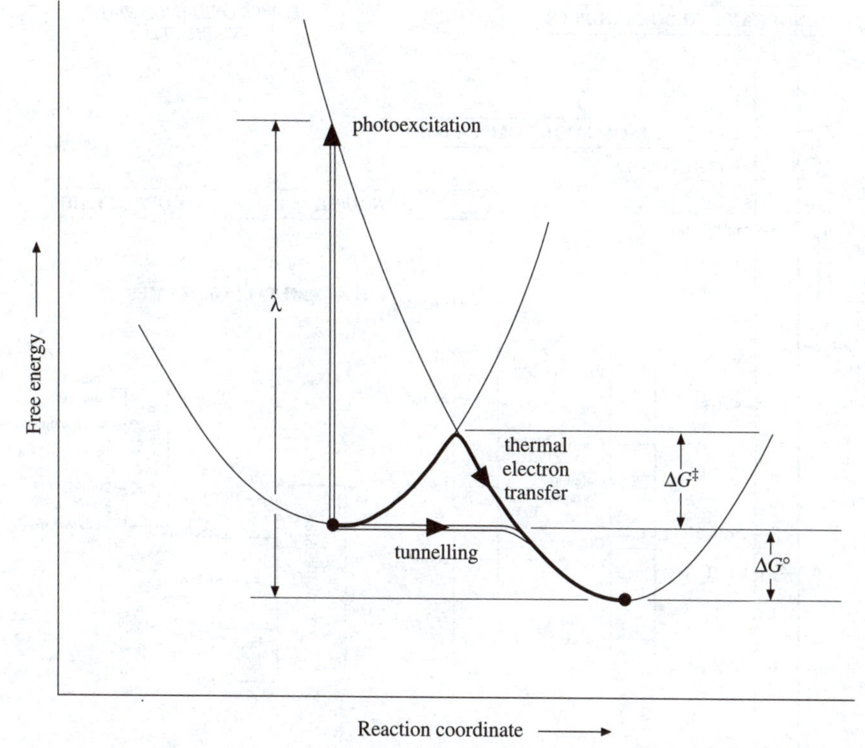

Fig. 9.22 Thermal, photoactivated, and tunnelling routes for electron transfer (λ is the nuclear reorganization energy, see text).

$$[Co(phen)_3]^{3+} + [Ru(NH_3)_6]^{2+} \rightleftharpoons [Co(phen)_3]^{2+} + [Ru(NH_3)_6]^{3+}$$

with k_{11} and k_{22} for exchange in the $[Co(phen)_3]^{3+/2+}$ and $[Ru(NH_3)_6]^{3+/2+}$ systems. Then the various electron exchange rate constants and the driving force (ΔG°_{12}, represented here by the equilibrium constant K_{12}; $\Delta G^\circ_{11} = \Delta G^\circ_{22} = 0$ for self-exchange, of course) are related, provided that ΔG°_{12} is not too large (the theory has been found to be valid up to at least 0.84 V driving force, for some U^{3+}aq reductions[112]), by

$$k_{12} = (k_{11}k_{22}K_{12})^{1/2}w$$

Here w is a work term representing the algebraic sum of the energies expended or liberated in bringing the reactants together in the self-exchange and cross-reactions. Clearly w is related to the association constants, especially for a cross-reaction involving charges of opposite sign, and may be much affected by the solvent.[113] K_{12} ($\equiv \Delta G^\circ_{12}$) can sometimes be measured directly, for example using UV–visible spectroscopy, but is usually obtained from redox potential data ($\Delta G^\circ = -n\mathscr{F}E^\circ$). Values of E° for the aqua-ion couples M^{3+}/M^{2+}aq are available in standard textbooks; values of

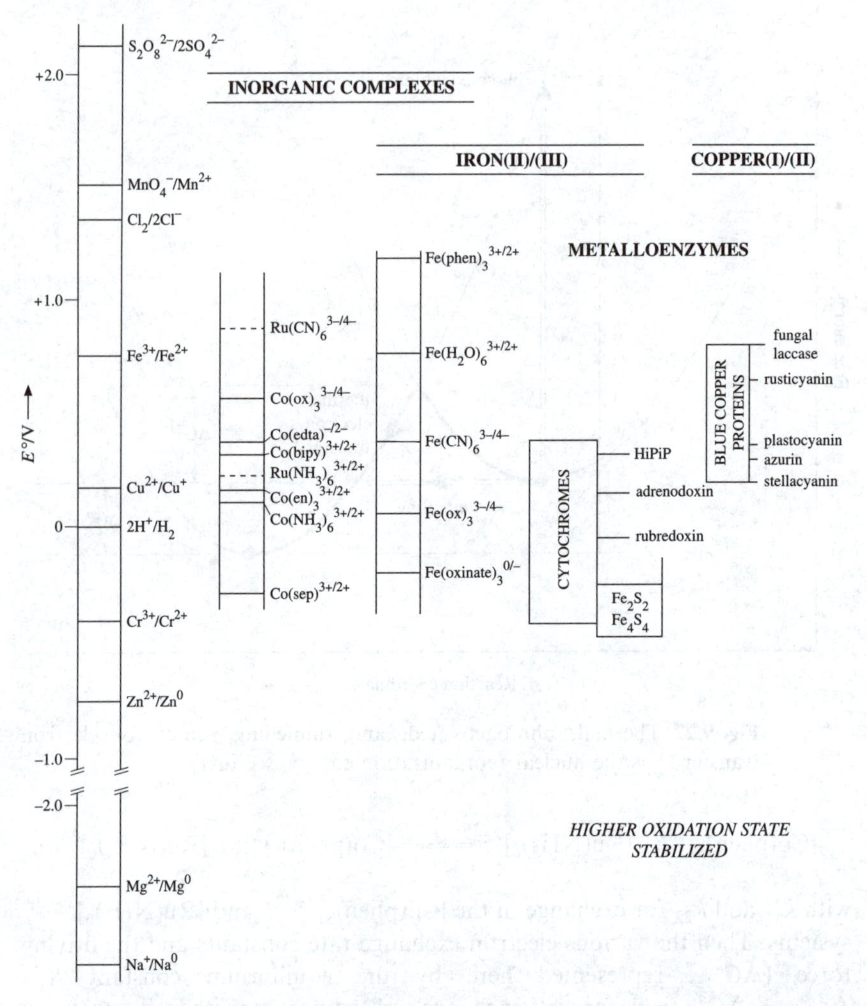

Fig. 9.23 Ligand effects on redox potentials—comparisons between iron and copper metalloenzymes, inorganic iron complex couples, and cobalt and ruthenium complex couples. The left-hand column gives a varied selection of inorganic redox couples to provide context for the values for the complex and metalloenzyme couples.

$E°$ for complexes take more finding. It is important to remember that complexation may have a large effect on redox potentials. Thus, for example, $E°$ values for iron(III)/iron(II) in a whole range of complexes span over $1\frac{1}{2}$ V, nearly half the total usable range in aqueous solution. Some ligand effects on this Fe^{3+}/Fe^{2+} couple are shown diagrammatically in Fig. 9.23, which also shows ranges of $E°$ values for iron- and copper-containing metalloproteins, and puts these values into their general inorganic context

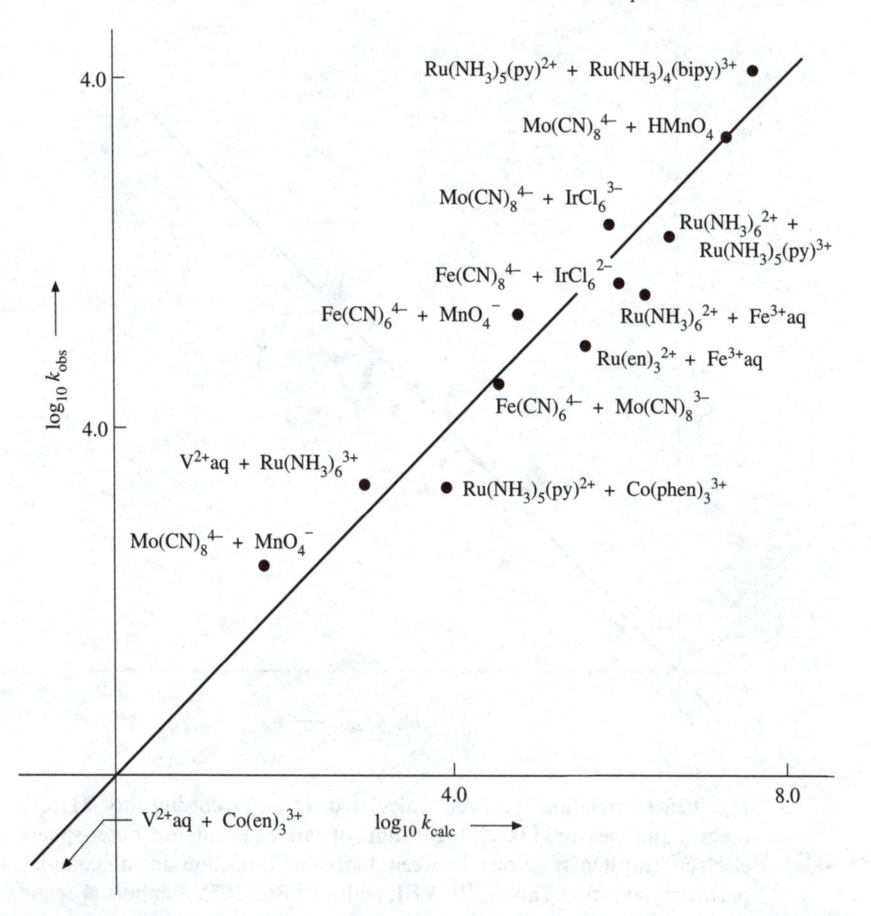

Fig. 9.24 Correlation between calculated (k_{calc}, according to Marcus theory) and measured (k_{obs}) logarithms of rate constants for outer-sphere electron transfer reactions between pairs of transition metal complexes (data from Table 1 of Ref. 107).

(left-hand column). Redox potentials for bio-derivatives of these two metals cover a range of nearly $1\frac{1}{2}$ V, thanks to tailoring of the appropriate metalloproteins. The copper(I)/(II) proteins cover the upper part of the range, the iron(II)/(III), whether cytochromes or iron–sulphur proteins, the lower part—with a generous overlap in the middle of the range (around 200–500 mV).

The rate constant k_{12} is generally measured directly, by conventional or fast-reaction techniques. Self-exchange rate constants are discussed in the following section—sometimes these are available by direct measurement, in other systems they are obtained by means of a Marcus analysis.

The Marcus–Hush theory has proved extremely valuable in correlating, rationalizing, explaining, and predicting. We illustrate its range of applic-

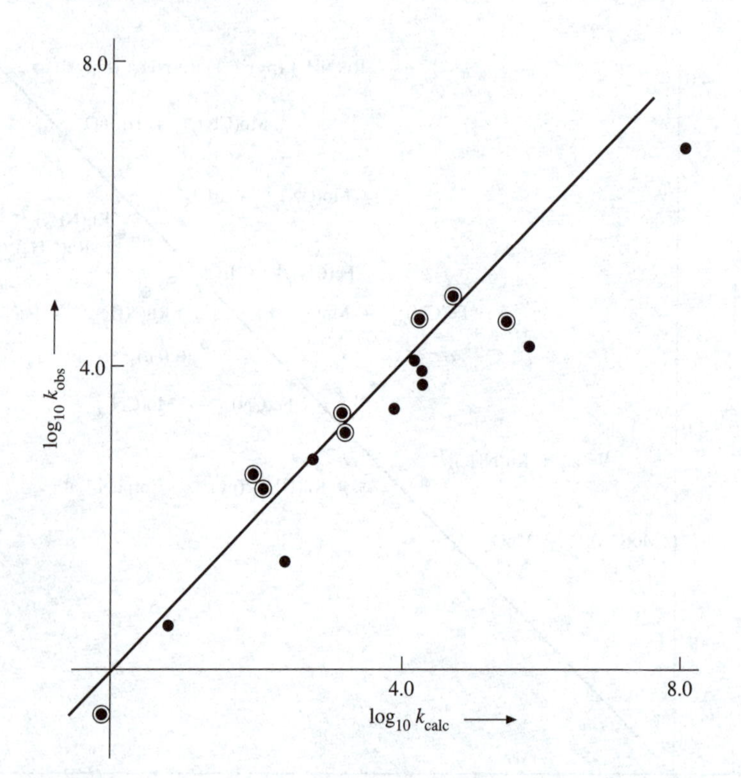

Fig. 9.25 Correlation between calculated (k_{calc}, according to Marcus theory) and measured (k_{obs}) logarithms of rate constants for outer-sphere electron transfer reactions between biorganic/transition metal complex partners (data from Tables VII, VIII, and X of Ref. 107). Symbols ● relate to various forms of cytochrome *c*, ◉ to copper proteins (azurin or plastocyanin).

ability in Figs 9.24–9.26. The first shows how well calculated rate constants agree with actual values for a number of purely inorganic systems. The others show similar agreement for reactions involving the oxidation or reduction of a bioinorganic substrate by an inorganic complex (Fig. 9.25) and for reactions between two bioinorganic species (Fig. 9.26). The theory applies equally well in organometallic chemistry,[114] and to electron transfer at electrodes.[115] The latter may be considered an extreme or limiting case, with the operation of inner- or outer-sphere mechanisms depending on the nature of the electrode. The theory has been successfully extended into the realms of organic chemistry,[116] for atom and methyl group transfer as well as for electron transfer.[117] There are cases where Marcus theory appears to work for inner-sphere electron transfer, but probably there is coincidental cancelling of opposing or interfering effects in what are generally closely related groups of reactions. On the other hand, there are a number of outer-sphere systems where there are large deviations from Marcus theory. These

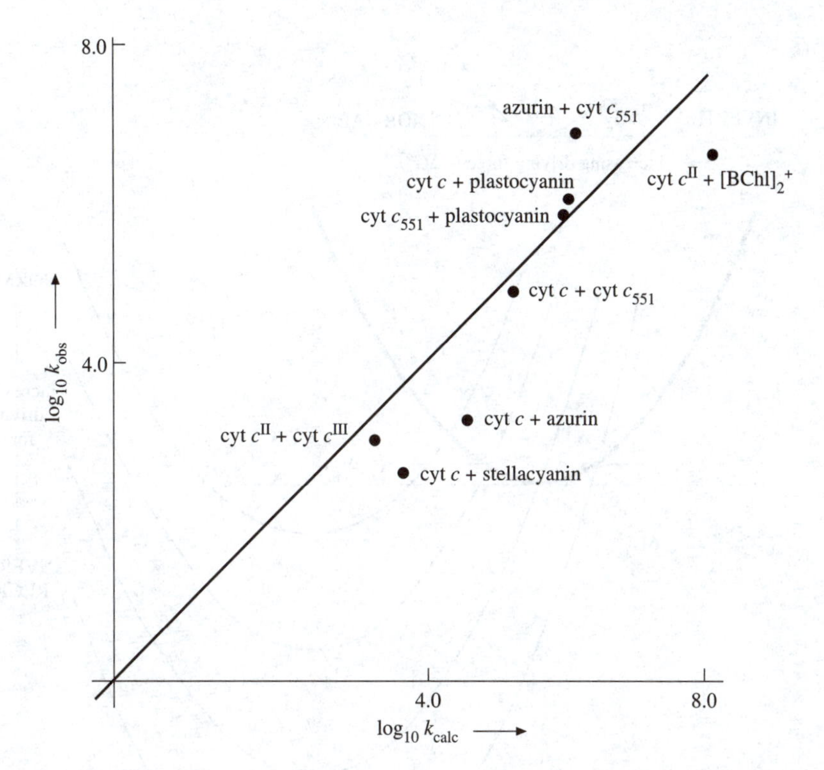

Fig. 9.26 Correlation between calculated (k_{calc}, according to Marcus theory) and measured (k_{obs}) logarithms of rate constants for outer-sphere electron transfer reactions between pairs of bioinorganic species (data from Tables VIII and IX of Ref. 107; BChl = bacteriochlorophyll).

often involve cobalt complexes, where the high-spin ↔ low-spin changeover are almost invariably involved and the possibility of non-adiabatic behaviour are often invoked as explanations.[118]

The expectation of a linear free-energy plot (LFER; $\Delta G^{\ddagger}$ against ΔG°) having a slope of 0.5, mentioned earlier in this section, applies to reactions with small or moderate driving forces ($-\Delta G^{\circ}$). As the driving force increases, such plots will start to curve, then to level off, and eventually to change direction. In other words, beyond a certain driving force, the point where $-\Delta G^{\circ}$ is equivalent to the nuclear reorganization energy, electron transfer rate constants begin to *decrease* as the driving force increases—the so-called 'inversion region'.[119] The reason underlying this sequence of behaviour is illustrated in Fig. 9.27. Curvature of LFER plots of this type beyond a value of $-\Delta G^{\circ}$ of around 70–80 kJ mol^{-1} (about 0.8 V) was reported many years ago, but experiments on reactions with driving forces up to -230 kJ mol^{-1} failed to reach an inversion point.[120] There has been much discussion of the breakdown of the kinetic–thermodynamic correlation at large negative ΔG° values, but experimental demonstrations proved

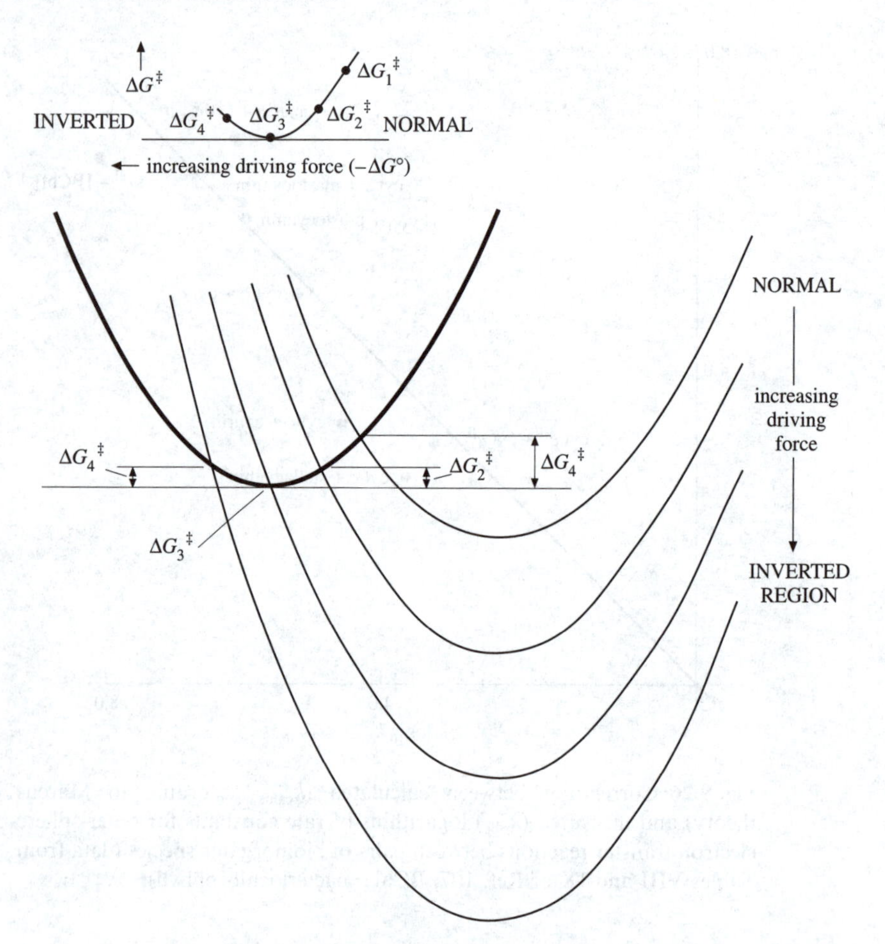

Fig. 9.27 The changeover from the normal behaviour of increasing rate constant with increasing driving force to the inverted behaviour of decreasing rate constant with increasing driving force at high values of the latter.

very hard to come by. The first examples related to very fast electron transfer in organic chemistry,[121] but there are examples from inorganic systems,[122] and several from bioinorganic chemistry[123] (see also Section 9.6.5 below).

9.5.4 Self-exchange reactions

These are amongst the simplest electron transfer reactions, involving only the movement of an electron from an ion or complex in one oxidation state to the analogous ion or complex in an oxidation state one unit higher. Examples are:

$$M^{2+}aq + {}^*M^{3+}aq = M^{3+}aq + {}^*M^{2+}aq$$

$$Fe(cp)_2 + {}^*Fe(cp)_2{}^+ = Fe(cp)_2{}^+ + {}^*Fe(cp)_2$$

$$MnO_4{}^- + {}^*MnO_4{}^{2-} = MnO_4{}^{2-} + {}^*MnO_4{}^-$$

$$Fe(CN)_6{}^{4-} + {}^*Fe(CN)_6{}^{3-} = Fe(CN)_6{}^{3-} + {}^*Fe(CN)_6{}^{4-}$$

In most cases such reactions are outer-sphere in character, but there are examples where the inner-sphere mechanism operates. Most examples involve pairs of charged species, so solvation, ion-pairing effects, and work terms are all often important in determining reactivities, and certainly have to be borne in mind in detailed discussions. This is, of course, particularly true for such cases as the hexacyanoferrate system[124] above, where $z_A z_B = 12$, but also affects the other much-studied system $MnO_4{}^{-2/-}$.[125] Only in such special cases as the ferrocene/ferrocinium exchange, where one of the reactants is uncharged, are solvent and ion-pairing effects sometimes negligible—and even here one needs to bear in mind the possibility of ion-pairing between the charged reactant and its gegenion.

Self-exchange kinetics have a long history, dating back to an examination of Pb^{II}/Pb^{IV} exchange in 1920. The ready availability of isotopes in the aftermath of the Second World War led to considerable activity in the study of isotopic exchange reactions in the late 1940s and early 1950s. The techniques available then restricted experimental investigations to reactions with what by today's standards are rather long half-lives. The advent of NMR and its attendant possibilities of obtaining kinetic information from signal coalescence and line width variation opened up the field to studies of self-exchanges with half-lives down to milliseconds. There are also often chemical factors which preclude the direct measurement of self-exchange rate constants. In such cases, the Marcus treatment described in the preceding section, linking self-exchange electron transfer rate constants with rate constants for each reductant/oxidant pair, can come into its own. Where self-exchange rate constants can be measured directly, the validity of the Marcus theory can be assessed. In other situations it is common practice to derive self-exchange rate constants via cross-reaction rate constants; a number of examples will be encountered in the ensuing discussion. When all the required kinetic and equilibrium data are available, then this approach to calculating experimentally inaccessible self-exchange rate constants can give reliable estimates. But the reader should be warned that some estimates of self-exchange rate constants using Marcus theory involve assumptions or approximations which may result in an uncertainty or error of several powers of ten in the final value (cf. $O_2/O_2{}^-$ and $SO_2/SO_2{}^{-\cdot}$ in Sections 9.7.6.3 and 9.7.6.4 below). An example of the Marcus analysis approach to the estimation of self-exchange rate constants is set out in Table 9.20.[126] Agreement between the four estimates in the final column is quite respectable, indeed the range is minuscule when compared with the

Table 9.20 Estimation of the self-exchange rate constant for the $[Co(tmen)_3]^{3+/2+}$ (tmen = tetramethylethane-1,2-diamine) couple from rate constants for reduction of the 3+ complex by $M^{2+}aq$.

			Rate constants ($dm^3\ mol^{-1}\ s^{-1}$)		
	$E°$(V)	$\log K_{12}$	$M^{2+/3+}aq$ self-exchange k_{11}	cross-reaction k_{12}	self-exchange for complex k_{22}
$Cr^{2+}aq$	−0.41	2×10^{-7}	11.7	0.010	1
$Eu^{2+}aq$	−0.38	4×10^{-4}	11.2	0.49	4
$V^{2+}aq$	−0.26	10^{-2}	9.0	0.66	20
$Ru^{2+}aq$	+0.21	20	1.3	0.0056	9

whole range of electron transfer rate constants. Readers wishing to study a system where Marcus analysis of self-exchange and cross-reactions runs into difficulties may like to read about the problems associated with the ascorbate system[127] or with copper$^{I/II}$[128].

In practice, recourse to the Marcus approach is frequently unavoidable, in view of the restrictions on systems in which self-exchange rate constants can be measured directly. Isotopic labelling can only provide kinetic data for relatively slow reactions—i.e. for self-exchange reactions with half-lives considerably longer than separation times. The techniques of NMR and, to a much lesser extent, ESR permit the estimation of rate constants for much more rapidly reacting systems. However, the line-broadening techniques used in such experiments can only give kinetic information over a relatively limited range of time constants—and there is the further restriction that the properties of the nuclei involved must be appropriate for these resonance techniques. Table 9.21[129] gives an idea of the range of techniques which can be used, and of the range of reactivities observed, for the copper(I)/(II) couple.

Measured or estimated self-exchange rate constants are available for many hundred, perhaps several thousand, reactions. A generous selection of such reactions have been documented[130]—with activation parameters $\Delta H^\ddagger$ and $\Delta S^\ddagger$ given for many. We shall discuss several groups of ions and complexes in this section, but before dealing with specific groups of compounds certain common and key features may be mentioned. Perhaps the most fundamental factor is the Frank–Condon principle. Normally the geometry of the species in the upper and lower oxidation states is different. Metal–ligand bond lengths are generally shorter in the higher oxidation state (Table 9.22), while for copper(I)/(II) there is a change in favoured coordination environment from tetrahedral copper(I) to tetragonally distorted octahedral copper(II). As electrons move very much more rapidly

Table 9.21 Range of experimental techniques used, and rate constants obtained, for electron exchange between copper(I) and copper(II) complexes (at or close to 298.2 K; in water or acetonitrile solution).

Ligand	Technique	$k_{11}/dm^3\ mol^{-1}\ s^{-1}$
chloride[a]	^{63}Cu NMR(T_2)	5×10^7
azurin	^{1}H NMR; $^{63/65}$Cu ESR	$1–100 \times 10^4$ [b]
stellacyanin	$^{63/65}$Cu ESR	1×10^5
$H_{-2}Aib_3$[c]	^{1}H NMR	5.5×10^4
2,9-Me_2phen	Marcus	2×10^4
various[d]	^{1}H NMR(T_1, T_2)	$1–130 \times 10^2$
phen	Marcus	43
[14]-ane-S_4	^{1}H NMR	4
water	Marcus	1×10^{-5}

[a] In conc. HCl.
[b] Over the range 277–319 K.
[c] $Cu^{II}[H_{-2}Aib_3]^- =$

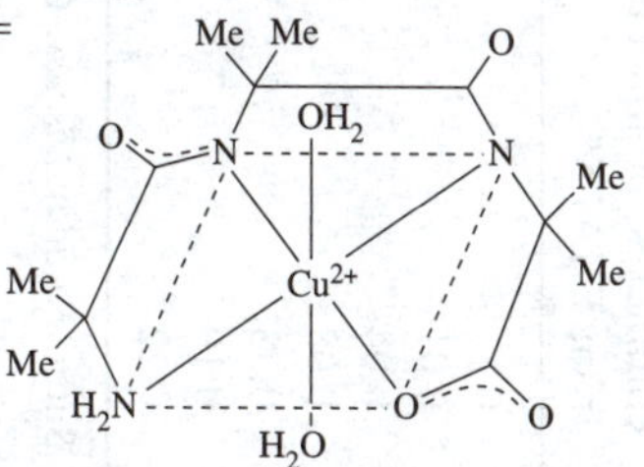

[d] N_4 and N_2S_2 macrocycles.

than nuclei, there must be some rearrangement of atoms before and/or after the actual electron transfer and therefore there will be an energy barrier arising from the required changes in bond lengths or angles. The importance of this Frank–Condon barrier has been widely discussed for many years, with this discussion not unclouded by changes in bond lengths dictated by improved crystal structure determinations, especially for the key complexes $[Co(NH_3)_6]^{n+}$.[131] Metal–ligand vibrational frequencies can give an idea of bond strengths and thus of the work which may be required to stretch or compress bonds to attain the appropriate geometry for electron transfer. Tables 9.22 and 9.23 contain values for self-exchange rate constants, to show their relation to the bond length changes reported in these tables. There are generally good correlations, though with a couple of cases of dramatic non-conformance.

The nature of the ligands is of prime importance in several respects. Their size affects the distance over which the electron has to move; their saturated or unsaturated nature affects the ease with which the electron can traverse this distance. Ligand nature also exerts a large influence over the redox potential of the metal couple, and thus the driving force for electron transfer. The presence of a potential bridging ligand may affect not just the rate but also the mechanism—hexaaqua and hexaammine complexes

Table 9.22 Metal–ligand bond distance changes and their relation to self-exchange rate constants.

	(Average) metal ion–ligand bond lengths (Å); in crystals from X-ray diffraction [a]		Difference between metal ion–ligand bond lengths (Å) in the 2+ and 3+ complexes; in aqueous solution from EXAFS [a]	Self-exchange rate constants $k_{11}(s^{-1})$
	2+	3+		
$Fe(phen)_3^{n+}$	1.97	1.97	0	1.3×10^7 [b]
$Fe(CN)_6^{n-}$	1.90	1.93	0.01	2×10^4
$Ru(NH_3)_6^{n+}$	2.14	1.93	0.04	6.6×10^3
$Ru(H_2O)_6^{n+}$	2.12	2.03	0.08	20
$Fe(H_2O)_6^{n+}$	2.12	1.99	0.13	1.1
$Co(phen)_3^{n+}$	2.13	1.93	0.19	12
$Co(NH_3)_6^{n+}$	2.16	1.97	0.22	$\sim 10^{-7}$

[a] From B. S. Brunschwig, C. Creutz, D. H. Macartney, T. -K. Sham, and N. Sutin, *Faraday Discuss. Chem. Soc.*, 1982, **74**, 113, and references cited therein.
[b] At 276 K.

Table 9.23 Relation between self-exchange rate constants, k_{11} ($dm^3\ mol^{-1}\ s^{-1}$), and changes in metal–ligand bond distances, Δ (Å), for series of 2,2-bipyridyl and sarcosine complexes.

	Δ	k_{11}		Δ	k_{11}
$Ru(bipy)_3^{3+/2+}$	0	4×10^8	$Ru(sar)^{3+/2+}$	0.01	6×10^5
$Fe(bipy)_3^{3+/2+}$	0	3×10^8	$Fe(sar)^{3+/2+}$	0.04	7×10^5
$Cr(bipy)_3^{3+/2+}$	0.10	1×10^9	$Ni(sar)^{3+/2+}$	0.09	4×10
$Ni(bipy)_3^{3+/2+}$	0.12	1.5×10^3	$Mn(sar)^{3+/2+}$	0.11	39
$Co(bipy)_3^{3+/2+}$	0.19	~ 10	$Co(sar)^{3+/2+}$	0.19	2

react by the outer-sphere mechanism perforce, but hydroxo and chloro species, for instance, may react very much more rapidly by an inner-sphere route.

9.5.4.1 Aqua-cations

Measurements or estimates have been made of self-exchange rate constants for a number of $M^{(n+1)+}/M^{n+}$ aqua-ion pairs. In a number of cases values are dubious, sometimes with estimates from different laboratories spanning several orders of magnitude, or large discrepancies between measured values and Marcus-derived estimates. Such divergences usually arise from differing assumptions about geometry, redox potentials, or adiabaticity being fed into Marcus derivations. There are also problems arising from the fact that Co^{3+} and Mn^{3+} oxidize water, and that $Cu^{+}aq$ is unstable with respect to disproportionation. A selection of rate constants and activation parameters which are believed to be fairly reliable is given in Table 9.24. Negative activation entropies and volumes are consistent with bimolecular transition states for electron transfer.

Table 9.24 Rate constants, k_{11} ($dm^3\ mol^{-1}\ s^{-1}$), and activation parameters, $\Delta H^{\ddagger}$ ($kJ\ mol^{-1}$), $\Delta S^{\ddagger}$ ($J\ K^{-1}\ mol^{-1}$), and $\Delta V^{\ddagger}$ ($cm^3\ mol^{-1}$), for electron exchange between aqua-cation pairs, in aqueous solution at 298.2 K.[a]

		k_{11}	$\Delta H^{\ddagger}$	$\Delta S^{\ddagger}$	$\Delta V^{\ddagger}$		k_{11}
sp-block	$Tl^{3+/+}$	7×10^{-5}					
d-block	$V^{3+/2+}$	1×10^{-2}	55	−105			
	$Cr^{3+/2+}$	2×10^{-6}					
	$Fe^{3+/2+}$	1.1	46	−88	−11	$Ru^{3+/2+}$	20
	$Co^{3+/2+}$	5	44	−92			
	$Cu^{2+/+}$	5×10^{-7}				$Ag^{2+/+}$	2×10^9
f-block	$Eu^{3+/2+}$	*ca* 10^{-5}					
	$Ce^{4+/3+}$	$\leq 6 \times 10^{-2}$				$U^{4+/3+}$	$\sim 6 \times 10^{-5}$

[a] Sources for the data given here may be traced via Tables 2.7 and 2.9 of Ref. 2.

Calculations of electron transfer and exchange rate constants in systems involving aqua-cations are complicated by work terms and, in many cases, by hydrolysis equilibria. For $M^{3+/2+}$ exchange, the most commonly studied, $z_A z_B = 6$, while for $M^{4+/3+}$ $z_A z_B$ is impossibly high at 12. Moreover, most M^{3+}aq cations, at least outside the *f*-block, have pK_a values between 1 and 5, while M^{4+}aq cations can have pK_a values below 1. Unless electron exchange kinetics are monitored in appropriately acidic conditions there will be significant, even dominant, reaction via a facile inner-sphere hydroxo-bridged path. Obviously, rate constant measurements over an appropriate pH range can give kinetic data for both the M^{3+}aq and MOH^{2+}aq paths; both for iron and for chromium the latter path has a rate constant about 10^6 times larger. Although this and other observations suggest that hydroxo complexes react by the inner-sphere mechanism and aqua complexes by outer-sphere electron transfer, such hypotheses have rarely been properly established. Indeed this is not always an easy task, although for instance in the case of the $Ru(H_2O)_6^{3+/2+}$ couple a combination of ^{17}O and ^{99}Ru NMR experiments has provided not only good values for k and for $\Delta H^{\ddagger}$ and $\Delta S^{\ddagger}$, but also strong evidence for the operation of the outer-sphere mechanism.[132] In the much-studied iron(III)/(II) system, activation volumes of -11.1 and $+0.8$ cm^3 mol^{-1} for the Fe^{3+}aq/Fe^{2+}aq and $FeOH^{2+}$aq/Fe^{2+}aq paths suggest outer-sphere and hydroxide-bridged inner-sphere mechanisms, respectively.[133]

9.5.4.2 *Cobalt(III)/(II)*

Some idea of the variety of systems for which self-exchange rate constants have been determined or estimated, and of the enormous range of rate constants, can be gleaned from Table 9.25.[134] It must be admitted that there are considerable problems in relation to two of the key cobalt(III)/(II) couples. The case of the aqua-ions has already been mentioned—the propensity of Co^{3+} to oxidize water precludes direct measurement, and there are questions over the validity of applying Marcus's theory. In the case of the $[Co(NH_3)_6]^{3+/2+}$ couple, estimates for the self-exchange rate constant have oscillated wildly over four decades since the original experiments using cobalt-60-labelled complexes. At least it is clear that the mechanism is outer-sphere, and that there is no hydroxide-dependent path. But difficulties stem from the substitution lability of the cobalt(II) complex, a conceivable contribution from $[Co(NH_3)_5(H_2O)]^{3+}$ even at high ammonia concentrations, from the possibility of oxygen oxidizing cobalt(II) species, and from the possible presence of catalysing impurities. Discussion of this system has also been clouded by uncertainties over cobalt–nitrogen bond lengths. For many years Co–N distances, derived from 1926 powder diffraction data, of 1.9 and 2.5 Å were quoted for the 3+ and 2+ complexes; the big difference indicated a very large reorganization energy on electron transfer. The most recent and reliable bond distances, given in Table 9.22 above, still indicate a change of 0.205 Å and thus a considerable reorganization barrier. The

Table 9.25 Kinetic parameters for self-exchange for cobalt(III)/(II) complex couples, measured or estimated at 298.2 K in aqueous solution.

Complex couple	k ($dm^3\ mol^{-1}\ s^{-1}$)	$\Delta H^{\ddagger}$ ($kJ\ mol^{-1}$)	$\Delta S^{\ddagger}$ ($J\ K^{-1}\ mol^{-1}$)	$\Delta V^{\ddagger}$ ($cm^3\ mol^{-1}$)
$[Co([9]aneS_3)_2]^{3+/2+}$ [a]	1.5×10^5	30	−45	−5
$[Co(terpy)_2]^{3+/2+}$	4×10^2			
$[Co(phen)_3]^{3+/2+}$	12	21	−156	
$[Co(bipy)_3]^{3+/2+}$	5.7	31	−127	
$[Co(sep)]^{3+/2+}$} [b]	5	41	−95	−6
$[CoW_{12}O_{40}]^{5-/6-}$	0.72	18	−184	
nitrogen macrocycles	10^{-2} to 10^{-5}			
$[Co(en)_3]^{3+/2+}$	7.7×10^{-5}	59	−132	−20
$[Co(NH_3)_6]^{3+/2+}$	$\sim 10^{-7}$			
$[Co(tmen)_3]^{3+/2+}$	8.5×10^{-8}			

[a] There have been significant disagreements over kinetic parameters for this self-exchange reaction, but there is no doubt that it is much the fastest Co^{III}/Co^{II} self-exchange.
[b] Rate constants are available for more than a dozen encapsulated systems—see text. They range from 0.024 to 4500, being affected by cage geometry and the nature of the donor atoms and of cage substituents; there is no correlation with, e.g., redox potentials.

change of spin involved in this electron transfer must also make a significant contribution to the overall activation barrier, but this is probably rather less than the reorganization energy associated with change in the cobalt–nitrogen distance.

None of these problems apply to $[Co(en)_3]^{3+/2+}$, where again rather slow electron exchange can be attributed in the main to reorganization energy. Substitution of four methyl groups per ethane-1,2-diamine ligand reduces the rate constant by 10^3. This marked decrease in rate constant is attributed to steric factors—the oxidized and reduced forms have to be docked carefully, keeping various methyl groups out of each other's way, in order for the cobalt centres to be in reasonable proximity. In fact, the two complexes have to approach along a mutual trigonal axis. The determination of the self-exchange rate for this $[Co(tmen)_3]^{3+/2+}$ couple provides a good example of the application of the Marcus approach, with consistent estimates from a range of different metal-ion reductants.[126]

Electron exchange rate constants for aza-macrocyclic ligand complexes[135] cover quite a range of reactivity, attributable to variation in redox potential and in reorganization energy. The latter depends considerably on ligand bulk and flexibility as well as on bond length changes. There is clearly a balance of several factors, for, for instance, triaza-macrocyclic ligands permit faster electron exchange ($0.19\ dm^3\ mol^{-1}\ s^{-1}$ for the $Co([9]aneN_3)_2^{3+/2+}$ couple) than tetraaza-macrocycles ($\sim 10^{-4}$ to $10^{-5}\ dm^3\ mol^{-1}\ s^{-1}$), and indeed than linear triaza-ligands, while self-exchange rate constants for rigid porphyrin systems are relatively fast (0.06–$20\ dm^3\ mol^{-1}\ s^{-1}$). Much faster self-exchange in trithia- ($\sim 10^5\ dm^3\ mol^{-1} s^{-1}$ for the $Co([9]aneS_3)_2{}^{3+/2+}$ couple) than in triaza- systems suggests that bond length changes and consequent reorganization energy changes dominate this comparison[136]—both Co^{II} and Co^{III} forms are low-spin for the trithia- ligands, but there is a spin change on oxidation or reduction for the triaza- ligands.

As stated towards the start of this section, the substitution lability of cobalt(II) complexes causes problems in obtaining reliable self-exchange rate constants. The best way to circumvent this problem is to encapsulate the cobalt atom, so as to make its escape from the surrounding donor atoms very difficult—see the entries for $[Co(sep)]^{n+}$ in Table 9.15. Kinetic parameters have been determined for a number of self-exchange reactions of encapsulated cobalt complexes (**50**); data for $[Co(sep)]^{3+/2+}$ are included in Table 9.25.

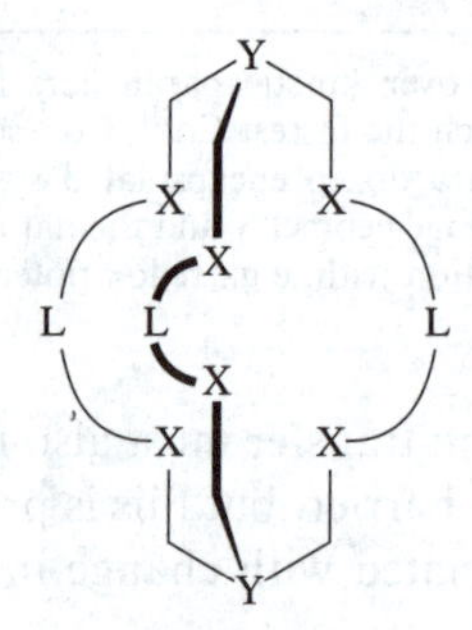

X = NH or S; Y = N or CR {R = H, Me, Cl, NO_2, $NH_3{}^+$}
L = CH_2CH_2, $CH_2CH_2CH_2$, or 1,2-C_6H_4

50
sep has X = NH; Y = N; L = CH_2CH_2
sar has X = NH; Y = CH; L = CH_2CH_2

Rate constants for more than a dozen cage systems have been determined[78,137]—they range from 2.4×10^{-2} to $2.8 \times 10^4\ dm^3\ mol^{-1}\ s^{-1}$. There is no simple correlation with, e.g., redox potentials. Reactivities are affected by cage geometry and flexibity, the nature of the donor atoms and of cage substituents, bond length changes, and solvation effects. Electron exchange is easier than in analogous linear (garland) ligand systems. Comparison of self-exchange rate constants for $Co^{III/II}$ macrocycle couples with those for their $Ni^{III/II}$ and $Cu^{III/II}$ analogues reveals the $Co^{III/II}$ systems to be much the slowest, reflecting the higher reorganization energies required.[138]

9.5.4.3 *Around the Periodic Table*

It is difficult to give detailed or extensive pictures of reactivity trends going across or down the Periodic Table. Limited data on some first-row transition element complexes show a vaguely consistent pattern for classical

Table 9.26 Trends in self-exchange rate constants ($dm^3\ mol^{-1}\ s^{-1}$) for Group VII and VIII complexes.[a,b]

	$[M(bipy)_3]^{3+/2+}$			
	in CD_3CN	in water	$[M(CN)_6]^{3-/4-}$	$[M(sar)]^{3+/2+}$
Fe	4×10^6	3×10^8	230 (0 °C)	6×10^3
Ru	8×10^6	4×10^8	$8\text{–}21 \times 10^3$	120×10^3
Os	19×10^6	20×10^8		

	$[M(dpme)_3]^{2+/+}$
Tc	2×10^6
Re	4×10^6

[a] In water at 25 °C except where indicated otherwise.
[b] Ligand formulae: sar—see **50** in the text; dmpe = $Me_2PCH_2CH_2PMe_2$.

ligands, but reactivity differences are very much less marked on going to bis-cyclopentadienyl complexes. The situation is more satisfactory when one considers trends down the Periodic Table. Table 9.26 shows increases in reactivity on going from iron to ruthenium to osmium and from technetium to rhenium. However, there is an enormous difference in reactivity on going from the [9]aneN_3, **51**, complexes of nickel(II)/(III) to the palladium(II)/(III) analogues. The nickel complexes have almost exactly the same geometry in the two oxidation states, but a huge reorganization is required in the palladium case. This difference is revealed as an almost 10^9 times smaller rate constant in the palladium system.

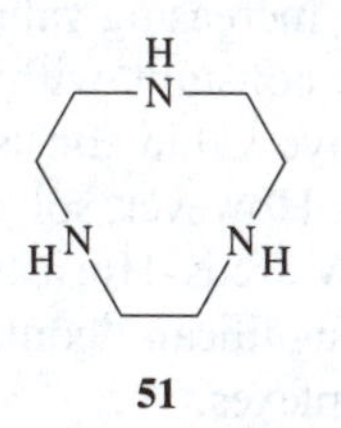

51

It is interesting to do a diagonal comparison between ruthenium and cobalt for the 2+ and 3+ oxidation states. Such comparisons have occasionally been used to check on Marcus estimates.[139] A modest correlation between values for the various 3+/2+ couples for the two metals is apparent from Table 9.27; the most striking thing is the considerably smaller range of values for the larger element.

The intuitively expected decrease in rate constant with increasing distance betweeen the metal centres is shown for ruthenium$^{3+/2+}$ redox reactions involving a variety of ammine, amine, pyridine, and cage ligands. There is a good correlation of log k with the reciprocal of the metal–metal separation

Table 9.27 Comparison of self-exchange rate constants, k_{11} ($dm^3\ mol^{-1}\ s^{-1}$), for cobalt and ruthenium couples.

	k_{11} (Co^{III}/Co^{II})	k_{11} (Ru^{III}/Ru^{II})	k_{11} (Ru)/k_{11}(Co)
$M(bipy)_3{}^{3+/2+}$	5.7	4×10^8	10^8
$M(sar)^{3+/2+}$	2.1	6×10^5	3×10^5
$M([9]aneN_3)_2{}^{3+/2+}$	0.2	5×10^4	3×10^5
$M(en)_3{}^{3+/2+}$	8×10^{-5}	3×10^4	4×10^8
$M(NH_3)_6{}^{3+/2+}$	10^{-7}	10^3	10^{10}

Table 9.28 Dependence of kinetic parameters for self-exchange reactions on size for manganese hexaisocyanide complexes $[Mn(CNR)_6]^{2+/+}$, in acetonitrile solution.

Ligand	k ($dm^3\ mol^{-1}\ s^{-1}$)	$\Delta H^\ddagger$ ($kJ\ mol^{-1}$)	$\Delta S^\ddagger$ ($J\ K^{-1}\ mol^{-1}$)
MeNC	2.1×10^7	8	−82
EtNC	2.0×10^6	15	−74
Me_2HCNC	2.6×10^5	20	−74
Me_3CNC	6.5×10^4	23	−78

over a range of 4 units of log k (0.07–0.15 Å).[140] The role of complex size and metal separation is clearly shown in the kinetic parameters for self-exchange in the manganese hexaisocyanide systems $[Mn(CNR)_6]^{2+/+}$ (Table 9.28).

As the bulk of the alkyl group R increases the self-exchange rate constant drops steadily, with this decrease reflecting steadily increasing values for the activation enthalpy. Activation entropies remain constant within experimental uncertainty, as usual at a markedly negative value consistent with bimolecular processes between like-charged species. However, solvent effects on reactivity in these systems do not conform to Marcus–Hush theoretical predictions, so work terms probably do make significant contributions, perhaps varying according to the sizes of the complexes.

Self-exchange rate constants have been measured or estimated for several polynuclear species, including rhodium(II) acetate,[141] a triangular oxotriruthenium complex,[142] some Fe_4S_4 and Mo_4S_4 clusters,[143] and a dodecatungstophosphate.[144] As Table 9.29 indicates, these are generally quite large. In most cases the metal centres are fairly close to the surface, so metal–metal separation barriers to electron transfer are not inordinately high. Thus for the oxotriruthenium couple $\Delta H^\ddagger = 19\ kJ\ mol^{-1}$ and $\Delta S^\ddagger = -29\ J\ K^{-1}\ mol^{-1}$. The equilibrium constant for pre-association of the two oxotriruthenium species may be about 7 (Eigen–Fuoss) or 0.14 (statistical), indicating that k_{et} is either about 2×10^7 or about $8 \times 10^8\ s^{-1}$ for the actual electron transfer.

Table 9.29 Self-exchange rate contants for bi- and polynuclear species.

Couple	Technique	$k\,(dm^3\,mol^{-1}\,s^{-1})$	Conditions
$[Rh_2(OAc)_4]^{+/0}$	NMR	2.9×10^5	water
	NMR	5.3×10^4	CD_3CN
	Marcus	3.0×10^5	CD_3CN
$[Ru_3O(OAc)_6(py)_3]^{+/0}$	NMR	1.1×10^8	CD_2Cl_2
$[Fe_4S_4(SR)_4]^{3-/2-}$	NMR	10^6 to 10^7	CD_3CN
$[Mo_4S_4(H_2O)_{12}]^{5+/4+}$	Marcus	7.6×10^2	water
$[Mo_4S_4(edta)_2]^{4-/3-}$	Marcus	1.5×10^7	water
$[Mo_4S_4(edta)_2]^{3-/2-}$	Marcus	8×10^5	water
$[PW_{12}O_{40}]^{4-/3-}$	NMR	1.1×10^5	water
$[PW_{12}O_{40}]^{5-/4-}$	NMR	1.6×10^2	water

9.5.4.4 Bioinorganic systems

Some self-exchange rate constants are available, mainly for iron and copper bioinorganic species (Table 9.30). Some have been obtained directly from NMR (iron; copper) or ESR (copper) experiments, others calculated using the Marcus equations. In the case of pseudoazurin from *Achromobacter cycloclastes* there is very good agreement between NMR and Marcus estimates;[145] in the case of rubredoxin agreement is not good.[146] There is a wide range of values for the iron systems. Electron exchange is very fast

Table 9.30 Self-exchange rate constants for bioinorganic iron and copper systems.

	$k\,(dm^3\,mol^{-1}\,s^{-1})$		$k\,(dm^3\,mol^{-1}\,s^{-1})$
IRON		**COPPER**	
Haems	10^6 to 10^{10} [a]	Azurin	2.4×10^6
Cytochromes		Stellacyanin	1×10^5
cytochromes *c*	10^2 to 10^4 [b]	Plastocyanins[d]	$\sim 10^3$ to 10^5
cytochrome c_{551}	10^6 to 10^7	Amicyanin	8.5×10^4
cytochrome *f*	5.0×10^5	Pseudoazurin	3×10^3
cytochrome b_5	2.6×10^3		
Iron–sulphur proteins			
rubredoxin	1.6×10^5		
several Fe_4S_4	10^3 to 10^4 [c]		
ferredoxin (parsley)	2×10^2		
HiPiP (*C. vinosum*)	~ 1		

[a] Most are within the range 5–10 × 10^7 $dm^3\,mol^{-1}\,s^{-1}$).
[b] There is a direct NMR determination of 1.2×10^3 $dm^3\,mol^{-1}\,s^{-1}$.
[c] Rates for electron exchange for simple $[FeS_4(SR)_4]^{2-/3-}$ pairs are faster, with *k* between 10^6 and 10^7 $dm^3\,mol^{-1}\,s^{-1}$ (cf. preceding table).
[d] Source-dependent (see text).

where there is a facile route through the protein to an exposed efficient electron transfer site, as for haem proteins (cf. Section 9.6.5). It is much less rapid for cytochromes, except for cytochrome c_{551}. The rapid self-exchange for this cytochrome may be due to its small size and to a favourable work term. Electron exchange is much slower when the metal centre is well buried in the protein, as in the case of the Fe_4S_4 unit in HiPiP of *Chromatium vinosum*. The electron transfer site can be probed through replacement of appropriate amino acid residues. Thus if valine-8 of rubredoxin is replaced by glutamate, thereby introducing a negative charge close to the cysteine residues (cys-9, cys-42) at the active site, the self-exchange rate constant decreases by nearly a hundred times.

Electron exchange in the copper systems[147] is usually fast; there appears to be a much smaller range of rate constants than for the iron proteins. In plastocyanin, azurin, and pseudoazurin the immediate environment of the copper consists of two donor nitrogens and two donor sulphurs, from two histidines, cysteine, and methionine. The geometry around the copper is distorted tetrahedral, with a particularly long bond to methionine-S. In azurin there is also a glycine-O almost as close to the copper as the distant methionine-S. Such arrangements represent compromises between the tetragonal geometry favoured by copper(II) and the tetrahedral geometry favoured by copper(I). Similarly, it is a compromise between the preference of copper(II) for nitrogen donors and of the softer copper(I) for sulphur donors. These systems provide examples of the *entatic state*,[148] where the metal centre is constrained in an environment intermediate between initial state and transition state, thereby reducing the activation barrier. There are considerable differences in rate constants for plastocyanins from different sources. This is hardly surprising, for the effective charges differ considerably, from 7−/6− for parsley plastocyanin to 1+/2+ for plastocyanin from *Anabaena variabilis*.[149]

9.6 Inner-sphere and outer-sphere reactions

9.6.1 Summary

A detailed comparison of the two electron transfer routes is given in Fig. 9.28, which is intended to show the much greater complication of the inner-sphere route, but also to emphasize features which are common to both mechanisms. In particular, initial outer-sphere association of oxidant and reductant is an important feature of both pathways. This pre-association may be reflected in activation entropies and volumes, especially when marked hydration changes accompany association. This has already been illustrated for iodide reduction of hexachloroiridate(IV) in the previous chapter (Section 8.2.4). The other important parallel to draw is between the first two steps of the inner-sphere path and the Eigen–Wilkins mechanism for complex formation (Section 7.5.1). In Taube's classic inner-sphere

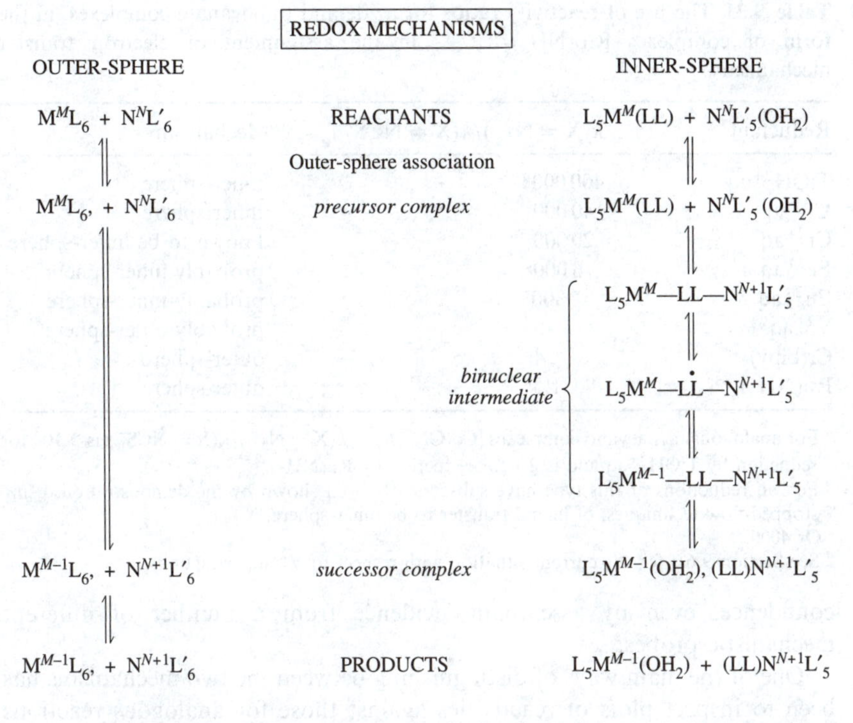

Fig. 9.28 Detailed comparison of the inner-sphere and outer-sphere electron transfer mechanisms.

system (Section 9.3.1), the replacement of one water in the Cr^{2+}aq reductant by $[Co(NH_3)_5Cl]^{2+}$ is in no way different from replacement of water in M^{n+}aq by chloride unencumbered by a $-Co(NH_3)_5{}^{3+}$ moiety. The key role of substitution in the inner-sphere path is illustrated in the discussion of V^{2+}aq reductions below (Section 9.6.3).

9.6.2 Distinguishing between these mechanisms

In cases where there is incontrovertible evidence for simultaneous electron and ligand transfer, it is clear that the mechanism is inner-sphere. In cases where electron transfer is very much faster than substitution required for inner-sphere electron transfer, the outer-sphere mechanism must be operative. Fortunately, the mechanisms of a large number of electron transfer reactions involving transition metal complexes can be deduced with confidence from these simple criteria. Unfortunately, the mechanisms of a large number of other electron transfer reactions involving transition metal complexes cannot be established by these simple criteria. Indeed, in several cases it has proved very difficult to establish the mechanism with any

Table 9.31 The use of reactivity ratios for azide and thiocyanate complexes, in the form of complexes $[Co(NH_3)_5X]^{2+}$ [a], in the assignment of electron transfer mechanisms.

Reductant	$k(X = N_3^-)/k(X = NCS^-)$	Mechanism
$TiOH^{2+}aq$	400 000[a]	inner-sphere
$U^{3+}aq$	40 000	inner-sphere
$Cr^{2+}aq$	20 000	known to be inner-sphere
$Fe^{2+}aq$	3 000	probably inner-sphere[b]
$Eu^{2+}aq$	300[c]	probably inner-sphere
$V^{2+}aq$	40	probably outer-sphere[d]
$Cr(bipy)_3^{2+}$	4	outer-sphere
$Ru(NH_3)_6^{2+}$	1.5[a]	outer-sphere

[a] For analogous pentacyano complexes $[Co(CN)_5X]^{3-}$, $k(X = N_3^-)/k(X = NCS^-)$ is $> 10^3$ for reduction by $TiOH^{2+}aq$ and is 2.1 for reduction by $Ru(NH_3)_6^{2+}$.
[b] $Fe^{2+}aq$ reductions of this type have subsequently been shown by the demonstration, using stopped-flow techniques, of ligand transfer to be inner-sphere.
[c] Or 4000.
[d] See Section 9.6.3 for the current situation with respect to $V^{2+}aq$ reductions.

confidence, even by assembling evidence from a number of different mechanistic probes.

One of the main ways of distinguishing between the two mechanisms has been to inspect plots of reactivities against those for analogous reactions involving indisputably outer-sphere processes involving such substitution-inert species as $[Fe(CN)_6]^{4-}$, $[Fe(phen)_3]^{2+}$, $[Ru(NH_3)_6]^{2+}$, $[Co(sep)]^{3+}$, or $[IrCl_6]^{2-}$. This is in effect assessment through linear free-energy plots—particularly useful when the non-conformance of some systems gives a clear indication of a difference in mechanism. Another approach is to compare, or contrast, kinetic parameters for reduction with those for complex formation (see the following section on $V^{2+}aq$ reductions). Comparison of reactivities for analogous azide and thiocyanate complexes has proved useful, based on the expectation that azide should be the more efficient bridging group in inner-sphere electron transfer but that there should be very little difference in reactivities for outer-sphere electron transfer (Table 9.31).[150] A similar approach uses ratios of rate constants for reduction by $V^{2+}aq$ and $Cr^{2+}aq$, as in the suggestion of outer-sphere reduction of rubredoxin by these cations.[151] In view of the fact that $V^{2+}aq$ sometimes reduces by the inner-sphere, sometimes by the outer-sphere mechanism (see Section 9.6.3, below), this seems to be a criterion to be used with considerable circumspection. Nonetheless, judicious comparisons of linear free-energy plots involving $V^{2+}aq$, $Cr^{2+}aq$, and $[Ru(NH_3)_6]^{2+}$ have proved informative.[152]

Other criteria suggested for distinguishing between the two pathways have proved less useful. Isotope effects (H/D) for aqua ligands are so small that their contribution has been very limited. Likewise, oxygen isotopic fractionation factors, the ratios of rate constants for reduction of $[Co(NH_3)_5(H_2{}^{16}O)]^{3+}$ to those for reduction of $[Co(NH_3)_5(H_2{}^{18}O)]^{3+}$, are

Table 9.32 Assignment of mechanism for iron(III) oxidations from activation volumes.

Mechanism (established)	Reaction	$\Delta V^{\ddagger}$ ($cm^3\ mol^{-1}$)	Deduced mechanism
OUTER-SPHERE ⟶	$Co(en)_3^{3+/2+}$	−20	
	$Fe(H_2O)_6^{3+/2+}$	−12	⟶ OUTER-SPHERE
	$Fe(OH)(H_2O)_5^{2+}/Fe(H_2O)_6^{2+}$	+1	⟶ INNER-SPHERE
INNER-SPHERE ⟶	$Cr(OH)(H_2O)_5^{2+}/Cr(H_2O)_6^{2+}$	+4	
INNER-SPHERE ⟶	$Co(NH_3)_5X^{2+}/Fe(H_2O)_6^{2+}$	+8 to +14	

Table 9.33 Mechanisms of electron transfer between aqua-cations and inorganic complexes.

	Inner-sphere	Outer-sphere
Reductants	Cr^{2+} Fe^{2+} Cu^{+}	Ti^{3+}
	⟵ V^{2+} [a] ⟶	
Oxidants	$FeOH^{2+}$	Fe^{3+}

[a] See Section 9.6.3.

all so close to unity (they are 1.021 for outer-sphere reduction by $[Ru(NH_3)_6]^{2+}$, 1.056 for inner-sphere reduction by Cr^{2+}aq) that differences are of dubious statistical significance. Absolute values of activation parameters can give ambiguous or even erroneous indications, generally due to problems with recognizing or quantifying hydration contributions. However, the determination of activation volumes for iron(III) oxidations proved useful in assigning electron transfer mechanisms for oxidation by Fe^{3+}aq and by $FeOH^{2+}$aq, as outlined in Table 9.32.[153] It has been suggested that Marcus analyses of activation volumes, rather than of the customary rate constants, may serve to distinguish between inner- and outer-sphere reactions.[154]

An overall summary of the areas of operation of the inner-sphere and outer-sphere mechanisms for reduction and oxidation involving metal ions is given in Table 9.33.

9.6.3 Reduction by vanadium(II)

V^{2+}aq is a particularly interesting and relevant reductant in the present context, since it may act as an inner-sphere or outer-sphere reductant, depending on the relation between the electron transfer rate and the rate of water exchange at this aqua-cation.[155] Fast electron transfer reactions have perforce to be outer-sphere, but many take place at rates comparable with substitution rates at V^{2+}aq and may therefore be outer- or inner-sphere in character. There are also a number of very slow redox reactions whose rate-limiting electron transfer step may again be outer- or inner-sphere. The relative lability of vanadium(III) complexes means that it can be difficult to confirm the operation of the inner-sphere mechanism by characterizing the immediate product of electron transfer. Nonetheless, such key intermediates have been detected in some systems.

The situation is summarized in Table 9.34,[156] which starts with kinetic data and estimates for substitution[157] in the top section. The first group of electron transfer systems is confidently believed to involve inner-sphere electron transfer, since not only have intermediates been detected but also the activation parameters are very similar to those for substitution. Admittedly the cationic complexes have electron transfer rate constants rather larger than estimates for rate-limiting substitution, but the complexes are quite large, so electrostatic repulsion between them and V^{2+}aq will be less than those between Ni^{2+}aq and the small positively charged entering groups from which these estimates were made. Again the somewhat slower than expected electron transfer rates for the pentacyanocobaltates may be attributed to their relatively large size. Certainly these anionic complexes react more rapidly with V^{2+}aq than their cationic cousins. The similarity of kinetic parameters for V^{2+}aq reduction of $[Co(NH_3)_5(OAc)]^{2+}$ to those for reduction of the complexes in the preceding group strongly suggest that reduction is inner-sphere here too. On the other hand, the differences between these kinetic parameters and those for reduction of $[Co(NH_3)_5Cl]^{2+}$ and $[Co(NH_3)_5Y]^{3+}$ ($Y = H_2O$, py, imid, dmf, or nicotinamide) suggest outer-sphere electron transfer for these latter complexes. Furthermore, the similarity of kinetic parameters for this group of complexes to those for the indubitably outer-sphere reductant $[Co(NH_3)_6]^{3+}$ strongly support outer-sphere electron transfer in these systems. The ruthenium and iron oxidants in the final group in Table 9.34 react so much more rapidly with V^{2+}aq than this aqua-ion undergoes substitution that they too must oxidize V^{2+}aq by outer-sphere electron transfer.

The cobalt(III)-oxalate complexes illustrate the balance between inner- and outer-sphere pathways. The ternary ammine-oxalate complexes, and their close relation $[Co(en)_2(ox)]^+$, oxidize V^{2+}aq by inner-sphere electron transfer, but $[Co(ox)_3]^{3-}$ oxidation goes almost exclusively by the outer-sphere route. However, the balance is sufficiently close in this case for very minor ($\sim 2\%$) but nonetheless significant inner-sphere electron transfer to have been detected.[158]

Finally, we mention a novel mechanism for reduction by V^{2+}aq. Rate constants for reaction with alkyl and aryl radicals ($R^{\cdot}$), at between 1 and 6×10^5 dm^3 mol^{-1} s^{-1}, are much too high for normal inner-sphere electron transfer, but a sort of S_N2 analogue has been proposed. In this variant, inner-sphere character is preserved by postulating an increase in coordination at the vanadium, the proposed transient intermediate being $[V(H_2O)_6R]^{2+}$.[159]

Table 9.34 Electron transfer mechanisms for reductions by V^{2+}aq.

Reaction	k ($dm^3\ mol^{-1}\ s^{-1}$)	$\Delta H^{\ddagger}$ ($kJ\ mol^{-1}$)	$\Delta S^{\ddagger}$ ($J\ K^{-1}\ mol^{-1}$)
SUBSTITUTION			
X^{3-} (estimated)[a]	~1000		
NCS^-	24	67	+5
bipy, phen, terpy	0.3–3.0	(~60)	
water exchange	1.6	68	+23
L^+; L^{2+} (estimated)[a]	~1; ~0.1		
ELECTRON TRANSFER			
Inner-sphere—intermediate detected			
$Cr(H_2O)_5(SCN)^{2+}$	16	52	−48
$Co(NH_3)_5(SCN)^{2+}$	30	69	+25
$Co(NH_3)_5(C_2O_4H)^{2+}$	12	51	−54
$Co(NH_3)_4(C_2O_4)^+$	45	51	−41
cis-$Co(en)_2(N_3)^{2+}$	33		
$Co(CN)_5X^{3-}$[b]	112, 140		
Inner-sphere—probably, though no intermediate detected			
$Co(NH_3)_5F^{2+}$	4	46	−77
$Co(NH_3)_5(OAc)^{2+}$	1.2	49	−8
$Co(CN)_5X^{3-}$[c]	120–280		
H_2O_2[d]	15.4		
Outer sphere—probably			
$Co(NH_3)_5Cl^{2+}$	10	31	−121
$Co(NH_3)_5Br^{2+}$	30		
$Co(NH_3)_5I^{2+}$	127		
$Co(NH_3)_5Y^{3+}$[e]	0.24, 0.53	34	−134
$Co(C_2O_4)_3{}^{3-}$	20 000	9	−132
Outer sphere—almost certainly			
$Ru(NH_3)_5X^{2+}$[f]	1300 to 5100	16	~ −135
$Ru(H_2O)_5Cl^{2+}$	1900		
Fe^{3+}aq	18 000		
$Co(NH_3)_6{}^{3+}$	0.004	38	−167

[a] Estimated by analogy with Ni^{2+} k_f trends (Table 7.14).
[b] X = N_3, SCN.
[c] X = Cl, Br, I.
[d] See Section 9.7.5.1.
[e] Y = py, H_2O.
[f] X = Cl, Br, OAc.

9.6.4 Similarities

Rate constants for intermolecular electron transfer are often remarkably similar to those for intramolecular electron transfer within a binuclear inner-sphere precursor analogue. Thus k_{et} within the $[Co(NH_3)_5(H_2O)]^{3+}$, $[Fe(CN)_6]^{4-}$ ion-pair is $0.19\,s^{-1}$ (*v.s.*), while within the original cobalt (III)–iron(II) precursor $[(H_3N)_5Co(nta)Fe(H_2O)_5]^{2+}$ it is $0.08\,s^{-1}$ (Section 9.4.4.5). A closer comparison is possible for some pyrazine-2,6-dicarboxylato (pzdc, **52**) complexes in ruthenium(II)–cobalt(III) systems. Thus k_{et} for electron transfer within the $[(NC)_5Fe(pzdcH)]^{4-}$,$[Co(dien)(pzdc)]^+$ ion-pair is, at $0.039\,s^{-1}$, six times greater than in the inner-sphere precursor complex $[(NC)_5Fe^{II}(pzdc)Co^{III}(dien)]^{2-}$, while k_{et} within the $[(H_3N)_5Ru(pzdcH)]^{2+}$, $[Co(dien)(pzdc)]^+$ ion-pair is seven times greater than that in $[(H_3N)_5Ru^{II}(pzdc)Co^{III}(dien)]^{3+}$.[160] Similar behaviour has been established for fast electron transfer. The second-order rate constant for reaction of $[Ru(phen)_2(py)Cl]^{2+}$ with $[Ru(bipy)_2(py)Cl]^+$, in acetonitrile is $4.9 \times 10^7\,dm^3\,mol^{-1}\,s^{-1}$. Using a Fuoss estimate of 0.6 for K_{OS} allows k_{et} for the outer-sphere electron transfer to be estimated as $8 \times 10^7\,s^{-1}$. For electron transfer within the inner-sphere analogue, **53**, k_{et} is between 5×10^7 and $10 \times 10^7\,s^{-1}$. In both reactions, the ruthenium–ruthenium separation is

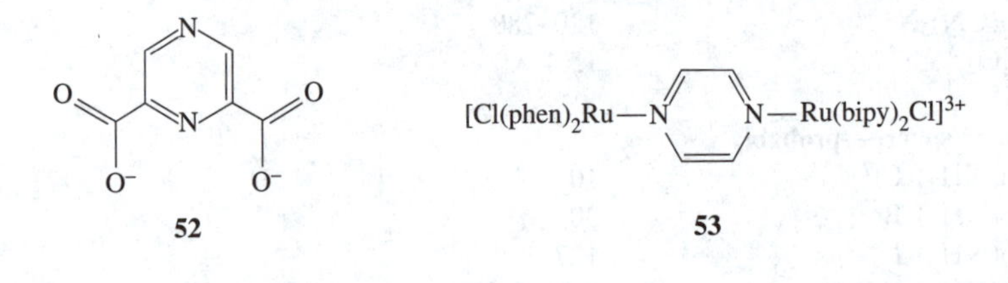

52 **53**

close to 13 Å.[161] If it is necessary to estimate k_{et} for very fast electron transfer within an ion-pair from IVCT absorption then the numerous approximations involved, detailed in Section 9.4.4.5 in connection with inner-sphere transfer, must be borne in mind. Direct comparison of estimates of k_{et} within the ion-pair $[Ru(NH_3)_5(py)]^{3+}$, $[Fe(CN)_6]^{4-}$ from kinetic results and from spectroscopic data reveals a discrepancy that amounts to a factor of a thousand.[162] It is safer to restrict comparisons to information gained by the same techniques. It is probably fairly safe to say that the almost identical frequencies of the absorption maxima of the IVCT bands of the ion-pair $[Co(NH_3)_6]^{3+}$, $[Ru(CN)_6]^{4-}$ and for the inner-sphere precursor $[(H_3N)_5Co^{III}{-}CN{-}Ru^{II}(CN)_5]^-$ indicate very similar rate constants for ruthenium → cobalt electron transfer in these two systems.[163]

We have mentioned $^*[Ru(bipy)_3]^{2+}$ as a reductant a couple of times earlier on in this section. It seems likely that electron transfer reactions of this and other photoexcited states will figure much more prominently in discussions of redox reactions in the near future.[164]

9.6.5 Bioinorganic electron transfer

There are a number of ways of monitoring or estimating the rate at which an electron moves across the organic material of a metalloprotein, or other metal-containing biological material, between metal and periphery. Inner-sphere, outer-sphere, and intramolecular systems may usefully be studied.

It is possible to carry out inner-sphere reduction by the use of Cr^{2+}aq, as has been done for, *inter alia*, plastocyanin, azurin, and ferrioxamine B, or, better, $Cr[15\text{-aneN}_4]^{2+}$aq.[165] The latter reduces $Fe^{III}{}_2S_2$-ferredoxin first to the mixed valence $Fe^{III}Fe^{II}S_2$, then to the $Fe^{II}{}_2S_2$ form. That the first electron transfers by the inner-sphere route is indicated by the product, in which chromium(III) is coordinated to a tyrosine residue of the $Fe^{III}Fe^{II}S_2$. This $Fe^{III}Fe^{II}S_2$-tyr-Cr^{III} intermediate then reacts with a second $Cr[15\text{-aneN}_4]^{2+}$aq by the outer-sphere mechanism to give the $Fe^{II}{}_2S_2$-tyr-Cr^{III} final product.[166]

We have already discussed the importance of steric and orientation factors in outer-sphere redox reactions of biochemical interest (Section 9.5.2.2), particularly the use of ionic strength effects on rate constants to estimate the effective charge on the electron transfer site of a metalloprotein. From this information, and additional information from pH or site-blocking effects on reactivity,[95–97] it may be possible to pinpoint the electron transfer site. Then one can deduce, either from X-ray crystallography or from models, the distance over which the electron has to travel, and the chemical make-up of the organic matrix through which it travels. A range of substitution-inert complexes is available (see Table 9.15 and the start of Section 9.5.2.2), so it is possible to select reductant or oxidant, anion or cation, and redox potential as required.

We have recalled the important pre-association step in the previous paragraph; it is now time to consider the electron transfer. In cases where the dependence of rate constant on concentration is curved (cf. Fig. 9.18), it is possible to derive both the ion-pairing constant and the rate constant for the subsequent electron transfer. The results of such analyses are shown, for oxidation of a range of metalloproteins by inorganic complexes, in Table 9.35 (cf. Table 9.17, in Section 9.5.2.1, for pairs of inorganic complexes). There is, unfortunately, no clear pattern to the somewhat scattered selection of k_{et} values in Table 9.35. Less information is available for inorganic complexes reducing bioinorganic species, but saturation kinetics have been reported for hexacyanoferrate(II) reduction of cytochrome c^{III} ($K_{OS} = 15$ or $400\ dm^3\ mol^{-1}$; $k_{et} = 100$ or $330\ s^{-1}$). If the reader is willing to estimate K_{OS} values for reactions whose kinetics do not exhibit saturation behaviour, then further k_{et} values can be estimated from published tabulations[167] of k_2 values ($k_2 = K_{OS}k_{et}$).

So far we have mentioned only simple inorganic partners, but it is also possible to carry out outer-sphere redox reactions between pairs of metalloproteins.[168] Thus rate constants have been measured for oxidation of rubredoxin by azurin and by cytochrome c_{551}.[146] It had been hoped that

Table 9.35 Association constants [K_{OS}(molar scale)] and electron transfer rate constants [$k_{et}(s^{-1})$] for electron transfer reactions between bioinorganic species and inorganic complexes.[a]

	Cytochromes			Rubredoxin		Ferredoxin[b]		Plastocyanin	
		K_{OS}	k_{et}	K_{OS}	k_{et}	K_{OS}	k_{et}	K_{OS}	k_{et}
$[(NC)_5Fe(CN)Co(CN)_5]^{5-}$	cyt f	4100	122						
$[Fe(CN)_6]^{3-}$	cyt c	~300[c]	~10[c]						
$[Co(NH_3)_5Cl]^{2+}$						~200	~2300		
$[Co(NH_3)_6]^{3+}$	cyt b_5	600	0.075			1000	19.2		
$[Co(en)_3]^{3+}$						600	2.7		
$[Co(phen)_3]^{3+}$								167	17.9
$[Co(terpy)_2]^{3+}$				470	515				
$[Ru(NH_3)_6]^{3+}$				4300	240				
$[Pt(NH_3)_6]^{4+}$	cyt b_5	14 800	0.080			21 000	3.3		
$[(H_3N)_5CoNH_2Co(NH_3)_5]^{5+}$	cyt b_5	16 600	3.8			26 400[d]	214[d]		

[a] At 298 K unless otherwise stated.
[b] For parsley ferredoxin—markedly different values are obtained with ferredoxins from other sources.
[c] There is rather poor agreement between several different published sets of values for this reaction.
[d] At 280 K.

such a simple outer-sphere electron transfer could be used in a Marcus analysis of cross-reactions to obtain a good estimate of the self-exchange rate constant (k_{22}) for the rubredoxin-$Fe^{III/II}$ couple. Unfortunately, this analysis yielded a value for k_{22} very different from that measured directly by NMR (Table 9.35 above). Similar disappointment ensued in an analogous treatment of redox reactions of rubredoxin with classical inorganic redox partners. There is clearly an extra factor operating here to complicate the Marcus analysis—possibly there is more than one electron transfer site on rubredoxin.[169] Haemoglobin and cytochrome b_5 are physiological redox partners, forming a very stable ($K = 3.4 \times 10^5$) association complex. Induced electron transfer between zinc-haemoglobin and cytochrome b_5 may therefore be assumed to take place within this complex, presumably from the haem edge of one component to the adjacent haem edge in the other. Electron transfer takes place with a rate constant of $8 \times 10^3\ s^{-1}$, over a distance of approximately 7 Å.[170] If it is required to have the two metalloproteins covalently linked, this can be achieved by the use of carbodiimide, to give a bonded unit with the components in a similar geometry and orientation to the normal outer-sphere associated pair.[171] Distance effects on inner- and outer-sphere electron transfer rate constants can thus be compared for pairs of metalloproteins.

There is always some uncertainty as to the exact geometry of reductant and oxidant in an outer-sphere electron transfer reaction. There are advantages in designing a redox system in which the two partners are bonded to each other, as in the carbodiimide link mentioned in the previous paragraph. Electron transfer then takes place intramolecularly in a well-defined binuclear species. The preparation and study of such species may take advantage of the particularly convenient aspect of ruthenium chemistry that it is substitution-inert in both the 2+ and the 3+ oxidation state. The metalloprotein is reacted with a ruthenium complex of the type $[Ru(NH_3)_5X]^{n+}$, so that the bioinorganic substrate displaces the ligand X, to give the Ru moiety bonded to the protein, usually at an imidazole-nitrogen donor site in a histidine residue.[172] After the preparation of the metallated substrate there remains the necessity of excitation in order to promote electron transfer. This can be effected in several ways, for example by photoexcitation, by an energetic radical such as methyl viologen or $CO_2^{-\cdot}$, or, most commonly, by the use of $^*[Ru(bipy)_3]^{2+}$. If the metalloprotein structure and the binding site have been established, electron transfer takes place across a well-defined distance and through a matrix of known composition.[173] The ruthenium will retain its ammine ligands through the 2+/3+ oxidation states involved. An example is provided by the estimate of k_{et} somewhere between 20 and 80 s^{-1}, over a distance of about 15 Å, in a ferricytochrome *c* derivative with $Ru^{II}(NH_3)_5$ attached at histidine-33.[174] Similarly, $k_{et} = 2\ s^{-1}$ for transfer over > 10 Å in $Ru^{II}(NH_3)_5$-labelled copper azurin.[175] Rate constants for electron transfer in ruthenated metalloproteins depend on the driving force, the distance the electron has to travel, and the nature of the organic matrix through which it has to pass.

The last-named effect can be probed through comparative studies of ruthenated mutants, as has been carried out for three ruthenated cytochrome b_2 mutants.[176] An alternative strategy to ruthenation is to utilize caged cobalt complexes, where the encapsulating ligand forces substitution-inertness on the cobalt(II) as well as the cobalt(III) state. This approach has been used for cytochrome *c* derivatization—a carboxylate side chain is used here to anchor the cobalt derivative to the metalloprotein. Electron transfer rate constants here range from 1.0 to 3.2 s^{-1}, remarkably little variation over distances ranging from about 14 to 20 Å.[177]

Electron transfer in ruthenated species should be amenable to treatment by the Marcus theory, especially if the structure of the ruthenated metalloprotein has been established. By suitable engineering of the ruthenating attachment and selection of the metalloprotein, systems can be designed in which there is a very large driving force for electron transfer. It should thus be possible to attain the so-called Marcus-inverted region, where increasing the driving force actually leads to a decrease in k_{et} (Section 9.5.3 above).[107] Attempts to do this have been only partially successful.[178] Thus, for example, for a series of cytochrome *c* derivatives ruthenated by groups of the $-Ru(phen)_2(CN)$ type, rate constants are in the narrow range between 2 and 5×10^5 s^{-1} for driving forces from -1.3 V right up to -1.9 V. Although there is not the significant decrease of k_{et} with increasing driving force forecast for this region, at least there is not the increase that characterizes the normal (i.e. pre-inversion) region.

Inner- and outer-sphere electron transfer kinetics can be compared via intramolecular electron transfer in a ruthenated derivative versus intermolecular electron transfer between metalloprotein and $[Ru(NH_3)_6]^{n+}$. This can be illustrated by cytochrome *c*.[179] Kinetic parameters for outer sphere electron transfer between cyt c^{III} and $[Ru(NH_3)_6]^{2+}$ are $k_2 = 3.8 \times 10^4$ $dm^3\,mol^{-1}\,s^{-1}$, $\Delta H^{\ddagger} = 12$ kJ mol^{-1}, and $\Delta S^{\ddagger} = -120$ J $K^{-1}mol^{-1}$. The horse heart cytochrome *c* can be ruthenated with $Ru(NH_3)_5$ at histidine-33 to give cyt c^{III}–Ru^{III}. Intramolecular electron transfer can then be initiated by excitation, with $CO_2^{-\cdot}$ or $*[Ru(bipy)_3]^{2+}$, to cyt c^{III}–Ru^{II}. Activation parameters are very similar ($\Delta H^{\ddagger} = 15$ kJ mol^{-1}, $\Delta S^{\ddagger} = -160$ J K^{-1} mol^{-1}) to those for the outer-sphere analogue; rate constant comparison is complicated by the difference in units and the need to allow for the contribution to k_2 from the (unknown) outer-sphere association constant. Overall there appears to be some sort of qualitative correlation between k_{et} and distance, at least for ruthenated metalloproteins,[180] especially if one makes allowances for perturbing features in several of the systems studied.

Metalloproteins such as haemoglobin and Fe_8S_8 ferredoxins contain more than one redox-active site, so that with suitable modification and excitation it is possible to monitor intramolecular electron transfer kinetics. Thus, for instance, photoexcitation of zinc–iron hybrid haemoglobins gives metal $\rightarrow$ metal electron transfer, with $k_{et} = 100$ s^{-1} at ambient temperature (9 s^{-1} at 170 K).[181] Outer-sphere reduction of such multi-site substrates may indicate reduction of more than one site and may also involve intramolecular electron transfer. Cytochrome cd_1 from nitrite reductase (*Pseudomonas aeruginosa*) has four redox-active centres, of two types (ccd_1d_1), in the one protein molecule. Reduction with $[Fe(edta)]^{2-}$ gives rapid reduction ($k \sim 0.01$–$0.02\,dm^3\,mol^{-1}\,s^{-1}$) of the accessible cyt *c* sites, but the well-buried cyt d_1 sites undergo reduction only slowly by electron transfer from reduced cyt *c*. A rate constant $k_{et} \sim 0.3\,s^{-1}$ at 298 K is sluggish even for the distance of about 14 Å involved. Electron transfer may be slow here because it is coupled to conformational changes.[182] Alternatively, the slowness may be due to the fact that the *c* and d_1 haems in each subunit are oriented perpendicular to each other—in metalloproteins where haem units are mutually parallel electron transfer is often rapid.[183] The relative effects of distance and orientation twist have been examined for electron transfer across a series of phenylene spacer groups in iron–zinc bis-porphyrins. Increasing distance caused only slight attenuation, with k_{et} still as high as $10^9\,s^{-1}$ for transmission across three coplanar benzene rings. The dominant factor in decreasing k_{et} with number of phenylene spacers is the twist angle between adjacent benzene rings.[67] Theoretical treatments of electron transfer rates should be more satisfactory for intramolecular transfers between donor and acceptor sites within a given molecule. The Marcus approach has been assessed and documented in relation to Co^{II}-protoporphyrin-myoglobin and -haemoglobins, where the theory has been checked out for systems of known geometry, after which it can legitimately be used to estimate $Co \rightarrow Fe$ electron transfer distances from measured k_{et} values.[184] There are some intramolecular electron transfers in metalloproteins which are very much faster than theory predicts—indeed very much faster than those mentioned so far in this section. A superexchange mechanism has been developed in attempts to cope with such claims as $k_{et} > 10^{11}\,s^{-1}$ over a distance of 17 Å. The nature of the protein matrix is a key element here, as indeed it is for the less fast electron transfers discussed above.[185]

There may be two routes to a given redox-active centre, with the possibility of balancing easier access, or more favourable association, at one site against easier through-protein transfer from a less favourable electron-transfer site elsewhere on the periphery of the metalloprotein (Fig. 9.29(a)). Comparisons of reactivities in relation to charges on oxidants, inhibition by complex ions, and pH effects on reactivities and inhibition, indicate that there are at least two electron transfer sites on the surface of HiPiP from *Chromatium vinosum*.[186] In the case of plastocyanin there appears to be slower electron transfer from the acidic site than from the hydrophobic site. This may be attributable to the longer distance the electron has to travel through the protein in the former case.[187] The site and mode of interaction may be affected by modest modification of reactants. Thus the respective kinetic patterns for reaction of copper(II)-plastocyanin with triplet-state zinc myoglobin and zinc cytochrome *c* indicate marked differences in interaction geometries (Fig. 9.29(b) and

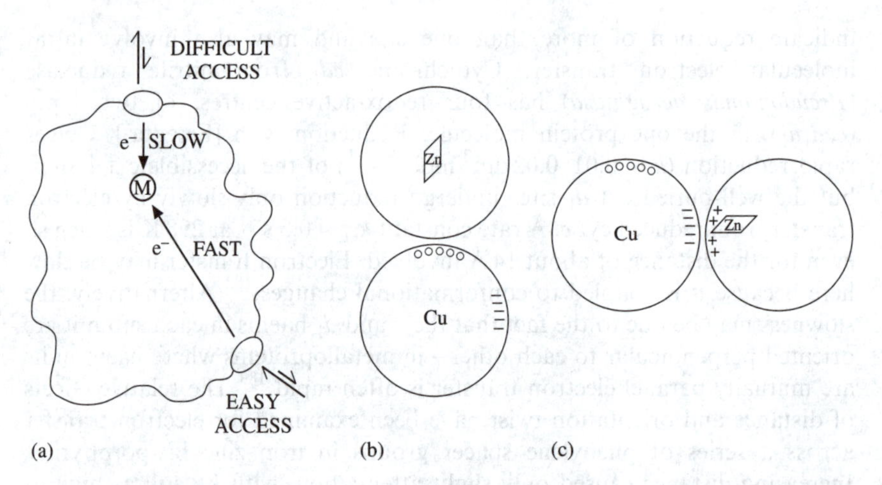

Fig. 9.29 (a) Possible competition between electron transfer via (i) an easy access site a long distance from the metal centre of a metalloprotein or (ii) a nearby difficult access site. (b) Mutual orientation of zinc myoglobin and copper(II) plastocyanin prior to electron transfer. (c) Mutual orientation of zinc cytochrome *c* and copper(II) plastocyanin prior to electron transfer.

(c)). The partially exposed haem in the Zn-cyt *c* is associated with a net positive patch on its surface which associates with a negative area on the plastocyanin, whereas the myoglobin derivative, with its haem buried and no charged patches, appears to interact at a hydrophobic patch elsewhere on the surface of the plastocyanin.[188] Ionic strength effects on rate constants for electron transfer between yeast cytochrome *c* and H_2O_2-oxidized cytochrome *c* peroxidase were interpreted as indicating non-optimal matching of electron transfer sites. Here charge effects or steric factors may be causing the reactants to dock in such an orientation that the electron transfer sites are not in close proximity, leading to a reduction in the electron transfer rate.[189] These last two examples illustrate the use of rate constants to probe electron transfer sites, bringing us back full circle to the start of Section 9.5.2.2.

We have said practically nothing so far about activation parameters $\Delta H^\ddagger, \Delta S^\ddagger$, and $\Delta V^\ddagger$ for electron transfer in bioinorganic systems. There have been a number of informative high-pressure kinetic studies of electron transfer kinetics in systems of bioinorganic relevance.[190] If reactions with only small free-energy changes are selected, then it is possible to determine activation volumes for both the forward and reverse reactions, and to measure the overall volume change, thereby obtaining a detailed volume profile. This has been achieved for reactions of cytochrome *c* with the cobalt-diimine complexes $[Co(phen)_3]^{n+}$, $[Co(bipy)_3]^{n+}$, and $[Co(terpy)_2]^{n+}$. These redox reactions are well behaved, both kinetically and in the sense that they conform well to the Stranks and Swaddle extension of the Marcus–Hush theory.[191] In each case the partial molar volume of the transition state is very close to the mean of the volumes of reactants and products, as is also the case for some ruthenium-ammine-pyridine/cytochrome *c* reactions.[193] In these ruthenium/cytochrome *c* reactions electrostriction effects may contribute significantly, and there is the additional complication that the application of pressure may make the organic matrix slightly more compact, reducing by a small but significant amount the distance over which the electron has to be transferred.[194]

Apparent activation energies for electron transfer within the zinc–iron haemoglobins and the $Ru^{II}(NH_3)_5$-cytochrome and -azurin compounds mentioned in earlier paragraphs are close to zero over certain temperature ranges. These and other bioinorganic examples of approximately zero, and sometimes temperature-dependent, activation energies at low temperatures have parallels in classical inorganic chemistry. These include the V^{IV}/V^{V} derivative included in Table 9.13 and Turnbull's Blue (at $< 50\,K$). Such activation energy behaviour has been ascribed to electron tunnelling. Early calculations of k_{et} for tunnelling, which emerged from an investigation of cytochrome *c* and bacteriochlorophyll, suggested that fairly rapid electron transfer was possible even over longer distances (e.g. $\sim 100\,s^{-1}$ at 74 Å!) for barrier heights of a few $kJ\,mol^{-1}$.[195]

9.7 Redox reactions involving *sp*-block species

9.7.1 Introduction

So far in this chapter we have devoted almost all our attention to complexes of the *d*-block transition elements. This reflects the great interest in, and extent of understanding of, redox reactions of such species. However, there are a large number of redox reactions which involve *sp*-block compounds and ions, often of considerable importance and interest. Unfortunately, the mechanisms of redox reactions of relatively simple species are often themselves not simple, involving several steps and transient intermediates. It is not possible to present a concise general overview and summary of this remarkably diverse field, so in this section we shall deal, often briefly, with a selection of systems. These will involve simple oxygen-containing oxidants (oxides, oxoanions, and dioxygen itself), aqua-ions of the *sp*-block (e.g. $Tl^{3+}aq$), a variety of radicals (inorganic and organic), and solvated electrons. Very often *sp*/*sp* and, particularly, *sp*/*d* or *f* redox reactions are complicated by mismatch of the number of electrons transferred. Whereas *d*- and *f*-block species tend to prefer one-electron transfers, oxidants and reductants from the *sp*-block generally prefer to undergo two-electron transfers. Examples include Tl^{3+}/Tl^{+}or Sn^{4+}/Sn^{2+} for metals and such pairs and sequences as $SO_3{}^{2-}/SO_4{}^{2-}$, $NO_2{}^{-}/NO_3{}^{-}$, or Cl^-, ClO^-, $ClO_2{}^-$, $ClO_3{}^-$, and $ClO_4{}^-$ for non-metals. The problem of one- versus two-electron tranfer can also crop up even for *sp*/*sp* reactions, as for example in the chlorite plus nitrite reaction, or indeed any reaction involving $NO_2/NO_2{}^-$.

Table 9.1 gave some examples of reductants which could act as one-electron *or* two-electron donors. We shall deal with this topic of non-complementary reactions in the following section, then move on to our selection of other redox reactions involving *sp*-block species.

9.7.2 Non-complementary reactions

The usual textbook example of such a reaction is that of thallium(III) oxidation of iron(II). The relevant redox potentials are $E^\circ(Tl^{3+}/Tl^{+}) = +1.26\,V$ and $E^\circ(Fe^{3+}/Fe^{2+}) = +0.77\,V$, so there is a respectable driving force for this reaction. Nevertheless, rates are relatively slow, and the mechanism operating for this reaction was deduced many years ago. The rate law under normal conditions is simple second-order:

$$-d[Fe^{II}]/dt = k[Fe^{II}][Tl^{III}]$$

and thus not indicative of the detailed mechanism. The key evidence was furnished by a series of experiments probing the kinetic consequences of addition of the respective products. It was found that addition of thallium(I) had no significant effect, but that addition of iron(III) caused inhibition:

$$-d[Fe^{II}]/dt = \frac{2k_1k_3[Tl^{III}][Fe^{II}]^2}{k_2[Fe^{III}]+k_3[Fe^{II}]}$$

where the subscripts correspond with those in the mechanism set out below.

For the reaction $2Fe^{II} + Tl^{III} = 2Fe^{III} + Tl^{I}$ (where M^{III} may be M^{3+} or MOH^{2+} depending on pH) initial one-electron transfer would give an unstable thallium(II) intermediate

$$Fe^{II} + Tl^{III} = Fe^{III} + \mathbf{Tl^{II}}$$

or initial two-electron transfer to give the stable thallium(III) product would generate unstable iron(IV):

$$Fe^{II} + Tl^{III} = \mathbf{Fe^{IV}} + Tl^{III}$$

In either case, a second electron transfer step completes the reaction sequence. Both Tl^{II} and Fe^{IV} are normally unstable oxidation states, but both are acceptable as transient intermediates. The effects of added products show unequivocally that the mechanism involves one-electron steps, with Tl^{II} as intermediate:

$$Fe^{II} + Tl^{III} \underset{k_2}{\overset{k_1}{\rightleftharpoons}} Fe^{III} + Tl^{II}$$

$$Fe^{II} + Tl^{II} \underset{k_3}{\longrightarrow} Fe^{III} + Tl^{III}$$

Although the above system provides the almost inescapable example, there are a number of analogous examples, as, for instance, in thallium(III) oxidation of vanadium(IV), where inhibition by vanadium(V) but not by thallium(I) again indicates that the likely mechanism is that of two consecutive one-electron transfers. Further examples of non-complementary reactions involving reactions between an *sp*-block and a *d*-block species which are believed to take place by consecutive one-electron steps are given in Table 9.36.[198,199]

In a sense the d^8 transition metal reductants platinum(0), platinum(II), and gold(I) have a formal similarity to *sp*-block reductants, for each of these needs to lose two electrons to reach the next higher stable oxidation state. These reductants also often react by two one-electron transfers, with a transient platinum(III) or gold(II) intermediate, e.g.

$$[Pt(NH_3)_4]^{2+} + 2Fe^{3+} + 2Br^- = trans\text{-}[Pt(NH_3)_4Br_2]^{2+} + 2Fe^{3+}$$

and, complementarily, platinum(IV) oxidation of $[Ru(NH_3)_6]^{2+}$. Platinum(IV) oxidation of vanadium(II), which could easily give vanadium(IV) by a single two-electron transfer, also proceeds by two distinct one-electron transfers. Platinum(II) and gold(I) reductions[200] are complicated by the change in coordination number accompanying the redox process, but the (relatively rare) nickel(IV) oxidations go from octahedral Ni^{IV} to octahedral nickel(II), yet again by two one-electron transfers. In this connection it must be admitted that roughly as many nickel(III) complexes are known as nickel(IV), so the intermediacy of nickel(III) here is not too surprising.

Chromium chemistry provides a number of interesting examples of non-complementary redox reactions involving changes of oxidation state of up to four units. Chromium(VI) to chromium(III) is, of course, a very common situation, in organic as well as in inorganic chemistry. Evidence is gradually accumulating that the three electrons involved are, especially for organic reductants, transferred one by one. Thus, for example, ESR evidence indicates the intermediacy of chromium(V) species for oxalate as reductant, as in the common laboratory preparation of $K_3[Cr(ox)_3]$ from potassium dichromate.

Complementarily, in cerium(IV) oxidation of chromium(III) complexes it is possible to detect chromium(IV) intermediates when these are stabilized by appropriate tetraaza-macrocycles, e.g. **54**. Once one has chromium(V), for example, stabilized as in **55**, then kinetic patterns for reduction by, for example, iodide, iron(II), or titanium(III) indicate the intermediacy of

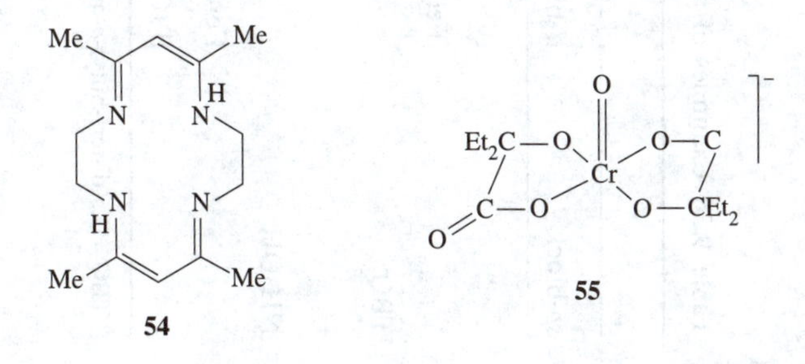

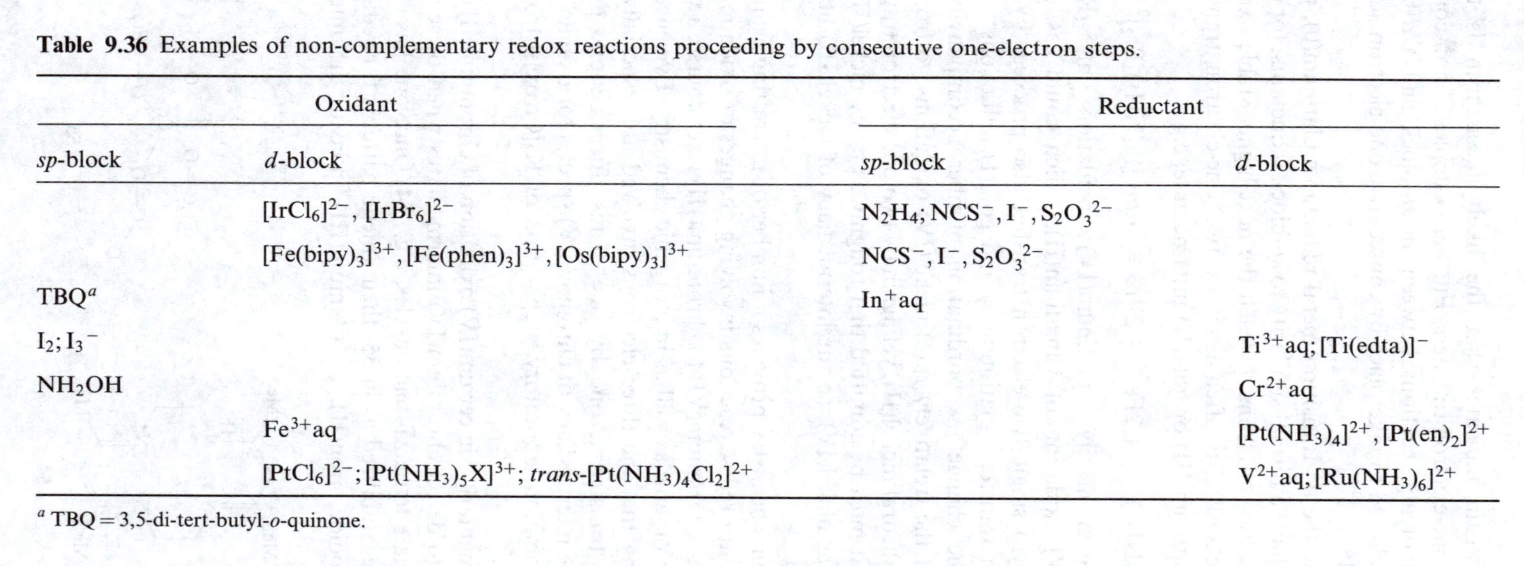

Table 9.36 Examples of non-complementary redox reactions proceeding by consecutive one-electron steps.

Oxidant		Reductant	
sp-block	*d*-block	*sp*-block	*d*-block
	$[IrCl_6]^{2-}$, $[IrBr_6]^{2-}$	N_2H_4; NCS^-, I^-, $S_2O_3^{2-}$	
	$[Fe(bipy)_3]^{3+}$, $[Fe(phen)_3]^{3+}$, $[Os(bipy)_3]^{3+}$	NCS^-, I^-, $S_2O_3^{2-}$	
TBQ[a]		In^+aq	
I_2; I_3^-			Ti^{3+}aq; $[Ti(edta)]^-$
NH_2OH			Cr^{2+}aq
	Fe^{3+}aq		$[Pt(NH_3)_4]^{2+}$, $[Pt(en)_2]^{2+}$
	$[PtCl_6]^{2-}$; $[Pt(NH_3)_5X]^{3+}$; *trans*-$[Pt(NH_3)_4Cl_2]^{2+}$		V^{2+}aq; $[Ru(NH_3)_6]^{2+}$

[a] TBQ = 3,5-di-tert-butyl-*o*-quinone.

chromium(IV) species. The reduction of chromium(VI) by glutathione has proved particularly revealing, for a combination of NMR, ESR, HPLC, and kinetic experiments have led to the characterization of relatively long-lived chromium(V) and chromium(IV) species en route to the chromium(III) product.[201]

Thus far we have produced a number of examples of multi-electron redox reactions proceeding by discrete one-electron steps. There is, however, evidence for two-electron and even four-electron transfer in certain areas of chromium chemistry. Chromium(IV) is involved in one of the more convincing demonstrations of simultaneous two-electron transfer. The oxidation of chromium(II) by certain two-electron oxidants gives significant amounts of binuclear chromium(III) products, either $Cr^{III}-(\mu\text{-}OH)_2-Cr^{III}$ or $Cr^{III}-(\mu\text{-}O)-Cr^{III}$. Such binuclear products are most likely to arise from comproportionation involving a chromium(IV) intermediate:[202]

$$Cr^{II} + 2e^- \longrightarrow Cr^{IV}; \quad Cr^{II} + Cr^{IV} \longrightarrow (Cr^{III})_2$$

The sole four-electron transfer involves the reduction of chromium(V) by hydroxylamine, but as the product contains NO coordinated to chromium the assignment of an oxidation state of +1 to the metal, and thus of the transfer of four electrons, must be considered not altogether unarguable.[203]

9.7.3 Transition metal oxidants and reductants

9.7.3.1 Aqua-ion oxidants

The main question here is, as ever, whether electron transfer takes place by an inner-sphere or an outer-sphere pathway. The former involves formation of a complex between the aqua-cation and the reductant followed by electron transfer from this coordinated ligand to the metal-ion oxidant:

$$M^{n+}aq + L \underset{k_b}{\overset{k_f}{\rightleftharpoons}} ML^{n+} \xrightarrow{k_{et}} \text{products}$$

The kinetic pattern observed depends on the relative values of k_f, k_b, and k_{et}; various possibilities are set out in Table 9.37. If the concentration of the complex ML is small throughout the reaction period, then application of the steady-state treatment gives

$$k = \frac{k_f k_{et}}{k_b + k_{et}}$$

and thence the various relations given in Table 9.37. There are three further complications. The first is that many aqua-ion oxidants have charge 3+ or 4+, and thus there will be a pH/rate profile reflecting the $M^{n+}aq \rightleftharpoons MOH^{(n-1)+}aq + H^+aq$ equilibrium. The second is that a 4+ aqua-ion oxidant in the necessarily relatively high concentration of acid required to minimize hydrolysis to MOH^{3+} will in most cases be mainly in the form of a complex with the anion of this added acid—e.g. cerium as

Table 9.37 Kinetic patterns for oxidations by aqua-cations.

Relation between rate constants	Expected pattern
(a) k_f, k_b both $\ll k_{et}$	Rapid equilibrium formation of ML^{n+} with rate-limiting electron transfer therein k_{obs} vs [L] with $k_{obs} = Kk_{et}$
(b) $k_b \sim k_{et}$	It may be possible to detect the intermediate, and possibly estimate K and k_{et} from k_{obs} vs [L]
(c) $k_{et} \gg k_b$	The rate is determined by k_f

$CeSO_4^{2+}$. The third is that the complex ML^{n+} may be a dead-end species (see Section 9.4.6 above) whose formation competes with, rather than leads to, redox chemistry.

Table 9.38 shows some examples of reactions of the type discussed in the present section where there is good evidence either for or against inner- or outer-sphere electron transfer. Evidence in favour of an inner-sphere mechanism is easier to obtain, especially when the complex ML^{n+} persists long enough to be detected spectrophotometrically. Indeed, in some cases, many of them involving iron(III) oxidation of what are potentially good ligands, e.g. cysteine, organic thiols, or thiocarboxylato compounds such as thiomalate, it is possible to see the colour changes corresponding to the formation of the intermediate complex and its subsequent disappearance consequent on electron transfer.[204]

9.7.3.2 Werner complexes

Substitution-inert transition metal complexes are much used for carrying out one-electron oxidations or reductions of *sp*-block species. Suitable reductants include hexaammineruthenium(II), hexacyanoferrate(II), and cobalt(II) sepulchrate; suitable oxidants include hexacyanoferrate(III) and hexahalogenoiridate(IV) anions and tris-diimineiron(III) cations. All of these undergo simple outer-sphere one-electron redox reactions, leaving

Table 9.38 Examples of aqua-cation oxidations of simple inorganic and organic species.

Oxidant	Inorganic reductants	Organic reductants
Evidence[a] for intermediates:		
thallium(III)	hypophosphite (s,k)	oxalate (s,k)
iron(III)	thiosulphate (s,t)	catechol (s,k)
cobalt(III)	chloride (s)	malic acid (s)
		alcohols (k)
cerium(IV)	phosphorous acid (k)	acetic acid (s,k)
		ethanol (s,k)
No evidence for intermediates:		
silver(II)	dithionate	
iron(III)	hydrazine	acetoin
cobalt(III)	thiocyanate	thiourea
cerium(IV)		benzaldehyde

[a] Evidence: s = spectrophotometric, k = kinetic, t = equilibrium constant.

the complicated chemistry to the various *sp*-block species as has been described for, for example, $IrCl_6^{2-}$, $IrBr_6^{2-}$, and $Fe(bipy)_3^{3+}$ oxidations of such *sp*-block species as iodide, thiocyanate, nitrite, and sulphite.[198]

9.7.3.3 Aqua-ion reductants

V^{2+}aq, Cr^{2+}aq, Eu^{2+}aq, Mo^{3+}aq, Ti^{3+}aq, and U^{3+}aq are all strongly reducing cations, Fe^{2+}aq mildly reducing. Kinetic studies of their reduction of a variety of *sp*-block species are mentioned in several of the the remaining sections of this chapter. As they are all substitution-labile, inner-sphere electron transfer is always a feasible pathway. It has been demonstrated for a number of Cr^{2+}aq reductions, as for example in the reduction of I_2, where the slow second step

$$[Cr(H_2O)_5I]^{2+} \xrightarrow{H_2O} [Cr(H_2O)_6]^{3+} + I^-$$

can be monitored and confirmed. Chromium(II) reduction of hydrogen peroxide is also inner-sphere, at least when conducted in the presence of suitable organic anions,[205] as is titanium(III) reduction of pyruvate.[206]

Most of the above aqua-cation reductants are one-electron transfer agents, but both Cr^{2+} and V^{2+} can act as two-electron, or more exactly oxygen-atom, transfer agents. Such behaviour is facilitated by the possibility of $CrOCr^{4+}$aq and VOV^{4+}aq intermediates ($M^{II}OM^{IV} \rightarrow M^{III}O^{III}M$ in both cases). It has been detected for, e.g., Cr^{2+} reduction of hydrogen peroxide.

Cations such as Cu^+aq, Cr^{2+}aq, and U^{3+}aq react rapidly with dioxygen and are also sufficiently strong reductants to reduce water. The kinetics of these reactions are often not simple; indeed, aqueous solutions of Cu^+ are metastable and their spontaneous disproportionation may take hours, especially if dioxygen is rigorously excluded. The fortunate corollary of this is that it has proved possible to monitor the kinetics of a number of $Cu^{+}aq$ reductions of transition metal complexes. The even less stable monovalent cation reductants $Mn^{+}aq$, $Co^{+}aq$, $Ni^{+}aq$, $Zn^{+}aq$, and $Cd^{+}aq$ are mentioned briefly in Section 9.7.6.4 below.

9.7.4 Oxoanions and peroxoanions[207]

Some *sp*-block oxoanions can act as oxidants (nitrate, chlorate, perchlorate), some as reductants (nitrite, sulphite, dithionite). A few oxoanions can act in either capacity, depending on the circumstances. Thus perchlorate is normally an oxidant, but in dilute solutions of chromate in concentrated perchloric acid the chromium(VI) is reduced to chromium(III), in the form of Cr^{3+}, $CrCl^{2+}$, and $CrCl_2^{+}$.[208] Trioxodinitrate, $HN_2O_3^{-}$, can reduce or oxidize haemoproteins, depending on conditions,[209] and it has been suggested that nitrate may reduce thallium(III).[210] Peroxoanions generally show the expected oxidizing powers, but they too can occasionally act, at least formally, as reductants. Thus peroxonitrite[211] usually acts as an oxidant, e.g. of bromide or hydroxylamine, but occasionally as a reductant, e.g. of permanganate. There is a wide range of kinetic and mechanistic behaviour to reflect this variety of chemistry. We shall start with three oxoanion reductants, and then deal with a wider range of oxidants. It is perhaps worth pointing out here that, particularly in the realms of sulphur chemistry, redox mechanisms may involve an initial dissociative step in which a binuclear reactant such as dithionite, $S_2O_4^{2-}$, or peroxodisulphate, $S_2O_8^{2-}$, gives the mononuclear species which are the actual reductants or oxidants. In the limit, such mechanisms give rate laws half-order in the binuclear reductant/oxidant, zero-order in the other reactant—several examples are cited below.

9.7.4.1 Dithionite, sulphite, and thiosulphate reductions

Dithionite, commonly employed in the form of its sodium salt, is a strong reductant. Its redox potential may be as low as −1.12 V, depending on conditions, particularly pH. It also has an unusually long, and weak, sulphur–sulphur bond (2.39 Å—cf. 2.15 Å in potassium dithionate), and is therefore prone to react by initial dissociation to $SO_2^{-\cdot}$ radicals, which are the actual reductants. Rapid disproportionation in acid solution

$$2S_2O_4^{2-} + H_2O = 2HSO_3^{-} + S_2O_3^{2-}$$

may also complicate kinetic studies.

Dithionite reduction of hexacyanoferrate(III) under conditions where the dithionite is present in stoichiometric amounts or in slight excess follows the simple rate law

$$\text{rate} = k[\text{Fe(CN)}_6^{3-}][\text{S}_2\text{O}_4^{2-}]^{1/2}$$

indicating solely reduction by $SO_2^{-\cdot}$ radicals. In the presence of a large excess of dithionite, the rate law is

$$\text{rate} = k[\text{Fe(CN)}_6^{3-}][\text{S}_2\text{O}_4^{2-}]^{1/2} + k'[\text{Fe(CN)}_6^{3-}][\text{S}_2\text{O}_4^{2-}]$$

indicating that now reduction by undissociated dithionite also plays a significant role.[212] The situation is similar, with both reduction pathways significant under appropriate conditions, for a number of inorganic complexes,[213] including $[Fe(CN)_5(4,4'\text{-bipy})]^{2-}$, $[Ru(NH_3)_5(\text{imid})]^{3+}$, $[Co(\text{terpy})_2]^{3+}$, and $[Co(NH_3)_5Cl]^{2+}$. However, only the $SO_2^{-\cdot}$ path has so far been observed for the majority of $[Fe^{III}(CN)_5L]^{n-}$ and $[Co^{III}(NH_3)_5X]^{n+}$ oxidants, whereas only the $S_2O_4^{2-}$ path was reported for reduction of $[Fe(\text{edta})]^-$ and $[Mn(\text{cydta})]^-$. Both pathways contribute to dithionite reduction of copper(II)-azurin from *P. aeruginosa* and of horse heart ferricytochrome *c*, but again only the $SO_2^{-\cdot}$ path has so far been reported for reduction of ferredoxins, of horse heart metmyoglobin, and of manganese(III) and cobalt(III) myoglobins.[214] Reduction rates for a series of myoglobin derivatives followed the order[215]

$$\text{imid} \gg CN^- > SCN^- \gg N_3^- \gg F^-$$

reflecting the positions of these ligands in the spectrochemical and nephelauxetic series. In fact reduction of the fluoro derivative is so slow that it probably undergoes aquation prior to reduction. Both $S_2O_4^{2-}$ and $SO_2^{-\cdot}$ generally act as outer-sphere reductants.

Sulphite reductions of oxoanions are often simple oxygen-transfer reactions, as has already been mentioned for hypochlorite and for nitrite, and is also the case for the halogens, thallium(III), and ferrate.[216] One-electron reduction is also possible, though not observed, for nitrite,[217] where NO_2^- to NO_2 is, rarely for oxoanions, feasible. Vanadium(V), chromium(VI), and manganese(VII) are reduced by concurrent one- and two-electron paths, giving dithionate and sulphate, respectively. Sulphite reductions may also have considerably more complicated mechanisms. We mentioned in Section 9.1 how permanganate oxidation of sulphite can give both dithionate and sulphate solely by one-electron transfers. Sulphite (sulphur dioxide) reductions of peroxomonosulphate,[218] of a nickel(III)-cyclam complex,[219] and in a range of environmentally important processes[220] also involve complex mechanisms. Sulphite reduction of the gold(III) complexes $[AuCl_4]^-$, *trans*-$[Au(CN)_2Cl_2]^-$, and *trans*-$[Au(CN)_2Br_2]^-$ occurs through halide-bridged two-electron transfer:

$$[\text{Au}^{\text{III}}\text{X}_2\text{Y}_2]^- + \text{SO}_3^{2-} \longrightarrow [\text{Au}^{\text{I}}\text{Y}_2]^- + \text{X}^- + \text{XSO}_3^-$$

The XSO_3^- rapidly reacts further to give halide plus bisulphate.[221]

Thiosulphate is usually oxidized to tetrathionate, by metal complexes as by iodine in volumetric analysis, by consecutive one-electron steps:

$$2S_2O_3^{2-} - 2e^- \longrightarrow S_4O_6^{2-}$$

$[IrX_6]^{2-}$ (X = Cl, Br), $[Fe(bipy)_3]^{3+}$, and $[Os(X\text{-}phen)_3]^{3+}$ (X = 5-Cl, 5,6-Me_2, 4,7-Me_2) provide typical examples. Marcus treatment of these cross-reactions has given a value of 2.3×10^5 dm^3 mol^{-1} s^{-1} for the $S_2O_3^{2-}$/$S_2O_3^{-\cdot}$ self-exchange rate constant. Such relatively fast electron exchange is consistent with the expected small difference in geometry between $S_2O_3^{2-}$ and $S_2O_3^{-\cdot}$.[222]

Reduction of the substitution-labile Fe^{3+}aq involves formation of the thiosulphato complex $Fe(S_2O_3)^+$aq prior to redox reaction with a second thiosulphate ion, a common mechanism for Fe^{III} oxidations of potential ligands (see Table 9.38 in Section 9.7.3.1). In the reaction with nitrite there is also a key intermediate formed by association between oxidant and reductant.[223] The rate law is

$$\text{rate} = k_3[H^+][HNO_2][S_2O_3^{2-}] + k_2[HNO_2]^2$$

The first, k_3, term is attributed to formation of the O_3SSNO^- intermediate via NO^+:

$$H^+ + HNO_2 \xrightleftharpoons{\text{fast}} NO^+ + H_2O$$

$$NO^+ + S_2O_3^{2-} \xrightarrow{\text{slow}} O_3SSNO^-$$

The second, k_2, term arises from rate-limiting formation of N_2O_3, which reacts quickly with thiosulphate to give the same O_3SSNO^- intermediate:

$$N_2O_3 + S_2O_3^{2-} \xrightarrow{k_3} O_3SSNO^- + NO_2^-$$

The reaction of thiosulphate with peroxodisulphate is zero-order with respect to the thiosulphate (the rate-limiting step is the generation of $SO_4^{-\cdot}$ radicals from the peroxodisulphate—see the next section).

In the $[Mn(acac)_3]$/thiosulphate system the initial electron transfer step is just one step in a cycle which results overall in reduction of the manganese(III) by acetylacetone—thiosulphate is not consumed but does play a key part as an essential catalyst.[224]

9.7.4.2 Oxoanion oxidations

The simplest mechanisms for redox reactions involving oxoanion or peroxoanion oxidants are electron transfer or the transfer of an oxygen atom, which is formally two-electron transfer. Which mechanism operates depends on such factors as the bond strength to the potentially transferable oxygen atom and the stability of the initial product of one-electron transfer. The first factor can be illustrated by comparing perchlorate with perbromate; oxygen atom transfer takes place from the latter but not the former, because the halogen–oxygen bond is much stronger in perchlorate. In the *sp*-block, where stable oxidation states tend to be two units apart, initial one-electron transfer is liable to yield a transient radical intermediate. Oxygen

Table 9.39 Examples of anion–anion redox reactions which involve oxygen atom transfer.

$ClO^- + Br^-$	$ClO_3^- + NO_2^-$	$HSO_5^- + NO_2^-$
$ClO^- + ClO_2^-$	$BrO_3^- + TeO_3^{2-}$	
$ClO^- + NO_2^-$		
$ClO^- + SO_3^{2-}$ [a]		

[a] $NO_2 + SO_3^{2-}$ also goes by oxygen atom transfer, $HClO + SO_3^{2-}$ probably involves Cl^+ transfer (see the end of Section 9.7.4.3).

atom transfer may therefore be favoured, as was demonstrated many years ago by labelled-oxygen studies of such reactions as the chlorate/sulphite reaction mentioned in the previous section, and similar reactions such as[225]

$$XO^- + NO_2^- = X^- + NO_3^- \quad (X = Cl, Br)$$

Several examples of oxoanion–oxoanion reactions proved, or believed, to involve oxygen atom transfer are given in Table 9.39; reactions in which an oxoanion transfers an oxygen atom to various types of substrate are mentioned in the following section.

Perchlorate oxidations of several low-valent transition metal ions, including V^{2+}aq, V^{3+}aq, Ti^{3+}aq, and Mo^{3+}aq, have been studied as a byproduct of the long-recognized inadvisability of using perchlorates to control ionic strength in kinetic studies of these reductants.[226] Interestingly, perchlorate oxidation of the particularly potent reductants Cr^{2+}aq and Eu^{2+}aq is extremely slow. These reactions clearly have high activation barriers rather than high thermodynamic barriers[227] ($E°(Cr^{3+}/Cr^{2+}) = -0.42$ V; $E°(Eu^{3+}/Eu^{2+}) = -0.35$ V; cf., e.g. $E°(V^{3+}/V^{2+}) = -0.26$ V). This marked behavioural difference has been attributed to the stability of various VO, TiO, and MoO entities—there are no stable chromium or europium analogues—and thus to the possibility of O-atom transfer as the favoured route for metal-ion reduction of perchlorate, though an alternative orbital explanation has been offered. There are marked differences in oxidation mechanisms between perbromates and perchlorates. Thus perbromate oxidation of $[Fe(bipy)_3]^{2+}$ and $[Fe(Xphen)_3]^{2+}$ cations is zero-order in perbromate; rate-limiting ligand dissociation is followed by rapid perbromate reaction with the product(s) of dissociation.[228] Oxygen transfer from perbromate is also possible, indeed easier than from perchlorate due to weaker bonding in the former. Oxygen transfer is easier still for the yet more weakly bound oxygens of periodate. Both periodate and bromate oxidize polycyanometallates, presumably by one-electron transfer in view of the substitution-inertness of complexes of this type. Kinetic parameters have been obtained for, e.g., bromate oxidation of octacyanoniobate(III), $[Nb(CN)_8]^{5-}$.[229] Here an activation entropy of $-139\ J\ K^{-1}\ mol^{-1}$ is consistent with an initial bimolecular reaction; the reduced bromate species then initiates a sequence of steps through various bromine oxoanions en route to the final product, bromide. This sequence, and the analogous sequencefrom perchlorate through chlorate and chlorite to hypochlorite ($ClO_4^- \rightarrow ClO_3^- \rightarrow ClO_2^- \rightarrow ClO^-$), both testify to the readiness of halate anions to react by oxygen transfer.[230] Hypochlorites are powerful (but unglamorous—their uses are in such applications as bleaches and disinfectants) oxidants. It has long been believed that the large number of known hypochlorite oxidations (nitrite to nitrate, sulphite to sulphate, cyanide to cyanate and on to nitrate and carbonate, halides to halates) all occur by oxygen transfer, but there are indications that some hypochlorite oxidations, for example of sulphite and of cyanide, may involve Cl^+ transfer (cf. end of Section 9.7.4.4).[231] Oxygen transfer from hypobromite to hypo-phosphite

$$O_3PPO_3^{4-} + BrO^- \longrightarrow O_3POPO_3^{4-} + Br^-$$

can hardly be simple, since the hypobromite oxygen is in the terminal position in the pyrophosphate product.[232]

Nitrates are well known for their oxidizing properties, but there seem to have been singularly few studies of mechanisms of oxidation by nitrate in aqueous solution. It is interesting to the inorganic chemist to note the proposal that, in the form of concentrated nitric acid, nitrate oxidizes cerium(III) in the shape of the NO_2^+ cation, the familiar nitrating agent of organic chemistry.

It is convenient at this point to digress to transition metal oxoanions, whose chemistry in many respects parallels that of their *sp*-block analogues. They can also participate in oxygen transfer reactions, as in permanganate oxidation (at high pH) of cyanide to cyanate. Indeed, the earliest established oxygen transfer involved the *d*-block oxide OsO_2 as donor, giving perchlorate from chlorate.[233] However, for transition metal oxoanions one-electron transfer can be an available and favourable option, since *d*-block elements often have oxidation states one unit apart, as in $MnO_4^{-/2-}$. It should be borne in mind that several important oxoanions of *d*-block elements give final products several units of oxidation state lower, for example chromate or dichromate usually ends up as chromium(III), permanganate as manganese(II). Such multi-electron transfers perforce involve several steps and transient intermediates (Section 9.7.2). There is also the additional complication of induction periods in a number of reactions, as for example in the familiar situation of permanganate titrations of oxalate. Here manganese(II) is a catalyst, so the reaction speeds up considerably once a significant amount of manganese(II) has been produced.

The Group 8 (VIII) oxoanions FeO_4^{2-}, RuO_4^{2-}, and RuO_4^-, and their close relatives RuO_4 and OsO_4, are powerful oxidants. OsO_4 is much used in organic chemistry, where its oxidation of alkenes, $R_2C{=}CR_2$, to diols, $R_2C(OH){-}C(OH)R_2$, involves two-oxygen transfer by way of esters, **56**, which are often isolable. Its enthusiasm for oxidizing organic matter, readily producing osmium metal in the presence of an excess of reductant, has led to

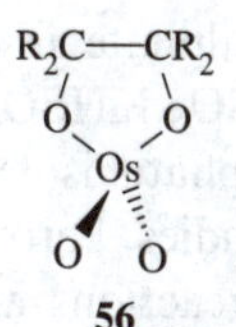

its use in biology as a stain. OsO_4 also transfers oxygen to phosphines, RuO_4 to sulphides. The ruthenium and iron oxoanions have similar properties, but have been much less studied. A mechanism of oxygen transfer has been proposed for ferrate oxidation of sulphite.[216] Oxygen transfer from ferrate is unlikely to be rate-limiting in slowish redox reactions, since the half-life for oxygen exchange with ferrate is about one minute. Ferrate oxidations of alcohols, aldehydes, diethyl sulphide, dimethyl sulphoxide, and 1,4-thioxane follow a three-term rate law, namely:[234]

$$\text{rate} = k_1[FeO_4^{2-}] + k_2[FeO_4^{2-}]^2 + k_3[FeO_4^{2-}][\text{reductant}]$$

with k_1, k_2, and k_3 all pH-dependent. The first two terms correspond to those in the rate law for oxidation of water by ferrate; the third term is attributed to bimolecular formation of ferrate esters or analogues. Readers interested in the mode of action of ruthenates as oxidants of organic substrates may enter the primary literature through the account of ruthenate oxidations of cinnamate.[235]

9.7.4.3 Peroxoanion oxidations

There are a number of peroxoanions of *sp*-block elements, of a range of stabilities and of oxidizing powers. Some are well- and long-established (O–O bond distances[236] are given in Table 9.40), others are unstable, others are reasonably well-characterized intermediates, and yet others may be no more than figments of the imagination. Probably the best known, and certainly the most studied by kineticists, is peroxodisulphate. Peroxomonosulphates are difficult to prepare—only the Cs^+, Rb^+, K^+ (and a K^+/NH_4^+

Table 9.40 Comparison of O–O bond distances, $d_{O\text{-}O}$, in peroxoanions with those in dioxygen species.

Peroxoanions	$d_{O\text{-}O}$(Å)	Dioxygen species	$d_{O\text{-}O}$ (Å)
Peroxodisulphate in $K_2S_2O_8$, $(NH_4)_2S_2O_8$	1.48, 1.50	O_2^{2-}	1.49[a]
Peroxomonosulphate in $KHSO_5H_2O$	1.46	O_2^-	1.26
Peroxodiphosphate in $[Na_2(H_2O)_9]_2[P_2O_8]$	1.49	O_2	1.21
Perborate in $Na_2[B_2(O_2)_2(OH)_4]6H_2O$	1.48	O_2^+	1.12

[a] 1.45 Å in solid H_2O_2.

double salt) have been obtained pure. Peroxomonosulphate is normally supplied as the triple salt 'oxone'®, which is $2KHSO_5.KHSO_4.K_2SO_4$. Peroxodiphosphate is fairly stable; peroxomonophosphate is less stable and is generally generated *in situ* for solution kinetics studies. Peroxonitrites have been studied;[211] they are intermediates in such reactions as that of nitrous acid with chlorite.[237] Sodium perborate is a stable compound containing a genuine peroxoanion.[238] Some percarbonates are genuine peroxo species,[239] others are carbonate with hydrogen peroxide of crystallization.[240] Dilute solutions of perborates or percarbonates consist of borate or carbonate plus hydrogen peroxide (or HO_2^-), but concentrated solutions contain significant concentrations of the respective peroxoanions, as shown by such spectroscopic probes as ^{11}B NMR and Raman spectroscopy.[241] Dissolved sodium perborate is about 50% dissociated to borate plus hydrogen peroxide in a 0.1 mol dm^{-3} solution. Both perborates and percarbonates are commercially available, supplied to act as solid sources of hydrogen peroxide for incorporation into household detergents, dental-care products, stain removers, and similar products. Perborates, percarbonates, and peroxomonosulphate are steadily replacing chlorine-based bleaches in a variety of industrial applications. The peroxoanions are generally less powerful bleaches than hypochlorite, but their environmental advantages are becoming increasingly attractive. Mechanisms of peroxoanion, and indeed of peroxide, oxidations may be radical or non-radical in nature—the role of nucleophilic attack at peroxide-oxygen in the latter was reviewed early in the development of this area by the key pioneer.[242]

(i) *Peroxodisulphate.* Whereas mononuclear peroxoanions can usually react by oxygen-atom transfer, e.g.

$$HSO_5^- + NO_2^- \rightarrow HSO_4^- + NO_3^-$$

some binuclear peroxoanions may not have this simple oxygen transfer option. Thus the most studied of these species, peroxodisulphate, acts only occasionally as an oxygen atom source. Rather it prefers to react by electron transfer or oxygen–oxygen bond fission:

$$S_2O_8^{2-} \rightarrow S_2O_8^{-\cdot} + e^-$$

$$S_2O_8^{2-} \rightarrow 2SO_4^{-\cdot} \text{ or } S_2O_8^{2-} \rightarrow SO_4^{2-} + SO_4^{-\cdot} + e^-$$

Although the redox potential for the peroxodisulphate anion is very high, $E°(S_2O_8^{2-}/2SO_4^{2-}) = 2.01$ V, the requirement for breaking the fairly strong oxygen–oxygen bond (cf. relevant bond distances in Table 9.40 above) means that most peroxodisulphate oxidations take place slowly. This is not an unmixed blessing, for the very numerous kinetic studies of peroxodisulphate oxidations from all over the world have produced a vast array of kinetic data and complicated mechanistic schemes, but not a comparable level of understanding of such systems. However, it is possible to distinguish several main types of mechanism and kinetic pattern for peroxodisulphate oxidations, which will be summarized in the following paragraphs.[243]

(a) Rate-limiting electron transfer:

$$\text{reductant} + S_2O_8^{2-} \longrightarrow \text{reductant}^+ + S_2O_8^{-\cdot}$$
$$\downarrow$$
$$SO_4^{2-} + SO_4^{-\cdot} \text{ etc.}$$

Such a mechanism gives the second-order rate law

$$\text{rate} = k_2[\text{reductant}][S_2O_8^{2-}]$$

This rate law and mechanism apply to oxidation of, *inter alia*, $[Nb(CN)_8]^{5-}$,[229] $[Co(cydta)]^{2-}$,[244] $[Ru(NH_3)_5(pz)]^{2+}$, $[Ru(NH_3)_4(bipy)]^{2+}$, and excited-state $*[Ru(bipy)_3]^{2+}$. Even dinuclear $[(H_3N)_5Ru(\mu\text{-}pz)Ru(NH_3)_5]^{4+}$ reacts by consecutive one-electron transfers, despite the apparent attraction of the possibility here of a two-electron transfer to two-electron oxidant $S_2O_8^{2-}$. These ruthenium(II) complexes, and also a number of iron(II)- and osmium(II)-diimine complexes, conform to Marcus theory.[245] Rate-limiting single-electron transfer is also assumed for the peroxodisulphate oxidations of species such as iodide and hexacyanoferrate(II), beloved by physical chemists who delve into charge product, cation nature, salt, and solvent effects on reactivities (cf. previous chapter, Section 8.3).[246]

(b) Oxygen atom transfer, as in peroxodisulphate oxidation of cysteine coordinated to cobalt(III),[247] may also follow the simple second-order rate law given in (a) above. It has been demonstrated, through ^{18}O isotope effects on ^{31}P NMR spectra, that the oxidation of PPh_3 to $OPPh_3$ by $S_2O_8^{2-}$ does involve transfer of a peroxodisulphate oxygen atom to the reductant, through rate-determining attack of the PPh_3 at one of the peroxidic oxygens.[248]

(c) Rate-limiting dissociation of peroxodisulphate,[249,250] with rapid substrate oxidation by the radicals produced:

$$S_2O_8^{2-} \xrightarrow{\text{slow}} 2SO_4^{-\cdot};\ SO_4^{-\cdot} + \text{reductant} \xrightarrow{\text{fast}} \text{intermediates/products}$$

The rate law for such a scheme is simply

$$\text{rate} = k_1[S_2O_8^{2-}]$$

with k_1 *identical* for all reductants. This mechanism operates for peroxodisulphate oxidation of, for example, ammonia, sulphite, Tl^+aq, Mn^{2+}aq, and $[Co(edta)]^{2-}$.[251] In strongly acidic media there is the possibility of dissociation to give peroxomonosulphate as a reactive intermediate:[252]

$$S_2O_8^{2-} + H_2O \xrightarrow{H^+} HSO_5^- + HSO_4^-$$

(d) Rate-determining dissociation of the substrate to give one or more products which are oxidized by peroxodisulphate much more rapidly than is the original substrate. Now the rate is independent of peroxodisulphate concentration, the rate law being

$$\text{rate} = k_d[\text{reductant}]$$

with k_d identical to the rate constant for dissociation of the species involved. The bis-maleonitriledithiolate complex of cobalt(II), $[Co(mnt)_2]$, illustrates this situation.

These are only the four kinetically simple options—many peroxodisulphate oxidations obey more complicated rate laws. Thus, for example, oxidation of aminoanthraquinone-bridged binuclear ruthenium complexes, here abbreviated to diRu, follows the rate law

$$-d[\text{diRu}]/dt = 2k_{et}K_{ip}[\text{diRu}][S_2O_8{}^{2-}]/\{1 + k_{ip}[S_2O_8{}^{2-}]\}$$

because rapid pre-equilibrium formation of an ion-pair precedes electron transfer.[253] The redox transition state is bimolecular, as in (a) above. Several low-spin iron(II) complexes follow the rate law

$$-d[\text{Fe(Xphen)}_3{}^{2+}]/dt = k_1[\text{Fe(Xphen)}_3{}^{2+}] + k_2[\text{Fe(Xphen)}_3{}^{2+}][S_2O_8{}^{2-}]$$

indicating concurrent dissociative and electron transfer paths. Values of k_1 equal separately established k_d values for the respective complexes. The balance between the two pathways reflects the electronic effects of the ligand substituents—when $X = NO_2$ the dissociative path dominates, when $X = Me_2$ or Me_4, dissociation of the complex is very much slower and the k_2 term dominates.[254] A complementary two-term rate law applies to the oxidation of bromide[250] and of nitrite:[255]

$$\text{rate} = k_1[S_2O_8{}^{2-}] + k_2[NO_2{}^-][S_2O_8{}^{2-}]$$

In this case it is the dissociation of $S_2O_8{}^{2-}$ into $SO_4{}^{-\cdot}$ radicals which occurs in parallel with the bimolecular oxidation. In view of the apparently different behaviours of $[Co(edta)]^{2-}$ and $[Co(cydta)]^{2-}$ (see (a) and (c) above), peroxodisulphate oxidation of this type of cobalt(II)-polyaminocarboxylate complex might also be found to obey an analogous two-term rate law if rate constants were to be obtained over a sufficient peroxodisulphate concentration range. In the nitrite reaction, and in several similar systems, the radical pathway assumes increasing importance as the temperature rises, since the dissociation of $S_2O_8{}^{2-}$ is characterized by a particularly high activation energy.

Many of the reactions which involve radical intermediates have complicated rate laws due to the variety of intermediates and lifetimes.[256] The peroxodisulphate oxidation of phosphite starts with the production of $SO_4{}^{-\cdot}$ radicals, which then give $HPO_3{}^{-\cdot}$ and $^{\cdot}OH$, which undergo further reactions to give a multistage mechanism and non-simple rate law.[257] The radicals $N_3{}^{-\cdot}$ and $O_3SN(O)SO_3{}^{2-\cdot}$ are proposed intermediates in the reaction with azide.[258] Peroxodisulphate oxidation of alcohols is slow—indeed solvent effects on relatively rapid redox kinetics can be studied in alcohol–water mixtures. The radical chain mechanisms involved give complicated rate laws, with that for methanol claimed to approximate to

$$\text{rate} = k[\text{MeOH}]^{1/2}[S_2O_8{}^{2-}]^{3/2}$$

under certain conditions. For the reaction with hydrogen peroxide, the overall order varies between one-half and three-halves, according to conditions.

Peroxodisulphate oxidations can often be greatly accelerated by the presence of a metal-ion catalyst, often added as a silver or iron salt. These presumably operate by catalytic cycles involving $Ag^{+/2+}$ (or perhaps $Ag^{+/3+}$) or $Fe^{2+/3+}$. Iron(II) induces oxidation of arsenic(III) by peroxodisulphate; copper(II) catalyses peroxodisulphate oxidation of oxalate. There can be serious problems with traces of transition metal ions catalysing peroxodisulphate dissociation/oxidation and causing misleading kinetic results and leading to erroneous mechanistic deductions.

As both p*K*s of $H_2S_2O_8$ correspond to strong acid behaviour, the oxidant is the $S_2O_8{}^{2-}$ anion over a wide pH range, with rate constants being essentially independent of pH over the range $3 < \text{pH} < 14$. In strongly acid solution there is a pH-dependent path involving acid-catalysed hydrolysis to $HSO_5{}^-$; at hydroxide concentrations above about molar there is a base hydrolysis term with rate $\propto [OH]^n$ with $n \gg 1$.

(ii) *Peroxomonosulphate.*[259] This is also strongly oxidizing ($E°(HSO_5{}^-, 2H^+/HSO_4{}^{2-}, H_2O) = 1.84$ V)[260], though not as powerful as peroxodisulphate. In contrast to the latter, there is a marked effect of pH on reactivities—the first dissociation constant of H_2SO_5 is similar to that of sulphuric acid, and therefore of no relevance to the normal run of kinetic studies in solution. However pK_2 is 9.4,[261] therefore interpretations of pH dependences of rate constants may have to consider both $HSO_5{}^-$ and $SO_5{}^{2-}$ as oxidants. Thus, for example, $HSO_5{}^-$ oxidizes azide, to nitrous oxide, about a hundred times more rapidly than $SO_5{}^{2-}$. ^{18}O-labelling experiments confirm an oxygen atom transfer mechanism.[262] There may also be acid–base behaviour of the reductant to bear in mind—$N_3{}^-$ is oxidized very much more readily than HN_3.

Peroxomonosulphate oxidizes chloride, bromide, and iodide by simple second-order kinetics.[263] It also oxidizes aldehydes according to an analogous rate law:

$$\text{rate} = k_2[\text{RCHO}][HSO_5{}^-]$$

Again oxygen atom transfer takes place, the products being the respective carboxylic acids. The mechanism here, and for peroxomonosulphate oxidation of anilines, of glycine, and of alicyclic ketones, is of nucleophilic attack by the reductant at the peroxo-oxygen of $HSO_5{}^-$.[264] Peroxomonosulphate oxidation of $Fe^{2+}aq$ and of $Ti^{3+}aq$ is autocatalytic, but HSO_5^- acts as a simple oxygen transfer agent to metalloproteins.[265]

Two consecutive oxygen atom transfers are involved (evidence provided by ^{18}O-tracer experiments) in the conversion of coordinated thiosulphate in $[Co(NH_3)_5(S_2O_3)]^+$ into coordinated $S_2O_5^{2-}$.[266] The first step is equilibrium formation of an intermediate (int), with the rate law providing an example of observed rate constants being the *sum* of forward and reverse rate constants for opposing reactions:

$$+d[\text{int}]/dt = \{k_{-1} + k_2[HSO_5^-]\}[Co(NH_3)_5(S_2O_3)^+]$$

Peroxomonosulphate also oxidizes thiosulphate and 4-substituted thiophenolates ($-SC_6H_4X$) coordinated to ruthenium(III) in two oxygen atom transfer steps, the latter giving first coordinated sulphenate (**57**), then coordinated sulphinate (**58**). This time the first step obeys a simple second-order rate law.[267]

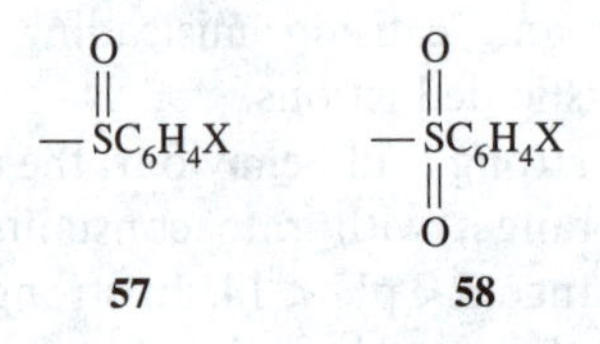

Peroxomonosulphate itself normally decomposes by a polar (i.e. non-radical) mechanism, following the rate law:

$$\text{rate} = k[HSO_5^-][SO_5^{2-}]$$

This function corresponds to the observed rate maximum at pH 9.3, the pK_2 for H_2SO_5.[268] Decomposition can be induced, e.g. by the presence of potential one-electron transfer catalysts such as $Fe^{3+/2+}$, $Ti^{4+/3+}$ (cf. above), $Ce^{4+/3+}$ or $Ag^{2+/+}$, or photochemically, to give radicals.[259] Under such conditions its mechanisms of oxidation of substrates may be radical rather than oxygen-transfer in nature.

(iii) *Peroxophosphates.*[269] Peroxodiphosphate (see Table 9.40 above for relevant bond distance information) oxidizes a variety of organic substrates, including aldehydes and ketones, according to a second-order rate law and thus presumably a rate-limiting electron transfer mechanism analogous to (a) above for peroxodisulphate. However, reaction of peroxodiphosphate with $[Fe(bipy)_3]^{2+}$ and with $[Fe(phen)_3]^{2+}$, as with iodide or malachite green, occurs solely through rate-determining dissociation, while there is no apparent reaction with the even more substitution-inert $[Fe(bipy)_2(CN)_2]$ and $[Fe(bipy)(CN)_4]^{2-}$. Peroxodiphosphate appears a less effective oxidant than peroxodisulphate for these low-spin iron(II) complexes.[270] Peroxomonophosphate[271] oxidizes halides,[263] hydroxylamine, hypophosphite, $[Co(edta)]^{2-}$,[272] aldehydes, and amines (including anilines) in second-order processes, but it oxidizes azo compounds such as methyl orange according to a rate law zero-order in organic substrate. For oxidations both by peroxodiphosphate and by peroxomonophosphate, pH has a marked effect on rate constants. Acidity constants are $pK_1 \sim 0, pK_2 \sim 0.5, pK_3 = 5.2, pK_4 = 7.7$ for $H_4P_2O_8$, and $pK_1 \sim 0.5$ or 1.1, $pK_2 = 5.5, pK_3 \sim 13$ for H_3PO_5.[273] In almost all peroxophosphate oxidations reactivity decreases with increasing pH, indicating reactivity orders[271,274]

$$H_4P_2O_8 > H_3P_2O_8^- > H_2P_2O_8^{2-} > HP_2O_8^{3-} > P_2O_8^{4-};$$
$$H_3PO_5 > H_2PO_5^- > HPO_5^{2-} > PO_5^{3-}$$

If the reductant exhibits acid–base behaviour within the pH range studied, then the pH-rate profile will also reflect this, as detailed for, for instance, peroxomonophosphate oxidation of amino acids.[275]

Peroxomonophosphate is generally felt to be a more rapid oxidizer than peroxodiphosphate. Thus, for instance, in the reaction of peroxodiphosphate with nitrite, with hydrazine, or with antimony(III) it is peroxomonophosphate produced by hydrolysis[273] of the peroxodiphosphate which is the actual oxidant.[276]

Oxidation of an aqua-cation may involve complex formation prior to electron transfer. This seems to be the case for the reactions of Ce^{3+}aq[277] and of Mn^{2+}aq[278] with peroxodiphosphate, though the latter reaction does not follow the simple second-order rate law which might be expected for rate-limiting complex formation (cf. the Eigen–Wilkins mechanism of Section 7.5.1).

(iv) Extensions and comparisons. Peroxocarboxylates are occasionally employed as oxidants of inorganic species. They can transfer oxygen atoms to inorganic complexes such as iron/edta species.[279]

Relative reactivities of various peroxoanion oxidants can be directly compared, inorganic/inorganic or inorganic/organic, in only a very few cases. There are some reactivity comparisons in a paper on the reaction of V^{2+}aq with peroxides $Me_2RC.O_2H$ (R = Me, Et, or Bz), where the kinetic parameters correspond to substitution at V^{2+}aq and thus indicate inner-sphere oxidation.[280] There are limited data on some peroxide, peroxomono- and peroxodisulphate, and peroxodiphosphate oxidations of titanium(III), iron(II), and copper(I) species,[281] and of organic substrates,[282] but we find ourselves unable to present our hoped-for wide-ranging correlatory tabulation.

9.7.4.4 Oxygen atom and halogen cation transfer

A number of the oxidation reactions mentioned in the previous section proceed through simple oxygen atom transfer between oxoanions (Table 9.39), or to coordinated sulphur. Such a mechanism also applies to a number of other redox reactions, involving a variety of species, ranging from simple species of the *sp*-block of the Periodic Table through transition

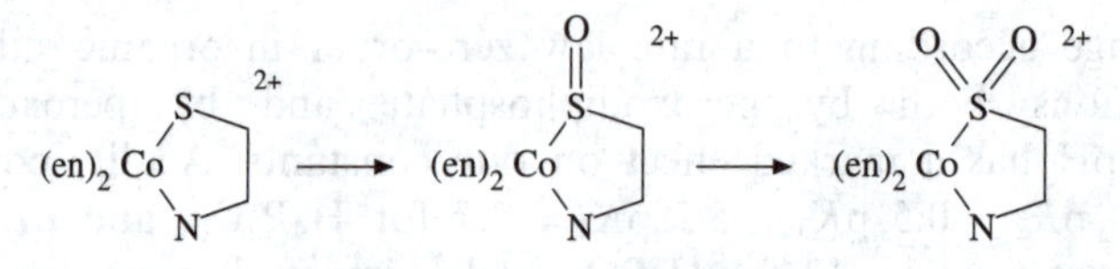

Fig. 9.30 Oxidation of cobalt(III)–coordinated sulphur-donor ligands by oxygen atom transfer.

metal oxoanions and oxohalide anions[283] to bioinorganic and organometallic substrates.[284] As in the previous section, we shall include some relevant *d*- and *f*-block chemistry here alongside the *sp*-block chemistry which is the main subject of Section 9.7.

A variety of oxidants convert phosphines into phosphine oxides and sulphides into sulphoxides to sulphones:

$$R_3P \longrightarrow R_3PO;$$

$$R_2S \longrightarrow R_2SO \longrightarrow R_2SO_2$$

Such oxidants include peroxodisulphate (Section 9.7.4.3), OsO_4 and RuO_4, oxohalide anions, such as $CrOCl_4^-$,[285] $VOCl_4^-$, and $TcOCl_4^-$,[286] and dioxo complexes containing $Mo^{VI}O_2$[287] or $Ru^{III}O_2$ units. A kinetic/thermodynamic correlation (log k versus E°) has been established for oxidation of PPh_3 by a series of *cis*-[$Mo^{IV}O_2(R_2dtc)_2$] complexes.[288] Oxygen transfer can also take place to compounds of this type when coordinated, as for example in the oxidation of sulphides to sulphoxides on ruthenium and in the sequence from cobalt chemistry shown in Fig. 9.30, where coordinated sulphide is oxidized successively to coordinated sulphenate and to coordinated sulphinate. The mechanisms of these coordinated-ligand reactions may be viewed as S_N2 reactions of the coordinating sulphur; kinetic parameters ($k, \Delta H^\ddagger$, and $\Delta S^\ddagger$) for twenty-six reactions of this type have been collated.[289] These oxidations of coordinated sulphur, bonded to cobalt(III) or ruthenium(III), may be achieved by a variety of oxidants, including hydrogen peroxide, periodate, peroxomono- and peroxodisulphate, and peroxo complexes of vanadium.[266,267,289,290,291]

Conversely, compounds of the type R_2SO, R_3AsO, PhIO, and pyridine- and pyrazine-*N*-oxides can act as oxygen-donor oxidants to appropriate reducing substrates, e.g.[292]

$$V^{2+}aq + O\text{—}NC_4H_4N \longrightarrow V^{2+}aq + NC_4H_4N,$$

$$Tc^{III}(dithiolate)(py)_2 + O\text{—}NC_4H_4N \longrightarrow Tc^{V}O(thiolate)_2 + 3\,NC_5H_5$$

and the generation of oxohalides such as $VOCl_3$ or $VOCl_4^-$ from PhIO and the appropriate halide. Even simple oxides of nitrogen can act as oxygen-donor oxidants to certain organometallic compounds, e.g.[293]

$$(\eta^5\text{-}C_5Me_5)_2VCl_2 + NO \longrightarrow (\eta^5\text{-}C_3Me_5)_2VOCl_2 + \tfrac{1}{2}N_2$$

$$(\eta^5\text{-}C_5Me_5)_2Hf(H)Ph + N_2O \longrightarrow (\eta^5\text{-}C_5Me_5)_2HfOPh \text{ and } (\eta^5\text{-}C_5Me_5)_2Hf(H)(OPh)$$

while phosphorus pentoxide transfers an oxygen to PCl_3 in the industrial manufacture of $POCl_3$. In practically all the above cases the mechanism of oxygen atom transfer appears to be straightforward.

Sometimes a transition metal complex is needed as a catalyst for oxygen atom transfer. Examples of such catalysts include $MeReO_3$, which is conveniently water-soluble and an efficient oxygen transfer agent in the hydrogen peroxide/bromide reaction,[294] the $[MoO(CN)_5]^{3-}$ anion, which effects oxygen transfer from dimethyl sulphoxide or N_2O to triphenyl phosphine,[295] and metalloporphyrins, which can mediate oxygen transfer from peroxomonosulphate in a variety of situations from simple oxoanion chemistry through to DNA cleavage.[296] Oxomolybdenum complexes are important oxygen transfer intermediaries in certain bioinorganic and model systems, and in organometallic chemistry and homogeneous catalysis. Thus, for example, xanthine oxidase and $Mo^{VI}(O)(S)$ models transfer oxygen to xanthine en route to uric acid, and $MoO(Et_2dtc)_2$ acts as an oxygen atom transfer agent to alkenes.[297] $[RuO(Me_4en)(terpy)]^{2+}$, $[RuO(edta)]^-$, and [Ru(edta)(NO)] are also claimed to be excellent oxygen atom transfer agents to a variety of alkenes, including cyclohexene and styrene.[298] Vanadium is a particularly versatile element in relation to oxygen transfer, in view of the stability of species V^{n+}, VO^{2+}, and VO_2^{+}.[299]

Intramolecular oxygen transfer reactions are also known, as in the reaction of $[U(dmso)_8]^{4+}$ with nitrite. This proceeds through rapid formation of a dinitro intermediate $[U(NO_2)_2(dmso)_n]^{2+}$, which undergoes two intramolecular oxygen transfers from these nitro ligands to the metal to give $[UO_2(dmso)_5]^{2+}$.[300] Intermolecular oxygen transfers from MnO_2 and PbO_2 to U^{4+} to give UO_2^{2+} were reported many years ago.

Some reactions which appear likely candidates for oxidizing substrates by oxygen atom transfer have in recent years been shown to act rather as positive halogen oxidants. Thus hypochlorous acid oxidation and hypobromite oxidation of sulphite, and hypochlorite oxidation of iodide, cyanide, and sulphite, may involve initial attack by Cl^+ or Br^+ at the reductant.[231,291,301] Similarly, the reactions of chloramine, NH_2Cl, with Br_2, Br_3^-, HOBr, and BrO^- are believed to involve Br^+ transfer, facilitated by nucleophilic attack by the lone pair of the nitrogen.[302] Atom transfer also plays a central role in self-exchange redox reactions of photochemically produced $M(\eta^5\text{-}C_5H_5)(CO)_3$ radicals in halide-containing media, where the reaction pair comprises $M(\eta^5\text{-}C_5H_5)(CO)_3^-$ and $MX(\eta^5\text{-}C_5H_5)(CO)_3$ ($M = Mo, W; X = Cl, Br, I$).[303]

9.7.5 Hydrogen peroxide and oxygen

9.7.5.1 Hydrogen peroxide as oxidant

This leads on logically from the previous section, for in alkaline solution hydrogen peroxide exists as the HO_2^- anion, a special case of the peroxoanions just covered. Nevertheless, its ease of protonation ($pK_a(H_2O_2) = 11.8$) means that uncharged H_2O_2 is comparably important. Both HO_2^- and H_2O_2 are relatively weak oxidants. As in the case of peroxoanion oxidations, there are a number of possible mechanisms and rate laws for oxidation by hydrogen peroxide (H_2O_2/HO_2^-). These include the cases set out in the following paragraphs.

Firstly, there are many simple bimolecular/second-order oxidations, as for example for copper(I)-diimine complexes[304] and for the often very slow oxidation of cobalt(II) complexes, e.g. of $[Co(sep)]^{2+}$.[305]

$$\text{e.g.} \quad -d[Cu(X\text{-phen})_2{}^+]/dt = k_2[Cu(X\text{-phen})_2{}^+][H_2O_2]$$

As would be expected for a simple electron transfer, the rate constant k_2 depends markedly on the electron-withdrawing or -releasing properties of the substituent X. Although the first step is a straightforward one-electron transfer from the copper(I), the mechanism becomes more complicated thereafter, with the detailed role of intermediate hydroxyl radicals not clearly established.[306]

It seems probable that labile aqua-cations such as V^{2+}aq, Fe^{2+}aq, and U^{3+}aq (and complexes such as $Fe(edta)^{2-}$aq and $Ti(edta)^-$aq) react by one-electron transfer, probably in species of the type $[M(H_2O)_5(H_2O_2)]^{n+}$, $[U(H_2O)_n(HO_2)]^{2+}$, $[M(edta)(H_2O_2)]^{n-}$, or $[M(edta)(HO_2)]^{n-}$, with the hydroxyl radical playing a key role.[307] The rate constant for oxidation of V^{2+}aq by H_2O_2 is 15.4 $dm^3\ mol^{-1}\ s^{-1}$ (at 298 K), which is within the range expected for inner-sphere oxidation of this cation (Section 9.6.3) and thus implicates $[V(H_2O)_5(H_2O_2)]^{2+}$ as an intermediate.[308] The Cr^{2+}/H_2O_2 reaction is more complicated, with product evidence for a significant two-electron transfer route—the detection of $[Cr(\mu\text{-}OH)_2Cr]^{4+}$ and $[Cr(\mu\text{-}O)Cr]^{4+}$ strongly indicates the intermediacy of transient Cr^{IV} species. Specific added anion effects show that some chromium(II) complexes react more rapidly than Cr^{2+}aq itself; inner-sphere electron transfer is indicated in these cases.[202,205] Hydrogen peroxide oxidation of the unstable Cu^+aq may also involve two-electron oxidation, with the intermediacy of copper(III).[309] Peroxo complexes of transition metals can themselves be efficient oxidants, though oxidation of sulphur(IV) by peroxotitanium(IV) actually involves rapid oxidation of HSO_3^- by H_2O_2 subsequent to rate-determining dissociation of the peroxo complex. Direct oxidation by the peroxo complex is more commonly seen for transition metals with many *d* electrons.[310]

Simple second-order reactions are also observed for H_2O_2 oxidation of a number of organic substrates. In this area it is often found that HO_2^- is a much better oxidant than H_2O_2 itself for dyes such as erythrosin, **59**. In this type of reaction, the substrate undergoes nucleophilic attack by the HO_2^-, as also, for example, in the oxidation of nitriles to amides and of azobenzene (**60**) to azoxybenzene (**61**).

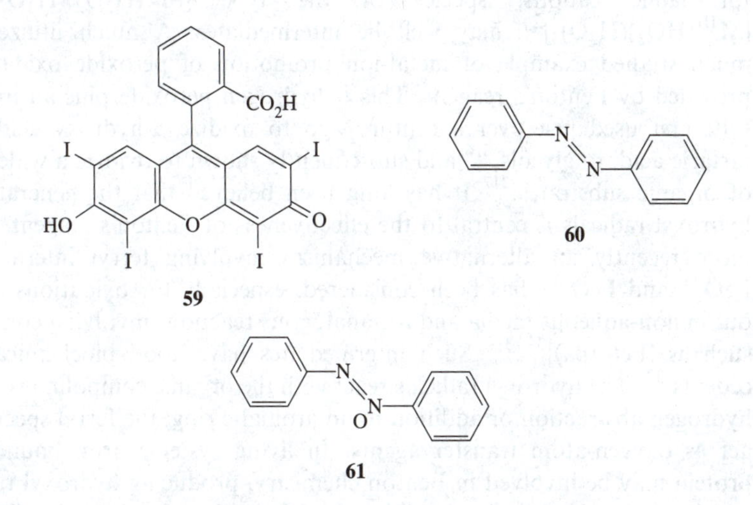

Secondly, when there is a large kinetic barrier to electron transfer from the metal ion in a complex, reaction may proceed via rate-limiting dissociation followed by more rapid oxidation of products of dissociation. This is the case for, e.g., reaction of hydrogen peroxide with low-spin iron(II) complexes such as those with diimine ligands. Thus for $[Fe(X\text{-}phen)_3]^{2+}$ cations, the rate law is

$$-d[Fe(X\text{-}phen)_3{}^{2+}]/dt = k_1[Fe(X\text{-}phen)_3{}^{2+}]$$

with the respective values for k_1 equalling those for aquation—i.e. aquation and reaction with H_2O_2 share the same dissociative rate-limiting step.[311] Similarly, oxidation of carbon disulphide by hydrogen peroxide occurs through rate-limiting oxidation of hydrolysis products from the carbon disulphide.[312] Similarly, oxidation of pentacyanoferrates(II) such as $[Fe^{II}(CN)_5(imid)]^{3-}$ involves rate-determining loss of imidazole, in the limiting dissociative (*D*) mechanism characteristic of this class of complex (See Chapter 4, Section 4.7.5.2), with competition between incoming hydrogen peroxide and leaving imidazole for the intermediate. One-electron transfer, with the production of hydroxyl radicals, follows.[313]

Thirdly, oxidation of a metal complex may involve ligand rather than metal-ion oxidation. Such reactions generally involve oxygen atom transfer, as in oxidation of coordinated dimethyl sulphide to coordinated dimethyl sulphoxide in

$$[Ru(bipy)(terpy)(Me_2S)]^{2+} \xrightarrow{H_2O_2} [Ru(bipy)(terpy)(Me_2SO)]^{2+}$$

In this case there is preferential ligand oxidation because the diimine ligands particularly favour the $t_{2g}{}^6$ ruthenium(II) state.[314] Hydrogen peroxide reactions are notoriously susceptible to catalysis by traces of transition metal cations, $M^{n+}aq$, which can mediate electron transfer.[315] Especially for labile cations, species of the type $[M^{II}(H_2O_2)(H_2O)_5]$ or $[M^{III}(HO_2)(H_2O)_5]^{2+}$ may well be intermediates. A much utilized and much studied example of metal-ion promotion of peroxide oxidation is provided by Fenton's reagent. This is hydrogen peroxide plus an iron(III) salt, first used just over a century ago to oxidize α-hydroxy acids, e.g. tartaric acid, to glycols,[316] and subsequently shown to oxidize a wide range of organic substrates.[317] It has long been believed that the generation of hydroxyl radicals is central to the effectiveness of Fenton's reagent,[318] but more recently an alternative mechanism involving ferryl intermediates FeO^{2+} and FeO^{3+}, has been considered, especially for oxidations carried out in non-aqueous media and for analogous reactions involving complexes such as $[Fe(dtpa)]^{3-}$.[319] Such intermediates have good biochemical precedents.[320] The hydroxyl radicals react with the organic compounds by, e.g., hydrogen abstraction or addition to an aromatic ring; the ferryl species can act as oxygen-atom transfer agents. In living systems, iron bound to a protein may be involved in Fenton chemistry, producing hydroxyl radicals which react with amino acid residues in the vicinity of the iron-binding site, thus impairing enzyme activity. There is a similar pattern to copper/hydrogen peroxide chemistry, with evidence favouring two-electron Cu^{III}/Cu^{I} cycles in some situations, one-electron reactions with hydroxyl radical intermediates in others.[321]

Finally, in special cases there is the possibility of oscillating reactions, as for example in the reaction of hydrogen peroxide with sulphide, S^{2-}.[322]

9.7.5.2 *Hydrogen peroxide as reductant*

Reduction of $[M(bipy)_3]^{3+}$ and $[M(phen)_3]^{3+}$ (M = Fe, Ru, Os, Ni), and of $[W(CN)_8]^{3-}$, appears to proceed by simple one-electron outer-sphere transfer. Rate constants for the polypyridyl complexes permitted an estimate of between 0.01 and 0.1 $dm^3\ mol^{-1}\ s^{-1}$ to be made for the self-exchange rate constant for the HO_2/HO_2^- couple, by the usual Marcus approach.[323] For the octacyanotungstate(V) it is suggested that electron transfer takes place within a $[W(CN)_8]^{3-}$, H_2O_2 associated species.[324] In the case of $[Mn(bipy)_2]^{3+}$, electron transfer is believed to take place in a peroxide complex $[Mn(bipy)_2(H_2O_2)(H_2O)]^{3+}$ or, more plausibly, $[Mn(bipy)_2(HO_2)(H_2O)]^{2+}$.[325] Purple $[Fe(edta)(HO_2)(OH)]^{3-}$ is the key intermediate in the Fe/edta-catalysed decomposition of hydrogen peroxide.[326] In these two cases, as for several aqua-cations $M^{n+}aq$, coordination of the peroxide to the metal centre is a prerequisite to electron transfer. The reduction of $[Ag(OH)_4]^-$ also involves coordination of the peroxide, giving $[Ag(OH)_4(O_2)]^{3-}$, and an Ag^{II} intermediate—in other words there are two one-electron transfers.[327] For $[Ag(bipy)_2]^{2+}$ loss of one bipy ligand precedes the redox step, though it is not clear whether peroxide coordination to the silver is necessary here too.[328] Finally, and differently, reaction of hydrogen peroxide with $[Fe(CN)_6]^{3-}$ involves initial attack of a peroxo species at coordinated carbon[329]—cf. analogous oxygen transfer to sulphur coordinated to cobalt or ruthenium from a variety of peroxo species (Sections 9.7.4.4, 9.7.5.1, and 10.2.5).

9.7.5.3 Oxidation by dioxygen

Hydrogen peroxide is often a well-behaved oxidant in terms of kinetics and mechanism—dioxygen less so. Again there are a variety of kinetic and mechanistic patterns, with a corresponding variety of simple and complicated rate laws. There is also the occasional example of oscillating behaviour, as in the benzaldehyde/bromide/cobalt(II)/oxygen system.[322] There are still a number of unanswered questions in this area, as has recently been discussed with particular relevance to biological oxidation mechanisms.[330]

There are a number of oxidations which follow a simple second-order rate law and appear to involve an initial one-electron transfer—and thus the intermediacy of O_2^-. Such reactions include those with $[Co(sep)]^{2+}$,[331] with $[Fe(CN)_5(N_3)]^{3-}$, and with some ammine complexes of ruthenium(II) . The much-studied, and recently reviewed,[332] reactions of copper(I)-diimine complexes also involve the intermediacy of O_2^-, but here there often seem to be deviations from simple second-order behaviour. Indeed, oxidation of copper(I) complexes of 2,2-bipyridyl or of (substituted) 1,10-phenanthroline has been claimed[304] to follow a third-order rate law:

$$\text{rate} = k[\text{Cu}^{\text{I}}]^2[\text{O}_2]$$

In reactions of several cobalt(II) complexes the O_2^- remains coordinated in the cobalt(III) product. Oxidations of alkyl complexes of chromium(II), $[Cr(H_2O)_5R]^{2+}$

$$[\text{Cr(H}_2\text{O)}_5\text{R}]^{2+} + \text{O}_2 = [\text{Cr(H}_2\text{O)}_5(\text{O}_2\text{R})]^{2+}$$

are chain reactions.[333] For the particular case of $R = CHMe_2$, the rate law is three-halves order, namely

$$\text{rate} = k[\text{Cr(H}_2\text{O)}_5\text{R}^{2+}]^{3/2}$$

The rate-limiting step is the generation of the $^{\cdot}CHMe_2$ radical; dioxygen does not appear in the rate law, since it is consumed by rapid reaction with $^{\cdot}CHMe_2$.

The reaction with nitric oxide produces nitrite in aqueous solution, nitrogen dioxide in the gas phase. Despite product differences, the rate law is third-order in water, in carbon tetrachloride, and in the gas phase:[334]

$$-\text{d}[\text{O}_2]/\text{d}t = k[\text{NO}]^2[\text{O}_2]$$

Kinetics and mechanisms of dioxygen reacting with sulphur(IV) species, including sulphur dioxide, under catalysed and uncatalysed conditions, have been fully described and discussed elsewhere.[220]

Oxidation of aqua-cations, such as Fe^{2+}aq, V^{2+}aq, V^{3+}aq, and Mo^{3+}aq, by dioxygen is generally kinetically complicated but generally involves an inner-sphere redox step.[335] Oxidation of iron(II) complexes is somewhat less complicated, though products, stoichiometry, and kinetics depend on the nature of the ligands. Whereas oxidation of the azidopentacyanoferrate(II) anion involves one-electron transfer (cf. above), oxidation of $[Fe(4,7\text{-dihydroxyphen})_3]^{2+}$ may, thanks to the unusual properties of this ligand, involve two-electron transfer with transient iron(IV). Oxidation of porphyrin complexes generally involves the coordination of the O_2 to the iron (cf. iridium(I) in Chapter 10), with subsequent formation of $Fe(\mu\text{-}O_2)Fe$-containing dimers. Oxidation of iron(II) complexes of ligands of the edta, ida type involves a four-step sequence, with the second step, O_2^- oxidation of Fe^{II}, allegedly rate-limiting. Oxidation of some iron(II) complexes involves only ligand oxidation, the diimine produced stabilizing the iron in the 2+ state, even though Fe^{III} intermediates are involved:[336]

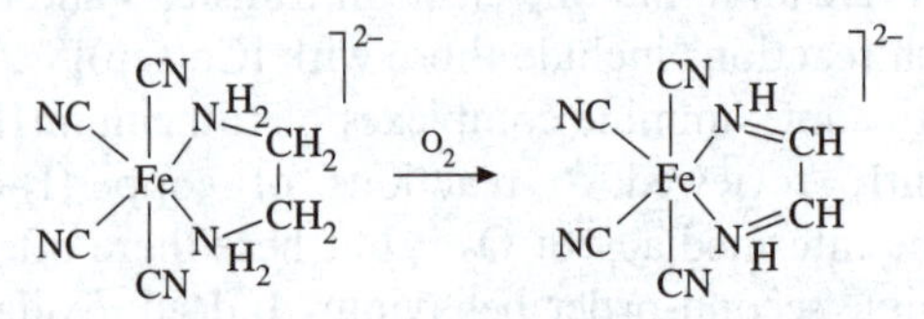

These examples lead us to haemoglobin and myoglobin, where of course the uptake of dioxygen does not involve conversion of the iron into iron(III) (cf. Chapter 10.3.1).

9.7.5.4 Ozone[337]

Ozone has a very short lifetime in neutral and weakly alkaline solution but has a half-life of several hours in strong alkali, above about 8 mol dm^{-3} hydroxide. It can react by initial one-electron transfer or oxygen atom transfer. The primary step in thermal decomposition is $O_3 \rightarrow O_2 + O$, for which the activation energy is relatively high, about 80 kJ mol^{-1}.[338] Electron transfer takes place in the exchange reactions O_2^-/O_3 and HO_2^-/O_3,[339] of which the former, being a radical reaction, is much the faster. Reaction with the bis-terpyridyliron(II) cation also involves electron transfer as the first step:

$$[Fe(terpy)_2]^{2+} + O_3 = [Fe(terpy)_2]^{3+} + O_3^{-\cdot}$$

Ozone oxidizes a number of metal complexes, for example acetylacetonates and dithiocarbamates, with clean kinetics, but with the ultimate production of a bewildering range of products.

The oxygen atom transfer mechanism operates in the reaction of ozone with the hydroxyl radical:

$$O_3 + {}^{\bullet}OH \longrightarrow O_2 + {}^{\bullet}O_2H$$

with chloride,[340] where the rate-determining step is

$$O_3 + Cl \longrightarrow O_2 + ClO^-$$

with Fe^{2+}, where an FeO^{2+} intermediate has been detected,[341] and, presumably, with cyanide ion. Oxidation of the mercurous ion, $Hg_2{}^{2+}aq$, has a more complicated mechanism,[342] since the rapid disproportionation equilibrium $2Hg_2{}^{2+} \rightleftharpoons Hg^{2+} + Hg^0$ plays a significant part. In fact two oxygen atom transfer reactions feature in ozone oxidation of $Hg_2{}^{2+}$ in aqueous acid:

$$Hg_2{}^{2+} + O_3 \longrightarrow Hg^{2+} + HgO\,(\rightarrow Hg^{2+} + H_2O) + O_2$$

and

$$Hg^0 + O_3 \longrightarrow HgO + O_2$$

$Hg_2{}^{2+}aq$ can undergo oxidation by one-electron ($[Fe(phen)_3]^{3+}$; $[Ru(bipy)_3]^{3+}$; Ag^{II}) or two-electron ($BrO_3{}^-$) paths. Clearly ozone prefers the latter, as do Tl^{III} and Mn^{III}, where the redox step is two-electron oxidation of Hg^0.

9.7.6 Radicals

The intermediacy of transient inorganic and organic radicals has been mentioned several times in this Chapter, especially in connection with non-complementary reactions (Section 9.7.2) and with oxidations by various oxoanions and peroxoanions (Section 9.7.4). Some radicals have only a very transient existence in solution, others are sufficiently long-lived to be considered as reagents—NO and NO_2 (and most transition metal ions!) provide familiar examples of stable radicals, as indeed do the organic radical traps purchasable from suppliers of chemical reagents. In this section we shall be concerned with the more reactive radicals, both inorganic and organic, and will try to give some sort of general picture of their reactivities. There are two compilations of rate constants[343] which give a wealth of values for reactions of a couple of dozen radicals with a variety of inorganic substrates. Before getting into details of various types of radical, it is interesting to note here the range of redox potentials shown by some relatively persistent free radicals. Values are given in Fig. 9.31,[198] which also includes some redox potentials for standard couples for comparison.

9.7.6.1 Hydrogen atoms

Hydrogen atoms can act as strong reductants or as oxidants, depending on circumstances. A selection of rate constants for hydrogen atoms acting as oxidants of aqua-metal ions and complexes appears in Table 9.41. Hydrogen atoms react with each other, and with dioxygen, very rapidly ($k_2 = 2 \times 10^{10}$ dm^3 mol^{-1} s^{-1} in each case).[344] Table 9.41 shows a very wide

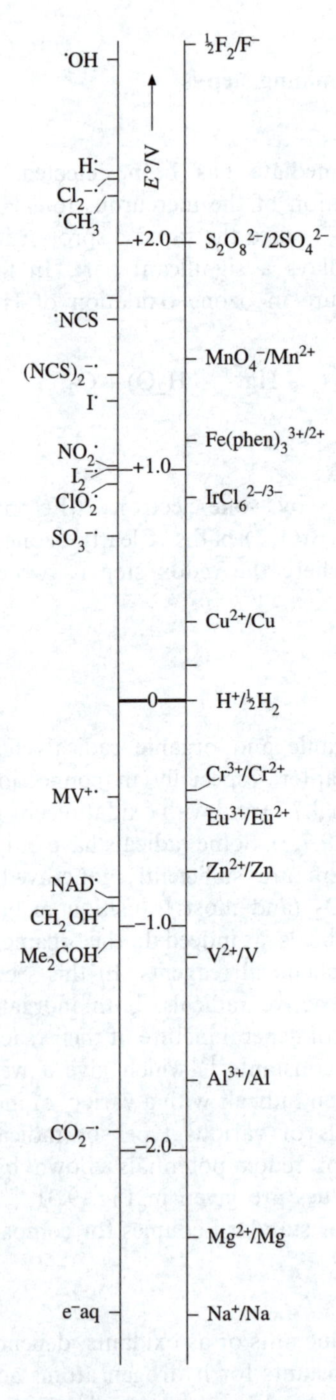

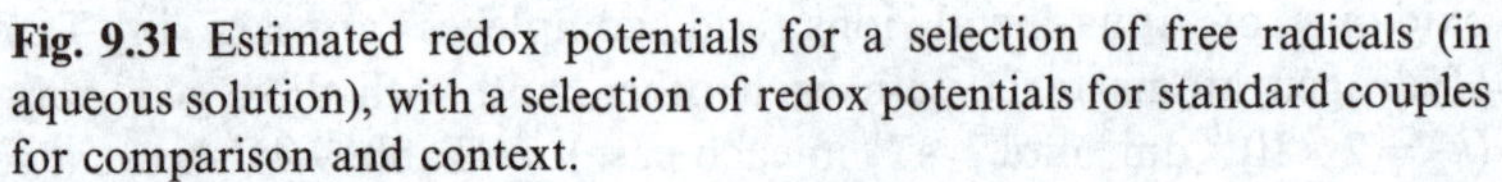
Fig. 9.31 Estimated redox potentials for a selection of free radicals (in aqueous solution), with a selection of redox potentials for standard couples for comparison and context.

Table 9.41 Second-order rate constants, k_2 ($dm^3\ mol^{-1}\ s^{-1}$), for oxidation of inorganic species by hydrogen atoms.

	k_2		k_2		k_2		k_2
Ag^+	1.1×10^{10}	Cr^{2+}	1.5×10^9			$Co(NH_3)_5I^{2+}$	3.3×10^{10}
Tl^+	1.2×10^8	Mn^{2+}	2.5×10^7			$Co(NH_3)_5Br^{2+}$	1.4×10^{10}
		Fe^{2+}	1.6×10^7	$Fe(CN)_6{}^{4-}$	4×10^7	$Co(NH_3)_5(NCS)^{2+}$	6.3×10^9
Sn^{2+}	$\sim 8 \times 10^{10}$	Co^{2+}	$< 10^3$			$Co(NH_3)_5Cl^{2+}$	1.6×10^9
		Ni^{2+}	$< 10^5$	$Ni(CN)_4{}^{2-}$	1.8×10^{10}	$Co(NH_3)_5(NO_2)^{2+}$	1.6×10^8
Zn^{2+}	$< 10^5$	Cu^{2+}	6×10^8			$Co(NH_3)_5(CN)^{2+}$	6.1×10^7
Cd^{2+}	$< 10^5$					$Co(NH_3)_5F^{2+}$	$\leq 2 \times 10^6$
Hg^{2+}	2.5×10^{10}	Ti^{3+}	4.2×10^7			$Co(NH_3)_6{}^{3+}$	$\leq 10^5$

range of reactivities, with the slowest rates corresponding to the least easily oxidized aqua-cations. The reactions with Cr^{2+}aq and Fe^{2+}aq are inner-sphere—these are the two cases where the product metal ion is sufficiently inert for convincing demonstration of ligand transfer. Oxidation of the hydroxide, formate, and thiocyanate anions takes place moderately slowly, with $k(OH^-) = 1.8 \times 10^7$, $k(HCO_2^-) = 2.3 \times 10^7$, and $k(NCS^-) = 3 \times 10^8$ $dm^3 mol^{-1} s^{-1}$. The activation energy for the reaction with hydoxide ion, which produces hydrated electrons, is 25 kJ mol^{-1}.

Hydrogen atoms react very rapidly with reducible oxoanions such as chromate, dichromate, or permanganate. Rate constants for such reactions are of the order of 10^{10} $dm^3 mol^{-1} s^{-1}$. Reduction of complexes such as $[Os(NH_3)_5Cl]^{2+}$ and $[Fe(CN)_6]^{3-}$ is almost as rapid, with $k \sim 5 \times 10^9$ $dm^3 mol^{-1} s^{-1}$, but of aqua-ions such as Fe^{3+}aq and Tl^{3+}aq rather slower (*ca* 10^8 $dm^3 mol^{-1} s^{-1}$). Faster reduction of $[Fe(CN)_6]^{3-}$ than of Fe^{3+}aq reflects stabilization of Fe^{II} by cyanide, just as very much faster oxidation of $[Ni(CN)_4]^{2-}$ than of Ni^{2+}aq (cf. Table 9.41) reflects stabilization of Ni^{III} by cyanide—an interesting example of the stabilization of lower and of higher oxidation states by the invaluable cyanide ligand! The final column of Table 9.41 shows the range of rate constants for reactions of classic pentaamminecobalt(III) complexes with hydrogen atoms. These rate constants are determined by a balance between Co–X and H–X bond strengths in what is effectively an inner-sphere mechanism.[345]

9.7.6.2 *Hydroxyl*

Hydroxyl radicals and hydrated electrons (Section 9.7.7) are primary products from pulse radiolysis or X-ray, γ-ray, or neutron irradiation of water:

$$H_2O \longrightarrow HO^{\cdot} + H^+ + e^-aq$$

Hydroxyl radicals may also be formed in oxidation of water by such strong oxidants as Co^{3+}aq. The hydroxyl radical is a strong oxidant, strong enough to remove a hydrogen from carbon in almost any organic compound. Rate constants for oxidation by hydroxyl radicals (Tables 9.42, 9.43, and 9.44[346]) almost all lie within the rather short range of 10^8–10^{10} $dm^3 mol^{-1} s^{-1}$. Considering the uncertainties in many reported values, it is unwise to attempt to identify trends. All that can be said with confidence is that these processes are very fast, often approaching the diffusion-controlled limit. Indeed there are reports of apparent rate constants for reactions with some organic substrates considerably exceeding the limiting value forecast from simple diffusion theory. Activation energies for reactions involving the hydroxyl radical are, as one would expect, very low—commonly in the range 3–7 kJ mol^{-1}. There is a marked correlation between rate constants for reactions of aqua-cations with hydroxyl radicals and with hydrogen atoms, for a number of *d*-block cations, for the borderline Zn^{2+}aq and Cd^{2+}aq, and for Tl^+aq.

Table 9.42 Second-order rate constants for oxidation of selected simple substrates by hydroxyl radicals in aqueous solution at (or very near) 25 °C and pH 7.

Reductant	k_2 ($dm^3\ mol^{-1}\ s^{-1}$)	Reductant	k_2 ($dm^3\ mol^{-1}\ s^{-1}$)
Hydroxyl	6×10^9	Tryptophan	1.3×10^{10}
Ozone	3×10^9	*N*-Methylimidazole	1.2×10^{10}
Iodide	9×10^9	Riboflavin	1.2×10^{10}
Chloride	4.3×10^9	2,2′-Bipyridyl	5.7×10^9
Carbonate	4.2×10^8	ROH	*ca* 10^9
Formate	*ca* 5×10^9	DMSO	7.0×10^9
Cyanate	5×10^7	Dopamine	6×10^9
HSO_4^-	8×10^5	MV^{2+}	2.5×10^8
HSO_3^-	5×10^9		
Thiosulphate	7.8×10^9		
HSO_5^-	2.9×10^8		

The hydroxyl radical can react by a variety of mechanisms. A number of M^{2+}aq and M^{3+}aq cations may react by hydrogen abstraction, namely

$$M^{n+}aq + {}^{\cdot}OH \longrightarrow M(H_2O)_5(OH)^{n+} + H_2O$$

Hydrogen abstraction may also occur in reactions of hydroxyl with protonated species of the $HSO_5^{-\cdot}$ and $HPO_5^{2-\cdot}$ type, and from coordinated hydroxide:[347]

$$[Fe^{III}(OH)(P_2O_7)_2]^{6-} + {}^{\cdot}OH \longrightarrow [Fe^{IV}O(P_2O_7)_2]^{6-} + H_2O.$$

It is difficult to distinguish between the hydrogen abstraction mechanism and the classical inner- and outer-sphere electron transfer mechanisms, just as it is difficult to distinguish between inner- and outer-sphere mechanisms (cf. Section 9.6.2). The inner-sphere mechanism probably operates for labile aqua-cations, such as Ag^+aq, Tl^+aq, Sn^{2+}aq, Cr^{2+}aq, Cu^{2+}aq, and Eu^{2+}aq.[348] The establishment of an activation volume of $+0.7\,cm^3\ mol^{-1}$ for reaction of Cu^{2+}aq[349] with hydroxyl radicals is consistent with such rate-determining ligand interchange. In the reaction with $PtCl_4^{2-}$, it has been proposed[350] that initial attack at the platinum leads to formation of a square pyramidal platinum(III) intermediate, which rearranges rapidly to form a second short-lived platinum(III) intermediate, which is trigonal bipyramidal (OH axial).

In contrast to these reactions involving interaction of the central metal with the hydroxyl radical, it does seem likely that the simple outer-sphere electron transfer mechanism operates for hydroxyl reactions with inert aqua-ions such as Cr^{3+}aq and for substitution-inert complexes such as hexacyanoferrate(II) and hexachloroiridate(III). There are also a number of reactions of metal complexes where the initial site of attack by hydroxyl is believed to be at a coordinated ligand. Examples include its reactions with a number of complexes of the $[Fe(bipy)_3]^{2+}$ type, where reaction ($k_2 = 2$–$17 \times 10^9\ dm^3\ mol^{-1}\ s^{-1}$) is about ten times slower at the coordinated than

Table 9.43 Second-order rate constants for oxidation of aqua-cations by hydroxyl radicals in aqueous solution at 25 °C.[a]

Reductant	k_2/dm^3 mol^{-1} s^{-1}	Reductant	k_2/dm^3 mol^{-1} s^{-1}	Reductant	k_2/dm^3 mol^{-1} s^{-1}
Ag^{+}aq	$\geq 3 \times 10^{9}$				
Tl^{+}aq	7.6×10^{9}				
Zn^{2+}aq	$< 5 \times 10^{5}$				
Cd^{2+}aq	$< 5 \times 10^{5}$				
Cr^{2+}aq	1.2×10^{10}			VO^{2+}aq	5.0×10^{8}
Mn^{2+}aq	$\geq 1.4 \times 10^{7}$				
Fe^{2+}aq	4.4×10^{8}				
Co^{2+}aq	8×10^{5}				
Ni^{2+}aq	$< 5 \times 10^{5}$	Pd^{2+}aq	5.1×10^{9}		
Cu^{2+}aq	3.5×10^{8} [b]				
		Ti^{3+}aq	3.0×10^{9}		
		Cr^{3+}aq	3.2×10^{8}		
Sm^{2+}aq	6.2×10^{9}	Ce^{3+}aq	$30/7.2 \times 10^{7}$		
Eu^{2+}aq	1.3×10^{9}	Pr^{3+}aq	4×10^{6}	NpO_2^{+}aq	4.5×10^{7}
Tm^{2+}aq	7×10^{9}			PuO_2^{+}aq	*ca* 4×10^{7} [d]
Yb^{2+}aq	3.2×10^{9}				
		U^{3+}aq	4.1×10^{8}	U^{4+}aq	8.6×10^{8}
		Pu^{3+}aq	1.8×10^{9} [c]	Pu^{4+}aq	*ca* 1×10^{8} [d]

[a] Data for the actinides from C. Lierse, K. H. Schmidt, and J. C. Sullivan, *Radiochim. Acta*, 1988, **44/45**, 71.
[b] Or *ca* 10^{10} (P. V. Bernhardt, G. A. Lawrance, and D. F. Sangster, *Polyhedron*, 1991, **10**, 1373).
[c] Predicted values for Np^{3+}aq, Am^{3+}aq, Bk^{3+}aq, and Cf^{3+}aq are all about 5×10^{8}.
[d] Predicted values.

Table 9.44 Second-order rate constants for oxidation of inorganic complexes by hydroxyl radicals in aqueous solution at (or very near) 25 °C and pH 7.

Reductant	k_2/dm^3 mol^{-1} s^{-1}	
$[Fe(CN)_6]^{4-}$	1.1×10^{10}	M^{II} variation; electron transfer
$[Ru(CN)_6]^{4-}$	5.7×10^9	
$[Os(CN)_6]^{4-}$	10×10^9	
$[Co(bipy)_3]^{2+}$	2.7×10^9	M^{II} variation; attack at ligand
$[Ru(bipy)_3]^{2+}$	8×10^9	
$[Os(bipy)_3]^{2+}$	*ca* 10^{10}	
$[Fe(CN)_6]^{4-}$	1.1×10^{10}	↓ mechanism change from electron transfer to attack at ligand
$[Fe(CN)_4(bipy)]^{2-}$	9.7×10^9	
$[Fe(CN)_2(bipy)_2]$	8.4×10^9	
$[Fe(bipy)_3]^{2+}$	4.1 or 80×10^8	
$[Fe(4,4'\text{-}Me_2bipy)_3]^{2+}$	17×10^9—faster than unsubstituted complex	
$[Co(ida)_2]^{2-}$	5.8×10^8	
trans-$[Cr(cyclam)(OH)_2]^+$	2.0×10^{10}	
$[Ni(cyclam)]^{2+}$	2×10^9	
$[Ni(NNNN)]^{2+}$ [a]	9.7–9.9$\times 10^9$	
$[IrCl_6]^{3-}$	1.3×10^{10}	
$[Au(CN)_2]^-$	4.7×10^9	

[a] NNNN = tetraza garland or macrocyclic ligands (details in P. V. Bernhardt, G. A. Lawrance, and D. F. Sangster, *Inorg. Chem.*, 1988, **27**, 4055).

at the free ligand,[351] and one path (there is a concurrent outer-sphere pathway here, *v.s.*) with $IrCl_6^{2-}$.[349] Attack at the coordinated organic ligand is also indicated for *trans*-$[Cr(cyclam)(OH)_2]^+$ and for *trans*-$[Cr(cyclam)F_2]^+$, since both react at the same very fast rate.[352] The final product from these chromium(III) complexes is a μ-peroxodichromium(III) species, formed via the intermediacy of chromium(IV).

The pK of the hydroxyl radical is 11.9, so that in alkaline media it is in the form O^-. This conjugate base form has been very much less studied but can react with comparable rapidity—thus rate constants for reaction with iodide and with hypoiodite are around 2×10^9 dm^3 mol^{-1} s^{-1}.[353] On the other hand, O^- reacts much more rapidly than hydroxyl with iodate; the presence of small amounts of the former vitiated earlier determinations of rate constants for the iodate/hydroxyl reaction.[354]

9.7.6.3 Superoxide

The hydroperoxyl radical $HO_2^{\cdot}$ can be generated by induced decomposition of hydrogen peroxide or by pulse radiolysis of acidic oxygen-saturated aqueous solutions. The superoxide radical is its conjugate base:

$$HO_2^{\cdot} \rightleftharpoons H^+ + O_2^{\cdot -}$$

The pK_a value for $HO_2^{\cdot}$ is between 4.8 and 4.9, so in neutral or alkaline aqueous media it is the anionic superoxide radical ion which is the relevant form. Redox potentials for O_2^- to O_2^{2-} and O_2^- to O_2 are such that O_2^- can act as oxidant or reductant, depending on substrate and circumstances. In organic chemistry it acts primarily as an aggressive nucleophile, with subsequent electron transfer.

Rate constants have been established for reaction of superoxide with more than three hundred organic and inorganic species,[355] ranging from simple compounds such as ozone to complexes such as cobalt(II) sepulchrates and to bioinorganic species such as copper-zinc superoxide dismutase. These rate constants cover a very wide range, from nearly 10^{10} dm^3 mol^{-1} s^{-1} for reaction with Cu^{2+}aq, about 10^9 dm^3 mol^{-1} s^{-1} for reaction with HOBr, ozone, or benzoquinone, and about 10^8 dm^3 mol^{-1} s^{-1} for reaction with $[Cu(arginine)]^+$, through 10^5 to 10^3 for a number of organic and bioorganic compounds, to 6×10^3 dm^3 mol^{-1} s^{-1} for ferrate(VI) and as low as 10^2 to 10^1 dm^3 mol^{-1} s^{-1} for phosphate esters. Superoxide can react very rapidly with bioinorganic compounds, for example with k_2 about 2×10^9 dm^3 mol^{-1} s^{-1} for CuZn superoxide dismutase, which is probably just as well in view of the need for the rapid removal of this very harmful radical from the human body. This rapid reaction is compared with rate constants for the reaction of superoxide radicals with other copper complexes in Table 9.45.[356] Most of these reactions are fast, apart from that with plastocyanin. The one reaction of superoxide which may be exceedingly slow is its self-exchange reaction with the dioxygen molecule. Many estimates for $k_{11}(O_2/O_2^-)$ have been made through Marcus cross-relation analyses. These range from 10^{-7} to 10^6 dm^3 mol^{-1} s^{-1}! Complications arising from the effects of very heavy hydration are often blamed for such uncertainty. An intermediate value between 1 and 10 dm^3 mol^{-1} s^{-1} seems to be favoured but is still slow for a reaction involving superoxide. Interestingly, this value is very similar to that estimated for O_3/O_3^- self-exchange.

Ligand effects on reactivity may best be illustrated from cobalt(III) chemistry. Table 9.45 shows that the nature of the ligand has a major effect, with much faster reaction when the coordinated ligand permits easy electron transfer between the metal and the periphery of the complex (diimine ligands).

9.7.6.4 Other inorganic radicals

There are two other important groups of inorganic radicals, one comprising halogen atoms and dihalogen anions, the other a variety of oxoanion-related species. Table 9.46 summarizes kinetic data for oxidations by dihalogen and dithiocyanogen anion radicals. Relative reactivities for these radicals are

$$k(Cl_2^{-\cdot}) > k(Br_2^{-\cdot}) \sim k((NCS)_2^{-\cdot}) > k(I_2^{-\cdot})$$

Table 9.45 Rate constants ($dm^3\ mol^{-1}\ s^{-1}$) for reactions of superoxide radicals with copper (II) and cobalt(III) species.

Cu^{2+}aq	*ca* 4×10^9	$[Co(phen)_3]^{3+}$	$> 10^4$
CuZnSOD[a]	2.3×10^9	$[Co(terpy)_2]^{3+}$	7×10^4
Cu^{2+}-ammines	2×10^8 to 2×10^9	$[Co(NH_3)_6]^{3+}$	31
Cu^{2+}-formates	$2–8 \times 10^8$	$[Co(en)_3]^{3+}$	24
Cu^{2+}-aa[b]/peptides	10^5 to 5×10^8 [c]	$[Co(chxn)_3]^{3+}$ [d]	16
Plastocyanin	$< 10^6$	$[Co(sep)]^{3+}$ [e]	*ca* 1

[a] SOD = superoxide dismutase.
[b] aa = amino acids.
[c] S. Goldstein, G. Czapski, and D. Meyerstein, *J. Am. Chem. Soc.*, 1990, **112**, 6489; 1×10^8 for $[Cu(arginine)]^+$, but 3.3×10^9 for $[Cu(arginineH)]^{2+}$—presumably due to more favourable electrostatics for the latter.
[d] chxn = 1,2-diaminocyclohexane.
[e] The encapsulated complex cobalt(III) sepulchrate (formula **50**); reaction with $[Co(sep)]^{2+}$ is very much faster (5×10^7—A. Bakac, J. H. Espenson, I. I. Creaser, and A. M. Sargeson, *J. Am. Chem. Soc.*, 1983, **105**, 7624).

Table 9.46 Rate constants ($10^{-8}k/dm^3\ mol^{-1}\ s^{-1}$) for oxidation by dihalogen and related radicals (aqueous solution, 25 °C).[a]

	$Cl_2^{-\cdot}$	$Br_2^{-\cdot}$	$I_2^{-\cdot}$	$(NCS)_2^{-\cdot}$
Tryptophan		14		3
NCS^-	14	16		
SO_3^{2-}		2.2		2.0
N_3^-	5	1.2		
HSO_3^-		0.63		0.034
NO_2^-		0.17		0.030
BPh_4^-		0.2		0.02
V^{2+}aq	20	15	1.4	
Cr^{2+}aq	24	19	15	
Mn^{2+}aq	0.085	0.06		
Fe^{2+}aq	*ca* 0.1[b]	0.03	0.04	
Co^{2+}aq	0.014			
U^{3+}aq	42	34	12	
$[Co(sep)]^{3+}$		140	50	
$[Fe(CN)_6]^{4-}$	1.8	0.27	0.024	0.24
$[Mo(CN)_8]^{4-}$	0.6	0.05		0.035
$[W(CN)_8]^{4-}$	3.0	0.58	0.037	0.71

[a] Data from G. S. Laurence and A. T. Thornton, *J. Chem. Soc., Dalton Trans.*, 1974, 1142; L. C. T. Shoute, Z. B. Alfassi, P. Neta, and R. E. Huie, *J. Phys. Chem.*, 1991, **95**, 3238; S. Padmaja, P. Neta, and R. E. Huie, *Int. J. Chem. Kinet.*, 1993, **25**, 445.
[b] For details see text.

Interestingly, $Cl_2^{-\cdot}$ reacts very much more slowly than chlorine atoms, $Cl^{\cdot}$. The latter reacts with rate constants in the range 10^8–10^{10} $dm^3\,mol^{-1}\,s^{-1}$, of the order of a thousand times faster than $Cl_2^{-\cdot}$, with Fe^{2+}aq and with a variety of organic substrates.[357] In relation to the reaction of azide with $Br_2^{-\cdot}$, it has been suggested that the $Br_2^{-\cdot} \rightleftharpoons Br^- + Br^{\cdot}$ reaction might make a significant contribution to the overall mechanism,[358] which might complicate kinetics and interpretation in some other $X_2^{-\cdot}$ reactions. Application of the Marcus free-energy analysis to appropriate self-exchange and cross-reactions involving a variety of inorganic couples has led to estimates of rate constants of 9×10^4 $dm^3\,mol^{-1}\,s^{-1}$ for both $Cl_2^{-\cdot}/Cl_2$ and $I_2^{-\cdot}/I_2$, 85 $dm^3\,mol^{-1}\,s^{-1}$ for $Br_2^{-\cdot}/Br_2$, and $\sim 4 \times 10^4$ $dm^3\,mol^{-1}\,s^{-1}$ for $N_3^{-\cdot}/N_3^-$ self-exchange (k_{11} for $(NCS)_2^{-\cdot}/(NCS)_2^-$ does not seem to be available).

The main interest in the dihalide radical reactions resides in their mechanisms of oxidation of aqua-cations. These have been established through comparisons of oxidation rate constants with those for water exchange under comparable conditions. Thus for V^{2+}aq, where reaction with dihalide-anion radicals is very much faster than water exchange, the outer-sphere mechanism for electron transfer must surely operate—as it must also for oxidation of the substitution-inert hexa- and octacyano-metallates. On the other hand, inner-sphere electron transfer seems much more likely for reactions of dihalide-anion radicals with Mn^{2+}aq, Co^{2+}aq, and, for $I_2^{-\cdot}$ and $Br_2^{-\cdot}$, Cr^{2+}aq. It has been suggested that the reactions of Cr^{2+}aq and of Fe^{2+}aq with $Cl_2^{-\cdot}$are at the outer-sphere/inner-sphere border, and indeed for Fe^{2+}aq the observed rate constant has been dissected into the components for the two pathways (outer-sphere electron transfer about twice as fast as inner-sphere). There is little information on ligand effects on reactivity. Comparison of $[Fe(CN)_6]^{4-}$ with Fe^{2+}aq reveals a surprisingly small effect on replacing coordinated water by cyanide, though admittedly there is a similarly small difference between rate constants for reactions of these species with hydroxyl radicals. However, Co^{2+} encapsulated in the sepulchrate ligand is oxidized very much more readily than Co^{2+}aq. In this latter case the very much greater stability of $[Co(sep)]^{3+}$ than of Co^{3+}aq is presumably the explanation.

Turning to oxoanion-related radicals, there is a lot of kinetic information on $SO_2^{-\cdot}$, $SO_4^{-\cdot}$,[359] $CO_3^{-\cdot}$,[360] $CO_2^{-\cdot}$, and $NO_3^{\cdot}$.[361] The $SO_2^{-\cdot}$ radical featured in Section 9.7.4.1 in connection with dithionite reduction mechanisms—indeed, most rate constants for reactions with $SO_2^{-\cdot}$ have been derived from kinetic studies of dithionite reductions. The $SO_2^{-\cdot}$ radical can also be generated electrochemically from sulphur dioxide in such solvents as acetonitrile or dimethylformamide.[362]

Unfortunately, different investigators have used different sets of substrates for their kinetic studies of different radicals, so direct comparison of reactivities and reactivity trends is very difficult. Such indirect comparisons as can be hazarded suggest that that they all have rather similar reactivities. This is certainly the case for the rapid reactions of $SO_4^{-\cdot}$ and $NO_3^{\cdot}$ with a

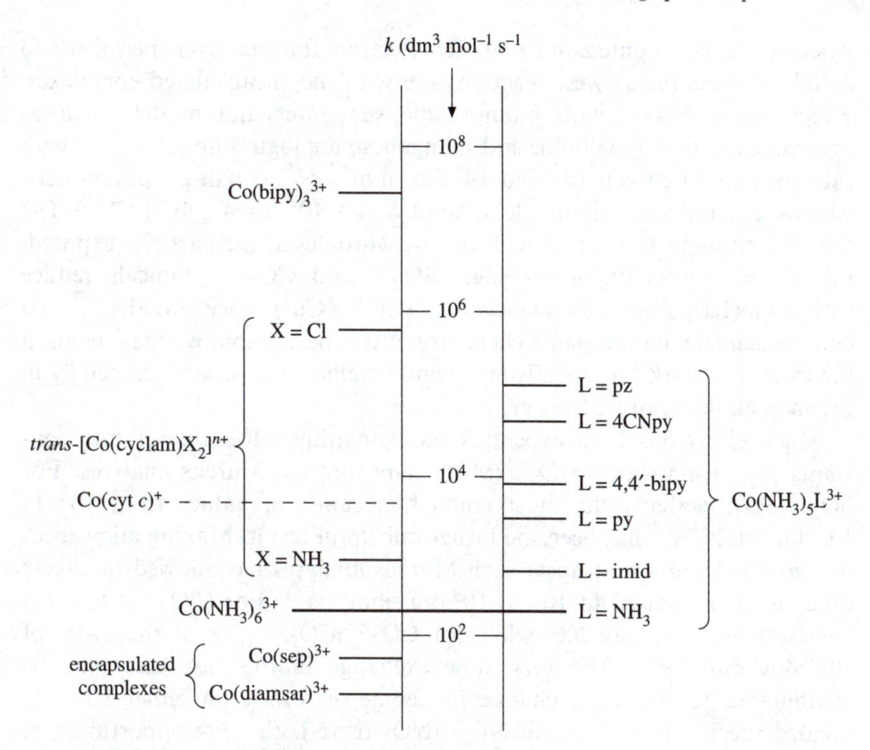

Fig. 9.32 Rate constants for $SO_2^{-\cdot}$ reduction of cobalt(III) complexes.

number of lanthanide and actinide aqua-cations and oxo-cations. Also they, and both $CO_2^{-\cdot}$ and $NO_3^{\cdot}$, react much more slowly with Co^{2+}aq and Ni^{2+}aq than with most other aqua-cations. $CO_2^{-\cdot}$ reacts very rapidly with Cr^{2+}aq and Cu^{2+}aq, as does $SO_4^{-\cdot}$ with Fe^{2+}aq and Ce^{3+}aq,[363] in all cases to give a metal-radical complex as initial product. These reactions are facile as water replacement in Fe^{2+}aq and Ce^{3+}aq is fast, and in Cr^{2+}aq and Cu^{2+}aq extremely fast (Chapter 7). Both $CO_2^{-\cdot}$ and $SO_4^{-\cdot}$ react very rapidly with $[Ru(bipy)_3]^{2+}$, presumably by outer-sphere electron transfer since the product from the $SO_4^{-\cdot}$ reaction is simply SO_4^{2-}.[364]

Ligand effects on reactivity have been established for $SO_2^{-\cdot}$ reduction of several pentacyanoferrates $[Fe^{III}(CN)_5L]^{n-}$ and of a number of cobalt(III) complexes.[365] As the rate constants for the pentacyanoferrates(III) are all similar, at just over 10^8 dm^3 mol^{-1} s^{-1}, it is more fruitful to consider the cobalt(III) complexes (Fig. 9.32). Rate constants range from 117 for $[Co(NH_3)_6]^{3+}$ through, for example, 188 for $[Co(NH_3)_5(imid)]^{3+}$ and 7000 for $[Co(NH_3)_5(4,4'\text{-bipy})]^{3+}$ to 10^5 dm^3 mol^{-1} s^{-1} for $[Co(NH_3)_5(pz)]^{3+}$. The encapsulated complexes $[Co(sep)]^{3+}$ and $[Co(diamsar)]^{3+}$ react slowly, with rate constants of 70 and 26 dm^3 mol^{-1} s^{-1}, respectively; $[Co(bipy)_3]^{3+}$ reacts rapidly, with $k_2 = 2 \times 10^7$ dm^3 mol^{-1} s^{-1}. As for the dihalide-radical anions discussed above, the most rapid reaction is with the diimine complex, where there is a conjugated path for electron transfer from periphery to metal, whereas the slowest reactions are with the encapsulated complexes. Reactions of $SO_2^{-\cdot}$ with bioinorganic substrates and models such as cytochromes, metmyoglobin, and manganese analogues are also fast, with rate constants between 10^4 and 10^8 dm^3 mol^{-1} s^{-1}; cobalt-porphyrin derivatives are reduced slightly less rapidly (4×10^3 to 4×10^4).[214,366] The $SO_2^{-\cdot}$ is thought to interact with the cytochromes at the partially exposed, or at least accessible, haem edge. $SO_2^{-\cdot}$ and $CO_2^{-\cdot}$ radicals reduce ruthenium(III)-copper(II)-stellacyanin ($Cu^{II} \rightarrow Cu^{I}$) very rapidly.[367] To complement the bioinorganic chemistry in this paragraph, we may mention the extensive work on radicals in organometallic chemistry,[368] especially in organometallic photochemistry.[369]

Much effort has been expended on estimating self-exchange rate constants (k_{11}) for various $MO_2^{-\cdot}/MO_2$ pairs through Marcus analyses. For $SO_2^{-\cdot}/SO_2$, perhaps the most studied, a range of values from 10^2 to 10^9 dm^3 mol^{-1} s^{-1} has been the rather unhelpful result. Making allowances for various factors that upset such Marcus analyses has allowed the likely range to be narrowed to 10^3 to 10^4 dm^3 mol^{-1} s^{-1}. For $ClO_2^{-\cdot}/ClO_2$, k_{11} has been estimated as 200, while for $CO_2^{-\cdot}/CO_2$, k_{11} is of the order of 10^{-6} dm^3 mol^{-1} s^{-1}. The very slow exchange in this last case may be attributable to the large change in geometry forced on either strongly bonded species. In the case of NO_2^-/NO_2 there is the rare opportunity of measuring k_{11} directly, since both components are stable entities. Such a direct measurement, by competition kinetics (NO_2 plus H_2O versus NO_2), gave a value of 580 dm^3 mol^{-1} s^{-1}, well above Marcus-derivation estimates of 0.01 to 5 dm^3 mol^{-1} s^{-1}. The suggestion here is that there may well be an inner-sphere path for electron transfer which is considerably easier than the outer-sphere path assumed in the Marcus calculations.[370] Of course this possibility also exists for the other $MO_2^{-\cdot}/MO_2$.

Finally, we shall make a brief mention of some low oxidation state aqua-ions, which are only arguably radicals but are conveniently mentioned here in view of their method of generation and their chemistry. We refer to the M^+aq cations, where M = e.g. Zn, Cd, Mn, Co, or Ni.[371] These species are generated by pulse radiolysis of the respective M^{2+}aq ions. They decay rapidly in solution by $M^+aq + M^+aq$, but if a cation M'^{2+}aq is present, they may also react by $M^+aq + M'^{2+}aq$, for which reactions some kinetic data are available. Both Zn^+aq and Cd^+aq reduce $[Cr(phen)_3]^{3+}$ with a rate constant of 1.7×10^9 dm^3 mol^{-1} s^{-1} (at 295 K), necessarily by outer-sphere electron transfer.[372] As always for phen complexes, electron transfer can be very fast as the delocalized nature of the ligands provides a facile path linking the central metal to the periphery of the complex. These M^+aq cations may also be useful reagents. An example of relevance to Section 9.4.4.1 above is the use of Cd^+aq to effect one-electron reduction of binuclear cobalt(III) complexes to give mixed-valence $Co^{III}Co^{II}$ complexes.[373]

Table 9.47 Rate constants, k_2 ($dm^3\ mol^{-1}\ s^{-1}$), for reactions with organic radicals.

	k_2		k_2
$Cr^{2+}aq + H$	1.5×10^9	$CH_2CH_3 + IrCl_6^{2-}$	3×10^9
CH_2CHO	3.5×10^8	$Fe(phen)_3^{3+}$	1×10^9
CH_2CO_2H	2.5×10^8	$Fe(bipy)_3^{3+}$	9×10^8
CH_3	2.2×10^8	$Cr(bipy)_3^{3+}$	2×10^7
CH_2CH_3	1.9×10^8		
CH_2OH	1.6×10^8	$CMe_2OH + Co(NH_3)_5Br^{2+}$	3.0×10^8
CH_2CH	8.5×10^7	$Co(NH_3)_5Cl^{2+}$	4.0×10^7
cyclo-C_5H_9	8×10^7	$Co(sep)^{3+}$	4.6×10^6
CMe_2OH	5.1×10^7	$Co(NH_3)_6^{3+}$	4.3×10^5
$CHMeOEt$	3.4×10^7	$Co(en)_3^{3+}$	1.7×10^5
$Ni(cyclam)^{2+} + CH_3$	1.1×10^9	$CH_2CH_3 + Co(NH_3)_5(N_3)^{2+}$	3.7×10^7
CH_2CH_3	2.3×10^8	$Co(NH_3)_5(SCN)^{2+}$	1.4×10^7
$CH_2CH_2CH_3$	1.7×10^8	$Co(NH_3)_5Br^{2+}$	3×10^6 [a]
cyclo-C_5H_9	1.0×10^7	$Co(NH_3)_5Cl^{2+}$	2×10^4
$CHCMe_2$	6.2×10^6		

[a] $Ru(NH_3)_5Br^{2+}$: 2×10^7.

9.7.6.5 *Organic radicals*

There is an extensive literature on the kinetics and mechanisms of reactions of inorganic ions and complexes with organic radicals.[202,356] These can act as oxidants or reductants—thus the CMe_2OH radical can both oxidize Eu^{2+}aq and reduce Eu^{3+}aq.[374] Table 9.47 gives several trends of rate constants for oxidation and for reduction by organic radicals, for variation of radical in the left-hand column, for variation of inorganic substrate for a given radical in the right-hand column.[375] There are many more rate constants in the literature for Cr^{2+}aq, series of values for, *inter alia*, Cu^{2+}aq[376] and U^{3+}aq,[377] and values for a number of other aqua-cations. A number of mechanisms are possible for reactions between aqua-cations and organic radicals. Mechanisms may vary according to the precise nature of the reactants, and indeed there often seem to be concurrent paths in this type of reaction. There may be hydrogen abstraction, as demonstrated for a number of organometallic compounds (e.g. metal hydride compounds such as $CoH(CO)_4$ and $FeH(CO)_2(\eta^5\text{-}C_5H_5)$[378]). This pathway has been suggested to operate for Eu^{2+}aq and V^{2+}aq. Otherwise there are several variants of the normal inner-sphere and outer-sphere mechanisms, often with some doubt as to whether transient intermediates, MR^{n+}aq, may be involved. Inner-sphere pathways seem likely for labile Cr^{2+}aq, Cu^{2+}aq, and U^{3+}aq; a variant of the standard inner-sphere path has been proposed for the less labile V^{2+}aq (see Section 9.6.3 above). Activation volumes for replacement of coordinated water from several transition metal binary and ternary aqua complexes follow a similar pattern to those for water exchange and complex formation (Chapter 7). Reported activation volumes range from between −4 and +3 $cm^3\,mol^{-1}$ for Mn^{II} and Fe^{II}, through +2 to +6 for Co^{II} to +4 for Ni^{II}, as $Ni(cyclam)^{2+}aq$, strongly suggesting that ligand interchange plays a key role, in other words that the inner-sphere mechanism is operative.[379]

The inner-sphere mechanism is also believed, on the basis of product evidence, to operate for reactions of a number of pentaammine-cobalt(III) complexes with the ethyl radical. The $^{\cdot}CMe_2OH$ radical is believed to reduce the metal ion directly when it reacts with complexes of the $[Co(NH_3)_5X]^{2+}$ type, but to reduce chromium complexes $[Cr(H_2O)_5X]^{2+}$ by initial reduction of the ligand, followed by rapid but kinetically distinct intramolecular electron transfer.[380] In the case of copper(II) complexes, oxidation to a copper(III) species may be followed by reductive elimination to give copper(I) as final product. The kinetics of reactions of copper(II) complexes with alkyl radicals may be complicated—in the case of $[Cu(gly)_2]$ reacting with methyl radicals, decomposition of the methyl-copper(III)-glycinate intermediate gives methane and ethane, with some concomitant intramolecular ligand oxidation. Interpretation of the kinetics of the often very fast reactions of complexes of the $[Fe(phen)_3]^{3+}$ type[381] is also not straightforward, as there may be attack at the coordinated ligand in parallel with outer-sphere electron transfer.

9.7.7 Reduction by solvated electrons

Solvated electrons can readily be generated in liquid ammonia; hydrated electrons[382] can be generated, though with much more difficulty, in aqueous solution. They are very reactive species, though their lifetime is sufficiently long for their visible absorption spectrum to have been established (λ_{max} = 720 nm) and for the kinetics of a large number of their reactions to have been monitored by appropriate fast-reaction techniques.[343] Activation enthalpies are generally low, in the region of 12–17 $kJ\,mol^{-1}$ (higher, at 20 $kJ\,mol^{-1}$, for reaction with water[383] and with the hydroxyl radical[384]), though non-Arrhenius behaviour has been described for a number of their reactions.[385] The redox potential of the hydrated electron has been estimated as −2.7 V, which is comparable with those for the alkali metals (cf. Fig. 9.31). A Livingston diagram (Section 8.3.1.3) summarizing ionic strength effects on rate constants for reduction of H^+, Ag^+, O_2, NO_2^-, and $[KFe(CN)_6]^{2-}$ by hydrated electrons is nicely consistent with their 1− charge.[386]

9.7.7.1 Simple compounds; oxoanions

Table 9.48 includes some rate constants for reduction of simple inorganic species by hydrated electrons. These, and rate constants for reaction with other simple species such as a hydrogen atom or a second electron, or compounds such as NO, CS_2, CCl_4, and I_2, are around the value expected

Table 9.48 Second-order rate constants ($10^{-8}k_2$/dm^3 mol^{-1} s^{-1}) for some reductions by hydrated electrons at 25 °C.

	$10^{-8}k_2$		$10^{-8}k_2$		$10^{-8}k_2$		$10^{-8}k_2$
O_3	360	BrO_3^-	21	$S_2O_8^{2-}$	80	$4\text{-}O_2NC_6H_4CH_2CO_2^-$	180
H^+	210	IO_3^-	77	$S_2O_3^{2-}$	1		
O_2	150	IO_4^-	111	$HO^{\cdot}$	300		
H_2O_2	120	ReO_4^-	130	$HSO_5^{-\cdot}$	84	ferrocenyl$(CH_2)_nCO_2^-$ [a]	~300
N_2O	91	TcO_4^-	130	$HPO_5^{2-\cdot}$	4		

[a] $n = 2, 3$, or 4.

for diffusion-controlled reactions, namely ~ 10^{10} dm^3 mol^{-1} s^{-1}. However, for simple anions electrostatic repulsion is beginning to have a significant effect by the time a charge of 2 (thiosulphate; the $HPO_5^{2-\cdot}$ radical) has been reached. Table 9.48 also shows that rate constants for large organic and organometallic carboxylates still approximate to diffusion-control. Two reactions which are much slower than diffusion-controlled are those with ammonia and with water. Details of these, and of some of the exceptionally fast elementary processes involving hydrated electrons and water-derived species, have been elucidated.[344,387]

9.7.7.2 *Aqua-ions*

Table 9.49 lists rate constants, mainly culled from standard textbooks,[388] for reactions of hydrated electrons with aqua-cations. The reactions with alkali metal cations and with Mg^{2+}aq are extremely slow, consistent with the great resistance of these ions to reduction. Once formed, the alkali metal atoms react very quickly ($k_2 = ca\ 1.5 \times 10^9$ dm^3 mol^{-1} s^{-1} for Na, Cs) with solvent water:[389]

$$2\text{alk}^0 + 2H_2O \longrightarrow 2\text{alk}^+ + H_2 + 2OH^-$$

Rate constants for *d*- and *f*-block aqua-cations reflect characteristics of their particular electron configurations. Many of the lanthanide cations react relatively slowly, reflecting their reluctance to adopt the 2+ state in aqueous solution, but Sm^{3+}aq, Eu^{3+}aq, and Yb^{3+}aq are eager to react and thus attain, or approach, the half-filled or filled f^7 or f^{14} configurations. The rate constants for reduction of these three cations correlate with the respective $Ln^{3+/2+}$ redox potentials. Rate constants for reduction of *d*-block aqua-cations also reflect electron configurations and redox potentials. Rate constants for reduction of Cd^{2+}aq by hydrated electrons have been determined over the wide temperature range of 20–250 °C, enabling an estimate of 15.5 kJ mol^{-1} to be made for this reaction.[390] This value is within the normal range mentioned above.

Table 9.49 Second-order rate constants ($10^{-8}k_2/dm^3\ mol^{-1}\ s^{-1}$) for reduction of aqua-cations by hydrated electrons at 25 °C.

Na^+	K^+		Cs^+			Ag^+		Tl^+	
0.0002	0.0003		0.0002			360		300	
Cr^{2+}	Mn^{2+}		Fe^{2+}	Co^{2+}		Ni^{2+}		Cu^{2+}	Zn^{2+}
420	0.8		3.5	120		220		300	15
									Cd^{2+}
									520
	Mg^{2+}				Sn^{2+}		Pb^{2+}		
	< 0.001				34[a]		390		
Al^{3+}	In^{3+}		Cr^{3+}						
20	560		600						
Y^{3+}									
2									
La^{3+}	Pr^{3+}	Nd^{3+}	Sm^{3+}	Eu^{3+}	Gd^{3+}	Tb^{3+}	Er^{3+}	Tm^{3+}	Yb^{3+}
3.4	2.9	5.9	250	610	5.5	0.17	0.7	30	430
Am^{3+}	Bk^{3+}	Cf^{3+}							
16	11	30							
Th^{4+}									
190									

[a] At pH 11.

9.7.7.3 *Complexes; bioinorganic species*

Table 9.50[343,372] shows a selection of rate constants for reduction of a number of chromium(III) complexes by hydrated electrons. Such reduction is favoured by the fact that chromium(II) is generally a fairly stable, if strongly reducing, state in aqueous media. The main feature of Table 9.50 is the decrease in rate constant down each column as the charge on the complex goes from 3+ towards 3−. Table 9.51[391] contains a collection of rate constants for reduction of assorted metal complexes by hydrated electrons, to illustrate reactivity trends. Comparisons of values for anionic and cationic complexes give further illustrations of the importance of

Table 9.50 Second-order rate constants ($10^{-8}k/dm^3\ mol^{-1}\ s^{-1}$) for the reduction of chromium(III) complexes by hydrated electrons in aqueous solution at 25 °C.

$Cr^{3+}aq$	600	$Cr(phen)_3{}^{3+}$	770
$Cr(en)_3{}^{3+}$	530	$Cr(phen)_2(ox)^+$	380
$Cr(en)_2Cl_2{}^+$	710	$Cr(phen)(ox)_2{}^-$	160
$Cr(edta)^-$	260		
$Cr(CN)_6{}^{3-}$	150		
$CrF_6{}^{3-}$	140		

Table 9.51 Second-order rate constants ($10^{-8}k_2$/dm^3 mol^{-1} s^{-1}) for reduction of complexes by hydrated electrons in aqueous solution at 25 °C.

Edta complexes		Cyano-complexes		Ammines		Other complexes	
Metal(I) complexes:							
$Ag(edta)^{3-}$	16	$Ag(CN)_2^-$	15	$Ag(NH_3)_2^+$	800		
Metal(II) complexes—square-planar:							
		$Pd(CN)_4^{2-}$	20			$PdCl_4^{2-}$	120
		$Pt(CN)_4^{2-}$	32			$PtCl_4^{2-}$	120
Metal(II) complexes—tetrahedral:							
		$Zn(CN)_4^{2-}$	1.8				
		$Cd(CN)_4^{2-}$	1.4				
Metal(II) complexes—octahedral:							
$Mn(edta)^{2-}$	< 0.02	$Mn(CN)_6^{4-}$	50				
$Co(edta)^{2-}$	5.1	$Fe(CN)_6^{4-}$	< 0.001				
$Ni(edta)^{2-}$	1.0	$Ru(CN)_6^{4-}$	very slow			$Ni(NNNN)^{2+}$ [a]	
$Zn(edta)^{2-}$	< 0.02						
$Cd(edta)^{2-}$	3.9					$Cu(sar)^{2+}$	700
Metal(III) complexes:							
$Al(edta)^-$	0.3						
$Cr(edta)^-$	260	$Cr(CN)_6^{3-}$	150			[b]	
		$Fe(CN)_6^{3-}$	30				
$Co(edta)^-$	290	$Co(CN)_6^{3-}$	36	$Co(NH_3)_6^{3+}$	900	$Co(NO_2)_6^{3-}$	580
				$Ru(NH_3)_6^{3+}$	700		

[a] Details of the various garland and macrocyclic ligands NNNN, and of rate constants, may be found in P. V. Bernhardt, G. A. Lawrence, and D. F. Sangster, *Inorg. Chem.*, 1988, **27**, 4055.
[b] See Table 9.50 for a range of chromium(III) complexes.

charge—see for instance the ammine versus cyanide or edta comparisons for silver(I) and cobalt(III). Diffusion-controlled rate constants have been estimated as 7.2×10^{10} for a 3+ metal complex, 1.6×10^9 for a 3− complex. Despite the dominant importance of charge, other ligand properties are of significance. Thus the hydrated electron is keen to react with chloride ligands, as shown by the palladium(II) and platinum(II) entries in Table 9.51, as well as the $Cr(en)_3^{3+}$ and $Cr(en)_2Cl_2^+$ entries in Table 9.44. The first two columns of Table 9.51 also give some idea of how reactivities vary with the nature of the metal in series of complexes with a common ligand. The main factor is the stability of the lower oxidation state produced—in the particular examples in Table 9.45 the trends for the two ligands do not match well because cyanide can be much more effective in stabilizing lower oxidation states of metals, edta higher.

For all the complexes in Table 9.50 and the ammine complexes in Table

9.51, rate constants are $> 10^{10}$ dm^3 mol^{-1} s^{-1}. Rate constants for a wide variety of other cationic metal complexes, for example cobalt(III)-ammines, nickel(II)-tetraza-macrocycles, encapsulated complexes of copper(II) and of rhodium(III), diimine complexes of ruthenium(II) and of rhodium(III), and binuclear ruthenium and cobalt complexes, are also all $> 10^{10}$ dm^3 mol^{-1} s^{-1}. Presumably with these very high rate constants the electron interacts first with one of the ligands before transferring to the metal centre. It is not usually possible to demonstrate this, but for the special case of ruthenium-polypyridyl (LL) complexes good evidence has been obtained for the initial production of radical-ligand species of the $[Ru(LL)_2(LL^{\cdot})]^{n+}$ type postulated in other contexts.[392]

Rate constants for reactions of a number of bioinorganic species have been measured. For methaemoglobin, iron(II) protoporphyrin, and two ferredoxins rate constants are between 60 and 340×10^8 dm^3 mol^{-1} s^{-1}, maintaining the consistent picture of very fast rates wherever the lower oxidation state is not too unstable.

9.7.7.4 Envoi

The preceding sections have dealt with hydrated electrons, species which may be considered to be in the realm of physics, or at least of physical chemistry. We have mentioned their interactions with ordinary inorganic species, with bioinorganic substrates, and with organometallic compounds. A little earlier we considered, albeit very briefly, reactions involving organic radicals. The closing pages of this chapter have thus covered a wide range of chemistry, centred on classical inorganic compounds, complexes, and ions, reflecting, we hope, the wide coverage of this book as a whole.

References

1. R. D. Cannon, *Electron Transfer Mechanisms*, Butterworths, London, 1980.
2. A. G. Lappin, *Redox Mechanisms in Inorganic Chemistry*, Ellis Horwood, Chichester, 1994.
3. J. Bolton, N. Mataga, and G. McLendon (eds), *Advances in Chemistry Series 228: Electron Transfer in Inorganic, Organic, and Biological Systems*, VCH, Weinheim, 1992.
4. M. A. Fox, *Chem. Rev.*, 1992, **92**, 365.
5. H. Taube, *Electron Transfer Reactions of Complex Ions in Solution*, Academic Press, New York, 1970.
6. C. Creutz, *Progr. Inorg. Chem.*, 1983, **30**, 1.
7. J. F. Endicott, K. Kumar, T. Ramasami, and F. P. Rotzinger, *Progr. Inorg. Chem.*, 1983, **30**, 141.
8. A. Haim, *Progr. Inorg. Chem.*, 1983, **30**, 273.
9. H. Taube, *Angew. Chem. Int. Edn. Engl.*, 1984, **23**, 329.

10. T. J. Meyer and H. Taube, in *Comprehensive Coordination Chemistry*, Vol. 1, eds G. Wilkinson, R. D. Gillard, and J. A. McCleverty, Pergamon, Oxford, 1987, Chapter 7.2.
11. T. Ernst and M. Cyfert, *Z. Phys. Chem. (Leipzig)*, 1987, **268**, 175.
12. K. Srinivasan and J. Rocek, *J. Am. Chem. Soc.*, 1978, **100**, 2789.
13. N. Rajasekar, R. Subramanian, and E. S. Gould, *Inorg. Chem.*, 1982, **21**, 4110.
14. H. Taube, *Chem. Rev.*, 1952, **50**, 69.
15. H. Taube and H. Myers, *J. Am. Chem. Soc.*, 1954, **76**, 2103.
16. A. Haim, *Progr. Inorg. Chem.*, 1983, **20**, 273.
17. R. G. Wilkins, *Kinetics and Mechanism of Reactions of Transition Metal Complexes*, 2nd edn, VCH, Weinheim, 1991, p. 272.
18. A. Haim, *J. Am. Chem. Soc.*, 1966, **88**, 2324.
19. R. D. Cannon and J. Gardiner, *J. Am. Chem. Soc.*, 1970, **92**, 3800; R. D. Cannon and J. Gardiner, *Inorg. Chem.*, 1974, **13**, 390; R. J. Balahura and A. L. Johnston, *Inorg. Chem.*, 1983, **22**, 3309.
20. C. S. Alexander and R. J. Balahura, *Inorg. Chem.*, 1994, **33**, 1399.
21. A. P. Szecsy and A. Haim, *J. Am. Chem. Soc.*, 1982, **104**, 3063; A. P. Szecsy and A. Haim, *Inorg. Chem.*, 1982, **21**, 247.
22. J. H. Espenson and J. P. Birk, *J. Am. Chem. Soc.*, 1965, **87**, 3280.
23. V. S. Srinivasan, A. N. Singh, K. Wieghardt, N. Rajasekar, and E. S. Gould, *Inorg. Chem.*, 1982, **21**, 2531.
24. E. S. Gould, *Accounts Chem. Res.*, 1985, **18**, 22.
25. H. A. Boucher, G. A. Lawrance, A. M. Sargeson, and D. F. Sangster, *Inorg. Chem.*, 1983, **22**, 3482; W. F. Hollaway, V. S. Srinivasan, and E. S. Gould, *Inorg. Chem.*, 1984, **23**, 2181; H. Spiecker and K. Wieghardt, *Inorg. Chem.*, 1977, **16**, 1290.
26. A. Vogler and H. Kunkely, *Angew. Chem. Int. Edn. Engl.*, 1980, **19**, 221.
27. G. V. Buxton and R. M. Sellers, *Coord. Chem. Rev.*, 1977, **22**, 195; K. D. Whitburn, M. Z. Hoffman, M. G. Simic, and N.V. Brezniak, *Inorg. Chem.*, 1980, **19**, 3180.
28. J. Ferguson, E. R. Krausz, and M. Maeder, *J. Phys. Chem.*, 1985, **89**, 1852.
29. N. E. Katz, S. L. Mecklenburg, D. K. Graff, Pinyung Chen, and T. J. Meyer, *J. Phys. Chem.*, 1994, **98**, 8959.
30. D. B. Brown (ed.), *Mixed Valence Compounds*, Reidel, Dordrecht, 1980.
31. Review: R. J. Crutchley, *Adv. Inorg. Chem.*, 1994, **41**, 273.
32. Razak bin Ali, M. J. Blandamer, J. Burgess, P. Guardado, and F. Sanchez, *Inorg. Chim. Acta*, 1987, **131**, 59; P. Guardado and R. van Eldik, *Inorg. Chem.*, 1990, **29**, 3477.
33. A. M. Sargeson, *Chem. Brit.*, 1979, 23.
34. D. E. Richardson and H. Taube, *J. Am. Chem. Soc.*, 1983, **105**, 40; D. E. Richardson and H. Taube, *Inorg. Chem.*, 1981, **20**, 1278.
35. D. E. Richardson and H. Taube, *Coord. Chem. Rev.*, 1984, **60**, 107.
36. J. E. Sutton, P. M. Sutton, and H. Taube, *Inorg. Chem.*, 1979, **18**, 1017.
37. See Tables I, III, IV, and V of Ref. 31, and Table XV on p. 322 of P. A. Lay and W. D. Harman, *Adv. Inorg. Chem.*, 1991, **37**, 219.
38. For example, V. M. Gooden, T. P. Dasgupta, N. R. Gordon, and G. G. Sadler, *Inorg. Chim. Acta*, 1998, **268**, 31.
39. L. O. Spreer, C. B. Allan, D. B. McQueen, J. W. Otvos, and M. Calvin, *J. Am. Chem. Soc.*, 1994, **116**, 2187.

40. A.-C. Ribou, J.-P. Launay, K. Takahashi, T. Nihira, S. Tarutani, and C. W. Spangler, *Inorg. Chem.*, 1994, **33**, 1325.
41. J. D. Petersen, W. R. Murphy, R. Sahai, K. J. Brewer, and R. R. Ruminski, *Coord. Chem. Rev.*, 1985, **64**, 261.
42. L. O. Spreer, Aiping Li, D. B. MacQueen, C. B. Allan, J. W. Otvos, M. Calvin, R. B. Frankel, and G. C. Papaefthymiou, *Inorg. Chem.*, 1994, **33**, 1753.
43. W. H. Morrison and D. N. Hendrickson, *J. Chem. Phys.*, 1973, **59**, 380.
44. M.-H. Desbois and D. Astruc, *Organometallics*, 1989, **8**, 1841; M.-H. Desbois, D. Astruc, J. Guillin, and F. Varret, *Organometallics*, 1989, **8**, 1848; M.-H. Desbois, D. Astruc, J. Guillin, F. Varret, A. X. Trautwein, and G. Villeneuve, *J. Am. Chem. Soc.*, 1989, **111**, 5800.
45. V. Boekelheide, *Pure Appl. Chem.*, 1986, **58**, 1; P. Jutzi, U. Siemeling, A. Müller, and H. Bögge, *Organometallics*, 1989, **8**, 1744.
46. W. E. Geiger, N. van Order, D. T. Pierce, T. E. Bitterwolf, A. L. Rheingold, and N. D. Chasteen, *Organometallics*, 1991, **10**, 2403.
47. J. W. Merkert, W. E. Geiger, M. N. Paddon-Row, A. M. Oliver, and A. L. Rheingold, *Organometallics*, 1992, **11**, 4109.
48. R. Winter, D. T. Pierce, W. E. Geiger, and T. J. Lynch, *J. Chem. Soc., Chem. Commun.*, 1994, 1949.
49. K. Weighardt, I. Tolksdorf, and W. Herrmann, *Inorg. Chem.*, 1985, **24**, 1230.
50. M. F. Moore and D. N. Hendrickson, *Inorg. Chem.*, 1985, **24**, 1236.
51. M. B. Robin and P. Day, *Adv. Inorg. Chem. Radiochem.*, 1967, **10**, 247; R. J. H. Clark, *Chem. Soc. Rev.*, 1984, **13**, 219.
52. W. K. Bratton and F. A. Cotton, *Inorg. Chem.*, 1970, **9**, 789.
53. U. Fürholz, S. Joss, H. B. Bürgi, and A. Ludi, *Inorg. Chem.*, 1985, **24**, 943.
54. J. B. Vincent, H.-R. Chang, K. Folting, J. C. Huffman, G. Christou, and D. N. Hendrickson, *J. Am. Chem. Soc.*, 1987, **109**, 5703.
55. M. Sorai and D. N. Hendrickson, *Pure Appl. Chem.*, 1991, **63**, 1503.
56. For example, R. C. Long and D. N. Hendrickson, *J. Am. Chem. Soc.*, 1983, **105**, 1513.
57. K. Y. Wong and P. N. Schatz, *Progr. Inorg. Chem.*, 1981, **28**, 369; U. Fürholz, H.-B. Bürgi, F. E. Wagner, A. Stebler, J. H. Ammeter, E. Krausz, R. J. H. Clark, M. J. Stead, and A. Ludi, *J. Am. Chem. Soc.*, 1984, **106**, 121; L. Dubicki, J. Ferguson, and E. R. Krausz, *J. Am. Chem. Soc.*, 1985, **107**, 179; E. Kraus and A. Ludi, *Inorg. Chem.*, 1985, **24**, 939; S. Joss, H. B. Bürgi, and A. Ludi, *Inorg. Chem.*, 1985, **24**, 949; V. Petrov, J. T. Hupp, C. Mottley, and L. C. Mann, *J. Am. Chem. Soc.*, 1994, **116**, 2171.
58. See p. 288 of Ref. 31, also Ref. 53.
59. A. Yeh and A. Haim, *J. Am. Chem. Soc.*, 1985, **107**, 369.
60. Adapted from A. Ludi, *Inorg. Chem.*, 1985, **24**, 953.
61. F. Fagalde and N. E. Katz, *Polyhedron*, 1994, **14**, 1213.
62. T. J. Meyer, *Chem. Phys. Lett.*, 1979, **64**, 417; N. Sutin, *Accounts Chem. Res.*, 1982, **15**, 275; L. A. A. de Oliveira and A. Haim, *J. Am. Chem. Soc.*, 1982, **104**, 3363; A. Yeh and A. Haim, *J. Am. Chem. Soc.*, 1985, **107**, 369.
63. A. Yeh and A. Haim, *J. Am. Chem. Soc.*, 1985, **107**, 369.
64. M. J. Weaver, *Chem. Rev.*, 1992, **92**, 463; M. N. Paddon-Row, *Accounts Chem. Res.*, 1994, **27**, 18.
65. A. Vassilian, J. F. Wishart, B. van Hemelryck, H. Schwarz, and S. S. Isied, *J. Am. Chem. Soc.*, 1990, **112**, 7278.
66. D. T. Pierce and W. E. Geiger, *Inorg. Chem.*, 1994, **33**, 373.

67. A. Helms, D. Heiler, and G. McLendon, *J. Am. Chem. Soc.*, 1992, **114**, 6227.
68. S. S. Isied and A. Vassilian, *J. Am. Chem. Soc.*, 1984, **106**, 1732.
69. K. D. Jordan and M. N. Paddon-Row, *Chem. Rev.*, 1992, **92**, 395.
70. J. R. Miller, L. T. Calcaterra, and G. L. Closs, *J. Am. Chem. Soc.*, 1984, **106**, 3047; S. Larsson, *Chem. Phys. Lett.*, 1982, **90**, 136.
71. N. I. Al-Shatti, M. A. Hussein, and Y. Sulfab, *Inorg. Chim. Acta*, 1985, **99**, 129.
72. W. S. Melvin and A. Haim, *Inorg. Chem.*, 1977, **16**, 2016.
73. G. C. Seaman and A. Haim, *J. Am. Chem. Soc.*, 1984, **106**, 1319.
74. Review: E. D. German, *Rev. Inorg. Chem.*, 1983, **5**, 123.
75. J. Halpern, R. J. Legare, and R. Lumry, *J. Am. Chem. Soc.*, 1963, **85**, 680; P. Hurwitz and K. Kustin, *Inorg. Chem.*, 1964, **3**, 823.
76. G. J. Kavarnos and N. J. Turro, *Chem. Rev.*, 1986, **86**, 401; R. A. Krause, *Struct. Bonding*, 1987, **67**, 1; H. Riesen and E. Krausz, *Comments Inorg. Chem.*, 1995, **18**, 27; C. D. Clark and M. Z. Hoffman, *Coord. Chem. Rev.*, 1997, **159**, 359.
77. J. R. Schoonover, G. F. Strouse, K. M. Omberg, and R. B. Dyer, *Comments Inorg. Chem.*, 1995, **18**, 165.
78. I. I. Creaser, L. R. Gahan, R. J. Geue, A. Launikonis, P. A. Lay, J. D. Lydon, M. G. McCarthy, A. W.-H. Mau, A. M. Sargeson, and W. H. F. Sasse, *Inorg. Chem.*, 1985, **24**, 2671.
79. T. J. Meyer, *Progr. Inorg. Chem.*, 1983, **30**, 389.
80. M. G. Segal and R. M. Sellers, *J. Chem. Soc., Faraday Trans. I*, 1982, **78**, 1149.
81. V. S. Srinivasan, C. A. Radlowski, and E. S. Gould, *Inorg. Chem.*, 1981, **20**, 3172.
82. D. Gaswick and A. Haim, *J. Am. Chem. Soc.*, 1971, **93**, 7347; A. Haim, *Pure Appl. Chem.*, 1983, **55**, 89; J. C. Curtis and T. J. Meyer, *J. Am. Chem. Soc.*, 1978, **100**, 6284.
83. Y. Kaizu, T. Mori, and H. Kobayashi, *J. Phys. Chem.*, 1985, **89**, 332.
84. D. A. Geselowitz, A. Hammershøi, and H. Taube, *Inorg. Chem.*, 1987, **26**, 1842; A. G. Lappin and R. A. Marusak, *Coord. Chem. Rev.*, 1991, **109**, 125; R. M. L. Warren, A. G. Lappin, and A. Tatehata, *Inorg. Chem.*, 1992, **31**, 1566, and references therein.
85. D. P. Martone, P. Osvath, and A. G. Lappin, *Inorg. Chem.*, 1987, **26**, 3094; S. E. Schadler, C. Sharp, and A. G. Lappin, *Inorg. Chem.*, 1992, **31**, 51.
86. H. Yoneda, U. Sakaguchi, and H. Nakazawa, *Bull. Chem. Soc. Jpn*, 1987, **60**, 2283.
87. R. M. L. Warren, K. J. Haller, A. Tatehata, and A. G. Lappin, *Inorg. Chem.*, 1994, **33**, 227.
88. From Tables 2.2–2.4 of Ref. 2; Ref. 82; A. J. Miralles, A. P. Szecsy, and A. Haim, *Inorg. Chem.*, 1982, **21**, 697; E. Kremer, G. Cha, M. Morkevicius, M. Seaman, and A. Haim, *Inorg. Chem.*, 1984, **23**, 3028; P. L. Gaus, R. M. Benefield, H. Hicsasmaz, B. Perry, J. Villanueva, C. E. Flack, V. B. Pett, S. A. Demoulini, J. C. Faulhaber, D. E. Lynch, and J. W. Bacon, *Inorg. Chim. Acta*, 1988, **141**, 61.
89. L. A. A. de Oliviera, E. Giesbrecht, and H. E. Toma, *J. Chem. Soc., Dalton Trans.*, 1979, 236.
90. S. W. Barr, K. L. Guyer, T. T-T. Liu, and M. J. Weaver, *J. Electrochem. Soc.*, 1984, **131**, 1626.
91. F. A. Armstrong, R. A. Henderson, and A. G. Sykes, *J. Am. Chem. Soc.*, 1980, **102**, 6545.

92. M. A. Augustin, S. K. Chapman, D. M. Davies, A. G. Sykes, S. H. Speck, and E. Margoliash, *J. Biol. Chem.*, 1983, **258**, 6405; G. Tollin, G. Cheddar, J. A. Watkins, T. E. Meyer, and M. A. Cusanovich, *Biochemistry*, 1984, **23**, 6345.
93. J. D. Rush, J. Lan, and W. H. Koppenol, *J. Am. Chem. Soc.*, 1987, **109**, 2679; G. Cheddar, T. E. Meyer, M. A. Cusanovich, C. D. Stout, and G. Tollin, *Biochemistry*, 1989, **28**, 6318.
94. Parsley plastocyanin also has an effective charge of about −7, while that of parsley ferredoxin is as high as −18. An effective charge of this magnitude at an electron transfer site seems highly unlikely.
95. D. Beoku-Betts and A. G. Sykes, *Inorg. Chem.*, 1985, **24**, 1142.
96. D. Beoku-Betts, S. K. Chapman, C. V. Knox, and A. G. Sykes, *Inorg. Chem.*, 1985, **24**, 1677.
97. S. A. White and S. K. Chapman, *Rec. Trav. Chim.*, 1987, **106**, 295.
98. A. G. Sykes, *J. Biol. Chem.*, 1983, **258**, 6400; G. D. Armstrong, J. A. Chambers, and A. G. Sykes, *J. Chem. Soc., Dalton Trans.*, 1986, 775.
99. M. A. S. Aquino and A. G. Sykes, *J. Chem. Soc., Dalton Trans.*, 1994, 683.
100. Joo-Sang Lim, M. A. S. Aquino, and A. G. Sykes, *Inorg. Chem.*, 1996, **35**, 614.
101. G. Dulz and N. Sutin, *Inorg. Chem.*, 1963, **2**, 917; J. D. Miller and R. H. Prince, *J. Chem. Soc. (A)*, 1966, 1370.
102. C. O. Adedinsewo and A. Adegite, *Inorg. Chem.*, 1979, **18**, 3597.
103. R. A. Marcus, *Ann. Rev. Phys. Chem.*, 1964, **15**, 155; *Nouv. J. Chim.*, 1987, **11**, 79; *Pure Appl. Chem.*, 1997, **69**, 13.
104. N. S. Hush, *Trans. Faraday Soc.*, 1961, **57**, 557.
105. See p. 195 of Ref. 1.
106. N. Sutin, *Accounts Chem. Res.*, 1982, **15**, 275; N. Sutin, *Progr. Inorg. Chem.*, 1983, **30**, 441.
107. R. A. Marcus and N. Sutin, *Biochim. Biophys. Acta*, 1985, **811**, 265.
108. M. D. Newton, *Chem. Rev.*, 1991, **91**, 767.
109. R. J. Campion, N. Purdie, and N. Sutin, *Inorg. Chem.*, 1964, **3**, 1091; D. P. Rillema, J. F. Endicott, and R. C. Patel, *J. Am. Chem. Soc.*, 1972, **94**, 394.
110. See, e.g., K. Tsukahara and R. G. Wilkins, *Inorg. Chem.*, 1985, **24**, 3399.
111. For a somewhat fuller treatment than we have room for here, see R. G. Wilkins, *Kinetics and Mechanism of Reactions of Transition Metal Complexes*, VCH, Weinheim, 1991, Section 5.4; A. G. Lappin, *Redox Mechanisms in Inorganic Chemistry*, Ellis Horwood, Chichester, 1994, Chapter 2; for a comprehensive treatment see Chapter 6 of R. D. Cannon, *Electron Transfer Reactions*, Butterworths, London, 1980.
112. M. K. Loar, M. A. Sens, G. W. Loar, and E. S. Gould, *Inorg. Chem.*, 1978, **17**, 330.
113. M. J. Weaver and G. E. McManis, *Accounts Chem. Res.*, 1990, **23**, 294.
114. J. K. Kochi, *Angew. Chem. Int. Edn. Engl.*, 1988, **27**, 1227.
115. T. Iwasita, W. Schmickler, and J. W. Schultze, *Ber. Bunsenges. Phys. Chem.*, 1985, **89**, 138; W. R. Fawcett and M. Opallo, *Angew. Chem. Int. Edn. Engl.*, 1994, **33**, 2131.
116. L. Eberson, *Adv. Phys. Org. Chem.*, 1982, **18**, 79.
117. J. A. Dodd and J. I. Brauman, *J. Am. Chem. Soc.*, 1984, **106**, 5356.
118. For example, T. Ramasami and J. F. Endicott, *Inorg. Chem.*, 1984, **23**, 3324.
119. See, e.g., Section 6.5.5 of Ref. 1; P. Suppan, *Top. Curr. Chem.*, 1992, **163**, 95; and Ref. 107.

120. For example, M. R. Hyde, R. Davies, and A. G. Sykes, *J. Chem. Soc., Dalton Trans.*, 1972, 1838; A. Ekstrom, A. B. McLaren, and L. E. Smythe, *Inorg. Chem.*, 1975, **14**, 2899.
121. I. R. Gould and S. Farid, *Accounts Chem. Res.*, 1996, **29**, 522.
122. See Section 5.2.6 of J. Burgess and E. Pelizzetti, *Progr. React. Kinet.*, 1992, **17**, 1.
123. Relevant references are cited in G. A. Mines, M. J. Bjerrum, M. G. Hill, D. R. Casimiro, I.-J. Chang, J. R. Winkler, and H. B. Gray, *J. Am. Chem. Soc.*, 1996, **118**, 1961.
124. See, e.g., H. Takagi and T. W. Swaddle, *Inorg. Chem.*, 1992, **31**, 4669.
125. L. Spiccia and T. W. Swaddle, *Inorg. Chem.*, 1987, **26**, 2265.
126. P. Hendry and A. Ludi, *Helv. Chim. Acta*, 1988, **71**, 1966.
127. C. Creutz, *Inorg. Chem.*, 1981, **20**, 4449; N. Kagayama, M. Sekiguchi, Y. Inada, H. D. Takagi, and S. Funahashi, *Inorg. Chem.*, 1994, **33**, 1881.
128. N. E. Meagher, K. L. Juntunen, C. A. Salhi, and D. B. Rorabacher, *J. Am. Chem. Soc.*, 1992, **114**, 10411; D. B. Rorabacher, N. E. Meagher, K. L. Juntunen, P. V. Robandt, G. H. Leggett, C. A. Salhi, B. C. Dunn, R. R. Schroeder, and L. A. Ochrymowycz, *Pure Appl. Chem.*, 1993, **65**, 573.
129. C. A. Koval and D. W. Margerum, *Inorg. Chem.*, 1981, **20**, 2315; S. Knapp, T. P. Keenan, Xiaohua Zhang, R. Fikar, J. A. Potenza, and H. J. Schugar, *J. Am. Chem. Soc.*, 1990, **112**, 3452.
130. See pp. 336–337 of Ref. 10.
131. J. M. Newman, M. Binns, T. W. Hambley, and H. C. Freeman, *Inorg. Chem.*, 1991, **30**, 3499.
132. P. Bernhard, L. Helm, A. Ludi, and A. E. Merbach, *J. Am. Chem. Soc.*, 1985, **107**, 312.
133. W. H. Jolley, D. R. Stranks, and T. W. Swaddle, *Inorg. Chem.*, 1990, **29**, 1948.
134. D. Geselowitz and H. Taube, *Adv. Inorg. Bioinorg. Mech.*, 1982, **1**, 391; J. M. Newton, M. Binns, T. W. Hambley, and H. C. Freeman, *Inorg. Chem.*, 1991, **30**, 3499; M. D. Newton, *J. Phys. Chem.*, 1991, **95**, 30.
135. H.-J. Küppers, A. Neves, C. Pomp, D. Ventur, K. Wieghardt, B. Nuber, and J. Weiss, *Inorg. Chem.*, 1986, **25**, 2400; D. Ventur, K. Wieghardt, B. Nuber, and J. Weiss, *Z. Anorg. Allg. Chem.*, 1987, **551**, 33; S. Lee, A. Bakac, and J. H. Espenson, *Inorg. Chem.*, 1990, **29**, 2480.
136. S. Chandrasekhar and A. McAuley, *Inorg. Chem.*, 1992, **31**, 480.
137. P. Osvath, A. M. Sargeson, B. W. Skelton, and A. H. White, *J. Chem. Soc., Chem. Commun.*, 1991, 1036; T. M. Donlevy, L. R. Gahan, and T. W. Hambley, *Inorg. Chem.*, 1994, **33**, 2668; R. J. Geue, A. Höhn, S. F. Ralph, A. M. Sargeson, and A. C. Willis, *J. Chem. Soc., Chem. Commun.*, 1994, 1513.
138. K. Kumar, F. P. Rotzinger, and J. F. Endicott, *J. Am. Chem. Soc.*, 1983, **105**, 7064.
139. R. M. Nielson and S. Wherland, *J. Am. Chem. Soc.*, 1985, **107**, 1505.
140. P. Bernhard and A. M. Sargeson, *Inorg. Chem.*, 1988, **27**, 2582.
141. M. A. S. Aquino, D. A. Foucher, and D. H. Macartney, *Inorg. Chem.*, 1989, **28**, 3357.
142. J. L. Walsh, J. A. Baumann, and T. J. Meyer, *Inorg. Chem.*, 1980, **19**, 2145.
143. P. W. Dimmock, J. McGinnis, Bee-Lean Ooi, and A. G. Sykes, *Inorg. Chem.*, 1990, **29**, 1085.
144. M. Kozik and L. C. W. Baker, *J. Am. Chem. Soc.*, 1990, **112**, 7604.

145. C. Dennison, T. Kohzuma, W. McFarlane, S. Suzuki, and A. G. Sykes, *J. Chem. Soc., Dalton Trans.*, 1994, 437.
146. Sang-Choul Im, Hua-Yun Zhuang-Jackson, T. Kohzuma, P. Kyritsis, W. McFarlane, and A. G. Sykes, *J. Chem. Soc., Dalton Trans.*, 1996, 4287.
147. Review of blue copper proteins: A. G. Sykes, *Adv. Inorg. Chem.*, 1991, **36**, 377.
148. B. L. Vallee and R. J. P. Williams, *Proc. Natl. Acad. Sci. USA*, 1968, **59**, 498; J. J. R. Fraústo da Silva and R. J. P. Williams, *The Biological Chemistry of the Elements*, Clarendon Press, Oxford, 1991, p. 182.
149. D. G. A. Harshani de Silva, D. Beoku-Betts, P. Kyritis, K. Govindaraju, R. Powls, N. P. Tomkinson, and A. G. Sykes, *J. Chem. Soc., Dalton Trans.*, 1992, 2145.
150. A. Haim, *Progr. Inorg. Chem.*, 1983, **30**, 273 (see Section IV.C); O. Oyetunji, O. Olubuyide, and J. F. Ojo, *Bull. Chem. Soc. Jpn*, 1990, **63**, 601.
151. C. A. Jacks, L. E. Bennett, W. N. Raymond, and W. Lovenberg, *Proc. Natl. Acad. Sci. USA*, 1974, **71**, 1118.
152. J. C. Chen and E. S. Gould, *J. Am. Chem. Soc.*, 1973, **95**, 5539; M. R. Hyde, R. S. Taylor, and A. G. Sykes, *J. Chem. Soc., Dalton Trans.*, 1973, 2730.
153. D. R. Stranks, *Pure Appl. Chem.*, 1974, **38**, 303.
154. M. R. Grace, H. Takagi, and T. W. Swaddle, *Inorg. Chem.*, 1994, **33**, 1915.
155. Y. Ducommun, D. Zbinden, and A. E. Merbach, *Helv. Chim. Acta*, 1982, **65**, 1385.
156. Data have been taken from R. G. Wilkins, *Kinetics and Mechanism of Reactions of Transition Metal Complexes*, 2nd edn, VCH, Weinheim, 1991, pp. 260, 271; A. G. Lappin, *Redox Mechanisms in Inorganic Chemistry*, Ellis Horwood, Chichester, 1994, p. 132.
157. K. Kustin and J. Swinehart, *Progr. Inorg. Chem.*, 1970, **13**, 107; S. F. Lincoln and A. E. Merbach, *Adv. Inorg. Chem.*, 1995, **42**, 1.
158. B. Grossmann and A. Haim, *J. Am. Chem. Soc.*, 1971, **93**, 6490.
159. J. H. Espenson, A. Bakac, and Jeong Hyun Kim, *Inorg. Chem.*, 1991, **30**, 4830.
160. A. Neves, W. Herrman, and K. Wieghardt, *Inorg. Chem.*, 1984, **23**, 3435.
161. T. J. Meyer, *Chem. Phys. Lett.*, 1979, **64**, 417.
162. L. A. A. de Oliviera and A. Haim, *J. Am. Chem. Soc.*, 1982, **104**, 3363.
163. A. Vogler and J. Kisslinger, *Angew. Chem. Int. Edn. Engl.*, 1982, **21**, 77.
164. T. J. Meyer, *Progr. Inorg. Chem.*, 1983, **30**, 389.
165. I. K. Adzamli, R. A. Henderson, J. D. Sinclair-Day, and A. G. Sykes, *Inorg. Chem.*, 1984, **23**, 3069.
166. Sang-Choul Im, Kin-Yu Lam, Meng-Chay Lim, Bee-Lean Ooi, and A. G. Sykes, *J. Am. Chem. Soc.*, 1995, **117**, 3635.
167. F. Armstrong, *Adv. Inorg. Bioinorg. Mech.*, 1982, **1**, 65.
168. G. McLendon and R. Hake, *Chem. Rev.*, 1992, **92**, 481.
169. Sang-Choul Im and A. G. Sykes, *J. Chem. Soc., Dalton Trans.*, 1996, 2219.
170. K. P. Simolo, G. L. McLendon, M. R. Mauk, and A. G. Mauk, *J. Am. Chem. Soc.*, 1984, **106**, 5012.
171. L. M. Peerey and N. M. Kostić, *Biochem.*, 1989, **28**, 1861, and references therein.
172. Review: J. R. Winkler and H. B. Gray, *Chem. Rev.*, 1992, **92**, 369.
173. Reviews: B. M. Hoffman, M. J. Natan, J. M. Nocek, and S. A. Wallin, *Struct. Bonding*, 1991, **75**, 85; M. J. Therien, J. Chang, A. L. Raphael, B. E. Bowler, and H. B. Gray, *Struct. Bonding*, 1991, **75**, 109.
174. S. S. Isied, G. Worosila, and S. J. Atherton, *J. Am. Chem. Soc.*, 1982, **104**, 7659.

175. N. M. Kostić, R. Margalit, Chi-Ming Che, and H. B. Gray, *J. Am. Chem. Soc.*, 1983, **105**, 7765.
176. E. Lloyd, K. Chapman, S. K. Chapman, Zhi-Shen Jia, Meng-Chay Lim, N. P. Tomkinson, G. A. Salmon, and A. G. Sykes, *J. Chem. Soc., Dalton Trans.*, 1994, 675.
177. D. W. Conrad, Hui Zhang, D. E. Stewart, and R. A. Scott, *J. Am. Chem. Soc.*, 1992, **114**, 9909.
178. These can be traced through references cited in Ref. 123 above.
179. J. R. Winkler, D. G. Nocera, K. M. Yocom, E. Bordignon, and H. B. Gray, *J. Am. Chem. Soc.*, 1982, **104**, 5798.
180. See, e.g., Figs 8 and 9 of Ref. 172.
181. S. E. Peterson-Kennedy, J. L. McGourty, and B. M. Hoffman, *J. Am. Chem. Soc.*, 1984, **106**, 5010.
182. S. A. Schichman and H. B. Gray, *J. Am. Chem. Soc.*, 1981, **103**, 7794.
183. N. W. Makinen, S. A. Schichman, S. C. Hill, and H. B. Gray, *Science (Washington, DC)*, 1983, **222**, 929.
184. E. Zahavy and I. Willner, *J. Am. Chem. Soc.*, 1996, **118**, 12 499.
185. J. P. Allen, G. Feher, T. O. Yeates, H. Komiya, and D. C. Rees, *Proc. Natl. Acad. Sci. USA*, 1987, **84**, 5730; M. Bixon, J. Jortner, M. E. Michel-Beyerle, and A. Ogrodnik, *Biochim. Biophys. Acta*, 1989, **977**, 273.
186. I. K. Adzamli, D. M. Davies, C. S. Stanley, and A. G. Sykes, *J. Am. Chem. Soc.*, 1981, **103**, 5543.
187. A. G. Sykes, *Chem. Soc. Rev.*, 1985, **14**, 283.
188. Jun Cheng, Jian S. Zhou, and N. M. Kostić, *Inorg. Chem.*, 1994, **33**, 1600.
189. J. T. Hazzard, G. McLendon, M. A. Cusanovich, and G. Tollin, *Biochem. Biophys. Res. Commun.*, 1988, **151**, 429.
190. R. van Eldik and C. D. Hubbard, *New J. Chem.*, 1997, **21**, 825, and references therein.
191. T. W. Swaddle, *Can. J. Chem.*, 1996, **74**, 631.
192. M. Meier and R. van Eldik, *Inorg. Chim. Acta*, 1994, **225**, 95; M. Meier and R. van Eldik, *Chem. Eur. J.*, 1997, **3**, 39.
193. B. Bänsch, M. Meier, P. Martinez, R. van Eldik, S. Chang, J. Sun, S. S. Isied, and J. F. Wishart, *Inorg. Chem.*, 1994, **33**, 4744; M. Meier, J. Sun, R. van Eldik, and J. F. Wishart, *Inorg. Chem.*, 1996, **35**, 1569.
194. M. Meier, R. van Eldik, I.-J. Chang, G. A. Mines, D. S. Wuttke, J. R. Winkler, and H. B. Gray, *J. Am. Chem. Soc.*, 1994, **116**, 1577.
195. D. DeVault and B. Chance, *Biophys. J.*, 1966, **6**, 825; D. DeVault, J. H. Parkes, and B. Chance, *Nature*, 1967, **215**, 642.
196. Review: R. D. Cannon, *Electron Transfer Reactions*, Butterworths, London, 1980; Chapter 3.2 provides not only a thorough treatment of a range of non-complementary reactions and mechanisms but also references to reviews of this subject.
197. K. G. Ashurst and W. C. E. Higginson, *J. Chem. Soc.*, 1953, 3044.
198. W. K. Wilmarth, D. M. Stanbury, J. E. Byrd, H. N. Po, and Chee-Peng Chua, *Coord. Chem. Rev.*, 1983, **51**, 155.
199. D. M. Stanbury, *Inorg. Chem.*, 1984, **23**, 2879; A. Adegite and J. F. Iyun, *Inorg. Chem.*, 1979, **18**, 3602; A. Bakac, J. L. Simunic, and J. H. Espenson, *Inorg. Chem.*, 1990, **29**, 1090; T. A. Annan, R. K. Chadha, P. Doan, D. H. McConville, B. R. McGarvey, A. Ozarowski, and D. G. Tuck, *Inorg. Chem.*, 1990, **29**, 3936; G. Nord, *Comments Inorg. Chem.*, 1992, **13**, 221.

200. A. Peloso, *J. Chem. Soc., Dalton Trans.*, 1979, 1160; C. S. Glennon, T. D. Hand, and A. G. Sykes, *J. Chem. Soc., Dalton Trans.*, 1980, 19.
201. R. N. Bose, S. Moghaddas, and E. Gelerinter, *Inorg. Chem.*, 1992, **31**, 1987.
202. J. H. Espenson, *Progr. Inorg. Chem.*, 1983, **30**, 189.
203. N. Rajasekar, R. Subramanian, and E. S. Gould, *Inorg. Chem.*, 1982, **21**, 4110.
204. For example, K. J. Ellis and A. McAuley, *J. Chem. Soc., Dalton Trans.*, 1973, 1533; R. F. Jameson, W. Linert, A. Tschinkowitz, and V. Gutmann, *J. Chem. Soc., Dalton Trans.*, 1988, 943; R. F. Jameson, W. Linert, and A. Tschinkowitz, *J. Chem. Soc., Dalton Trans.*, 1988, 2109.
205. W. Gaede and R. van Eldik, *Inorg. Chem.*, 1994, **33**, 2204.
206. J. Konstantatos, E. Vrachnou-Astra, and D. Katakis, *Inorg. Chim. Acta*, 1984, **85**, 41.
207. T. A Turney, *Oxidation Mechanisms*, Butterworths, London, 1965, Chapters 5 and 6; A. G. Sykes, *Kinetics of Inorganic Reactions*, Pergamon, Oxford, 1966; D. Benson, *Mechanisms of Inorganic Reactions in Solution*, McGraw-Hill, London, 1968, Chapter 4. References to some of the unreferenced facts quoted in this section, and to much related material, can be traced through these books.
208. K. E. Collins, C. Archundia, and C. H. Collins, *Quim. Nova*, 1983, **6**, 164.
209. M. P. Doyle, S. N. Mahapatro, R. D. Broene, and J. K. Guy, *J. Am. Chem. Soc.*, 1988, **110**, 593.
210. J. W. Gryder and M. C. Dorfman, *J. Am. Chem. Soc.*, 1961, **83**, 1254.
211. Review: J. O. Edwards and R. C. Plumb, *Progr. Inorg. Chem.*, 1994, **41**, 599.
212. C. W. J. Scaife and R. G. Wilkins, *Inorg. Chem.*, 1980, **19**, 3244.
213. D. Pinnell and R. B. Jordan, *Inorg. Chem.*, 1979, **18**, 3191; Z. Bradic and R. G. Wilkins, *J. Am. Chem. Soc.*, 1984, **106**, 2236; R. J. Balahura and M. D. Johnson, *Inorg. Chem.*, 1987, **26**, 3860.
214. D. O. Lambeth and G. Palmer, *J. Biol. Chem.*, 1973, **248**, 6095; P. Hambright, S. Lemelle, K. Alston, P. Neta, H. H. Newball, and S. Di Stefano, *Inorg. Chim. Acta*, 1984, **92**, 167; R. Langley, P. Hambright, K. Alston, and P. Neta, *Inorg. Chem.*, 1986, **25**, 114.
215. R. P. Cox and M. R. Hollaway, *Eur. J. Biochem.*, 1977, **74**, 575.
216. M. D. Johnson and J. Bernard, *Inorg. Chem.*, 1992, **31**, 5140.
217. R. E. Huie and P. Neta, *J. Phys. Chem.*, 1986, **90**, 1193.
218. E. A. Betterton and M. R. Hoffmann, *J. Phys. Chem.*, 1988, **92**, 5962.
219. D. E. Linn, S. D. Rumage, and J. L. Grutsch, *Int. J. Chem. Kinet.*, 1993, **25**, 489.
220. C. Brandt and R. van Eldik, *Chem. Rev.*, 1995, **95**, 119; V. Lepentsiotis, F. F. Prinsloo, R. van Eldik, and H. Gutberlet, *J. Chem. Soc., Dalton Trans.*, 1996, 2135.
221. J. Berglund and L. I. Elding, *Inorg. Chem.*, 1995, **34**, 513.
222. R. Sarala, S. B. Rabin, and D. M. Stanbury, *Inorg. Chem.*, 1991, **30**, 3999; R. Sarala and D. M. Stanbury, *Inorg. Chem.*, 1992, **31**, 2771.
223. M. S. Garley and G. Stedman, *J. Inorg. Nucl. Chem.*, 1981, **43**, 2863.
224. R. Banerjee, R. Das, and A. K. Chakraburtty, *J. Chem. Soc., Dalton Trans.*, 1991, 987.
225. J. Halperin and H. Taube, *J. Am. Chem. Soc.*, 1950, **72**, 3319; J. Halperin and H. Taube, *J. Am. Chem. Soc.*, 1952, **74**, 375, 380; J. O. Edwards, *Chem. Rev.*, 1952, **50**, 455; M. Anbar and H. Taube, *J. Am. Chem. Soc.* 1958, **80**, 1073.
226. E. F. Hills, C. Sharp, and A. G. Sykes, *Inorg. Chem.*, 1986, **25**, 2566.
227. Bo-Ying Liu, P. A. Wagner, and J. E. Earley, *Inorg. Chem.*, 1984, **23**, 3418.

228. A. M. Kjaer and J. Ulstrup, *Inorg. Chem.*, 1982, **21**, 3490.
229. B. Sieklucka and D. H. Macartney, *Transition Met. Chem.*, 1996, **21**, 200.
230. Review of halate oxidations: R. C. Thompson, *Adv. Inorg. Bioinorg. Mech.*, 1986, **4**, 65.
231. B. S. Yiin and D. W. Margerum, *Inorg. Chem.*, 1990, **29**, 1942; C. M. Gerritsen and D. W. Margerum, *Inorg. Chem.*, 1990, **29**, 2757.
232. B. V. L. Potter, *J. Chem. Soc., Chem. Commun.*, 1986, 21.
233. K. A. Hofmann, *Chem. Ber.*, 1912, **45**, 3329.
234. R. Bartzatt, A. Tabatabai, and J. Carr, *Synth. React. Inorg. Met.-Org. Chem.*, 1985, **15**, 1171; R. L. Bartzatt and J. Carr, *Transition Met. Chem.*, 1986, **11**, 116; R. Bartzatt and J. Carr, *Transition Met. Chem.*, 1986, **11**, 414; J. F. Read, K. D. Boucher, S. A. Mehlman, and K. J. Watson, *Inorg. Chim. Acta*, 1998, **267**, 159.
235. D. G. Lee and S. Helliwell, *Can. J. Chem.*, 1984, **62**, 1085.
236. R. D. Powell and A. C. Skapski, *Inorg. Chim. Acta*, 1988, **148**, 15.
237. S. S. Emeish and K. E. Howlett, *Can. J. Chem.*, 1980, **58**, 159.
238. M. A. A. F. de C. T. Carrondo and A. C. Skapski, *Acta Cryst.*, 1978, **B34**, 3551.
239. D. P. Jones and W. P. Griffith, *J. Chem. Soc., Dalton Trans.*, 1980, 2526.
240. J. M. Adams and R. G. Pritchard, *Acta Cryst.*, 1977, **B33**, 3650.
241. R. Pizer and C. Tihal, *Inorg. Chem.*, 1987, **28**, 3639; J. Flanagan, D. P. Jones, W. P. Griffith, A. C. Skapski, and A. P. West, *J. Chem. Soc., Chem. Commun.*, 1986, 20; J. Flanagan, W. P. Griffith, R. D. Powell, and A. P. West, *J. Chem. Soc., Dalton Trans.*, 1989, 1651.
242. J. O. Edwards, in *Peroxide Reaction Mechanisms*, ed. J. O. Edwards, Interscience, New York, 1962, p. 67.
243. Review: D. A. House, *Chem. Rev.*, 1962, **62**, 185.
244. M. P. Pujari and P. Banerjee, *Transition Met. Chem.*, 1983, **8**, 91.
245. U. Fürholz and A. Haim, *Inorg. Chem.*, 1987, **26**, 3243.
246. For example, A. D. Pethybridge and J. E. Prue, *Progr. Inorg. Chem.*, 1972, **17**, 327 (see Section VI.B.4); A. Rodriguez, F. Sanchez Burgos, J. Burgess, and C. Carmona, *Can. J. Chem.*, 1990, **68**, 926; C. Carmona, F. Sanchez, A. Rodriguez, and J. Burgess, *J. Chem. Soc., Faraday Trans.*, 1990, **86**, 3731; A. Rodriguez, C. Carmona, E. Munoz, F. Sanchez, and J. Burgess, *Transition Met. Chem.*, 1991, **16**, 535, and references therein.
247. O. Vollárová and J. Benko, *J. Chem. Soc., Dalton Trans.*, 1983, 2359.
248. S. Srinivasan and K. Pitchumani, *Int. J. Chem. Kinet.*, 1982, **14**, 1315; J. M. Risley and R. L. Van Etten, *Int. J. Chem. Kinet.*, 1984, **16**, 1167.
249. E. J. Behrmann and J. O. Edwards, *Rev. Inorg. Chem.*, 1980, **2**, 179.
250. W. K. Wilmarth, N. Schwartz, and C. R. Giuliano, *Coord. Chem. Rev.*, 1983, **51**, 243.
251. K. Ohashi, M. Matsuzama, E. Hamano, and K. Yamamoto, *Bull. Chem. Soc. Jpn.*, 1976, **49**, 2240.
252. S. Fronaeus, *Acta Chem. Scand.*, 1986, **40A**, 572.
253. V. M. Gooden, T. P. Dasgupta, N. R. Gordon, and G. G. Sadler, *Inorg. Chim. Acta*, 1998, **268**, 31.
254. J. Burgess and R. H. Prince, *J. Chem. Soc. (A)*, 1966, 1772; J. Burgess and R. H. Prince, *J. Chem. Soc. (A)*, 1970, 2111.
255. R. E. Ball, A. Chako, J. O. Edwards, and G. Levey, *Inorg. Chim. Acta*, 1985, **99**, 49.

256. Full details for several of the systems mentioned here can be found in, e.g., Chapter 5 of D. Benson, *Mechanisms of Inorganic Reactions in Solution*, McGraw-Hill, London, 1968.
257. E. Ben-Zvi, *J. Phys. Chem.*, 1963, **67**, 2698.
258. D. Rehorek and E. G. Janzen, *Z. Chem.*, 1985, **25**, 100.
259. C. Marsh and J. O. Edwards, *Progr. React. Kinet.*, 1989, **15**, 35.
260. J. S. Price, I. R. Tasker, E. H. Appelman, and P. A. G. O'Hare, *J. Chem. Thermodyn.*, 1986, **18**, 923.
261. D. L. Ball and J. O. Edwards, *J. Am. Chem. Soc.*, 1956, **78**, 1125.
262. R. C. Thompson, P. Wieland, and E. H. Appelman, *Inorg. Chem.*, 1979, **18**, 1974.
263. D. H. Fortnum, C. J. Battaglia, S. R. Cohen, and J. O. Edwards, *J. Am. Chem. Soc.*, 1960, **82**, 778; F. Secco and M. Venturini, *J. Chem. Soc., Dalton Trans.*, 1976, 1410.
264. R. Renganathan and P. Maruthamuthu, *Int. J. Chem. Kinet.*, 1986, **18**, 49.
265. A. Robert and B. Meunier, *New J. Chem.*, 1988, **12**, 885.
266. M. D. Johnson and R. J. Balahura, *Inorg. Chem.*, 1988, **27**, 3104.
267. M. D. Johnson and D. Nickerson, *Inorg. Chem.*, 1992, **31**, 3971.
268. D. F. Evans and M. W. Upton, *J. Chem. Soc., Dalton Trans.*, 1985, 1151.
269. Review: I. I. Creaser and J. O. Edwards, *Top. Phosphorus Chem.*, 1972, **7**, 379.
270. D. K. Mishram, P. Parasher, and P. D. Sharma, *J. Indian Chem. Soc.*, 1990, **67**, 948; J. Burgess and B. Shraydeh, *Polyhedron*, 1992, **11**, 2015.
271. A. K. Panda, S. N. Mahapatro, and G. P. Panigrahi, *J. Org. Chem.*, 1981, **46**, 4000.
272. P. S. Swaroop, K. A. Kumar, and P. V. K. Rao, *Transition Met. Chem.*, 1991, **16**, 416.
273. M. M. Crutchfield, in *Peroxide Reaction Mechanisms*, ed. J. O. Edwards, Interscience, New York, 1962, p. 41.
274. Y. Ogata, K. Tomizawa, and T. Morikawa, *J. Org. Chem.*, 1979, **44**, 352; G. P. Panigrahi and A. K. Panda, *Int. J. Chem. Kinet.*, 1983, **15**, 989.
275. G. P. Panigrahi and R. C. Paiccha, *Int. J. Chem. Kinet.*, 1991, **23**, 345.
276. S. Kapoor and Y. K. Gupta, *J. Inorg. Nucl. Chem.*, 1977, **39**, 1019.
277. P. Parasher, D. K. Mishra, and P. D. Sharma, *Oxid. Commun.*, 1992, **15**, 99.
278. P. Sivaswaroop, K. A. Kumar, and P. V. K. Rao, *React. Kinet. Catal. Lett.*, 1989, **39**, 33.
279. P. N. Balasubramanian and T. C. Bruice, *J. Am. Chem. Soc.*, 1986, **108**, 5495.
280. H. P. Kim, J. H. Espenson, and A. Bakac, *Inorg. Chem.*, 1987, **26**, 4090.
281. B. C. Gilbert and J. K. Stell, *J. Chem. Soc., Perkin Trans. II*, 1990, 1281; B. C. Gilbert and J. K. Stell, *J. Chem. Soc., Faraday Trans.*, 1990, 3261.
282. See several of the references cited in the perphosphate section above.
283. Review: W. P. Griffith, *Coord. Chem. Rev.*, 1970, **5**, 459.
284. Review: R. H. Holm, *Chem. Rev.*, 1987, **87**, 1401 gives a comprehensive, wide-ranging, and fully referenced (545 citations) treatment of oxygen transfer reactions and mechanisms.
285. K. F. Miller and R. A. D. Wentworth, *Inorg. Chim. Acta*, 1979, **36**, 37.
286. Jun Lu, A. Yamano, and M. J. Clarke, *Inorg. Chem.*, 1990, **29**, 3483.
287. S. A. Roberts, C. G. Young, C. A. Kipke, W. E. Cleland, K. Yamanouchi, M. D. Carducci, and J. H. Enemark, *Inorg. Chem.*, 1990, **29**, 3659, and references therein.

288. K. Onoura, Y. Kato, K. Abe, A. Iwase, and H. Ogino, *Bull. Chem. Soc. Jpn*, 1991, **64**, 3372.
289. O. Vollarova and J. Benko, *Curr. Top. Soln. Chem.*, 1994, **1**, 107.
290. I. K. Adzamli and E. Deutsch, *Inorg. Chem.*, 1978, **12**, 1366; A. F. Ghiron and R. C. Thompson, *Inorg. Chem.*, 1990, **29**, 4457.
291. R. C. Troy and D. W. Margerum, *Inorg. Chem.*, 1991, **30**, 3538
292. N. de Vries, A. G. Jones, and A. Davison, *Inorg. Chem.*, 1989, **28**, 3728.
293. G. A. Vaughan, P. B. Rupert, and G. L. Hillhouse, *J. Am. Chem. Soc.*, 1987, **109**, 5538.
294. J. H. Espenson, O. Pestovsky, P. Huston, and S. Staudt, *J. Am. Chem. Soc.*, 1994, **116**, 2869.
295. H. Arzoumanian, D. Nuel, and J. Sanchez, *J. Mol. Catal.*, 1991, **65**, L9.
296. Review: B. Meunier, *New J. Chem.*, 1992, **16**, 201.
297. K. G. Moloy, *Inorg. Chem.*, 1988, **27**, 677.
298. C. Ho, Chi-Ming Che, and Tai-Chu Lau, *J. Chem. Soc., Dalton Trans.*, 1990, 967; M. M. Taqui Khan, D. Chatterjee, R. R. Merchant, A. Bhatt, and S. S. Kumar, *J. Mol. Catal.*, 1991, **66**, 289; M. M. Taqui Khan, K. Venkatasubramanian, Z. Shirin, and M. M. Bhadbhade, *J. Chem. Soc., Dalton Trans.*, 1992, 1031.
299. Yiping Zhang and R. H. Holm, *Inorg. Chem.*, 1990, **29**, 911.
300. Yoon-Yul Park, Y. Ikeda, M. Harada, and H. Tomiyasu, *Chem. Lett.*, 1991, 1329.
301. K. D. Fogelman, D. M. Walker, and D. W. Margerum, *Inorg. Chem.*, 1989, **28**, 986, and references therein.
302. M. Gazda and D. W. Margerum, *Inorg. Chem.*, 1994, **33**, 118.
303. C. Creutz, Jeong-Sup Song, and R. M. Bullock, *Pure Appl. Chem.*, 1995, **67**, 47.
304. S. Goldstein and G. Czapski, *Inorg. Chem.*, 1985, **24**, 1087.
305. F. Pina, M. Ciano, L. Moggi, and V. Balzani, *Inorg. Chem.*, 1985, **24**, 844.
306. G. R. A. Johnson and N. B. Nazhat, *J. Am. Chem. Soc.*, 1987, **109**, 1990.
307. D. Golub, H. Cohen, and D. Meyerstein, *J. Chem. Soc., Dalton Trans.*, 1985, 641; C. R. Johnson, T. K. Myser, and R. E. Shepherd, *Inorg. Chem.*, 1988, **27**, 1089.
308. H. P. Kim, J. H. Espenson, and A. Bakac, *Inorg. Chem.*, 1987, **26**, 4090.
309. G. R. A. Johnson, N. B. Nazhat, and R. A. Saadalla-Nazhat, *J. Chem. Soc., Chem. Commun.*, 1985, 407; G. R. A. Johnson, N. B. Nazhat, and R. A. Saadalla-Nazhat, *J. Chem. Soc., Faraday Trans. I*, 1988, **84**, 501.
310. A. Bakac, J. H. Espenson, and J. A. Janni, *J. Chem. Soc., Chem. Commun.*, 1994, 315; R. C. Thompson, *Inorg. Chem.*, 1986, **25**, 184.
311. J. Burgess and R. H. Prince, *J. Chem. Soc. (A)*, 1970, 2111.
312. Y. G. Adewuyi and G. R. Carmichael, *Environ. Sci. Technol.*, 1987, 170.
313. M. L. Bowers, D. Kovacs, and R. E. Shepherd, *J. Am. Chem. Soc.*, 1977, **99**, 6555.
314. M. J. Root and E. Deutsch, *Inorg. Chem.*, 1985, **24**, 1464.
315. See, e.g., D. G. Bray and R. C. Thompson, *Inorg. Chem.*, 1994, **33**, 905.
316. H. Fenton, *J. Chem. Soc.*, 1894, **65**, 899.
317. J. H. Mertz and W. A. Waters, *J. Chem. Soc.*, 1949, 2427; *Encyclopaedia of Reagents for Organic Synthesis*, Vol. 4, ed. L. A. Paquette, Wiley, Chichester, 1995, p. 2738.
318. M. G. Evans, *J. Chem. Soc.*, 1947, 266; C. Walling, *Accounts Chem. Res.*, 1975, **8**, 125; C. Walling and K. Amarnath, *J. Am. Chem. Soc.*, 1982, **104**, 1185.

319. S. Rahhal and H. W. Richter, *J. Am. Chem. Soc.*, 1988, **110**, 3126.
320. J. E. Frew and P. Jones, *Adv. Inorg. Bioinorg. Mech.*, 1986, **4**, 1.
321. M. Masarwa, H. Cohen, D. Meyerstein, D. L. Hickman, A. Bakac, and J. H. Espenson, *J. Am. Chem. Soc.*, 1988, **110**, 4293; T. Ozawa and A. Hanaki, *J. Chem. Soc., Chem. Commun.*, 1991, 330.
322. I. R. Epstein and K. Kustin, *Struct. Bonding.*, 1984, **56**, 1.
323. D. H. Macartney, *Can. J. Chem.*, 1986, **64**, 1936.
324. B. Sieklucka and A. Samotus, *Transition Met. Chem.*, 1996, **21**, 226.
325. M. P. Heyward and C. F. Wells, *J. Chem. Soc., Faraday Trans. I*, 1988, **84**, 815.
326. K. C. Francis, D. Cummins, and J. Oakes, *J. Chem. Soc., Dalton Trans.*, 1985, 493.
327. E. T. Borish and L. J. Kirschenbaum, *J. Chem. Soc., Dalton Trans.*, 1983, 749.
328. M. P. Heyward and C. F. Wells, *J. Chem. Soc., Dalton Trans.*, 1981, 1863;
329. D. R. Eaton and M. Pancratz, *Can. J. Chem.*, 1985, **63**, 793.
330. R. J. P. Williams, *Chem. Scripta*, 1988, **28A**, 5.
331. I. I. Creaser, R Geue, J. M. Harrowfield, A. J. Herlt, A. M. Sargeson, M. R. Snow, and J. Springborg, *J. Am. Chem. Soc.*, 1982, **104**, 6016.
332. K. D. Karlin, S. Kaderli, and A. D. Zuberbühler, *Accounts Chem. Res.*, 1997, **30**, 139.
333. D. A. Ryan and J. H. Espenson, *J. Am. Chem. Soc.*, 1982, **104**, 704.
334. H. H. Awad and D. M. Stanbury, *Int. J. Chem. Kinet.*, 1993, **25**, 375.
335. E. F. Hills, P. R. Norman, T. Ramasami, and A. G. Sykes, *J. Chem. Soc., Dalton Trans.*, 1986, 157.
336. V. L. Goedken, *J. Chem. Soc., Chem. Commun.*, 1972, 207.
337. F. Westley, *NBS Spec. Publ. (U.S.)*, 1983, **665**, 1, gives 164 relevant references.
338. K. Sehestad, H. Corfitzen, J. Holcman, C. H. Fischer, and E. J. Hart, *Environ. Sci. Technol.*, 1991, **25**, 1589.
339. H. Tomiyasu, H. Fukutomi, and G. Gordon, *Inorg. Chem.*, 1985, **24**, 2962.
340. L. R. B. Yeatts and H. Taube, *J. Am. Chem. Soc.*, 1949, **71**, 4100.
341. T. Løgager, J. Holcman, K. Sehestad, and T. Pedersen, *Inorg. Chem.*, 1992, **31**, 3523.
342. W. J. McElroy and J. Munthe, *Acta Chem. Scand.*, 1991, **45**, 254.
343. P. Neta, R. E. Huie, and A. B. Ross, *J. Phys. Chem. Ref. Data*, 1988, **17**, 1027–1284; G. V. Buxton, C. L. Greenstock, W. P. Helman, and A. B. Ross, *J. Phys. Chem. Ref. Data*, 1988, **17**, 513–886.
344. B. Hickel and K. Sehested, *J. Phys. Chem.*, 1985, **89**, 5271.
345. J. Halpern and J. Rabani, *J. Am. Chem. Soc.*, 1966, **88**, 699.
346. J. A. Howard, in *Advances in Free-Radical Chemistry*, ed. G. H. Williams, Logos Press, 1972, Chapter 2 (see pp. 52–58).
347. J. D. Melton and B. H. J. Bielski, *Radiat. Phys. Chem.*, 1990, **36**, 725.
348. N. Selvarajan and N. V. Raghavan, *J. Chem. Soc., Chem. Commun.*, 1980, 336.
349. H. Cohen, R. van Eldik, M. Masarwa, and D. Meyerstein, *Inorg. Chim. Acta*, 1990, **177**, 31.
350. G. E. Adams, R. K. Broszkiewicz, and B. D. Michael, *Trans. Faraday Soc.*, 1968, **64**, 1256.
351. N. M. Dimitrijević and I. Mićić, *J. Chem. Soc., Dalton Trans.*, 1982, 1953; A. C. Maliyackel, W. L. Waltz, J. Lilie, and R. J. Woods, *Inorg. Chem.*, 1990, **29**, 340.
352. O. Mønsted, G. Nord, and P. Pagsberg, *Acta Chem. Scand.*, 1987, **41A**, 104.
353. G. V. Buxton and R. M. Sellers, *J. Chem. Soc., Faraday Trans. I*, 1985, **81**, 449.
354. S. P. Mezyk and A. J. Elliot, *J. Chem. Soc., Faraday Trans.*, 1994, **90**, 831.

355. B. H. J. Bielski, D. E. Cabelli, R. L. Arudi, and A. B. Ross, *J. Phys. Chem. Ref. Data*, 1985, **14**, 1041.
356. G. V. Buxton and R. M. Sellers, *Coord. Chem. Rev.*, 1977, **22**, 195; K. Tsukahara and R. G. Wilkins, *Inorg. Chem.*, 1985, **24**, 3399.
357. G. G. Jayson, B. J. Parsons, and A. J. Swallow, *J. Chem. Soc., Faraday Trans. I*, 1973, **69**, 1597; B. C. Gilbert, J. K. Stell, W. J. Peet, and K. J. Radford, *J. Chem. Soc., Faraday Trans. I*, 1988, **84**, 3319.
358. M. R. DeFelippis, M. Faraggi, and M. H. Klapper, *J. Phys. Chem.*, 1990, **94**, 2420.
359. W. J. McElroy and S. J. Waygood, *J. Chem. Soc., Faraday Trans.*, 1990, **86**, 2557; Z. B. Alfassi, S. Padmaja, P. Neta, and R. E. Huie, *Int. J. Chem. Kinet.*, 1993, **25**, 151; S. Padmaja, Z. B. Alfassi, P. Neta, and R. E. Huie, *Int. J. Chem. Kinet.*, 1993, **25**, 193.
360. R. E. Huie and P. Neta, *Int. J. Chem. Kinet.*, 1991, **23**, 541.
361. Y. Katsumura, P. Y. Jiang, R. Nagaishi, T. Oishi, K. Ishiguri, and Y. Yoshida, *J. Phys. Chem.*, 1991, **95**, 4435; M. Exner, H. Herrmann, and R. Zellner, *Ber. Bunsenges. Phys. Chem.*, 1992, **96**, 470.
362. D. Knittel, *Monatsh. Chem.*, 1986, **117**, 359.
363. B. G. Ershov, E. Janata, M. Michaelis, and A. Henglein, *J. Phys. Chem.*, 1991, **95**, 8996.
364. R. H. Baker, J. Lilie, and M. Grätzel, *J. Am. Chem. Soc.*, 1982, **104**, 422; M. D'Angelantonio, Q. G. Mulazzini, M. Venturi, M. Ciano, and M. Z. Hoffman, *J. Phys. Chem.*, 1991, **95**, 5121.
365. R. J. Balahura and M. D. Johnson, *Inorg. Chem.*, 1987, **26**, 3860.
366. D. M. Davies and J. M. Lowther, *J. Chem. Soc., Chem. Commun.*, 1986, 385; K. Tsukahara, *Rec. Trav. Chim.*, 1987, **106**, 291.
367. M. P. Jackman, J. McGinnis, R. Powls, G. A. Salmon, and A. G. Sykes, *J. Am. Chem. Soc.*, 1988, **110**, 5880; O. Farver and I. Pecht, *Inorg. Chem.*, 1990, **29**, 4855.
368. Review: D. R. Tyler, *Progr. Inorg. Chem.*, 1988, **36**, 125.
369. Review: C. R. Bock and E. A. Koerner von Gustorf, *Adv. Photochem.*, 1977, **10**, 221.
370. D. M. Stanbury, M. M. deMaine, and G. Goodloe, *J. Am. Chem. Soc.*, 1989, **111**, 5496.
371. J. H. Baxendale, J. P. Keene, and D. A. Stott, *Chem. Commun.*, 1966, 715.
372. G. A. Lawrance and D. F. Sangster, *J. Chem. Soc., Dalton Trans.*, 1987, 1425.
373. P. Natarajan and N. V. Raghavan, *J. Chem. Soc., Chem. Commun.*, 1980, 268.
374. S. Muralidharan and J. H. Espenson, *Inorg. Chem.*, 1984, **23**, 636.
375. D. G. Kelly, J. H. Espenson, and A. Bakac, *Inorg. Chem.*, 1990, **29**, 4996; D. G. Kelly, A. Marchaj, A. Bakac, and J. H. Espenson, *J. Am. Chem. Soc.*, 1991, **113**, 7583.
376. J. C. Scaiano, W. J. Leigh, and G. Ferraudi, *Can. J. Chem.*, 1984, **62**, 2355.
377. D. Golub, H. Cohen, and D. Meyerstein, *J. Chem. Soc., Dalton Trans.*, 1985, 641.
378. D. C. Eisenberg, C. J. C. Lawrie, A. E. Moody, and J. R. Norton, *J. Am. Chem. Soc.*, 1991, **113**, 4888.
379. R. van Eldik, H. Cohen, A. Meshulam, and D. Meyerstein, *Inorg. Chem.*, 1990, **29**, 4156; R. van Eldik, H. Cohen, and D. Meyerstein, *Inorg. Chem.*, 1994, **33**, 1566.

380. A. Bakac, V. Butkovic, J. H. Espenson, R. Marcec, and M. Orhanovic, *Inorg. Chem.*, 1991, **30**, 481.
381. K. L. Rollick and J. K. Kochi, *J. Am. Chem. Soc.*, 1982, **104**, 1319; J. Grodkowski, P. Neta, C. J. Schlesener, and J. K. Kochi, *J. Phys. Chem.*, 1985, **89**, 4373.
382. E. J. Hart and M. Anbar, *The Hydrated Electron*, Wiley–Interscience, New York, 1970.
383. E. M. Fielden and E. J. Hart, *Trans. Faraday Soc.*, 1968, **64**, 3158.
384. A. J. Elliot and D. C. Ouellette, *J. Chem. Soc., Faraday Trans.*, 1994, **90**, 837.
385. G. V. Buxton and S. R. Mackenzie, *J. Chem. Soc., Faraday Trans.*, 1992, **88**, 2833.
386. D. N. Hague, *Fast Reactions*, Wiley–Interscience, London, 1971, p. 97.
387. Y. Gauduel, S. Pommeret, A. Migus, and A. Antonetti, *J. Phys. Chem.*, 1989, **93**, 3880.
388. Values also from Ref. 356; values for actinide cations may be traced through J. C. Sullivan, K. H. Schmidt, L. R. Morss, C. G. Pippin, and C. Williams, *Inorg. Chem.*, 1988, **27**, 597, and references therein.
389. T. Telser and U. Schindewolf, *J. Phys. Chem.*, 1986, **90**, 5378.
390. H. Shiraishi, Y. Katsumura, D. Hiroishi, K. Ishigure, and M. Washio, *J. Phys. Chem.*, 1988, **92**, 3011.
391. Most of the data in this table have been taken from M. Anbar and E. J. Hart, *J. Phys. Chem.*, 1965, **69**, 973.
392. B. J. Parsons, P. C. Beaumont, S. Navaratnam, W. D. Harrison, T. S. Akasheh, and M. Othman, *Inorg. Chem.*, 1994, **33**, 157.

10 Activation, addition, insertion, and catalysis

Chapter contents

10.1 Introduction

There are a number of important reactions—even more than listed in the chapter title—which do not fit into the two main reaction types, substitution and redox, which have dominated this book so far. In this final chapter we discuss some of these reactions, and, towards the end of the chapter, show how various combinations operate in a selection of homogeneous catalysis cycles.

Starting materials, products, and catalysts have been highlighted in several of the figures in Section 10.7, which depict catalytic cycles. The convention adopted is to show starting materials in boxes, products in boxes with background shading, and catalysts or catalyst precursors on a shaded background. A small square in a potential ligand position indicates a vacant coordination site. Curved arrows show the direction of operation of the cycle to obtain the required products—they should not necessarily be taken to imply that the reactions shown are irreversible.

10.2 Reactions of coordinated ligands[1,2,3]

10.2.1 Introduction

The organometallic case of reaction at coordinated carbonyl has already been discussed in Chapter 6 (Section 6.4). Coordination of organic compounds to metal centres can have a profound effect on their reactivity, in nucleophilic and electrophilic addition, substitution, and abstraction reactions.[4] This has been much studied for dienes and dienyls,[5] and for arenes[6] modified by the presence of such moieties as $M(CO)_3$ (M = Cr, Mo, Mn). Two well-studied examples (Y^- = carbanion; Nu^- = e.g., H^- or OR^-) are:

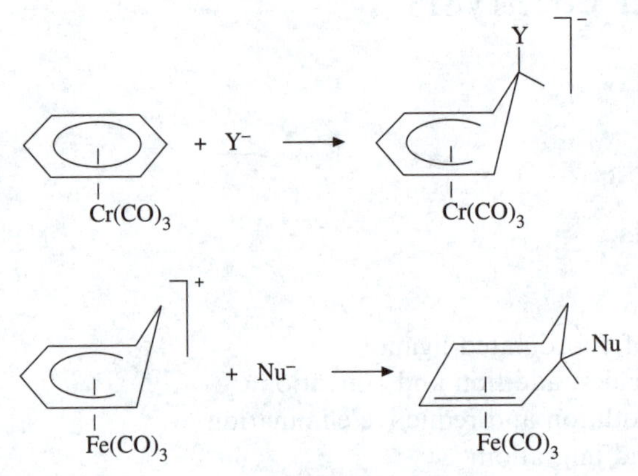

Such reactions are often very stereospecific—the above two reactions almost always give *exo*-products. A generous selection of kinetic data and a detailed discussion of possible mechanisms are available.[7] Another well-tilled organometallic field has been that of nucleophilic attack at allyl compounds coordinated to palladium, while the Wacker process (Section 10.7.9) involves nucleophilic attack at palladium-bonded alkene.

In the present section we deal mainly with reactions which take place at the coordinated ligand of classical Werner complexes and of a few bioinorganic species. There is a link between these reactions and the organometallic examples already mentioned, provided by reactions of coordinated ligands with azide. In Section 6.4 we discussed the reaction of Group 6 metal hexacarbonyls with azide:

$$M(CO)_6 + N_3^- = [M(NCO)(CO)_5]^- + N_2$$

Azide similarly reacts with coordinated nitriles in complexes $[Co(NH_3)_5(NCR)]^{3+}$, this time without extrusion of dinitrogen, to give complexes **1**. As for the hexacarbonyl plus azide reaction, the cobalt complexes react with azide according to simple second-order kinetics.

Werner himself recognized the importance of ligand reactions, especially of coordinated thiocyanate,[8] but they were thereafter much neglected until

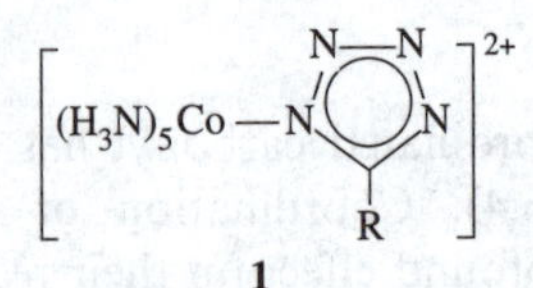

1

the past twenty or thirty years. The renaissance of coordinated-ligand chemistry was sparked by studies of reactions of ammine and amine complexes, often of nickel(II), with organic carbonyl compounds.[9] Often the chemistry involved is the same for the ligand in the coordinated as in the free state, but also very often the reaction takes place much more rapidly when the species is coordinated. In other circumstances coordination of the molecule may change its chemistry, sometimes permitting reactions to take place that are not possible for the free ligand. In similar vein, a number of organic reactions take place more cleanly when one or both are coordinated to an appropriate metal ion. An example of this is provided by Schiff base formation from glyoxal and methylamine. This goes in low yield, producing very impure product, in the absence of a metal-ion catalyst, but in the presence of Fe^{2+} the required Schiff base is formed in high yield; it may be efficiently removed from the iron by the addition of cyanide:

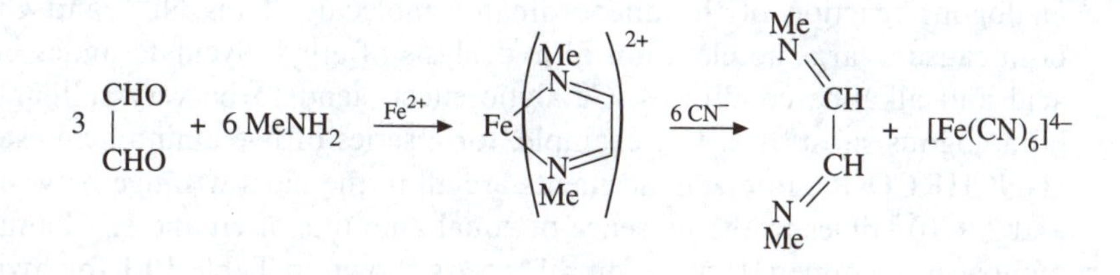

Conversely, metal ions in other circumstances can promote hydrolysis of coordinated Schiff base ligands. The Schiff base **2** is stabilized by coordination to iron(II), but hydrolyses about 10^5 times more rapidly when coordinated to copper(II) than in the free state.[10] We should also mention cases where the ligand is very unstable in the free state but is stabilized sufficiently on coordination for its chemistry to be studied. Again an example can be provided from this area of iron chemistry, where the simplest diimine, **3**, can only be prepared coordinated to, and stabilized by, iron(II) or ruthenium(II):[11]

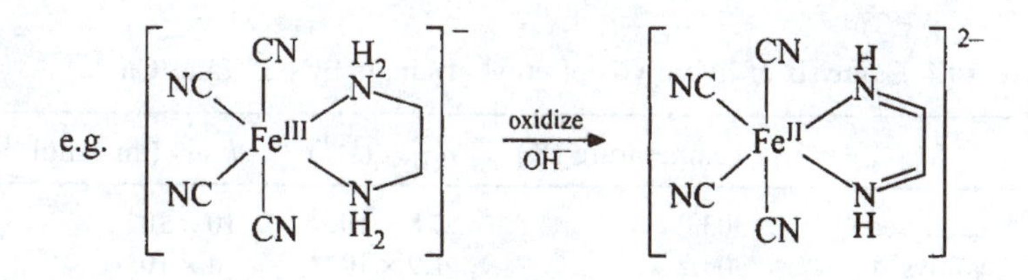

The stabilization of cyclobutadienes by coordination to nickel(0) provides another example, this time from organometallic chemistry.

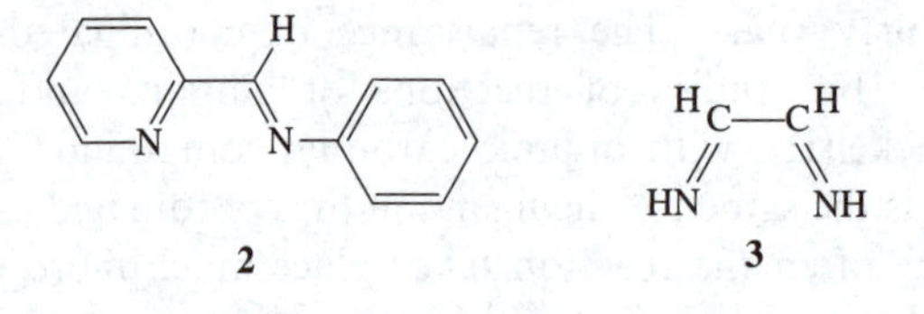

10.2.2 Substitution

10.2.2.1 Nucleophilic substitution at organic ligands

Examples of *nucleophilic substitution* processes at coordinated ligands include aquation and base hydrolysis of coordinated esters, amides, peptides, or nitriles:

$$\text{e.g. } [Co(NH_3)_5(RCO_2R')]^{3+} + OH^- \longrightarrow [Co(NH_3)_5(RCO_2)]^{2+} + R'OH$$

Such nucleophilic substitutions often occur much more rapidly than the analogous reaction of the uncoordinated molecule. Thus Ni^{2+} and Cu^{2+} both cause a large acceleration of hydrolysis of ethyl glycinate under both acid and alkaline conditions. Catalytic effects tend to be very similar for homologous substrates. For example, for a series of five amino acid esters, $H_2NCHRCO_2R'$, rate enhancements are all in the narrow range between 1 and 2×10^3 times in the presence of equal amounts of ethane-1,2-diamine-monoacetatocopper(II), $[Cu(edma)]^+$.[12] As shown in Table 10.1 for hydrolysis and base hydrolysis of ethyl glycinate,[13] and more fully in Table 10.2 for a greater range of cations and substrates, different cations have significantly different effects. However, values for the metal-catalysed rate constants fall within a relatively narrow band compared with the difference between them and the value for the free ligand. This difference is probably largest for nitrile hydrolysis to produce the respective carboxamide. Rate constants for the case of acetonitrile are given in Table 10.3,[14] which shows a steady small decrease down the sequence $Co^{III} \rightarrow Rh^{III} \rightarrow Ir^{III}$ and a somewhat larger increase going to Ru^{III}. The dramatic comparison is with free acetonitrile, with hydrolysis of acetonitrile coordinated to $Ru(NH_3)_5{}^{2+}$

Table 10.1 Catalysis of hydrolysis of ethyl glycinate by Ni^{2+} and Cu^{2+}.

	Temperature (K)	k_{H_2O} (s^{-1})	k_{OH^-} ($dm^3\ mol^{-1}\ s^{-1}$)
Co^{2+}-catalysed	303.2	3.8×10^{-4}	10×10^3
Ni^{2+}-catalysed	303.2	1.9×10^{-4}	4×10^3
Cu^{2+}-catalysed	298.2	0.3×10^{-4}	76×10^3
Zn^{2+}-catalysed	303.2	3.3×10^{-4}	23×10^3
Uncatalysed	298.2	[a]	0.64

[a] Less than, probably very much less than, $10^{-6}\ s^{-1}$ (see R. P. Bell and B. A. W. Coller, *Trans. Faraday Soc.*, 1964, **60**, 1087).

Table 10.2 Effects of varying metal-ion nature on catalysis of base hydrolysis; all rate contants at 298.2 K.

Coordin-ated ligand:	glyOMe [a]	egda [b]	L-Me cyst [c]
	k_{OH^-} ($dm^3\ mol^{-1}\ s^{-1}$)	k_{OH^-} ($dm^3\ mol^{-1}\ s^{-1}$)	k_{OH^-} ($dm^3\ mol^{-1}\ s^{-1}$)
Mn^{2+}		4	
Fe^{2+}		39	
Co^{2+}	19	10	
Ni^{2+}	52	4	10.8
Cu^{2+}	460	218	
Zn^{2+}	35	66	4.7
Cd^{2+}		2	3.5
Hg^{2+}			2.3
Pb^{2+}		283	5.7
$La^{3+} \rightarrow Lu^{3+}$		$115 \rightarrow 3450$	

[a] glyOMe = methyl glycinate, in the ternary complexes M(nta)(glyOMe) (nta = nitrilotriacetate).
[b] egta = ethyl glycinate N,N-diacetate.
[c] L-Me cyst = L-methyl cysteinate, in complexes $M(L\text{-Mecyst})_2$.

Table 10.3 Rate constants for hydrolysis of coordinated acetonitrile in aqueous solution at 298.2 K.

	k_{OH^-} ($dm^3\ mol^{-1}\ s^{-1}$)
Free MeCN	1.6×10^{-6}
$[Co(NH_3)_5(MeCN)]^{3+}$	3.4
$[Rh(NH_3)_5(MeCN)]^{3+}$	1.0
$[Ir(NH_3)_5(MeCN)]^{3+}$	0.23
$[Ru(NH_3)_5(MeCN)]^{3+}$	220

just over 10^8 times faster than that of free acetonitrile. The former reaction is over within a second, whereas a physical chemist interested in kinetics in binary aqueous solvent mixtures can safely keep dilute solutions of sodium hydroxide in acetonitrile–water mixtures for long periods without worrying about hydrolysis of the organic component. The same cannot be said about hydroxide solutions in dimethylformamide–water. Base hydrolysis of dimethylformamide is speeded up by coordination to cobalt(III), but the kinetics of base hydrolysis of $[Co(NH_3)_5(dmf)]^{3+}$ are complicated by hydrolysis of the complex to $[Co(NH_3)_5(OH)]^{2+}$ plus free dimethylformamide in parallel with base hydrolysis of coordinated dimethylformamide.

Normally accelerations of the type described in the preceding paragraphs can be attributed to the positive charge on the adjacent metal ion polarizing the coordinated ligand, making the site of nucleophilic attack effectively

Table 10.4 Effect of coordination and protonation on rate constants for bromination of aniline.

	k_2 ($dm^3\ mol^{-1}\ s^{-1}$)
$C_6H_5NH_2$	3×10^{10}
$[Co(en)_2Cl(C_6H_5NH_2)]^{2+}$	0.14
$C_6H_5NH_3^+$	3×10^{-7}

more positive. Conversely, electrophilic attack may be greatly retarded by metal ion coordination, though not as dramatically as by protonation (Table 10.4). When acid-catalysed nucleophilic substitution involves a basic nucleophile, as is so often the case, then a metal ion, being more selective than the proton, may simultaneously allow high concentrations of both the catalyst and the nucleophile.

It should be borne in mind that the uncharged moiety $-Cr(CO)_3$ can also cause large rate accelerations in hydrolyses of aromatic molecules coordinated to the chromium(0). Coordination of $-M(CO)_3$ to an aromatic ring increases the electron density around the ring, making the arene protons more acidic, and facilitating nucleophilic attack. Electronic and bonding factors must also come into play in these systems. It may be commented that rate accelerations produced by labile metal ions are easy enough to observe, but in such situations it is difficult to establish the corresponding mechanism. Inert complexes may also produce similarly large rate accelerations, and now there is a much better chance of establishing mechanism. The problem here is often the synthesis of the required complex, especially when isotopic tracer techniques are to be used to track reaction moieties. However, the synthesis of a number of cobalt(III) complexes of the $[Co(en)_2(LL)]^{n+}$ type, with LL = ester or amide, and subsequent investigation of the course of (base) hydrolysis of coordinated LL, has shown the operation of the expected mechanisms of metal-ion-promoted nucleophilic attack (Fig. 10.1). This type of complex permits comparison of relative susceptibilities of coordinated esters, amides, and peptides to base hydrolysis (Table 10.5). Despite the reactivity differences, the coordination of the $Co(en)_2^{3+}$ moiety produces a million-fold acceleration of base hydrolysis both for the ester and for the peptide shown in this table.[15]

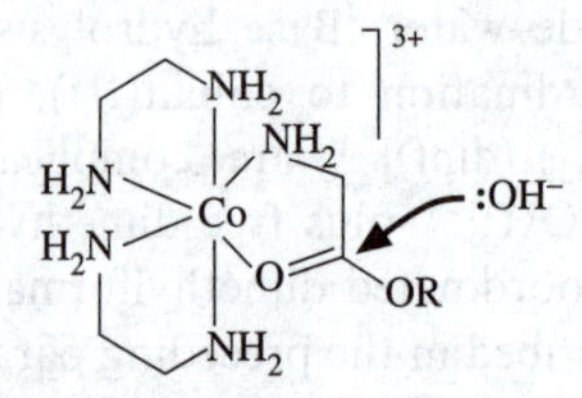

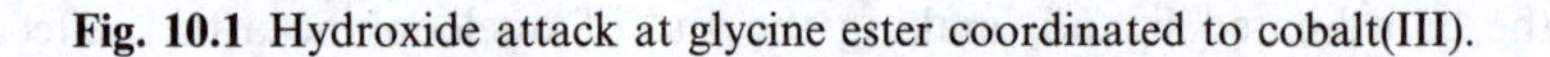
Fig. 10.1 Hydroxide attack at glycine ester coordinated to cobalt(III).

Table 10.5 Reactivity comparisons for base hydrolysis of ester, amide, and peptide derivatives of glycine, free and coordinated to cobalt(III), at 298.2 K.

Coordinated	k_2 (dm^3 mol^{-1} s^{-1})	Free ligand	k_2 (dm^3 mol^{-1} s^{-1})
$[Co(en)_2(glyO^iPr)]^{3+}$	1.5×10^6	glyOEt	0.6
$[Co(en)_2(glyNH_2)]^{3+}$	25		
$[Co(en)_2(glyNHMe)]^{3+}$	1.6		
$[Co(en)_2(glyNMe_2)]^{3+}$	1.1		
$[Co(en)_2(glyglyO)]^{2+}$	2.6	$glyglyO^-$	$\sim 6 \times 10^{-6}$

The electron-withdrawing properties of cobalt(III) may also be important in facilitating intramolecular nucleophilic attack, while its substitution-inertness permits the characterization of intermediates and products. Figure 10.2 shows two examples of cobalt(III)-promoted intramolecular cyclizations. In both cases the cobalt(III) facilitates proton loss from coordinated nitrogen in an initial rapidly established equilibrium. In example (a), the proton is lost from the organic ligand, and the deprotonated nitrogen

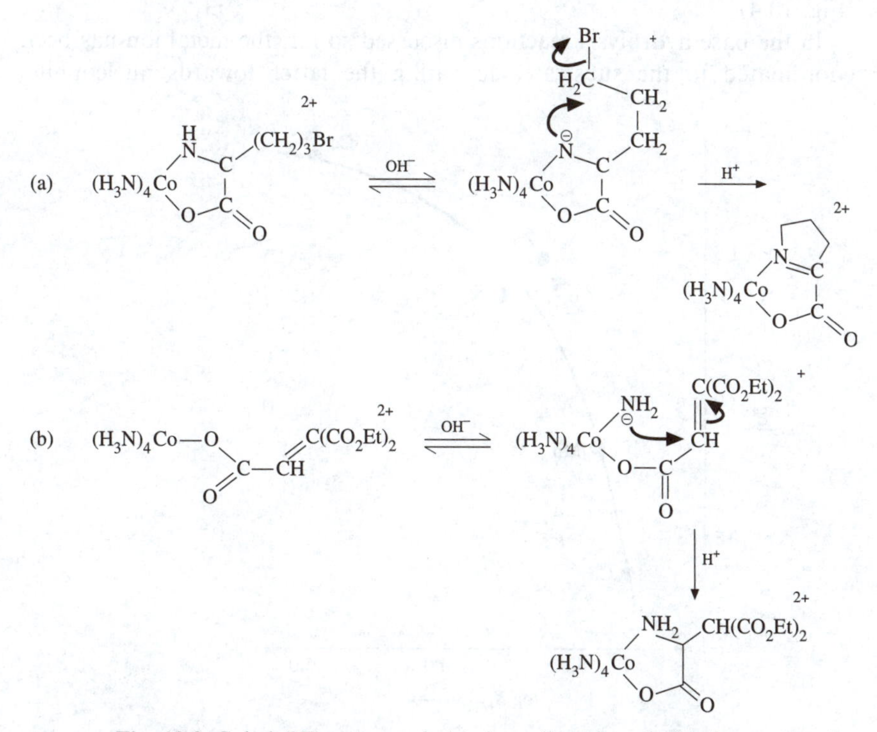

Fig. 10.2 Cobalt(III)-promoted (a) intra-ligand and (b) intramolecular cyclizations.

attacks at carbon to give an intra-ligand cyclization. In example (b), proton loss is from an ammine ligand, whose deprotonated nitrogen then attacks at one of the alkene-carbons of the adjacent organic ligand to give the chelate shown. The kinetics of cobalt(III)-enhanced base hydrolysis were established for the two $-CH_2CO_2Et$ ester groups in this coordinated chelating ligand.[16] Reversible proton loss from coordinated ammonia is also important in the reaction of $[Ru(NH_3)_6]^{3+}$ with thiosulphate and thiophosphate, which attack at the deprotonated nitrogen of the conjugate base $[Ru(NH_3)_5(NH_2)]^{2+}$—thiosulphate gives the sulphamato complex $[Ru(NH_3)_5(NHSO_3)]^+$ as product.[17]

The importance of intermediate complex formation in catalysis by labile metal ions is generally probed by seeking a correlation between rate constants and stability constants. The data for decarboxylation of acetonedicarboxylate ($^-O_2CCH_2COCH_2CO_2{}^-$) and of oxaloacetate[18]

$$^-O_2CCOCH_2CO_2H \longrightarrow {}^-O_2CCOCH_3 + CO_2$$

show a strong, if curved (Fig. 10.3 for the latter), correlation. There is the additional feature of *keto* ↔ *enol* tautomerism of the coordinated ligand to consider; in fact the enol form is the more reactive.[19] The Lewis acid behaviour of the metal ion, which is coordinated to the carbonyl and adjacent carboxylate, facilitates the departure of the carbon dioxide (Fig. 10.4).

In the base hydrolysis reactions discussed so far, the metal ion has been coordinated to the substrate, activating the latter towards nucleophilic

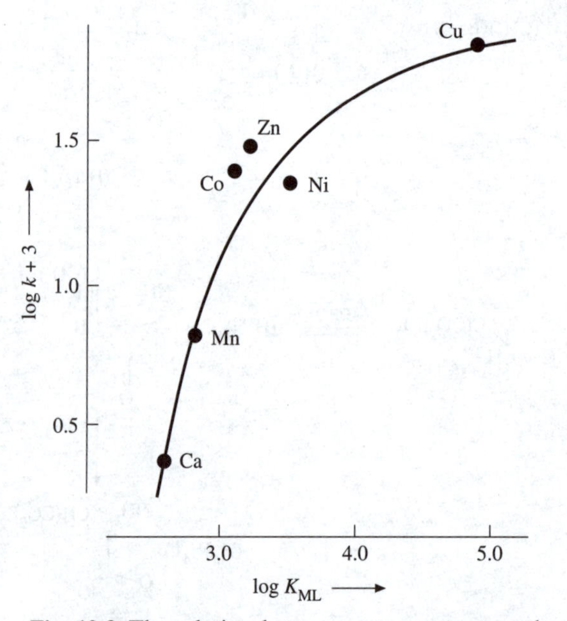

Fig. 10.3 The relation between rate constants and stability constants for metal-ion-promoted decarboxylation of oxaloacetate.

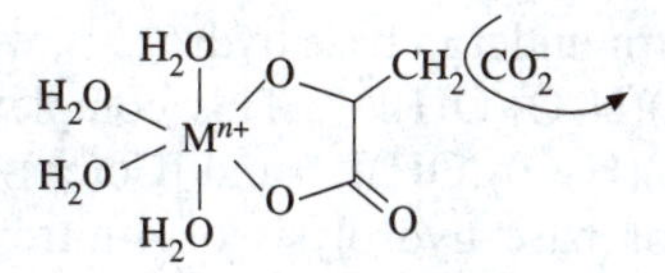

Fig. 10.4 Facilitation of carbon dioxide liberation from oxalatoacetate by coordination to a metal ion.

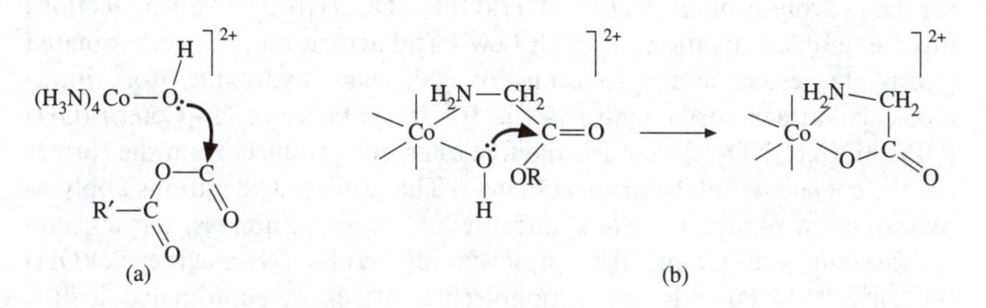

Fig. 10.5 (a) Intermolecular and (b) intramolecular nucleophilic attack by coordinated hydroxide.

attack by hydroxide. An alternative and complementary approach is to coordinate the hydroxide to a metal ion, as in $[M(NH_3)_5(OH)]^{2+}$ (M = Co, Rh, Ir, Ru), and use this complexed hydroxide as a nucleophile. One such reaction is shown in Fig. 10.5(a), with an example of an intramolecular variant in Fig. 10.5(b). Such intramolecular activation of hydroxide, analogous to that of deprotonated ammine (Fig. 10.2(b)), has been demonstrated for hydrolysis of coordinated esters, amides, and nitriles, with rate accelerations up to 10^{11} reported.[20]

10.2.2.2 Nucleophilic substitution—phosphate esters

Coordinated phosphate esters have provided another area of intensive kinetic and mechanistic study, generally in complexes with cobalt(III) and, where even greater substitution-inertness is required at the metal centre, iridium(III). Thus $[Ir(NH_3)_5(tmp)]^{2+}$ hydrolyses in basic solution to $[Ir(NH_3)_5(dmp)]^{2+}$ (tmp, dmp = trimethyl, dimethyl phosphate) and methanol, but $[Co(NH_3)_5(tmp)]^{2+}$ hydrolyses with cobalt–oxygen bond fission, to give $[Co(NH_3)_5(OH)]^{2+}$ and tmp.[21] For phosphate esters, as for the organic substrates mentioned above, metal-ion enhancement of base hydrolysis may also be effected by activating the hydroxide by coordination to, e.g., cobalt(III). Thus the complex $[Co(N\{CH_2CH_2CH_2NH_2\}_3)(H_2O)(OH)]^{2+}$ accelerates base hydrolysis of ATP by a million times, which may be not entirely unconnected with the unusually rapid rate of water exchange or replacement in this complex.[22] Normally cobalt(III)-promoted hydrolysis of ATP gives ADP and phosphate, but there are cobalt(III) complexes which catalyse the breakdown of ATP to AMP plus pyrophosphate.[23] Cyclic AMP can in turn undergo base hydrolysis, which is accelerated by 10^8 times by $[Co(trien)(H_2O)(OH)]^{2+}$. This complex, and closely related species such as $[Co(dien)(H_2O)_2(OH)]^{2+}$ and $[Co(en)_2(H_2O)(OH)]^{2+}$, are also effective catalysts for base hydrolysis of 4-nitrophenyl phosphate and related esters.[24]

Intramolecular attack by coordinated hydroxide at phosphorus has been demonstrated, by the establishment of the rate–pH profile and ^{18}O-labelling, for the hydrolysis of *cis*-$[Co(en)_2(OH)(OPO_2\{OC_6H_4NO_2\})]$. Such reactions involve double activation, through Lewis acid activation of the coordinated phosphate ester and provision of adjacent hydroxide for intramolecular attack. Ester hydrolysis is 10^5 times faster in *cis*-$[Co(en)_2(OH)(OPO_2\{OC_6H_4NO_2\})]$ than for the free ester; the products from the former are the chelate **4** and 4-nitrophenolate.[25] The same considerations apply to hydrolysis of phosphate esters, and for hydrolysis of amides, nitriles, and carboxylate esters, in the presence of *cis*-$[Co(N_4\text{-macrocycle})(OH)(H_2O)]^{3+}$.[26] Metal-activated intramolecular attack of coordinated hydroxide at adjacent coordinated phosphate is also the likely mechanism for catalysis by labile copper complexes such as $[Cu(tmen)(OH)(H_2O)]^+$ or $[Cu(bipy)(OH)(H_2O)]^+$.[27] As M^{2+}aq ions of bioinorganic importance, both from the *d*-block and the *sp*-block, are all labile it is very difficult to be sure of their detailed role in phosphate ester hydrolysis. It has therefore been suggested that, *faute de mieux*, Pt^{II} complexes might be used to model these 2+ cations (cf. Chapter 3, Section 3.3.3.5).[28]

There is analogous intramolecular attack, this time by coordinated $-NH_2$, in base hydrolysis of the iridium complexes $[Ir(NH_3)_5L]^{2+}$, with L = 4-nitrophenylphosphate or ethyl 4-nitrophenylphosphate. Now there is a rapidly established pre-equilibrium to give the reactive conjugate base, and subsequent reactions after the intramolecular attack at phosphorus (Fig. 10.6).[29] Iridium(III) is actually considerably less effective than cobalt(III) in the systems mentioned in this paragraph, but, as already stated, the much higher barriers to substitution in its complexes ensure predominant, or exclusive, reaction at the coordinated ligands in these base hydrolyses.

High activation also may be achieved by coordinating the phosphate ester to one cobalt centre, the hydroxide to another. This type of double activation has been demonstrated for, e.g., $[Co(tn)_2(H_2O)(OH)]^{2+}$ catalysis of hydrolysis of coordinated inorganic phosphate ligands in $[Co(en)_2(P_2O_7)]^-$ and $[Co(tacn)(P_3O_{10}H_2)]$. The base hydrolysis of coordinated acetonitrile, **5**, is the intramolecular equivalent, with the acetonitrile and hydroxide bonded to two different but linked cobalt centres.[30] There is a similar situation in which hydroxide on one cobalt attacks at coordinated adenosine triphosphate on an adjacent cobalt, but this time part of the ATP also acts as the bridging ligand linking the two activating cobalt centres. There are several examples of intermolecular analogues where the substrate and hydroxide are in two separate cobalt complexes.

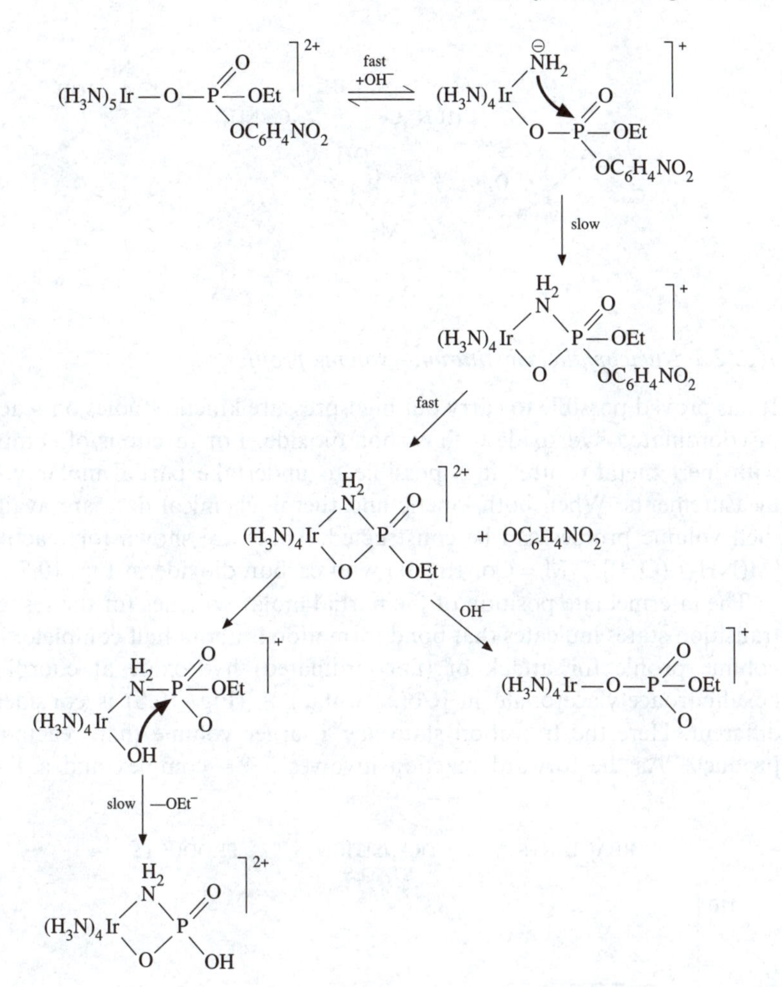

Fig. 10.6 Iridium(III)-catalysed base hydrolysis of coordinated 4-nitrophenylphosphate.

There is much information on the catalysis of hydrolysis of ATP and other nucleoside triphosphates by labile M^{2+}aq ions,[31] and on the role of metal ions in enzymatic hydrolysis of phosphate esters.[32] Mechanisms are harder to establish at a labile metal centre. However, there is, for example, good evidence that the very large rate enhancements, of 10^{13}-fold, produced by cerium(IV), added as $[Ce(NO_3)_6]^{2-}$, of hydrolysis of 3′,5′-cyclic monophosphates of adenosine and guanosine operate through intramolecular attack of coordinated hydroxide on adjacent coordinated phosphate ester.[33] Urease provides another example of double-metal activation, with nickel-activated hydroxide attacking at nickel-activated urea, **6**.

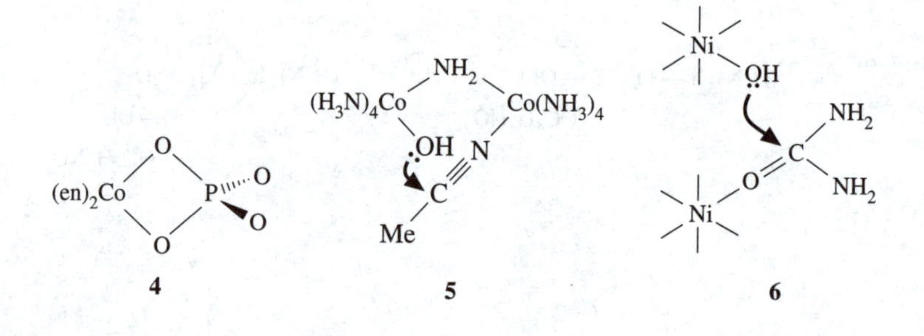

10.2.2.3 Nucleophilic substitution—volume profiles

It has proved possible to carry out high-pressure kinetic studies on reactions of coordinated hydroxide with carbon dioxide. For reactions of complexes with inert metal centres it is possible to undertake partial molar volume measurements. When both kinetic and thermochemical data are available, then volume profiles can be constructed. These are shown for reaction of $[M(NH_3)_5(OH)]^{2+}$ (M = Co, Rh, Ir) with carbon dioxide in Fig. 10.7.

The intermediate position of the partial molar volumes for the respective transition states indicates that bond formation is about half complete.[34] The volume profile for attack of (uncoordinated) hydroxide at coordinated hexafluoroacetylacetonate in $[Co(en)_2(hfac)]^{2+}$ (Fig. 10.8) is considerably different. Here the transition state has a larger volume than reactants or products. As the forward reaction involves a 2+ complex and a 1− ion

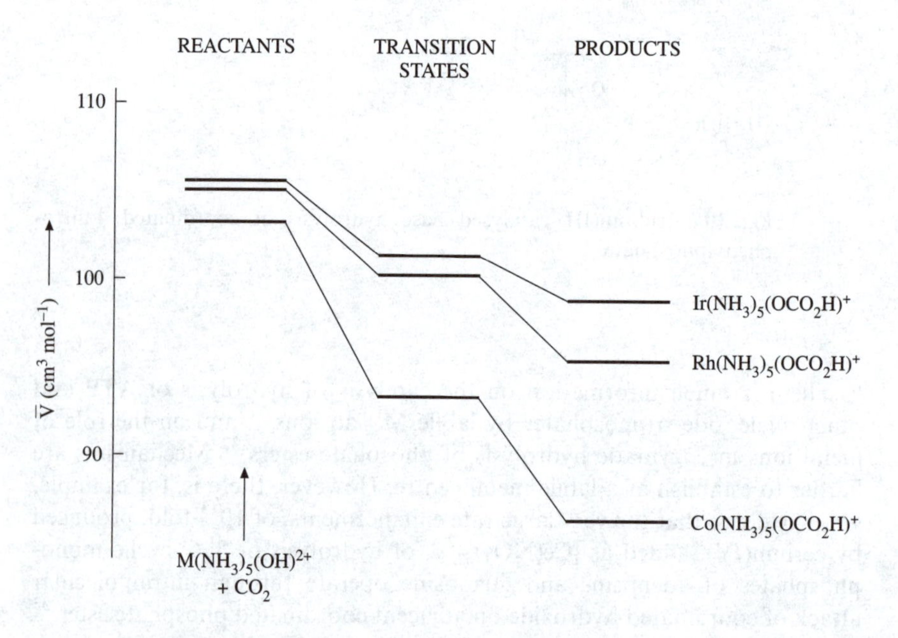

Fig. 10.7 Volume profiles for reaction of $[M(NH_3)_5(OH)]^{2+}$ with carbon dioxide in aqueous solution at 298.2 K.

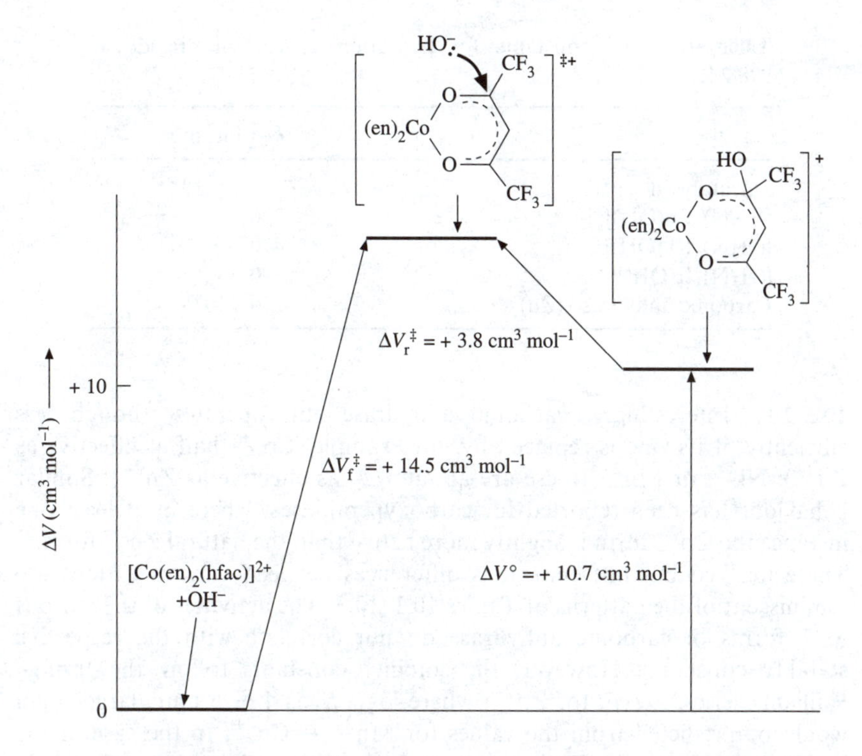

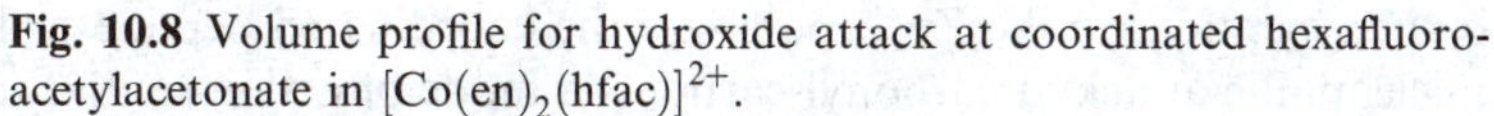
Fig. 10.8 Volume profile for hydroxide attack at coordinated hexafluoroacetylacetonate in $[Co(en)_2(hfac)]^{2+}$.

combining to give a transition state, and indeed a product, with only a 1+ net charge, the relatively large increase in partial molar volume on forming the transition state can be attributed to release of electrostricted water into bulk solvent (cf. Section 8.2.3). The small positive activation volume for the reverse reaction, where there is no change in charge, is consistent with extension of the C–OH bond on transition-state generation.[35]

10.2.2.4 Bioinorganic reactions

Bioinorganic examples of metal-ion activation of a coordinated ligand include carbonic anhydrase and the carboxypeptidases. In the former case the Zn^{2+} probably activates coordinated hydroxide to attack at the substrate's carbon, a reaction analogous to the attack by coordinated hydroxide as described in Section 10.2.2.2 above. Table 10.6 compares the enormous catalytic effect of carbonic anhydrase (in its zinc form)[36] and the much smaller, but still large, effects of $[M(NH_3)_5(OH)]^{2+}$ complexes.[37] The nature of the metal in the hydroxo complexes has only a small effect, as observed in a different context earlier. The trend $Co^{III} \rightarrow Rh^{III} \rightarrow Ir^{III}$ is the opposite of that of these metal centres on coordinated amide (cf. Section

Table 10.6 Rate constants for hydration of carbon dioxide at 298.2 K.

	k_2 (dm^3 mol^{-1} s^{-1})
Uncatalysed	6.7×10^{-4}
$[(H_3N)_5Co(OH)]^{2+}$	220
$[(H_3N)_5Rh(OH)]^{2+}$	470
$[(H_3N)_5Ir(OH)]^{2+}$	590
Carbonic anhydrase (Zn)	10^7–10^8

10.2.2.1). Interestingly, carbonic anhydrase still operates, though less efficiently, if its zinc is replaced by, for example, Co^{2+} (half as effective as Zn^{2+}), Ni^{2+}, or Mn^{2+} (these are about 1/20 as effective as Zn^{2+}). Similar behaviour has been reported for carboxypeptidases, where in at least one instance the Co^{2+} form is slightly more active than the natural Zn^{2+} form.[38] These marked but not enormous differences between various cations are reminiscent of the patterns of Tables 10.1–10.3. The activities of the various M^{2+} forms of carbonic anhydrase do not correlate with the respective stability constants. However, the stability constants follow the Irving–Williams series, *except* for Zn^{2+}, where $\log_{10} K_1$ is 4 or 5 units larger than would be predicted from the values for $Mn^{2+} \rightarrow Cu^{2+}$. In the case of the carboxypeptidases, the Zn^{2+} polarizes the coordinated peptide to facilitate nucleophilic attack at carbonyl-carbon—a mode of action similar to that in the reactions discussed at the start of this section.

This section is linked to our earlier discussions of organic substrates (Sections 10.2.2.1 and 10.2.2.2) through ever-popular model system studies. Modelling carbonic anhydrase has proved difficult, as obvious species such as $[Zn(N_4\text{- or }N_3\text{-macrocycle})(H_2O)]^{2+}$ or their hydroxo equivalents do not work—in contrast to the *cis*-$[Co(N_4\text{-macrocycle})(OH)(H_2O)]^{3+}$/phosphate-ester systems mentioned in Section 10.2.2.2. Limited success has been achieved by complexing zinc, or cobalt, with the semi-encapsulating ligand derived from 1,3,5-triaminocyclohexane and cinnamaldehyde,[39] and by using μ-hydroxo binuclear complexes.[40] Efforts are being made to develop a unified mechanistic approach to the development of artificial metalloenzymes with realistic catalytic turnover characteristics. The approach has been applied, *inter alia,* to hydrolysis of esters, amides, nitriles, and phosphate esters at cobalt, copper, and zinc centres.[41]

10.2.2.5 Oxoanion complexes

To return to classical inorganic systems, there is a group of complexes of inert metal centres which undergo aquation unexpectedly rapidly. This group comprises a number of oxoanion complexes, mainly of chromium(III) and of cobalt(III). Isotopic-labelling studies have shown that, in these cases where aquation occurs very much faster than for normal complexes of these metals, it is not the metal–oxygen bond to the coordinated oxoanion but the coordinated oxygen to the oxoanion's central atom bond that breaks. The nitrogen–oxygen bond in, say, nitrate is not weak, but is less reluctant to break than the chromium–nitrate bond in $[Cr(H_2O)_5(ONO_2)]^{2+}$.[42] Activation energies for hydrolysis of the nitrite complexes $[M(NH_3)_5(ONO)]^{2+}$, M = Cr, Co, Rh, are all the same within experimental uncertainty (66–68 kJ mol^{-1}) and are significantly less than that for hydrolysis of $[Co(NH_3)_5(NO_2)]^{2+}$ (75 kJ mol^{-1}).[43] This is consistent with nitrogen–oxygen rather than metal–oxygen bond breaking for the *O*-nitrito-complexes. Other complexes where hydrolysis involves intra-ligand rather than metal–oxygen bond fission include $[Cr(NH_3)_5(OSO_2)]^{2+}$ and $[Co(NH_3)_5(OAsO_3)]^{2+}$; a recent example where both aquation and formation are coordinated-ligand reactions is provided by the molybdate complex $[Co(NH_3)_5(OMoO_3)]^{+}$.[44]

10.2.2.6 Electrophilic substitution

Electrophilic substitution, the converse of nucleophilic substitution, involves attack by an electron-seeking reagent at a negatively charged site in a molecule or complex. In fact any S_N2 reaction is S_E2 if looked at from the complementary viewpoint—mercury(II)-catalysed aquation of $[Co(NH_3)_5Br]^{2+}$ is S_E2 attack of the Hg^{2+} at the bromide, but is equally S_N2 attack of the coordinated bromide at Hg^{2+}aq. This class of metal-ion-catalysed aquations—both soft/soft as in this $Co^{III}Br/Hg^{2+}$ case, or in complementary hard/hard systems of the $M^{N}F/Zr^{IV}$ type—have been mentioned in Chapter 4 (see Section 4.5.1).

Probably the most investigated group of electrophilic substitutions at ligands has been that of halogenation of coordinated β-diketones, whose chelate rings behave in many respects as aromatic systems. Bromination of tris-acetylacetonato-chromium(III) was described in 1925. Since then, halogenation of coordinated β-diketones, by elemental halogen or, better, by *N*-halogenosuccinimide, nitration (by dinitrogen tetroxide), formylation,[45] and other similar reactions have been examined, especially for complexes of the substitution-inert Cr^{3+}, Co^{3+}, and Rh^{3+} ions.[3,46] Substituents in the coordinated acetylacetone can have major effects on reactivity—halogenation of coordinated dipivaloylmethane is very much slower than of coordinated acetylacetone, thanks to steric hindrance by the t-butyl groups in the former. Nucleophilic substitution at coordinated β-diketones is extremely difficult, but for coordinated 3-bromoacetylacetone, as in products of bromination of $[M(acac)_3]$, nucleophilic replacement of bromide by thiophenolate is actually easier than electrophilic substitution.

Although this field is dominated by β-diketone complexes, there have been studies, sometimes including kinetic work, on electrophilic substitution in other coordinated ligands. These have included, for example, bromination of coordinated 8-hydroxyquinoline and of coordinated imidazole. In relation to bromination of imidazole, the dramatic activation by coordination of $[Co(NH_3)_5]^{3+}$ towards electrophilic attack has been compared quantitatively with activation by methylation; bromination of the complex **7** and is a thousand times faster than that of the *N*-methyl analogue **8**.[47]

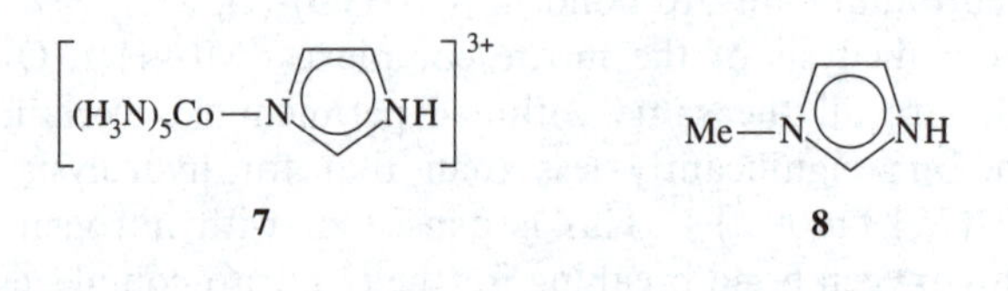

10.2.3 Orientation and site blocking

By bringing reagents together in the coordination shell of the metal ion, modes of combination which are otherwise statistically or chemically unlikely can be promoted. This category includes *template reactions*,[48] which produce cyclic oligomers in a wide variety of cases where, in the absence of a metal-ion template, either no reaction or uncontrolled polymerization would occur. A classic example is that of diethylenetriamine and 2,6-diacetylpyridine, which react in the presence of Ni^{2+} to give the expected macrocyclic Schiff base coordinated to the metal ion, **9**. The best syntheses of many of the macrocyclic ligands which appear at various junctures in this book are very often template reactions. The building up of catenanes and knots on metal-ion centres such as Cu^{2+} provides a recent example where the synthesis of exotic materials has been optimized through a knowledge of mechanism.[49] Examples of template reactions from organometallic chemistry appear later in this chapter. One way of controlling template reactions is to limit the number of coordination sites available on the metal ion by blocking certain coordination positions by firmly bonded ligands—the best examples come from organometallic chemistry (*v.i.*). Template reactions can also be important in bioinorganic chemistry, as in the use of nickel(II) to complete the macrocyclic ring of corrin.[50]

The variety of mechanisms possible can be illustrated for the simple case of synthesis from the appropriate metal ion and the two ligand components, schematically:†

$$M^{n+} + L + L' = M(LL')^{n+}$$

† The ligand is often formed from its components with the elimination of a water molecule—this has, in the interests of simplicity of representation, been ignored in most of the equations that follow, as has proton elimination from such Schiff base ligands as contain the potential donor oxygen atom in a hydroxyl group.

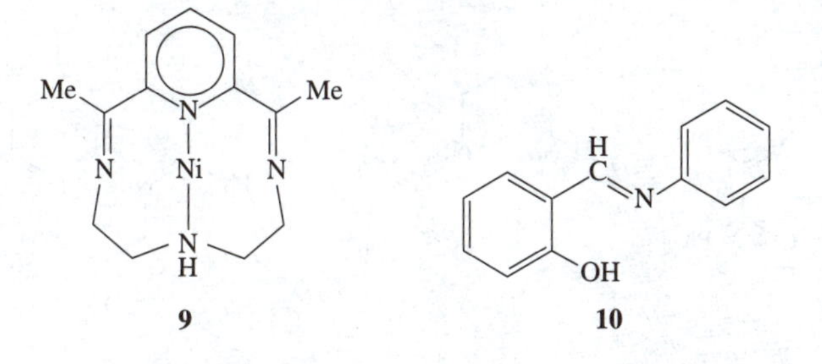

Thus for a salicylideneaniline complex

$$M^{n+} + 3o\text{-}C_6H_4(CHO)(OH) + 3PhNH_2 + 3OH^- = M(\mathbf{10})_3 + 6H_2O$$

Such reactions are often loosely termed 'template reactions', but it is usually not clear whether the actual mechanism for chelate ring formation is a true template reaction

$$M^{n+} + L + L' \rightarrow M\begin{matrix}\diagup L \\ \diagdown L'\end{matrix} \rightarrow M\begin{matrix}\diagup L \\ \diagdown L'\end{matrix}\Big)$$

i.e. reaction between the two components after they have *both* become coordinated to the metal ion, or attack by one ligand at the other after the latter has become coordinated to, and activated by, the metal ion:

$$M^{n+} + L \rightarrow ML^{n+} \xrightarrow{+L'} M\begin{matrix}\diagup L \\ \diagdown L'\end{matrix}\Big)$$

or simply by reaction between the ligand components followed by reaction of the Schiff base thus generated with the metal ion:

$$L + L' \rightarrow LL' \xrightarrow{+M^{n+}} M\begin{matrix}\diagup L \\ \diagdown L'\end{matrix}\Big)$$

In some systems the template step appears to follow non-template organic chemistry. Thus benzil and 1,3-diaminopropane have long been known to react in the presence of cobalt(II) acetate to give a tetraaza-macrocycle, but some years after the characterization of the product it was shown that the intermediate **11** was isolable, which in the final stage could react with the second molecule of the diamine only on a metal ion such as Co^{2+}. Subsequently it was demonstrated that the very similar reaction of the unsymmetrical diketone PhCOCOMe with 1,3-diaminopropane in the presence of M^{2+}, M = Fe, Co, or Ni, involved initial condensation to give the Schiff base **12**, which then dimerized stereospecifically on the metal ion to give **13** with the pairs of phenyl and methyl *trans* to each other across the ring.[51]

10.2.4 Ligand migration and insertion

These reactions involve both the ligand and the metal to which it is bonded. As important examples are all organometallic, discussion of this class of reactions is dealt with later in the chapter, in Section 10.5.

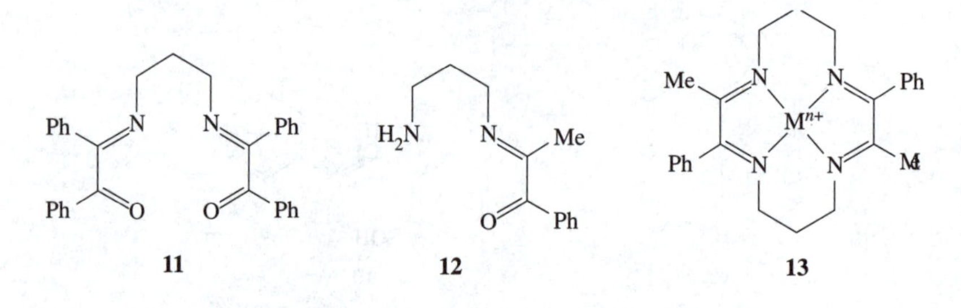

10.2.5 Electron transfer

It could be claimed that all inner-sphere redox reactions (Section 9.4) are reactions of coordinated ligands, but there are also numerous examples where ligand oxidation leaves the central metal ion and some or all donor atoms unaffected. The earliest example is probably Werner's demonstration of oxidation of the coordinated thiocyanate in $[Co(NH_3)_5(NCS)]^{2+}$ to ammonia, which he established in the process of showing the thiocyanate in $[Co(NH_3)_5(NCS)]^{2+}$ to be *N*-bonded to the cobalt. Oxidation of this complex by hydrogen peroxide obeys the simple second-order rate law

$$[Co(NH_3)_5(NCS)]^{2+} + H_2O_2 = k_2[Co(NH_3)_5(NCS)^{2+}][H_2O_2]$$

(k_2 is acid-dependent).[52] More recently, a number of examples involving oxygen transfer to coordinated organic thio-ligands have been established. Thus oxidation of coordinated mercaptoacetate by hydrogen peroxide, peroxodisulphate, or periodate results in transfer of first one oxygen and then, with periodate, of a second to the donor sulphur atom:

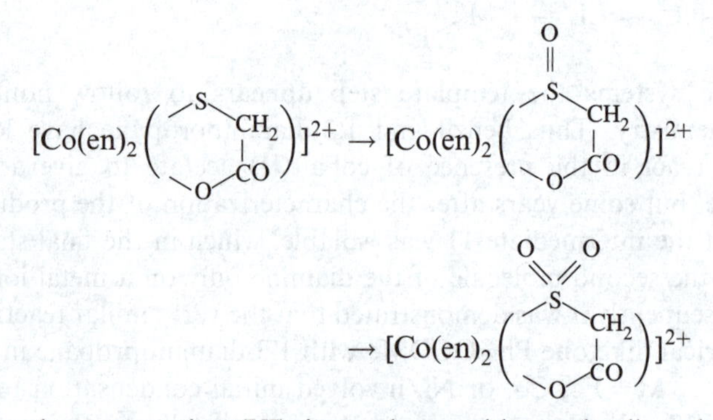

However, when neptunium(VI) is used as oxidant, the ligand is instead oxidized to thiooxalate; an analogous product is obtained from cerium(IV) oxidation of the chromium(III) complex:

$$[M(en)_2(S\text{–}CH_2\text{–}CO\text{–}O)]^{2+} \rightarrow [M(en)_2(S\text{–}CO\text{–}CO\text{–}O)]^{2+}$$

There is a lot of fascinating, and not yet fully elucidated, chemistry involved in such reactions as these, and coordinated-ligand oxidations in similar complexes such as $[Co(en)_2(SCH_2CH_2NH_2)]^{2+}$, where four different ligand-oxidized products may be obtained by the use of different oxidizing agents.[53] The reaction of $[Co(NH_3)_5(S_2O_3)]^+$ with peroxomonosulphate, SO_5^{2-}, is simpler. This takes place by two successive nucleophilic attacks of peroxo-oxygen at sulphur transferring two oxygen atoms to the coordinated thiosulphate to give $[Co(NH_3)_5(S_2O_5)]^+$.[54]

The metal can act as a source or sink of electrons and thereby serve to reduce or oxidize a coordinated ligand. Thus the ruthenium(III) ammine complex $[Ru(NH_3)_6]^{3+}$ reacts with biacetyl to give a ruthenium(II)-diimine complex as final product:

$$[Ru(NH_3)_6]^{3+} + MeCOCOMe \xrightarrow{OH^-} \left[(H_3N)_4Ru(HN{=}C(Me)C(Me){=}NH)\right]^{2+}$$

The redox step is presumably driven by the particularly high stability of the $t_{2g}^{\ 6}$-diimine grouping. In similar vein, $[Ru(NH_3)_6]^{3+}$ reacts with nitriles to give as final products the ruthenium(II) complexes $[Ru(NH_3)_5(NCR)]^{2+}$.[55] The metal centre in the iron(III) complex $[Fe(CN)_4(en)]^-$ is actually *reduced* by dioxygen, in basic solution, to iron(II), while the 1,2-ethanediamine ligand is oxidized to the diimine HN=CHCH=NH. The product, $[Fe(CN)_4(HN{=}CHCH{=}NH)]^{2-}$, is particularly stable, as the $t_{2g}^{\ 6}$ configuration of the iron(II) permits synergic π-back bonding (metal-to-ligand charge transfer) to the HN=CHCH=NH to reinforce the usual σ-bonding from the nitrogen donors. 1,2-Ethanediamine coordinated to ruthenium(II) also readily undergoes oxidation to coordinated HN=CHCH=NH, again stabilized by the $t_{2g}^{\ 6}$ Ru^{2+}.[11]

The dual ligand-activation (or template) and electron transfer function of metal ions is one of their major roles in many bioinorganic systems, particularly metalloenzymes. At other times the metal ion can act as the transmitter of electrons and facilitate electron transfer between coordinated ligands. The change of a pair of electrons in the central atom from a non-bonding to a bonding role can formally reduce a single molecule to two donor fragments upon which further chemistry might be carried out. This is an alternative way of looking at the chemistry involved in *oxidative addition.* The reverse process, *reductive elimination*, can combine two donor fragments in the same complex to form one molecule, generally quickly lost from the metal centre, with two electrons going into a non-bonding orbital on the metal ion. These processes occur in *sp*-block chemistry as well as in the chemistry of classical *d*-block coordination compounds, but their main importance is in organometallic chemistry. The conventional approaches to oxidative addition and reductive elimination are set out in Section 10.4, in the context of the organometallic reactions which dominate the central portion of this chapter.

10.3 Small molecules: addition and activation

Section 10.2 dealt with the activation of coordinated ligands; this section deals with the preliminary step of coordination of small molecules to metal centres, and their consequent activation[56] to onward reaction. Most of the molecules considered are themselves strongly bonded, for example dioxygen, dinitrogen, carbon monoxide and dioxide, and alkanes. Therefore either a considerable input of energy or an appropriate catalyst is required to enable them to indulge in bond-breaking reactions. Interaction with a suitably complexed metal ion may provide sufficient modification of their properties for reaction to occur under mild conditions. Both bioinorganic and organometallic systems have been much studied in respect to both the addition of small molecules and their activation.[57] Organometallic examples will appear frequently in the subsequent section on homogeneous catalysis (Section 10.7).

There are several ways in which a small molecule can approach and bind to a metal centre. Thus for reaction with nitrogen or oxygen there is a choice of bonding mode, η^1 or η^2, in the product. The former mode, namely M–Y–Y , is possible thanks to the lone pairs on these molecules. The latter mode may be represented as either

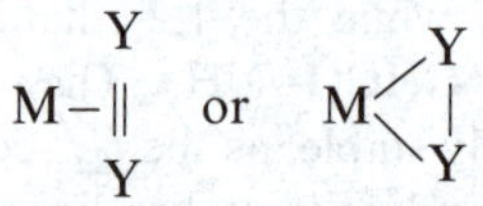

$$\mathrm{M}-\overset{\mathrm{Y}}{\underset{\mathrm{Y}}{\|}} \quad \text{or} \quad \mathrm{M}\begin{matrix}\diagup\mathrm{Y}\\ \quad|\\ \diagdown\mathrm{Y}\end{matrix}$$

The choice between these two representations for η^2-attachment also applies to alkenes and alkynes (to which of course the η^1-mode is inapplicable), and indeed to carbon dioxide, ketones, and almost any other molecule with π-bonding. The choice of bonding model has implications as to metal oxidation states and ligand charges, as will be discussed in individual cases in the following sections.

10.3.1 Oxygen[58–61]

Complexes of the dioxygen molecule play a vital role in bioinorganic chemistry, and have spawned an extensive chemistry of model oxygen carriers.[62] Dioxygen complexes also feature in organometallic chemistry. Here dioxygen is normally found η^2-bonded, as in the products of reaction with platinum(0) and iridium(I) complexes, e.g.:

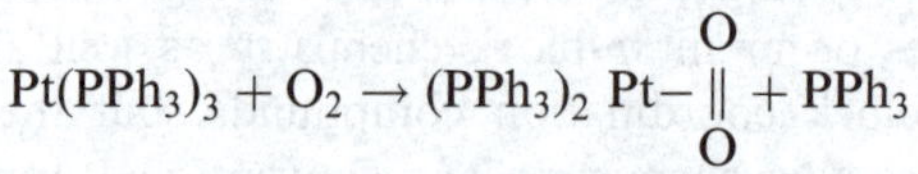

$$\mathrm{Pt(PPh_3)_3} + \mathrm{O_2} \rightarrow \mathrm{(PPh_3)_2\ Pt}-\overset{\mathrm{O}}{\underset{\mathrm{O}}{\|}} + \mathrm{PPh_3}$$

On the other hand, dioxygen is normally η^1-bonded in its complexes with bioinorganic molecules, and when added to first-row transition metal complexes in oxidation state 2+ and higher. The question arises as to

Table 10.7 The assignment of formal oxidation states in dioxygen complexes from oxygen–oxygen bond distances.

Dioxygen species	O–O bond distance (Å)	Dioxygen complex	O–O bond distance (Å)	Formal oxidation state of metal
O_2 molecule	1.21	Myoglobin.O_2	1.21	2+
		Haemoglobin.O_2	1.22	2+
		[Fe(ppp)(Me imid)(O_2)] [c]	1.22	2+
Superoxide, O_2^-	1.28 [a]	[Co(bzacen)(py)(O_2)]	1.26	3+
		$[Cr(O_2)_4]^{3-}$ [d]	1.41	5+
		$[Rh(diphos)_2(O_2)]^+$	1.42	3+
		$[Pt(PPh_3)_2(O_2)]$	1.45	2+
		$[Rh(AsPhMe_2)_4(O_2)]^+$	1.46	3+
		$[IrCl(CO)(PPh_2Et)_2(O_2)]$ [e]	1.47	3+
Peroxide, O_2^{2-}	1.49 [b]	$[Ir(PMe_2Ph)_4(O_2)]^+$	1.49	3+
		$[IrI(CO)(PPh_3)_2(O_2)]$	1.51	3+
		$[Ir(dppe)_2(O_2)]^+$	1.52	3+
		$[Mo(O_2)_4]^{2-}$	1.55	6+
		$[Ir(diphos)_2(O_2)]^+$	1.63	3+

[a] In KO_2.
[b] In BaO_2.
[c] ppp = picket fence porphyrin.
[d] O–O distances range from 1.31 to 1.44 Å for a series of chromium–N-donor–O_2 complexes (A. F. Wells, *Structural Inorganic Chemistry*, 5th edn, Clarendon Press, Oxford, 1984, pp. 503–4).
[e] The early report of O–O = 1.30 Å in $[IrCl(CO)(PPh_3)_2(O_2)]$ should be treated with some suspicion.

whether the M–O–O product should be regarded as a complex of peroxide, O_2^{2-}, superoxide, O_2^-, or simply the uncharged O_2 molecule.[63] The oxidation state of the metal is a corollary to this—thus for the binding of dioxygen to a metal(II) complex in these three forms, the formal oxidation states of the metal in the product are M^{IV}, M^{III}, or M^{II}, respectively. This matter is not altogether clear-cut, but it is possible to obtain indications of formal oxidation states from such properties as oxygen–oxygen bond distances (see Table 10.7)[59,64] and $\nu_{O\text{–}O}$ in vibrational spectra of these species. However, the spread of oxygen–oxygen distances for the peroxide entries in Table 10.7, and the effects of spectator ligands on $\nu_{O\text{–}O}$ in a series of dioxygen adducts of Vaska lookalikes (Table 10.8),[65] counsel caution in using such evidence. The reversibility or irreversibility of oxygen uptake provides another, qualitative, indicator of bonding and strength of interaction in this type of adduct. The relation of reversibility to $\nu_{O\text{–}O}$ is shown in Table 10.8; in similar vein, short O–O distances are associated with reversible uptake. Oxygen–oxygen distances, $\nu_{O\text{–}O}$ values, and (ir)reversibility of oxygen uptake all give some information about electron distribution and thus about consequent activation possibilities for the coordinated dioxygen molecule.

Iron is in the Fe^{II} form in the oxygen adducts of haemoglobin and of

Table 10.8 Ligand effects on O–O stretches[a] for dioxygen adducts of *trans*-$[IrX(CO)(PPh_3)_2]$.

X	ν_{O-O} (cm^{-1})	
Me	827	irreversible oxygen uptake
C≡CPh	835	
SH	845	
I	850	
NCS	855	reversible oxygen uptake
NCO	855	
Cl	855	

[a] ν_{O-O} in the dioxygen molecule is at 1580 cm^{-1}, in O_2^- at 1097 cm^{-1} (in LiO_2), and in O_2^{2-} at 802 cm^{-1} (in Li_2O_2).

myoglobin, but Fe^{III} in all but the most carefully designed models. Oxygenation of cobalt(II) complexes generally gives products which are formally superoxide complexes of cobalt(III), though a small number of peroxocobalt(III) and of cobalt(II) complexes with η^2-dioxygen are known.[59] Oxygenation of manganese complexes gives products which, although still not as well characterized as iron and cobalt analogues, may best be regarded as peroxomanganese(IV) species.[66] Structures of a number of superoxo and peroxo complexes of chromium, molybdenum, and tungsten have been documented;[67] there are η^2-peroxide complexes of a further dozen or so *d*-block elements, and of uranium, known. This group of peroxide complexes have O–O distances within the range 1.43–1.51 Å.

There is a considerable body of kinetic data on the binding of oxygen to, and its release from, model and bioactive iron complexes (summarized extremely briefly in Table 10.9).[68–70] Such kinetic data are often obtained from photolysis experiments, either starting with the dioxygen adduct or from a carbon monoxide adduct in the presence of dioxygen.[71] The dependence of rate constants on ligand nature has been explored in detail for iron-porphyrins. A range of rate constants of at least 10^4-fold has been attained by various modifications, generally engineered to prevent the iron(II) being oxidized to iron(III) by the oxygen. We are in the land of tulips (adamantyl petals), picket fences, strapped and capped haems, and other fanciful, descriptive, and wildly non-systematic nomenclature—one review contains more than four pages filled with their elaborate formulae, amply illustrating the strategies needed to prevent oxidation of the metal to iron(III) in potential model oxygen carriers.[69]

The existence of four haem centres in haemoglobin, their co-operativity, and the reversibility of oxygen addition made precise and detailed kinetic studies and analysis of its oxygenation difficult, but individual rate constants are now available for addition to the four iron atoms. The problems with haemoglobin kinetics have been well documented in the proceedings of a

Table 10.9 Rate constants for the reaction of natural and synthetic iron oxygen carriers with dioxygen.[a]

Substrate	k_2 ($dm^3\ mol^{-1}\ s^{-1}$)
Myoglobin (sperm whale)	1.5×10^7 [b]
Haemoglobin	$0.3–7 \times 10^7$ [c]
Haemerythrin	$0.7–8 \times 10^7$ [d]
Synthetic oxygen carriers	$10^5–7 \times 10^8$ [e]

[a] Rate constants at 293 or 298 K; natural carriers in water, models generally in toluene.
[b] Horse myoglobin has a very similar rate constant, 1.4×10^7, and activation parameters $\Delta H^\ddagger = 20\ kJ\ mol^{-1}$ and $\Delta S^\ddagger = -150\ J\ K^{-1}\ mol^{-1}$.
[c] This range covers rate constants for the four iron centres in haemoglobins from various sources.
[d] This range covers rate constants for a number of haemerythrins, from monomer to octamers.
[e] The majority of the nearly a hundred values, for some modified myoglobins and haemoglobins as well as for synthetic complexes, that have been reported lie within the range $1–30 \times 10^7$.

conference session devoted to these reactions.[72] The inorganic kineticist will be intrigued by the fact that such mercury compounds as methylmercury(II) hydroxide and the diuretic mersalyl (**14**), though not mercury(II) chloride, bound to –SH groups near the active site actually increase the affinity of

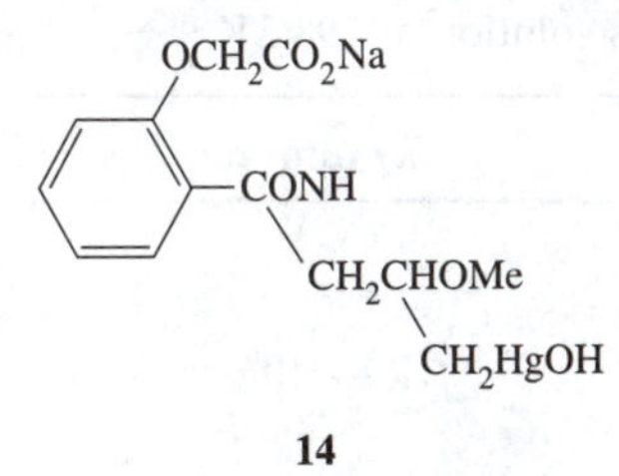

14

haemoglobin for oxygen.[73] Myoglobin, with its single iron centre, is more tractable. Haemerythrin, which in its native form is an octamer with four di-iron centres, exhibits co-operativity in some of its forms, but here co-operativity is much less marked than for haemoglobin. Rate constants for reactions of myoglobin[68,69] and of haemerythrins[70,74] with dioxygen are of similar magnitude to those for haemoglobin, and indeed of many synthetic iron-containing oxygen carriers (Table 10.9).

Oxygen uptake by certain classes of cobalt(II) complexes has attracted interest for many years,[75] in the hope that a reversible oxygen carrier system could be developed. Interest seems to centre on this property, to the virtual exclusion of interest in the cobalt activating the dioxygen to further reaction. Indeed, the activation of coordinated dioxygen which encourages it to bind to a second cobalt, either as bridging superoxide or peroxide, i.e.

$$[Co(L)(H_2O)_2]^{2+} + O_2 \rightleftharpoons [Co(L)(H_2O)(O_2)]^{2+}$$

$$\overset{[Co(L)(H_2O)_2]^{2+}}{\rightleftharpoons} [(H_2O)(L)Co(\mu - O_2)Co(L)(H_2O)]^{2+}$$

provides complications in establishing kinetic and thermodynamic properties. There are some kinetic data available, both on the addition of oxygen to such complexes and on its dissociation from the adducts. Table 10.10 shows a selection of rate constants for addition of oxygen to cobalt(II) complexes of polydentate nitrogen-donor ligands. The macrocyclic complexes react considerably more rapidly than the complexes of open-chain ligands. The products of these oxygenation reactions have oxygen–oxygen bond distances either in the range 1.26–1.36 Å or in the range 1.45–1.47 Å. The former may be regarded as superoxo complexes of cobalt(III), the latter as peroxo derivatives. However, ν_{O-O} frequencies often differ significantly from the values expected.[76]

For both iron and cobalt there has been considerable interest in tailoring the dimensions of the cavity adjacent to the metal into which the dioxygen molecule is to fit. The design of such lacunar complexes has been reviewed, along with their electrochemical and ESR properties, and possible mechanisms of dioxygen uptake.[77]

Table 10.10 Rate constants [a] for the reaction of cobalt(II) complexes with dioxygen in aqueous solution at 298.2 K

Complex	k_2 (dm^3 mol^{-1} s^{-1})
[Co(cyclam)]$^{2+}$ (structure: Co with four NH donors in 14-membered macrocycle)	1.2×10^7
(structure: Co with four NH donors in hexamethyl-substituted macrocycle, Me groups shown)$^{2+}$	5.0×10^6
$[Co(NH_3)_5(H_2O)]^{2+}$	2.5×10^4
$[Co(trien)(H_2O)_2]^{2+}$	2.5×10^4
$[Co(tetren)(H_2O)]^{2+}$	10^3

[a] The activation volume for binding dioxygen to $[Co(cyclam)]^{2+}$ is close to zero (Menglin Zhang, R. van Eldik, J. H. Espenson, and A. Bakac, *Inorg. Chem.*, 1994, **33**, 130).

Copper is another metal of importance for transport of dioxygen in living systems, with haemocyanin a ubiquitous dioxygen carrier for invertebrates.[78] The key inorganic feature of the haemocyanins consists of a pair of copper(I) ions. Haemocyanins from different species contain different numbers of such sub-units—the tarantula has 24 sub-units grouped in four entities—but the chemistry of dioxygen addition is $Cu_2{}^{I} + O_2 \rightleftharpoons Cu_2{}^{II}O_2$, with the product containing μ^2-η^2,η^2-peroxide (O–O = 1.4 Å),[79] and each copper having three histidine nitrogens attached, in all cases. Other bio-important copper-containing enzymes utilizing oxygen include tyrosinase, cytochrome *c* oxidase, and a range of oxygenases, but these do not involve simple dioxygen addition. Returning to the haemocyanins, there is some kinetic information on dioxygen take-up. It is known that there is significant co-operativity, and it is believed that a mononuclear $Cu \cdot O_2$ π-complex is an intermediate. There are some kinetic data[80] on model systems.[81] In particular, a significant substituent effect, of about 25-fold in uptake rate constant in going from nitro through t-butyl to fluoro, has been established for a system containing dinucleating ligands with pyridyl and amine-donor sites. Activation entropies are, as expected, negative for all these complexes; activation enthalpies show a surprisingly wide range, from 6 kJ mol^{-1} for the nitro-substituted complex to 29 kJ mol^{-1} for the fluoro. These substituent effects are ascribed, at least in part, to the effect of the substituent on the Cu–Cu distance. As this decreases, the rate constant for dioxygen uptake increases, tending towards the value for haemocyanin (6×10^7 dm^3 mol^{-1} s^{-1} for the monomer, in aqueous solution[82]). In a related study, kinetic and thermodynamic properties of mononuclear model complexes have been investigated in an attempt to increase understanding of the dinuclear models and thence of the real thing.

This section has so far been concerned mainly with the uptake of dioxygen—with the essential prerequisite rather than with the results of activation. We shall therefore close with a system which provides an example of dioxygen activation, and draws together its organic/organometallic and bioinorganic aspects. It is well established that bleomycin can activate dioxygen in the presence of certain iron-containing species. Thus, for instance, β-methylstyrene can be oxidized by dioxygen activated by bleomycin bonded to Fe^{III}/H_2O_2, Fe^{III}/PhIO, or Fe^{II}/O_2 to a mixture of products.[83] In general such oxidizing mixtures oxidize alkenes to a mixture of products, especially their epoxides, and can in turn result in oxidative cleavage:

$$R^1HC{=}CHR^2 \longrightarrow R^1HC\overset{O}{\frown}CHR^2 \longrightarrow R^1CHO + R^2CHO$$

10.3.2 Nitrogen

The dinitrogen molecule usually bonds end-on, i.e. η^1, to metal centres, in classical coordination complexes, in organometallic derivatives such as

$[Ni(PEt_3)_3(N_2)]$ and $[Mo(PMe_3)_5(N_2)]$,[84] and in the myriad inorganic models for nitrogen fixation. There are also a few η^2 and bridged complexes, and a number of matrix-isolated complexes of uncertain structure, such as $[NiCl_2(N_2)]$ and $[M(N_2)_n]$ with M = Ni, Pd, Pt. The N–N bond distance in the original coordination complex $[Ru(NH_3)_5(N_2)]^{2+}$ is 1.09 Å, only very slightly shorter than in the dinitrogen molecule, suggesting retention of most of the triple-bond character in the complex. There are a number of similar transition metal complexes containing dinitrogen bonded to, *inter alia*, cobalt, technetium, rhenium, chromium, molybdenum, and tungsten. The complexes of these last two metals are the most important, for they have proved the most amenable for inclusion in nitrogen fixation sytems. The coordination of dinitrogen activates it to electrophilic attack, particularly by the proton in the first step in reduction to coordinated –N=NH, =N–NH_2, =NH, and other species en route to the desired hydrazine and ammonia (Section 10.7.5.2). Coordinated nitrogen is also activated with respect to reaction with acyl halides (to give coordinated –N=NCOR), alkyl halides (radical reactions), and to nucleophilic attack by, for example, lithium or magnesium alkyls.

10.3.3 Carbon monoxide and isonitriles

Carbon monoxide is an effective, and potentially lethal, competitor for haemoglobin. The stronger affinity of haemoglobin and myoglobin for CO than for oxygen is actually a boon in the context of inorganic mechanisms, for it simplifies the kinetic analysis from that for reversible reactions to that for reactions going to completion. This is particularly helpful for haemoglobin, where there is the added complication of four potentially reactive iron centres.[85] The reaction sequence for interaction between haemoglobin and carbon monoxide was set up more than seventy years ago,[86] but there was a thirty-year gap before the development of photolytic and stopped-flow techniques permitted the establishment of the respective rate constants.[87] These reflected co-operativity dramatically, for whereas the four rate constants for the successive additions of four carbon monoxide molecules would on statistical grounds be expected to be in the ratio 4 : 3 : 2 : 1, the observed rate constants were 4 : 6 : 2 : 80.[88] The relevance of isonitriles at this point is that the kinetics of their addition to iron in haem species have also been studied successfully, permitting the investigation of steric effects, as have those for nitric oxide.[89]

Kinetics of haemoglobin or of myoglobin binding to small molecules have generally been studied by monitoring the products of flash dissociation of the appropriate adduct, or of the carbon monoxide adduct in the presence of a large excess of oxygen. In the case of the carbon monoxide adduct, recombination kinetics could be studied advantageously by time-resolved infrared spectroscopy.[90] Some kinetic results are summarized in Table

Table 10.11 Rate constants for the addition of small molecules to haem derivatives in aqueous media at temperatures between 293.2 and 298.2 K.

	k_2 (dm^3 mol^{-1} s^{-1})			
	+CO	+O_2	+RNC	
Myoglobin (sperm whale)	5×10^5	1.5×10^7	MeNC	1.0×10^5
			^{n}BuNC	3.7×10^4
			^{n}HxNC	2.5×10^4
			TMIC [a]	2.3×10^2
Haemoglobin	6×10^6	$0.3–7 \times 10^7$		
Chelated protohaem [b]	4×10^6	2.6×10^7	^{n}BuNC	2×10^8
			TMIC	2×10^8
Synthetic haem	8.2×10^6	3.5×10^7	RNC [c]	$1.4–2.1 \times 10^8$

[a] TMIC = tosylmethyl isocyanide.
[b] Oversimple model compound.
[c] R groups are methyl, ethyl, n-propyl, n-butyl, n-amyl, n-hexyl, iso-propyl and t-butyl.

10.11.[59,72,90,91] This compares rate constants for addition of carbon monoxide, oxygen, and isonitriles to myoglobin, haemoglobin, and simple model compounds. The isonitrile results illustrate the importance of the structure of the adding molecule for addition to myoglobin, with a marked decrease in rate constant as bulk increases and approach to the active site becomes hindered. Hindrance can also be induced by use of viscous solvents, such as aqueous glycerol or glycol.[92] Uncluttered model compounds react very rapidly even with bulky isonitriles.[93] Table 10.11 also gives the range covered by the four individual rate constants for addition of oxygen to the four individual iron centres (*R* and *T* states; α- and β-chains) of haemoglobin. They are of similar magnitude to that for myoglobin, and to that for reaction of haemoglobin with carbon monoxide. The question of how myoglobin and haemoglobin distinguish between carbon monoxide and oxygen has been addressed by the design of a series of mutant myoglobins and haemoglobins, and a kinetic assessment of the detailed way in which selectivity works in these systems.[94]

Whereas the addition of carbon monoxide is a matter of considerable bioinorganic interest, its activation is rather in the realms of organometallic chemistry. Examples include the water gas shift reaction, hydroformylation, and the Monsanto process for acetic acid manufacture, all of which appear in Section 10.7.

10.3.4 Carbon dioxide; sulphur dioxide; nitrogen dioxide

There has recently been considerable interest in carbon dioxide activation[95] in or by organometallic systems, both in relation to synthesis of organic compounds from this very simple and (too) abundantly available feedstock and in relation to its removal from the atmosphere. Of course bioinorganic activation, in the sense of hydration by carbonic anhydrase (see Section 10.2.2.4) and of photosynthesis, has been around for a very long time and been much studied by biological chemists.

Both η^1- and η^2-bonding modes are well established for mononuclear complexes, while carbon dioxide can also act as a μ_2 and a μ_3 bridging ligand. Protonation of η^1-CO_2, the favoured bonding mode to, e.g., Co^I in tetraaza-macrocyclic derivatives, Rh^I, and Ir^I, gives coordinated carboxylate; there is only a little qualitative kinetic information on subsequent extrusion of carbon dioxide ($M-CO_2H \rightarrow M-H + CO_2$). Fixation by insertion is a possibility (cf. Section 10.5.5.1); alkenes, alkynes, conjugated dienes, allenes, and strained carbocyclic rings can all be carboxylated with appropriate metal promotion. Cyclic carbonates (useful solvents) and polycarbonates are also accessible. Indeed, carbon dioxide may undergo nucleophilic attack at carbon, electrophilic attack at oxygen, or electron transfer, providing a large number of possibilities for metal-promoted activation or fixation.[96] These organometallic approaches are being complemented by studies and modelling of biological fixation at a number of metal centres, ranging from iron–sulphur proteins and ruthenium complexes to aluminium porphyrins.[97]

Many of the oxides and oxoanions of nitrogen and of sulphur are serious environmental contaminants. The lower oxidation state species are fairly reactive, but it may be desirable to activate them for more rapid conversion into less undesirable species. The commonest approach is that of oxidation by dioxygen, in which catalysis by traces of metal ions which are effective in redox cycling is important.[98]

10.3.5 Alkenes and alkynes

In contrast to most of the small molecules in this section, namely H_2, O_2, N_2, CO, CO_2, and alkanes, alkenes and alkynes are, like NO_2 and SO_2, quite reactive molecules. Nonetheless, they may become much more reactive on coordination to an appropriate metal centre. Ethene is unreactive towards nucleophiles such as $[(MeO_2C)_2CH]^-$, but this anion reacts at room temperature with ethene coordinated to, and activated by, iron in **15**.

The addition of an alkene or an alkyne to a low oxidation state transition metal centre with a low coordination number may be regarded as a simple addition, or to be oxidative in nature:[99]

$$M^N + C{=}C \rightarrow M^{N+2}\!\left\langle\begin{matrix} C \\ | \\ C \end{matrix}\right. \quad \textit{or} \quad M^N - \begin{matrix} C \\ \| \\ C \end{matrix}$$

The former metallocyclopropane representation is probably the closer approximation for, e.g., C_2F_4, $C_2(CF_3)_4$, or $C_2(CN)_4$ (tcne), the latter for, e.g., C_2H_4 or, with appropriate alkene to alkyne modification, C_2H_2.[100]

Where any particular compound lies on the path linking the two extreme representations may be assessed in several ways, which are often in tolerable rather than good agreement. Thus, for example, infrared spectroscopy has been used, often monitoring $\upsilon_{C=O}$ for an adjacent carbonyl ligand.[101] NMR spectroscopy has proved informative, as in studies of adducts [Ni(alkene){P(O-*o*-tolyl)$_3$}$_2$].[102] Alternatively, bond distances and bond angles for the coordinated alkene or alkyne can be used to assess whether the C–C bond approximates to a single or double bond or whether the carbon atoms appear nearer to sp^2 or sp^3 hybridization. Thus, the geometry of the alkyne complex derived from Pt(PPh$_3$)$_2$ and diphenylacetylene, **16**, indicates that it approximates to an alkene complex of platinum(II). Similarly, the C–C bond distance for coordinated tetracyanoethene, at 1.51 Å, and the almost tetrahedral geometry of the coordinated carbon atoms, indicate that [IrBr(CO)(PPh$_3$)$_2$(tcne)] is best regarded as an iridium(III) complex. Indeed, the metal–tcne geometry here, and in an analogous nickel–tcne complex, is remarkably similar to that of the epoxide **17**.[103] However, it should be recognized that there is a gradation of bond distances in alkyne and alkene complexes. Values range from 1.26 to 1.33 Å for the former (C–C = 1.20 Å in ethyne, 1.34 Å in ethene), while the C–C distance in coordinated ethene is 1.37 Å in Zeise's salt, **18**, but 1.47 Å (C–C = 1.54 Å in ethane) in its dimethylamine analogue, **19**.

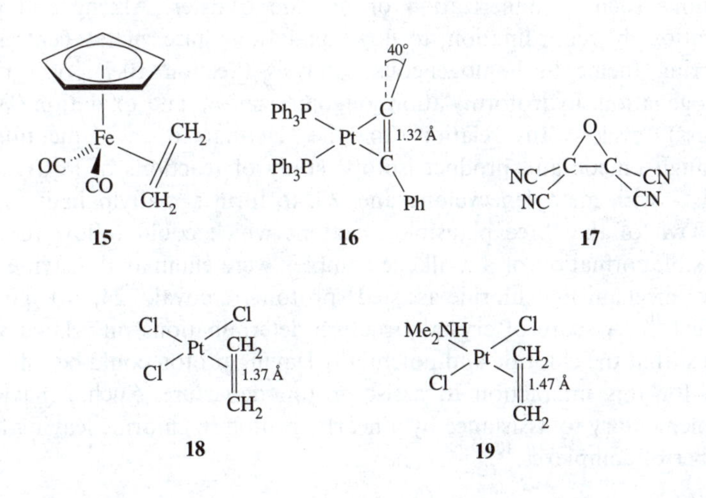

The geometry of coordinated dienes again reflects single/double-bond character. This varies as one goes from left to right across the transition metals, as indicated by **20** and **21**, which show the relevant bond distances in the coordinated segments of [Zr(cp)$_2$(η^4-dimethylbutadiene)] and [Fe(CO)(η^4-cyclohexadiene)], respectively. Bonding differences are also reflected in the metal–carbon distances. The complex [Rh(cp)(C$_2$H$_4$)(C$_2$F$_4$)] has two alkenes separately attached to the metal. Here there is a kinetic reflection of the difference in strength of bonding. The coordinated ethene shows rotation in low-temperature NMR spectra, with an activation barrier of 57 kJ mol^{-1}, whereas the coordinated C_2F_4 shows no sign of rotation even at 100 °C.[104]

The main mechanistic question in relation to the formation of alkene and alkyne complexes is whether their initial approach to the metal centre is end-on (η^1) or broadside-on (η^2). Several pieces of evidence indicate that, especially for electrophilic alkenes such as C_2F_4, the favoured route is η^2-approach. Thus the very small solvent effects on rate constants disfavour the polar transition state expected for η^1-attack, while the stereospecificity of reaction with the alkene **22** favours η^2-attack.[105] Once an alkene or

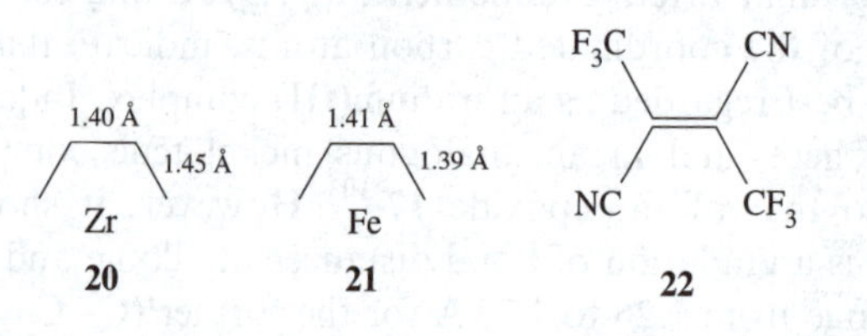

alkyne is coordinated, it may undergo a variety of reactions. Both electrophilic and nucleophilic attack are possible, as are intramolecular reactions such as dimerization or hydride transfer. Alkene and alkyne activation by coordination to low oxidation state metal centres is a recurring theme in homogeneous catalysis (Section 10.7), occurring in hydrogenation, hydroformylation, oligomerization, and oxidation (Wacker process) cycles. In relation to the last-named, we mention the thorough kinetic and product isotope study of reactions of $PdCl_2$ and of $PdCl_4^{2-}$ with methylenecyclohexane, **23**, to form a π-allylpalladium product. Two of the three plausible reactions which could follow the initial reversible formation of a π-alkene complex were eliminated, leaving as the likely mechanism chloride-assisted proton removal, **24**, to give the product.[106] A score of crystal structure determinations on related species suggest that the chloride and potentially leaving proton could be sufficiently close for this interaction to assist proton departure. Such behaviour is complementary to assistance by a nearby proton in chloride leaving from a number of complexes.[107]

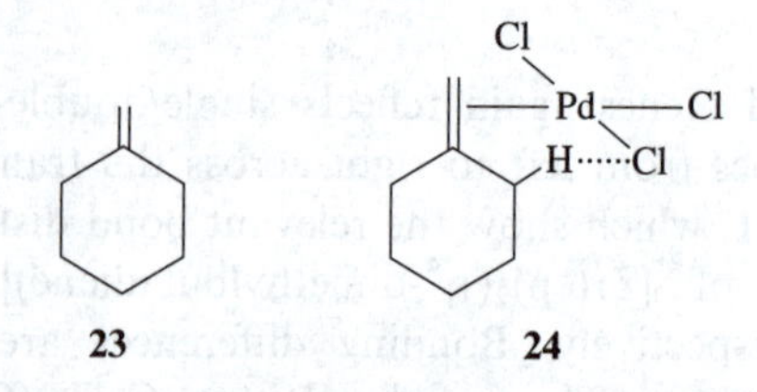

We mentioned the activation of aromatic hydrocarbons to nucleophilic attack upon coordination to such moieties as $-Cr(CO)_3$ in Section 10.2.1.

In such systems benzene is η^6-bonded to the metal. Benzene, and substituted derivatives, can also coordinate η^2 to a metal, as in $[Os(NH_3)_5(\eta^2\text{-}C_6H_6)]^{2+}$, where it behaves as an activated alkene. Thus, for instance, this η^2-coordinated benzene can be reduced to η^2-coordinated cyclohexene under much milder conditions than free benzene.[108]

10.3.6 Alkanes; hydrogen

It is not possible for alkanes or for dihydrogen to form end-on-bonded complexes, as they lack the requisite lone pairs. Structures of the type shown as **25** are also ruled out by the lack of π-bonding in these molecules. Furthermore, it is not possible to draw a conventional all-single-σ-bonded structure of the type shown as **26** for a three-centre adduct of dihydrogen or of an alkane. Maintaining electron-pair bonds in these adducts requires them to be formulated as products of oxidative addition. However, there are a number of compounds containing units whose geometry approximates to **26**, and such entities are likely intermediates for oxidative addition. In view of the dominance of oxidative addition in respect to these species, they will be discussed in that context (Section 10.4.1.2). Their importance is their role in activation, for the dihydride or alkylhydride products are reactive compounds, which may undergo further reaction, particularly reductive elimination. In this way alkanes and dihydrogen are activated to the formation of products or, in appropriate cases, to forming feedstocks for homogeneous catalysis cycles (Section 10.7). In almost every case, activation of alkanes is with respect to C–H bond breaking. C–C bonds are much harder to break, indeed are only susceptible to rupture in strained hydrocarbons such as cyclopropane. This does undergo C–C bond fission, to give **27**, on reaction with $PtCl_2$. Recent studies of activation of dihydrogen, methane, ethane, and ethene in such media as supercritical carbon dioxide, xenon, fluoroform, and hydrocarbons have proved promising in respect both of synthesis and of the elucidation of mechanisms.[110]

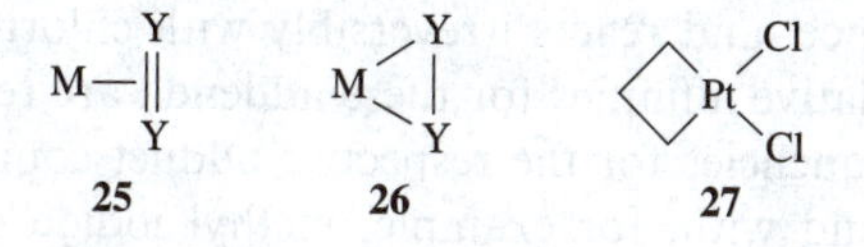

10.4 Oxidative addition and reductive elimination[111]

10.4.1 Introduction

10.4.1.1 Oxidative addition—general[112]

Oxidative addition is a special case of the addition reactions discussed in the preceding section. It is the reaction between two compounds involving bond fission in one and an increase in the formal oxidation state and coordination number of the central atom of the other on bonding the two fission fragments to it:

$$M^N L_n + XY \longrightarrow M^{N+2} L_n(X)(Y)$$

The change in oxidation state of the central atom is formally achieved by regarding the ligands X and Y to be both in anionic form, namely X^- and Y^-, in the product. There are a number of simple examples in the chemistry of the *p*- and *d*-block elements, e.g.

$$PCl_3 + Cl_2 \longrightarrow PCl_5$$

$$SnBr_2 + Br_2 \longrightarrow SnBr_4$$

$$PtCl_4^{2-} + Cl_2 \longrightarrow PtCl_6^{2-}$$

the formation of Grignard reagents

$$Mg + RX \longrightarrow RMgX$$

and, for complexes, e.g.

$$Sn^{II}(\text{maltolate})_2 + I_2, \longrightarrow Sn^{IV}(\text{maltolate})_2I_2$$

In each of these cases the increase in oxidation state is clear to all right-thinking chemists—we are avoiding the equivocal and ambiguous cases mentioned in Section 10.3.

The main interest in *mechanisms* of oxidative addition resides in slightly more complicated examples, particularly where these are model reactions for, or components of, homogeneous catalysis processes (cf. Section 10.7). One of the most studied *d*-block complexes has been Vaska's compound, the square-planar iridium(I) complex *trans*-$[IrCl(CO)(PPh_3)_2]$.[101] This complex reacts with molecules such as oxygen and sulphur dioxide to give adducts which dissociate quite readily, reacts with molecules such as hydrogen chloride, methyl iodide, and tetrafluoroethene to give stable adducts which dissociate with reluctance, and reacts irreversibly with chlorine and bromine. Interestingly, the relative affinities for these addenda are reflected in the carbonyl stretching frequencies for the respective adduct complexes. Reaction of Vaska's compound with, for example, methyl iodide gives a product whose octahedral stereochemistry, inertness, and other properties all indicate it to be a complex of iridium(III):

$$\textit{trans}\text{-}[IrCl(CO)(PPh_3)_2] + MeI \longrightarrow [IrMeClI(CO)(PPh_3)_2]$$

Analogous comments apply to the almost equally well-studied rhodium(I) complex $RhCl(PPh_3)_3$, sometimes called Wilkinson's complex, and to a number of other similar transition metal complexes. Sometimes the product reacts further, as in the following oxidative addition followed by insertion (ligand migration; Section 10.5):[113]

$$\textit{trans}\text{-}[RhCl(CO)(PR_3)_2] + MeI \longrightarrow [RhMeClI(CO)(PR_3)_2]$$

$$[RhMeClI(CO)(PR_3)_2] \longrightarrow [Rh(COMe)ClI(PR_3)_2]$$

Kinetic data (rate constants and activation parameters) were reported for complexes of several phosphines, PR_3. Acetylrhodium(III) complexes of this type, and higher homologues, are also available by direct oxidative addition:[114]

$$[RhCl(PMe_3)_3] + RCHO \longrightarrow [Rh(COR)ClH(PMe_3)_3]$$

As is clear from the previous paragraph, oxidative addition is an important reaction for square-planar d^8 complexes. Obviously its importance is restricted to lower oxidation states such as rhodium(I), iridium(I), palladium(II), and platinum(II). Oxidative addition is the main reaction of the first two, whereas for Pd(II) and Pt(II) substitution is much more important than oxidative addition. By the time we reach the 3+ oxidation state, e.g. gold(III) or copper(III), the central metal is itself a strong oxidant.

Oxidative addition is also of importance for tetrahedral d^{10} complexes, such as those of nickel(0) and of platinum(0), and for a number of five-coordinate compounds:

e.g. $Ru(CO)_5 + I_2 \longrightarrow \textit{cis}\text{-}[Ru(CO)_4I_2]$

In this and a number of similar cases the restriction to octahedral coordination for ruthenium(II) in the product means that one of the carbonyl ligands in the starting compound has to be eliminated to permit both iodine atoms (iodide ligands) to be coordinated in the product. For five-coordinate complexes with at least one ligand that leaves fairly easily, or can be photolytically removed (e.g. $[Ir(\eta^5\text{-}C_5Me_5)(H)_2(PMe_3)]$[115]), oxidative addition can take place at a four-coordinate intermediate:

$$ML_4L' \rightleftharpoons \quad \text{or} \quad \xrightarrow{h\nu} ML_4 + L'$$

$$ML_4 + RX \longrightarrow M(R)(X)L_4$$

Addition of hydrogen halides, halogens, or electrophiles can proceed by reaction with X^+ to give an octahedral species, which may undergo substitution to give the final product. Thus the first step in the reaction of $[Ir^I(CO)(dppe)_2]$ with HBF_4 is the oxidative addition of H^+ to give $\textit{cis}\text{-}[Ir^{III}(H)(CO)(dppe)_2]^{2+}$.[116] Similarly, reaction of $[Os(CO)_3(PPh_3)_2]$ with hydrohalic acids proceeds through addition of H^+ to give a cationic six-coordinate hydride intermediate, while reaction of this same osmium(0) complex with bromine or iodine involves addition of X^+ to the osmium to give $[OsX(CO)_3(PPh_3)_2]X$, with slow onward reaction to $[Os(CO)_2(PPh_3)_2(X)_2]$ and carbon monoxide.[117]

For elements favouring high cordination numbers, it has occasionally proved possible to oxidatively add to an octahedral substrate without concomitant ligand loss:[118]

$$\text{e.g.}\quad [Tc(diars)_2Cl_2]^+ + Cl_2 \longrightarrow [Tc(diars)_2Cl_4]^+$$

A couple of examples of oxidative addition to simple compounds of low coordination number were given at the start of this section. Oxidative addition to complexes and to organometallic compounds where the central element has a coordination number of two or three are also known, for example in the chemistry of gold(I) and of tin(II). Thus the transient dialkenyltin(II) $Sn(CPh{=}CPh_2)_2$ reacts with n-butyl bromide to give the tin(IV) product $Sn(CPh{=}CPh_2)_3{}^nBu$, bromide reacting with $Li(CPh{=}CPh_2)$ remaining from the initial generation of the dialkenyltin(II) from $SnCl_2$.[119]

In general, oxidative addition becomes less prevalent as one moves to the left-hand side of the *d*-block. Examples involving rhenium and molybdenum appear later in this section, while reactions of Cr^{2+} aq with organic radicals R· to give $Cr^{III}R^{2+}$[120] have already appeared in Section 9.7.6.5. In Group 4, $Zr(C_5H_5)_2$ and $Zr(C_5H_5)(PR_3)_2{}^+$ oxidatively add alkyl halides.[121,122] Overall, oxidative addition is a process of major importance in the organometallic chemistry of many of the *d*-block elements, of modest importance in the chemistry of several of the *f*- and *p*-block elements, but only of significance in one aspect of bioinorganic chemistry. This solitary case is that of vitamin B_{12}, where oxidative addition to cobalt(I) is central to the methylations which produce such essential compounds as methionine and such undesirable species as the methylmercury cation.

Just as a wide variety of transition metal complexes and *p*-block compounds can act as substrates for oxidative addition, so there is a considerable range of compounds which can oxidatively add, including even hydrocarbons[123] and water[124] (Table 10.12). In the case of water, the remarkable species $[Ir(H)(OH)(PMe_3)_4]^+$, whose identity was confirmed by

Table 10.12 Examples of compounds which oxidatively add to low-valence transition metal centres.[a]

Compounds whose addition involves bond breaking:	Compounds which add without any bond breaking:
$M + YZ \rightarrow M<^{Y}_{Z}$	$M + YZ \rightarrow M - \begin{matrix} Y \\ \Vert \\ Z \end{matrix}$
HX	O_2; SO_2; CS_2
PH_3; AsH_3; SbH_3	$F_2C{=}CF_2$; $RC{\equiv}CR'$
RX;[b] RCOX; R_3EX; SnX_4	$(CF_3)_2CO$
RHgX	RNCS; RNCO
H_2[c]	

[a] R = alkyl, perfluoroalkyl, or aryl; X = Cl, Br, or I; E = Si, Ge, or Sn.
[b] Including X = H, i.e. alkanes.
[c] See the following Section (§10.4.1.2) for a discussion of the extent of bond breaking in (oxidative) addition of the dihydrogen molecule.

neutron diffraction, was obtained.[125] Recent examples include addition of *trans*-[IrCl(CO(PPh$_3$)$_2$] to the currently fashionable buckminsterfullerenes (C_{60} and C_{70}),[126] and of iodine, phosphorus, and ferrocene to C_{60}.[127] Whether the latter products ($C_{60}(I_2)_2$; $C_{60}(P_4)_2$; $C_{60}\{Fe(cp)_2\}_2$—especially the last) are oxidative adducts or just inclusion compounds is not altogether clear. In some cases (left-hand column of Table 10.12) a bond in the adding molecule is broken, with the two parts effectively adding as separate ligands. Usually this type of oxidative addition results in an octahedral product of *trans* stereochemistry, but on occasion the product is the *cis* isomer, as for addition of PH_3 to *trans*-[IrX(CO)(PEt$_3$)$_2$],[128] or a mixture of *cis* and *trans* isomers, as in the addition of carboxylic acids to Vaska's compound.[129] In other cases (right-hand column) no bonds in the adding molecule break on bonding to the metal; here the adding molecule generally has a π-bond available for sideways coordination. Questions of bonding and metal oxidation state involved in this type of adduct have been considered earlier in this chapter (Section 10.3).

10.4.1.2 Dihydrogen, alkanes, and agostic interactions

In an earlier section (Section 10.3.5) we pointed out that simple bonding theories predict that addition of dihydrogen or of alkanes should be oxidative. Nonetheless, it is apparent, especially from structural information from X-ray and neutron diffraction studies, that addition of dihydrogen to a low oxidation state four- or five-coordinate metal centre does not by any means always give an octahedral *cis*-dihydride product, **28**. Often the product is better represented as **29**, with significant H–H interaction indicated by a short distance between the hydrogens. Structural evidence is supported by the persistence of $\nu_{H\text{-}H}$ in vibrational spectra of adducts, and by NMR evidence from, e.g., relaxation times (T_1).[130] The introduction to a theoretical treatment of the problem[131] is a convenient source of references; an extensive account of dihydrogen complexes provides context.[132] The first example of a dihydrogen compound demonstrated to contain an apparently intact H–H bond, as in **29**, was [W(H$_2$)(CO)$_3$(PCy$_3$)$_3$].[133] Vibrational spectra showed bands for $\nu_{H\text{-}H}$ (infrared; 2690 cm^{-1}), $\nu_{H\text{-}D}$ (infrared; 2360 cm^{-1}), and $\nu_{D\text{-}D}$ (a weak, broad Raman band at ~1900 cm^{-1}). Neutron diffraction indicated a D–D separation in this compound of 0.84 Å, little longer than the H–H distance of 0.74 Å in the free dihydrogen molecule. The H–H distance in [Fe(H)$_2$(H$_2$)(PEtPh$_2$)$_3$] is 0.82 Å; other examples have somewhat longer H–H distances. Some adducts appear to occupy intermediate positions between the two alternatives. Thus, for instance, a low-temperature neutron diffraction study[134] of [ReH$_7$\{(4MeC$_6$H$_4$)$_3$P\}$_2$] established H···H distances of 1.36 Å, midway between the distances expected for the dihydride (**28**) and η^2-dihydrogen (**29**) formulations. Other evidence indicates marked H···H interaction; this rhenium compound, and the cationic osmium complex *trans*-[Os(H)(H$_2$)(Et$_2$PCH$_2$CH$_2$PEt$_2$)$_2$]$^+$, probably mark the effective crossover from **29** to **28**. A detailed NMR study of another osmium(II) complex, $[Os(H)(Cl)(\eta^2\text{-}H_2)(CO)(P^iPr_3)_2]$ and its deutero-analogue, indicated a short H–H distance, of 0.8 Å, rapid spinning of the η^2-$H_2(D_2)$ ($\Delta H^\ddagger \sim 5\,\text{kJ mol}^{-1}$), and very slow exchange of hydrogen ($\Delta H^\ddagger = 73\,\text{kJ mol}^{-1}$) between the η^2-H_2 and the hydride ligand.[135]

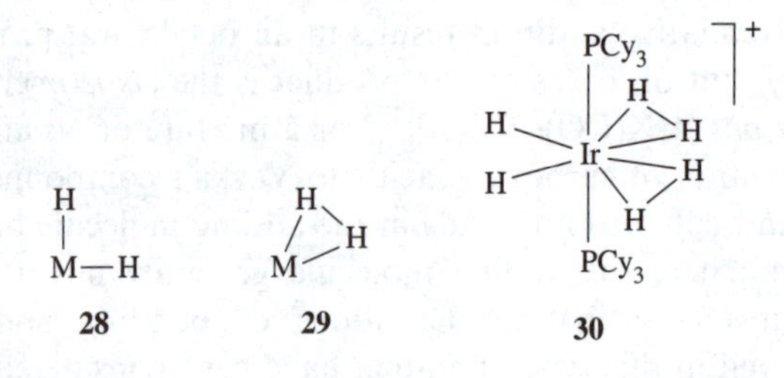

Both types of interaction may be present in the same compound, as in the osmium(II) complexes mentioned above and in the cation **30**. The latter is generated by protonation of the complex $[IrH_5(PCy_3)_2]$, which contains, like several of these η^2-dihydrogen complexes, the particularly bulky triscyclohexylphosphine ligand. The cation **30** contains iridium in oxidation state 3+, which is much more plausible than the 7+ required for reformulation as a hexahydride complex.[136] The formally eight-coordinate hydrides $[ML_4H_4]$ are probably just that for M = Mo and W, but may be mixed hydride–dihydrogen species $[ML_4(H)_2(H_2)]$ for M = Fe and Ru. The precise nature of the bonding may be affected by spectator ligands, as explored through *ab initio* calculations for iridium-hydrogen complexes,[137] while it has been suggested that the two forms might coexist in equilibrium for certain tungsten and palladium compounds.[138]

The rate law for the reversible addition of dihydrogen to such rhodium(I) complexes as $[Rh(dppe)_2]^+$, dpppe = $PhCH_2CH_2CH_2Ph$, takes the expected form:

$$-d[Rh(dpppe)_2{}^+]/dt = k_2[Rh(dpppe)_2{}^+][H_2] - k_{-1}[Rh(dpppe)_2H_2{}^+].$$

Values of k_2 depend greatly on the nature of the rhodium(I) complex, with, for instance, reaction of $[RhCl(PPh_3)_2(C_6H_6)]$ being at least 10^4 times faster than of $[RhCl(PPh_3)_3]$. Addition to iridium(I), in Vaska's compound, is somewhat slower than to rhodium(I), and shows the small solvent effect to be expected for broadside-on (η^2) incorporation of the dihydrogen into the transition state.[139] The H/D kinetic isotope effect on this reaction is consistent with this, implying a reactant-like, i.e. early, transition state.[140] There was originally some confusion over the stereochemistry of addition of dihydrogen—not in relation to the uniformly *cis* arrangement of the two hydrogen atoms, but to the remainder of the product. This confusion arose from isomerization subsequent to oxidative addition:

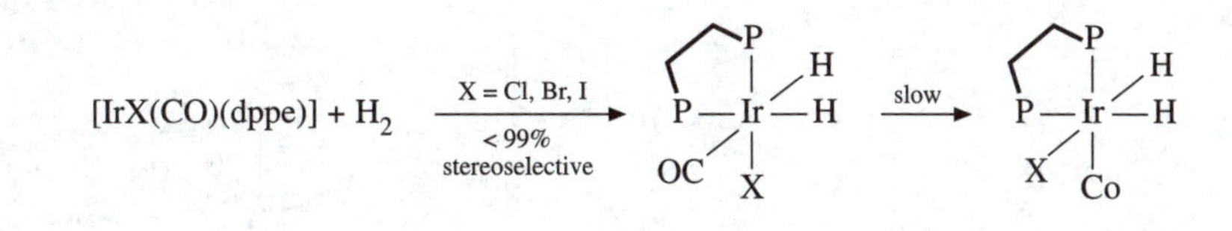

The kinetics of the second step indicated that bimolecular dihydride transfer took place concurrently with the expected reductive elimination–oxidative addition path. The former path contributed a term in the rate law second-order in the initial iridium(III) product, the latter a term first-order in this complex.[141] Oxidative addition of dihydrogen stripped down to its bare essentials is represented by a study of addition to iron atoms in low-temperature dihydrogen/noble gas matrices. Here there was no evidence for the reaction proceeding through FeH plus H atoms, but rather η^2-approach to an early transition state[142]—a conclusion consistent with calculations on the hypothetical reaction of $[Pt(PH_3)_2]$ with dihydrogen, and with the conclusions reached in the study of the reaction of dihydrogen with Vaska's compound (*v.s.*).

There is some information on the replacement of dihydrogen from its addition compounds. Thus a rate constant for its replacement from $(\eta^5\text{-}C_5H_5)Mn(CO)_2(\eta^2\text{-}H_2)$ by dinitrogen, to give $(\eta^5\text{-}C_5H_5)Mn(CO)_2(\eta^1\text{-}N_2)$, can be estimated from published repeat-scan spectra. This substitution was monitored in supercritical xenon, where the half-life is about half an hour at room temperature. Supercritical xenon offers the great advantage, both in relation to kinetics and to characterization, of giving minimal interference with vibrational spectra of solutes.[143]

Central to the question of alkane activation is the concept of *agostic interactions*,[144] in which a carbon–hydrogen bond acts as a ligand to a transition metal, e.g. manganese, iron, ruthenium, osmium, palladium, or tungsten, sharing an electron pair (**31**; cf. the intact C–H bond with the intact H–H in **29** above[145]). Such bonding is in a sense hydrogen bonding C–H$\cdots$M and may be compared with B–H–B bonding in electron-deficient boranes. It can be inferred from short metal–hydrogen distances in, for example, the formally sixteen-electron complex **32** and the alkyl compound **33** ($\widehat{PP}$ = dmpe, $Me_2PCH_2CH_2PMe_2$). Agostic interactions may also be inferred from bond angles, as in **34**. The significance in relation to reaction mechanisms is that this type of interaction may well precede hydrogen transfer in orthometallation reactions and in alkyl loss. If an orthometallation situation is set up for a complex of an element that is unable to undergo oxidative addition, then it can be possible to isolate and characterize the agostic species. Thus interaction between an ortho-hydrogen in a phenyl ring and the metal has been shown in the product of the reaction of $[Sm(\eta^5\text{-}C_5Me_5)_2(thf)_2]$ with azobenzene.[146]

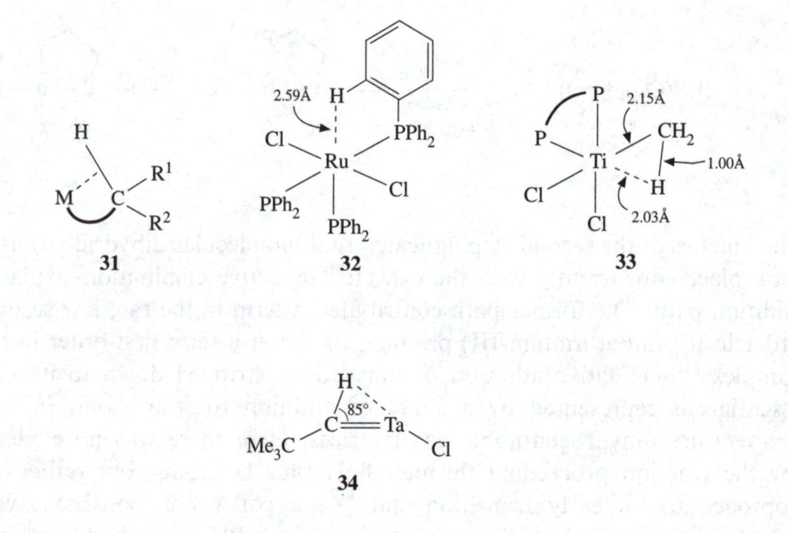

10.4.1.3 *Intramolecular oxidative addition*

This provides a variant on the oxidative addition theme. It is often encountered for triphenyl phosphine complexes of metals in low oxidation states. Here the C–H bond ortho to C–P in one of the ligand phenyl groups is broken, with the carbon and hydrogen, as hydride, bonding to the iridium, as shown in **35**.[147] This so-called orthometallation makes the coordination about the iridium octahedral, and its oxidation state 3+.

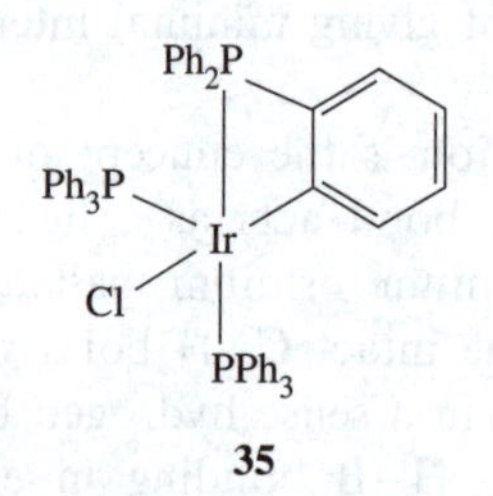

10.4.1.4 *Reductive elimination*

The main importance of the reverse reaction to oxidative addition, namely reductive elimination, is as the step that normally follows oxidative addition, both in synthetic applications and in catalytic cycles (Section 10.7). The potential of the combination of oxidative addition with reductive elimination as a preparative method or a catalysis component may be gauged from a simple sequence such as:

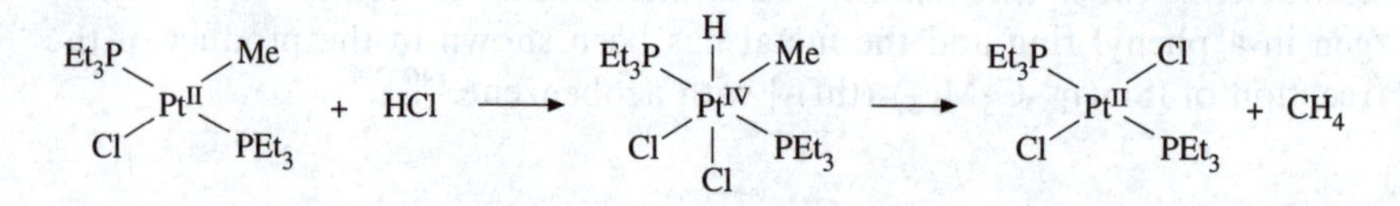

Several factors relevant to the kinetics and mechanism of such oxidative addition/reductive elimination cycles can be illustrated by reference to a study of the reaction of $[Pt(neopentyl)_2(PEt_3)_2]$ with dihydrogen. The bulkiness of the neopentyl ligands plays a key role—one phosphine ligand has to dissociate before the dihydrogen can interact with the metal centre. In the absence of added phosphine, PEt_3 loss is the rate-determining step, there is no H/D kinetic isotope effect, and the order with respect to dihydrogen is zero. In the presence of added PEt_3 the kinetics reflect the reversibility of the first step, while the rate-determining step comes later in the reaction sequence, probably at the stage of neopentane loss from the platinum.[148]

The complementary sequence of reductive elimination followed by oxidative addition has appeared in platinum chemistry, in that base hydrolysis of *trans*-$[Pt(CN)_4Br_2]^{2-}$ has been claimed to proceed in this manner, through a platinum(II) intermediate.[149] There have also been a number of examples of reactions between pairs of complexes in which oxidative addition and reductive elimination are coupled, for instance:[150]

$$[Pt(CN)_4]^{2-} + [AuCl_4]^- \longrightarrow \textit{trans}\text{-}[Pt(CN)_4Cl_2]^{2-} + [AuCl_2]^-$$

$$\textit{trans}\text{-}[PtCl_4(PEt_3)_2] + \textit{trans}\text{-}[IrCl(CO)(PEt_3)_2] \longrightarrow \textit{trans}\text{-}[PtCl_2(PEt_3)_2] + IrCl_3(CO)(PEt_3)_2$$

In several reactions of the latter kind it has been shown that phosphine exchange or loss takes place much more slowly, and therefore does not affect the oxidative addition/reductive elimination coupling. Reactions of this type are believed to proceed by two-electron transfer through bridging chloride.

Mechanisms of oxidative addition have been much studied (Section 10.4.2, immediately following). Reductive elimination has, for good reasons, been much less studied than oxidative addition—as will become apparent in Section 10.4.3. The paucity of information on mechanisms of reductive elimination is to be regretted, especially in this economics-driven age, since it has been observed that it is the reductive elimination step in the appropriate catalytic cycle which is the most profit-determining in commercial applications.

10.4.2 Mechanisms of oxidative addition

10.4.2.1 Introduction; rate laws

Most oxidative additions to square-planar complexes follow second-order kinetics

$$-\mathrm{d[complex]}/\mathrm{d}t = +\mathrm{d[product]}/\mathrm{d}t = k_2\mathrm{[complex][oxidant]}$$

suggesting bimolecular reaction. This rate law evidence is supported by generally negative values for activation entropies and uniformly negative activation volumes, ranging from -13 to $-44\ \mathrm{cm^3\ mol^{-1}}$, for oxidative addition to rhodium(I)[151] and to iridium(I) complexes.[152] Activation entropies and volumes vary with the nature of the solvent, since transition states are often polar, and with the nature of the adding molecule. If oxidative addition does not go to completion, then a rate law appropriate to a reversible reaction is obeyed. Thus for reaction of *trans*-$[RhCl(CO)(PR_3)_2]$ with tetracyanoethylene,[153]

$$-d[RhCl(CO)(PR_3)_2]/dt = k_2[RhCl(CO)(PR_3)_2][tcne] - k_{-1}[RhCl(CO)(PR_3)_2(tcne)]$$

There is an interesting contrast between rhodium and iridium in the rate laws for reaction of the respective compounds *trans*-$[MCl(CO)(PR_3)_2]$ with methyl iodide. In the case of rhodium the system attains equilibrium, obeying a two-term rate law of the type shown just above, but for iridium reaction goes to completion and the rate law is the simple one-term second-order type shown at the start of this paragraph.

Oxidative additions to five-coordinate complexes often follow somewhat more complicated rate laws reflecting reversible ligand loss from the starting complex. For oxidative addition of a molecule XY to $[Ir(H)(CO)(PPh_3)_3]$, the mechanism

$$[Ir(H)(CO)(PPh_3)_3] \underset{k_{-1}}{\overset{k_1}{\rightleftharpoons}} [Ir(H)(CO)(PPh_3)_2] + PPh_3$$

$$[Ir(H)(CO)(PPh_3)_2] + XY \xrightarrow{k_2} [Ir(H)(X)(Y)(CO)(PPh_3)_2]$$

is reflected in the rate law:

$$\text{rate} = \frac{k_1k_2[Ir(H)(CO)(PPh_3)_3][XY]}{k_{-1}[PPh_3]+k_2[XY]}$$

Reversible CO loss from $Rh(\eta^5\text{-}C_5H_5)(CO)_2$ has been demonstrated, and the resulting transient $Rh(\eta^5\text{-}C_5H_5)(CO)$ detected, in gas-phase photolysis.[154] Reversible ligand loss may also occur from four-coordinate complexes, with the reaction of $Ni(CO)_2(PPh_3)_2$ with cyanogen involving parallel CO or PPh_3 loss.[155] Kinetics of oxidative addition to, as of substitution at, $Pt(PPh_3)_4$ and $Pt(PPh_3)_3$ are complicated by possibilities of parallel associative and ligand-loss pathways.[156]

The limited information from kinetics needs to be supplemented by stereochemical and other evidence to gain a fuller understanding of mechanisms, for at least three distinct pathways operate for various oxidative additions to low oxidation state metal centres. These three involve nucleophilic attack at the metal, concerted three-centre reaction, and radical pathways. There are also a number of variants, for example initial attack by chloride at the iridium in the reaction of Vaska's compound or of cyclooctadiene complexes $[Ir(cod)L_2]^+$ with HCl,[157] iodide catalysis of oxidative addition to $[Ir(cod)(phen)]^+$,[158] initial attack by H^+ or X^+ (as mentioned in the previous section), or the template reaction involving a binuclear intermediate for addition of benzonitrile to the nickel complex $[Ni(LL)_2]$, where LL is the long-chain two-phosphorus-donor ligand $Et_2P(CH_2)_4PEt_2$.[159]

10.4.2.2 Nucleophilic attack by the metal

The nucleophilic substitution mechanism was the first to be established.[160] It involves S_N2 attack by the metal at, for example, carbon in an alkyl halide, e.g.

$$Ir^{I}: \curvearrowright \overset{\displaystyle R}{\underset{\displaystyle I}{CH_2}} \rightarrow [Ir^{III}(I)(CH_2R)]^{\ddagger}$$

If the carbon atom is chiral, then inversion at this centre accompanies the oxidative addition. The establishment of such inversion for iridium(I) attack at, e.g., $MeCHBrCO_2Et$, and for a series of palladium(0) reactions,[161] provided the main evidence for the operation of this mechanism. The *trans* stereochemistry of the product indicates that the initial nucleophilic attack by the Ir^I is followed by more rapid bond breaking in the organic species, with one moiety migrating to the *trans* position on the iridium. The five-coordinate intermediates implied by this mechanism have been observed by low-temperature NMR spectroscopy for, e.g., addition of methyl iodide or of benzyl bromide to $[PtL_2Me_2]$[162] and addition of CD_3I to $PdMe_2(bipy)$.[163] This mechanism is also believed to operate for oxidative addition to rhodium(I) and to cobalt(I). The cobalt case is bioinorganically relevant in relation to the chemistry of vitamin B_{12s}. ^1H NMR studies have demonstrated the key feature of inversion at carbon for addition of appropriately substituted cyclohexanes to model complexes of the $[Co^I(dmgH)_2(py)]^-$ type.

The observed trends in rate constants for a typical series of reactions, namely between *trans*-$[IrX(CO)L_2]$ and RI, proceeding by this mechanism are:

$$R = \text{adamantyl} < \text{cyclohexyl} < \text{secondary alkyl} < Et < Me$$

$$X = Cl < Br \sim \text{tosylate} < I$$

$$L = P(C_6H_4\text{-}4Cl)_3 < PPh_3 < P(4\text{-tolyl})_3$$

Such trends are entirely consistent with expectations for S_N2 attack of Ir^I at carbon of the various groups R, as is that for oxidative addition to nickel(0) complexes[164]

$$Ni(cod)_2 < Ni(PR_3)_2(\text{alkene}) < Ni(PAr_3)_4 < Ni(PR_3)_4$$

and hydrogen/deuterium kinetic isotope effects on oxidative addition of methyl iodide and of methyl fluorosulphonate to Vaska's compound.[165]

10.4.2.3 Three-centre mechanisms

Oxidative addition by some simple diatomic molecules may proceed through a three-centre transition state, e.g.

$$Pt^0 + O_2 \rightarrow \left[Pt\begin{matrix} \diagup O \\ \diagdown | \\ O \end{matrix} \right]^{\ddagger}$$

or involve such a species as a product-like intermediate formed through a linear transition state $[Pt \cdots O{-}O]^{\ddagger}$. The product of such a three-centre mechanism for addition to a square-planar or tetrahedral complex will clearly give a *cis* octahedral product; the nucleophilic attack mechanism outlined in the previous section generally gives a *trans* product, as shown in Fig. 10.9. Thus the reaction of $[Ir(\eta^5\text{-}C_5Me_5)(PMe_3)]$, photochemically generated from $[Ir(H)_2(\eta^5\text{-}C_5Me_5)(PMe_3)]$, with hydrocarbons RH gives the *cis* product **36**, presumably through the transition state **37**, rather than the *trans* product expected from the nucleophilic attack mechanism.[115] The latter pathway is, of course, ruled out here by the pentamethylcyclopentadienyl ligand occupying three *cis* sites in octahedral Ir^{III}. Three-centre reactions can be very fast, especially when, as in this case and for instance that of reaction of $[Ru(dmpe)_2]$ with dihydrogen,[166] the low-valence reactant is a high-energy product of photolytic generation from a stable precursor.

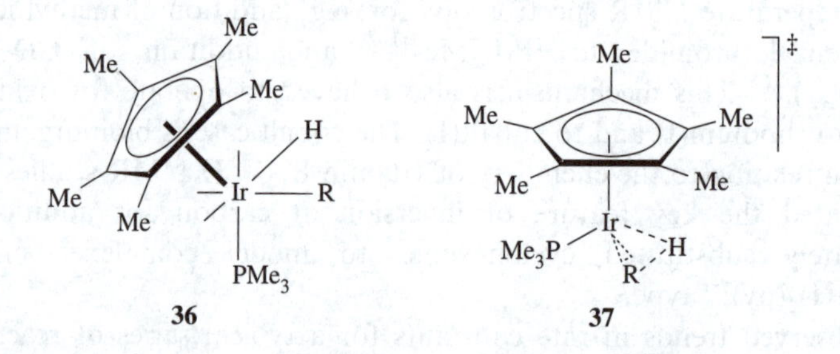

The use of a suitable optically active compound in the three-centre reaction should in principle indicate its operation by retention of configuration, in contrast to the inversion at carbon characteristic of nucleophilic attack by a low-valence metal (cf. above).

10.4.2.4 Radical mechanisms

Radical mechanisms are also possible, especially for alkyl and aryl halides at which nucleophilic substitution is difficult, and if the organic intermediates involved are relatively stable radicals, e.g.

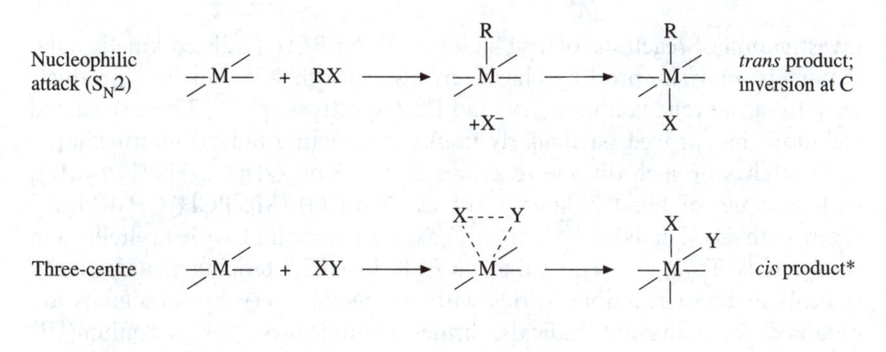

* The initially formed *cis* isomer may subsequently isomerize to the *trans* form; three-centre additions may (where relevant) be stereoselective

Fig. 10.9 Stereochemical consequences of oxidative addition.

$$Pt^0 + RX \longrightarrow Pt^IX + R\cdot (\text{attack of Pt at halogen})$$

$$Pt^0 + RX \longrightarrow Pt^I + R\cdot + X^- (\text{electron transfer from Pt})$$

If the radicals R· are very reactive they do not escape from the reaction site, but if they are less reactive then they may react with some other centre than the initially immediately adjacent Pt^I, and thereby generate a radical-chain mechanism. Participation by radicals is often probed by spin trapping. The favourability of a radical path also depends on the (in)stability of the one-electron-transfer oxidation state of the metal, e.g. Pd^I, Ir^{II}, Au^{II}, Pt^{III}. The use of ^tBuNO and of 2,3,5,6-Me_4C_6HNO, along with a rigorous set of control experiments, permitted the demonstration of radical and radical-chain mechanisms for addition of several alkyl and acyl halides to platinum(0) and platinum(II) complexes. A platinum(I) complex was trapped and identified in the reaction of $Pt(C_2H_4)(PPh_3)_2$ with ethyl iodide, while strong evidence for the intermediacy of the $Ph_3C\cdot$ radical was obtained in the reaction of $Pt(PPh_3)_3$ with Ph_3CCl. As expected, there was no evidence for a significant radical pathway for similar reactions involving rhodium(I) and iridium(I).[167] A radical-chain mechanism seems to operate for oxidative addition to dirhodium(II) complexes, such as that of photoinduced addition of alkyl iodides to the $[Rh_2(\text{diisocyanopropane})_4]^{2+}$ cation,[168] and of benzyl bromide to the octaethylporphyrin complex $[Rh_2(\text{oep})_2]$.[169] In the latter case chain propagation involves both $Bz^\cdot$ and $Rh(\text{oep})^\cdot$ radicals; there is an $[Rh_2(\text{oep})_2]^{1/2}$ term in the rate law.

The operation of a radical mechanism may be probed by a number of techniques and approaches other than trapping. A particularly thorough investigation of reactions of aryl halides with $Ni(PEt_3)_4$ utilized kinetic data, Hammett analysis, product characterization (both Ni^I and Ni^{II} products were found), cyclic voltammetry, and ESR spectroscopy.[170] The last-named technique has proved particularly useful in detecting radical intermediates, as in studies of such diverse reactions as those of $Zr(\eta^5\text{-}C_5H_5)(PPh_2Me)_2$ with a range of alkyl halides,[171] of *cis*-$[Mo(CO)_2(Me_2PCH_2CH_2PMe_2)_2]$ again with alkyl halides,[172] and of Vaska's compound with aryltellurium compounds Te_2R_2.[173] The value of ESR in characterizing metal-centred radicals in these reactions varies with the metal—very broad signals are obtained for tellurium radicals, firmer identification for zirconium(III) species.

For the addition of a molecule RX, one may expect,[174] on the basis of knowledge from organic chemistry on relative stabilities of various types of organic radicals, a radical mechanism to be increasingly likely for:

$$R = \text{primary alkyl} < \text{secondary} < \text{tertiary} < \text{benzyl or allyl}$$

$$X = RSO_3 < Cl < Br < I$$

These sequences may be compared with those set out for nucleophilic attack by low-valence metals in Section 10.4.2.2 above. The stereochemical consequence of a radical mechanism is racemization, in contrast to inversion in the S_N2 mechanism.

10.4.2.5 Diagnosis of mechanism; activation parameters

It is difficult to provide general guidelines for collating and forecasting mechanisms of oxidative addition. Some indication of mechanism for various adding molecules may be obtained from reactivity trends (cf. Sections 10.4.2.2 and 10.4.2.4 above). The trend with varying R for addition of RX is diagnostic; the expected trends with variation of halide X for nucleophilic attack and radical mechanisms are the same. Stereochemical, trapping, and ESR experiments have proved informative in choosing between the three main mechanisms; values of activation parameters have not.[175] Long ranges of $\Delta H^\ddagger$ and $\Delta S^\ddagger$ values have been reported for reactions involving (oxidative) addition to *trans*-$[IrCl(CO)(PPh_3)_2]$ (Table 10.13), with values for the presumably three-centre addition of dioxygen and of $CF_3C(CN){=}C(CN)CF_3$ lying between the extremes for S_N2 attack, by methyl iodide and methyl fluorosulphonate.[176] Indeed, there can be large variations in $\Delta H^\ddagger$ and $\Delta S^\ddagger$ merely on modifying spectator phosphine ligands (see Table 10.14 for $\Delta H^\ddagger$ examples).[177] It should be added that there are also large ranges of values in similar sets of ΔH° values, for example from -88 to $-185\ \text{kJ mol}^{-1}$ for reaction of alkyl iodides with iridium(I) compounds. Both steric and electronic effects may affect $\Delta H^\ddagger$ and $\Delta S^\ddagger$. It is thus not possible to specify ranges of $\Delta H^\ddagger$ and $\Delta S^\ddagger$ values characteristic of the three types of mechanism. It has been claimed that $\Delta S^\ddagger$ values within the range -150 to $-250\ \text{J K}^{-1}\ \text{mol}^{-1}$ are suggestive of the operation of the S_N2 mechanism. A cursory glance at Table 10.13 suggests

Table 10.13 Range of activation parameters for oxidative addition to *trans*-$[IrCl(CO)(PPh_3)_2]$, in benzene solution.

Compound added	$\Delta H^\ddagger$ (kJ mol^{-1})	$\Delta S^\ddagger$ (J K^{-1} mol^{-1})
CH_3I	23	−213
$CF_3C(CN){=}C(CN)CF_3$	30	−163
$4\text{-}YC_6H_4SH$ [a]	33–42	−113 to −155
H_2	45	−96
O_2	55	−88
CH_3OSO_2F	71	−46

[a] Y = H, F, Cl, Br, Me, OMe, NO_2.

Table 10.14 Effect of variation of phosphine ligands on activation enthalpies for oxidative addition of *trans*-$[IrCl(CO)(PR_3)_2]$.

PR_3	X	$\Delta H^\ddagger$ (kJ mol^{-1})	
		+ MeI	+ H_2
PPh_3		29	
PEt_2Ph		41	
$P{-}C_6H_4{-}X$	Me	58	18
	Cl	62	41
	F	71	49

that this is hardly a safe guide, but it is possible to make a vague generalization that S_N2 reactions tend to have more negative activation entropies than three-centre reactions. It is not surprising that activation parameters are not diagnostic for the various mechanisms. 'End-on' intermediates, transients, or transition states of the form of **38** could occur at an early stage on the reaction path of any of the three mechanisms (Fig. 10.10). The determination of the structure, **39**, of the initial product of reaction of I_2 with the appropriate platinum(II) compound provides a model for just such an intermediate.[178] Such end-on species will be polar, so a three-centre addition could show similar solvent effects on reactivity to an S_N2 addition if the three-centre intermediate or transition

$[M{\cdot\cdot}X{-}X]^\ddagger$

38

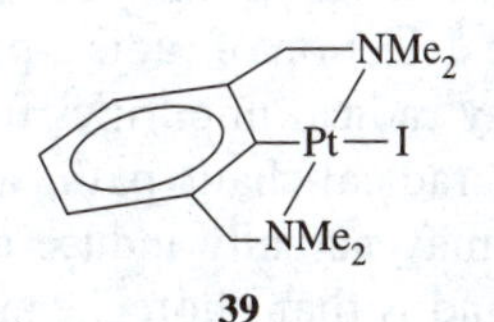

39

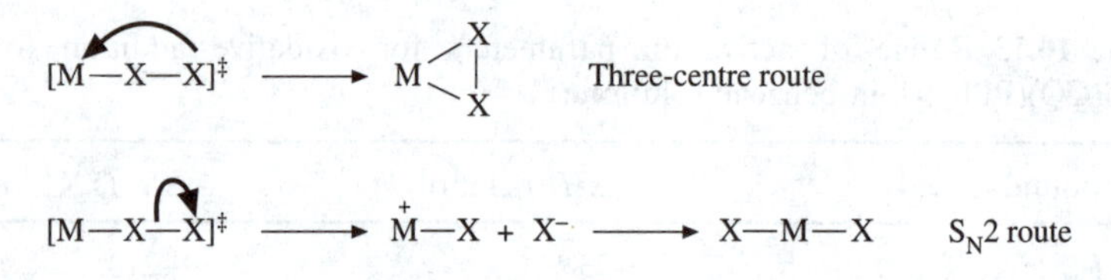

Fig. 10.10 Two possible fates for intermediate **38**.

state was formed through initial approach of one end of the adding molecule to the metal. Readers interested in delving further into solvent effects on reactivity in oxidative addition reactions, and assessing for themselves how informative these effects may be, could consult publications on, for example, [Rh(β-diketonate)(CO)(PPh_3)],[179] $[Rh(CO)_2I_2]^-$,[180] and *trans*-[IrCl(CO)$(PPh_3)_2$] plus MeI,[160,181] and *trans*-[IrX(CO)$(PR_3)_2$] plus various C=C and C≡C compounds.[182] The first three provide examples of S_N2 addition, the last of several indubitably three-centre additions. Activation enthalpies are generally much less than the appropriate bond-dissociation energies for the molecule being added. Thus, for instance, $\Delta H^{\ddagger}$ is about 40 kJ mol^{-1} for addition of H_2 to *trans*-[IrCl(CO)$(PPh_3)_2$] or to [RhCl$(PPh_3)_3$], though the bond-dissociation energy of the dihydrogen molecule is 430 kJ mol^{-1}. However, this comparison, though suggestive, cannot be taken as firm evidence for the concerted (three-centre transition state) mechanism.

Alkenes necessarily add *cis*, presumably through a three-centre transition state. Dioxygen and dihydrogen molecules generally add by such a mechanism, but it may well be that dihydrogen adds in a *trans* fashion to [IrH(CO)$(PPh_3)_3$],[183] and there is kinetic evidence for dioxygen attacking the iridium end-on in the five-coordinate iridium(I) complex [IrI(cod)(phen)].[184] Concerted (three-centre transition state) addition has also been claimed for the reaction of acid chlorides with [RhCl$(PPh_3)_3$].[185] Alkyl halides usually add via the S_N2[186] or, more rarely, radical routes; the balance between these mechanisms may be quite fine for several metal centres. n-Alkyl halides add to platinum(II) by the S_N2 mechanism, but isopropyl iodide adds by a radical-chain mechanism.[187] Small changes in the non-reacting ligands may also tip the balance,[185] as in oxidative addition of $PhCHFCH_2Br$ to *trans*-[IrCl(CO)$(PR_3)_2$]. This goes completely by the radical mechanism for $PR_3 = PMe_3$ or PMe_2Ph, but the radical route only contributes 10% to overall reaction when $PR_3 = PMePh_2$.[174]

We end this section with two further cautionary points relating to the establishment of oxidation mechanisms. The first is the usual pair of warnings in relation to the detection of radical pathways—that it is essential to remove all traces of any radical impurities from the reagents, especially any that might induce a radical-chain path, and that the addition of a radical trap to a system may actually induce an otherwise non-operative radical pathway. The second is that photochemical reactions subsequent to the oxidative addition step may scramble the ligands in the product, giving misleading stereochemical information. This situation has been described for, e.g., oxidative addition of bromine to *trans*-$[PtCl_2(PEt_3)_2]$. When this reaction is carried out in the dark, all-*trans*-$[PtBr_2Cl_2(PEt_3)_2]$ is obtained, but in light a series of compounds $[PtBr_xCl_{4-x}(PEt_3)_2]$ is obtained.[188]

10.4.2.6 *Oxidative coupling*

This class of reaction is closely related to oxidative addition, in that the oxidation state and coordination number of the metal involved both increase by two, but in oxidative coupling this is achieved by adding two molecules to the metal, coupling them as they bond. Thus in the classic case, two alkene molecules couple to give a metallocycle:

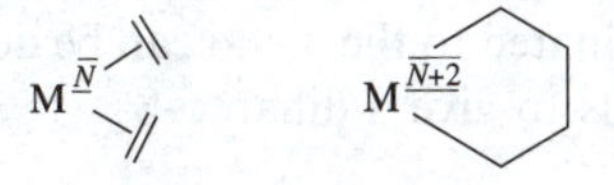

A *p*-block example is provided by the oxidative addition and coupling of two hexafluoroacetone molecules at phosphorus(III) to give dioxophospholanes **40**.[189]

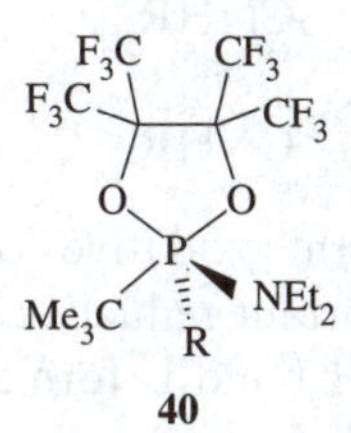

10.4.3 Reductive elimination[190]

Reductive elimination is the converse of oxidative addition. The starting material generally loses two ligands and drops two oxidation states, e.g.:

$$Me_2SI_2 \longrightarrow Me_2S + I_2$$

$$cis\text{-}[PtHMe(PPh_3)_2] \longrightarrow [Pt(PPh_3)_2] + CH_4$$

$$[(cp)IrRH(PMe_3)] \longrightarrow [(cp)Ir(PMe_3)] + RH$$

$$cis\text{-}[AuMe_2(CO_2Me)(PPh_3)] \longrightarrow [Au(CO_2Me)(PPh_3)] + C_2H_6$$

$$\searrow [Au(Me)(PPh_3)] + MeCO_2Me$$ [191]

In this last example there is a mixture of products; in some systems product distribution depends on the nature of non-eliminated ligands. Thus reductive elimination from **41** gives a mixture of $SiMeEt_3$ and CH_4 when R = Et, but solely methane when R = OEt or Ph.[192]

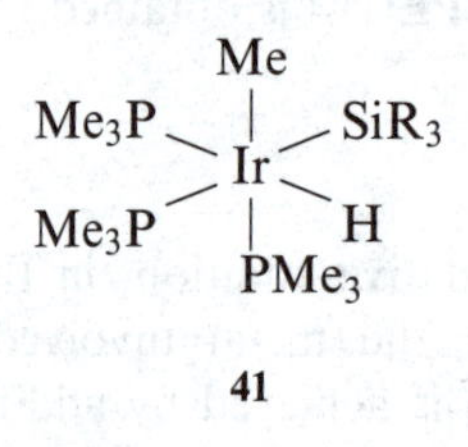

The reactions listed above are sometimes called *reductive coupling*, though it might be felt preferable to reserve this term to cases where an intermediate containing the coupled ligands still coordinated to the metal can be detected, as in the coupling of two sulphide ligands to give a dihapto-$S_2{}^{2-}$:

$$\mathrm{Mo^{VI}}\begin{matrix} \diagup S \\ \diagdown S \end{matrix} \rightarrow \mathrm{Mo^{IV}}\begin{matrix} \diagup S \\ | \\ \diagdown S \end{matrix}$$

or[193]

$$\left[\mathrm{X(RNC)_4W^{IV}}\begin{matrix} \diagup CNR \\ \diagdown CNR \end{matrix}\right]^+ \rightarrow \left[\mathrm{X(RNC)_4W^{II}}\begin{matrix} \diagup CNHR \\ \| \\ \diagdown CNHR \end{matrix}\right]^+$$

Such reductive coupling may be compared with oxidative coupling, described in Section 10.4.2.6 above. The most important reductive elimination reactions are those in which a C–C or a C–H bond is formed, these reactions being driven by the strengths of the bonds formed.

The small molecules eliminated in this class of reaction include aliphatic and aromatic hydrocarbons, tetraalkylsilanes, alcohols, ketones, esters, and others. Most of the reactions displayed above take place on heating, but the second occurs even at −25°C. There is often a marked difference in the tendency to undergo reductive elimination between rhodium(III) species and their iridium(III) analogues. Sometimes this is advantageous, in that the complete sequence $Rh^{I} \rightarrow Rh^{III} \rightarrow Rh^{I}$ may take place for rhodium, providing product and, at times, potential for homogeneous catalysis, whereas for iridium only the first step, $Ir^{I} \rightarrow Ir^{III}$, may take place, providing a model intermediate for the rhodium system. An example of this is provided by intramolecular oxidative addition involving coordinated triphenylphosphine (Section 10.4.1.3). $[IrCl(PPh_3)_3]$ gives the stable octahedral iridium(III) species **35**, but $[RhMe(PPh_3)_3]$ undergoes intramolecular oxidative addition to give **42**, the rhodium(III) analogue of **35**, which undergoes reductive elimination with loss of methane to give the rhodium(I) compound **43** as final product.

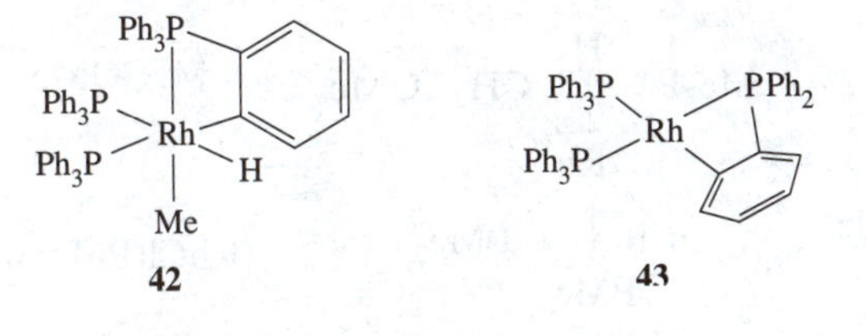

Reductive elimination often occurs quite rapidly, which means that the study of kinetics and mechanisms of such reactions can be difficult. This is especially so for homogeneous catalysis cycles (Section 10.7), where rate-limiting oxidative addition may make subsequent rapid reductive elimination inaccessible to normal kinetic methods. Thus there have been rather few kinetic and mechanistic studies of reductive elimination. An early and classic study established that reductive elimination from *fac*-$[PtMe_3X(PR_3)_2]$ gave simply ethane and $[PtMeX(PR_3)_2]$, via phosphine loss to generate a five-coordinate intermediate. However, reductive elimination from other alkyl-halide-platinum(II) complexes may give appreciable amounts of alkyl halide alongside the desired dialkyl major product. This is the case, for example, for $[PtMe_3I(dppe)]$, which gives methyl iodide as well as ethane. Again reaction proceeds through a five-coordinate intermediate, which in this case was shown to be the same for both paths.[194] The generality of initial phosphine dissociation is supported by the kinetic demonstration of the operation of this mechanism, this time in a $Pt^{II} \rightarrow Pt^0$ system, in thermal reductive elimination from *cis*- and *trans*-$[PtMe(SiMe_3)(PMePh_2)_2]$.[195] Five-coordinate intermediates have also been proposed in reactions involving charged leaving groups.[163] Thus the mass-law retardation observed on adding iodide to $[Pd(Me)_3(bipy)I]$ is attributed to the initial equilibrium

$$[Pd(Me)_3(bipy)I] \rightleftharpoons [Pd(Me)_3(bipy)]^+ + I^-$$

followed by loss of ethane, presumably via intramolecular carbon–carbon bond formation

$$[Pd(Me)_3(bipy)]^+ \longrightarrow [Pd(Me)(bipy)]^+ + C_2H_6$$

and recombination with the liberated iodide

$$[Pd(Me)(bipy)]^+ + I^- \longrightarrow [Pd(Me)(I)(bipy)]$$

A chemically (^1H and ^{31}P NMR) and kinetically well-characterized example of reductive elimination is the second stage of the following sequence. The first step shows the method of generating the substrate of interest, which was in fact the first pre-RH-elimination *cis*-[Rh(H)(R)] species to be isolated:

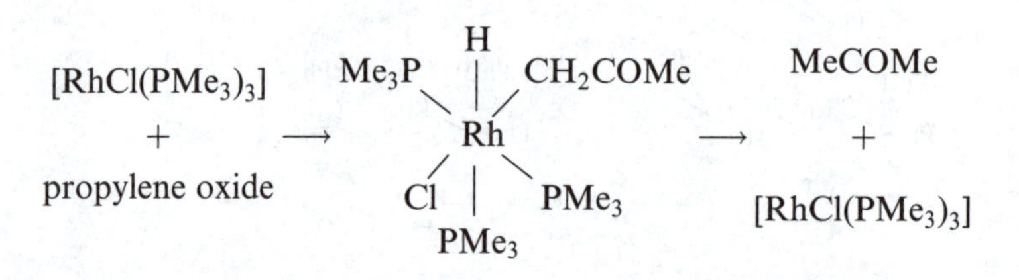

This reductive elimination has a half-life of about four hours in deuterobenzene solution at 25 °C, with activation parameters $\Delta H^{\ddagger} = 102$ kJ mol^{-1} and $\Delta S^{\ddagger} = +22$ J K^{-1} mol^{-1}. Results from kinetic, deuterium-labelling, and PMe_3-addition experiments are consistent with an intramolecular mechanism for the elimination of acetone from, once again, a five-coordinate species produced by initial loss of a PMe_3 ligand from the rhodium(III) compound.[196] Similar mechanisms have been proposed for the reductive elimination of alkanes from square-planar palladium(II) and gold(III) complexes, for example of ethane from *cis*-$[PdMe_2(PPh_3)_2]$ or from $[AuMe_3(PPh_3)]$. These reactions have been shown to be intramolecular from the lack of crossover products on reacting mixtures of, e.g., $[AuMe_3(PPh_3)]$ and $[AuMe_2Et(PPh_3)_3]$.[197]

The examples cited so far all go by intramolecular mechanisms. Some intermolecular reductive elimination reactions are known. These proceed through the formation of binuclear intermediates. A simple example is provided by the extrusion of hydrogen from hydridocarbonyls of manganese and of cobalt, e.g.

$$2Co(H)(CO)_4 \longrightarrow Co_2(CO)_8 + H_2$$

A full kinetic study of the reaction

$$(4\text{-}MeOC_6H_4CH_2)Mn(CO)_5 + Mn(H)(CO)_5 \longrightarrow 4\text{-}MeOC_6H_4Me + Mn_2(CO)_{10}$$

established the rate law

$$d[(4\text{-}MeOC_6H_4CH_2)Mn(CO)_5]/dt = \{k[(4\text{-}MeOC_6H_4CH_2)Mn(CO)_5][Mn(H)(CO_5]\}/\{k'[CO] + k''[Mn(H)(CO)_5]\}$$

for this reaction (in non-polar solvents). From this and other evidence a concerted (one-centre) mechanism was deduced.[198] Labelling has also proved useful in establishing these intermolecular mechanisms. Thus deuterium labelling ruled out simple intramolecular elimination of H_2 from $[Os(H)_2(Me)(CO)_4]$ or of CH_4 from $[Os(H)(Me)(CO)_4]$. An intermolecular mechanism is indicated by the labelling experiments, though not by the first-order kinetics established for these reactions. The first-order behaviour suggests rate-limiting carbon monoxide loss from one molecule of substrate prior to a bimolecular elimination step.[199] All these intermolecular reductive elimination reactions involve at least one hydride ligand, which prompted the suggestion that hydride-bridged intermediates M·····H·····M are probably involved.

10.5 Insertion and migration

10.5.1 Overview[200]

A considerable variety of small molecules can insert into a variety of bonds. The most important of the small molecules are alkenes, alkynes, and carbon monoxide,[201] but a wide range of other species can insert. Some of these other inserting molecules are detailed in the following sections. The most important bonds into which the small molecules are inserted are metal–hydrogen and metal–carbon. Insertion into particularly strong metal–carbon bonds such as $M{-}CF_3$, $M{-}C_6F_5$, and $M{-}CH_2CN$ is difficult to achieve. The chief importance of the molecules and bonds just mentioned is, of course, in relation to homogeneous catalysis (Section 10.7)—ligand insertion and migration comprise one of the most important functions of transition metal catalysts. Indeed, the insertion of carbon monoxide into organic molecules is currently one of the most important uses of transition metal complexes in organic synthesis, preparative or catalytic. The transition metals concerned are distributed around the *d*-block, including, for example, iron, cobalt, nickel, palladium, and zirconium.

Insertion and migration occur once the ligand or fragments are incorporated into the coordination shell of the metal ion. In general, one ligand changes its mode of bonding to the metal, generating a highly reactive site to which another ligand from the coordination shell may move. This is generally preceded, accompanied, or followed by the entry of another ligand from the vicinity, to make up the coordination number and stabilize the product.

An insertion reaction is often just that, but there are some important examples which are rather more complicated than appears at first sight. The most studied is that of the 'insertion' of carbon monoxide. This was the first insertion reaction to be convincingly demonstrated,[202] in the classic case of the formation of acetylmanganese pentacarbonyl from the reaction of methylmanganese carbonyl and carbon monoxide:

$$Mn(Me)(CO)_5 + CO = Mn(COMe)(CO)_5$$

However, this reaction in fact involves migration of a ligand as well, as will be detailed in Section 10.5.2 below.

Crabtree[203] distinguishes between 1,1 and 1,2 insertions:

$$\text{1,1-INSERTION}\quad \begin{matrix} X \\ | \\ M{-}A{=}B \end{matrix} \longrightarrow M{-}A\begin{matrix} \diagup X \\ \diagdown\!\!\diagdown B \end{matrix}$$

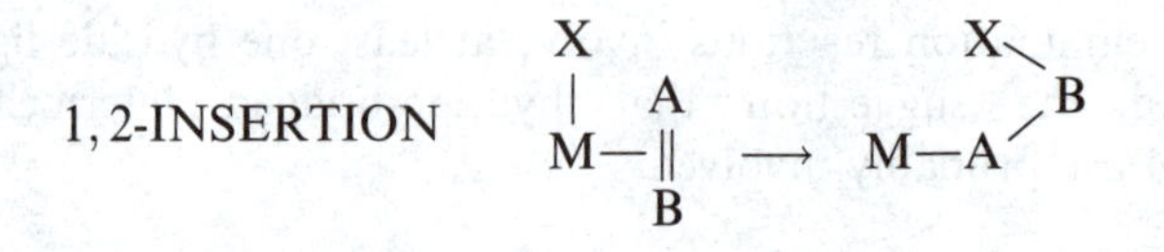

Ethene gives only 1,2-insertions, carbon monoxide only 1,1-insertions. Indeed one can generalize that η^1 ligands normally insert 1,1, η^2 ligands 1,2. Sulphur dioxide is the only common small molecule that inserts by 1,1- or 1,2-routes.

The reverse reaction, in which, for example, a σ-bonded alkyl ligand transfers a hydrogen atom to the metal and becomes bonded as a π-complex (generally resulting in subsequent loss of the alkene from the metal), is β-elimination, which is dealt with in Section 10.6. Insertion and β-elimination complement each other in a manner similar to oxidative addition and reductive elimination.

10.5.2 Carbon monoxide[204]

Carbon monoxide inserts into metal–alkyl bonds to give metal acyl derivatives, as in the long-established and much studied example

$$(OC)_5MnMe + CO \longrightarrow (OC)_5Mn(COMe)$$

cited in the preceding section, and in the post-oxidative addition reaction of rhodium mentioned in Section 10.4.1:

$$[RhMeClI(CO)(PR_3)_2] \longrightarrow [Rh(COMe)ClI(PR_3)_2]$$

Experiments on the manganese system using ^{13}CO as reagent gave **44** as product, showing that the mechanism was not one of simple insertion, but that group migration also took place. Fortunately there was no rapid subsequent scrambling to confuse this result! If the ^{13}CO is used as one of the ligands, i.e. if **45** is treated with (unlabelled) carbon monoxide, then the isomer distribution of the labelled product confirms that the methyl group has indeed migrated.[205] The central role of migration has also been established, albeit slightly less satisfactorily, by using PPh_3 as a signpost, and, for five-coordinate species, by stereochemical studies of insertion into the iron and ruthenium compounds $[M(CO)_2(PMe_3)_2X]$ (X = I, CN).[206]

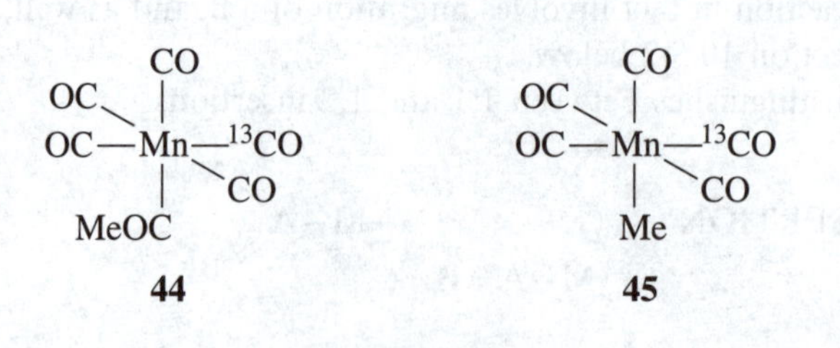

Carbon monoxide 'insertion' reactions may also be promoted by a variety of ligands (L), such as phosphines or t-butyl isocyanide:

$$(OC)_5MnR + L \longrightarrow (OC)_4Mn(COR)L$$

In general they take place by alkyl (methyl) migration, with reaction rate constants increasing, though only to a small extent, as the nucleophilicity of L increases.[207] Typical activation parameters are the ranges of $\Delta H^{\ddagger}$ between 50 and 60 kJ mol^{-1} and of $\Delta S^{\ddagger}$ between -40 and -80 J K^{-1} mol^{-1} reported for PPh_3-promoted insertion into the Rh–R bond of $[Rh(CO)(C_5Me_5)(R)I]$.[208] Electron-releasing groups R facilitate, electron-withdrawing retard, migration,[209] with rate constants decreasing about 10^4-fold on going from $R = CH_2Me$ or CH_2Et to $R = CO_2H$. Steric effects also affect reactivity, as has been established for $[Mn(CH_2C_6H_{5-n}Me_n)(CO)_5]$, for $[Fe(\eta^5\text{-}C_5H_5)(CO)(CH_2\text{-}cyclo\text{-}C_6H_{11})(dmso)]$, and for $[Mo(\eta^5\text{-}C_5H_5)(CH_2C_6H_5)(CO)_3]$ reacting with phosphines of varying bulkiness.[210] Steric effects on reactivity are relatively small but do correlate (albeit very non-linearly) with cone angles. The promoting ligand L in the above equation can be a donor solvent, such as acetonitrile in an investigation of insertion kinetics for complexes $[Mn(CO)_5R]$ containing a variety of substituted benzyl groups R. This study established Hammett substituent and reaction parameters σ and ρ for this sytem, and steric effects resulting from *ortho*-substitution in the benzyl group.[211]

Solvent effects on reactivity for methyl migration are usually large. There has been disagreement as to whether this signifies a role for transient intermediates containing solvent bonded to metal or merely large solvation effects in the initial and/or transition states.[212] The former seems more plausible for, e.g., phosphine-promoted methyl migration in $[Mo(\eta^5\text{-}C_5H_5)(Me)(CO)_3]$ than in $[Mn(Me)(CO)_5]$. However, there is some fairly convincing infrared and ^1H NMR evidence that has been presented for the intermediacy of a solvento species $[Mn(COR)(CO)_4(solv)]$, even for $[Mn(R)(CO)_5]$, when the alkyl migration reaction is carried out in good donor solvents such as DMSO, DMF, acetonitrile, or pyridine.[213] Non-negligible effects of solvent bulk also lend some support to direct solvent–metal bonding in intermediates in migration reactions of $[Mn(CH_2C_6H_{5-n}Me_n)(CO)_5]$. The methyl migration mechanism operates in donor and polar solvents, but there is strong kinetic evidence for an alternative, very slow, bimolecular pathway in very non-polar solvents. This has been established for rhodium(III) complexes of the type $[Rh(\eta^5\text{-}C_5Me_5)(R)(I)(CO)]$ as well as for the standard pentacarbonylmanganese substrates. When R = carboxylate, η^2-bonding permits a stable intermediate, revealed from saturation kinetic behaviour.[214] The affinity of transition metals to the left of the Periodic Table, and in the *f*-block, for oxygen means that η^2-acyls may be obtained more easily for these elements than for metals with many *d* electrons, e.g.:[215]

$$(\eta^5\text{-}C_5Me_5)_2Zr\begin{matrix}\diagup R\\ \diagdown R\end{matrix} \xrightarrow{CO} (\eta^5\text{-}C_5Me_5)_2Zr\begin{matrix}\diagup C(R){=}O \leftarrow\\ \diagdown R\end{matrix}$$

Also in the left-hand region of the *d*- and *f*-blocks,[216] the often difficult insertion into a metal–hydrogen bond takes place quite readily, giving an η^2-formyl product. Again this is presumably facilitated by the formation of the particularly favoured metal–oxygen bond, e.g:

$$(\eta^5C_5Me_5)_2Th\begin{matrix}\diagup H\\ \diagdown OR\end{matrix} \xrightarrow{CO} (\eta^5\text{-}C_5Me_5)_2Th\begin{matrix}\diagup C(H){=}O \leftarrow\\ \diagdown OR\end{matrix}$$

The migration mechanism also appears to operate for the great majority of carbonyl 'insertion' reactions of square-planar complexes.[217]

In contrast to all these 'insertions', which are actually migrations, real non-migratory insertion of carbon monoxide, into iron–carbon bonds,[218] has been claimed for octaethylporphyrin complexes [Fe(oep)R] (Fe^{III} with R = n-alkyl, Fe^{II} with R = t-butyl) and in the following reaction (cf. the η^2-acyl intermediate here with the product in the previous equation):

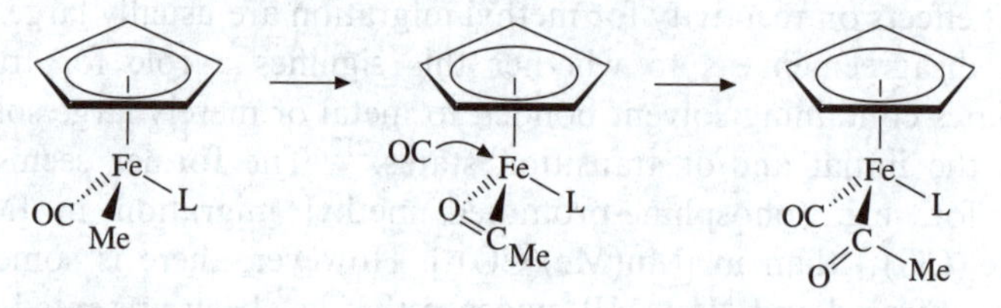

10.5.3 *Isocyanides*

Although there is a close similarity between the insertion of an isocyanide, RNC, and of carbon monoxide, the former has been very much less studied. Insertion of RNC into M–H is easier than CO into M–H.[219] An early example of RNC insertion into a metal–carbon bond involved $[Ni(cp)(R')(PPh_3)]$; parallel insertion and substitution gave **46** as product.[220]

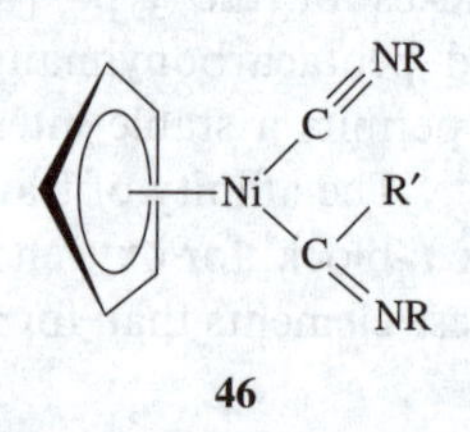

46

Later studies showed that the products of isocyanide insertion depended both on the nature of the isocyanide and on the nature of the transition metal substrate. Thus, for instance, in the reaction of $[Fe(Me)X(CO)_2(PMe_3)_2]$ with such compounds as cyclohexyl and t-butyl isocyanides, when X = I ionic products were obtained, but when X = CN there was no ionization of X^- in the course of the reaction.[221] In contrast to carbon monoxide insertion, isonitriles can perform multiple insertions. So, for example, Pt–$(RNC)_n$–C can be obtained from Pt–C, whereas Mn–$(CO)_n$–R species with $n > 2$ are unattainable.

As in the case of carbon monoxide insertion (previous section), insertion into metal–carbon bonds involving metals towards the left of the Periodic Table may give an η^2 product, e.g.:[222]

$$(\eta^5\text{-}C_5H_5)_3U{-}R \xrightarrow{R'NC} (\eta^5\text{-}C_5H_5)_3U{-}\overset{\displaystyle CR}{\underset{\displaystyle NR'}{\|}}$$

Isonitriles, and nitriles, insert readily into metal–carbene bonds, for example in chromium[223] and tungsten carbenes. Large negative activation entropies and volumes for insertion of electron-rich triple-bonding systems, for example dipropylcyanamide, into chromium- and tungsten-carbene bonds indicate considerable bond formation in the rate-determining step of the proposed three-stage mechanism. Further details will be found in Section 10.5.4, in connection with analogous alkyne insertions.[224] A large negative activation entropy, $-100\ J\ K^{-1}\ mol^{-1}$, has also been found for reaction of the diene complex [Ru(cp)(PhC=CHCH=CPh)Br], **47**, with t-butyl isocyanide to give the η^4-cyclopentadiene derivative **48**. This value, and the simple second-order rate law for the reaction, were interpreted as suggesting that the initial step in the insertion reaction was rate-determining associative attack by the isocyanide at the metal.[225]

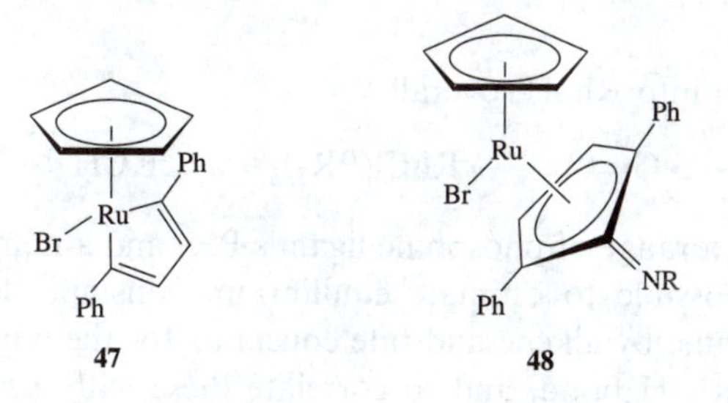

10.5.4 *Alkenes and alkynes*

10.5.4.1 *Alkenes → metal–hydrogen*

Insertion of alkenes into metal–hydrogen bonds is a common and important reaction, which can be promoted by various incoming groups, even dinitrogen.[226] Insertion of alkenes into metal–carbon[227] is less common but is very important in relation to homogeneous catalysis of alkene polymerization (Section 10.7.3). This M–H versus M–C difference can be illustrated by ethene insertion in **49**, which takes place for R = H, but not for R = Me or Ph[228]:

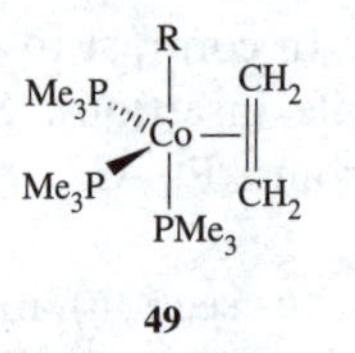

49

Although insertion of alkenes into a wide range of metal–hydrogen bonds (e.g. into Co–H, Rh–H, Ir–H, Pt–H, Ru–H, Mo–H, Zr–H) has been reported,[229] there have been rather few studies of the detailed mechanisms of such reactions. The product of insertion is generally the *syn* isomer, though sometimes this subsequently isomerizes to the *anti* form—probably through an η^2-bonded intermediate.

Alkene insertion into the Rh^I–H bond in *trans*-$[RhH(C_2H_4)(P^iPr_3)_2]$ has been shown[230] to be intramolecular in an intermediate with the H and alkene ligands mutually *cis*, as one would expect. The transition state is identical with that for the reverse, elimination, reaction (details in Section 10.6 below). Ease of reaching such a transition state is clear from structure **50**, established by X-ray diffraction (cf. **49** with R = H). A detailed kinetic

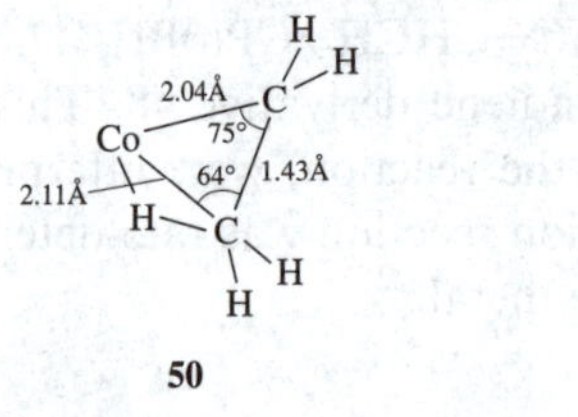

50

study of alkene insertion into Rh–H, overall

$$RhH_2Cl(PR_3)_3 + {>}C{=}C{<} \longrightarrow RhCl(PR_3)_3 + {>}CHCH{<}$$

has been carried out for a range of phosphine ligands PR_3 and a number of alkenes.[231] It proved possible to estimate equilibrium constants for the replacement of a phosphine by alkene and rate constants for the migration of the alkene into the Rh–H bond, and to correlate these with electronic effects of the R groups and of the groups in the alkenes. Thus, for example, migration rate constants increase with increasing electron-donating power of substituent in alkenes 4-X-$C_6H_4CH{=}CH_2$. As the substituent effect on the alkene-for-phosphine equilibrium constant follows the opposite trend, there is compensation and the overall range of reactivity is small. Steric and electronic effects on alkene insertion have been examined in detail for insertion into niobium- and tantalum-hydride bonds.[232]

Dienes may also insert, for example into the manganese–hydrogen bond of $[MnH(CO)_5]$ and into the iron–hydrogen bond of $[Fe(cp)(CO)_2H]$. The rate law for the latter reaction is first-order in iron compound and first-order in diene; added carbon monoxide has no effect on rates. There is an inverse primary isotope (H/D) effect. The kinetic data are best accommodated by a radical mechanism in which the rate-determining step is hydrogen abstraction from the diene to give an allyl radical, $H_3CCR{=}\dot{C}R'CH_2$.[233] A radical mechanism is also believed to operate for the $[MnH(CO)_5]$ reaction.

10.5.4.2 Alkenes → metal–carbon[204]

Palladium systems are widely used for alkene insertions into metal–carbon bonds. Unfortunately, for the mechanistically minded though not for the synthetic chemist, the intermediates usually rearrange very quickly and thus cannot be isolated and characterized. However, in the particular case of $[PdCl(C_6F_5)(cod)]$, which has been fully characterized by X-ray structural analysis, insertion proceeds slowly, presumably due to constraints imposed by the cyclooctadiene ligand. The half-life is of the order of half an hour, so here it has proved possible to monitor the insertion process, using ^{19}F NMR spectroscopy to determine rate constants.[234]

Kinetic studies of ethene insertion into a rhodium–carbon bond, in the second stage of the sequence **51** → **52** → **53**, gave activation parameters of $E_a = 80\ \mathrm{kJ\ mol^{-1}}$ and $\Delta S^{\ddagger} = +14\ \mathrm{J\ K^{-1}\ mol^{-1}}$ for the insertion step, which follows dissociation of the triflate group from the rhodium.[235]

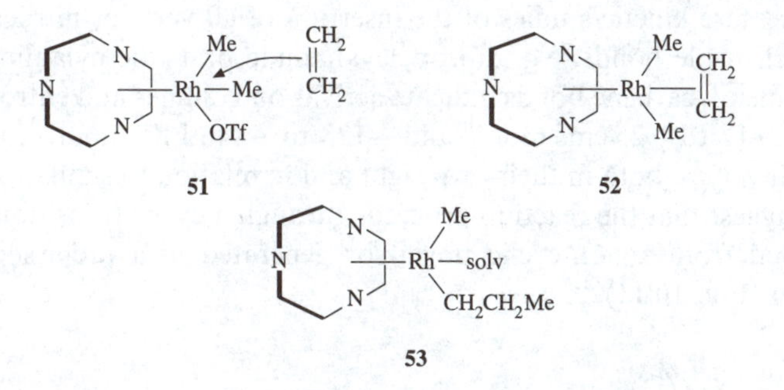

Functionalized alkenes studied include tetracyanoethene, tcne, which probably can insert into the metal–hydrogen bond in $IrH(CO)(PPh_3)_3$ and certainly inserts into metal–carbon in $Fe(cp)(CO)_2(CH_2C_6H_5)$. In the latter case, Fe ← N- and Fe ← C-bonded isomers are both formed.[236]

10.5.4.3 Alkynes

The important alkynes in relation to insertion[227] are activated molecules such as $F_3CC{\equiv}CCF_3$ and $MeO_2CC{\equiv}CCO_2Me$. Insertion is generally *syn*, conserving stereochemistry where appropriate. Details of insertion stereochemistry have been established for, *inter alia*, ruthenium- and

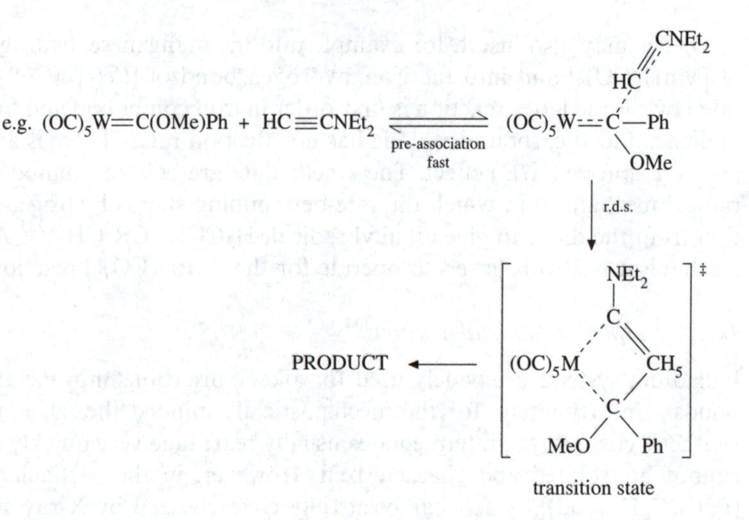

Fig. 10.11 Mechanism of insertion of the electron-rich alkyne diethylaminoethyne into a tungsten–carbene bond.

rhenium-hydride bonds.[237] Kinetic studies of alkyne insertion into the ruthenium–hydrogen bond of $[Ru(CO)H(NCMe)_2(PPh_3)_2]^+$ showed the expected second-order rate law.[238] Palladium(0)-catalysed insertion of alkynes, and of allenes, into tin–tin or tin–silicon bonds is generally regio- and stereospecific.[239]

High-pressure kinetic studies of the insertion of alkynes, or nitriles, with electron-rich triple bonds, e.g. dipropylcyanamide or 1-(diethylamino)propyne, into metal–carbene bonds indicate activation volumes and entropies in the ranges -17 to $-25\ cm^3\ mol^{-1}$ and -128 to $-146\ J\ K^{-1}\ mol^{-1}$, respectively. Such values, both in their own right and in relation to similar organic systems, suggest that the reaction proceeds through a cyclic transition state, itself derived from a non-cyclic precursor generated in a pre-association equilibrium (Fig. 10.11).[224]

10.5.4.4 Comparisons

Comparisons of alkene and alkyne reactivity, into metal–hydrogen and into metal–carbon bonds, are available for complexes $[Sc(\eta^5\text{-}C_5Me_5)_2R]$, R = H or alkyl. Reaction is too fast to follow for R = H; rate constants follow the sequence R = Me > Et < ^{n}Pr for insertion into Sc–R. The slightly slower reaction of the ethyl compound was attributed to modest ground-state stabilization through agostic interaction involving β-hydrogen. Alkyne insertion is characterized by similar rate constants to alkene insertion. The activation entropy of $-150\ J\ K^{-1}\ mol^{-1}$ for $H_3CC{\equiv}CCH_3$ insertion indicates a very ordered transition state, presumably four-centred (**54**). As reaction with alkynes is stoichiometric, with alkenes catalytic, the transition state for the former can usefully act as a model for the latter. The discussion of these reactions touches on agostic interactions, and on elimination from alkyl products, thus drawing together several topics from this section and Section 10.6.[240]

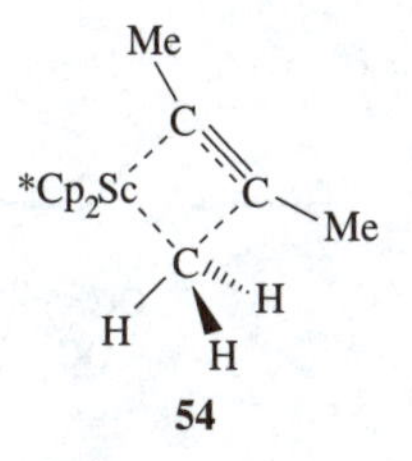

10.5.5 Other compounds

10.5.5.1 Carbon dioxide[97] and carbon disulphide

Carbon dioxide insertion is a reaction of great importance to the use, or recycling, of this compound as a feedstock for synthesis, preparative or catalytic, of organic compounds. The electrophilicity of carbon dioxide facilitates its insertion into M–H, M–O, M–N, and M–C bonds,[95,241] though there are marked variations in ease of insertion. Thus many examples are known of carbon dioxide insertion into metal–alkyl bonds, fewer into metal–alkoxide bonds.[242] Photoactivation may encourage insertion, as in the reaction of $[Ti(cp)_2(Me)_2]$ with carbon dioxide, the first example of photoinsertion of carbon dioxide into a metal–carbon bond. At appropriate wavelengths, a methyl is converted into an acetate ligand at around room temperature; thermal insertion requires 80 °C. The precise products vary with the wavelength of irradiation, which is hardly surprising, since at least five different photoreactions of $[Ti(cp)_2(Me)_2]$ have been reported.[243]

The normal mechanism of carbon dioxide insertion into a metal–carbon bond is I_a (S_N2), with retention of configuration at the α-carbon.[244] Kinetics, stereochemistry, and mechanisms of carbon dioxide and carbon monoxide insertion have been compared for reactions involving a number of metals, including tungsten,[245] ruthenium,[246] and rhodium.[247]

Insertion of carbon dioxide into metal–hydrogen bonds gives formates, as in the reactions giving $[NiH(O_2CH)(PCy_3)_2]$,[248] $[MoH(O_2CH)(dppe)_2]$,[249] and $[U(O_2CH)(C_5H_4SiMe_3)_3]$.[250]

Insertion of carbon dioxide into tungsten–oxygen bonds has been studied in detail for the substrate $[W(CO)_5(OPh)]^-$,[251] where insertion takes place much less slowly than into tungsten–carbon in $[W(CO)_5(Me)]$. Added carbon monoxide has no effect on the rate of insertion of carbon dioxide into the W–OPh bond, no $W(CO)_6$ being formed in such experiments. The W–OPh bond must therefore remain unbroken during the course of the insertion, which takes place by electrophilic attack by the carbon monoxide at a lone pair of the phenoxide-oxygen. Relative effects of phosphorus-

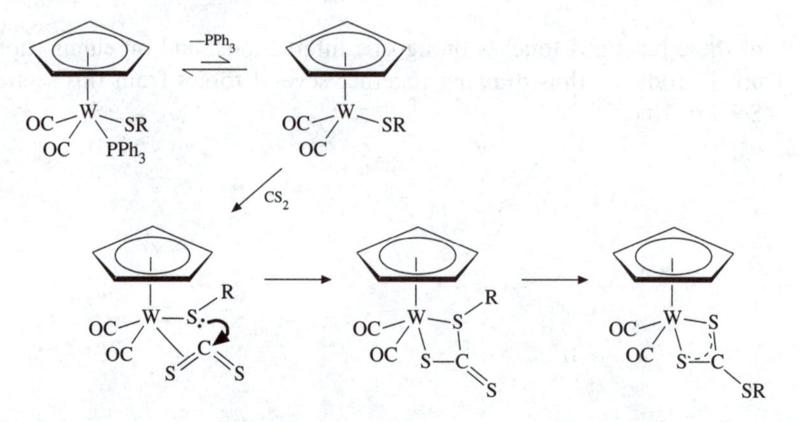

Fig. 10.12 Mechanism of carbon disulphide insertion into $[W(cp)(CO)_2(PPh_3)(SR)]$.

donor spectator ligands were also assessed, for the closely related anions $[M(CO)_4(OPh)L]^-$. Here the order of decreasing rate constants is $L = P(OMe)_3 > PMe_3 > PPh_3$.

Examples of CS_2 and of COS insertion are much less common. A mechanism has been proposed for insertion of sulphur into the rhodium–hydrogen bond of $[RhH(PPh_3)_4]$ on reaction with COS. η^2-Addition of the COS to the metal is followed by hydride transfer to the sulphur, promoting C–S bond cleavage to give the final product *trans*-$[Rh(SH)(CO)(PPh_3)_2]$.[252] Carbon disulphide inserts into the platinum–oxygen bond of platinum(II) methoxides, the palladium–oxygen bond of $[PdMe\{OCH(CF_3)Ph\}(dpe)]$, and both palladium–carbon and palladium–phosphorus bonds of *trans*-$[PdMeI(PMe_3)_2]$ and the $[PdMe(PMe_3)_3]^+$ cation.[253] Kinetics and product characterization for the insertion of carbon disulphide into the tungsten–sulphur bond of $[W(cp)(CO)_2(PPh_3)(SR)]$ indicate preliminary reversible loss of the PPh_3 and a final ligand reorganization step to give the thioxanthate product.[254] The mechanism is set out in Fig. 10.12. For a final, and simpler, example of carbon disulphide insertion we cite its reaction with $MeZnNEt_2$ to give the dithiocarbamate complex $MeZnS_2CNEt_2$.

Relative reactivities of CS_2, COS, and CO_2 have been established for their respective insertions into the metal–oxygen bonds of $[M(CO)_5(OPh)]$, M = Cr, W. The reactivity order is $CS_2 > COS > CO_2$, rate constants probably spanning about one order of magnitude.[251]

10.5.5.2 Sulphur dioxide

The kinetics and mechanism of sulphur dioxide insertion into metal (*sp*-block and *d*-block) to carbon bonds have been studied for many years.[255] There is a marked difference between sulphur dioxide insertion and carbon monoxide insertion, arising from the greater nucleophilicity of the former. Insertion of sulphur dioxide into a metal–alkyl bond takes place by an S_E2

mechanism, with the SO_2 attacking at the rearside of the alkyl group, to give an *O*-bonded sulphinate species that usually isomerizes to the thermodynamically more stable *S*-bonded form:

$$M-R \rightarrow M-O-S\begin{matrix}\nearrow O \\ \searrow R\end{matrix} \rightarrow M-\overset{O}{\underset{O}{\overset{\|}{\underset{\|}{S}}}}-R$$

This applies for, e.g., Fe–R, Ru–R,[256] Ni–R[257]—and to insertion into the tungsten–tin bond in the $[(OC)_5W\text{–}SnMe_3]^-$ anion.[258]

Inversion takes place at the carbon when SO_2 inserts into Co–C in cobaloximes[259], though SO_2 insertion into Co–C in cobalamins (under photochemical conditions) takes place by a free-radical chain mechanism.[260]

In contrast to these numerous examples of insertion into metal–carbon bonds, sulphur dioxide is conspicuously reluctant to insert into metal–hydrogen bonds.[261]

10.5.5.3 Miscellaneous

There is an assortment of other small molecules which have been shown to participate in insertion reactions. We give a few examples here, both to indicate the variety and to provide appropriate references for the interested reader to follow up. Our selection includes NO (into transition metal–alkyl and –aryl bonds),[262] O_2 (into Co–C,[263] Ni–C,[264] and Ti–, Zr–, and Hf–C[265]), N_2 (from diazonium compounds into Rh–H, Pt–H),[266] P_4 (into Zr–P),[267] $GeCl_2$,[268] and $SnCl_2$ (into Mn–Cl, Re–Cl).[269] Most of the references cited provide at least some mechanistic insight. Insertion of a metal-containing species into, for example, carbon–carbon or carbon–hydrogen bonds is generally oxidative addition of the organic compound to the metal (see Section 10.4 above). The $GeCl_2$ and $SnCl_2$ reactions mentioned above are oxidative additions at the germanium and tin centres, but insertions with respect to the other reactants.

10.6 Elimination

The extrusion of an organic compound from an organometallic species to leave a metal-hydride species is termed an elimination reaction, denoted α-, β-, γ-, etc. depending on the number of carbon atoms between the metal and the hydrogen in question in the starting compound:

$$M\text{–}[C_n]\text{–}H \longrightarrow M\text{–}H + [C_n]$$

For $n = 1$ elimination is difficult (hence the possibility of preparing, even though with difficulty, such compounds as tungsten hexamethyl and the $[WMe_8]^{2-}$ anion[270]), though cases are known, e.g. in tantalum and tungsten chemistry (where photochemical activation is required).[271] Examples are also known of eliminations for compounds with $n=3$, 4, etc., but the most important case is when $n = 2$, β-elimination. This is the reverse of the 1,2-insertion discussed in Section 10.5 above:

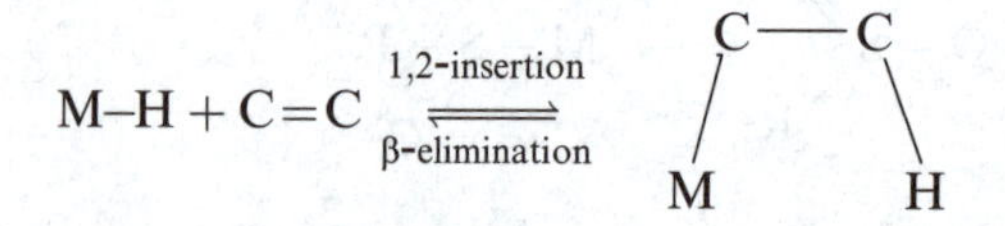

The direction of spontaneous reaction depends on the nature of the compounds involved. In most systems of interest reaction proceeds essentially 100% in one direction or the other, but there are a few cases documented where an equilibrium position can be attained from either side. Examples include

$$\text{Cl—PtL}_2\text{—H} + \text{H}_2\text{C=CH}_2 \rightleftharpoons \text{Cl—PtL}_2\text{—CH}_2\text{CH}_3$$

with L = an alkyl or aryl phosphine,[272] and the $Co(H)(PMe_3)_3/C_2H_4$ system mentioned in Section 10.5.4.1 above.

β-Elimination and insertion have a common planar four-centre transition state (microscopic reversibility), **55**. In the case of the $[Sc(\eta^5\text{-}C_5Me_5)_2R]$ system discussed in Section 10.5.4.4 above, a detailed kinetic study, including H/D isotope effects, indicated a transition state of this type with partial charges as shown in **56**.[240] Evidence in support of such transition states has been provided by, for example, the demonstrating of the close approach of a β-hydrogen to the titanium in **57**. Titanium is a favourable choice of metal here, as it is more willing than elements to its right in the Periodic Table to adopt a coordination number of eight. In **57** the titanium

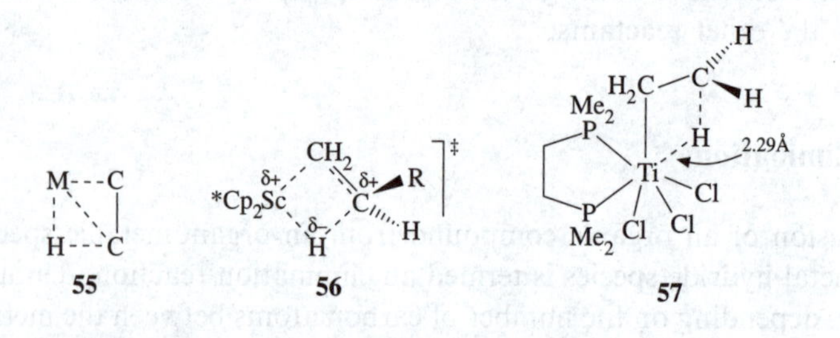

to hydrogen distance (note the advantage of using the relatively light titanium to maximize the chances of locating this hydrogen in the X-ray structure determination) is 2.29 Å, considerably less than the sum of the van der Waals radii (Ti ~ 1.9, H = 1.20 Å).[273] A similar strong agostic interaction between metal and β-hydrogen has been demonstrated by ^{1}H and ^{13}C

NMR spectroscopy as well as by X-ray structure determination for the cobalt-ethyl complex **58**. The strength of the cobalt–β-hydrogen interaction can be gauged from the report that the ethyl group is rigid on the NMR time-scale at −80 °C.[274] Barriers to enantiomer interconversion in $[M(L)(Et)(C_2H_4)]^+$, for M = Co, Rh, and L = η^5-C_5H_5 or η^5-C_5Me_5, must be low, as there is fast interconversion, probably with associative activation, even at low temperatures. This indicates that interconversion goes via terminal hydrides, as in **59** → **60** + L, and suggests that energy differences between terminal and agostic hydrogens must be small. Species of these types are well on the way to being models for the transition states for β-elimination, and for alkene insertion.[275]

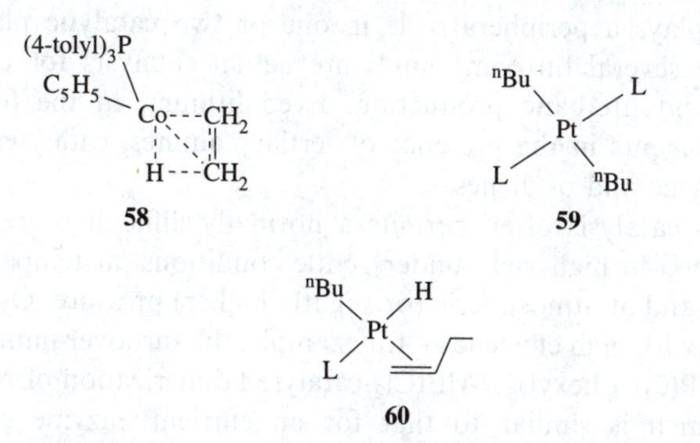

These reactions may also be viewed as intramolecular oxidative addition followed by dissociation, if one writes, e.g.,

M H ⟶ M—H rather than M H ⟶ M—H

for the initial stage. It is advantageous to have, or to be able to generate by facile ligand loss, a vacant coordination site on the metal to accept the hydride. The fifth and sixth coordination sites on square-planar complexes cannot fulfil this role—one of the phosphines has to dissociate from **61** before β-elimination can take place. Such constraints explain why β-elimination takes place around 10^4 times more slowly from compounds of the type **62** than from their analogues **61**.[276]

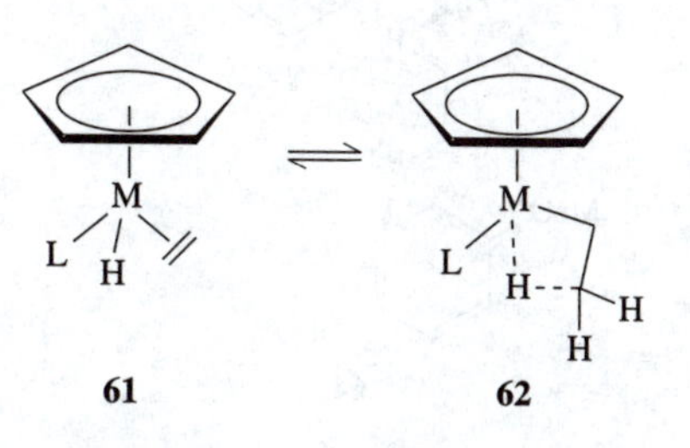

10.7 Homogeneous catalysis[277]

10.7.1 General principles

Many metal complexes, particularly complexes of transition metals in low oxidation states, catalyse a variety of organic reactions, such as hydrogenation, isomerization, dimerization and oligomerization, and carbonylation, in solution. The roles of the transition metals in such catalysis provide the main subject matter of this section. There also are a number of systems in which *sp*-block metal compounds act as catalysts, or at least vital catalyst components. The part played by alkyl aluminium halides in the Ziegler–Natta process is dealt with in Section 10.7.3.2 below. Tin, in the form of the $-SnCl_3$ ligand, plays a peripheral role in one or two catalytic platinum complexes, while several tin compounds are actual catalysts for epoxide polymerization and urethane production. Even lithium, in the form of organolithium reagents in the presence of tertiary amines, catalyses polymerization of ethene and of dienes.

Homogeneous catalysis often permits a normally difficult or reluctant reaction to proceed in high yield under gentle conditions, at temperatures close to ambient and at atmospheric (or slightly higher) pressure. Often the catalysts operate with high efficiency—for example, the turnover number for the $Ni(C_3H_5)Br\{P(cyclohexyl)_3\}/AlEtCl_2$-catalysed dimerization of propene is *ca* 60 000, which is similar to that for an efficient enzyme such as catalase.[278] They can be stereoselective—catalysts with appropriate chiral ligands can be used in asymmetric syntheses with enantiomeric excesses as high as 99%.[4,279,280] The first chiral phosphine to be used successfully was **63** (cyclohexylanisylmethylphosphine, or 'camp'), soon followed by the more effective chelate dipamp, **64**. In complexes of these ligands chirality resides in the phosphorus atoms. It was realized that chirality could equally well reside in carbon centres, with diop, **65**, quickly becoming popular as it can be synthesized relatively easily from tartaric acid. More recently, some tripodal tris-phosphane ligands[281] have been developed to expand the range of ligands for the production of chiral compounds, while the synthesis of

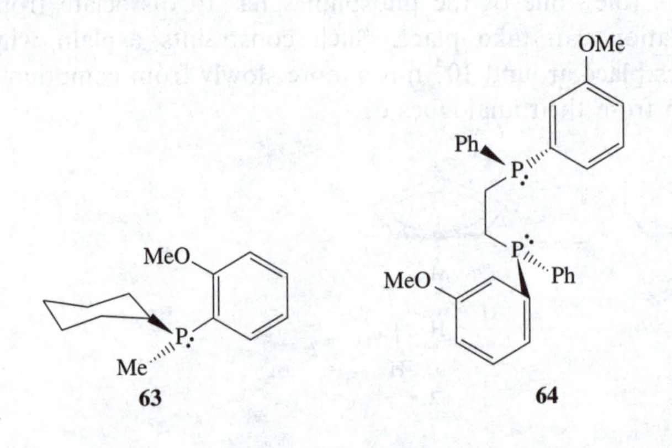

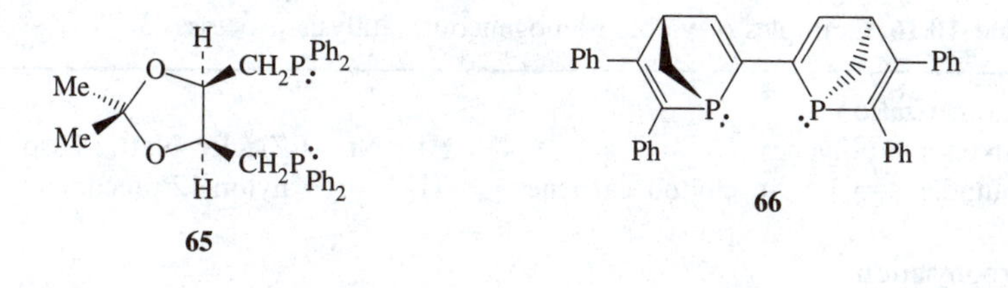

bipnor, **66**, makes available a chiral ligand which is effectively non-racemizable since the donor-phosphorus atoms are in bridgehead positions.[282]

A great deal of the research interest in homogeneous catalysis stems from the range of useful organic compounds that can be generated under mild conditions—homogeneous catalysis has considerable industrial potential. After the initial flush of enthusiasm it was realized that the undoubted advantages of homogeneous catalysis were balanced by some serious drawbacks (Table 10.15).[283] Nonetheless, since about 1950 a series of processes has been successfully developed to exploit its advantages.[284,285] Table 10.16 gives some commercially viable examples. As so often, the empirical optimization of such processes tended to precede, rather than be guided by, detailed laboratory work on the respective reaction mechanisms. However, as will be seen in the following sections, the mechanisms of a number of these important processes are now fairly well understood. Some asymmetric syntheses have proved of commercial value, for example the large-scale (in pharmaceutical terms) production of L-dopa (3,4-dihydroxy-phenylalanine), using rhodium(I)-dipamp catalysts,[279] and the production

Table 10.15 Advantages and disadvantages of homogeneous and heterogenous catalysis.

	Homogeneous	Heterogeneous
Efficiency of catalyst use	All metal centres active	Only surface sites active
Conditions	Gentle	Often high temperature and/or high pressure required
Separation of product from catalyst	Difficult [a]	Usually easy
Catalyst recovery	Difficult [a]	Straightforward
Establishment of reaction mechanism	Kinetic studies and rate laws usually informative	Often difficult

[a] Current examination of biphasic reaction systems, for example using sulphonated phosphine ligands to confer water solubility on the catalyst (cf. hydroformylation section), or using fluorous biphase conditions (cf. hydroboration section), and of such solvent media as supercritical carbon dioxide (cf. opening section), may lead to amelioration of these difficulties.

Table 10.16 Examples of viable homogeneous catalysis processes.[a]

Oligomerization		
alkenes → polyenes	Al[b]; Ni	Ziegler–Natta; Esso; Shell
butadiene → 1,5,9-cyclododecatriene	Ti	nylon-12 precursor
Carbonylation		
methanol → acetic acid	Rh	Monsanto; now BP
alkenes → carboxylic acids	Ni	Reppe
Hydroformylation		
alkenes → alcohols or aldehydes	Co; Rh	Oxo reaction [c]
Hydrocyanation		
butadiene → adiponitrile	Ni	nylon-66 precursor
Oxidation		
ethylene → acetaldehyde	Pd	Wacker; Hoechst
propene → propylene oxide	Mo	polyurethanes; polyesters [d]
p-xylene → terephthalic acid	Co; Mn	Sabatier
cyclohexane → adipic acid	Co; Mn	nylon-12 precursor

[a] In deference to worldwide industrial relevance, rather than to current UK academic conformance to political correctness in respect of nomenclature, we generally use long-established trivial names in this section.
[b] Also (more importantly) Al/Ti.
[c] At least eighteen companies use, or have used, hydroformylation processes.
[d] For foams and plastics, respectively.

of (–)-menthol using $[Rh(binap)L_2]^+$ as catalyst (dipamp = **64**; binap = **67**). The mechanism of the latter process has been thoroughly probed by NMR spectroscopy (^{1}H and ^{31}P) and kinetic and labelling studies.[286] In relation to kinetics and mechanisms of reaction sequences which give chiral products, it is interesting to note the concept of a selectivity-determining step complementing the familiar notion of a rate-determining step.

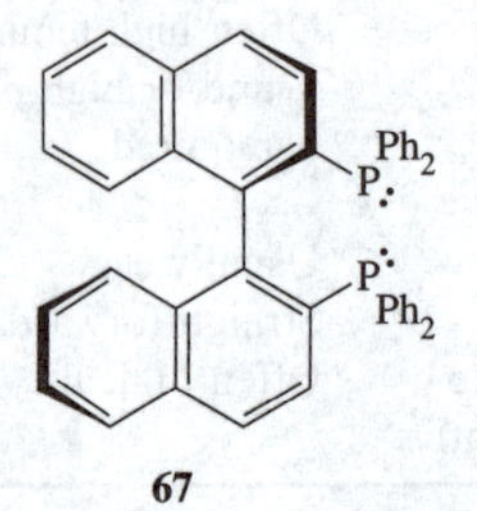

67

The overall general pattern for most homogeneous catalysis processes involves a number of individual reactions drawn from the reaction types discussed in this and earlier chapters. These include oxidative addition and reductive elimination (Section 10.3), insertion (Section 10.5), and reactions of coordinated ligands (Section 10.2). Dissociation, needed to get the active catalyst at the start and to liberate the product from the catalytic metal centre, has been discussed at several points in the chapters on substitution mechanisms (Chapters 2–4 and, particularly relevant, Chapter 6). In the subsequent sections of this chapter we shall discuss a number of types of catalysed reactions, showing how the overall process is built up from the individual reactions. Wherever possible we shall show how the individual steps are combined into a catalytic cycle. But before turning to the detailed build-up for selected reactions we need to deal with some general matters. These include the general nature of complexes used as homogeneous catalysts, ligand dissociation, and interaction of species of reduced coordination number with substrates.

Many homogeneous catalysts are phosphine[287] or carbonyl compounds. The advantages of an *in situ* carbonyl ligand are obvious for a carbonylation or hydroformylation reaction. Phosphine ligands have the advantage that they can be tailored to vary steric and electronic effects, and to modify the solubilities of their metal complexes. Thus sulphonation can confer water solubility, as mentioned in relation to hydroformylation below, as also can guanidinium groups[288] or the use of ligands of the $P(CH_2OH)_3$ type.[289] Perfluoroalkyl substituents can lead to complexes which are freely soluble in supercritical carbon dioxide,[290] which is at the time of writing a reaction medium of considerable interest, for instance for catalytic hydrogenation.[291] The advantages of nitrogen-donor ligands have been extolled,[292] but their complexes are only occasionally encountered as homogeneous catalysts. The catalyst may, especially if it is four-coordinate, start its cycle by (oxidatively) adding one of the reactants. On the other hand, the potential catalyst may first need to lose a phosphine, nitrogen donor, or carbon monoxide ligand to give an active catalyst, i.e. a species of reduced coordination number that is ready to accept one of the reactants. There is kinetic evidence for a catalytically active three-coordinate intermediate $[Ir(H)(CO)(PPh_3)]$ being formed from the precursor $[Ir(H)(CO)(PPh_3)_2]$.[293] There is also evidence for reversible phosphine loss from tetrakis-phosphine-nickel(0) complexes, as for example prior to oxidative addition of HCN in hydrocyanation. Equilibrium constants for such phosphine loss are given in Table 10.17.[294] They are presumably determined by inductive and steric (cone angle) factors. Rate constants for reversible phosphine loss are not often known, but for the familiar catalyst $[Rh(PPh_3)_3Cl]$ it is known that the rate constant for phosphine loss is $0.31\ s^{-1}$ at 303 K.[295] It has also been established that there is intramolecular exchange between the two types of coordinated phosphine, *cis* and *trans* to the chloride, and that this process is more rapid ($k = 22\ s^{-1}$ at 297 K). The situation with regard to the need for phosphine loss from $[Rh(PPh_3)_3Cl]$ is complicated, as oxidative addition to give a complex of higher coordination number is also a possible initial step. This matter is discussed in relation to catalytic hydrogenation of alkenes in Section 10.7.5.1 below.

Table 10.17 Equilibrium constants for $NiL_4 \rightleftharpoons NiL_3 + L$ in benzene at 298.2 K.

Ligand L	*K* (molar scale)
$P(OEt)_3$	$< 10^{-10}$
$P(O\text{-}4\text{-tolyl})_3$	6×10^{-10}
$P(O^iPr)_3$	2.7×10^{-5}
$P(O\text{-}2\text{-tolyl})_3$	4.0×10^{-2}

Phosphine loss can be facilitated by appropriate complex design. In the extreme, the presence of bulky phosphines, as in $[Ru(H)(Cl)(CO)(P^iPr_3)_2]$, gives an unusually low coordination number, here five rather than the normally expected six for this type of complex. The disadvantage here is that approach to the active site is hindered by the bulky phosphines. Complex **68** is to be preferred; the relatively large bulk (cone angle) of the PPh_3 ligands combined with the strong *trans* effect of the hydride facilitates loss of the PPh_3 *trans* to the hydride without significant obstruction to incoming reactants. Thus this PPh_3 can be replaced by CO, ^{t}BuNC, or pyridine to give model catalytic intermediates.

In practice, the species generated by reversible ligand loss are generally either stabilized by interactions which attempt to maintain the metal's preferred coordination number, as in **69**, or have easily-displaced solvent

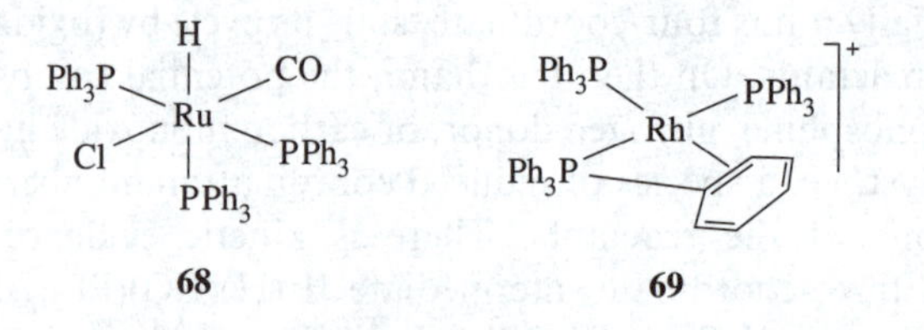

molecules in the 'vacant' site, as for example in the model complex $[Rh(PPh_3)_3(MeCN)]^+$.[296] The presence of two coordinated solvent molecules and rhodium–ligand distances have been established by EXAFS for the intermediate $[Rh(dipamp)(MeOH)_2]$ (cf. L-dopa production above; dipamp = **64**).[297] This thorough study compared rhodium to ligand-donor atom distances obtained by EXAFS of solutions, EXAFS of solid samples, and X-ray crystallography. The great advantage of EXAFS is that this technique can be used to probe the geometry of the immediate environment of the metal both in solution and in any solid material obtained from the system, or indeed in model compounds.[298] The most straightforward application is to use EXAFS to compare the geometry around the metal in the intermediate in solution with geometry established by X-ray diffraction in a single crystal of a model compound. This approach was in fact used for the $[Rh(PPh_3)_3(MeCN)]^+$ species mentioned above; EXAFS results from solution studies were interpreted with the aid of the crystal structure of its BF_4^- salt. X-ray photoelectron spectroscopy, or ESCA, is claimed to be easier to use than EXAFS, but more difficult to interpret with confidence. Interested readers may learn of its assessment in relation to the mechanism of catalysis by $[RhCl(PPh_3)_3]$ and of its use in probing Ziegler–Natta catalysis in the reference cited.[299]

Once one has a vacant site at the metal centre, or a site containing a weakly bonded solvent molecule, then this can become occupied by a reactant molecule. The existence of equilibria of the type $L_nM + \text{alkene} \rightleftharpoons L_nM(\text{alkene})$ has been demonstrated for, for example, interaction of bis(tricyclohexylphosphine)platinum(0) with ethene. This interaction was studied in solution by 2-D ^{31}P NMR; X-ray diffraction indicated weak platinum–ethene bonding in the solid state.[300] There is a large body of thermodynamic data on equilibria involving association of alkynes and, especially, alkenes with metal centres.[301] Much of this refers to silver(I) so is relevant but not central to homogeneous catalysis. Data for such centres as platinum(II), palladium(II), and nickel(0) are, unfortunately, not generally for systems which lend themselves to shedding light directly on steps in the homogeneous catalysis systems discussed in this book. Nonetheless, it is interesting to see the variation in equilibrium constant with the number and nature of substituents for silver(I)/alkene interactions (Table 10.18), and to compare equilibrium constants for interaction of Vaska's iodide with, on the one hand, ethene and ethyne, and, on the other, ethene and perfluoroethene (Table 10.19). The steric factors which are so important in determining stability constants for coordination of bulky alkenes are also important in directing the approach of dihydrogen to the active centre in asymmetric hydrogenation.

A detailed kinetic study of reactions of $[Ru(H)(Cl)(PPh_3)_3]$ with substrates such as cycloheptatriene, dienes, and dimethyl maleate showed that the predominant reaction pathway involved the initial loss of one phosphine, followed by reaction of the reactive four-coordinate species thus generated with the organic substrate. However, there was some evidence for a minor contribution from direct attack of the alkene at the ruthenium,

Table 10.18 Equilibrium constants for the association of alkenes with silver(I).

Alkene	K	Alkene	K
$H_2C{=}CH_2$	17.5		
$D_2C{=}CD_2$	19.7	cyclopentene	10.2
$H_2C{=}CHR$ (R = Me → n-C_5H_{11})	3–9		
cis-MeHC=CHMe	4.9		
trans-MeHC=CHMe	1.6		
$Me_2C{=}CHMe$	1.1	$H_2C{=}CHCH{=}CH_2$	4.5
$Me_2C{=}CMe_2$	0.34		

Table 10.19 Equilibrium constants, and enthalpies, for association of alkenes and ethyne with Vaska's iodide, *trans*-$[IrI(CO)(PPh_3)_2]$.

Alkene/Alkyne	*K* (20 °C)	*K* (30 °C)	$\Delta H°$ (kJ mol^{-1})
$H_2C{=}CH_2$	2.7	1.1	−50
$F_2C{=}CF_2$		>100	
$HC{\equiv}CH$	1.2		−39

followed by an insertion reaction of the six-coordinate $[RuClH(PPh_3)_3(alkene)]$ produced.[302] Similar pathways are believed to operate for $[Rh(PPh_3)_3Cl]$ catalysis, with the added complication of a possible third route involving catalysis by the $[(Ph_3P)_2Rh(\mu\text{-}Cl)_2Rh(PPh_3)_2]$ dimer. In this system, as in others, the relative contributions of competing routes can be varied by varying concentrations of reactants, especially of added phosphine.

Dissociation is again important in that the product must drop off the metal, or the active catalyst will not be regenerated to go round the cycle again! Rhodium and iridium form a very useful pair of metals in relation to efficient catalysis and mechanistic studies. The different reactivity time-scales characteristic of these elements often, happily, result in a given rhodium complex acting as a good catalyst, but its iridium analogue assisting in the diagnosis of mechanism. The considerably greater reluctance of iridium–ligand bonds to break may give relatively long-lived intermediates or even halt the catalytic cycle, with the iridium product then providing a stable model for the transient rhodium intermediate. This approach may be exemplified by just one example from the field of asymmetric hydrogenation, where stable enamide and hydridoalkyl complexes of iridium were prepared to model their reactive rhodium counterparts in the catalytic cycle. The absolute configuration of one iridium complex was established, by NMR spectroscopy, to provide a stereochemical guide to the rhodium-based catalysis.[303]

Many catalysts are versatile, promoting a variety of reactions. Some such rhodium catalysts are given in Table 10.20. Some idea of the versatility of $[RhCl(PPh_3)_3]$ is afforded by the observation that as long ago as 1981 it was possible to devote a complete review, with more than 650 references, to this one catalyst.[304] This catalyst has also been the subject of a number of theoretical studies, in which energy profiles have been calculated for plausible intermediates and transition states.[305] Our main mention of mechanisms of catalysis by $[RhCl(PPh_3)_3]$ can be found in the section on hydrogenation (Section 10.7.5.1). Ranges of examples similar to those in Table 10.20 could be given for catalysts based on other metals—for instance platinum carbonyl derivatives are useful catalysts for, *inter alia*, hydrogenation, carbonylation, hydroformylation, and the water gas shift reaction.[306] The varied aspects of organonickel chemistry in dimerization and oligomer-

Table 10.20 Some versatile homogeneous catalysts.

Reaction type	Typical substrates		
	$RhCl(PPh_3)_3$	$Rh_4(CO)_{12}$ $Rh_6(CO)_{16}$	$[Rh(CO)_2X]_2$ $[Rh(CO)_2I_2]^-$
Isomerization	butenes; allyls		allyl aromatics; allylamines
Polymerization	alkenes; alkynes		butadiene
Hydrogenation	(cyclo)alkenes; steroids	aldehydes; butadiene	
Reduction		nitrobenzene	
Carbonylation	methanol	methanol	methanol; alkynes
Hydroformylation	alkenes	alkenes	alkenes
Oxidation	1-alkenes; alkylbenzenes; anthracene	cyclohexanol; carbon monoxide; triphenyl phosphine	ethanol

ization of alkenes, cyclo-oligomerization of alkynes, and asymmetric catalysis have been well reviewed.[307] Different catalysts may give different products from the same substrate, as in the isomerization and cyclization reactions shown in Fig. 10.13.[308] Readers interested in preparing catalysts for kinetic and mechanistic studies could profitably consult the extensive chapter in a wide-ranging handbook on organic syntheses using transition metal complexes which details the preparation and use of some two dozen transition metal complexes commonly used as homogeneous catalysts.[309] Changing the emphasis from catalysts to processes, it is difficult to give guidelines as to preferred catalysts for specific reaction types, but Table 10.21 gives some very rough generalizations. For catalysis by palladium compounds, catalysts, processes, and mechanisms are comprehensively documented on a CD-ROM,[310] whose database can be searched by, *inter alia*, authors, reagents, products, or mechanisms.

Thus far we have dealt mainly with mononuclear compounds as catalysts, but there has been significant interest in binuclear and polynuclear homogeneous catalysts.[311] We mentioned platinum carbonyl clusters[312] briefly above, and included two rhodium carbonyl clusters in Table 10.20. They activate hydrogen and catalyse such processes as hydrogenation and hydroformylation.[306] The main interest has been in platinum/gold/phosphine/carbonyl clusters,[313] with the inclusion of a few phosphines often helping the catalytic properties by relatively easy reversible loss giving active sites. However, a number of other types of polynuclear species have been assessed. These have included a selection of polyoxoanions of molybdenum and tungsten,[314–316] for example the familiar $[SiW_{12}O_{40}]^{4-}$ anion, whose properties can be tailored and optimized by some molybdenum-for-tungsten

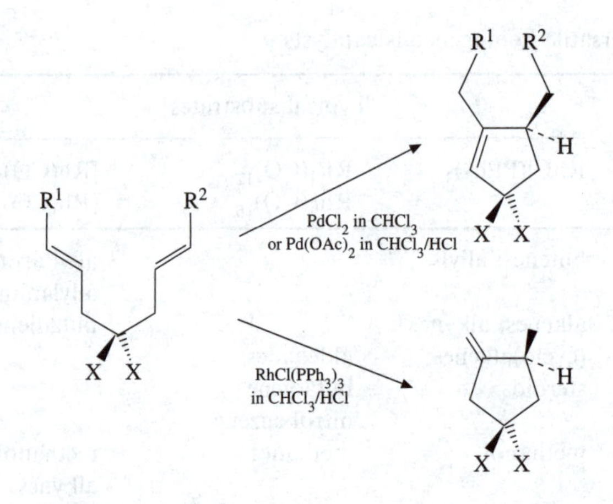

Fig. 10.13 Effect of catalyst nature on products of cyclization (X = e.g. COMe or CO_2Et).

replacement, namely $[SiW_{12-x}Mo_xO_{40}]^{4-}$.[317] For reactions bringing two species together there may well be advantages in using a catalyst with two adjacent metal centres. For such a catalyst it is only necessary to generate one active site per metal, rather than two at the same metal. Moreover, with appropriate mutual orientations of active sites in a polynuclear catalyst it should be possible to control the detailed stereochemistry of polymerization of alkenes, in other words to control the isotactic (all tertiary carbons having the same configuration)/syndiotactic (carbon atoms with alternating configuration) nature of, e.g., polypropylene. The advantages of binuclear catalysts have been illustrated by, e.g., dirhodium catalysts for hydroformylation[318] and dicopper complexes in biologically relevant hydrolyses.[319] Despite the promised advantages of bi- and polynuclear catalysts, there has been a general lack of enthusiasm for them, with opinion divided as to whether they offer any significant advantages.[198]

Table 10.21 Preferred catalysts for reactions involved in homogeneous catalysis.

Reaction	Catalyst
C–H activation	Iridium (rhodium) complexes [a]
C–C activation	Palladium complexes [a]
C=C activation	Early transition metal complexes
Hydroformylation	Rhodium (cobalt) complexes
Hydrocyanation	Nickel/phosphite complexes
Asymmetric catalysis	It is preferable to have the chirality at carbon rather than at coordinating phosphorus.

[a] Generally in the form of phosphine (phosphite) complexes.

The main question in regard to most cluster catalysis mechanisms is whether the intact cluster is the catalyst, or whether the cluster is merely the precursor for smaller catalytically active moieties, particularly coordinatively unsaturated mononuclear fragments.[320] Sometimes, as in the case of metal–metal-bonded catalysts such as $Co_2(CO)_{10}$ or rhodium(II) dimers, binuclear species are known to be stable precursors of mononuclear active catalysts. It is often difficult to devise and execute experiments which provide incontrovertible evidence that the catalytically active species is the intact cluster, or at least retains the original nuclearity. In one of the first studies to provide convincing evidence for whole-cluster catalysis, μ^4-PPh groups were used to bind the cobalt cluster **70** (L = CO) tightly together, and the effects of varying L (CO; PR_3) on rates and selectivity assessed.[321] Appropriate control experiments with monomeric and dimeric species provided good evidence that alkene hydrogenation by $[Fe(\eta^5\text{-}C_5H_5)(CO)]_4$ does involve whole-cluster catalysis. Cluster versus monomer comparisons for hexarhodium clusters containing optically active phosphines reveal differences in optical characteristics that confirm whole-cluster catalysis here too, though there are indications that the catalytically active form for surface-supported $Rh_6(CO)_{16}$ is binuclear.[322] The structure of one of the products isolated from reaction of t-butylacetylene with $Ru_3(CO)_{12}$ contains a three-alkyne unit interacting with all three ruthenium atoms. It chelates the central ruthenium, as shown in **71**, and is also π-bonded to the other two ruthenium atoms.[323] Here, and in the dirhodium hydroformylation case mentioned above, the bi- or trinuclear nature of the catalyst is intrinsic.

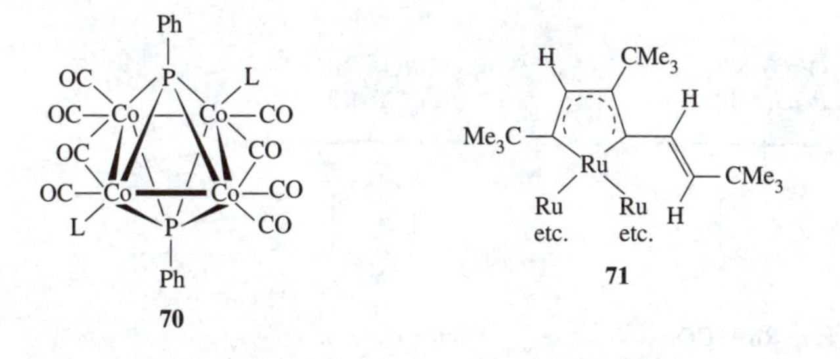

The other main mechanistic interest in polynuclear catalysts lies in the possibility of forging a link between homogeneous and heterogeneous catalysis.[324] Efforts in this direction have involved such species as polynuclear carbonyls and polynuclear nickel isocyanides.[320,325] Polynuclear carbonyls can, of course, be very large molecules, with peripheries approximating to a small section of the surface of a solid catalyst.[326] A recently advertised example was the $[Ni_{38}Pt_6(CO)_{48}H]^-$ anion, while $[Pt_{38}(CO)_{44}]^{2-}$ has been claimed to be almost as large as the finest particles preparable from bulk solids. The sequence of catalysts from mononuclear to dinuclear to polynuclear clusters to surfaces has been reviewed succinctly in relation to selective catalytic hydrogenation of alkynes and aromatic hydrocarbons,[327] and at length.[328] Examples of kinetic and mechanistic studies of cluster catalysis will be found in the sections on hydrogenation (Section 10.7.5) and hydroformylation (Section 10.7.8.1) below.

The use of supported catalysts provides another link between homogeneous and heterogeneous catalysis. A number of low-valence metal catalysts have successfully been derivatized and attached, physically or chemically, to alumina, silica, or polystyrene. Towards the heterogeneous end of the range, mechanistic studies have been carried out on, for example, hydrogenation promoted by platinum or palladium electrodes, or catalysed by these metals, rhodium, and ruthenium adsorbed on alumina.[329] The need to ionize adsorbed hydrogen in the proposed mechanisms separates them from mechanisms of catalysis of homogeneous hydrogenation. Moving from uncomplexed metal(0) towards more conventional homogeneous catalysts, rhodium acetate has been intercalated in a mica (i.e. silicate) matrix to give catalytically active rhodium(I).[330] The general approach for derivatization and chemical bonding into the support may be illustrated by the example shown in **72**.[331] Copolymerization of vinylpyridine with divinylbenzene gives the organic backbone with pendant pyridyl groups which can be coordinated to rhodium(I); the rhodium has the usual sort of ligands occupying the other three coordinated positions. Polymers with pendant benzyldiphenylphosphine groups can coordinate a range of low oxidation state metal centres, **73**. To link this approach with the efforts mentioned elsewhere to confer water solubility on catalysts, a derivative of **72** has been prepared with sulphonated phosphine ligands. Supported

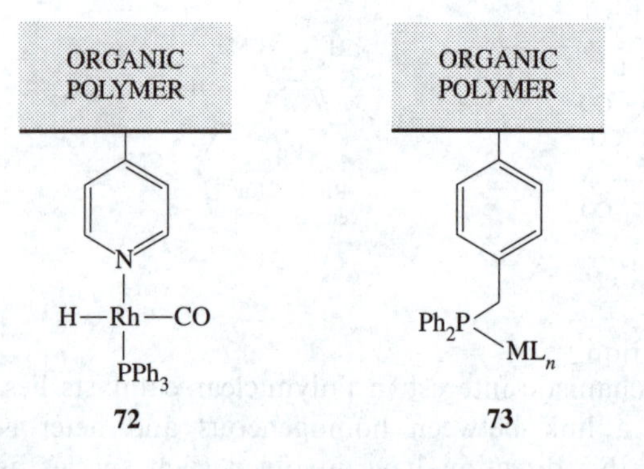

catalysts can be more active and more selective than analogous normal homogeneous catalysts, which themselves tend to be more selective than heterogeneous catalysts. Supported homogeneous catalysts are also, obviously, easier to separate from products. However, as so often, these advantages are offset by other problems, particularly severe technical difficulties caused by the supporting materials. The use of heteropolyanion catalysts provides other ways in which the homogeneous/heterogeneous division is blurred. Such catalysts may absorb such solvents as water or alcohols, so that the apparently heterogeneous catalysis actually takes place in a pseudo-liquid medium at the surface.[315] In similar vein, catalysis has been established in colloids and in microemulsions. There are times when it is not certain whether a system involves homogeneous or heterogeneous catalysis, indeed it may not be right to take a black-and-white stance on this matter. Metals in colloidal form, or extremely finely divided in micelles or microemulsions, may bridge the two. Colloidal platinum with its particles covered with chiral molecules shows good selectivity and very high catalytic activity in asymmetric hydrogenation.[332] We shall allude to this homogeneous or heterogeneous question again, in relation to hydrosilylation (Section 10.7.4.2) and to the Fischer–Tropsch reaction (Section 10.7.5.3). Some of the differences between homogeneous and heterogeneous catalysis have been summarized earlier in this chapter (Table 10.15)—it may be that polynuclear, supported, colloidal, microemulsion-stabilized, or other such catalyst forms might bridge the gap and combine the advantages of both.

The preceding paragraphs have dealt with gross differences of medium and environment. Restricting attention to homogeneous catalysis in solution, there are still significant effects on yields, stereochemistry, and rates produced by the more modest environmental changes wrought by change of solvent. These result from a variety of solute–solvent and solute–solute interactions, i.e. from solvation and ion-pairing, respectively. Solvent effects may be attributed to such factors as coordination to active sites, solvent polarity, hydrogen bonding, and ion-pairing.[333]

We should perhaps end this section by commenting that mechanistic hypotheses and speculations are rife in the field of homogeneous catalysis, but that the number of really firmly established mechanisms is still disappointingly small.

10.7.2 Alkene isomerization and metathesis

10.7.2.1 Isomerization

The alkene isomerization process most commonly carried out under the influence of a homogeneous catalyst is that of double-bond migration—which can equally be viewed as hydrogen migration, e.g.

$$CH_3CH_2CH{=}CH_2 \longrightarrow CH_3CH{=}CHCH_3$$

Alkene isomerizations catalysed by $CoH(CO)_4$ have long been believed to involve the intermediacy of alkylcobalt intermediates, as in the sequence **74** → **75** → **76**.[334] An analogous reaction, but this time involving migration of the double bond from C=C to C=O, as in the conversion of allyl alcohol into propionaldehyde, is also $CoH(CO)_4$ -catalysed.[335] The catalytic cycle is shown in Fig. 10.14. The mechanism involves addition of the π-bond of the allyl alcohol to the active catalyst, a 1,2-hydride shift, and the separation of the unstable α,β-ene-ol isomerization product. Deuterium labelling gave important mechanistic evidence—the use of $CoD(CO)_4$ as catalyst gave exclusively CH_2DCH_2CHO. This alkyl-intermediate mechanism operates for catalysts which have a hydride ligand and one (potential) vacant site. Strong support for the intermediacy of an alkyl-metal complex is provided by the observation that if the perfluoroalkene $F_2C{=}CF_2$ is reacted with $[RhH(CO)(PPh_3)_3]$, stable $[Rh(CF_2CF_2H)(CO)(PPh_3)_2]$ is obtained. Moreover, transfer of hydride specifically to the β-carbon of the alkyl ligand has been demonstrated in this case, as required for the mechanism shown in Fig. 10.14.[336]

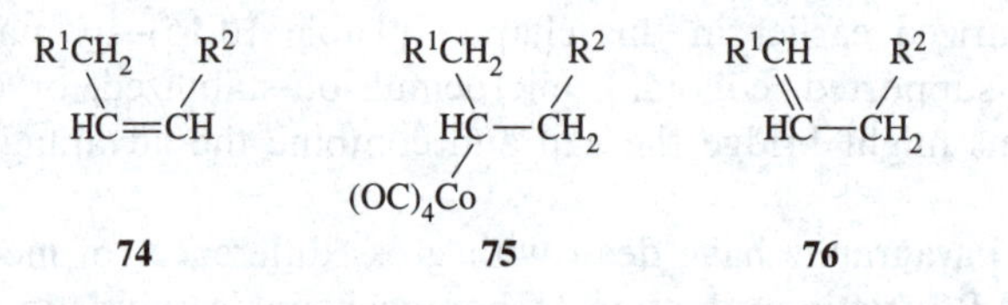

The alkyl-intermediate mechanism is not feasible if there is no hydride on the metal, but if there is a possibility of having two vacant sites, then an allyl-intermediate route (Fig. 10.15) may be followed. A cluster carbonyl can act as precursor; for example there is evidence that $Fe_3(CO)_{12}$ may undergo fragmentation to give transient $Fe(CO)_3$, which would have two vacant (or perhaps lightly solvated) sites as it is two ligands (four electrons) short of the stable eighteen-electron mononuclear carbonyl $Fe(CO)_5$. The key steps in this cycle are the addition and dissociation of starting alkene and isomerized product, and the intermediate intramolecular oxidative addition/reductive elimination alkene → allyl → alkene sequence. The structure of the compound $Zr(allyl)_3(\eta^5\text{-}C_5H_5)$ is of interest in this connection, since it contains three different geometries for the metal–allyl interactions (Fig. 10.16). These three forms may be taken to model allyl binding in substrates, intermediates, and transition states; interchange between the three types of ligand is rapid on the NMR time-scale even at low temperatures.[337]

Double-bond migration reactions are often complicated by parallel *cis* ↔ *trans* (*Z* ↔ *E*) isomerization, which can result in a mixture of products. The mechanism of catalysis of *cis* ↔ *trans* isomerization has been studied in its own right for, e.g., dialkyl maleates, catalysed by the ruthenium(II) compound $Ru(H)(Cl)(CO)(PPh_3)_3$.[338] One of the cycles believed to be involved is shown in Fig. 10.17—the other involves a binuclear catalyst but is of similar form.

A rather different type of isomerization is that of quadricyclene, **77**, to produce its more stable valence isomer the alkene norbornadiene, **78**.[339] This can be carried out under conditions of photochemical activation in the

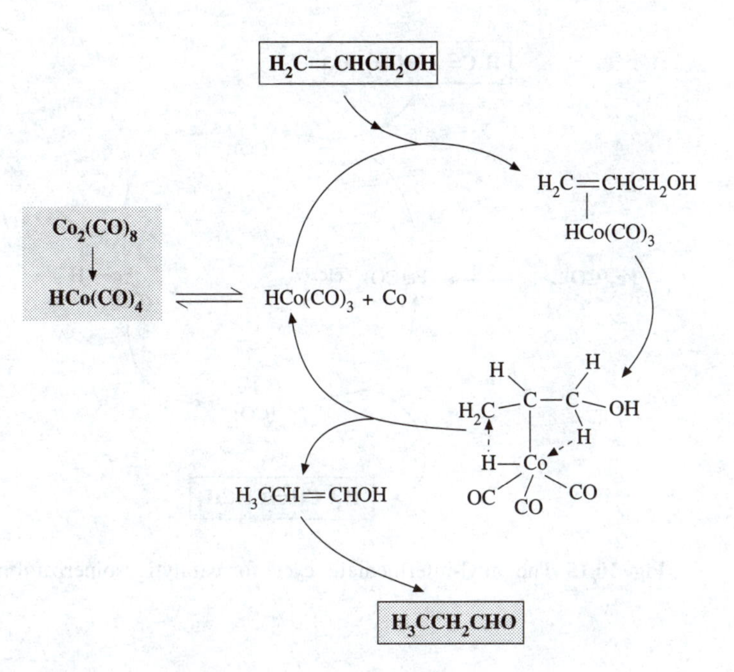

Fig. 10.14 Reaction cycle for alkene isomerization using $[CoH(CO)_4]$ as catalyst. Starting materials, products, and catalysts have been highlighted in several of the figures in Section 10.7 which depict catalytic cycles. The convention adopted is to show starting materials in boxes, products in boxes with background shading, and catalysts or catalyst precursors on a shaded background. A small square in a potential ligand position indicates a vacant coordination site. Curved arrows show the direction of operation of the cycle to obtain the required products—they should not necessarily be taken to imply that the reactions shown are irreversible.

presence of $[Ru(bipy)_3]^{2+}$ and the methylviologen cation, so presumably involves redox steps in the catalytic cycle.

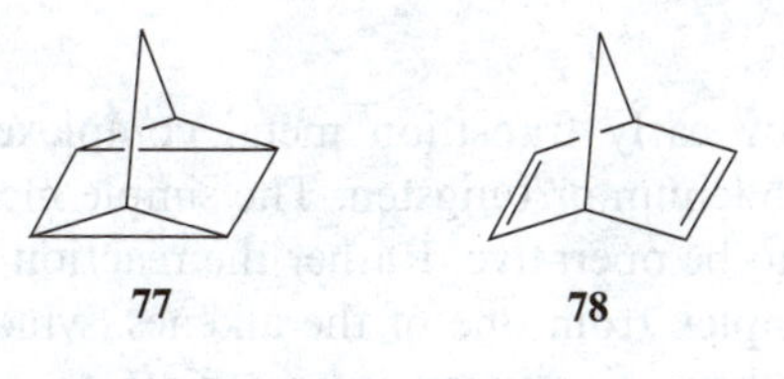

10.7.2.2 Metathesis

A reaction closely related to the above isomerizations is that of alkene metathesis, e.g.

$$R_2C{=}CR_2 + R'_2C{=}CR'_2 \longrightarrow 2R_2C{=}CR'_2$$

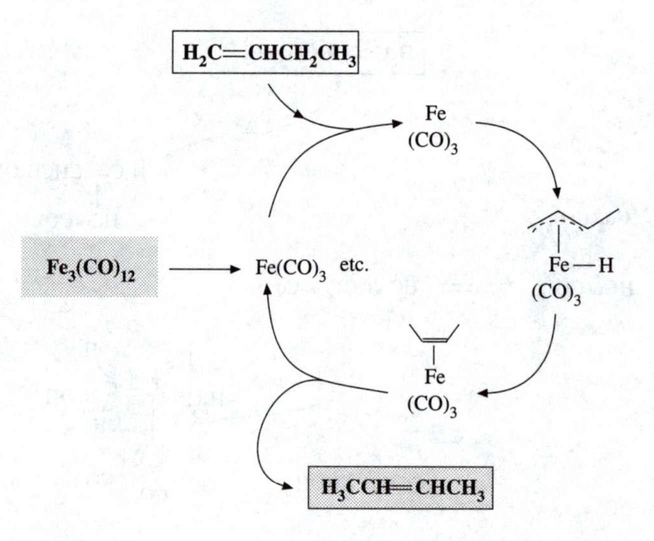

Fig. 10.15 The 'allyl-intermediate' cycle for catalytic isomerization.

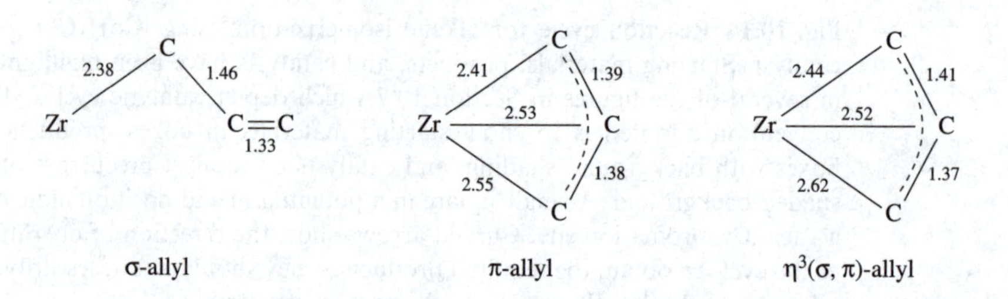

Fig. 10.16 The three modes of zirconium–allyl bonding in $[Zr(allyl)_3(cp)]$; the central carbon atom in the σ-bonded ligand is at least 3 Å from the metal.

Such reactions are catalysed by early transition metal complexes, often alkoxide derivatives[340] of molybdenum or tungsten. The simple mechanism shown in Fig. 10.18 seems not to be operative. Rather the reaction involves the formation of a carbene complex from one of the alkenes, which reacts with a molecule of the other alkene to give an intermediate containing a metallocyclobutane ring (Fig. 10.19). The intermediates involved here are much more plausible than a cyclobutane ring wrapped too tightly around the central metal as in the Fig. 10.18 intermediate. Distinction between the two mechanisms has been made in several product distribution studies, generally based on metathetical reactions involving a cycloalkene plus a linear alkene.[294] μ-Carbene complexes of the type shown as **79** (M = Fe or

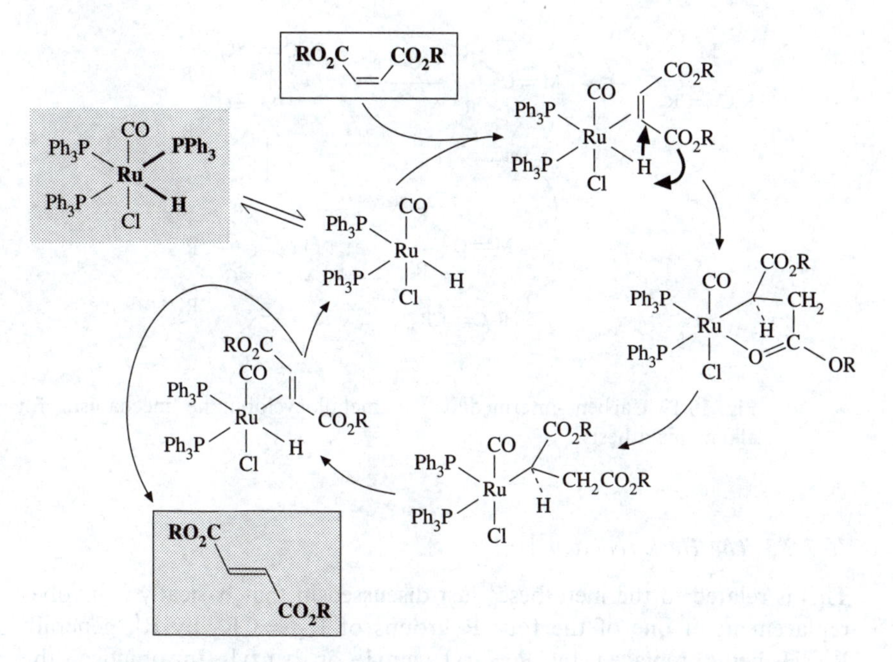

Fig. 10.17 Ruthenium(II)-catalysed isomerization of dialkyl maleates.

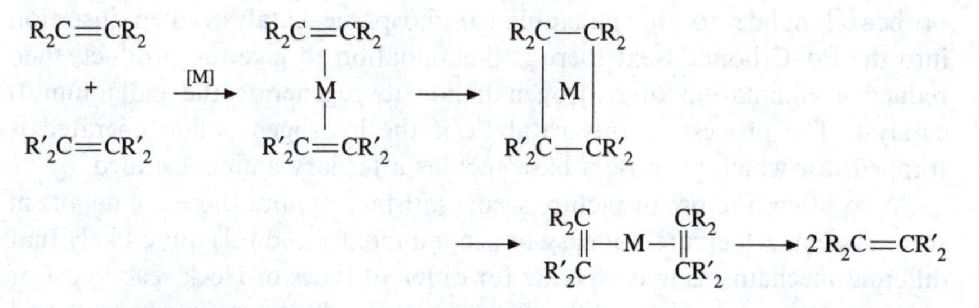

Fig. 10.18 Hypothetical simple four-centre mechanism for alkene metathesis.

Ru), where the combination of σ- and π-bonding to adjacent metal atoms is equivalent to two of the intermediates in the Fig. 10.19 cycle, might be involved in alkene metathesis at binuclear centres.[341]

79

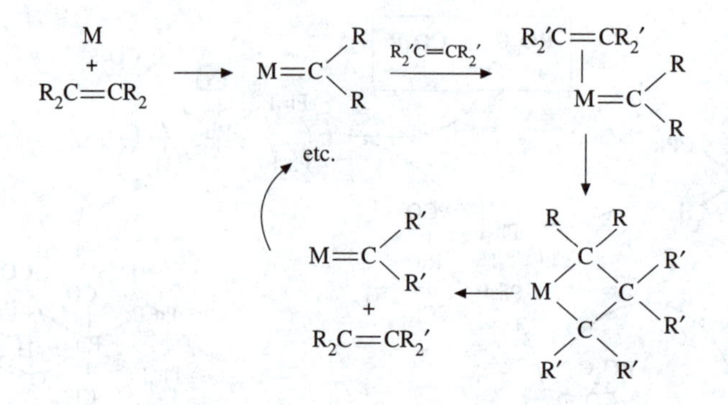

Fig. 10.19 Carbene-intermediate or metallocyclobutane mechanism for alkene metathesis.

10.7.2.3 The Heck reaction[342]

This is related to the metatheses just discussed in that basically it involves replacement of one of the four R groups of $R_2C{=}CR_2$ by R′, generally R = H being replaced by R = aryl, vinyl, or benzyl. In practice, this type of reaction has proved very versatile in synthetic organic chemistry, providing good synthetic routes to a range of compound types.[343,344] Heck reactions are catalysed by palladium compounds. The normally accepted mechanism (Fig. 10.20) involves oxidative addition of the aryl, vinyl, or benzyl halide to the palladium(0)-phosphine catalyst, then insertion into the Pd–C bond. Next there is β-elimination to give the product, then reductive elimination of hydrogen halide to regenerate the palladium(0) catalyst. The process is only catalytic if the hydrogen halide liberated is trapped, for which purpose a base such as a tertiary amine is added.

As so often, the broad picture seems satisfactory, but there are details in some systems which are not easy to accommodate, and it is quite likely that different mechanisms may operate for different types of Heck reaction. For example, it is particularly difficult to persuade chloroarenes to react, and considerable effort has been expended to develop suitable catalysts.[344] These consist of a mixture of, e.g., palladium(II) acetate and tris(*o*-tolyl)phosphine. The active catalyst is dimeric; it has been isolated and characterized as **80**. It seems to operate in a different manner from that detailed in the previous paragraph. There is no evidence for the involvement of palladium(0) in the catalytic cycle. Rather the key reaction appears to be oxidative addition to palladium(II) to give a palladium(IV) metallacycle. An analogous reaction has been proposed for aryl coupling ('Suzuki coupling'), where again a Pd^{II}/Pd^{IV} cycle seems much more likely than a Pd^0/Pd^{II} cycle. The latest contribution to the study of the nature and mechanism of Heck reactions relates to the demonstration that palladacycles are efficient catalysts for vinylation of aryl halides.[345]

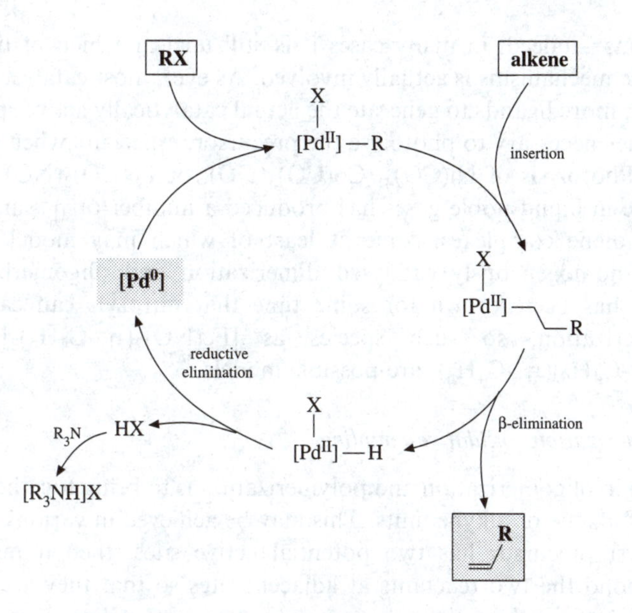

Fig. 10.20 Catalytic cycle for the Heck reaction.

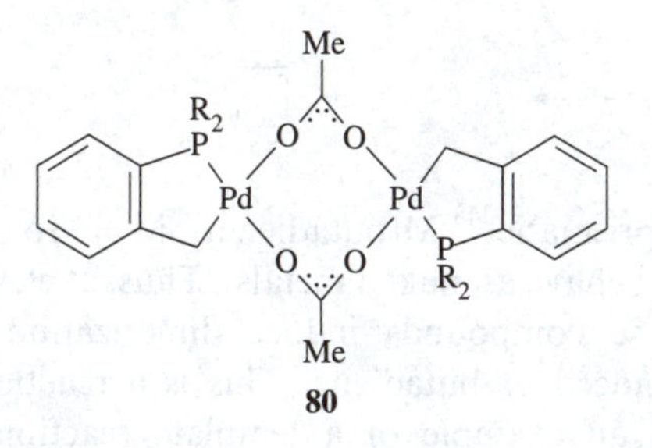

10.7.3 Dimerization, oligomerization, and polymerization

Catalytic dimerization, oligomerization, and polymerization are industrially important for converting cheap two-to-four carbon alkenes into valuable higher alkenes, both linear for use in the manufacture of detergents and branched for use in petrol. Homogeneous catalysts are the key to the preparation of the ubiquitous polythene, polypropylene, polyvinyl chloride, and PTFE under mild conditions. The industrially important processes usually involve assembly of alkenes, but there are analogous reactions for alkynes, which sometimes provide useful complementary information and model systems.

Numerous catalysts have been investigated and employed for this type of reaction, ranging from simple halides and carbonyls of molybdenum and tungsten[346] through the usual range of transition metal carbonyl/phosphine complexes to aluminium alkyls, alone or mixed with, e.g., titanium trichloride. Needless to say, this variety of catalysts operates by a variety of mechanisms—indeed, in many cases it is still unclear which of two or more possible mechanisms is actually involved. As ever, most catalysts have to lose one or more ligands to generate the actual catalytically active species. It is sometimes necessary to photolyse the precursor, especially when this is a carbonyl. Photolysis of $Ni(CO)_4$, $Co(CO)_3(NO)$, or $Fe(CO)_2(NO)_2$ with 1,3-butadiene in liquid noble gases has produced a number of η^2- and η^4-bonded butadiene complexes, some at least of which may model intermediates in homogeneously catalysed dimerization and oligomerization reactions. It has been known for some time that nitrosyls can catalyse alkene dimerization, so such species as $[Fe(NO)_2(\eta^2\text{-}C_4H_6)_2]$ and $[Fe(NO)_2(\eta^2\text{-}C_4H_6)(\eta^4\text{-}C_4H_6)]$ are possible models.[347]

10.7.3.1 Dimerization; oxidative coupling

The first step in oligomerization and polymerization is to bring together and join a pair of alkene or alkyne units. This may be achieved in various ways. If the catalyst precursor has two potential active sites, then it may be possible to bond the two reactants at adjacent sites so that they are close enough, and sufficiently activated, to combine to give a linear or cyclic product, e.g.:

$$\equiv + \equiv \xrightarrow{M} M(\equiv)_2 \longrightarrow \text{butadiene or cyclobutadiene}$$

Compounds such as the trigonal prismatic[348] $M(butadiene)_3$, M = Mo or W, are useful in this respect, for they behave as 'naked metals'. Thus, at elevated temperatures and pressures these compounds induce dimerization and polymerization of ethene, and indeed of butadiene. This is a reaction of coordinated ligands, specifically an example of a template reaction (cf. Section 10.2.3). The binding and joining processes may involve change in oxidation state for the central metal, in which case the reaction is one of *oxidative coupling*. Such a reaction occurs on adding two alkyne molecules to the iron(0) complex $Fe(dad)_2$, where dad is the Schiff base diimine ligand **81** derived from glyoxal and cyclohexylamine. The substituted butadiene produced is a good chelating ligand for $t_{2g}{}^6$ iron(II), forming a metallocyclopentadiene ring, as in the X-ray-characterized compound **82**.[349] Similar (substituted) butadiene/bidentate phosphine species are known for nickel(II).[350] Analogous reactions of alkenes yield metallocyclopentanes, as exemplified by the dimerization of ethene promoted by $[Ta(C_4H_9)(C_2H_4)_2(PMe_3)_2]$. The tantalacyclopentane species produced, **83**, reacts in a similar cycle with another ethene molecule to give **84**, indicating the possibility of catalytic polymerization[351] (see the discussion of the Ziegler–Natta process below; Section 10.7.3.2). As an example of an *sp*-block analogue, we might mention the generation of aluminocyclopentane, **85**, in low-temperature matrices.[352]

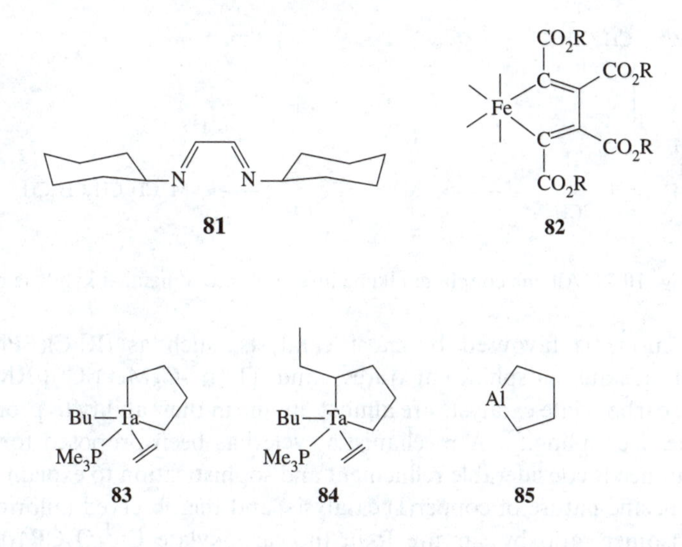

Two other aspects have been widely discussed for the individual coupling steps in dimerization and the growth of polymers on transition metal catalysts. The first is alkene insertion into a metal–alkyl bond:

$$\begin{array}{c} H_2C{=}CH_2 \\ | \\ M{-}R \end{array} \longrightarrow MCH_2CH_2R$$

This possibility is supported by the establishment of such insertion at, e.g., Rh–C[353] and Lu–C.[354] However, proof of the insertion of an alkene directly into a metal–alkyl bond is fairly rare, and insertion was felt by some to be prohibitively difficult for a facile homogenous catalysis system. To circumvent this, the alternative of alkene insertion into a metal–alkylidene bond, as shown in Fig. 10.21, was suggested. This is more complicated, involving hydrogen and electron transfer, but may well provide an attractive route where there is a large barrier to direct insertion. There is a close relation between this and the alkene metathesis mechanism discussed above. This route, like the examples of oxidative coupling mentioned earlier, involves two-electron oxidation state changes for the metal involved. They are thus plausible for many *d*-block elements, but not for, e.g., lanthanide catalysts—hence the demonstration of insertion of alkenes and alkynes into the lanthanide–methyl bond of $[Ln(\eta^5\text{-}C_5Me_5)_2]$ (Ln = Yb, Lu), alluded to above, is a particularly important indication of the insertion route, at least in lanthanide systems.

Unsymmetrical alkynes (RC≡CH) can dimerize in three ways:

- Head-to-tail, to give $RC{\equiv}CCR{=}CH_2$
- Head-to-head, to give $RC{\equiv}CCH{=}CHR$
- Cumulene-wise, to give $RCH{=}C{=}C{=}CHR$.

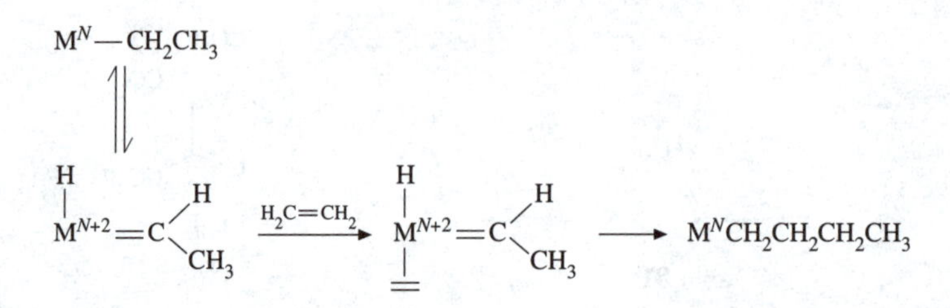

Fig. 10.21 Alkene coupling: alkene insertion into a metal–alkylidene bond.

The first mode is favoured by most catalysts, such as $[RhCl(PPh_3)_3]$, palladium acetate/phosphine mixtures, and $[Ti(\eta^5\text{-}C_5Me_5)_2Cl_2]$/RMgX. Copper(I) carboxylate catalysts are almost unique in their ability to promote head-to-head coupling.[355] A mechanistic cycle has been proposed for this system, but needs considerable refinement and sophistication to explain fully the regiospecific nature of copper(I) catalysis, and the observed tailoring of the E/Z isomer ratio by varying R in the carboxylate $Cu^I(O_2CR)$ used. Head-to-head coupling has also been achieved through the agency of a μ^3-benzyl complex of rhodium, where, unusually, intramolecular migration gives the enynyl unit RC≡C−C=CHR σ-bonded to the rhodium.[356]

Binuclear[357,358] complexes should, at least in principle, be useful dimerization or coupling catalysts if their active sites are disposed in an appropriate geometry. Binding of the dimer or developing polymer to adjacent metal centres, as in formula **79** in the alkene metathesis discussion above (Section 10.7.2.2), may model intermediates in the polymerization of alkynes catalysed by binuclear carbonyls $M_2(CO)_9$.[341] There are a number of species containing units of this and related types, including **86–89**, which are useful synthetically for carbon–carbon bond formation, and therefore may be involved in alkene or alkyne dimerization and polymerization when catalysed by such species as $Ru_2(CO)_9$ or $Ru_3(CO)_{12}$. $Fe_2(CO)_9$ can promote the formation of products containing the unusual unit **90** from silicon-containing di-ynes.[359]

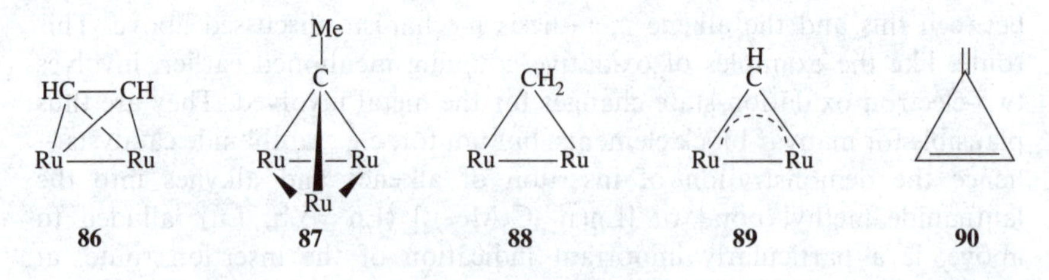

10.7.3.2 *Linear products—the Ziegler–Natta process*

The Ziegler–Natta system[360] is probably both the most important[361] and the least well-understood example of alkene polymerization.[362] Its aluminium/titanium catalysts and its mechanism are ill-defined and the process may well be microheterogeneous. In contrast to dinitrogen or paraffins, many alkenes (though not ethene) are fairly reactive, and the aim of catalysis here is often less one of activation, more one of control. This is especially true for polymerization of unsymmetrical alkenes such as propene. The initial observations of Ziegler that aluminium trialkyls reacted with ethene to produce oligomers were followed by the development of the so-called Ziegler–Natta catalysts. These incorporated an early transition metal halide with the *sp*-block alkyl/halide catalyst, and often nowadays contain further promoters to maximize reactivity, stereospecificity, and yield. As so often in the field of catalysis (as of inorganic pharmaceuticals), while people were still trying to work out the mechanism or even to identify the actual catalytic species, the process was producing polythene and polypropylene on a very large scale.

To start at the beginning, catalysis of dimerization of propene by tri-n-propylaluminium, in the absence of any transition metal compound, gives 2-methylpent-1-ene as the sole product. This is produced in a cycle (Fig. 10.22) involving insertion and subsequent β-elimination.[363] Moving on to Ziegler–Natta catalysts, the mechanism of their action probably involves a metal alkyl with a vacant site reacting with an alkene, which inserts into the metal–alkyl bond (cf. preceding section). An added Lewis acid may control competition between Al–C and Ti–C insertion.[364] Insertion leaves a vacant site for coordination of the next C=C. The key steps are shown in Fig. 10.23. There is a great deal of evidence to support a mechanism of a series of rapid sequential insertions into metal–carbon single bonds,[365] but several distinguished chemists have favoured an alternative carbene/alkylidene mechanism reminiscent of this type of mechanism for alkene metathesis (cf. Section 10.7.2.2 above). This route involves a 1,2-hydrogen shift from α-carbon and formation of carbene and metallacycle intermediates. A 1,2-hydrogen shift should be easy for an early transition metal, such as titanium or tantalum, with few or no *d* electrons. A general metal-based cycle is shown in Fig. 10.24.[366] Support for this mechanism is provided by the results of experiments with the tantalum alkylidene hydride complex $[Ta(CHCMe_3)(H)(PMe_3)_3I_2]$. This catalyses ethene polymerization, producing various free and coordinated polymeric species depending on whether or not extra PMe_3 is added. The results suggest that the initial complex reacts with ethene to give a tantalacyclobutane hydride, which gives an alkyl complex which in turn undergoes α-hydride elimination to give a new alkylidene hydride, which can react further to build up the polymer.[367]

There are no stereochemical complications when ethene is used as substrate, but unsymmetrical propene may polymerize in three distinct ways. The methyl groups of propene in the polypropylene product may be all in the same orientation (isotactic, which is commercially preferred), all alternating (syndiotactic), or random (atactic), as shown in Fig. 10.25. A third component, a Lewis base, may be added to the aluminium/titanium catalyst to improve stereospecificity for such unsymmetrical alkene substrates—making matters even less well understood mechanistically. However, it is apparent that the orientation of each incoming monomer in

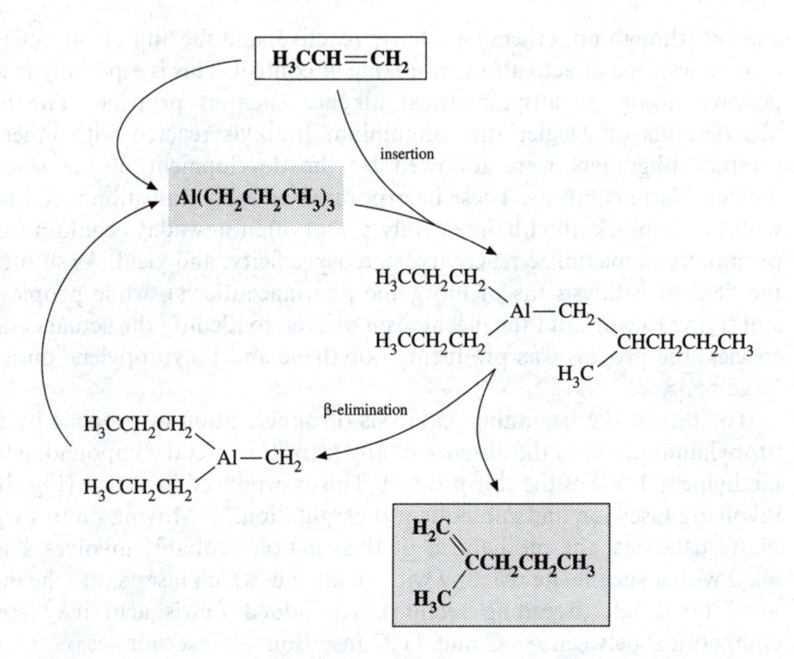

Fig. 10.22 Catalytic cycle for the production of 2-methylpent-1-ene from propene in the presence of tri-n-propylaluminium.

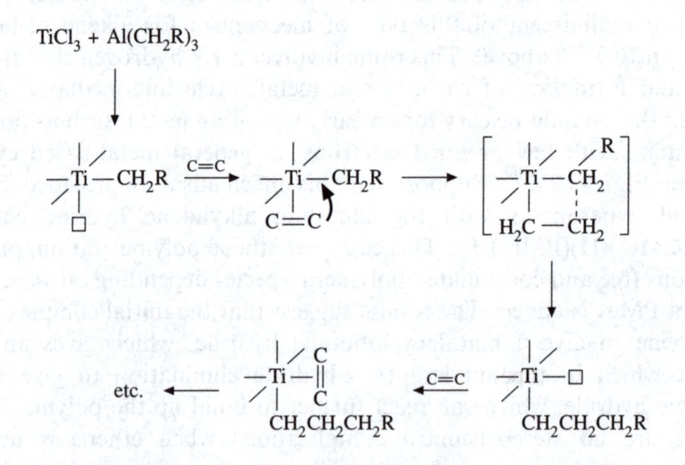

Fig. 10.23 Outline of the key steps in Ziegler–Natta catalysis.

relation to the growing polymer is determined by the steric requirements of the coordination shell of the metal ion to which they are both bound. Ligand tailoring to control the stereochemistry of polymerization has been achieved by the use of substituents in metallocene dihalide/aluminium alkyl catalysis,[368] which incidentally is probably truly homogeneous. Figure 10.26

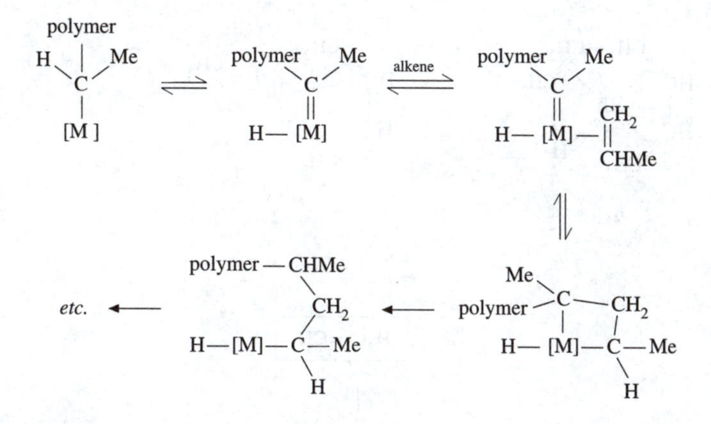

Fig. 10.24 The alkylidene hydride mechanism for alkene polymerization.

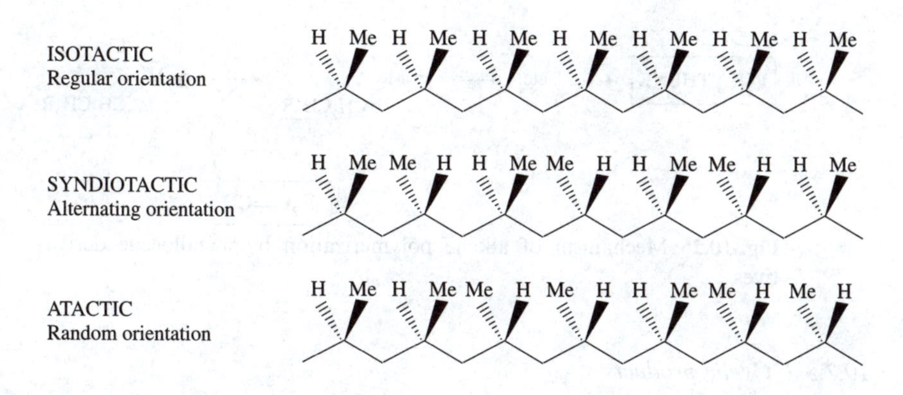

Fig. 10.25 Isotactic, syndiotactic, and atactic forms of polypropylene.

shows the assumed mechanism of vacant site creation, followed by π-binding of the alkene, then insertion. Sufficiently bulky groups attached to the cyclopentadienyl rings can impose orientation on the incoming and inserting alkene, giving isotactic or syndiotactic polymers, as required.

Catalysis of the addition of an alkene to a diene, sometimes termed codimerization, is also of interest and importance. Codimerization of butadiene with ethene may be catalysed by rhodium or nickel complexes. For both metal centres the mechanism involves formation of a reactive metal hydride species with one ethene and one butadiene ligand, e.g. **91**. Hydride transfer to the coordinated butadiene gives the π-crotyl species **92** en route to the required product, *trans*-1,4-hexadiene, a precursor for the production of synthetic rubber.[369]

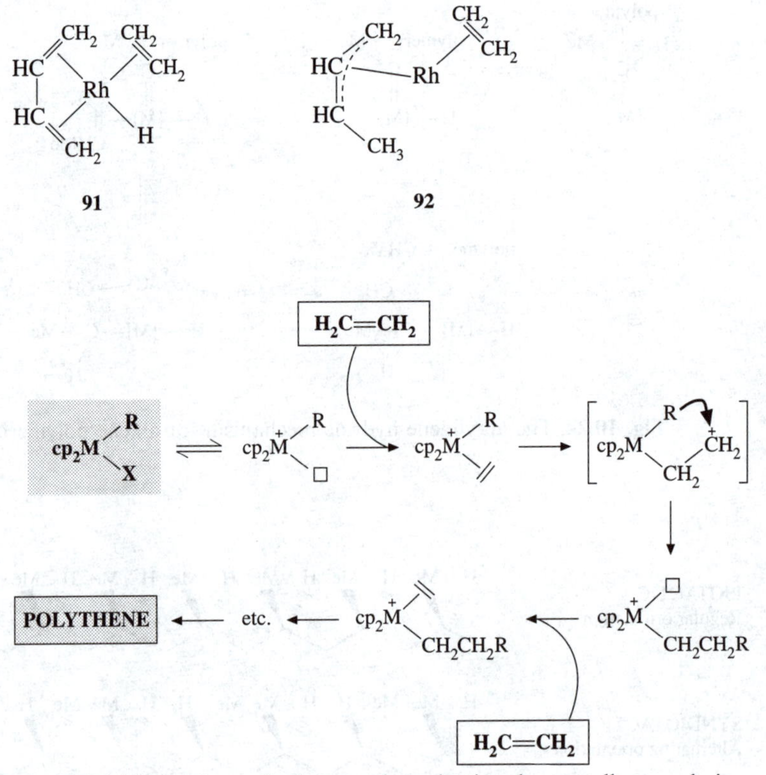

Fig. 10.26 Mechanism of alkene polymerization by metallocene derivatives.

10.7.3.3 Cyclic products

Cyclo-oligomerization involving alkynes was first developed by Reppe[370] using as catalysts nickel(II) cyanide or nickel(II) complexes with a range of bidentate ligands such as pentane-2,4-dione or salicaldehyde. The requirement was that the nickel remained high-spin octahedral (or tetrahedral) and that substitution was rapid (a reasonable requirement for a catalyst). The main reaction was the tetramerization of acetylene (ethyne):

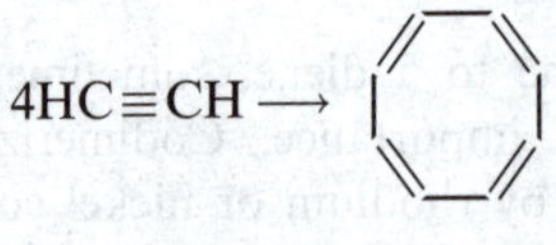

$$4HC\equiv CH \longrightarrow$$

This must have been a disappointment if the original intention had been to develop a facile synthesis of aromatics, but it is a simple and elegant route to cyclooctatetraene, a compound of considerable value to organometallic chemists. It has been thought[371] that four ethyne molecules bond to the nickel at the four available sites in, e.g., $Ni(CN)_2$ or $[Ni(acac)]^+$ and that the bonds rearrange simultaneously to give the sole product—stepwise oligomerization would lead to a mixture of products. The reaction can be modified, and the original purpose vindicated, by adding a competing ligand. Triphenyl phosphine (PPh_3 : Ni :: 1 : 1) inhibits the formation of cyclooctatetraene and benzene is formed instead. This is fully in keeping with the idea that only three coordination sites are now available. Bidentate inhibitors, such as $Ph_2PCH_2CH_2PPh_2$, 1,10-phenanthroline, or 2,2′-bipyridyl, prevent reaction altogether. Presumably, mixed complexes such as [Ni(acac)(phen)] would be square-planar, leaving the alkyne coordination sites *trans* to each other and therefore preventing combination.[372] This pattern is attractively neat and tidy, and the 'zipper-stepwise' mechanism for tetra- and trimerization has considerable support, especially from labelling studies using ^{13}C–^{12}C ethyne.[373] The simple connection between available sites and product, anyway not as complete as implied in the preceding sentences, does not apply in other systems. Thus, for example, oligomerization of alkynes in the presence of $[Ni(dad)_2]$ (dad = **74** above) gives specifically tetramers, while $Ni(CO)_4$ in the presence of triphenylphosphine gives rise to substituted benzenes. For these reactions, and also for Reppe's original reactions, it has been suggested that reaction proceeds through formation of dinuclear species containing bridging butadiene. An X-ray crystal structure determination of a likely intermediate lends credence to this theory.[374] The main attraction of butadiene-containing intermediates is that a combination of two such gives cyclooctatetraene without forming the more stable trimeric product benzene.

A number of compounds of other metals catalyse cyclo-oligomerization of alkynes.[375] Thus the Ziegler–Natta mixture of $TiCl_4$ and Et_2AlCl is effective,[376] as are cyclooctatetraene complexes of titanium or titanium trichloride for trimerization.[377] These titanium catalysts appear to operate, like the nickel catalysts mentioned above, by a concerted mechanism involving intermediates in which all three alkynes are simultaneously bonded to the metal. Cobalt, in the form of $[Co(\eta^5\text{-}C_5H_5)(CO)_2]$ or $[Co(\eta^5\text{-}C_5H_5)(C_2H_4)_2]$, can also catalyse a variety of [2 + 2 + 2] cycloadditions, building up organic molecules of complexity up to that of steroids or tetracyclic diterpenes.[378] The latter one-pot synthesis is actually a cascade process, with the [2 + 2 + 2] cycloaddition followed by a further [4 + 2] reaction, with the C–C bonds being created chemo-, regio-, and stereoselectively. Cobalt catalysis may involve metallacyclopentadiene intermediates, which give final products by subsequent alkyne insertion;[379] for niobium catalysis it has been proposed that the intermediate is an η^4-cyclobutadiene complex.[380] There are a number of more or less stable models for such complexes, e.g. $[Ni(\eta^4\text{-}C_4Me_4)(bipy)]$, $[Co(\eta^4\text{-}C_4H_4)(\eta^5\text{-}C_5H_5)]$, and $[Mo(\eta^4\text{-}C_4H_4)(CO)_3]$. Once the C_4-metal intermediate has been formed, the next question is the mechanism for addition of the third alkyne molecule. For the case of a metallacyclopentadiene intermediate, binding this third alkyne (**93**) may be followed by formation of either a metallacycloheptatriene ring or a species **94**. A crystal structure established[381] for a tantalum compound containing the unit **94** has been regarded as the characterization of a 'stable transition state' for generation of the complexed trimer.

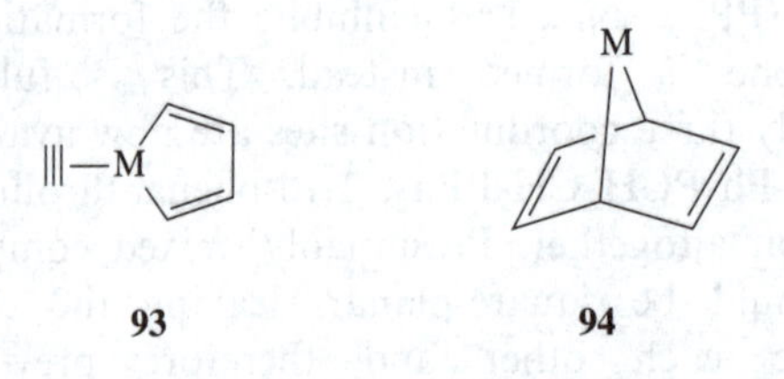

Whereas platinum forms stable alkyne complexes, palladium compounds tend to react with alkynes to produce a mixture of oligomers—except in such exceptional cases as di-t-butylacetylene, where the exceptional bulk of the t-butyl groups prevents further reaction and permits the isolation of a monoalkyne complex. Normally alkynes react initially with palladium dichloride by insertion into the Pd–Cl bond. The σ-vinylpalladium intermediate produced undergoes a further *cis* insertion to give a σ-butadienyl compound. The course of the remaining reaction depends on the nature of the alkyne. If there are bulky substituents, the σ-butadienyl species isomerizes to an η^4-C_4H_4 complex, analogous to the M(η^4-C_4H_4) species listed in the previous paragraph. Otherwise a further insertion is followed by ring closure to produce the trimeric product, an appropriately substituted benzene.[382]

This type of oligomerization is of considerable importance; suitable modification of the catalyst and the alkyne can lead to the synthesis of a wide range of conjugated products. Cyclodimerization and cyclotrimerization of, for instance, butadiene to give cyclooctadiene and cyclododecatriene (important as a precursor of nylon-12) can be catalysed by $TiCl_4/Et_2AlCl$ or by nickel(0) complexes.[383] π-Allyl nickel complexes are likely intermediates in a stepwise mechanism for the latter. Blocking coordination positions with suitable ligands leads to a wide variety of products. For example, blocking by phosphine or phosphite gives cyclooctadiene from butadiene. In most of this oligomerization and polymerization work, the devising of new catalysts savours more of craftsmanship than of science. The catalysts are none the worse for their parentage and do their jobs every bit as well as if they were to the manner born—aristocratic designer molecules of known composition, structure, and function.

10.7.4 Hydrocyanation, hydrosilylation, and hydroboration

10.7.4.1 Hydrocyanation[384]

The simplest hydrocyanation, of ethene to give ethyl cyanide (propionitrile), is catalysed by bis-(tri-*o*-tolylphosphite)nickel(0) in a catalytic cycle including the customary coordination and insertion stages. The overall reaction obeys a simple second-order rate law. This and the large negative activation entropy were taken as evidence in favour of reductive elimination of propionitrile from a five-coordinate nickel complex at the appropriate point in the cycle.[385]

Moving on to a larger substrate and a much more important reaction, homogeneous catalysis of the addition of hydrogen cyanide to butadiene is the key to the economic production of adiponitrile, the building block in the production of Nylon-66:

$$H_2C{=}CHCH{=}CH_2 + 2HCN \rightarrow NCCH_2CH_2CH_2CH_2CN$$

Asymmetric hydrocyanation has recently assumed importance in the pharmaceutical industry, since it can provide the required enantiomers of precursors for the widely used anti-inflammatory drugs naproxen and ibuprofen.

Hydrogen cyanide adds readily to activated alkenes, but for unactivated alkenes or dienes a transition metal catalyst is needed. The formation of adiponitrile is catalysed by triarylphosphite complexes of nickel(0), $Ni(P\{OAr\}_3)_4$. The best of these appears to be that with Ar = 4-tolyl, which has the optimal balance of steric bulk, which controls ease of dissociation from the nickel, and electron-withdrawing power, which is needed to assist the bond making between the substrates. The mechanism seems still not to be completely established—there appear to be perhaps twenty steps in the overall addition of two HCN molecules to butadiene. There have been problems in understanding how unwelcome isomerization of the monocyanoalkene intermediate is minimized and how regiospecificity of addition of the second HCN molecule is determined. However, it has been clear for a long time that the basic catalytic cycles are made up of a selection of the familiar steps of ligand dissociation, oxidative addition of HCN, and reductive elimination to couple the HCN to the butadiene.

Whereas HCN can readily oxidatively add, e.g. $Pd^0 + HCN = H{-}Pd^{II}{-}CN$, nitriles RCN tend to prefer to coordinate as ligands, M + RCN = M:NCR. The metal activates the ligand to reaction with alcohols, the basis of an efficient catalytic route for the production of esters and lactones.[386] The cycle is shown in Fig. 10.27. The metal remains in the same oxidation state throughout the cycle; its role is to activate the ligand to nucleophilic attack, not this time to undergo oxidative addition.

10.7.4.2 Hydrosilylation[387]

Hydrosilylation may not be the most industrially important example of homogeneous catalysis, but it does provide a versatile method for the preparation of a wide range of organosilicon compounds. The original Chalk–Harrod cycle[388] for platinum-catalysed hydrosilylation (outlined in Fig. 10.28) provides another illustration of oxidative addition–reductive

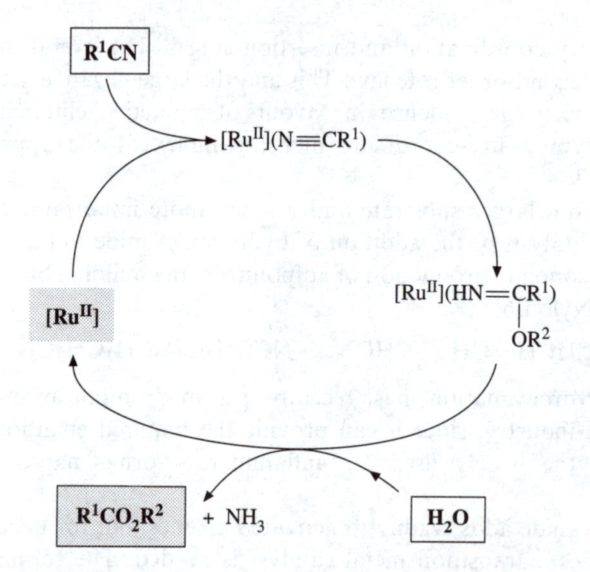

Fig. 10.27 Ruthenium-catalysed production of esters from nitriles.

elimination. The platinum can cycle between the 2+ and 0 or 4+ and 2+ oxidation states. In one variant,[389] the catalyst precursor is the binuclear species **95**, prepared from $[Pt(cod)_2]$ via $[Pt(Pcx_3)(C_2H_4)_2]$ (cx = cyclohexyl). Here the active platinum catalyst is generated by reaction with an alkene HC=CHR′ to give a mononuclear platinum(II) complex $[Pt(H)(SiR_3)(Pcx_3)$ (HC=CHR′)], which reacts with HC=CHR′ to start proceeding round the catalytic cycle. Figure 10.28 shows insertion into Pt–H, but it subsequently became apparent that insertion could also take place into the Pt–Si bond. The possibility of alkene insertion into a metal–silicon bond was decisively demonstrated[390] for the photochemical reaction

$$Fe(\eta^5\text{-}C_5H_5)(SiMe_3)(CO)_2 + H_2C{=}CH_2 \longrightarrow Fe(\eta^5\text{-}C_5H_5)(CH_2CH_2SiMe_3)(CO)_2$$

The operation of Chalk–Harrod cycles has been assumed for many years, but only recently has it proved possible to demonstrate the participation of the proposed reductive elimination step in platinum-catalysed hydrosilylation,[391] though an analogous reductive elimination at iron, from $[Fe(CO)_4(SiMe_3)(R)]$, had previously been established.[392] The compound *cis*-$[Pt(Me)(SiPh_3)(PMePh_2)_2]$ was characterized and shown to undergo reductive elimination. A rather limited kinetic study showed that added alkenes or alkynes produced small accelerations, but that added $PMePh_2$ inhibited the reaction. The latter observation implicated reversible phosphine dissociation as the first step. In general, particular attention needs to be paid to the precise nature of the catalyst in respect to the reductive elimination stage, in view of the demonstration that reductive elimination from the iridium(III) compound **96** produces 100% methane when R = OEt or Ph, but when R = Et both methane and Et_3MeSi are produced.[393]

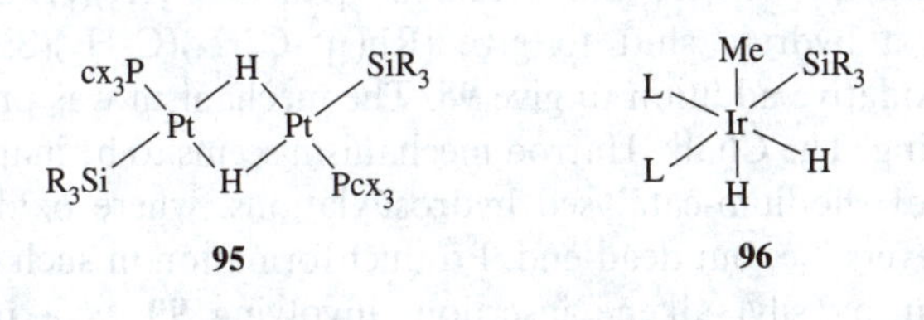

The role and detailed mechanism of oxidative addition and reductive elimination in hydrosilylation are now becoming clear, but though Si–H bond making and bond breaking have been quite extensively studied, very little is known about bond making and bond breaking involving silicon–carbon bonds, or indeed bonds between silicon and other *sp*-block elements. The reductive elimination stage, only recently documented in hydrosilylation, has been particularly little examined for these other silicon–element bonds.[394]

The difficulty of obtaining definitive evidence for the reductive elimination step, indeed the known reluctance of compounds of platinum and some other transition metals to undergo this reaction, has led to the search for other hydrosilylation mechanisms.[395] Some relevant iron pentacarbonyl

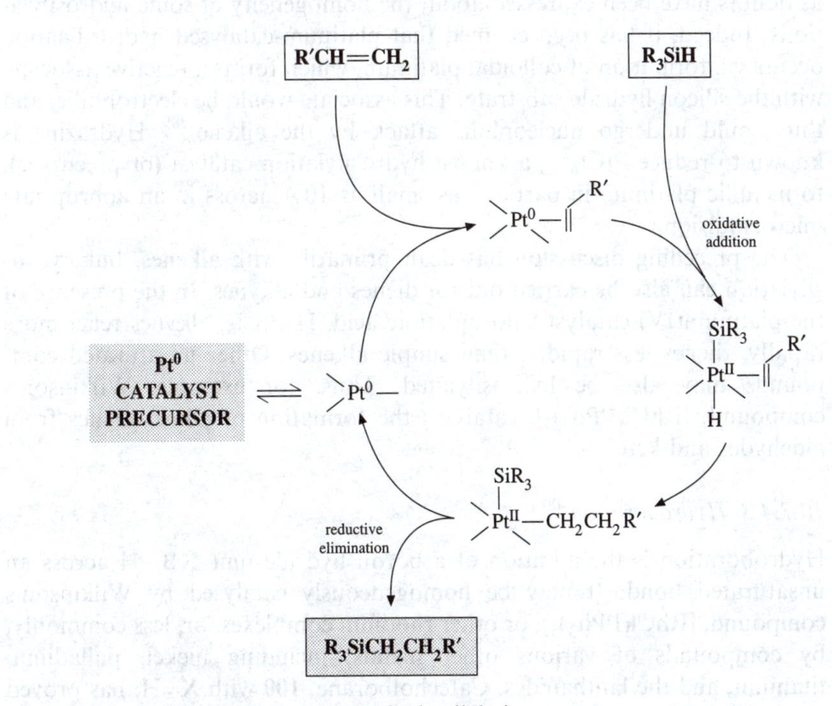

Fig. 10.28 Catalytic cycle for hydrosilylation.

chemistry led to the proposal of a cycle involving an intermediate, **97**, for cobalt carbonyl catalysis of hydrosilylation.[396] A similar cycle has been proposed for rhodium,[397] where the precursor [Rh(η^5-C_5H_5)(H)($SiEt_3$)(C_2H_4)] undergoes a hydride shift to give [Rh(η^5-C_5H_5)(C_2H_5)($SiEt_3$)], which undergoes oxidative addition to give **98**. The mechanism was probed by deuterium labelling. The Chalk–Harrod mechanism seems to be inapplicable to a number of rhodium-catalysed hydrosilylations, where oxidative addition is facile, reversible, but dead-end. Product formation in such cases is claimed to occur by silyl alkene insertion, involving **99** as a major intermediate.[398]

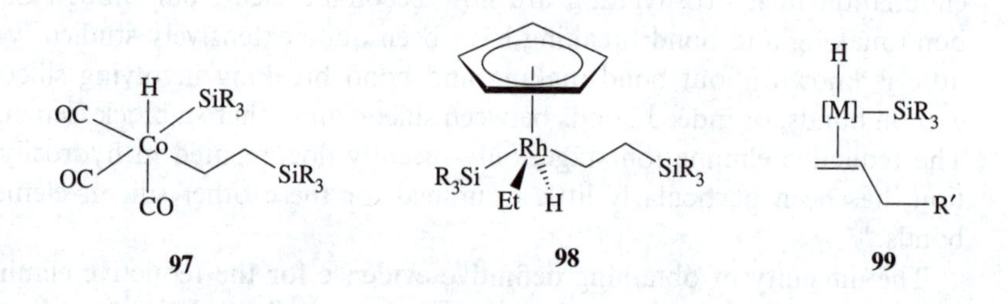

In the study of the [Rh(η^5-C_5H_5)(C_2H_5)($SiEt_3$)]-catalysed system mentioned above care was taken to show the reaction to be truly homogeneous, as doubts have been expressed about the homogeneity of some hydrosilylations. Indeed, it has been claimed that platinum-catalysed hydrosilylation occurs via formation of colloidal platinum, which forms a reactive associate with the silicon hydride substrate. This associate would be electrophilic, and thus could undergo nucleophilic attack by the alkene.[399] Hydrazine is known to reduce $PtCl_6^{2-}$, a known hydrosilylation catalyst (or precursor), to metallic platinum in particles as small as 10 Å across in an appropriate microemulsion.

The preceding discussion has dealt primarily with alkenes, but hydrosilylation can also be carried out for dienes and alkynes. In the presence of the platinum(IV) catalyst chloroplatinic acid, H_2PtCl_6, alkynes react more rapidly, dienes less rapidly, than simple alkenes. Other unsaturated compounds may also be hydrosilylated. Thus, for example, Wilkinson's compound, [RhCl(PPh_3)$_3$], catalyses the formation of alkoxysilanes from aldehydes and ketones.

10.7.4.3 Hydroboration[400]

Hydroboration is the addition of a boron hydride unit >B–H across an unsaturated bond. It may be homogeneously catalysed by Wilkinson's compound, [RhCl(PPh_3)$_3$], or other rhodium complexes, or, less commonly, by compounds of various other metals, including nickel, palladium, titanium, and the lanthanides. Catecholborane, **100** with X=H, has proved a popular substrate. It reacts with the active catalyst [RhCl(PPh_3)$_2$] to give

101, which can then pick up, for example, an alkene to give the intermediate **102**. Insertion into the rhodium–hydride bond gives alkyl complexes, in turn giving as major products **100** with X = $-CHRCH_3$ and/or $-CH_2CH_2R$.

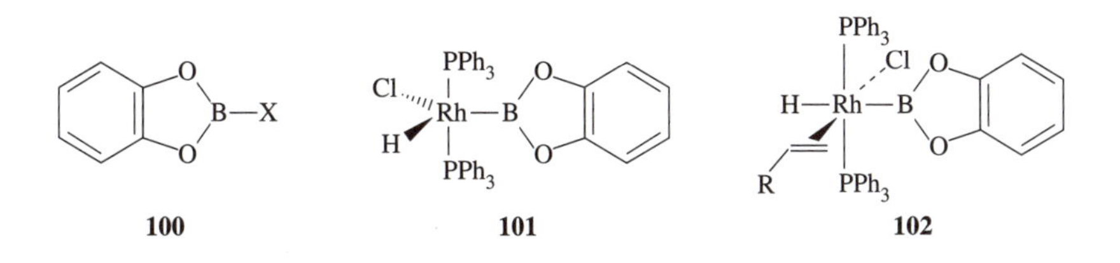

Hydroboration was selected as the system to demonstrate the advantages of using fluoroalkylphosphine ligands in homogeneous catalysis, specifically in the form of the rhodium(I) complex $[Rh(P\{CH_2CH_2(CF_2)_5CF_3\}_3)_3Cl]$.[401] The solubility of this complex in perfluoromethylcyclohexane, $C_6F_{11}CF_3$, permits solvent extraction of products and recycling of the catalyst solution—and thus provides both economic and environmental rewards.

10.7.5 Hydrogenation[402]

10.7.5.1 Organic substrates

The addition of hydrogen to alkenes is of considerable importance in a variety of processes, but it requires catalysis and quite often high pressures of hydrogen. Over the past few decades homogeneous catalysts have been developed that are clean, often selective, and by keeping their molecular integrity allow a study to be made of the detailed mechanism of their mode of action.[403] A typical catalyst, commercially available for many years now, is the rhodium(I) complex $RhCl(PPh_3)_3$. The mechanism of catalysis of hydrogenation by this compound demonstrates the combination of the various primary steps that have been discussed in earlier chapters, and earlier in this chapter, and provides examples of some of the complications that may beset the unequivocal establishment of mechanism in this area.

An early study[404] of hydrogenation kinetics established the rate law

$$\text{rate} = k[H_2][\text{alkene}][RhCl(PPh_3)_3]/\{1+k'[H_2]+k''[\text{alkene}]\}$$

Rate constants are given in Table 10.22; they correlate with the extent of rhodium–alkene interaction expected on electronic and, mainly, steric grounds. This rate law is similar to that established for oxidative addition to five-coordinate complexes, or to $Pt(PPh_3)_4$ (cf. Section 10.3.2). The rate of hydrogenation is reduced by an excess of triphenyl phosphine, suggesting pre-equilibrium ligand loss:

$$RhCl(PPh_3)_3 \rightleftharpoons RhCl(PPh_3)_2 + PPh_3$$

In practice it is likely that the rhodium in the $RhCl(PPh_3)_2$ has a molecule of

Table 10.22 Rate constants for alkene hydrogenation catalysed by $[RhCl(PPh_3)_3]$ in benzene solution at 298 K.

1-Dodecene	34	Styrene	93
1-Hexene	29	Cyclopentene	34
cis-4-Methyl-2-pentene	10	Cyclohexene	32
trans-4-Methyl-2-pentene	2	1-Methylcyclohexene	1

solvent weakly bonded, but easily replaceable, in the fourth coordination position (cf. Section 10.7.1 above). The catalyst can π-bond to a molecule of alkene, and undergo oxidative addition of dihydrogen. Hydride migration and reductive elimination of the alkane product complete the catalytic cycle, shown in Fig. 10.29. It is the reductive elimination step which drives these cycles, as it is the only non-reversible reaction. The intermediacy of, and reductive elimination from, *cis*-dihydroalkylrhodium species was first convincingly supported by the isolation of stable complexes **103** (R = Me or Ph), and confirmation that they do undergo reductive elimination to rhodium(I).[405] References to theoretical calculations and their forecasts of energy profiles for $RhCl(PPh_3)_3$ catalysis of alkene hydrogenation by this mechanism have already been cited in Section 10.7.1.[305] Intensive study over the years has revealed a number of complications. In the first place, the order of addition of dihydrogen and the alkene may vary with conditions. Secondly, (oxidative) addition may take place either to $[RhCl(PPh_3)_3]$ or to $[RhCl(PPh_3)_2]$—the former could give octahedral $[Rh(H)_2Cl(PPh_3)_3]$, the latter square-planar $[RhCl(PPh_3)_2(alkene)]$. If both routes were operative, the ratio would be concentration-sensitive. Thirdly, there is the possibility of involvement of the dimer $[(Ph_3P)_2Rh(\mu\text{-}Cl)_2Rh(PPh_3)_2]$. Fourthly, an NMR study of ligand exchange at $[RhCl(PPh_3)_3]$ and at $[Rh(H)_2Cl(PPh_3)_3]$ has shown that exchange processes at the latter take place on a sufficiently short time-scale possibly to affect homogeneous catalysis hydrogenation cycles. Specifically, the rate constant for phosphine exchange at the site *trans* to hydride is around $400\ s^{-1}$ at ambient temperatures.[295] Although $[RhCl(PPh_3)_3]$ is an excellent catalyst, life is easier for the kineticist if the closely related $[RhH(CO)(PPh_3)_3]$ is used.[294]

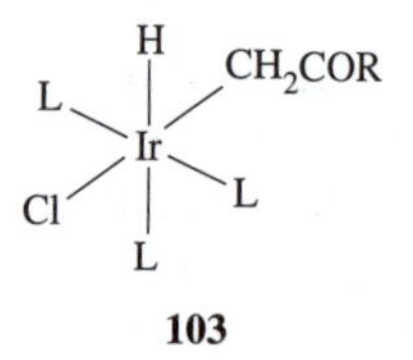

103

Asymmetric hydrogenation of alkenes can be achieved by the use of rhodium catalysts with chiral bidentate phosphine ligands (*PP), as in the ternary cyclooctadiene rhodium(I) cation $[Rh(cod)(*PP)]^+$. Very high

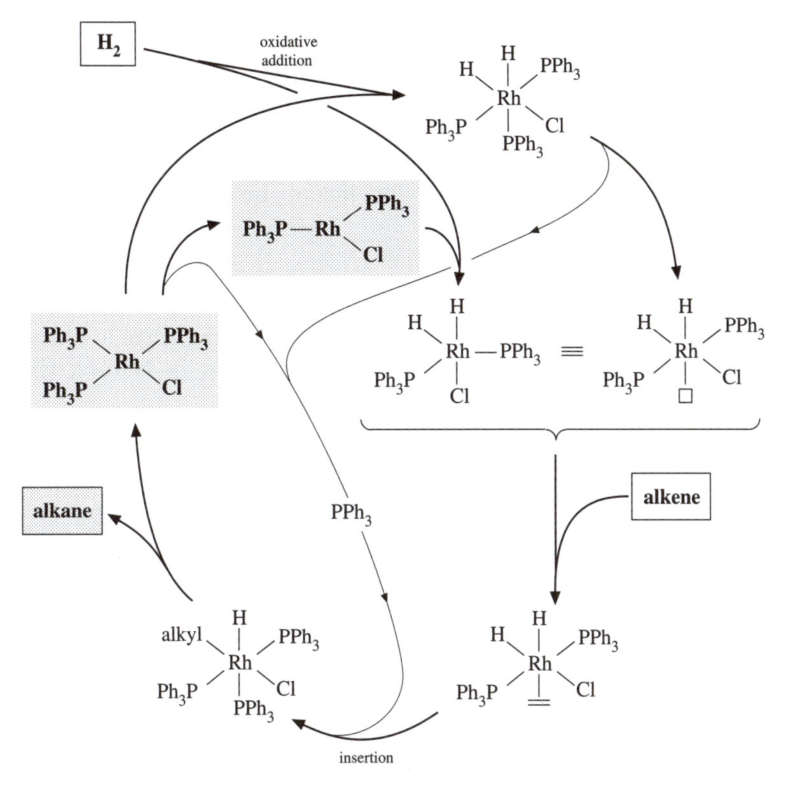

Fig. 10.29 Cycle for rhodium-catalysed alkene hydrogenation.

enantioselectivities can be achieved, both in the usual organic solvents such as methanol or hexane and in supercritical carbon dioxide. This last solvent has the advantages both of dissolving the catalyst (as an appropriate salt) and of dissolving a large amount of dihydrogen—not to mention its environmentally friendly (in the very small amounts involved here) properties.[406]

Iron pentacarbonyl catalysis requires photoactivation to generate the active catalyst, $[Fe(H)_2(CO)_3(H_2C{=}CH_2)]$ being the key intermediate.[407] Ruthenium(II) catalysts probably act differently from rhodium catalysts. The likely ligands would disfavour the ruthenium(IV) state, and to avoid this the mechanism probably involves heterolysis of dihydrogen. Pentacyanocobaltate would be expected to catalyse by yet a third mechanism, this time involving homolysis of dihydrogen:

$$2[Co(CN)_5]^{3-} + H_2 \longrightarrow 2[Co(H)(CN)_5]^{3-}$$

Alkene hydrogenation provides several well-established examples of cluster catalysis in which the cluster remains intact. $[Ru_4H_4(CO)_{12}]$ reacts by reversible CO loss followed by addition of H_2, insertion of ethene into one of the Ru–H bonds, and finally rate-limiting irreversible elimination of ethane. Here, as for catalysis by ruthenium(II) dimers, presumably the hydrogen and alkene have to be coordinated to the same metal atom to facilitate insertion of the alkene into a metal–hydrogen bond.[408] In $[Ru_3(CO)_{12}]$ catalysis of nitrobenzene reduction, isolation and characterization of intermediates showed the Ru_3 cluster to remain intact. Several intermediates contain a μ^3-NPh ligand linking the three ruthenium atoms.[409]

10.7.5.2 Nitrogen fixation[410]

The fixation of atmospheric nitrogen has been a challenge to chemists for a long time. It occurs under very mild conditions in Nature, though the overall reactions

$$N_2 + 6H_2O + 6e^- \longrightarrow 2NH_3 + 6OH^-$$

$$N_2 + 10H^+ + 8e^- \longrightarrow 2NH_4{}^+ + H_2$$

require a lot of energy, provided by Mg^{2+}-catalysed ATP hydrolysis. Military and agricultural requirements for ammonium salts and nitrates prompted the Haber–Bosch process for the direct combination of nitrogen and hydrogen, a very effortful and energy-intensive process which lies outside the scope of this book. At least half a century of research, whose intermittent progress has been chronicled frequently over the years,[411] into nitrogen fixation catalysed by inorganic species under mild conditions in aqueous media is beginning to bear fruit. There has been considerable success in characterizing nitrogenases, especially the stoichiometry and structure of their active centres, e.g. the $MoFe_7S_8$ cluster in *Azotobacter vinelandii*, and in elucidating mechanisms of nitrogen coordination to, and activation by, molybdenum in a variety of model complexes. Some models work well but approximate poorly to natural systems; a few are both close to the real thing and are moderately effective. They involve the initial formation of N_2 complexes, possibly with the dinitrogen bridging two metal centres, various $-NH_xNH_y$ ligands formed by reduction of the coordinated and activated N_2, and coordinated =NH remaining after N–N bond breaking and NH_3 liberation. Both a fully active and appropriate model system and a detailed mechanism for natural nitrogen fixation lie in the future.

10.7.5.3 Carbon dioxide and carbon monoxide

The most important reaction of carbon dioxide with hydrogen[412] is linked with water carbonylation in the *water gas shift reaction*:

$$CO_2 + H_2 \rightleftharpoons CO + H_2O$$

This equilibrium is discussed in the following section. The other simple hydrogenation reaction of carbon dioxide is the formation of formic acid:

$$CO_2 + H_2 \longrightarrow HCO_2H$$

This reaction was shown to be efficiently catalysed heterogeneously by Raney nickel in 1935; homogeneous catalysis was not established until forty years later. Phosphine/hydride or halide complexes of a number of second- and third-row transition elements from the right-hand side of the *d*-block are effective catalysts. The presence of base generally improves rate and efficiency, and indeed is in some cases obligatory. The choice of solvent is also relevant, with supercritical carbon dioxide often giving faster reaction than orthodox solvents such as tetrahydrofuran.[110,413]

There are two main candidates for the mechanism of formate formation. In the first, the carbon dioxide forms a complex with the metal, with the coordinated and thus activated carbon dioxide then reacting with hydrogen. While there are many examples of carbon dioxide acting as a ligand[414]—in a variety of bonding modes—there is little firm precedent for the subsequent hydrogenation. Nonetheless, it has been suggested that this route may operate for catalysts such as palladium chloride or Wilkinson's compound, $[RhCl(PPh_3)_3]$. In the second mechanism, the carbon dioxide inserts into a metal–hydrogen bond to give coordinated formate. This pathway has been demonstrated for the cycle involving the *cis*-$[Ru(bipy)_2(CO)H]^+$ cation[415] (though insertion of carbon dioxide into Co–H has been demonstrated to be most unlikely in a kinetic study of relevant reactions involving the cobalt(I) macrocyclic complex $[Co([14]\text{-diene})]^+$ [416]). While there is only one way of inserting carbon dioxide into Ru–H, there are at least four different ways of completing the catalytic cycle, variously involving reductive elimination of formic acid, hydrolysis of the formate complex to a hydroxide complex, addition of dihydrogen to the formate complex, and direct hydrogenolysis of the metal to formate-oxygen bond. These and other variants on the mechanism of formate formation have been described and reviewed in detail elsewhere.[412] Preparative and mechanistic aspects of homogeneous catalysis of hydrogenation of carbon dioxide in the presence of an organic substrate have also been much studied. Alkyl formates may thus be produced from alcohols or alkyl halides, formamides from amines, and diols from oxiranes. Carbon dioxide may also be fixed by unsaturated compounds, with the aid of appropriate transition metal catalysts.[417] Several nickel-containing species have been isolated, such as the chelate **104** and an X-ray structurally characterized acrylate-bridged binuclear intermediate.[418] Some mechanistic details have been documented for such reactions as the carboxylations of alkenes, alkynes, and conjugated dienes, and the insertion of carbon dioxide into strained carbocyclic rings to give lactones.[96]

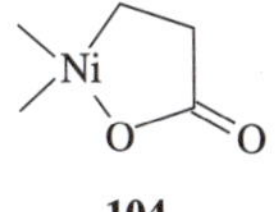

104

Hydrogenation of carbon monoxide[419,420] under mild conditions produces formaldehyde[421] and then methanol:

$$CO + H_2 \longrightarrow HCHO; \quad HCHO + H_2 \longrightarrow CH_3OH$$

The conversion of carbon monoxide into methanol has from time to time provoked considerable industrial and academic interest, with both homogeneous and heterogeneous processes planned, developed, and investigated mechanistically.[422] One homogeneous route involves the unusual soluble oxide catalyst hexamethyldisiloxane. Four cycles are involved, two linked pairs producing formaldehyde, then two further linked cycles hydrogenating the formaldehyde to give methanol.

Under forcing conditions, hydrogenation of carbon monoxide gives a mixture of liquid hydrocarbons containing some CHO compounds. The now defunct Fischer–Tropsch process converted synthesis gas ('syn gas'—see start of next section), made from coal, into fuel. The commercial process used heterogeneous catalysis, but there was considerable interest in developing a homogeneous process that would be more selective and give a cleaner product. Compounds of the early transition metals, e.g. $(\eta^5\text{-}C_5H_5)$/Zr compounds, and a number of clusters, e.g. $Ru_3(CO)_{12}$ and $Ir_4(CO)_{12}$, were investigated. Metal–formyl intermediates were presumed, with later proposals for the intermediacy of σ-vinyl and σ-alkenyl species.[423]

10.7.6 The water gas shift reaction

One of the important feedstocks for industrial organic chemistry is so-called synthesis gas, which is a mixture of carbon monoxide, carbon dioxide, hydrogen, and water. It is often desirable to change the CO : H_2 ratio, which can be achieved by the *water gas shift reaction*:

$$CO + H_2O \rightleftharpoons H_2 + CO_2$$

This, in the forward direction, is hydration of carbon monoxide, in the reverse direction hydrogenation of carbon dioxide (cf. preceding section). The reactions of the water–gas shift system are catalysed by a variety of transition metal compounds, including binary and ternary carbonyls of iron, ruthenium, rhodium, copper, tungsten, platinum, and palladium,[97,423] acting by a range of mechanisms.[424] Complexes of 2,2′-bipyridyl and of 1,10-phenanthroline also figure prominently in kinetic and mechanistic studies of catalysis of the water gas shift reaction.[425] The ruthenium complex $[RuCl(CO)(bipy)_2]^+$ links these two classes of catalyst, while the $[Ir(\eta^5\text{-}C_5Me_5)Cl(bipy)]^+$ cation and its phen analogue also link organometallic and classical Werner chemistry. All these catalysts require photoactivation. The non-organometallic complexes $[Ru(bipy)_2Cl_2]$, $[Rh(bipy)_2]^+$, and $[Rh(phen)_2]^+$ are also effective catalysts for the water–gas shift reaction. The ruthenium complex catalyses by the cycle shown in Fig. 10.30. This involves a series of octahedral $t_{2g}{}^6$ complexes, except for the rather dubious looking seven-coordinate precursor to dihydrogen liberation.

Returning from the excursion into almost classical coordination chemistry represented by $[Ru(bipy)_2Cl_2]$ as catalyst to catalysis by carbonyls, the

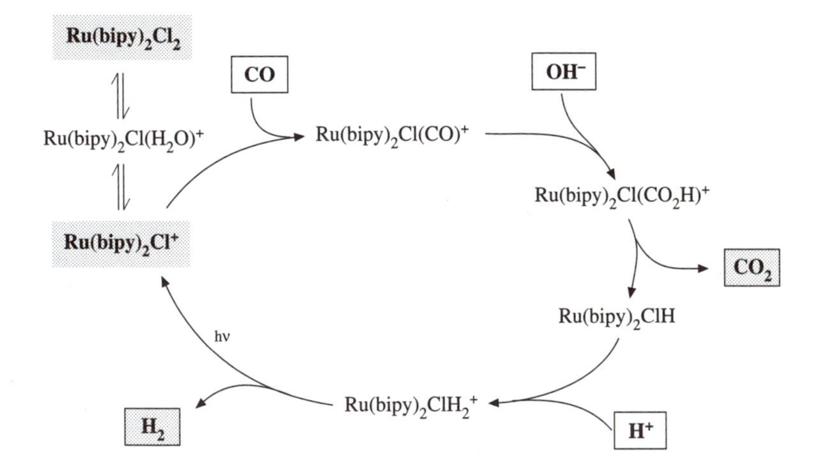

Fig. 10.30 The $[Ru(bipy)_2Cl_2]$-catalysed water gas shift reaction. Most steps in the cycle are reversible; it is driven in the $CO + H_2O \rightarrow CO_2 + H_2$ direction by the irreversibility of the CO_2-releasing step. The rate-limiting step is the photoactivated H_2-releasing step.

cycle for catalysis of hydration of carbon monoxide by iron pentacarbonyl is shown in Fig. 10.31. The key step is here, as in Fig. 10.30, that of attack of hydroxide at coordinated carbonyl, indicated by the rate law:

$$+\mathrm{d}[H_2]/\mathrm{d}t = k[Fe(CO)_5][OH^-]$$

As formate formed generally buffers the system, in a given experiment the rate law is simply:

$$+\mathrm{d}[H_2]/\mathrm{d}t = k'[Fe(CO)_5]$$

The rate is independent of the partial pressure of carbon monoxide, except that this must be above the threshold for preventing dissociation of the iron pentacarbonyl. The dihydride intermediate further round the cycle has been modelled in analogous iridium systems, where reaction stops at this stage as the iridium dihydride complexes do not lose their hydrogen readily.

The $Fe(CO)_5$-catalysed cycle provides an example of what has been termed the associative mechanism, which also operates for catalysis by the $[Rh(LL)\{P(aryl)_3\}]^+$ cations with LL = norbornadiene or cyclooctadiene. The effects of phosphine cone angle support the assignment of associative attack by hydroxide at coordinated carbonyl in the appropriate intermediates here.[426] In contrast to these, catalysis by the Group VI hexacarbonyls is dissociative in mechanism (Fig. 10.32). Here dissociation of a carbon monoxide ligand, often photo-induced, to give catalytically active $M(CO)_5$ is necessary. This, and the subsequent reaction of the $M(CO)_5$ with formate, are indicated by the form of the rate law for these systems:

$$+\mathrm{d}[H_2]/\mathrm{d}t = k''[M(CO)_6][HCO_2^-]/[CO]$$

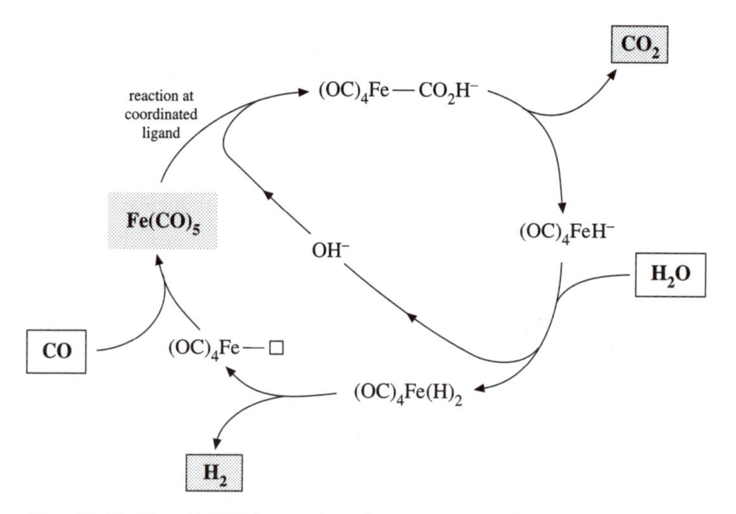

Fig. 10.31 The $Fe(CO)_5$-catalysed water gas shift reaction.

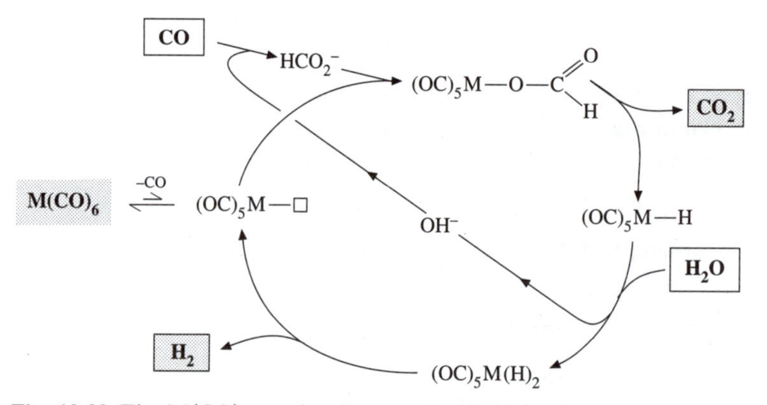

Fig. 10.32 The $M(CO)_6$-catalysed water gas shift reaction.

The platinum-catalysed cycle differs slightly but significantly from the $Fe(CO)_5$-catalysed cycle in that the platinum activates both water, whose loss of a proton is facilitated, and carbon monoxide. The rhodium-carbonyl-iodide cycle of Fig. 10.33, whose key feature is cycling between the rhodium(I) anion $[Rh(CO)_2I_2]^-$ and the rhodium(III) anion $[Rh(CO)_2I_4]^-$, is important in relation to rhodium-catalysed carbonylation of methanol, detailed in the following section. All these cycles are composed wholly of reversible reactions—of necessity, since the water gas equilibrium has to be shiftable in both directions—but have for simplicity been shown in Figs 10.30–10.33 operating in the water carbonylation direction. Thus far each mechanism has consisted of one cycle, but the long-known $K_2[PtCl_4]/SnCl_4$-catalysed system has been analysed in terms of two linked cycles, one processing the oxides of carbon, the other the hydrogen and water.[427]

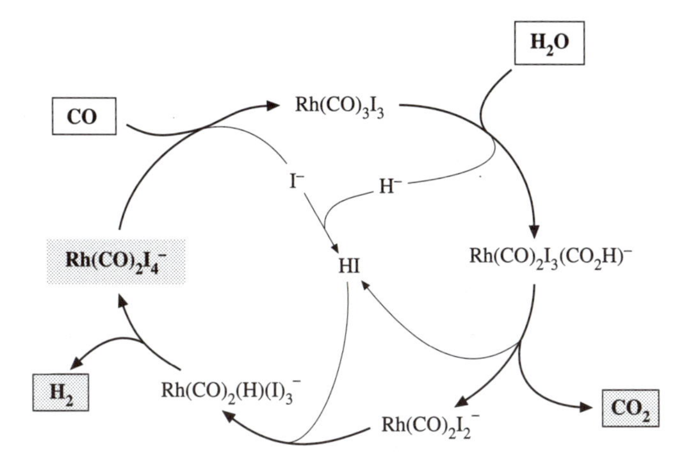

Fig. 10.33 The rhodium-carbonyl-iodide-catalysed water gas shift reaction.

Catalysis of the water gas shift reaction by carbonyl clusters has been studied. Kinetic and spectroscopic studies of the catalysis by $Ru_3(CO)_{12}$ indicate reversible separation of the trinuclear core, with concomitant addition of CO, into bi- and mononuclear carbonyls:

$$Ru_3(CO)_{12} + CO \rightleftharpoons [Ru_3(CO)_{13}] \rightleftharpoons Ru(CO)_5 + Ru_2(CO)_8$$

Both $Ru(CO)_5$ and $Ru_2(CO)_8$ are catalysts for the water gas shift reaction, though here the path involving the binuclear complex dominates. The mechanism of this path is analogous to that for the Group VI carbonyls, with rate-limiting associative attack of formate at ruthenium giving $[Ru_2(CO)_8(HCO)_2]$ as the first intermediate.[424] There is spectroscopic evidence for the subsequent intermediacy of a binuclear ruthenium carbonyl hydride cation.[428] Interest has recently been shown in kinetic and mechanistic details of the water gas shift reaction catalysed by $Ru_3(CO)_{12}$, and by mononuclear $Mo(CO)_6$, $W(CO)_6$, and $Fe(CO)_5$, in binary aqueous solvent media.[429]

10.7.7 Carbonylation[430,431]

The most important homogeneously catalysed carbonylation is that of methanol

$$CH_3OH + CO \longrightarrow CH_3CO_2H$$

Over a million tons of acetic acid (ethanoic acid) is produced annually by a process still generally known as the Monsanto process, even though BP bought the process and plant from Monsanto over a decade ago. Research groups in Monsanto, BP, and academe[432] have all contributed to the establishment of the mechanism of this process, which is now fairly well understood. The main kinetic and spectroscopic evidence may be summarized as follows.

- Reaction is first-order with respect to rhodium species.
- Reaction is first-order with respect to methyl iodide.
- Reaction is independent of CO presssure above about 3 bar. The rate law is thus:

$$\text{rate} = k[\text{Rh}^{\text{I}}][\text{CH}_3\text{I}].$$

- Infrared spectroscopy supports the intermediacy of *cis*-$[Rh(CO)_2I_2]^-$.

The deduced catalytic cycle contains examples of oxidative addition, reductive elimination, and insertion (i.e. methyl migration), as shown in the bottom half of Fig. 10.34. This homogeneous catalysis cycle, into which

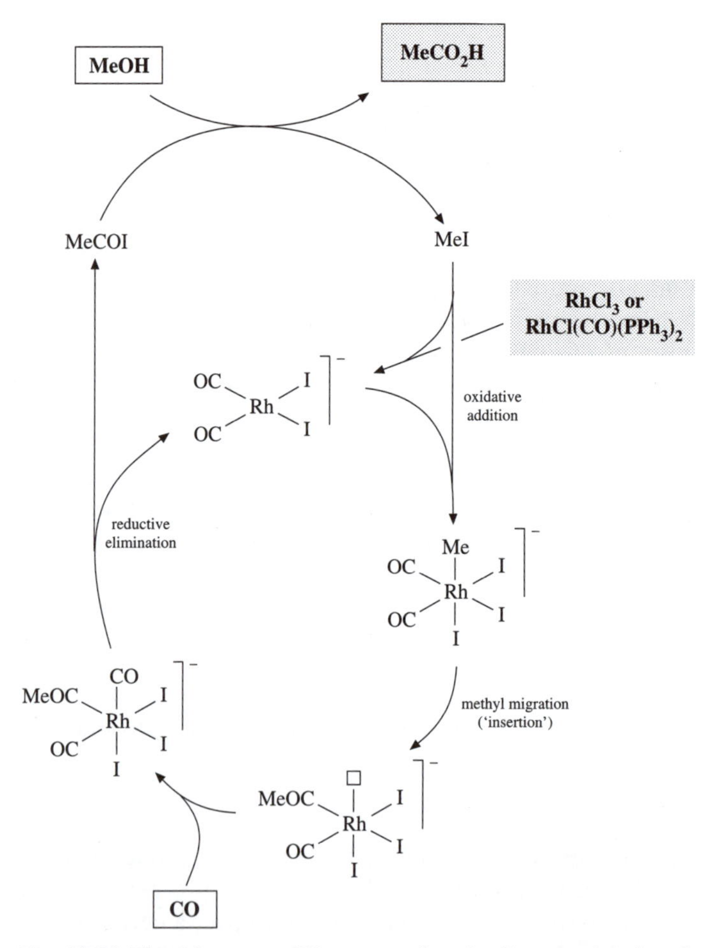

Fig. 10.34 The Monsanto–BP process for rhodium-catalysed carbonylation of methanol.

the carbon monoxide is fed, is coupled (top half of Fig. 10.34) with the reaction

$$CH_3OH + CH_3COI \longrightarrow CH_3CO_2H + CH_3I$$

to convert the starting material, methanol, into the product. There are, however, two complications to add to this simple central mechanism of two linked cycles. The first is that it is necessary to add a little HI with the rhodium(I) at the start, to generate some methyl iodide and thence the active catalyst *cis*-$[Rh(CO)_2I_2]^-$. The second is that it is necessary to have significant amounts (>10%) of water present to maintain a sufficient level of HI to stabilize the rhodium species involved. In fact the rhodium catalytic species are maintained by a water gas shift reaction—see Section 10.7.5.2 above.

It is interesting to compare and contrast the above rhodium catalytic system with its iridium analogue, featured in Fig. 10.35. Both involve oxidative addition, methyl migration, and reductive elimination, and in both systems ternary carbonyl anions play a key role—but there are important kinetic and mechanistic differences. Firstly, the oxidative addition of methyl iodide to the metal centre in the iridium cycle is over a hundred times faster than in the rhodium case. Secondly, stronger iridium–carbon bonding slows the methyl migration step dramatically, by nearly a million times compared with the rhodium system. These differences result in a shift of the rate-limiting step from oxidative addition of methyl iodide, in the rhodium cycle, to the dissociation of iodide and addition of carbon monoxide in the iridium cycle. There is a water gas shift reaction cycle coupled to the carbonylation cycle, and there is also some methane generation in a side reaction associated with these iridium cycles. These extras are included in Fig. 10.35 to show their relation to the main carbonylation process.

Carbonylation of alkenes is also of both academic and commercial interest. Catalysis of carbonylation of ethene by molybdenum hexacarbonyl has been the subject of a detailed kinetic, spectroscopic, and modelling investigation. The hexacarbonyl is, as usual, the precursor, the coordinatively unsaturated pentacarbonyl the active catalyst. Reaction is initiated with ethyl iodide, giving $Mo(CO)_5I$ and ethyl radicals. The latter react with $Mo(CO)_6$ to give very reactive molybdenum-based radicals which catalyse carbonylation of the ethene en route to the desired product, propionic acid or its anhydride.[433]

The mechanism of repeated carbonylation of alkenes to form polyketones has been much studied.[431,434] Alternate insertion of alkene and of carbon monoxide into palladium–carbon bonds has been shown to be much more plausible than the alternative route through palladium–carbene intermediates. Recent interest has centred on the synthesis of lactones and lactams, produced, for instance, by palladium-catalysed carbonylation of halide-containing alcohols, or by palladium/copper-catalysed hydrocarboxylation of alkenes. Carbonylation of methyl isocyanide can be achieved by reductive

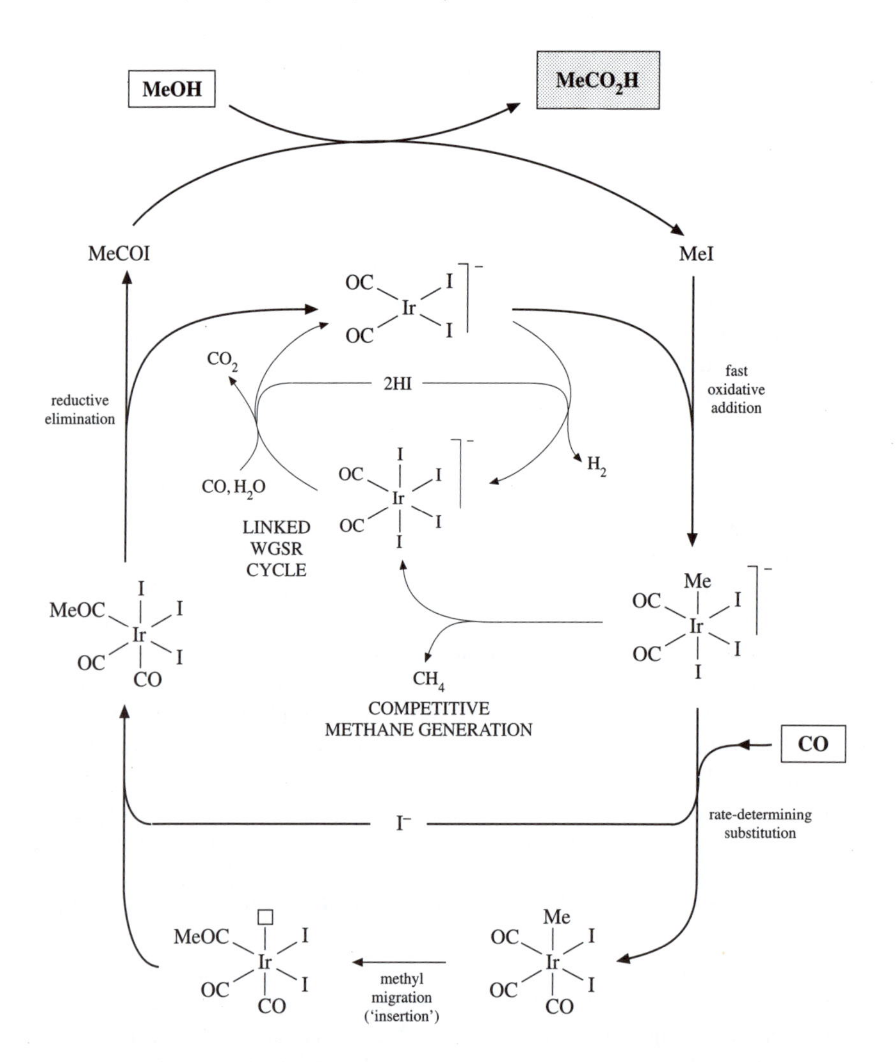

Fig. 10.35 Simplified iridium-catalysed carbonylation cycle, coupled with the associated water gas shift cycle and competing methane-generation cycle (see T. Ghaffar, H. Adams, P. M. Maitlis, G. J. Sunley, M. J. Baker, and A. Haynes, *J. Chem. Soc., Chem. Commun.*, 1998, 1023).

coupling at niobium.[435] The mechanism of production of the highly functionalized alkyne product, RR′NCCOR″, centres on assembling the reactants at the metal. The geometry of the structurally characterized intermediate contains the product coordinated as **105** rather than **106**—seven-coordinate niobium(III) rather than six-coordinate niobium(I). The C–C bond length and bond angles included in **105** also indicate a closer approximation to this metallacyclopropene representation than to the 2-alkyne representation **106**.

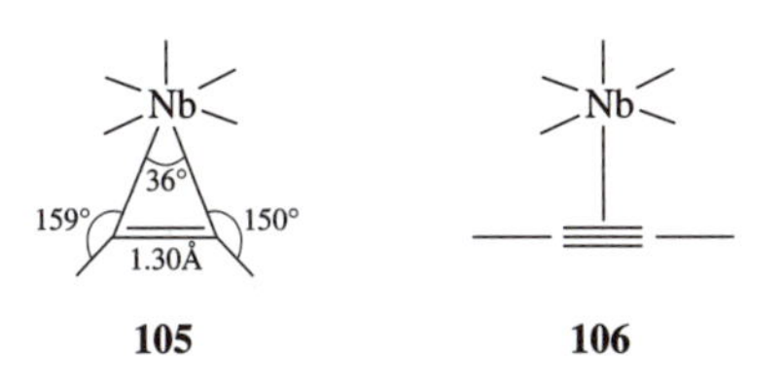

10.7.8 Hydroformylation and hydrocarboxylation

10.7.8.1 Hydroformylation[429,430,436]

This process, also known as the *oxo reaction*, involves the addition of hydrogen and carbon monoxide to an alkene

$$RCH{=}CH_2 + CO + H_2 \rightarrow RCH_2CH_2CHO + RCHMeCHO$$

and is one of the most widely used homogeneously catalysed reactions commercially (see Table 10.16 above). The original industrial application was developed by BASF some fifty years ago, using $Co_2(CO)_8$ as the catalyst precursor. This was treated with hydrogen to give $HCo(CO)_4$, which in turn could lose a carbonyl ligand to give the active catalyst $HCo(CO)_3$. The catalytic cycle (Fig. 10.36)[437] contains the customary steps—coordination of alkene, insertion, and so on—and indeed the initial stages are very similar to those in the alkene hydrogenation cycle shown in Fig. 10.29 (Section 10.7.5.1). An unsymmetrical alkene $RCH{=}CH_2$, for example propene or but-1-ene, can insert into the metal–carbon bond in two ways, Markownikov or anti-Markownikov, to give $(OC)_3CoCH_2CH_2R$ or $(OC)_3CoCHRCH_3$. These intermediates will, under kinetic control of the subsequent migration (co-insertion) step, give linear and branched aldehydes as

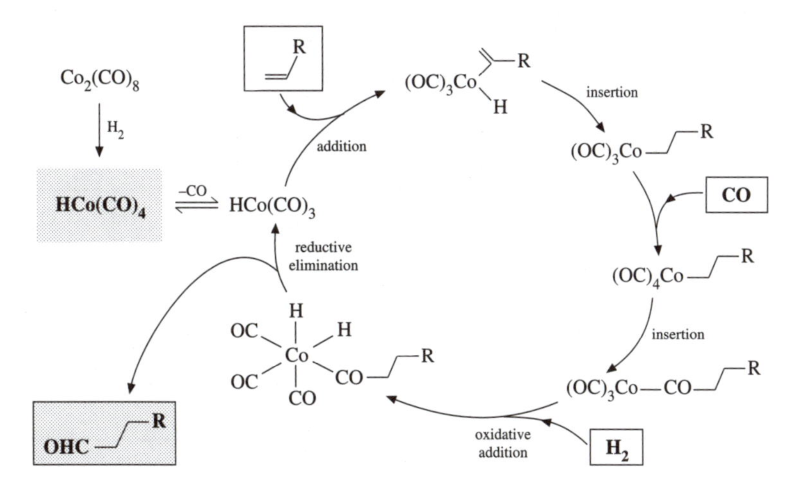

Fig. 10.36 Hydroformylation cycle with $Co_2(CO)_8$ catalysis.

products. This is a commercially important variable, since the linear aldehydes are the required starting materials for commercial use.

Catalysis of hydroformylation by cobalt carbonyl suffers from the dual disadvantages of needing relatively high temperatures and pressures. Rhodium catalysts operate under much milder conditions and, fortunately from the economic viewpoint, much more efficiently. Rhodium catalysts offer additional advantages. Their phosphine ligands can be tailored, both sterically and electronically, to optimize conditions and yields, and to maximize the linear: branched aldehyde product ratio. Thus $[Rh(H)(CO)(PPh_3)_3]$ at 25 °C and 1 atmosphere gives essentially just one product. The phosphine ligands may also be derivatized, e.g. with sulphonate substituents as in $P(3\text{-}C_6H_4SO_3Na)_3$,[438] to make them water-soluble. This has already been mentioned (Section 10.7.1) in connection with the polymer-supported catalyst **72**, which is a good promoter of hydroformylation.[331] It is also possible to build chiral centres into the catalysts by use of appropriate ligands (Section 10.7.1 again). Much research into enantioselective asymmetric hydroformylation[439] has been rewarded by efficient and profitable synthetic routes to such products as L-Dopa and (−)-menthol. The catalytic cycle for rhodium catalysis, for example by $[RhH(PEt_3)_3]$,[440] is believed to be similar to that shown in Fig. 10.36 for catalysis by $Co_2(CO)_8$, though in the case of $[RhCl(PPh_3)_3]$ there has been a suggestion of associative and dissociative pathways (cf. the discussion of the water–gas shift reaction above, Section 10.7.6). Relative hydroformylation rates for various alkenes seem to be similar (Table 10.23), though this statement has to be based on rather limited strictly comparable experimental data.

Compounds of several other metals also act as hydroformylation catalysts, but with very different speeds and efficiencies. We have already seen that rhodium catalysts are much better than cobalt carbonyl. The former act about three orders of magnitude more quickly than the latter, but cobalt carbonyl catalysis is of the order of a hundred times faster than catalysis by ruthenium complexes, which in turn operate several orders of

Table 10.23 Rate constants for hydroformylation of alkenes.

alkene	$[Co(H)(CO)_4]$ $10^5 k(s^{-1})$ [a]	$[Rh(H)(CO)(PPh_3)_3]$ (ml min^{-1}) [b]
1-pentane → 1-octene	110	3.5–3.8
2-hexene; 2-octene	30; 31	
2-pentene		0.15
2-methyl-1-pentene	13	0.06
2-methyl-2-pentene	8	

[a] At 110 °C.
[b] Gas uptake; at 25 °C.

magnitude more rapidly than their iron analogues. The platinum(II)–tin catalyst **107** brings together several strands, from Section 10.7.1 as well as from the present section. It combines the use of tin in the form of the $SnCl_3^-$ ligand, chiral chelating bis-phosphine ligands, asymmetric hydroformylation, homogeneous and polymer-supported modes, and pharmaceutical usefulness—at least one of the products has anti-inflammatory properties and is analgesically active.[441] A number of clusters, mainly carbonyl–phosphine or –phosphite compounds of rhodium, cobalt, and ruthenium, are effective hydroformylation catalysts.[320] There is good evidence that their cores remain intact around the respective catalytic cycles.

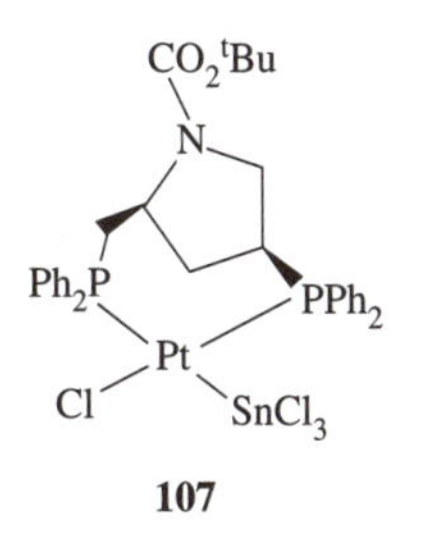

107

10.7.8.2 Hydrocarboxylation

Hydrocarboxylation, often termed the *Reppe reaction*, involves the addition of carbon monoxide and water to an alkene or alkyne:

$$C{=}C(\text{or } C{\equiv}C) + CO + H_2O \longrightarrow -CH_2CH_2CO_2H(\text{or } -CH{=}CHCO_2H)$$

The commercially important example is the conversion of acetylene into acrylic acid using a nickel bromide/copper halide catalyst. Despite the long-time importance of this type of reaction, the mechanism still appears not to be fully understood, though the chemistry at the nickel centre is almost certainly as depicted in Fig. 10.37. Alkene hydrocarboxylation in the presence of a rhodium catalyst has a cycle rather similar to that for the Monsanto/BP process, with the *cis*-$[Rh(CO)_2I_2]^-$ anion playing a central role.[442] Currently there is interest in hydrocarboxylation of alkenes in aqueous sulphuric acid using copper(I) or silver(I) carbonyl cations, e.g. $[Cu(CO)_3]^+$ or $[Ag(CO)_2]^+$, as catalysts. These reactions can be carried out at room temperature and atmospheric pressure, to give mixtures of carboxylic acids.[443] 1-Octene gives 2,2-dimethylheptanoic acid as main product—an understanding of the mechanism should assist with the tailoring of conditions to produce a single product.

10.7.9 Oxidation

The Wacker process for the oxidation of ethylene to acetaldehyde was of great importance for many years, indeed was one of the first examples of

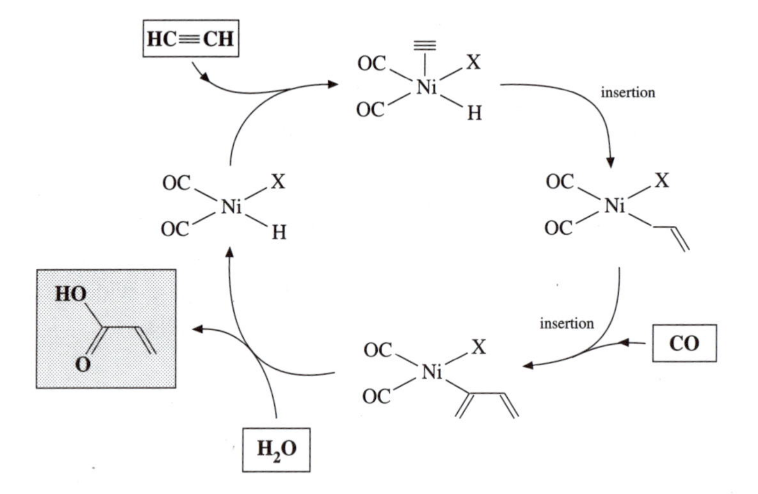

Fig. 10.37 The role of nickel in the Reppe hydrocarboxylation cycle.

industrially significant homogeneous catalysis. Both palladium and polyoxoanion catalysts were patented,[316] with the former developed for industrial use. The Wacker process was a key step in the production of acetic acid—now more generally produced by oxidation of methanol (Monsanto/BP process; Section 10.7.6). The catalytic cycle (Fig. 10.38) involves such components as nucleophilic substitution, reaction at a coordinated ligand, and elimination. The establishment of the rate law, namely

$$-\mathrm{d}[\mathrm{C_2H_4}]/\mathrm{d}t = k[\mathrm{C_2H_4}][\mathrm{PdCl_4}^{2-}][\mathrm{H^+}][\mathrm{Cl^-}]^2$$

and stereochemical studies helped considerably in the elucidation of the mechanism. Coordination of the alkene to palladium(II) activates it to nucleophilic attack by water, with concomitant formation of a $-CH_2CH_2OH$ ligand. Acetaldehyde is then probably formed via the η^2-vinyl intermediate shown (vinyl acetate is produced if the process is run in acetic acid rather than in the usual aqueous medium). On losing the acetaldehyde, the palladium(II) is reduced to an ill-defined palladium(0) state, reoxidized by coupling through copper(I)/(II) to oxygen. This copper/oxygen oxidation can be improved by further coupling to a heteropolyanion redox couple, as shown in Fig. 10.39, which provides a redox sequence of favourably small steps in redox potentials.[444] Oxidation of alkenes by oxygen atom transfer can be achieved in catalytic cycles based on [Ru(edta)(NO)]. This complex oxidizes 1-hexene to 2-hexanone and cyclohexene to its epoxide.[445]

Oxidation of water, to hydrogen peroxide or oxygen, has attracted attention recently. Catalysts for these multi-electron redox reactions include di- and polynuclear manganese and ruthenium species. The interest in the former stems from the connection with photosynthesis.[446] In the ruthenium systems the oxidation states of the metal range from 2+ to 6+, with a pair of

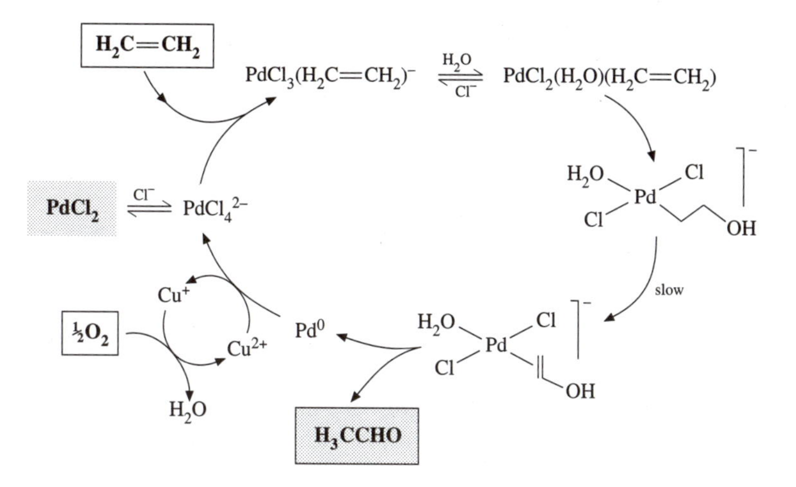

Fig. 10.38 The Wacker process.

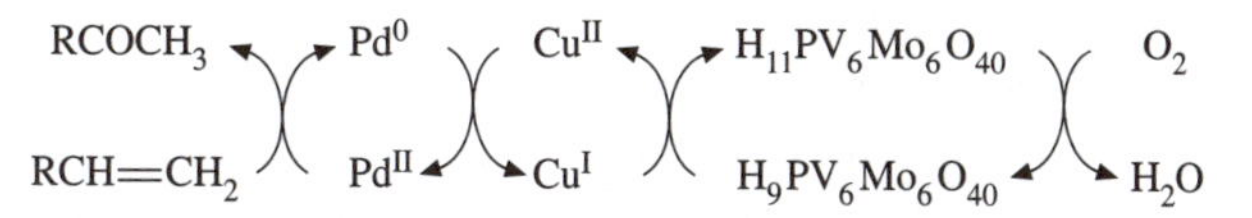

Fig. 10.39 Heteropolyanion redox couple linked to the Wacker process.

ligands taken variously from water, hydroxide and oxide. Cycle components range from $[Ru^{II}(bipy)_2(H_2O)_2]^{2+}$ through complexes of Ru^{III}, Ru^{IV}, and Ru^{V} to $[Ru^{VI}(bipy)_2(O)_2]^{2+}$.[447] In similar vein, an $Ru^{II}(H_2O)/Ru^{IV}{=}O$ couple in which the metal is also complexed by terpy and a diazabutadiene ligand has proved an effective mediator for catalysis of cerium(IV) oxidation of water to dioxygen.[448]

Photocatalytic oxidation of hydrocarbons is also subject to catalysis by transition metal complexes, for example of iron (where comparisons have been drawn with the mode of action of cytochrome P450) or of ruthenium.[449] In the case of the ruthenium-catalysed systems, comprising oxidation of alkenes by oxygen and of alkanes by hydrogen peroxide, free-radical mechanisms seem likely. The rate-determining step in these reactions is thought to be hydrogen atom abstraction. Oxidation of methane is a perenially challenging topic, in view of its unreactivity. Recent research has produced a rhodium trichloride/iodide catalyst for oxidation of methane by oxygen, in the company of carbon monoxide, to acetic acid.[450] Initial probing of the mechanism suggests that the oxidation does not proceed through methanol, despite the adumbration of the Monsanto/BP process by the rhodium-carbonyl-iodide combination.

Apart from a very few instances of catalysis by polynuclear oxoanions, almost all the catalysts mentioned in this section on homogeneous catalysis, certainly in Sections 10.7.1—10.7.8, have been of the carbonyl/phosphine low oxidation state type. Intermediate oxidation states have, not unexpectedly, made more of an appearance in this present section on oxidation. Our final reference is to a catalyst in the highest known oxidation state, 8+. Osmium tetroxide is well known as a powerful oxidant in organic chemistry, and this property also makes it a useful catalyst in organic oxidations by other oxidants, especially in alkene dihydroxylation.[451]

10.7.10 Further reading

We have been conscious during the preparation and writing of this chapter, particularly this final section on homogeneous catalysis, of the severe restrictions imposed by its length, or rather its brevity. To cover, even sketchily, the range from the borders with biology through to organometallic catalytic systems, from pure chemistry to industrially important processes, has the inevitable consequences of gross selectivity and of oversimplification. Readers who would like a somewhat fuller treatment, and a different perspective, but at roughly the same level, are recommended to read the appropriate chapters in the books by Atwood and Crabtree. The former is particularly appropriate in view of its strong emphasis on detailed kinetic results and mechanistic discussion. These authors cover essentially the whole field of our Chapter 10. For fuller and much more detailed treatments of specific areas, the relevant sections in *Comprehensive Coordination Chemistry* are recommended for reactions of coordinated ligands and for the activation of small molecules. A few entries in the *Encyclopaedia of Inorganic Chemistry* (Wiley) give useful kinetic and mechanistic information, especially that entitled 'Carbonylation Processes by Homogeneous Catalysis'. These entries are cited where appropriate in earlier sections of this chapter. The development of knowledge and understanding of kinetics and mechanisms in the area of organometallic chemistry and its applications in homogeneous catalysis can be followed, from 1960 up to the present, through the appropriate sections in the Chemical Society/Royal Society of Chemistry Specialist Periodical Reports on inorganic reaction mechanisms, and on the successor series (*Mechanisms of Inorganic and Organometallic Reactions*) from Plenum.

In relation to homogeneous catalysis, there are a number of books which go into more detail of preparative, synthetic, technical, and commercial aspects, as well as of the mechanisms with which we are primarily concerned. A selection is given, in chronological order, below. Some are dedicated to homogeneous catalysis, some have just one or two relevant chapters—the latter are often at a more appropriate level for our treatment, but the former need to be consulted for greater depth and detail. One or two of the books cited below are out of print but will be available in libraries. Indeed, in the current economic climate readers in academic institutions are more likely to find these slightly older classic texts on the shelves than the most up-to-date volumes!

C. Masters, *Homogeneous Transition-metal Catalysis*, Chapman & Hall, London, 1981.

R. S. Dickson, *Homogeneous Catalysis with Compounds of Rhodium and Iridium*, Reidel, Dordrecht, 1985.

A. W. Parkins and R. C. Poller, *An Introduction to Organometallic Chemistry*, Macmillan, London, 1986.

J. P. Collman, L. S. Hegedus, J. R. Norton, and R. G. Finke, *Principles and Applications of Organotransition Metal Chemistry*, 2nd edn, University Science Books/W. H. Freeman, Basingstoke, 1987.

P. Powell, *Principles of Organometallic Chemistry*, 2nd edn, Chapman & Hall, London, 1988.

Advances in Chemistry Series, No. 230: Homogeneous Transition Metal Catalyzed Reactions, eds W. R. Moser and D. W. Slocum, American Chemical Society, Washington, DC, 1992.

G. W. Parshall and S. D. Ittel, *Homogeneous Catalysis*, 2nd edn, Wiley, New York, 1992.

Ch. Elschenbroich and A. Salzer, *Organometallics*, 2nd edn, VCH, Weinheim, 1992.

R. H. Crabtree, *The Organometallic Chemistry of the Transition Metals*, 2nd edn, Wiley, New York, 1994.

J. D. Atwood, *Inorganic and Organometallic Reaction Mechanisms*, 2nd edn,VCH, Weinheim, 1997.

References

1. M. M. Jones, *Ligand Reactivity and Catalysis*, Academic Press, New York, 1968; R. P. Houghton, *Metal Complexes in Organic Chemistry*, Cambridge University Press, 1979; E. C. Constable, *Metals and Ligand Reactivity*, Ellis Horwood, Chichester, 1990; E. C. Constable, *Metals and Ligand Reactivity: An Introduction to the Organic Chemistry of Metal Complexes*, VCH, Weinheim, 1996.
2. Review: R. W. Hay, *Comprehensive Coordination Chemistry*, Vol. 6, eds G. Wilkinson, R. D. Gillard, and J. A. McCleverty, Pergamon, Oxford, 1987, Chapter 61.4. This review is thorough, fully referenced, and should be consulted for fuller details and further citations for topics discussed in this section.
3. Review: D. St. C. Black, *Comprehensive Coordination Chemistry*, Vol. 6, eds G. Wilkinson, R. D. Gillard, and J. A. McCleverty, Pergamon, Oxford, 1987, Chapter 61.1.4.1. This review provides some useful general chemical information on reactions of coordinated ligands.
4. R. H. Crabtree, *The Organometallic Chemistry of the Transition Metals*, 2nd. edn, Wiley, New York, 1994.
5. A. J. Pearson, in *Comprehensive Organometallic Chemistry II*, Vol. 12, eds E. W. Abel, F. G. A. Stone, and G. Wilkinson, ed. L. S. Hegedus, Pergamon, Oxford, 1995, Chapter 6.3.
6. M. F. Semmelhack, in *Comprehensive Organometallic Chemistry II*, Vol. 12, eds E. W. Abel, F. G. A. Stone, and G. Wilkinson, ed. L. S. Hegedus, Pergamon, Oxford, 1995, Chapter 9.1.

7. L. A. P. Kane-Maguire, E. D. Honig, and D. A. Sweigart, *Chem. Rev.*, 1984, **84**, 525.
8. A. Werner, *Z. Anorg. Chem.*, 1900, **22**, 91; *Chem. Ber.*, 1911, **44**, 876; *Liebigs Ann. Chem.*, 1919, **386**, 1.
9. N. F. Curtis, *Coord. Chem. Rev.*, 1968, **3**, 3.
10. R. W. Hay and K. B. Nolan, *J. Chem. Soc., Dalton Trans.*, 1976, 548.
11. V. Goedken, *J. Chem. Soc., Chem. Commun.*, 1972, 207; D. F. Mahoney and J. K. Beattie, *Inorg. Chem.*, 1973, **12**, 2561.
12. R. W. Hay and P. K. Banerjee, *J. Chem. Soc., Dalton Trans.*, 1980, 2452.
13. H. L. Conley and R. B. Martin, *J. Phys. Chem.*, 1965, **69**, 2914; J. E. Hix and M. M. Jones, *Inorg. Chem.*, 1966, **5**, 1863.
14. A. W. Zanella and P. C. Ford, *Inorg. Chem.*, 1975, **14**, 42, 700.
15. D. A. Buckingham, C. E. Davis, D. M. Foster, and A. M. Sargeson, *J. Am. Chem. Soc.*, 1970, **92**, 5571.
16. P. J. Lawson, M. G. McCarthy, and A. M. Sargeson, *J. Am. Chem. Soc.*, 1982, **104**, 6710; N. E. Dixon and A. M. Sargeson, *J. Am. Chem. Soc.*, 1982, **104**, 6716.
17. J. N. Armor and H. Taube, *Inorg. Chem.*, 1971, **10**, 1570.
18. E. Gelles, *J. Chem. Soc.*, 1956, 4736; E. Gelles and A. Salama, *J. Chem. Soc.*, 1958, 3683, 3689.
19. W. D. Covey and D. L. Leussing, *J. Am. Chem. Soc.*, 1974, **96**, 3860; N. V. Raghaven and D. L. Leussing, *J. Am. Chem. Soc.*, 1974, **96**, 7147.
20. D. A. Buckingham, D. M. Foster, and A. M. Sargeson, *J. Am. Chem. Soc.*, 1969, **91**, 4102; D. A. Buckingham, D. M. Foster, and A. M. Sargeson, *J. Am. Chem. Soc.*, 1970, **92**, 6151; D. A. Buckingham, F. R. Keene, and A. M. Sargeson, *J. Am. Chem. Soc.*, 1973, **95**, 5649; D. A. Buckingham, P. J. Morris, A. M. Sargeson, and A. Zanella, *Inorg. Chem.*, 1977, **16**, 1910.
21. P. Hendry and A. M. Sargeson, *J. Chem. Soc., Chem. Commun.*, 1984, 164.
22. R. M. Milburn, S. S. Massoud, and F. Tafesse, *Inorg. Chem.*, 1985, **24**, 2591.
23. F. Tafesse and R. M. Milburn, *Inorg. Chim. Acta*, 1987, **135**, 119.
24. Jik Chin and Xiang Zou, *J. Am. Chem. Soc.*, 1988, **110**, 223.
25. D. R. Jones, L. F. Lindoy, and A. M. Sargeson, *J. Am. Chem. Soc.*, 1983, **105**, 7327.
26. J. A. Connolly, Jung Hee Kim, M. Banaszczyk, M. Drouin, and Jik Chin, *Inorg. Chem.*, 1995, **34**, 1094.
27. J. Morrow and W. C. Trogler, *Inorg. Chem.*, 1989, **28**, 2330; R. W. Hay and N. Govan, *Polyhedron*, 1996, **15**, 2381.
28. R. N. Bose, N. Goswami, and S. Moghaddas, *Inorg. Chem.*, 1990, **29**, 3461.
29. P. Hendry and A. M. Sargeson, *Inorg. Chem.*, 1990, **29**, 97.
30. N. J. Curtis, K. S. Hagen, and A. M. Sargeson, *J. Chem. Soc., Chem. Commun.*, 1984, 1571.
31. H. Sigel, F. Hofstetter, R. B. Martin, R. M. Milburn, V. Scheller-Kratiger, and K. H. Scheller, *J. Am. Chem. Soc.*, 1984, **106**, 7935.
32. D. Gani and J. Wilkie, *Struct. Bonding*, 1997, **89**, 133.
33. J. Sumaoka, S. Miyama, and M. Komiyama, *J. Chem. Soc., Chem. Commun.*, 1994, 1755.
34. U. Spitzer, R. van Eldik, and H. Kelm, *Inorg. Chem.*, 1982, **21**, 2821.
35. Y. Kitamura and R. van Eldik, *Inorg. Chem.*, 1987, **26**, 2907.
36. I. Bertini, C. Luchinat, S. Mangani, and R. Pierattelli, *Comments Inorg. Chem.*, 1995, **17**, 1.

37. R. W. Hay, *Bio-inorganic Chemistry*, Ellis Horwood, Chichester, 1984, Chapter 4 (see p. 95 of the 1993 reprint).
38. J. E. Coleman and B. L. Vallee, *J. Biol. Chem.*, 1961, **236**, 2244.
39. B. Greener, M. H. Moore, and P. H. Walton, *J. Chem. Soc., Chem. Commun.*, 1996, 27; B. Greener and P. H. Walton, *J. Chem. Soc., Dalton Trans.*, 1997, 3733.
40. Y. Gultneh, Allwar, B. Ahvazi, D. Blaise, R. J. Butcher, J. Jasinski, and J. Jasinski, *Inorg. Chim. Acta*, 1996, **241**, 31.
41. J. Chin, *Accounts Chem. Res.*, 1991, **24**, 145.
42. Shieh-Jun Wang and E. L. King, *Inorg. Chem.*, 1980, **19**, 1506.
43. B. E. Crossland and P. J. Staples, *J. Chem. Soc. (A)*, 1971, 2853.
44. A. A. Holder and T. P. Dasgupta, *J. Chem. Soc., Dalton Trans.*, 1996, 2637.
45. T. Schirado, E. Gennari, R. Merello, A. Decinti, and S. Bunel, *J. Inorg. Nucl. Chem.*, 1971, **33**, 3417.
46. T. J. Cardwell and T. H. Lorman, *Inorg. Chim. Acta*, 1984, **85**, 1.
47. A. G. Blackman, D. A. Buckingham, and C. R. Clark, *J. Am. Chem. Soc.*, 1991, **113**, 2656.
48. Reviews: M. de S. Healy and A. J. Rest, *Adv. Inorg. Chem. Radiochem.*, 1978, **21**, 1; L. F. Lindoy, *The Chemistry of Macrocyclic Ligand Complexes*, Cambridge University Press, Cambridge, 1989; R. Hoss and F. Vögtle, *Angew. Chem. Int. Edn. Engl.*, 1994, **33**, 375. Ten more reviews on template reactions (not all dealing with kinetic and mechanistic aspects) can be traced through Table 13 on p. 19 of *Comprehensive Coordination Chemistry*, Vol. 7, eds G. Wilkinson, R. D. Gillard, and J. A. McCleverty, Pergamon, Oxford, 1987.
49. J.-C. Chambron, C. O. Dietrich-Buchecker, V. Heitz, J.-F. Nierengarten, J.-P. Sauvage, C. Pascard, and J. Guilhem, *Pure Appl. Chem.*, 1995, **67**, 233.
50. A. Eschenmoser, R. Scheffold, E. Bertelle, M. Pesaro, and H. Gschwind, *Proc. Roy. Soc. A*, 1965, **288**, 306.
51. D. S. Eggleston and S. C. Jackels, *Inorg. Chem.*, 1980, **19**, 1593.
52. K. Schug, M. D. Gilmore, and L. A. Olsen, *Inorg. Chem.*, 1967, **6**, 2180.
53. O. Vollárová and J. Benko, *Current Topics in Solution Chemistry*, 1994, **1**, 107.
54. M. D. Johnson and R. J. Balahura, *Inorg. Chem.*, 1988, **27**, 3104.
55. I. P. Evans, G. W. Everett, and A. M. Sargeson, *J. Chem. Soc., Chem. Commun.*, 1975, 139; K. Schug and C. P. Guengerich, *J. Am. Chem. Soc.*, 1975, **97**, 4135.
56. Review: A. Spencer, *Comprehensive Coordination Chemistry*, Vol. 6, eds G. Wilkinson, R. D. Gillard, and J. A. McCleverty, Pergamon, Oxford, 1987, Chapter 61.2.
57. Review: Chapter 12 of R. H. Crabtree, *The Organometallic Chemistry of the Transition Metals*, Wiley, New York, 1988, deals with the activation of small molecules in organometallic systems.
58. A. G. Sykes, *Adv. Inorg. Bioinorg. Mech.*, 1982, **1**, 121.
59. A. Bakac, *Progr. Inorg. Chem.*, 1995, **43**, 267.
60. T. G. Spiro (ed.), *Metal Ion Activation of Dioxygen*, Wiley, New York, 1980; A. E. Martell and D. T. Sawyer (eds), *Oxygen Complexes and Oxygen Activation by Transition Metals*, Plenum, New York, 1988.
61. Reviews: In 1994 *Chem. Rev.* devoted most of one issue to a series of reviews entitled *Metal Dioxygen Complexes—A Perspective*. Specific mention will be made of several of these at appropriate points in this section.

62. Reviews: F. Basolo, B. M. Hoffman, and J. A. Ibers, *Accounts Chem. Res.*, 1975, **8**, 384; G. McLendon and A. E. Martell, *Coord. Chem. Rev.*, 1976, **19**, 1; R. D. Jones, D. A. Summerville, and F. Basolo, *Chem. Rev.*, 1979, **79**, 139.
63. For a valiant attempt to present a unified picture of oxygen adducts, see L. Vaska, *Accounts Chem. Res.*, 1976, **9**, 175.
64. P. T. Cheng, C. D. Cook, S. C. Nyburg, and K. Y. Wan, *Can. J. Chem.*, 1971, **49**, 3772.
65. S. A. Cotton, *Chemistry of Precious Metals*, Blackie, London, 1997, p. 139.
66. V. L. Pecoraro, M. J. Baldwin, and A. Gelasco, *Chem. Rev.*, 1994, **94**, 807; M. Suzuki, T. Ishikawa, A. Harada, S. Ohba, M. Sakamoto, and Y. Nishida, *Polyhedron*, 1997, **16**, 2553.
67. M. H. Dickman and M. T. Pope, *Chem. Rev.*, 1994, **94**, 569.
68. E. Rose, B. Boitrel, M. Quelquejeu, and A. Kossanyi, *Tetrahedron Lett.*, 1993, **34**, 7267; S. David, B. R. James, D. Dolphin, T. G. Traylor, and M. A. Lopez, *J. Am. Chem. Soc.*, 1994, **116**, 6.
69. M. Momenteau and C. A. Reed, *Chem. Rev.*, 1994, **94**, 659.
70. A. L. Feig and S. J. Lippard, *Chem. Rev.*, 1994, **94**, 759.
71. Q. H. Gibson, *J. Physiol.*, 1956, **134**, 112.
72. Chien Ho (ed.), *Hemoglobin and Oxygen Binding*, Elsevier North Holland and Macmillan, London, 1982, see Session III.
73. A. F. Riggs and R. A. Wolbach, *J. Gen. Physiol.*, 1956, **39**, 585.
74. P. C. Wilkins and R. G. Wilkins, *Coord. Chem. Rev.*, 1987, **79**, 195; R. E. Stenkamp, *Chem. Rev.*, 1994, **94**, 715.
75. Key references can be traced through Section II of Ref. 58 and through C. Bianchini and R. W. Zoellner, *Adv. Inorg. Chem.*, 1996, **44**, 264.
76. See Table I of Ref. 59.
77. D. H. Busch and N. W. Alcock, *Chem. Rev.*, 1994, **94**, 585.
78. K. D. Karlin and Y. Gultneh, *Progr. Inorg. Chem.*, 1987, **35**, 219; K. A. Magnus, Hoa Ton-That, and J. E. Carpenter, *Chem. Rev.*, 1994, **94**, 727; N. Kitajima and Y. Moro-oka, *Chem. Rev.*, 1994, **94**, 737.
79. J. M. Brown, L. Powers, B. Kincaid, J. A. Larrabee, and T. G. Spiro, *J. Am. Chem. Soc.*, 1980, **102**, 4210.
80. K. D. Karlin, M. S. Nasir, B. I. Cohen, R. W. Cruse, S. Kaderli, and A.D. Zuberbühler, *J. Am. Chem. Soc.*, 1994, **116**, 1324; Dong-Heon Lee, Ning Wei, N. N. Murthy, Z. Tyeklár, K. D. Karlin, S. Kaderli, B. Jung, and A. D. Zuberbühler, *J. Am. Chem. Soc.* 1995, **117**, 12498.
81. Review: N. Kitajima, *Adv. Inorg. Chem.*, 1992, **39**, 1.
82. G. D. Armstrong and A. G. Sykes, *Inorg. Chem.*, 1986, **25**, 3135.
83. A. Suga, T. Sugiyama, M. Otsuka, M. Ohno, Y. Sugiura, and K. Maeda, *Tetrahedron*, 1991, **47**, 1191.
84. For example, R. Luck and R. H. Morris, *Inorg. Chem.*, 1984, **23**, 1489.
85. See Q. H. Gibson, p. 321 of Ref. 72.
86. G. S. Adair, *J. Biol. Chem.*, 1925, **63**, 529.
87. Q. H. Gibson, *J. Physiol.*, 1956, **134**, 112, 123.
88. Q. H. Gibson and F. J. W. Roughton, *Proc. Roy. Soc.*, 1957, **B146**, 206.
89. Q. H. Gibson and F. J. W. Roughton, *J. Physiol.*, 1955, **128**, 69 *P*.
90. A. J. Dixon, P. Glyn, M. A. Healy, P. M. Hodges, T. Jenkins, M. Poliakoff, and J. J. Turner, *Spectrochim. Acta*, 1988, **44A**, 1309.

91. T. G. Traylor, N. Koga, L. A. Deardurff, P. N. Swepston, and J. A. Ibers, *J. Am. Chem. Soc.*, 1984, **106**, 5132; T. G. Traylor, S. Tsuchiya, D. Campbell, M. Mitchell, D. Stynes, and N. Koga, *J. Am. Chem. Soc.*, 1985, **107**, 604.
92. E. F. Caldin and B. B. Hasinoff, *J. Chem. Soc., Faraday Trans. I*, 1975, **71**, 515; B. B. Hasinoff, *Can. J. Chem.*, 1978, **82**, 2630; see also p. 363 of Ref. 72.
93. J. S. Olson, R. E. McKinnie, M. P. Mims, and D. K. White, *J. Am. Chem. Soc.*, 1983, **105**, 1522.
94. B. A. Springer, S. G. Sligar, J. S. Olson, and G. N. Phillips, *Chem. Rev.*, 1994, **94**, 699.
95. Reviews: D. Walther, *Coord. Chem. Rev.*, 1987, **79**, 135; A. Behr, *Angew. Chem. Int. Edn. Engl.*, 1988, **27**, 661.
96. M. Aresta, E. Quaranta, and I. Tommasi, *New J. Chem.*, 1994, **18**, 133.
97. K. Tanaka, *Adv. Inorg. Chem.*, 1995, **43**, 409.
98. Review: N. Coichev and R. van Eldik, *New J. Chem.*, 1994, **18**, 123.
99. R. H. Crabtree and G. G. Hlatky, *Inorg. Chem.*, 1980, **19**, 571.
100. W. J. Louw, D. J. A. de Waal, T. I. A. Gerber, C. M. Demanet, and R. G. Copperthwaite, *Inorg. Chem.*, 1982, **21**, 1667.
101. L. Vaska, *Accounts Chem. Res.*, 1968, **1**, 335.
102. C. A. Tolman, A. D. English, and L. E. Manzer, *Inorg. Chem.*, 1975, **14**, 2353.
103. C. Elsenbroich and A. Salzer, *Organometallics: A Concise Introduction*, 2nd edn, VCH, Weinheim, 1992, Chapter 15.1.
104. R. Cramer, J. B. Kline, and J. D. Roberts, *J. Am. Chem. Soc.*, 1969, **91**, 2519.
105. M. Cooke, M. Green, and D. C. Wood, *Chem. Commun.*, 1968, 733.
106. D. R. Chrisope, P. Beak, and W. H. Saunders, *J. Am. Chem. Soc.*, 1988, **110**, 230.
107. For example, A. J. Poë and D. H. Vaughan, *Inorg. Chim. Acta*, 1967, **1**, 255; P. Cavoli, R. Graziani, U. Casellato, and P. Uguagtiati, *Inorg. Chim. Acta*, 1986, **111**, L35.
108. W. D. Harmann and H. Taube, *J. Am. Chem. Soc.*, 1988, **110**, 7906.
109. Reviews: G. W. Parshall, *Accounts Chem. Res.*, 1975, **8**, 113; A. E. Shilov and A. A. Shteinman, *Coord. Chem. Rev.*, 1977, **24**, 97; R. H. Crabtree, *Chem. Rev.*, 1985, **85**, 245; J. Halpern, *Inorg. Chim. Acta*, 1985, **100**, 41; I. P. Rothwell, *Polyhedron*, 1985, **4**, 177; B. A. Arndtsen, R. G. Bergman, T. A. Mobley, and T. H. Peterson, *Accounts Chem. Res.*, 1995, **28**, 154.
110. R. van Eldik and C. D. Hubbard, *New J. Chem.*, 1997, **21**, 825.
111. See R. H. Crabtree, *The Organometallic Chemistry of the Transition Metals*, Wiley, New York, 1988, Chapter 6, and J. D. Atwood, *Inorganic and Organometallic Reaction Mechanisms*, VCH, 1997, Chapter 5 for rather more detailed discussions at about the same level.
112. J. J. Eisch, *Pure Appl. Chem.*, 1984, **56**, 35; L. M. Rendina and R. J. Puddephat, *Chem. Rev.*, 1997, **97**, 1735.
113. J. C. Douek and G. Wilkinson, *J. Chem. Soc. (A)*, 1969, 2604.
114. D. Milstein, *Organometallics*, 1982, **1**, 1549.
115. A. A. Bengali, B. A. Arndtsen, P. M. Burger, R. H. Schultz, B. H. Weiller, K. R. Kyle, C. B. Moore, and R. G. Bergman, *Pure Appl. Chem.*, 1995, **67**, 281.
116. M. A. Lilga and J. A. Ibers, *Inorg. Chem.*, 1984, **23**, 3538.
117. J. P. Collman and W. R. Roper, *J. Am. Chem. Soc.*, 1966, **88**, 3504.
118. J. E. Fergusson and R. S. Nyholm, *Chem. Ind.*, 1960, 347; K. A. Glavan, R. Whittle, J. F. Johnson, R. C. Elder, and E. Deutsch, *J. Am. Chem. Soc.*, 1980, **102**, 2103.

119. C. J. Cardin, D. J. Cardin, R. J. Norton, H. E. Parge, and K. W. Muir, *J. Chem. Soc., Dalton Trans.*, 1983, 665.
120. J. H. Espenson, *Progr. Inorg. Chem.*, 1983, **30**, 189.
121. *Aldrichimica Acta*, 1985, **18**, 31.
122. G. M. Williams, K. I. Gell, and J. Schwartz, *J. Am. Chem. Soc.*, 1980, **102**, 3660.
123. J. K. Hoyano, A. D. McMaster, and W. A. G. Graham, *J. Am. Chem. Soc.*, 1983, **105**, 7190; R. A. Periana and R. G. Bergman, *Organometallics*, 1984, **3**, 508; A. H. Janowicz, R. A. Periana, J. M. Buchanan, C. A. Kovac, J. M. Stryker, M. J. Wax, and R. G. Bergman, *Pure Appl. Chem.*, 1984, **56**, 13.
124. D. Milstein, J. C. Calabrese, and I. D. Williams, *J. Am. Chem. Soc.*, 1986, **108**, 6387; P. K. Monaghan and R. J. Puddephat, *Organometallics*, 1984, **3**, 444.
125. R. C. Stevens, R. Bau, D. Milstein, O. Blum, and T. F. Koetzle, *J. Chem. Soc., Dalton Trans.*, 1990, 1429.
126. A. L. Balch, D. A. Costa, J. W. Lee, B. C. Noll, and M. M. Olmstead, *Inorg. Chem.*, 1994, **33**, 2071.
127. M. L. H. Green, *Pure Appl. Chem.*, 1995, **67**, 249.
128. E. A. V. Ebsworth and R. A. Mayo, *J. Chem. Soc., Dalton Trans.*, 1988, 477.
129. J. A. van Doorn, C. Masters, and C. van der Woude, *J. Chem. Soc., Dalton Trans.*, 1978, 1213.
130. R. H. Crabtree, *Accounts Chem. Res.*, 1990, **23**, 95.
131. P. J. Hay, *J. Am. Chem. Soc.*, 1987, **109**, 705.
132. Review: P. J. Jessop and R. H. Morris, *Coord. Chem. Rev.*, 1992, **121**, 155.
133. G. J. Kubas, R. R. Ryan, B. I. Swanson, P. J. Vergamini, and H. J. Wasserman, *J. Am. Chem. Soc.*, 1984, **106**, 451; G. J. Kubas, *Accounts Chem. Res.*, 1988, **21**, 120; G. J. Kubas, *Comments Inorg. Chem.*, 1988, **7**, 17.
134. L. Brammer, J. A. K. Howard, O. Johnson, T. F. Koetzle, J. L. Spencer, and A. M. Stringer, *J. Chem. Soc., Chem. Commun.*, 1991, 241.
135. V. I. Bakhmutov, J. Bertrán, M. A. Esteruelas, A. Lledós, F. Maseras, J. Modrego, L. A. Oro, and E. Sola, *Chem. Eur. J.*, 1997, **2**, 815.
136. R. H. Crabtree and M. Lavin, *J. Chem. Soc., Chem. Commun.*, 1985, 1661.
137. A. L. Sargent and M. B. Hall, *Inorg. Chem.*, 1992, **31**, 317.
138. G. J. Kubas, R. R. Ryan, and D. A. Wrobleski, *J. Am. Chem. Soc.*, 1986, **108**, 1339; G. A. Ozin and J. García-Prieto, *J. Am. Chem. Soc.*, 1986, **108**, 3099.
139. B. R. James and D. Mahajan, *J. Organomet. Chem.*, 1985, **279**, 31.
140. Peng Zhou, A. A. Vitale, J. San Filippo, and W. H. Saunders, *J. Am. Chem. Soc.*, 1985, **107**, 8049.
141. C. E. Johnson and R. Eisenberg, *J. Am. Chem. Soc.*, 1985, **107**, 3148; A. J. Kunin, C. E. Johnson, J. A. Maguire, W. D. Jones, and R. Eisenberg, *J. Am. Chem. Soc.*, 1987, **109**, 2963.
142. G. A. Ozin and J. G. McCaffrey, *J. Phys. Chem.*, 1984, **88**, 645.
143. S. M. Howdle and M. Poliakoff, *J. Chem. Soc., Chem. Commun.*, 1989, 1099.
144. M. Brookhart and M. L. H. Green, *J. Organometal. Chem.*, 1983, **250**, 395.
145. R. H. Crabtree, *Angew. Chem. Int. Edn. Engl.*, 1993, **32**, 789.
146. W. J. Evans, D. K. Drummond, S. G. Bott, and J. L. Atwood, *Organometallics*, 1986, **5**, 2389.
147. M. A. Bennett and D. L. Milner, *J. Chem. Soc., Chem. Commun.*, 1967, 581.
148. R. H. Reamey and G. M. Whitesides, *J. Am. Chem. Soc.*, 1984, **106**, 81.
149. Pongchan Chandayot and Yueh-Tai Fanchiang, *Inorg. Chem.*, 1985, **24**, 3535, and references therein.

150. S. Al-Jibori, C. Crocker, and B. L. Shaw, *J. Chem. Soc., Dalton Trans.*, 1981, 319; L. Drougge and L. I. Elding, *Inorg. Chem.*, 1987, **26**, 1073.
151. J. A. Ventes, J. G. Leipoldt, and R. van Eldik, *Inorg. Chem.*, 1991, **30**, 2207.
152. R. van Eldik, in *Inorganic High Pressure Chemistry: Kinetics and Mechanisms*, ed. R. van Eldik, Elsevier, Amsterdam, 1986, Chapter 3; see p. 194.
153. M.-A. Haga, K. Kawakami, and T. Tanaka, *Inorg. Chem.*, 1976, **15**, 1946.
154. E. P. Wasserman, C. B. Moore, and R. G. Bergman, *Science (Washington)*, 1992, **255**, 315.
155. M. Basato, B. Corain, G. Favero, and P. Rosano, *J. Chem. Soc., Dalton Trans.*, 1984, 2513.
156. A. Sen and J. Halpern, *Inorg. Chem.*, 1980, **19**, 1073.
157. T. V. Ashworth, J. E. Singleton, D. J. A. de Waal, W. J. Louw, E. Singleton, and E. van der Stok, *J. Chem. Soc., Dalton Trans.*, 1978, 340; R. G. Pearson and C. T. Kresge, *Inorg. Chem.*, 1981, **20**, 1878; and see Chapter 6.4 of Ref. 4.
158. D. J. A. de Waal, T. I. A. Gerber, and W. J. Louw, *Inorg. Chem.*, 1982, **21**, 1259.
159. G. Favero and A. Morvillo, *J. Organomet. Chem.*, 1984, **260**, 363.
160. P. B. Chock and J. Halpern, *J. Am. Chem. Soc.*, 1966, **88**, 3511.
161. J. K. Stille and K. S. Y. Lau, *Accounts Chem. Res.*, 1977, **10**, 434.
162. M. Crespo and R. J. Puddephat, *Organometallics*, 1987, **6**, 2548.
163. A. J. Canty, *Accounts Chem. Res.*, 1992, **25**, 83.
164. E. Uhlig and D. Walther, *Coord. Chem. Rev.*, 1980, **33**, 3.
165. P. J. Stang, M. D. Schiavelli, K. H. Chenault, and J. L. Breidigam, *Organometallics*, 1984, **3**, 1133.
166. R. N. Perutz, *Chem. Soc. Rev.*, 1993, **22**, 361.
167. T. L. Hall, M. F. Lappert, and P. W. Lednor, *J. Chem. Soc., Dalton Trans.*, 1980, 1448.
168. S. Fukuzumi. N. Nishizawa, and T. Tanaka, *Bull. Chem. Soc. Jpn*, 1983, **56**, 709.
169. R. S. Paonessa, N. C. Thomas, and J. Halpern, *J. Am. Chem. Soc.*, 1985, **107**, 4333.
170. T. T. Tsou and J. K. Kochi, *J. Am. Chem. Soc.*, 1979, **101**, 6319.
171. G. M. Williams and J. Schwartz, *J. Am. Chem. Soc.*, 1982, **104**, 1122.
172. J. A. Connor and P. I. Riley, *J. Chem. Soc., Dalton Trans.*, 1979, 1318.
173. R. T. Mehdi and J. D. Miller, *J. Chem. Soc., Dalton Trans.*, 1984, 1065.
174. J. A. Labinger and J. A. Osborn, *Inorg. Chem.*, 1980, **19**, 3230; J. A. Labinger, J. A. Osborn, and N. J. Coville, *Inorg. Chem.*, 1980, **19**, 3236.
175. R. G. Pearson and P. E. Figdore, *J. Am. Chem. Soc.*, 1980, **102**, 1541.
176. J. Burgess, M. J. Hacker, and R. D. W. Kemmitt, *J. Organomet. Chem.*, 1974, **72**, 121.
177. R. Ugo, A. Pasini, A. Fusi, and S. Cenini, *J. Am. Chem. Soc.*, 1972, **94**, 7364.
178. J. A. M. van Beek, G. van Koten, W. J. J. Smeets, and A. L. Speck, *J. Am. Chem. Soc.*, 1986, **108**, 5010.
179. S. S. Basson, J. G. Leipoldt, and J. T. Nel, *Inorg. Chim. Acta*, 1984, **84**, 167.
180. C. E. Hickey and P. M. Maitlis, *J. Chem. Soc., Chem. Commun.*, 1984, 1609.
181. H. Stieger and H. Kelm, *J. Phys. Chem.*, 1973, **77**, 290.
182. M. Walper and H. Kelm, *Z. Anorg. Allg. Chem.*, 1981, **483**, 225.
183. J. F. Harrod, G. Hamer, and W. Yorke, *J. Am. Chem. Soc.*, 1979, **101**, 3987.
184. W. J. Louw, T. I. A. Gerber, and D. J. A. de Waal, *J. Chem. Soc., Chem. Commun.*, 1980, 760.

185. K. S. Y. Lau, Y. Becker, F. Huang, N. Baenziger, and J. K. Stille, *J. Am. Chem. Soc.*, 1977, **99**, 5664; R. G. Pearson and P. E. Figdore, *J. Am. Chem. Soc.*, 1980, **102**, 1541.
186. R. G. Pearson and W. R. Muir, *J. Am. Chem. Soc.*, 1970, **92**, 5519.
187. R. H. Hill and R. J. Puddephat, *J. Am. Chem. Soc.*, 1985, **107**, 1218.
188. S. Al-Jibori, C. Crocker, W. S. McDonald, and B. L. Shaw, *J. Chem. Soc., Dalton Trans.*, 1981, 2589.
189. R. Francke, G.-V. Röschenthaler, R. di Giacomo, and D. Dakternieks, *Phosphorus Sulphur*, 1984, **20**, 107.
190. J. Halpern, *Accounts Chem. Res.*, 1982, **15**, 332.
191. S. Komiya, M. Ishikawa, and S. Ozaki, *Organometallics*, 1988, **7**, 2238.
192. M. Aizenberg and D. Milstein, *Angew. Chem. Int. Edn. Engl.*, 1994, **33**, 317.
193. E. M. Carnahan, J. D. Protasiewicz, and S. J. Lippard, *Accounts Chem. Res.*, 1993, **26**, 90.
194. M. P. Brown, R. J. Puddephat, and C. E. E. Upton, *J. Chem. Soc., Dalton Trans.*, 1974, 2457; K. I. Goldberg, Ji Yang Yan, and E. L. Winter, *J. Am. Chem. Soc.*, 1994, **116**, 1573.
195. F. Ozawa, T. Hikida, and T. Hayashi, *J. Am. Chem. Soc.*, 1994, **116**, 2844.
196. D. Milstein, *J. Am. Chem. Soc.*, 1982, **104**, 5226.
197. S. Komiya, T. A. Albright, R. Hoffmann, and J. K. Kochi, *J. Am. Chem. Soc.*, 1976, **98**, 7255; A. Gillie and J. K. Stille, *J. Am. Chem. Soc.*, 1980, **102**, 4933.
198. J. Halpern, *Inorg. Chim. Acta*, 1982, **62**, 31.
199. J. R. Norton, *Accounts Chem. Res.*, 1979, **12**, 139.
200. Reviews: A. Wojcicki, *Accounts Chem. Res.*, 1971, **4**, 344; A. Wojcicki, *Adv. Organomet. Chem.*, 1974, **12**, 31; G. K. Anderson and R. J. Cross, *Accounts Chem. Res.*, 1984, **17**, 67.
201. Review: B. C. Soderberg, in *Comprehensive Organometallic Chemistry II*, eds E. W. Abel, F. G. A. Stone, and G. Wilkinson, Pergamon, Oxford, 1995, Vol. 12, ed. L. S. Hegedus, Chapter 3.5, deals with insertion of carbon monoxide, alkenes, and alkynes into transition metal–alkyl bonds.
202. T. H. Coffield, J. Kozikowski, and R. D. Closson, *J. Org. Chem.*, 1957, **22**, 598.
203. See pp. 161–162 of Ref. 4.
204. Review: K. J. Cavell, *Coord. Chem. Rev.*, 1996, **155**, 209.
205. R. A. Henderson, *The Mechanisms of Reactions at Transition Metal Sites*, Oxford University Press, Oxford, 1993, pp. 80–81.
206. G. Cardaci, G. Reichenbach, G. Bellachioma, B. Wassink, and M. C. Baird, *Organometallics*, 1988, **7**, 2475.
207. I. S. Butler, F. Basolo, and R. G. Pearson, *Inorg. Chem.*, 1967, **6**, 2074.
208. M. Bassetti, G. J. Sunley, and P. Maitlis, *J. Chem. Soc., Chem. Commun.*, 1988, 1012.
209. J. N. Cawse, R. A. Fiato, and R. L. Pruett, *J. Organometal. Chem.*, 1979, **172**, 405; F. U. Axe and D. S. Marynick, *J. Am. Chem. Soc.*, 1988, **110**, 3728.
210. J. D. Cotton and R. D. Markwell, *Inorg. Chim. Acta*, 1982, **63**, 13; J. D. Cotton and R. D. Markwell, *Organometallics*, 1985, **4**, 937.
211. J. D. Cotton and R. D. Markwell, *J. Organometal. Chem.*, 1990, **338**, 123.
212. M. J. Wax and R. G. Bergman, *J. Am. Chem. Soc.*, 1981, **103**, 7028.
213. T. L. Bent and J. D. Cotton, *Organometallics*, 1991, **10**, 3156.
214. M. Bassetti, G. J. Sunley, F. P. Fanizzi, and P. M. Maitlis, *J. Chem. Soc., Dalton Trans.*, 1990, 1799; D. Monti, M. Bassetti, G. J. Sunley, and P. M. Maitlis, *Organometallics*, 1991, **10**, 4015.

215. T. Y. Meyer, L. R. Garner, N. C. Baenziger, and L. Messerle, *Inorg. Chem.*, 1990, **29**, 4045.
216. K. G. Moloy and T. J. Marks, *J. Am. Chem. Soc.*, 1984, **106**, 7051.
217. Review: G. K. Anderson and R. J. Cross, *Accounts Chem. Res.*, 1984, **17**, 67.
218. H. Brunner and H. Vogt, *Angew. Chem. Internat. Edn. Engl.*, 1981, **20**, 405; C. Gueutin, D. Lexa, M. Momenteau, and J.-M. Savéant, *J. Am. Chem. Soc.*, 1990, **112**, 1874.
219. See p. 168 of Ref. 4.
220. Y. Yamamoto, H. Yamazaki, and N. Hagihara, *J. Organomet. Chem.*, 1969, **18**, 189.
221. G. Bellachioma, G. Cardaci, and P. Zanazzi, *Inorg. Chem.*, 1987, **26**, 84.
222. A. Dormond, A. A. Elbouadili, and C. Moïse, *J. Chem. Soc., Chem. Commun.*, 1984, 749; P. Zanella, G. Paolucci, G. Rossetto, F. Benetollo, A. Polo, R. D. Fischer, and G. Bombieri, *J. Chem. Soc., Chem. Commun.*, 1985, 96.
223. D. C. Yang, V. Dragisich, W. D. Wulff, and J. C. Huffman, *J. Am. Chem. Soc.*, 1988, **110**, 307.
224. K. J. Schneider, A. Neubrand, R. van Eldik, and H. Fischer, *Organometallics*, 1992, **11**, 267.
225. M. O. Albers, D. J. A. De Waal, D. C. Liles, D. J. Robinson, and E. Singleton, *J. Organometal. Chem.*, 1987, **326**, C29.
226. A. Vigalok, H.-B. Kraatz, L. Konstantinovsky, and D. Milstein, *Chem. Eur. J.*, 1997, **3**, 253.
227. Review: W. W. Leong and R. C. Larock, in *Comprehensive Organometallic Chemistry II*, eds E. W. Abel, F. G. A. Stone, and G. Wilkinson, Pergamon, Oxford, 1995, Vol. 12, ed. L. S. Hegedus, Chapter 3.3.
228. H.-F. Klein, R. Hammer, J. Gross, and U. Schubert, *Angew. Chem., Int. Edn. Engl.*, 1980, **19**, 809.
229. K. Hiraki, N. Ochi, Y. Sasada, H. Hayashida, Y. Fuchita, and S. Yamanaka, *J. Chem. Soc., Dalton Trans.*, 1985, 873; L. Versluis, T. Ziegler, and Liangyou Fan, *Inorg. Chem.*, 1990, **29**, 4530; and references therein.
230. D. C. Roe, *J. Am. Chem. Soc.*, 1983, **105**, 7770.
231. J. Halpern and T. Okamoto, *Inorg. Chim. Acta*, 1984, **89**, L53.
232. B. J. Burger, B. D. Santarsiero, M. S. Trimmer, and J. E. Bercaw, *J. Am. Chem. Soc.*, 1988, **110**, 3134.
233. T. A. Shackleton and M. C. Baird, *Organometallics*, 1989, **8**, 2225.
234. E. G. Samsel and J. R. Norton, *J. Am. Chem. Soc.*, 1984, **106**, 5505; A. C. Albéniz, P. Espinet, Y. Jeannin, M. Philoche-Levisalles, and B. E. Mann, *J. Am. Chem. Soc.*, 1990, **112**, 6594.
235. Lin Wang and T. C. Flood, *J. Am. Chem. Soc.*, 1992, **114**, 3169.
236. W. H. Baddley and M. S. Fraser, *J. Am. Chem. Soc.*, 1969, **91**, 3661; S. R. Su, J. A. Hanna, and A. Wojcicki, *J. Organomet. Chem.*, 1970, **21**, P21.
237. G. E. Herberich and W. Barlage, *J. Organomet. Chem.*, 1987, **331**, 63; J. D. Vessey and R. J. Mawby, *J. Chem. Soc., Dalton Trans.*, 1993, 51.
238. J. López, A. Romero, A. Santos, A. Vegas, A. M. Echavarren, and P. Noheda, *J. Organomet. Chem.*, 1989, **373**, 249.
239. T. N. Mitchell, H. Killing, R. Dicke, and R. Wickenkamp, *J. Chem. Soc., Chem. Commun.*, 1985, 354.
240. B. J. Burger, M. E. Thompson, W. E. Cotter, and J. E. Bercaw, *J. Am. Chem. Soc.*, 1990, **112**, 1566.

241. D. J. Darensbourg and R. A. Kudaroski, *Adv. Organomet. Chem.*, 1987, **22**, 129; A. Behr, *Angew. Chem. Int. Edn. Engl.*, 1988, **27**, 661.
242. D. J. Darensbourg, K. M. Sanchez, and A. L. Rheingold, *J. Am. Chem. Soc.*, 1987, **109**, 290.
243. R. F. Johnston and J. C. Cooper, *Organometallics*, 1987, **6**, 2448.
244. D. J. Darensbourg and G. Grötsch, *J. Am. Chem. Soc.*, 1985, **107**, 7473.
245. D. J. Darensbourg, R. K. Hanckel, C. G. Bauch, M. Pala, D. Simmons, and J. N. White, *J. Am. Chem. Soc.*, 1985, **107**, 7463.
246. J. F. Hartwig, R. G. Bergman, and R. A. Andersen, *J. Am. Chem. Soc.*, 1991, **113**, 6499.
247. D. J. Darensbourg, G. Grötsch, P. Wiegreffe, and A. L. Rheingold, *Inorg. Chem.*, 1987, **26**, 3827.
248. D. J. Darensbourg, M. Y. Darensbourg, Lai Yoong Goh, M. Ludwig, and P. Wiegreffe, *J. Am. Chem. Soc.*, 1987, **109**, 7539.
249. T. Ito and T. Matsubara, *J. Chem. Soc., Dalton Trans.*, 1988, 2241.
250. J.-C. Berthet and M. Ephritikhine, *New J. Chem.*, 1992, **16**, 767.
251. D. J. Darensbourg, K. M. Sanchez, J. Reibenspies, and A. L. Rheingold, *J. Am. Chem. Soc.*, 1989, **111**, 7094.
252. T. R. Gaffney and J. A. Ibers, *Inorg. Chem.*, 1982, **21**, 2857.
253. Yong-Joo Kim, K. Osakada, K. Sugita, T. Yamamoto, and A. Yamamoto, *Organometallics*, 1988, **7**, 2182, and references therein.
254. A. Shaver, B. Soo Lum, P. Bird, and K. Arnold, *Inorg. Chem.*, 1989, **28**, 1900.
255. See, e.g., A. Wojcicki, *Accounts Chem. Res.*, 1971, **4**, 344 for early work in this area.
256. M. F. Joseph and M. C. Baird, *Inorg. Chim. Acta*, 1985, **96**, 229.
257. E. Wenschuh and R. Zimmering, *Z. Chem.*, 1988, **28**, 190.
258. D. J. Darensbourg, C. G. Bauch, J. H. Reibenspies, and A. L. Rheingold, *Inorg. Chem.*, 1988, **27**, 4203.
259. J. D. Cotton and G. T. Crisp, *J. Organomet. Chem.*, 1980, **186**, 137.
260. B. D. Gupta, M. Roy, M. Oberoi, and V. Dixit, *J. Organomet. Chem.*, 1992, **430**, 197.
261. L. K. Bell and D. M. P. Mingos, *J. Chem. Soc., Dalton Trans.*, 1982, 673.
262. C. J. Jones, J. A. McCleverty, and A. S. Rothin, *J. Chem. Soc., Dalton Trans.*, 1985, 405.
263. A. Sauer, H. Cohen, and D. Meyerstein, *Inorg. Chem.*, 1989, **28**, 2511.
264. A. Sauer, H. Cohen, and D. Meyerstein, *Inorg. Chem.*, 1988, **27**, 4578.
265. T. V. Lubben and P. T. Wolczanski, *J. Am. Chem. Soc.*, 1987, **109**, 424.
266. V. L. Frost and R. A. Henderson, *J. Chem. Soc., Dalton Trans.*, 1985, 2059; R. A. Henderson, *J. Chem. Soc., Dalton Trans.*, 1985, 2067.
267. E. Hey, M. F. Lappert, J. L. Atwood, and S. G. Bott, *J. Chem. Soc., Chem. Commun.*, 1987, 597.
268. A. N. Nesmeyanov, K. N. Anisimov, N. E. Kolobova, and F. S. Demisov, *Izvest. Akad. Nauk SSSR, Ser. Chim.*, 1968, 1419.
269. B. Alvarez, D. Miguel, J. A. Pérez-Martínez, V. Riera, and S. García-Granda, *J. Organomet. Chem.*, 1992, **427**, C33.
270. A. J. Shortland and G. Wilkinson, *J. Chem. Soc., Dalton Trans.*, 1973, 872; A. L. Galyer and G. Wilkinson, *J. Chem. Soc., Dalton Trans.*, 1976, 2235.
271. S. M. B. Costa, A. R. Dias, and F. J. S. Pina, *J. Chem. Soc., Dalton Trans.*, 1981, 314; M. R. Churchill, H. J. Wasserman, H. W. Turner, and R. R. Schrock, *J. Am. Chem. Soc.*, 1982, **104**, 1710.

272. J. Chatt and B. L. Shaw, *J. Chem. Soc.*, 1962, 5075.
273. Z. Dawoodi, M. L. H. Green, V. S. B. Mtetwa, and K. Prout, *J. Chem. Soc., Chem. Commun.*, 1982, 802.
274. R. B. Cracknell, A. G. Orpen, and J. L. Spencer, *J. Chem. Soc., Chem. Commun.*, 1984, 326.
275. G. M. Whitesides, J. F. Gaasch, and E. R. Stedronsky, *J. Am. Chem. Soc.*, 1972, **94**, 5258.
276. M. Brookhart, D. M. Lincoln, M. A. Bennett, and S. Pelling, *J. Am. Chem. Soc.*, 1990, **112**, 2691.
277. Section 10.7.10 contains a selection of books and reviews covering this subject.
278. B. Bogdanović, B. Spliethoff, and G. Wilke, *Angew. Chem. Int. Edn. Engl.*, 1980, **19**, 622.
279. W. S. Knowles, *Accounts Chem. Res.*, 1983, **16**, 106
280. For example, B. Bosnich, *Chem. Brit.*, 1984, 808; H. Kawano, Y. Ishii, T. Kodama, M. Saburi, and Y. Uchida, *Chem. Lett.*, 1987, 1311; J. M. Brown, *Chem. Brit.*, 1989, 25, 276; I. Ojima (ed.), *Catalytic Asymmetric Synthesis*, VCH, Weinheim, 1993.
281. M. J. Burk and R. L. Harlow, *Angew. Chem. Int. Edn. Engl.* 1990, **29**, 1462.
282. F. Robin, F. Mercier, L. Ricard, F. Mathey, and M. Spagnol, *Chem. Eur. J.*, 1997, **3**, 1365.
283. P. Powell, *Principles of Organometallic Chemistry*, 2nd edn, Chapman & Hall, London, 1988; J. Falbe and H. Bahrmann, *J. Chem. Ed.*, 1984, **61**, 961.
284. R. Whyman, in *Selected Developments in Catalysis*, ed. J. R. Jennings, SCI/Blackwell, Oxford, 1985, p. 128; F. J. Waller, *J. Mol. Catal.*, 1985, **31**, 123; G. W. Parshall and R. E. Potscher, *J. Chem. Ed.*, 1986, **63**, 189; A. Heaton (ed.), *An Introduction to Industrial Chemistry*, Blackie, Glasgow, 1996, Chapter 11.
285. The March 1986 issue of *J. Chem. Ed.* contains a section entitled 'Symposium on Industrial Applications of Organometallic Chemistry and Catalysis' (pp. 188–223).
286. S.-I. Inoue, H. Takaya, K. Tani, S. Otsuka, T. Sato, and R. Noyori, *J. Am. Chem. Soc.*, 1990, **112**, 4897.
287. E. C. Alyea and D. W. Meek (eds), *Catalytic Aspects Of Metal–Phosphine Complexes*, Adv. Chem. Ser., 1982, **196**; L. H. Pignolet (ed.), *Homogeneous Catalysis with Metal Phosphine Complexes*, Plenum, New York, 1983; F. H. Jardine, *Progr. Inorg. Chem.*, 1981, **28**, 63; F. H. Jardine, *Progr. Inorg. Chem.*, 1984, **31**, 265; M. C. Simpson and D. J. Cole-Hamilton, *Coord. Chem. Rev.*, 1996, **155**, 163.
288. L. S. Hegedus, *Coord. Chem. Rev.*, 1997, **159**, 129.
289. K. N. Harrison, P. A. T. Hoye, A. G. Orpen, P. G. Pringle, and M. B. Smith, *J. Chem. Soc., Chem. Commun.*, 1989, 1096.
290. S. Kainz, D. Koch, W. Baumann, and W. Leitner, *Angew. Chem. Int. Edn. Engl.*, 1997, **36**, 1628.
291. *Chem. Brit.*, 1997, **33**, 17.
292. A. Togni and L. M. Venanzi, *Angew. Chem. Int. Edn. Engl.*, 1994, **33**, 497.
293. D. J. A. de Waal, T. I. A. Gerber, and W. J. Louw, *J. Chem. Soc., Chem. Commun.*, 1982, 100.
294. J. D. Atwood, *Inorganic and Organometallic Reaction Mechanisms*, 2nd edn, VCH, Weinheim, 1997.

295. J. M. Brown, P. L. Evans, and A. R. Lucy, *J. Chem. Soc., Perkin Trans. II*, 1987, 1589.
296. G. Pimblett, C. D. Garner, and W. Clegg, *J. Chem. Soc., Dalton Trans.*, 1985, 1977.
297. B. R. Stults, R. M. Friedman, K. Koenig, W. Knowles, R. B. Greegor, and F. W. Lytle, *J. Am. Chem. Soc.*, 1981, **103**, 3235.
298. H. Bertagnolli and T. S. Ertel, *Angew. Chem. Int. Edn. Engl.*, 1994, **33**, 15.
299. P. G. Gassman, D. W. Macomber, and S. M. Willging, *J. Am. Chem. Soc.*, 1985, **107**, 2380.
300. H. C. Clark, G. Ferguson, M. J. Hampden-Smith, B. Kaitner, and H. Ruegger, *Polyhedron*, 1988, **7**, 1349.
301. R. J. Cvetanović, F. J. Duncan, W. E. Falconer, and R. S. Irwin, *J. Am. Chem. Soc.*, 1965, **87**, 1827; F. R. Hartley, *Chem. Rev.*, 1973, **73**, 163.
302. G. Alibrandi and B. E. Mann, *J. Chem. Soc., Dalton Trans.*, 1994, 951.
303. N. W. Alcock, J. M. Brown, A. E. Derome, and A. R. Lucy, *J. Chem. Soc., Chem. Commun.*, 1985, 575.
304. F. H. Jardine, *Progr. Inorg. Chem.*, 1981, **28**, 63.
305. For example, A. Dedieu, *Inorg. Chem.*, 1980, **19**, 375; A. Dedieu, *Inorg. Chem.*, 1981, **20**, 2803; N. Koga, C. Daniel, J. Han, X. Y. Fu, and K. Morokuma, *J. Am. Chem. Soc.*, 1987, **109**, 3455; N. Koga, C. Daniel, J. Han, X. Y. Fu, and K. Morokuma, *J. Am. Chem. Soc.*, 1988, **110**, 3773.
306. H. C. Clark and V. K. Jain, *Coord. Chem. Rev.*, 1984, **55**, 151.
307. G. Wilke, *Angew. Chem. Int. Edn. Engl.*, 1988, **27**, 186.
308. R. Grigg, T. R. B. Mitchell, and A. Ramasubbu, *J. Chem. Soc., Chem. Commun.*, 1980, 27.
309. H. M. Colquhoun, J. Holton, D. J. Thompson, and M. V. Twigg, *New Pathways for Organic Synthesis*, Plenum, New York, 1984, Chapter 9.
310. J.-L. Malleron and A. Juin, *Database of Palladium Chemistry: Reactions, Catalytic Cycles and Chemical Parameters*, Version 1.0, Academic Press, London, 1997.
311. B. C. Gates, L. Guczi, and H. Knözinger (eds), *Metal Clusters in Catalysis*, Elsevier, Amsterdam, 1986. This book contains the useful feature of a detailed index of cluster catalysts.
312. G. Longoni and P. Chini, *J. Am. Chem. Soc.*, 1976, **98**, 7225.
313. Review: L. H. Pignolet, M. A. Aubart, K. L. Craighead, R. A. T. Gould, D. A. Krogstad, and J. S. Wiley, *Coord. Chem. Rev.*, 1995, **143**, 219.
314. Review: C. L. Hill and C. M. Prosser-McCartha, *Coord. Chem. Rev.*, 1995, **143**, 407.
315. M. Misono, T. Okuhara, T. Ichiki, T. Arai, and Y. Kanda, *J. Am. Chem. Soc.*, 1987, **109**, 5535.
316. S. F. Davidson, B. E. Mann, and P. M. Maitlis, *J. Chem. Soc., Dalton Trans.*, 1984, 1223.
317. K. Nomiya, H. Saijoh, and M. Miwa, *Bull. Chem. Soc. Jpn*, 1980, **53**, 3719.
318. G. Süss-Fink, *Angew. Chem. Int. Edn.*, 1994, **33**, 67.
319. M. W. Göbel, *Angew. Chem. Int. Edn.*, 1994, **33**, 1141.
320. L. N. Lewis, *Chem. Rev.*, 1993, **93**, 2693.
321. C. U. Pittman, G. M. Wilemon, W. D. Wilson, and R. C. Ryan, *Angew. Chem. Int. Edn.*, 1980, **19**, 478.

322. A. K. Smith, F. Hugues, A. Theolier, J. M. Basset, R. Ugo, G. M. Zanderighi, J. L. Bilhou, V. Bilhou-Bougnol, and W. F. Graydon, *Inorg. Chem.*, 1979, **18**, 3104.
323. E. Sappa, A. M. Manotti Lanfredini, and A. Tiripicchio, *Inorg. Chim. Acta*, 1980, **42**, 255.
324. G. Henrici-Olivé and S. Olivé, *The Chemistry of the Catalysed Hydrogenation of Carbon Monoxide*, Springer, Berlin, 1984; W. A. Herrmann and B. Cornils, *Angew. Chem. Int. Edn. Engl.*, 1997, **36**, 1049.
325. E. L. Muetterties, E. Band, A. Kokorin, W. R. Pretzer, and M. G. Thomas, *Inorg. Chem.*, 1980, **19**, 1552; E. L. Muetterties and M. J. Krause, *Angew. Chem. Int. Edn. Engl.*, 1983, **22**, 135.
326. E. L. Muetterties, *Bull. Soc. Chim. Belges*, 1975, **84**, 959.
327. E. L. Muetterties, *Inorg. Chim. Acta*, 1981, **50**, 1.
328. E. L. Muetterties, T. N. Rhodin, E. Band, C. F. Bruker, and W. R. Pretzer, *Chem. Rev.*, 1979, **79**, 91.
329. K. Shimazu and H. Kita, *J. Chem. Soc., Faraday Trans. I*, 1985, **81**, 175.
330. T. J. Pinnavaia, R. Raythatha, J. G.-S. Lee, L. J. Halloran, and J. F. Hoffman, *J. Am. Chem. Soc.*, 1979, **101**, 6891.
331. N. Yoneda, Y. Nakagawa, and T. Mimami, *Catal. Today*, 1997, **36**, 357.
332. H. Bönnemann and G. A. Braun, *Chem. Eur. J.*, 1997, **3**, 1200.
333. E. Cesarotti, R. Ugo, and L. Kaplan, *Coord. Chem. Rev.*, 1982, **43**, 275.
334. P. Taylor and M. Orchin, *J. Am. Chem. Soc.*, 1971, **93**, 6504.
335. G. E. Coates, M. L. H. Green, P. Powell, and K. Wade, *Principles of Organometallic Chemistry*, Methuen, London, 1968, p. 235.
336. F. A. Cotton, G. Wilkinson, and P. Gaus, *Basic Inorganic Chemistry*, 2nd edn, Wiley, New York, 1987, pp. 646–647.
337. G. Erker, K. Berg, K. Angermund, and C. Krüger, *Organometallics*, 1987, **6**, 2620.
338. K. Hiraki, N. Ochi, H. Takaya, Y. Fuchita, Y. Shimokawa, and H. Hayashida, *J. Chem. Soc., Dalton Trans.*, 1990, 1679.
339. C. Kutal, C. K. Kelley, and G. Ferraudi, *Inorg. Chem.*, 1987, **26**, 3258.
340. Review: R. R. Schrock, *Polyhedron*, 1995, **14**, 3177.
341. A. F. Dyke, S. A. R. Knox, P. J. Naish, and G. E. Taylor, *J. Chem. Soc., Chem. Commun.*, 1980, 803.
342. R. F. Heck and J. P. Nolly, *J. Org. Chem.*, 1972, **37**, 2320; R. F. Heck, *Palladium Reagents in Organic Synthesis*, Academic Press, London, 1985.
343. J. P. Collman and L. S. Hegedus, *Principles and Applications of Organotransition Metal Chemistry*, University Science Books, Mill Valley, California, 1980, pp. 580–584.
344. W. A. Herrmann, C. Brossmer, K. Öfele, C.-P. Reisinger, T. Priermeier, M. Beller, and H. Fischer, *Angew. Chem. Int. Edn. Engl.*, 1995, **34**, 1844; M. Beller, H. Fischer, W. A. Herrmann, K. Öfele, and C. Brossmer, *Angew. Chem. Int. Edn. Engl.*, 1995, **34**, 1848.
345. W. A. Herrmann, C. Brossmer, C.-P. Reisinger, T. H. Riermeier, K. Öfele, and M. Beller, *Chem. Eur. J.*, 1997, **3**, 1357.
346. T. Masuda and T. Higashimura, *Accounts Chem. Res.*, 1984, **17**, 51.
347. G. E. Gadd, M. Poliakoff, and J. J. Turner, *Organometallics*, 1987, **6**, 391.
348. W. Gausing and G. Wilke, *Angew. Chem. Int. Edn. Engl.*, 1981, **20**, 186.

349. H. tom Dieck, L. Stamp, R. Diercks, and C. Müller, *Nouv. J. Chem.*, 1985, **9**, 289; H. tom Dieck, R. Diercks, L. Stamp, H. Bruder, and T. Schuld, *Chem. Ber.*, 1987, **120**, 1943.
350. K.-R. Pörschke, *Angew. Chem. Int. Edn. Engl.*, 1987, **26**, 1288.
351. J. D. Fellmann, R. R. Schrock, and G. A. Rupprecht, *J. Am. Chem. Soc.*, 1981, **103**, 5752.
352. J. H. B. Chemier, J. A. Howard, and B. Mile, *J. Am. Chem. Soc.*, 1987, **109**, 4109.
353. Lin Wang and T. C. Flood, *J. Am. Chem. Soc.*, 1992, **114**, 3169.
354. P. L. Watson, *J. Am. Chem. Soc.*, 1982, **104**, 337.
355. N. Balcioğlu, I. Uraz (Ünalan), C. Bozkurt, and F. Sevin, *Polyhedron*, 1997, **16**, 327.
356. H. Werner, R. Wiedemann, P. Steinert, and J. Wolf, *Chem. Eur. J.*, 1997, **3**, 127.
357. P. O. Nubel and T. L. Brown, *J. Am. Chem. Soc.*, 1984, **106**, 3474.
358. M. Green, P. A. Kale, and R. J. Mercer, *J. Chem. Soc., Chem. Commun.*, 1987, 375.
359. H. Sakurai, K. Hirama, Y. Nakadaira, and C. Kabuto, *J. Am. Chem. Soc.*, 1987, **109**, 6880.
360. K. Ziegler, E. Holzcamp, H. Breil, and H. Martin, *Angew. Chem.*, 1955, **67**, 426, 541; G. Natta, *Macromol. Chem.*, 1955, **16**, 213; G. Natta, *Angew. Chem.*, 1956, **68**, 393; W. A. Herrmann, *Angew. Chem. Int. Edn. Engl.*, 1988, **27**, 1297.
361. The Nobel Prize lectures give the inventor's stories : K. Ziegler, *Angew. Chem.*, 1964, **76**, 545; G. Natta, *Angew. Chem.*, 1964, **76**, 553.
362. E. P. Bierwagen, J. E. Bercaw, and W. A. Goddard, *J. Am. Chem. Soc.*, 1994, **116**, 1481.
363. A. W. Parkins and R. C. Poller, *An Introduction to Organometallic Chemistry*, Macmillan, Basingstoke, 1986, p. 229, and references therein.
364. N. S. Barta, B. A. Kirk, and J. R. Stille, *J. Organomet. Chem.*, 1995, **487**, 47.
365. J. Boor, *Ziegler–Natta Catalysts and Polymerizations*, Academic Press, New York, 1979.
366. K. J. Ivin, J. J. Rooney, C. D. Stewart, M. L. H. Green, and R. Mahtab, *J. Chem. Soc., Chem. Commun.*, 1978, 604.
367. H. W. Turner, R. R. Schrock, J. D. Fellmann, and S. J. Holmes, *J. Am. Chem. Soc.*, 1983, **105**, 4942.
368. Review: H. H. Brintzinger, D. Fischer, R. Mülhaupt, B. Rieger, and R. M. Waymouth, *Angew. Chem. Int. Edn. Engl.*, 1995, **34**, 1143.
369. R. Cramer, *Accounts Chem. Res.*, 1968, **1**, 186; C. A. Tolman, *Chem. Rev.*, 1977, **77**, 313.
370. J. W. Reppe, O. Schlichting, K. Klager, and T. Toepel, *Liebigs Ann. Chem.*, 1948, **560**, 1.
371. G. Wilke, *Pure Appl. Chem.*, 1978, **50**, 677.
372. G. N. Schrauzer and S. Eichler, *Chem. Ber.*, 1962, **95**, 550.
373. R. E. Colborn and K. P. C. Vollhardt, *J. Am. Chem. Soc.*, 1981, **103**, 6259.
374. R. Diercks, L. Stamp, J. Kopf, and H. tom Dieck, *Angew. Chem. Int. Edn. Engl.*, 1984, **23**, 893.
375. L. P. Yur'eva, *Russ. Chem. Rev.*, 1974, **43**, 48.
376. G. Wilke, *J. Organomet. Chem.*, 1980, **200**, 349.
377. M. E. E. Meijer-Veldman and H. J. de L. Meijer, *J. Organomet. Chem.*, 1984, **260**, 199.

378. K. P. C. Vollhardt, *Angew. Chem. Int. Edn. Engl.*, 1984, **23**, 539; K. P. C. Vollhardt, *Pure Appl. Chem.*, 1985, **57**, 1819; R. Boese, A. P. van Sickle, and K. P. C. Vollhardt, *Synthesis*, 1995, 1374; P. Cruciani, C. Aubert, and M. Malacria, *J. Org. Chem.*, 1995, **60**, 2664.
379. D. R. McAlister, J. E. Bercaw, and R. G. Bergman, *J. Am. Chem. Soc.*, 1977, **99**, 1666; K. P. C. Vollhardt, *Accounts Chem. Res.*, 1977, **10**, 1.
380. A. N. Nesmayanov, A. I. Gusev, A. A. Pasynskii, K. N. Anisimov, N. E. Kolobova, and Yu. T. Struchkov, *J. Chem. Soc., Chem. Commun.*, 1969, 739.
381. M. A. Bruck, A. S. Copenhaver, and D. E. Wigley, *J. Am. Chem. Soc.*, 1987, **109**, 6525.
382. P. M. Maitlis, *Accounts Chem. Res.*, 1976, **9**, 93; P. M. Maitlis, *J. Organomet. Chem.*, 1980, **200**, 161.
383. G. Wilke, *Angew. Chem. Int. Edn. Engl.*, 1988, **27**, 185.
384. A. L. Casalnuovo, R. J. McKinney, and C. A. Tolman, in *Encyclopaedia of Inorganic Chemistry*, Vol. 3, ed. R. B. King, Wiley, Chichester, 1994, p. 1428.
385. R. J. McKinney and D. C. Roe, *J. Am. Chem. Soc.*, 1985, **107**, 261.
386. T. Naota, Y. Shichijo, and S.-I. Murahashi, *J. Chem. Soc., Chem. Commun.*, 1994, 1359.
387. J. F. Harrod, in *Encyclopaedia of Inorganic Chemistry*, Vol. 3, ed. R. B. King, Wiley, Chichester, 1994, p. 1486.
388. A. J. Chalk and J. F. Harrod, *J. Am. Chem. Soc.*, 1965, **87**, 16.
389. M. Green, J. L. Spencer, F. G. A. Stone, and C. A. Tsipis, *J. Chem. Soc., Dalton Trans.*, 1977, 1519.
390. C. L. Randolph and M. S. Wrighton, *J. Am. Chem. Soc.*, 1986, **108**, 3366.
391. F. Ozawa, T. Hikida, and T. Hayashi, *J. Am. Chem. Soc.*, 1994, **116**, 2844.
392. K. C. Brinkman, A. J. Blakeney, W. Krone-Schmidt, and J. R. Gladysz, *Organometallics*, 1984, **3**, 1325.
393. M. Aizenberg and D. Milstein, *Angew. Chem. Int. Edn. Engl.*, 1994, **33**, 317.
394. U. Schubert, *Angew. Chem. Int. Edn. Engl.*, 1994, **33**, 419.
395. M. A. Schroeder and M. S. Wrighton, *J. Organomet. Chem.*, 1977, **128**, 345.
396. F. Seitz and M. S. Wrighton, *Angew. Chem. Int. Edn. Engl.*, 1988, **27**, 289.
397. S. Duckett and R. N. Perutz, *Organometallics*, 1992, **11**, 90.
398. S. H. Bergens, P. Noheda, J. Whelan, and B. Bosnich, *J. Am. Chem. Soc.*, 1992, **114**, 2128.
399. L. N. Lewis, *J. Am. Chem. Soc.*, 1990, **112**, 5998.
400. K. Burgess and W. A. van der Donk, in *Encyclopaedia of Inorganic Chemistry*, Vol. 3, ed. R. B. King, Wiley, Chichester, 1994, p. 1420.
401. J. J. Juliette, I. T. Horváth, and J. A. Gladysz, *Angew. Chem. Int. Edn. Engl.*, 1997, **36**, 1610.
402. B. R. James, *Homogeneous Hydrogenation*, Wiley, New York, 1973.
403. Review: Volume 415 of *Ann. N. Y. Acad. Sci.*, published in 1983, contains articles on structure and reactivity of hydride complexes of transition metals as well as on their role in homogeneous catalysis of carbon monoxide, alkenes, alkynes, and other organic substrates.
404. J. A. Osborn, F. H. Jardine, J. F. Young, and G. Wilkinson, *J. Chem. Soc. (A)*, 1966, 1711; F. H. Jardine, J. A. Osborn, and G. Wilkinson, *J. Chem. Soc. (A)*, 1967, 1574.
405. D. Milstein, *J. Am. Chem. Soc.*, 1982, **104**, 5227.
406. M. J. Burk, Shaoguang Feng, M. F. Gross, and W. Tumas, *J. Am. Chem. Soc.*, 1995, **117**, 8277.

407. D. M. Hayes and E. Weitz, *J. Phys. Chem.*, 1991, **95**, 2723.
408. Y. Doi, K. Koshizuka, and T. Keii, *Inorg. Chem.*, 1982, **21**, 2732; A. J. Lindsay, G. McDermott, and G. Wilkinson, *Polyhedron*, 1988, **7**, 1239.
409. S. Bhaduri, H. Khwaja, N. Sapre, K. Sharma, A. Basu, P. G. Jones, and G. Carpenter, *J. Chem. Soc., Dalton Trans.*, 1990, 1313.
410. W. Kaim and B. Schwederski, *Bioinorganic Chemistry: Inorganic Elements in the Chemistry of Life*, Wiley, Chichester, 1994, Chapter 11.2; J. J. R. Fraústo da Silva and R. J. P. Williams, *The Biological Chemistry of the Elements*, Clarendon Press, Oxford, 1991, Chapter 17.7.4.
411. For example, J. Chatt, J. R. Dilworth, and R. L. Richards, *Chem. Rev.*, 1978, **78**, 589; *New Trends in the Chemistry of Nitrogen Fixation*, ed. J. J. Chatt, Academic Press, New York, 1980; R. A. Henderson, G. J. Leigh, and C. J. Pickett, *Adv. Inorg. Chem. Radiochem.*, 1983, **27**, 197; T. A. Bazhenova and A. E. Shilov, *Coord. Chem. Rev.*, 1995, **144**, 69.
412. Review: P. G. Jessop, T. Ikariya, and R. Noyori, *Chem. Rev.*, 1995, **95**, 259.
413. P. G. Jessop, T. Ikariya, and R. Noyori, *Nature*, 1994, **368**, 231.
414. The reference cited at the start of this section lists no fewer than 16 reviews devoted to carbon dioxide complexes. For a recent review on coordination of carbon dioxide and its relevance to catalysis, see W. Leitner, *Coord. Chem. Rev.*, 1996, **153**, 257.
415. J. R. Pugh, M. R. M. Bruce, B. P. Sullivan, and T. J. Meyer, *Inorg. Chem.*, 1991, **30**, 86.
416. C. Creutz, H. A. Schwartz, J. F. Wishart, and N. Sutin, *J. Am. Chem. Soc.*, 1989, **111**, 1153.
417. Reviews: D. Walther, *Coord. Chem. Rev.*, 1987, **79**, 135; A. Behr, *Angew. Chem. Int. Edn. Engl.*, 1988, **27**, 661; T. Tsuda, *Gazz. Chim. Ital.*, 1995, **125**, 101.
418. R. Alvarez, E. Carmona, D. J. Cole-Hamilton, A. Galindo, E. Gutiérrez-Puebla, A. Monge, M. L. Poeda, and C. Ruíz, *J. Am. Chem. Soc.*, 1985, **107**, 5529; H. Hoberg, Y. Peres, C. Krüger, and Yi-Hung Tsay, *Angew. Chem. Int. Edn. Engl.*, 1987, **26**, 771.
419. G. Henrici-Olivé and S. Olivé, *The Chemistry of the Catalysed Hydrogenation of Carbon Monoxide*, Springer, Berlin, 1984.
420. J. R. Blackborow, R. J. Daroda, and G. Wilkinson, *Coord. Chem. Rev.*, 1982, **43**, 17.
421. D. R. Fahey, *J. Am. Chem. Soc.*, 1981, **103**, 136.
422. Review: R. J. Klingler and J. W. Rathke, *Progr. Inorg. Chem.*, 1991, **39**, 113.
423. P. M. Maitlis, I. M. Saez, N. J. Meanwell, K. Isobe, A. Nutton, A.Váquez de Miguel, D. W. Bruce, S. Okeya, P. M. Bailey, D. G. Andrews, P. R. Ashton, and I. R. Johnstone, *New J. Chem.*, 1989, **13**, 419.
424. R. B. King and F. Ohene, *Ann. N. Y. Acad. Sci.*, 1983, **415**, 135.
425. D. Choudhury and D. J. Cole-Hamilton, *J. Chem. Soc., Dalton Trans.*, 1982, 1885; D. Mahajan, C. Creutz, and N. Sutin, *Inorg. Chem.*, 1985, **24**, 2063; R. Ziessel, *J. Chem. Soc., Chem. Commun.*, 1988, 16.
426. A. Cabrera, J. Gómez-Lara, and M. Alcaráz, *New J. Chem.*, 1989, **13**, 103.
427. C.-H. Cheng and R. Eisenberg, *J. Am. Chem. Soc.*, 1978, **100**, 5968; M. Kubota, *Inorg. Chem.*, 1990, **29**, 574.
428. P. Yarrow, H. Cohen, C. Ungermann, D. Vandenberg, P. C. Ford, and R. G. Rinker, *J. Mol. Catal.*, 1983, **22**, 239.
429. F. Ungváry, *Coord. Chem. Rev.*, 1997, **160**, 129.

430. A review of available reviews can be found on p. 575 of G. G. Stanley, in *Encyclopaedia of Inorganic Chemistry*, Vol. 2, ed. R. B. King, Wiley, Chichester, 1994.
431. R. van Asselt, E. E. C. G. Gielens, R. E. Rülka, K. Vrieze, and C. J. Elsevier, *J. Am. Chem. Soc.*, 1994, **116**, 977.
432. For example, A. Haynes, B. E. Mann, D. J. Gulliver, G. E. Morris, and P. M. Maitlis, *J. Am. Chem. Soc.*, 1991, **113**, 8567; A. Haynes, B. E. Mann, G. E. Morris, and P. M. Maitlis, *J. Am. Chem. Soc.*, 1993, **115**, 4093; P. M. Maitlis, A. Haynes, G. J. Sunley, and M. J. Howard, *J. Chem. Soc., Dalton Trans.*, 1996, 2187.
433. J. R. Zoeller, E. M. Blakely, R. M. Moncier, and T. J. Dickson, *Catal. Today*, 1997, **36**, 227.
434. A. Sen, *Accounts Chem. Res.*, 1993, **26**, 303.
435. E. M. Carnahan and S. J. Lippard, *J. Am. Chem. Soc.*, 1990, **112**, 3230.
436. There are many reviews and an annual survey of hydroformylation—see, e.g., Refs 97, 429 and 430.
437. R. F. Heck and D. S. Breslow, *J. Am. Chem. Soc.*, 1961, **83**, 4023; L. Versluis, T. Ziegler, and L. Fan, *Inorg. Chem.*, 1990, **29**, 4530.
438. P. Escaffre, A. Thorez, and P. Kalck, *J. Chem. Soc., Chem. Commun.*, 1987, 146.
439. F. Agbossou, J.-F. Carpentier, and A. Mortreux, *Chem. Rev.*, 1995, **95**, 2485; S. Gladiali, J. C. Bayon, and C. Claver, *Tetrahedron: Asymmetry*, 1995, **6**, 1453.
440. J. K. MacDougall, M. C. Simpson, M. J. Green, and D. J. Cole-Hamilton, *J. Chem. Soc., Dalton Trans.*, 1996, 1161.
441. G. Parrinello and J. K. Stille, *J. Am. Chem. Soc.*, 1987, **109**, 7122.
442. D. C. Roe, R. E. Sheridan, and E. E. Bunel, *J. Am. Chem. Soc.*, 1994, **116**, 1163.
443. Y. Souma and H. Kawasaki, *Catal. Today*, 1997, **36**, 91.
444. E. Monflier, E. Blouet, Y. Barbaux, and A. Mortreux, *Angew. Chem. Int. Edn. Engl.*, 1994, **33**, 2100.
445. M. M. Taqui Khan, K. Venkatasubramanian, and M. M. Bhadbhade, *J. Chem. Soc., Dalton Trans.*, 1992, 1031.
446. W. Rüttinger and G. C. Dismukes, *Chem. Rev.*, 1997, **97**, 1.
447. S. W. Gersten, G. J. Samuels, and T. J. Meyer, *J. Am. Chem. Soc.*, 1982, **104**, 4029.
448. N. C. Pramanik and S. Bhattacharya, *Transition Met. Chem.*, 1997, **22**, 524.
449. L. Weber, R. Hommel, J. Behling, G. Haufe, and H. Hennig, *J. Am. Chem. Soc.*, 1994, **116**, 2400; A. S. Goldstein, R. H. Beer, and R. S. Drago, *J. Am. Chem. Soc.*, 1994, **116**, 2424.
450. M. Lin and A. Sen, *Nature*, 1994, **368**, 613.
451. B. B. Lohray, *Tetrahedron: Asymmetry*, 1992, **3**, 1317; see also Ref. 292.

Index

This index should be used in conjunction with the detailed list of contents at the start of the book, especially for locating general topics. Frequently occurring compounds and classes of compounds, such as cobalt and ruthenium ammines, iron diimines, and hexacyanoferrates have not been given index entries – it has been deemed preferable to devote the limited space available to less frequently mentioned species which might otherwise be difficult to trace. Similarly, headings such as rate constants and mechanisms do not appear, but there are detailed index entries for topics, such as photochemistry, rate laws, and exchange reactions, whose treatment is scattered throughout the book. In general names of people have only been indexed when they act as identifiers of compounds, reactions, equations, or parameters.